APPLIED COMPUTATIONAL FLUID DYNAMICS

Applied Computational Fluid Dynamics

Editor:

Olga Moreira, Ph.D.

ArclerPress

www.arclerpress.com

Applied Computational Fluid Dynamics

Editor: Olga Moreira, Ph.D.

708 3rd Avenue, 6th Floor
New York
NY 10017
United States of America
www.arclerpress.com

ISBN: 978-1-68094-354-2

Library of Congress Control Number: 2016950958

About the Editor

Olga Moreira, Ph.D.

Olga Moreira obtained her Ph.D. in Astrophysics from the University of Liege (Belgium) in 2010, her BSc. in Physics and Applied Mathematics from the University of Porto (Portugal). Her post-graduate travels and international collaborations with the European Space Agency (ESA) and European Southern Observatory (ESO) led to great personal and professional growth as a scientist. Currently, she is working as an independent researcher, technical writer, and editor in the fields of Mathematics, Physics, Astronomy and Astrophysics.

Contents

Section III: CFD Applications to Architecture Design and Optimization in Terms of Energy Consumption, Ventilation and Cooling Systems

List of Contributors

Victor Udoewa
George Washington University, USAID Development Engineer, AAAS Science & Technology Policy Fellow, AAAS, 2009-2011

Vinod Kumar
Mechanical Engineering, University of Texas at El Paso, USA

Adel Abbas-Bayoumi
Aerodynamic Strategies, Airbus, Avd. John Lennon, Getafe, Spain

Klaus Becker
Aerodynamic Strategies, Airbus, Airbusallee 1, 28199 Bremen, Germany

N. Kroll
Institute of Aerodynamics and Flow Technology, German Aerospace Center (DLR), Lilienthalplatz 7, 38108 Braunschweig, Germany

M. Abu-Zurayk
Institute of Aerodynamics and Flow Technology, German Aerospace Center (DLR), Lilienthalplatz 7, 38108 Braunschweig, Germany

D. Dimitrov
Institute of Aeroelasticity, German Aerospace Center (DLR), Go¨ttingen, Germany

T. Franz
Institute of Aerodynamics and Flow Technology, German Aerospace Center (DLR), Lilienthalplatz 7, 38108 Braunschweig, Germany

T. Fuhrer
Institute of Composite Structures and Adaptive Systems, German Aerospace Center (DLR), Braunschweig, Germany

T. Gerhold
Institute of Aerodynamics and Flow Technology, German Aerospace Center (DLR), Lilienthalplatz 7, 38108 Braunschweig, Germany

S. Go¨rtz
Institute of Aerodynamics and Flow Technology, German Aerospace Center (DLR), Lilienthalplatz 7, 38108 Braunschweig, Germany

R. Heinrich
Institute of Aerodynamics and Flow Technology, German Aerospace Center (DLR), Lilienthalplatz 7, 38108 Braunschweig, Germany

C. Ilic
Institute of Aerodynamics and Flow Technology, German Aerospace Center (DLR), Lilienthalplatz 7, 38108 Braunschweig, Germany

J. Jepsen
Institute of Air Transportation Systems, German Aerospace Center (DLR), Hamburg, Germany

J. Ja¨gersku¨pper
Institute of Aerodynamics and Flow Technology, German Aerospace Center (DLR), Lilienthalplatz 7, 38108 Braunschweig, Germany

M. Kruse
Institute of Aerodynamics and Flow Technology, German Aerospace Center (DLR), Lilienthalplatz 7, 38108 Braunschweig, Germany

A. Krumbein
Institute of Aerodynamics and Flow Technology, German Aerospace Center (DLR), Lilienthalplatz 7, 38108 Braunschweig, Germany

S. Langer
Institute of Aerodynamics and Flow Technology, German Aerospace Center (DLR), Lilienthalplatz 7, 38108 Braunschweig, Germany

D. Liu
Institute of Aerodynamics and Flow Technology, German Aerospace Center (DLR), Lilienthalplatz 7, 38108 Braunschweig, Germany

R. Liepelt
Institute of Aeroelasticity, German Aerospace Center (DLR), Go¨ttingen, Germany

L. Reimer
Institute of Aerodynamics and Flow Technology, German Aerospace Center (DLR), Lilienthalplatz 7, 38108 Braunschweig, Germany

M. Ritter
Institute of Aeroelasticity, German Aerospace Center (DLR), Go¨ttingen, Germany

A. Schwo¨ppe
Institute of Aerodynamics and Flow Technology, German Aerospace Center (DLR), Lilienthalplatz 7, 38108 Braunschweig, Germany

J. Scherer
Institute of Structures and Design, German Aerospace Center (DLR), Stuttgart, Germany

F. Spiering
Institute of Aerodynamics and Flow Technology, German Aerospace Center (DLR), Lilienthalplatz 7, 38108 Braunschweig, Germany

R. Thormann
Institute of Aeroelasticity, German Aerospace Center (DLR), Go¨ttingen, Germany

V. Togiti
Institute of Aerodynamics and Flow Technology, German Aerospace Center (DLR), Lilienthalplatz 7, 38108 Braunschweig, Germany

D. Vollmer
Institute of Aerodynamics and Flow Technology, German Aerospace Center (DLR), Lilienthalplatz 7, 38108 Braunschweig, Germany

J.-H. Wendisch
Institute of Aerodynamics and Flow Technology, German Aerospace Center (DLR), Lilienthalplatz 7, 38108 Braunschweig, Germany

Yun Jiang
UAV Research Institute, Northwestern Polytechnical University, 127 West Youyi Road, Xi'an 710072, China

Bo Zhang
UAV Research Institute, Northwestern Polytechnical University, 127 West Youyi Road, Xi'an 710072, China

Tao Huang
UAV Research Institute, Northwestern Polytechnical University, 127 West Youyi Road, Xi'an 710072, China

Antonio Viviani
Seconda Università degli Studi di Napoli, Aversa, Italy

Giuseppe Pezzella
Centro Italiano Ricerche Aerospaziali, Capua, Italy

Carsten Othmer
Vehicle Technology, Group Research, Volkswagen AG, 38436 Wolfsburg, Germany

Mulugeta Biadgo Asress
Department of Aeronautics, Faculty of Mechanical Engineering, University of Belgrade, Kraljice Marije 16, Belgrade, Serbia

Jelena Svorcan
Department of Aeronautics, Faculty of Mechanical Engineering, University of Belgrade, Kraljice Marije 16, Belgrade, Serbia

C Z Xia
School of Hydraulic, Energy and power Engineering Yangzhou University, Yangzhou, Jiangsu, P.R.China, 225009

L Cheng
School of Hydraulic, Energy and power Engineering Yangzhou University, Yangzhou, Jiangsu, P.R.China, 225009

Y N Shang
School of Hydraulic, Energy and power Engineering Yangzhou University, Yangzhou, Jiangsu, P.R.China, 225009

J R Zhou
School of Hydraulic, Energy and power Engineering Yangzhou University, Yangzhou, Jiangsu, P.R.China, 225009

F Yang
School of Hydraulic, Energy and power Engineering Yangzhou University, Yangzhou, Jiangsu, P.R.China, 225009

Y Jin
School of Hydraulic, Energy and power Engineering Yangzhou University, Yangzhou, Jiangsu, P.R.China, 225009

Vinicius C. Rispoli
Department of Electrical Engineering, University of Brasilia, Brasília, Brazil
UnB Gama College, University of Brasilia, Brasília, Brazil

Jon F. Nielsen
fMRI Laboratory, Biomedical Engineering Department, University of Michigan, Ann Arbor, USA

Krishna S. Nayak
Magnetic Resonance Engineering Laboratory, Ming Hsieh Department of Electrical Engineering, University of Southern California, Los Angeles, USA.

Joao L. A. Carvalho
Department of Electrical Engineering, University of Brasilia, Brasília, Brazil

Alberto Coccarelli
Wales Heart Research Institute, School of Medicine, Cardiff University, Cardiff, UK

Etienne Boileau
Zienkiewicz Centre for Computational Engineering, College of Engineering, Swansea University, Swansea, UK

Dimitris Parthimos
Wales Heart Research Institute, School of Medicine, Cardiff University, Cardiff, UK

Perumal Nithiarasu
Zienkiewicz Centre for Computational Engineering, College of Engineering, Swansea University, Swansea, UK

Yoshihito Ohhara
Department of Oral and Maxillofacial Surgery, Yokohama City University Graduate School of Medicine, 3-9 Fukuura, Kanazawa-ku, Yokohama 236-0004, Japan

Marie Oshima
Department of Interfaculty Initiative in Information Studies, The University of Tokyo, 7-3-1 Hongo, Bunkyo-ku, Tokyo 113-0033, Japan
Institute of Industrial Science, The University of Tokyo, 4-6-1 Komaba, Meguro-ku, Tokyo 153-8505, Japan

Toshinori Iwai
Department of Oral and Maxillofacial Surgery, Yokohama City University Graduate School of Medicine, 3-9 Fukuura, Kanazawa-ku, Yokohama 236-0004, Japan

Hiroaki Kitajima
Department of Oral and Maxillofacial Surgery, Yokohama City University Graduate School of Medicine, 3-9 Fukuura, Kanazawa-ku, Yokohama 236-0004, Japan

Yasuharu Yajima
Department of Oral and Maxillofacial Surgery, Yokohama City University Graduate School of Medicine, 3-9 Fukuura, Kanazawa-ku, Yokohama 236-0004, Japan

Kenji Mitsudo
Department of Oral and Maxillofacial Surgery, Yokohama City University Graduate School of Medicine, 3-9 Fukuura, Kanazawa-ku, Yokohama 236-0004, Japan

Absy Krdy
Department of Interfaculty Initiative in Information Studies, The University of Tokyo, 7-3-1 Hongo, Bunkyo-ku, Tokyo 113-0033, Japan
Institute of Industrial Science, The University of Tokyo, 4-6-1 Komaba, Meguro-ku, Tokyo 153-8505, Japan

Iwai Tohnai
Department of Oral and Maxillofacial Surgery, Yokohama City University Graduate School of Medicine, 3-9 Fukuura, Kanazawa-ku, Yokohama 236-0004, Japan

Stephan Rhode
Department of Mechanical Engineering, Karlsruhe Institute of Technology, 76131 Karlsruhe, Germany

Manosh C Paul
School of Engineering, University of Glasgow, Glasgow G12 8QQ, UK

Eckhard Martens
Department of Mechanical Engineering, University of Applied Sciences, 76133 Karlsruhe, Germany

Duncan F Campbell
Department of Oral & Maxillofacial Surgery, Queen Margaret's Hospital, Fife, UK.

Yu Chen
Department of Applied Mechanics, Sichuan University, Chengdu, 610065, China

Yan Xiong
School of Manufacturing Science & Engineering, Sichuan University, Chengdu, 610065

Wentao Jiang
Department of Applied Mechanics, Sichuan University, Chengdu, 610065, China

Fei Yan
Department of Applied Mechanics, Sichuan University, Chengdu, 610065, China

Meng Guo
School of Biological Science and Medical Engineering, Beihang University, Beijing, 100191, China.

Qingyuan Wang
Department of Applied Mechanics, Sichuan University, Chengdu, 610065, China

Yubo Fan
School of Biological Science and Medical Engineering, Beihang University, Beijing, 100191, China.

Sujoy Kumar Saha
Mechanical Engineering Department, Bengal Engineering and Science University, Shibpur, Howrah, West Bengal 711 103, India

Gian Piero Celata
ENEA Casaccia Research Centre, Institute of Thermal Fluid Dynamics, Office Building F-20, Via Anguillarese 301, S. M. Galeria, Rome 00123, Italy

Oladapo S. Akinyemi
Department of Mechanical Engineering, University of Louisiana at Lafayette, Lafayette, LA 70504, USA

Yucheng Liu
Department of Mechanical Engineering, Mississippi State University, Mississippi State, MS 39762, USA

Andrius Jurelionis
Faculty of Civil Engineering and Architecture, Kaunas University of Technology, Studentu str. 48, 51367 Kaunas, Lithuania

Demetri G. Bouris
School of Mechanical Engineering, National Technical University of Athens, 15780 Zografou, Greece

Jorge Flores-Velázquez
Department of Agricultural Engineering, University of Chapingo, Chapingo, México

Guillermo De la Torre-Gea
Department of Biosystems, School of Engineering Queretaro State University Querétaro, México

Enrique Rico-García
Department of Biosystems, School of Engineering Queretaro State University Querétaro, México

Irineo L. López-Cruz
Department of Agricultural Engineering, University of Chapingo, Chapingo, México

Abraham Rojano-Aguilar
Department of Agricultural Engineering, University of Chapingo, Chapingo, México

John Kaiser Calautit, Dominic O'Connor, Polytimi Sofotasiou and Ben Richard Hughes
Department of Mechanical Engineering, University of Sheffield, Sheffield S10 2TN, UK

Andrea Alaimo, Antonio Esposito, Antonio Messineo, Calogero Orlando and Davide Tumino
Faculty of Engineering and Architecture, Kore University of Enna, Cittadella Universitaria, 94100 Enna, Italy

R. Lanzafame
Department of Industrial Engineering, University of Catania, Viale A. Doria, 6, 95125 Catania, Italy

S. Maurol, M. Messina
Department of Industrial Engineering, University of Catania, Viale A. Doria, 6, 95125 Catania, Italy

B Huang
Faculty of Engineering, Kyushu Institute of Technology

K Yoshida
Faculty of Engineering, Kyushu Institute of Technology

T Kanemoto
Faculty of Engineering, Kyushu Institute of Technology

X H Duan
National Research Center of Pumps, Jiangsu University, Zhenjiang, Jiangsu 212013, China

F Y Kong
National Research Center of Pumps, Jiangsu University, Zhenjiang, Jiangsu 212013, China

Y Y Liu
National Research Center of Pumps, Jiangsu University, Zhenjiang, Jiangsu 212013, China

R J Zhao
National Research Center of Pumps, Jiangsu University, Zhenjiang, Jiangsu 212013, China

Q L Hu
National Research Center of Pumps, Jiangsu University, Zhenjiang, Jiangsu 212013, China

Xiana Romani Fernandez
Karlsruhe Institute of Technology, Campus South, Institute for Mechanical Process Engineering and Mechanics, Karlsruhe Germany

Lars Egmont Spelter
Karlsruhe Institute of Technology, Campus South, Institute for Mechanical Process Engineering and Mechanics, Karlsruhe Germany

Hermann Nirschl
Karlsruhe Institute of Technology, Campus South, Institute for Mechanical Process Engineering and Mechanics, Karlsruhe Germany

Roberto Meloni
Department of Astronautics, Electrical and Energy Engineering, University of Rome, "Sapienza", via Eudossiana 18, Rome 00184, Italy

Ali Hadi Ghawi
Department of Civil Engineering, Faculty of Civil Engineering, AL-Qadisyia University Iraq

Jozef Kriš
Department of Sanitary and Environmental Engineering, Faculty of Civil Engineering, Slovak University of Technology, Bratislava Slovakia

Maitraye Sen
Department of Chemical and Biochemical Engineering, Rutgers, The State University of New Jersey, Piscataway, NJ 08854, USA

Dana Barrasso
Department of Chemical and Biochemical Engineering, Rutgers, The State University of New Jersey, Piscataway, NJ 08854, USA

Ravendra Singh
Department of Chemical and Biochemical Engineering, Rutgers, The State University of New Jersey, Piscataway, NJ 08854, USA

Rohit Ramachandran
Department of Chemical and Biochemical Engineering, Rutgers, The State University of New Jersey, Piscataway, NJ 08854, USA

Masroor Mohajerani
Department of Chemical Engineering, Ryerson University, Toronto, Ontario Canada

MehrabMehrvar
Department of Chemical Engineering, Ryerson University, Toronto, Ontario Canada

Farhad Ein-Mozaffari
Department of Chemical Engineering, Ryerson University, Toronto, Ontario Canada

Preface

The same basic procedures are applied in all Computational Fluid Dynamics (CFD) implementations.

- Pre-processing - this procedure is prior to computational simulation and establishes the physical conditions (geometry definition), mesh size (volume division), hydrodynamic models and boundary conditions.
- Simulation - this is the actual computational experiment. It includes numerical discretization methods, turbulence models, linear algebraic equation systems and other scientific libraries.
- Post-processing - this is the final procedure when the computations are complete. This involves analysis and visualization of the resulting solution. Usually, the information is displayed in color graphics and, sometimes, the information is animated to show physical changes in time, follow a fluid particle or a streamline.

First, the book overviews these aforementioned procedures in some detail. Next, it reviews the applications CDF to the advances in transportation, construction, biomedical engineering, turbines and centrifuges design, as well as the modeling of processes that can impact environments such as pollutant emissions, water treatment, powder handling and degradation of organic materials. This book is divided into the following 5 sections:

- Section 1 - CFD applications in the designing, optimization and testing of transportation.
- Section 2 - CFD applications to cardiovascular hemodynamics and biomimetics.
- Section 3 - CFD applications to architecture design and optimization in terms of energy consumption, ventilation and cooling systems.
- Section 4 - CFD Applied to the designing and optimization of turbines and centrifuges
- Section 5 - CFD applications to the study of combustion, sedimentation, granulation, oxidation processes in manufacturing engineering.

Editor
Olga Moreira, Ph.D.

SECTION I: CFD APPLICATIONS IN THE DESIGNING, OPTIMIZATION AND TESTING OF TRANSPORTATION

CHAPTER 1

COMPUTATIONAL FLUID DYNAMICS

Victor Udoewa[1] and Vinod Kumar[2]

[1]George Washington University, USAID Development Engineer, AAAS Science & Technology Policy Fellow, AAAS, 2009-2011

[2]Mechanical Engineering, University of Texas at El Paso, USA

INTRODUCTION

Computational Fluid Dynamics (CFD) is the emerging field of fluid mechanics in which fluid flow problems are solved and analyzed using computational methods and numerical algorithms. In fluid mechanics, there are generally three routes of work in the field, three ways to conduct experiments. The first category is theoretical, or analytical, fluid mechanics. Theoretical fluid mechanics includes theorizing, manipulating and solving equations with pen and paper. The Navier-Stokes equation governing incompressible fluid flow is an example of theoretical fluid mechanics. Secondly, many engineers and physicist work in the area of experimental fluid mechanics. Experimental fluid mechanics involves conducting actual physical experiments and studying the flow and the effect of various disturbances, shapes, and stimuli on the flow. Examples include waves generated by pools, air flow studies in actual wind tunnels, flow through physical pipes, etc. Lastly, a growing number of engineers, mathematicians, computer scientists, and physicists work in the area of computational fluid dynamics (CFD). In CFD, you may still run an experiment of waves across water, an airplane in a wind tunnel, or flow through pipes, but now it is done through the computer Instead of actual, physical, 3D objects. A computer model is created, and computer programmers code the equations representing the physical laws that govern the flow of the molecules of fluid. Then the flow results (such as velocity and pressure) are output into files that can be visualized through pictures or animation so that you see the result just as you do with physical experiments.

In cases where an analytical, or theoretical, solution exists, CFD simulations and the mathematical models, which are coded in the computer program, are corroborated by comparison to the exact solutions. This comparative check is called validation. CFD is not yet to a point where solutions to problems are used without corroboration by existing, known, analytical or exact solutions when available. Validation is not to be confused with verification, however; validation is a check to make sure that the implemented, coded model accurately represents the conceptual, mathematical description and the solution intended to be modeled.

Still, there are many times when there is no analytical solution. In these cases, one often uses a computational approach. In such cases without a known solution, CFD is used to approximate a solution. Most often, CFD is used when a computational solution is faster, cheaper, or more convenient. Convenience may be due to time or safety or another reason. If we wanted to create a database of information about 3D flow around a cylinder for different types of fluids at varying speeds,

Citation: Victor Udoewa and Vinod Kumar (2012). Computational Fluid Dynamics, Applied Computational Fluid Dynamics, Prof. Hyoung Woo Oh (Ed.), InTech, DOI: 10.5772/28614.

CFD is quite helpful. It would take quite a long time to change the fluid in our flow container and clean the container for every type of fluid we desired. It would also take some time to change the speed of the flow. In this case, it's much faster to simulate it computationally. Then, anytime we wanted to change the speed of the flow, we simply would change one number in a computer input file. Or if we wanted to change viscosity and density for the fluid (representing a different fluid) we would just change the corresponding values in a computer input file. In this case, CFD is faster.

Now imagine if you were doing space defense work for a government, and you were asked to do fluid dynamics simulation of the combustion dynamics during an explosion or when a space shuttle launched (1). It would take immeasurable amounts of money to do test launches over and over as you studied the combustion dynamics of space shuttle thruster ignition. And it would take large amounts of money to test explosive devices, especially considering the damage they cause. In these cases, CFD is, again, quite helpful. The only costs in CFD are the time of creating a computer model, choosing the right mathematical model, coding it, and the power and computer time required to solve the equations. But it is plain to see that CFD is cheaper.

What if you were hired to help design material for the outfits of swimming athletes? Your company gave you the job of studying sharks and their agile ability to swim and maneuver through the water. So you start by trying to study the fluid dynamics around the shark skin (2). How convenient is it to locate sharks and place them in some type of testing container where you have probes and measuring devices located? How convenient is it to place probes on the body of the shark itself? How safe is it to work with the sharks in that manner? No, it's better to create a computer model of a shark and get the information for the shape, design, feel, and density of its skin and to use this information to run simulations. It is clear that CFD is more convenient in this situation. Sometimes a CFD simulation can be all three—safer, cheaper, and more convenient. Imagine a situation in which two paratroopers, jumping from both side doors of a military cargo aircraft, always crash into each other down below (3). In order to analyze the fluid dynamics of the problem to see what air flow forces are affecting the paratrooper paths, you would need to perform test jumps with paratroopers. However, that is potentially injurious and not safe. You would also have to rent the plane, pay for the rental by hour, hire the test pilot, and pay for all the equipment for the jump. That is expensive. Lastly, the organization of the use of the military aircraft and personnel and equipment takes many months, and it can take from 6 to 12 months to plan the test. In this case a CFD experiment is more convenient: faster, cheaper, and safer. Usually solving CFD problems involves three stages. First there is the pre-processing stage. In this stage, the geometric boundaries of the problem are defined. In 3D, a volume is created (in 2D, an area) over which the equation will be solved. This volume is broken into smaller units or cells creating a mesh (though there are meshless methods for computing CFD problems). This may be uniform or non-uniform. Along with constitutive equations, the particular equations are chosen for the problem in order to properly physically model the flow. These equations may be manipulated depending on the mathematical method being used. Boundary conditions are prescribed along the boundary. For time dependent problems, initial conditions are prescribed.

Secondly, the problem is solved numerically. At the least, we usually solve for velocity and pressure, but the list of unknowns can be longer depending on the mathematical model (equation) chosen for the physical situation. Other unknowns may include temperature, energy, and density. The numerical solution is usually computed iteratively for steady-state solutions. For time-dependent problems, a step in time is taken, and the equation is numerically solved again, eventually producing a solution for every time step. The final step in the CFD process is post-processing. In this stage, the solution is analyzed usually with the help of visualization and possibly animation for dynamic, or timedependent, problems. It is in this stage that CFD results are usually compared to any previous experimental results or known analytical, or theoretical, solutions. This comparison is usually called validation. Today, confidence in CFD is growing, but we have not arrived at the point of trusting CFD solutions without validation. Even if a particular model is validated, we still corroborate the results of a simulation with experimental or analytical results. Because of the hybrid nature of CFD, advances in CFD are usually made in three areas: computational and applied mathematics, mechanical/chemical engineering, and computer science/electrical engineering. Some researchers work on new theoretical, mathematical models creating new discretization methods (ways to discretize the problem in order to numerically solve it over the discrete units or cells), or turbulence models. They might publish in applied mathematics or computational mathematics journals. Others work on computer architecture (such as different types of supercomputers or computer clusters), coding techniques such as parallel programming, or speeding up the computational processes through faster mesh generation and mesh reordering. They might publish their results in electrical engineering or computer science journals. Lastly others might use CFD to concentrate on new insights in the engineering aspect of the problem such as the mechanics of bird flight or sharkskin-inspired speedo design for less water resistance, or resistance to blood flow inherent to certain veins. They might publish their results in engineering journals next to experimental or theoretical engineering results. Their focus is on the application more so than the math or computer science.

Pre-Processing

Pre-processing refers to the work that must be done prior to the actual computational experiment or simulation. This work can be reduced to four general areas: geometry definition, volume division, model choice and definition, and boundary condition definition. For the purposes of this article, we will ascribe the work of coding and the choice of computational implementation to the processing stage called simulation. Sometimes researchers refer to mesh generation as pre-processing in general because a mesh is generated when the volume is divided.

Geometry Definition

The first step is to define the computational domain of the problem. The purpose of this definition is to confine the problem to a finite space and limit the computation. It is true that a plane flying in the air has some residual effects on air flow patterns 1,000 miles away, but because of the negligible nature of those effects, we are relatively safe in looking at the effect of the plane on air flow within a reasonable vicinity of the plane thereby limiting our computational work and making the problem finite. It would be quite a task to compute the effect of the airplane on the air flow at a certain height around the entire earth. To limit the flow in this way usually requires defining a domain inside which we will compute the flow, outside of which we will not compute the flow. Usually the geometry of the domain is chosen to be a box of some sort, usually a rectangular prism in 3D or a rectangle in 2D. However, any closed shape may be chosen as long as the shape closes off an inside computational domain from an outside space in which computations will not take place. A closed space does not imply, however, an empty domain box. For example imagine that we are simulating intravenous blood flow (4) (5) around a cancerous growth. The vascular domain is modeled by a 3D cylindrical prism (our domain box), but we still have an object inside. In this case, our object is a semi-spherical cancerous growth on the surface of a wall of the vein. Traditionally, our domain must be totally closed, so the surface of the domain goes from the wall of the vein, joins the surface of the cancerous growth, and continues on the other side rejoining the wall of the vein creating a closed 3D space that does not go under the tumor but continues over the surface of it. Likewise, if we were calculating flow around a sphere (6), the domain box would be the outer half of the domain surface. The inner half of the surface would be the sphere inside the domain box. Just as in the cancer example, we are not calculating the flow inside the sphere, just as we were not calculating the blood flow in the tumor. But the tumor and sphere form part of the boundaries of the domain helping to close off the computational space in which we are interested in the velocity and pressure of the fluid. Remember, domain boxes may contain objects inside which no flow is calculated, but whose surface forms part of the surface of the domain helping to limit the computational space and better define where the fluid flows.

Volume Division

The second step is volume division or mesh generation. Why we must divide the volume is not obvious until one remembers that it is easier to solve a flow problem over a smaller area or volume than a larger one. So dividing the volume into smaller units transforms the large problem over the entire domain into a large number of smaller problems over smaller sub-domains. However, the real reason we divide the volume is because we seek to find, for example, the velocity and pressure of the fluid at various points throughout the domain volume or area. In order to do this, we fill the inner domain volume or area with nodespoints at which we will calculate, in this instance, velocity and pressure. Once we have filled the inner volume or area with nodes, we connect the nodes with edges (and sides in 3D) creating smaller sub-volume or sub-area elements. For example, if our computational box is a rectangular prism and we fill it with nodes, we can connect the nodes to create quadrilaterals or tetrahedrals. If the domain is a rectangle and we fill the rectangle with nodes, we can connect all the nodes to create small rectangles or triangles. This network of rectangles/triangles and quadrilaterals/tetrahedrals creates a mesh of nodes; a mesh has been generated. Remembering that the purpose of volume or area division is to create more manageable subvolumes or sub-areas, it behooves us to evenly space out the distribution of nodes. If we do not, we may find that there are large spaces (volumes or areas) with no nodes. This is problematic because it means some of the sub-volume or sub-area elements will still be large; though our goal is to make them small.

Our second goal is to make each sub-element evenly shaped. In 2D, evenness in both directions means that 2D rectangles tend towards squares or that our triangles are equilateral. In 3D, if evenness is desired in all 3 directions, our quadrilaterals tend toward cubes and the tetrahedrals tend to be equilateral. Oblong and unevenly shaped sub-elements also create spaces (areas or volumes) with fewer nodes than parts of the domain with evenly shaped subelements. In these cases, our second goal serves a similar purpose as our first goal: to divide the domain into smaller, more manageable sub-domains with evenly distributed nodes. An important aspect of mesh generation is choosing the appropriate size, or refinement, of the mesh sub-elements, such that important aspects of the flow are properly resolved. The general rule is that no fluid particle should advance through multiple sub-elements, cells, or units in one time step. Therefore our third goal is to increase the refinement (or the number of nodes) in areas or sub-volumes of increased fluid velocity or vorticity or any interesting fluid flow phenomena that you would like to capture computationally. This will allow us to visualize it later. So far, the realm of geometry definition and mesh generation fall into the computer science side of CFD. When generating the mesh by defining a geometry and dividing that geometric area or volume, one must decide if one will use uniform sub-elements or non-uniform subelements. Definitely in parts of the domain with special flow requiring increased refinement, the elements in those parts will not match the refinement of elements elsewhere. But in the general flow one can still choose a uniform, structured mesh or a non-uniform, unstructured mesh. The ability to create non-uniform meshes is important in CFD because of the physical nature of fluids to occupy and fill any void left unoccupied. When dealing with complex geometries and small nooks, crannies, and crevices of an automobile or a model of a city block, it helps to have unstructured, non-uniform meshes that allow for the modeler to create the best shapes to fit the 2D or 3D space (7). Likewise, the division of the volume or area in mesh generation requires the modeler to choose between quadrilaterals and rectangular prisms or triangles and tetrahedrals (there are other choices of shapes, as well, such as wedges and pyramids in 3D). Generally, quadrilaterals and rectangular prisms have a more accurate solution than triangles and triangular prisms, but there are ways to increase the accuracy of the latter. Because of the non-uniform and sharp geometries found in fluid problems by nature of fluids, triangles and triangular prisms work better geometrically for CFD applications. Normally, CFD researchers will utilize triangles and triangular prisms and then increase the number of interpolation points inside these elements so that no accuracy is lost. Interpolation points are points inside an element at which the solution is calculated. From these interpolation points, we can approximate the solution at any location inside an element.

All of the choices in dividing the volume and discretizing the mesh have the potential to introduce errors. Such errors, due to bad distribution of nodes or parts of the domain where the refinement is too low, are called discretization errors. These are errors that would disappear if we appropriately divided the volume or area or appropriately discretized the mesh.

Physical Model Definition

The first two steps dealt with the computer science side of CFD and there are many CFD engineers who work on geometric mesh discretization and mesh partitioning methods. Step three deals with the computational and applied mathematical side of CFD choosing the appropriate mathematical model. When preparing to model a certain fluid flow situation, one must decide which equation accurately describes the fluid flow one wishes to simulate. If no equation currently exists, the CFD engineer must do work in the theoretical side of CFD and formulate a new equation or a more specific equation for his or her specific fluid flow situation. If equations do exist, the CFD engineer must simply choose the correct equation for the fluid flow. This is not a trivial step as sometimes the same situation may require different mathematical models at different velocity regimes or different temperature regimes, for instance. So the specific parameters of the flow must be looked at in detail velocity, viscosity, density, pressure, etc. so that the correct equation is chosen.

A good example from fluid mechanics is the Navier-Stokes equation which is the basic or fundamental equation for fluid dynamics. If you remove viscosity from the equation, the Navier-Stokes equations become the Euler equations. Since all fluids have some amount of viscosity this approximation is important in flows in which the viscosity is negligible (8) such as sonic flows. A plane flying at sonic speeds will have air sliding past it, relatively, as if it had no viscosity. So the use of the Navier-Stokes equations also depends on the velocity of the flow, or more accurately the Reynolds number which governs the ratio of the kinematic forces to the inertial forces. You can still go further: if you remove vorticity from the Euler equations, you arrive at the full potential equations.

The point in this illustration is that choosing the correct mathematical model is important, sometimes difficult, and always specific to the flow situation. The choice of the mathematical model affects the unknown values you will compute. Some CFD simulations are really computational fluidothermodynamics because temperature and energy are calculated as well (9). For compressible flows, density is an unknown value and we would seek to solve for this value of density in the simulation. So the choice of mathematical modeling affects what unknowns we will compute. More accurately, the unknowns we want to compute in a given situation (along with other details about the flow situation) may help positively affect our choice of a mathematical model or the need to formulate a new one. Remember that errors can be introduced at this step as well. If a inappropriate or poorly approximating governing equation is chosen, this affects the final solution. If a governing solution is chosen or formulated for which no analytical solution or experimental solution exists, we lose the opportunity for validation to reduce errors. Any simplification in the model or any untrue assumptions the mathematical model uses introduces errors as well. All of these types of errors can be classified as physical approximation errors because they deal with the physical, mathematical model (not the geometric model).

Boundary Condition Definition

After a mathematical model is chosen to model the physical phenomenon, usually boundary conditions must be chosen. This is the fourth step of the pre-processing stage, and this step falls on the applied math side of CFD. Usually we deal with boundary-value problems which require values to be assigned along the boundary of the domain of the problem in order to solve the problem throughout the 2D or 3D space. For instance, in some problems, one may specify the value of the unknown on the boundary.

Imagine prescribing the value of the velocity of the fluid on the boundary. Such a boundary condition is called a Dirichlet or a direct boundary condition because you are setting the value of the unknown. If one specifies the value of a derivative of the unknown, the boundary condition is called a Neumann or natural boundary condition. For example in hydrodynamic flow around a submerged rock with moss, a CFD researcher would usually place the boundary condition of free-stream velocity on the entrance side and exit side of the domain box assuming that the entrance side is sufficiently upstream from the rock so as to still be undisturbed, and the exit side is sufficiently downstream that the flow conditions have returned to free-stream conditions. The prescription of free-stream velocity would be Dirichlet or direct boundary conditions. The same researcher might assume stress (a derivative of velocity) to be zero in the direction perpendicular to the side surfaces of the domain box. When she prescribes stress in that direction she is setting Neumann or natural boundary conditions in the direction perpendicular to the side surfaces of the domain box. Such boundaries are spatial boundaries. For time-dependent problems, there are temporal boundaries in a sense.

Time dependent problems require an initial condition, prescribed values for the unknowns set at the temporal start of the simulation. In the same hydrodynamic example, let us say we want to simulate the flow when a rock, half the size of the stationary, mossy rock, was thrown into the river passing next to the mossy rock and hitting the riverbed. To start the simulation we need to have the steady flow of the river around the mossy, stationary rock without the 2nd rock thrown in. Once we have computed this flow, we can use the values of velocity and pressure from this flow as initial conditions for a simulation of a moving 2nd rock that is falling to the bottom of the river. From there the simulation will march in time and use the flow results from the previous time step as initial conditions for the next time step.

SIMULATION

The second phase of CFD work is the actual simulation or the "processing" work once the pre-processing work is completed. However, there are still some pre-simulation decisions to be made. CFD work is done through computers which not only decreases the time it takes to perform calculations, but also increases the amount of calculations that can be done in a given time period. As computing power has increased over the years, CFD has been used to solve larger and larger problems.

Large problems, however, were traditionally reserved for supercomputers. Supercomputers are large computers made up of multiple computers or CPUs. A desktop or laptop computer could only handle so many calculations due to hard drive limitations on different types of computer memory. As computers in general become more advanced, not only has the memory capacity of supercomputers increased, but so has the memory of desktops and laptops increased. This has created a cycle where problems solved by supercomputers today are solved by desktop computers and laptop computers tomorrow. And the problems solved by desktop and laptop computers today were only solved by supercomputers yesterday. For example, historically, a simulation of flow past an automobile was done on supercomputers (10) (11).

Today one can create a model of an automobile and run a flow simulation of air flow past the automobile with one desktop or laptop machine. This example is one of many indicative of this ever-improving cycle. Besides memory the other limitation on computing ability in today's world is clock speed. CFD workers are dependent upon computer scientist researchers to continue to increase the clock speed of microprocessors. In general, the faster computers become, the faster is the speed CFD scientists and engineers can compute solutions to problems. And as the speed of computations increases, the time to do computations decreases, and CFD scientists and engineers can compute larger problems (as long as they have the memory capacity for the calculations and storage for the solution). Currently, as this article goes to press, the fastest machine in the world is the K computer which computes at 10 Petaflops (12). Flops are floating point operations per second, and the prefix peta means 10^15. Therefore, the K computer can compute 10^16 floating point operations per second. Supercomputers come in varying shapes and sizes. A small supercomputer may have 36-100 CPUs. A larger supercomputer could have 50,000 different nodes, or CPUs. For example, the K computer has 68,544 CPUs, each with 8 cores (octo-core) for a total of 548,352 cores (12). As well, today we have small supercomputing clusters, in which different CPUs are linked together to act as a supercomputer. Often one will find Linux clusters arranged in this way. Each node of a cluster can actually contain multiple processors itself acting as a single computer.

A computer or node with 2 processors is called a dual core machine or node. A computer or node with 4 processors is called a quad-core machine or node. Processors in a multicore machine or cluster can use memory in different ways. Some use a shared memory architecture. In this case, all processors can access all the memory because it is completely shared between all processors. Some may have distributed memory where each processor or node has access only to its own memory. Finally there are hybrid machines like the K machine. Each of the 68,544 CPUs has its own distributed memory. But inside each CPUs memory is a system of 8 cores that share memory. Historically supercomputers used vector-based architecture, but this created a niche market since codes for such machines could not simply be run on non-vector based machines like a desktop computer. Today laptop computers are very similar to supercomputers because many supercomputers use bus-based architecture which is a modified architecture allowing a desktop computer to run more than one processor like a quad core Linux machine (13). Because most CFD work is done on clusters or supercomputers, CFD programmers often learn parallel programming, a type of computer programming with instructions or directives for communication and transmission of information between processors/cores or nodes on a supercomputer or cluster. Parallel programming is especially important because the purpose of supercomputing is to divide the large problem into smaller pieces given to each node to process. However, in order to solve the larger problem, the nodes must communicate especially and specifically about border regions of the partitioned mesh.

Once the mesh unit and resolution are chosen in the pre-processing stage, the mesh is partitioned and a piece of the mesh is given to each processor or node. However, sometimes researchers will re-order the mesh for large problems. Inefficient mesh partitioning contributes increasing costs of calculations and time for larger and larger problems. To facilitate calculations, mesh re-ordering schemes seek to minimize communication. Communication is minimized most when each processor or node manages a contiguous portion of the mesh. In this case, each processor only shares geometric nodes on the borders of mesh portions with processors that work on neighboring portions of the mesh. Imagine the opposite situation in which mesh elements are randomly distributed. A processor might have to communicate with 8 other processors if 8 of the bordering elements lie on 8 distinct processors! In CFD work, communication can take more than 50% of the computational time depending on the specific problem and its size. So it is very important to minimize this as much as possible, to leave more computational work for actual mathematical computations.

Examples of mesh partitioning and re-ordering methods include MATLAB's MESHPART (14), METIS (15), and PARMETIS (16) (17). Ordering the mesh so that each CPU has access to cells or units that are connected to each other is important. We also re-order and partition to maintain proper load balance so that no CPU has more work than any other. All of these choices—mesh unit geometry, mesh resolution, partitioning, and order can affect the ability of each processor to solve the resulting algebraic system of equations. Therefore, researchers work on the parallelism of such computer codes. There are many parallel programming language directives such as OpenMP (18), and MPI (19) and languages such as Manticore (20) and NESL (21).

After the mesh partitioning and re-ordering scheme is chosen, the next step a CFD engineer takes is choosing a discretization method. The mesh partitioning lies mostly in the realm of computer science, but this next step of discretization lies in the area of computational and applied mathematics. The mathematical model and governing equation has been chosen, but the CFD scientists or engineer must choose how to discretize the solution of this model over the entire domain.

Numerical Discretization Methods

In computational and applied mathematics, there are different numerical discretization methods (22). Three popular methods are finite difference, finite volume, and finite volume methods. Each of these classes of methods contains many variations usually specific to an application area. Other methods include boundary element methods, higher-resolution methods, and meshless methods like spectral methods.

Finite Difference, Finite Volume, and Finite Element Methods

The finite difference method is probably the oldest of the main three (23). It lends itself quite well to orderly and structured geometries. It is not used as commonly as the finite element method or finite volume method, probably due to the geometric limitations on applications. Still, there are modern finite difference codes that employ overlapping grids and embedded boundaries allowing the use of the finite difference method for difficult or irregular geometries. However, it is the easiest method to code and is often taught first in courses that teach numerical discretization methods.

The finite volume method is a method in which the governing partial differential equation is solved over smaller finite control volumes (24). Since the governing equations are cast in a conservative manner over each control volume, the fluxes across the volumes are conserved In terms of tests, applications, validation, and literature, the most robust of all methods is the finite element method (25). The finite element method is a type of residual method in which a residual equation is weighted and integrated over the domain. Since the domain is broken into many elements, this integration actually takes places over each element in the mesh. The finite element method requires more memory than the finite volume method but is also more stable than the finite volume method.

There are other methods as well. Boundary elements methods include methods in which the boundary is meshed into separate sub-elements (26) (27). In 3D, a boundary element method domain would be a 2D surface. In 2D, a boundary element method domain would be represented as a 1D surface or edge. There are also immersed boundary methods to deal with situation in which elastic structures interact with fluid flows (28). There are numerous other discretization methods. Each of the above discretization methods can be used with functions of varying order. Every increase in the order of the functions used in the discretization method (from linear to quadratic, quadratic to cubic, from cubic to quartic, etc.) carries increased computational costs since higher order functions have more terms and more coefficients. In fact, higher order functions require more sample points or interpolation points to properly resolve them. Accuracy comes at a price. Still, such accuracy is sometimes warranted in cases where there are sharp gradients or shocks in velocity, pressure, density, or temperature, for instance. In these cases, some of the previous discretization schemes fail and introduce what we call "spurious oscillations" which are not actually physical but a byproduct of computational approximation. In these cases, CFD engineers may choose to use higher-order discretization methods or shock capturing methods such as Total Variation Diminishing (TVD) schemes (29), Essentially NonOscillating (ENO) schemes (30), and the Piecewise Parabolic Method (PPM) (31).

Spurious oscillations are an example of discretization errors mentioned earlier. They are also called numerical errors because they are not physical. Specifically, such numerical error resulting in spurious oscillations is often called dispersive error or dispersion. Truncation error is the type of numerical error resulting from the difference between the partial differential equation and the finite equation that we actually code.

Additionally, with CFD we can experience a third error—computer error. For example, for one calculation or one floating point operation, computer roundoff is usually negligible. However, when doing repeated calculations over and over in simulations dealing with billions of nodes at which we solve for multiple unknowns at each node, computer roundoff error can build.

Turbulence Models

Another area of CFD modeling is turbulence modeling. Turbulence modeling is a difficult yet still potentially fruitful area of CFD research because of the computational difficulties it poses. Turbulence is a complicated phenomenon that occurs over a wide range of time scales and length scales. This is where the trouble lies making it impossible to fully resolve turbulent phenomena when using most approaches. Because of the wide ranges of time periods over which turbulent periodicity manifests and the wide range of length scales on which turbulence acts, most CFD engineers and scientist choose to resolve a certain length scale and time scale range of turbulence and model the rest. In this case resolving turbulence means that we actually compute and calculate turbulent quantities like vorticity and velocities and pressures in turbulent regions. Modeling turbulence means that we add an expression into our equation, the effect of which is approximately to create the effect of fully resolved turbulence on our unknown values such as velocity and pressure.

As one increases the range of time scales and length scales over which the turbulence is resolved, one must increase the refinement of the simulation both in length (refinement of the mesh) and in time (temporal refinement—the size of the time steps). This increased refinement increases the computational costs (the number of equations to be solved and the time it takes to compute the entire simulation). If one tends toward the other end of the model-resolve spectrum models all turbulence of all length and time scales, the computational costs decrease but one loses accuracy in the simulation. CFD researchers usually tend to resolve a range of turbulent length and time scales and model the rest. Usually the range is related to the range of interest of the flow simulation. For instance, if someone is simulating gas dynamics in the inner-ballistics of a particular weapon, it would be best to resolve turbulence on the length range of the chamber inside the weapon and model anything much larger and much smaller. Likewise, any turbulence that occurs over a longer period than the time it takes to fire the weapon would have smaller effects on the simulation; it would be best to model turbulence on longer time scales and much shorter time scales. The most expensive turbulence scheme is Direct Numerical Simulation (DNS) (32). In DNS, all length scales of the turbulence are resolved and little or no modeling is done. It is not used in cases of extremely complex geometries which can create prohibitive expense in resolving the turbulence especially in special geometrically complex portions of the domain. Less expensive than DNS, the Large Eddy Simulation (LES) resolves turbulence on large length scales as the name suggests (33).

A model is used to represent sub-grid scale effects of turbulence. The computational cost of turbulence at small length scales is reduced through modeling. Reynolds-averaged Navier-Stokes (RANS) is the oldest approach to turbulence modeling and it is cheaper than LES. It involves solving a version of the transport equations with new Reynolds stresses introduced. This addition brings a 2nd order tensor of unknowns to be solved as well. Examples of RANS methods are K-ε methods (34), Mixing Length Model (35), and the Zero Equation model (35). The Detached-eddy simulation (DES) is a version of the RANS model in which portions of the grid use RANS turbulence modeling and portions of the grid (or mesh) use an LES model (36).

Since RANS is usually cheaper to implement than LES, DES is usually more expensive than using RANS throughout the entire domain or grid and usually cheaper than using LES throughout the entire grid. If the turbulent length scales fit within the grid dimensions or the particular portion of the grid is near a boundary or wall, a RANS model is used. However, when the turbulence length scale exceeds the maximum dimension of the grid, DES switches to an LES model. Care must be taken when creating a mesh over which DES will be used to model turbulence due to the switching between RANS and LES. Therefore thought must be given to proper refinement to minimize computation while maximizing accuracy (especially refinement near walls). DES itself does not utilize zonal functions; there is still one smooth function used across the entire domain regardless of the use of RANS or LES in certain regions. There are many more turbulence models including the coherent vortex simulation which separates the flow field turbulence into a coherent part and a background noise part, somewhat similar to LES (37). Many of the contributions today are coming through different versions of RANS models.

Linear Algebraic Equation System

Once the mathematical equation has been chosen in the pre-processing stage, the mesh has been partitioned and distributed, and the numerical discretization is chosen and coded, the last part of the computer program is to solve the resulting algebraic system for the unknowns. Remember in CFD we are calculating the unknowns (velocity and pressure, for instance) at all nodes in the domain. The algebraic system usually looks like Ax = b. Usually with many steady problems, Ax is a linear equation system. Therefore we can just invert the matrix A to find the vector x of unknown values. The problem is that for large computations, for instance a computation involving 500 million nodes, inverting the matrix A takes too long. Remember that the number of nodes does not necessarily equal the number of unknowns. If you're calculating 5 unknowns at every node, than you must multiply the number of nodes by the number of unknowns per node to get the value of actual number of unknowns and the length of the unknown vector x. So for such large problems, we solve the equation system iteratively. These systems are solved iteratively with iterations such as Newton or the Picard iteration. In many cases with unsteady flows, Ax represents a system of ordinary or partial differential equations. In this case, CFD researchers must choose either an implicit method or an explicit method to deal with the time integration.

An explicit method calculates the solution at a later time based on the current solution. Another way to rephrase that is that it calculates the current solution based on an earlier solution in time. An implicit method calculates the solution at a later time based on both the solution at a later time and the current solution. Rephrasing that, an implicit method calculates the current solution based on both the current solution and the solution at an earlier time. Once a method is chosen and the time derivatives have been expanded using an implicit or explicit method, Ax is usually a nonlinear algebraic system. As stated earlier, for large problems, it may be too time-consuming or too costly (computationally) to compute this directly by inverting A. Usually, then, the linear system is solved iteratively as well (38).

One must then choose which type of iterative solver to employ to find the unknown vector. Depending on the characteristics of this equation system, an appropriate solver is chosen and employed to solve the system. The major implication of all computational research in this area is that there is no one perfect iterative solver for every situation. Rather there are solvers that are better for certain situations and worse for others. When operating and computing with no information about the equation system, there are, however, general solvers that are quite robust at solving many types of problems though may not be the fastest to facilitate the specific problem one may be working on at the moment. A popular class of iterative solvers for linear systems includes Krylov subspace methods of which GMRES appears to be the most general and robust (38) (39).

Its robustness makes it a great choice for a general solver especially when a researcher has no specific information about the equation system she is asked to solve. GMRES minimizes residuals over successively larger subspaces in an effort to find a solution. Still, in recent years, the Multigrid method has shown better optimal performance than GMRES for the same problems, motivating many to use this method in place of GMRES. The Multigrid method minimizes all frequency components of the residual equally providing an advantage over conventional solvers that tend to minimize high-frequency components of the residual over low-frequency components of the residual. Operating on different scales, the Multigrid method completes in a mesh-independent number of iterations (40) (41).

CFD experts also employ methods to simplify the ability of a solver to solve an algebraic equation system by pre-conditioning the matrix. The point of preconditioning is to transform the matrix closer to the identity matrix which is the easiest matrix to invert. Though we rarely invert matrices for such large problems, matrices that are invertible or closer to being invertible are also easier to solve iteratively. Preconditioning is an art, involving the correct choice of preconditioner according to the type of linear system (just as in choosing the correct solver or iterative method). Examples include lumped preconditioners, diagonal preconditioners, incomplete LU, ConjugateGradient, Cholesky or block preconditioners such as block LU or Schwarz (38) (42) (43). The best preconditioners are those that take advantage of the natural structure of the algebraic system in order to facilitate the transformation of A to a matrix closer to the identity matrix, thereby facilitating the solving of the system. The condition number roughly measures at what rate the solution x will change with respect to a change in b. Low conditions numbers imply more ease in inverting and solving the system. Characteristics such as condition number and even definite-ness of systems become important in choosing a preconditioner.

Libraries

Many CFD programs require thousands of lines of code for one specific simulation. To avoid writing a different program for every application, researchers often write general programs and software packages that can be used for a variety of situations. This generality greatly increases the lines of code even more. To help, there are many CFD and computational solid dynamics (CSD) libraries and applied mathematics libraries for the solving of such differential equations. They include code libraries such as OFELI (44), GETFEM (45), OOFEM (46), deal.ii (47), fdtl (48), RSL (49), MOUSE (50), OpenFOAM (51), etc. Sometimes a CFD engineer may use a package like ANSYS Fluent (52) to do a simulation, and sometimes a CFD researcher may find that the software does not give him the freedom to do what he would like to do (usually for very specific research applications). In these cases, he can write his own code or piece code together using these libraries. The most general Finite Element Library is deal.ii which is written in C++. It contains great support but currently does not support triangles or tetrahedral.

Moving Problems

Moving problems are another class of CFD challenges. One can either use a body-fitted mesh approach (an approach where the mesh fits around the solid body of interest), an overlapping mesh method, or a meshless method. Body-fitted mesh methods must actual move the mesh elements or cells. Overlapping methods have the option of moving the mesh but capture the movement of interest in the overlapping regions. Meshless methods avoid the need to move the mesh as moving the mesh has the ability to introduce error. Usually when a mesh is moving the aspect ratio of the mesh elements must be checked. If the ratio goes beyond some limit, the mesh must be reordered and remade, and the solution from the old mesh must be projected to the new mesh in order to continue the simulation. Each of those steps has the ability to introduce error. Therefore it is best to limit mesh distortion or concentrate it in elements that have the ability to absorb the deformation without reaching the aspect ratio limit for the computation (7) (53). There are also numerous methods that avoid meshes. Due to the problems and errors that poor refinement and mesh distortion introduce, some people avoid mesh methods altogether. Some meshless methods (54) include spectral methods (55), the vortex method (meshless method for turbulent flows), and particle dynamics methods (56).

POST-PROCESSING

This is the final stage when the computations are complete. This involves taking the output files full of velocity, pressure, density, and similar. information at each node for each time step and displaying the information visually, hopefully with color. Many CFD engineers also animate the results so as to better show what happens in time or to follow a fluid particle or a streamline.

Engineering

Up to this point, all of the work we have spoken about deals with the computer science/electrical engineering part or the computational and applied mathematics part of CFD. The mechanical and chemical engineering part of CFD comes in the post-processing and analysis.

The interpretation of the visualized results is also another point at which error can be introduced. CFD scientists and engineers must therefore take care in not overextending the analysis and simultaneously not missing important implications or conclusions that can be drawn from it. Engineers also make sure to corroborate the results with physical experiments and theoretical analysis. Additionally, the choice of problem and the motivation can come from the engineering side of CFD at the very beginning before any preprocessing work. Finally, the engineering analysis of the phenomenon directs the feedback loop of the research work. At this point the CFD worker must decide what values to change, whether it should be run again, what parameters should be maintained at current values, etc. He must decide which parameters' effect should be tested. He must determine if the results make sense and how to properly communicate those results to other engineers, scientists, scientists outside of the field, lay persons, and policy makers.

CFD work has taken exciting directions and comes to bear in many ways in society. When the fluid is water, CFD workers study hydrodynamics. When the fluid is air, they study aerodynamics. When it is air systems, they study meteorology and climate change. When it is the expanding universe, astronomy and cosmology; blood, medicine; oil and fossil flows, geology and petroleum engineering. CFD workers work with zoologists studying the mechanics of bird flight, paleontologists studying fossils, and chemists studying mixing rates of various gases in chemical reactions and in cycles like the nitrogen cycle.

Because of the importance and presence of fluids everywhere, the interaction of fluids with structures throughout the real world, and the application of mechanics and chemical engineering everywhere, CFD remains important. Moreover, the computational tools CFD engineers use can be utilized to solve and help other computational fields. Since the advent and continual innovation of the computer, all scientific fields have become computational computational biology, computational chemistry, computational physics, etc. The same computational tools used in CFD can often be applied in other areas (predicting the weather involves a linear algebraic system for instance). This allows CFD engineers and scientists to move in and out of fields and bring their engineering analytical skills and critical reasoning to bear in other situations. CFD engineers and scientists are even used to make video games and animations look more physically realistic instead of simply artistic (57).

CLOSING

There are a host of other interesting areas in CFD such as two-phase or multi-phase flow (58), vorticity confinement techniques (similar to shock capturing methods) (59), probability density function methods (60), and fluid-structure interaction (FSI) (61). Other chapters in this book address those, so we will not talk specifically about them here. It is simply important to understand that as computing capability increases as computer scientists increase the clock speed of the microchip all the time, our ability to solve real-life problems increases. Problems involving two liquids or two phases occur all the time, and there are many instances of fluids interacting with deforming solids—in fact that is the general reality. So combining CFD with computational solid dynamics is an important partnership that lends itself well to solving the major challenges facing us in the 21st century.

REFERENCES

1. Phing, Anselm Ho Yen. Simulating Combustion Flow in a Rocket Chamber. Lulea: Lulea University of Technology, 2008.
2. A Study of Sharkskin and its Drag Reducing Mechanism. Freidmann, Elfriede, Portl, Julia and Richter, Thomas. Berlin: Springer-Verlag, 2010.
3. Three-Dimensional Aerodynamic Simulations of Jumping Paratroopers and Falling Cargo Payloads. Udoewa, Victor. 5, Reston, Virginia: AIAA, 2009, Vol. 46. 0021-8669.
4. Blood Flow in Arteries. Ku, David N. 1, Palo Alto, CA: Annual Reviews, 1997, Vol. 29.
5. Outflow boundary conditions for three-dimensional finite element modeling of blood flow and pressure in arteries. Vignon-Clementel, Irene E., et al., et al. 29-32, Stanford : Elsevier, 2006, Vol. 195.
6. Numerical investigation of transitional and weak turbulent flow past a sphere. Tomboulides, Ananias G. and Orszag, Steven A. Cambridge: Cambridge University Press, 2000, Vol. 416.
7. Mesh Generation and Update Techniques for 3D Aerodynamic Simulations. Udoewa, Victor. 7, s.l.: Wiley, 2009, Vol. 29.
8. Slip formulation for numerical simulations of jumping paratroopers. Udoewa, Victor. 7, Johannesburg: Academic Journals, 2009, Vol. 2.
9. Models and Finite Element Techniques for Blood Flow Simulation. Behr, Marek. s.l.: International Journal for Computational Fluid Dynamics, 2006, Vol. 20.
10. Sawley, Mark L. Numerical Simulation of the Flow around a Formula 1 Racing Car. EPFL Supercomputing Review.
11. Online. November 1, 1997.
12. Cited: July 13, 2011.. http://ditwww.epfl.ch/SIC/SA/publications/SCR97/scr9-page11.html.
13. Computational study of flow around a simplified car body. Guilmineau, Emmanuel. 6-7, Ottowa, Canada: Elevier, 2008, Vol. 96.
14. Top 500. Top 500 Supercomputing Sites. Top500.org.
15. Online. Top 500, June 1, 2011.
16. Cited: July 13, 2011.. www.top500.org/lists/2011/06.
17. Salisbury, David F. News Release. Stanford News Service.
18. Online. November 19, 1997.
19. Cited: July 13, 2011.. http://news.stanford.edu/pr/97/971119supercomp.html.
20. MESHPART a Matlab Mesh Partitioning and Graph Separator Toolbox. Cerfacs.
21. Online. Cerfacs, February 8, 2002.
22. Cited: July 13, 2011.. http://www.cerfacs.fr/algor/Softs/MESHPART/.
23. Karypis, George and Kumar, Vipin. METIS - Unstructured Graph Partitioning and Sparse Matrix Ordering System, Version 2.0. Minneapolis: s.n., 1995.
24. Karypis, George, Schloegel, Kirk and Kumar, Vipin. PARMETIS - Parallel Graph Partitioning and Sparse Matrix Ordering. Minneapolis: University of Minnesota, 2003.
25. Applying Parmetis To Structured Remeshing For Industrial CFD Applications. Laflamme, S., et al., et al. 1, s.l.: Spring: International Journal of High Performance Computing Applications, 2003, Vol. 17.
26. OpenMP. OpenMP News. OpenMP.
27. Online. OpenMP, July 9, 2011.
28. Cited: July 13, 2011.. http://openmp.org/wp/.
29. Pacheco, Peter S. Parallel Programming with MPI. San Francisco: Morgan Kaufmann Publishers, Inc., 1997.
30. Manticore: A heterogeneous parallel language. Fluet, Matthew, et al. Nice, France: DAMP 2007, ACM 2007, 2007.
31. NESL: A Parallel Programming Language. Computer Science at Carnegie Mellon University.
32. Online. Scandal Lab, August 1, 2005.
33. Cited: July 13, 2011.. http://www.cs.cmu.edu/~scandal/nesl.html.
34. Kaw, Autar and Kalu, E. Eric. Numerical Methods with Applications. s.l.: Lulu Selfpublished, 2008.
35. Oliver, Rubenkonig. The Finite Difference Method (FDM) - An introduction. Freiburg: Albert Ludwigs University of Freiburg, 2006.
36. Leveque, Randall. Finite Volume Methods for Hyperbolic Problems. Cambridge: Cambridge University Press, 2002.
37. Huebner, K. H., Thornton, E. A. and Byron, T. D. The Finite Element Method for Engineers. s.l.: Wiley Interscience, 1995.
38. Ang, W. T. A Beginner's Course in Boundary Element Methods. s.l.: Universal Publishers, 2007. 978-1581129748.
39. Beer, Gernot, Smith, Ian and Duenser, Christian. The Boundary Element Method with Programming: For Engineers and Scientists. s.l.: Springer, 2008. ISBN 978-3211715741.
40. The immersed boundary method. Peskin, C. S. s.l.: Acta Numerica, 2002, Vol. 11.
41. Wesseling, P. Principles of Computational Fluid Dynamics. s.l.: Springer-Verlag, 2001.
42. Uniformly High Order Accurate Essentially Non-Oscillatory Schemes III. Harten, A., et al., et al. s.l.: Journal of Computational Physics, 1987, Vol. 71.
43. The Piecewise parabolic Method (PPM) for Gasdynamical Simulations. Colella, P. and Woodward, P. s.l.: Journal of Computational Physics, 1984, Vol. 54.
44. Pope, S. B. Turbulent Flows. Cambridge : Cambridge University Press, 2000. ISBN 978- 0521598866.
45. Garnier, E., Adams, N. and Sagaut, P. Large eddy simulation for compressible flows. s.l.: Springer, 2009.

978-90-481-2818-1.

46. The Numerical Computation of Turbulent Flows. Launder, B. E. and Spalding, D. B. 2, s.l.: Computer Methods in Applied Mechanics and Engineering, 1974, Vol. 3.

47. Wilcox, David C. Turbulence Modeling for CFD (3 ed.). s.l.: DCW Industries, Inc., 2006. 978-1928729082.

48. Comments on the feasibility of LES for wing and on a hybrid RANS/LES approach. Spalart, P. R. Arlington, TX: 1st ASOSR CONERFENCE on DNS/LES, 1997.

49. Coherent Vortex Simulation (CVS), A Semi-Deterministic Turbulence Model Using Wavelets. Farge, Marie and Schneider, Kai. 4, s.l.: Flow Turbulence and Combustion, 2001, Vol. 66.

50. Saad, Y. Iterative Methods for Sparse Linear Systems, 2nd edition. s.l.: SIAM, 2003. ISBN 978- 0-89871-534-7.

51. GMRES: A generalized minimal residual algorithm for solving nonsymmetric linear systems. Saad, Y. and Schultz, M. H. s.l.: SIAM Journal of Scientific and Statistical Computing, 1986, Vol. 7.

52. Wienands, Roman and Joppich, Wolfgang. Practical Fourier analysis for multigrid methods. s.l.: CRC Press, 2005. 1584884924.

53. Trottenberg, U., Oosterlee, C. W. and Schüller, A. Multigrid. s.l.: Academic Press, 2001. ISBN 012701070X.

54. Trefethen, Lloyd N. and Demmel, James W. Numerical Linear Algebra. s.l.: SIAM, 1997. 0898713617.

55. Golub, Gene H. and Van Loan, Charles F. Matrix Computations 3rd Edition. s.l.: John Hopkins University Press, 1996. 0801854148.

56. OFELI. OFELI An Object Oriented Finite Element Library. OFELI.

57. Online. OFELI, June 1, 2011.

58. Cited: July 13, 2011.. http://www.ofeli.net/.

59. Renard, Yves; Pommier, Julien. GETFEM++. GNA.

60. Online. GETFEM, June 1, 2010.

61. Cited: July 13, 2011.. http://download.gna.org/getfem/html/homepage/.

62. OOFEM. OOFEM - free object oriented finite element solver. OOFEM.

63. Online. OOFEM, March 29, 2011.

64. Cited: July 13, 2011.. http://www.oofem.org/en/oofem.html.

65. Dealii. Dealii Homepage. Dealii.

66. Online. Dealii, January 9, 2011.

67. Cited: July 13, 2011.. http://www.dealii.org/.

68. Google. FDTL - Finite Difference Template Library. FDTL.

69. Online. Google, June 2, 2011.

70. Cited: July 13, 2011.. http://code.google.com/p/fdtl/.

71. Michalakes, John. A Runtime System Library for Parallel Finite Difference Models with Nesting. Argonne National Laboratory.

72. Online. May 18, 1995.

73. Cited: July 13, 2011.. http://www.mcs.anl.gov/~michalak/RSL/.

74. MOUSE team. MOUSE. USASK.

75. Online. University of Duisburg, January 7, 2010.

76. Cited: July 13, 2011.. http://homepage.usask.ca/~ijm451/finite/fe_resources/node563.html.

77. OpenFOAM. OpenFOAM - The Open Source CFD Toolbox. OpenFOAM.

78. Online. OpenFOAM, June 2, 2011.

79. Cited: July 13, 2011.. http://www.openfoam.com/.

80. Ansys. Features of Ansys Fluent. Ansys.

81. Online. Ansys, June 3, 2011.

82. Cited: July 13, 2011.. http://www.ansys.com/Products/Simulation+Technology/Fluid+Dynamics/ANSYS+FLUENT/Features.

83. Huang, Weizhang and Russell, Robert D. Adaptive Moving Mesh Methods. s.l.: Springer, 2011. 978-1-4419-7915-5.

84. Liu, G. R. Mesh Free Methods, 2nd ed. s.l.: CRC Press, 2009. 978-1-4200-8209-9.

85. Hesthaven, J., Gottlieb, S. and Gottlieb, D. Spectral methods for time-dependent problems. Cambridge: Cambridge University Press, 2007.

86. Belytschko, T. and Chen, J.S. Meshfree and particle methods. s.l.: John Wiley and Sons, Ltd., 2007. 0-470-84800-6.

87. Gourlay, Michael J. "Fluid Simulation for Video Games". s.l.: Intel Software Network, 2009.

88. Two phase flow in complex systems. Levy, Salomon. s.l.: Wiley, 1999.

89. Numerical Simulation of Vortical Flows Using Vorticity Confinement Coupled with Unstructured Adaptive Grid Refinement. Murayama, M. and Kato, T. 1, s.l.: Computational Fluid Dynamics Journal, 2001, Vol. 10.

90. Fox, Rodney. Computational methods for turbulent reacting flows. Cambridge: Cambridge University Press, 2003. 978-0-521-65049-6.

91. FOIST: Fluid–object interaction subcomputation technique. Udoewa, V. 9, s.l.: Wiley: Communications in Numerical Methods in Engineering, 2009, Vol. 25.

CHAPTER 2

AN INDUSTRIAL VIEW ON NUMERICAL SIMULATION FOR AIRCRAFT AERODYNAMIC DESIGN

Adel Abbas-Bayoumi[1] and Klaus Becker[2a]

[1] Aerodynamic Strategies, Airbus, Avd. John Lennon, Getafe, Spain

[2] Aerodynamic Strategies, Airbus, Airbusallee 1, 28199 Bremen, Germany

ABSTRACT

In Airbus view, one major objective for the aircraft industry is the reduction of aircraft development lead-time and the provision of robust solutions with highly improved quality. In that context it is important to exploit all opportunities provided by enhanced or new classes of numerical simulation tools, e.g. high fidelity multi-disciplinary Computational Fluid Dynamics (CFD) and powerful High Performance Computing (HPC) capabilities.

To help meet the challenge of superior product development it will finally be essential to numerically 'flight-test' a virtual aircraft with all its multi-disciplinary interactions in a computer environment and to compile all of the data required for the development and certification with guaranteed accuracy in a reduced time frame. Numerical simulation is foreseen to provide a tremendous increase in aircraft design efficiency and quality over the next decades. This concept is considered by Airbus as one of the long term main objectives for aircraft development.

Progress in HPC will essentially contribute to achieve this goal. Considerable changes of aircraft design processes and way of working will lead to significant reduction of development times while including more and more disciplines in the early phases of design activities in order to find an overall optimum aircraft design.

Aerodynamic Design deals with the development of outer shapes of an aircraft, optimizing for its performance, handling qualities and loads. A major ingredient to the design process is the numerical simulation of the external airflow. The capabilities to predict the flow not only near the design point but also under other challenging conditions in a given flight envelope is a prerequisite for optimization towards market requirements.

Since it began about 50 years ago, CFD has made important progress in terms of accuracy of the physical models, robustness and efficiency of the nonlinear solution algorithms and reliability of the overall prediction approach. This trend will continue over the next decades. In our view, along with the increasing capability to model and compute all major multi-disciplinary aspects of an aircraft, in the long term it will become possible to 'fly' and investigate the complete aircraft in the computer.

Currently numerical simulation provides good means to analyze the flow around the aircraft in detail, although the regime of flow separation onset up to maximum lift conditions is still not modelled accurately enough, nonlinearities and turbulence modelling for separated flows are still a major concern.

Citation: Adel Abbas-Bayoumi and Klaus Becker, "An industrial view on numerical simulation for aircraft aerodynamic design," DOI 10.1186/2190-5983-1-10.

It was not only the increase in HPC power that made more sophisticated Navier-Stokes solving enter the daily industrial design process. Better understanding and mathematical analysis of the system of Navier-Stokes equations led to more powerful algorithms, to more capable software and more comprehensive analysis of aircraft flows.

However, a lot work remains to be done. Next decade's goal will be to better exploit more accurate and efficient numerical formulations, advanced turbulence models and to achieve a fully flexible and automatic CFD capability that works in a fully adaptive manner, providing the best quality solution at minimum cost and time. This will lead to a complete change in the way future aircraft will be designed.

TODAY'S CFD IN AERODYNAMIC DESIGN

Today, the aircraft industry has the experience, best practices and up to date capabilities to conduct a lot of numerical simulation in its daily design and development work [1]. At the forefront, Aerodynamics is using a variety of CFD methods and tools, which essentially help to analyse global as well as local flow behaviour about simplified and complex aircraft configurations. Reynolds Averaged Navier-Stokes (RANS) methods including 2-equation turbulence modeling is the most widespread approach to tackle, with reasonable accuracy and best practices, even highly complex 3D take-off and landing configurations. Aerodynamic design as well as aerodynamic modelling of the aircraft is highly supported by these means.

In a wider sense, simulation is also approaching multiple interacting disciplines. Flexibility effects on aircraft aerodynamics and structural loads are in the direct scope of CFD simulations coupled to CSM (Computational Structural Mechanics) models. This area also extends to use flight control modules in order to simulate trimmed aircraft configurations or even full flight manoeuvres. This approach, however, suffers the same simulation drawbacks and requires very high computer resources.

The requirements on predictive capabilities have reached a level where full aircraft simulation is a must. Any judgment on design progress with respect to aircraft performance, handling qualities or loads can no longer be based on geometrical, physical or aerodynamic simplifications. Instead all potential interactions between aircraft components have to be taken into account. Previously favored linear superposition principles no longer yield the required accuracy and consistency of aerodynamic data. It has become necessary to account for full nonlinear effects, requiring the study of the aircraft globally, and not just as a sum of components. High fidelity RANS CFD has made a big step forward to help solve this prediction task, although the simulation of nonlinear flows and related turbulence modelling are still a major problem for accuracy and cost and considered as the main challenge for the future.

A further area where numerical simulation has already offered real benefit is design optimization. Although fast strategies to find the optimum for multi-disciplinary multipoint design in 3D are still under development, the aircraft industry already uses optimization algorithms for detailed design tasks. However, there is a need to further explore available optimization techniques since they represent a significant potential in enhancing design.

Figure 1 shows that CFD is used today on a wide variety of tasks in the aircraft development at Airbus. While essential external shape design activities are largely based on CFD there is a more moderate use only on topics dealing with increased local geometric complexity, thus requiring considerably more effort in the future. Limited, but growing, use of CFD can be found in areas that deal with highly complex geometries or need multi-disciplinary coupling, e.g. aero-thermics and aero-acoustics. Some examples may illustrate what has been achieved in the industrial context.

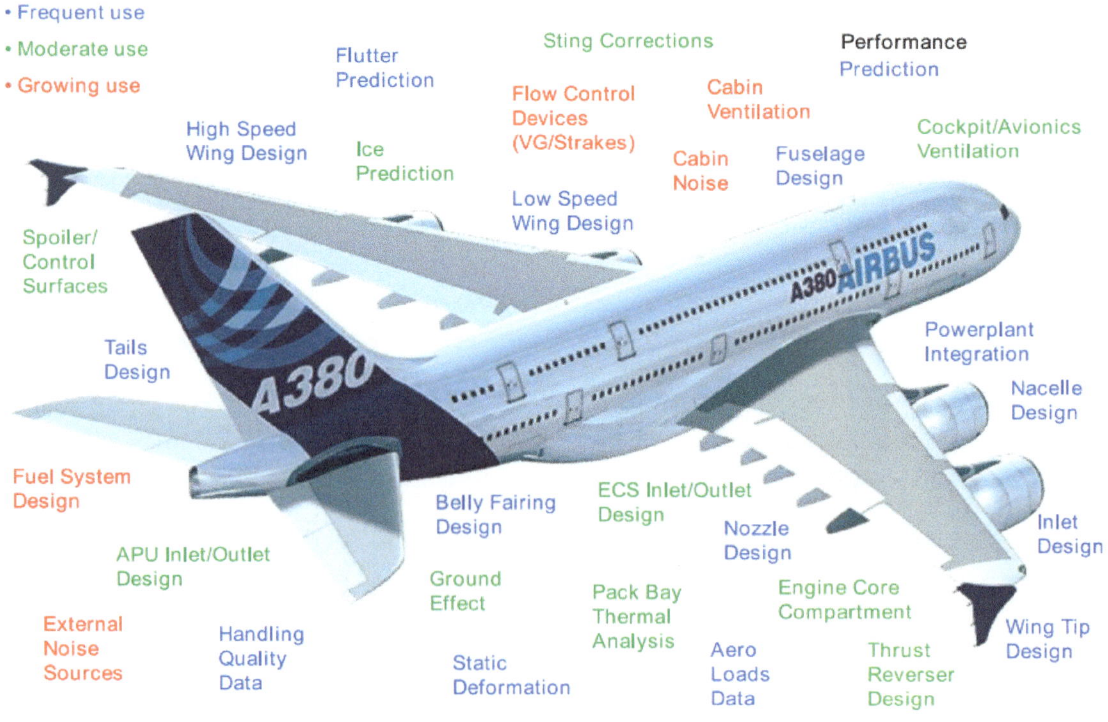

Figure 1: 2010 use of CFD at Airbus. CFD moved from an exploratory tool to a full flight physics production capability.

Prediction of Aerodynamic Performance

The aircraft design process is relying on continuously growing knowledge about the final product. Therefore detailed aerodynamic analysis is used to judge on progress with respect to aerodynamic and overall aircraft performance. CFD plays an increasing role in this business because it can deliver aerodynamic quantities with acceptable accuracy, at least in the cruise speed regime. This finally allows the optimization of the aircraft with respect to certain customer or market requirements like the typical mission and payload.

The average flight efficiency is measured as the fuel needed for a certain trip divided by the distance. This ratio can be expressed through some major aircraft parameters:

$$\frac{\text{TripFuel}}{\text{Distance}} \approx \frac{(\text{SFC} \cdot \text{Weight})}{\left(\text{Mach} \cdot \frac{\text{Lift}}{\text{Drag}}\right)}.$$

SFC is the specific fuel consumption of the engine. The cruise Mach number and the lift over drag ratio can be identified as aerodynamic contributions. Thus optimizing the fuel consumption aerodynamically means to design the aircraft for the highest L/D at a given cruise Mach number. Compared to the first A300 our today's aircraft are about 46% more efficient.

A typical example of what CFD can deliver is in providing the local air pressure distributions. Near the preferred point of aircraft operation, i.e. the design cruise Mach number and related cruise lift, the CFD results are very close to the experiment. Figure 2 shows a comparison of wing sectional pressure distributions between wind tunnel results (dotted line) and two different CFD codes (elsA [2] and TAU [3, 4]). Although both codes use different computational meshes (block-structured vs. hybrid-unstructured) there is hardly any difference between the two. Compared to the experimental values the pressure peaks, gradients, overall pressure levels, flow accelerations and decelerations are very well matched. Only the transonic shock position is not properly captured. However, the picture at the right shows that a major reason for this deviation, beside turbulence model effects, is the influence of the wind tunnel model support (sting). Including the sting in the simulation reduced the differences between wind tunnel and CFD results all over the wing. This effect was more pronounced on the outer wing, as shown on wing section 4 of Figure2.

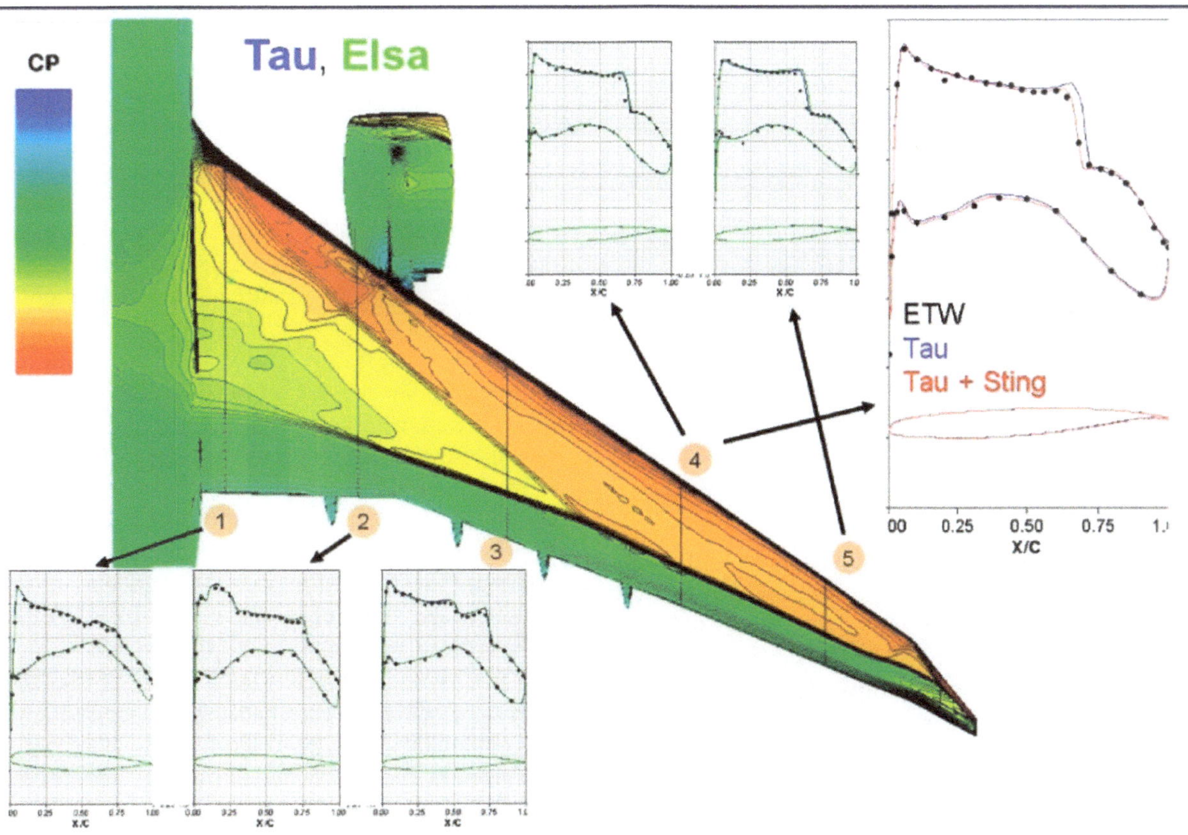

Figure 2: Aircraft cruise configuration transonic flow. Comparison of sectional wing pressure with experimental data; difference between codes and influence of model support.

Wind Tunnel Test Support

Wind tunnel experiments continue to be a major means to provide aerodynamic information. However, all specific modelling effects like for example, model size and simplification of geometrical details, wind tunnel walls and test support must be corrected in order to predict the free flight aircraft data. This correction process was formerly based on a number of corrections which were applied in a kind of linear summation. This follows the assumption that those effects are mainly independent and thus can be superimposed with only minor error. Today this is no longer feasible as nonlinearity has to be taken into account. This means that for a sufficiently accurate overall result more detailed local flow field corrections have to be applied. Specifically concerning the model support effect via CFD it is now possible to quantify even the influence on local flow (Figures 2 and 3). Figure 3[5] shows the local pressure differences between a CFD calculation for the aircraft mounted on a strut and the free aircraft. Obviously the strut has significant influence on the lower wing pressure distribution and even on the upper wing near the inboard leading edge.

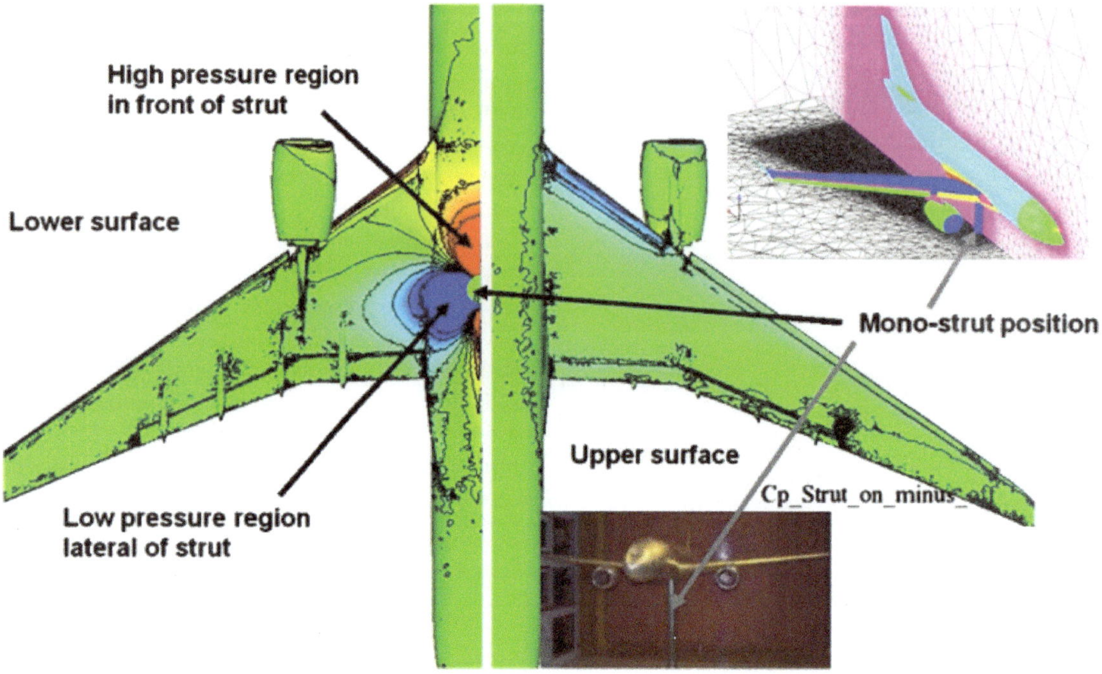

Figure 3: Mono-strut effect on aircraft flow in wind tunnel. Local strut effects on pressure - results of a CFD investigation.

MAIN ISSUES WITH NUMERICAL SIMULATION

There are a number of recent publications that provide a good overview on what numerical simulation has delivered to aircraft design and what challenges we are going to face, e.g. [6]. Some aspects are highlighted in the following sections.

Aircraft Models

Aircraft Design is based on principal models of flight, telling about the relations between basic geometry and configuration parameters used to define the wing, e.g., wing area, span, taper, bending, twist etc., and aerodynamic performance, aircraft manoeuvre and controllability qualities, and loads on the structure due to aerodynamic forces and moments. Such an aircraft, pre-defined according to the needs derived from target missions, has to be given optimized external shapes. Geometrical modelling is necessary to allow the designer to construct and modify aircraft components and shapes. Computer Aided Design (CAD) systems like the commercial CATIA software [7] essentially help to perform respective designs work. They provide mathematical descriptions of surfaces and ensure certain quality in terms of smoothness, curvature, and joints. All numerical simulation of the flow around the complete aircraft needs a watertight model with fitting components, properly prepared to enter the mesh generation process of CFD. The management of shapes during a design process and their assembly for a full aircraft are also tough tasks. We still lack automatic assembly due to imperfections of the CAD systems, non-conformal use of those tools by the designers and too strict requirements of the follow-on numerical mesh generation process and tools. Figure 4 shows the pylon intersection part of a wing in high lift configuration. Curved surfaces, 90 degree sharp corners, small gaps between components, and the complex flow characteristics to be represented make automatic grid generation a challenge and one of the subjects identified for further development. This is specifically true because we request a certain mesh and cell quality in order to diminish the numerical error being produced by the discretization process. It has been shown that for a given number of degrees of freedom regular hexahedral cells aligned with the principal flow provide the best results [8]. This explains the emphasis onto meshing based on these principles. Mesh adaptation based on error estimation and Chimera techniques are also considered as a way forward for improved mesh and solution quality.

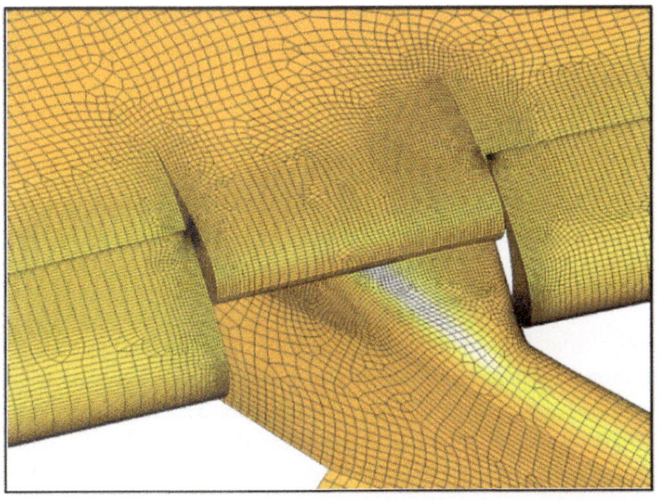

Figure 4: Surface mesh - pylon/wing intersection. SOLAR mesh for an aircraft in landing configuration.

Physical Models

Over a wide range of the flight envelope, i.e. the complete range of speed, longitudinal and lateral on-flow angles, flight levels and configuration variations, the flow about aircraft exhibits a smooth behavior. It can be predicted pretty well using the Navier Stokes equations [9] as basic physical model. However, physically relevant scales of the flow range from the order of kilometres (downstream wake effects) down to the order of microns (near wall turbulence) or even less. For a computational mesh resolving these scales would mean a mesh size of 10^9 points, which results in a nonlinear system of 10^{10} equations. Such a system is unrealistic to be solved on today's industrial computers, at least not at acceptable time and cost. This is also true for the semi-deterministic computations such as LES (Large Eddy Simulation) which on top need quite a big number of time steps to converge to sufficiently accurate statistics of turbulence. Therefore the smaller scale physical effects need to be modelled, e.g. by so-called turbulence models.

The Navier-Stokes equations comprise of 5 differential or integral equations, arising from the conservation laws of mass, momentum and energy. The open element in these equations is the so-called Reynolds stress tensor, which in 3 dimensions needs to correlate 9 entities - the Reynolds stresses - to the flow variables. By assuming an isotropic behavior of the fluid medium we end up with 6 quantities for which we seek additional equations. There are however no conservation relations known for a direct closure of the resulting system. Therefore these quantities are modelled using specific assumptions on the flow.

The development and calibration of such models depend on the flow phenomena that appear in the aircraft flight envelope. Figure 5 provides an overview of the flow conditions and effects that specifically appear at the borders of the envelope. Massively separated flows at high-lift low speed conditions, low local Mach number flows (low compressibility flow weekly coupled with the mean flow), strong nonlinearity at buffet boundary and shock boundary layer interaction and finally unsteady effects in separated flows are all situations where numerical simulation suffers low accuracy and very high cost and time.

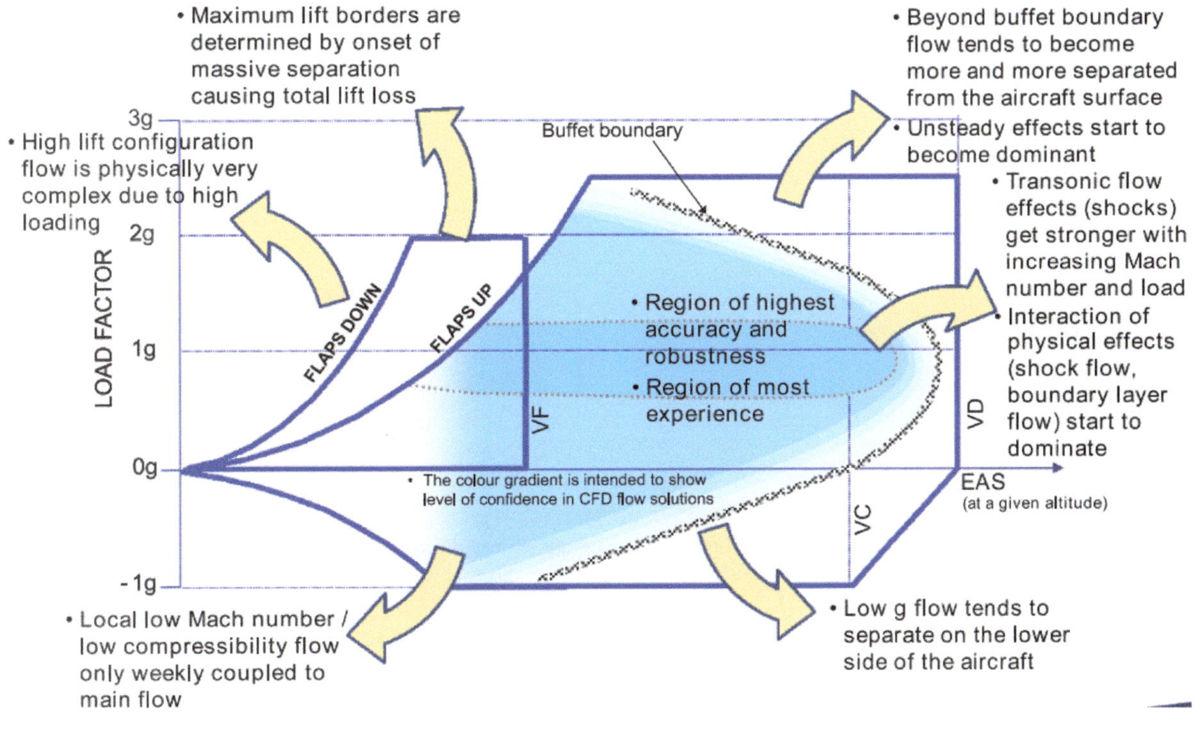

Figure 5: Flight envelope challenges on CFD. While CFD is widely developed for the cruise design regime it still faces essential challenges towards the borders of the flight envelope.

The effects of pressure, surface curvature and surface quality, viscosity and even temperature on local flow behaviour have to be taken into account. Increasing demands on accuracy have necessitated a move from 2-equation models to more sophisticated Reynolds-Stress models (RSM). The objective is to correctly predict with high accuracy all local flow phenomena for a wide range of flow parameters, mainly Mach, flow incidence and Reynolds numbers. For the sake of consistency and generality in prediction it is highly important to avoid any hard switch between turbulence models, depending on the flow conditions. The RSM model class seems to provide the best results for the whole flow regime.

With these considerations at hand, the flow stability is looked at as a separate phenomenon. So-called transition models have been developed to predict the location of transition from laminar to turbulent flow. These models more or less deal with the analysis of amplification factors of relevant modes natural to the flow. Once these factors have reached a certain threshold this is marked as transition location and the turbulence model can be activated.

A remaining difficulty is to predict the onset of flow separation. i.e. the location where the flow starts to detach from the surface. This physical effect is not fully understood up to now, however, designers need this information for reasons of safety, comfort, and handling qualities of the aircraft. Unfortunately the onset of separation is very sensitive to local properties of the surface (roughness, curvature, kinks, etc.) and the general properties of the flow (Reynolds number, energy content of the boundary layer, etc.). For the correct prediction of separated flow the turbulence model plays a major role and currently best practices on how to predict separation onset using such models is the preferred approach.

Even more important to aircraft design is the appearance of massive flow separation. As long as a flow separation is confined to a small area of the wing or tails only, it might be controllable and the pilot can counteract. However, when separation starts to cover large areas this normally drives the aircraft into a complete lift loss and thus a catastrophic situation. Thus massive separation prediction is directly coupled to the prediction of maximum lift properties of the aircraft, which is a limiting factor in take-off and landing performance. This prediction depends on the ability of the turbulence model to detect the local separation and to describe the extension of the separation up to the massive breakdown of the flow. Figure 6 provides a typical picture of such extension of separated flow areas on an aircraft model in the wind tunnel. For this high lift configuration at low Mach number, the local extension grows with increasing angle of attack.

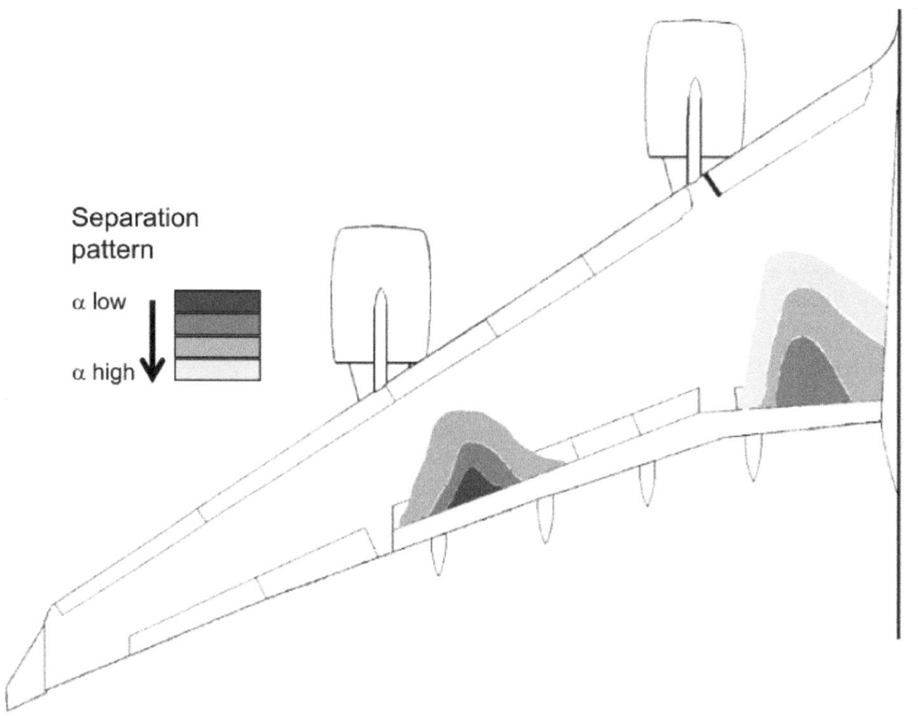

Figure 6: Typical successive growth of separation region on aircraft. Starting at the trailing edge of the wing separation increases with increasing angle of attack. Specifically the inboard region is prone to develop larger areas that finally extend to major parts of the wing surface.

Discretization and Numerical Approaches

CFD simulation in practical industrial application is mainly confined to maximum 2nd order approximations on computational meshes that are specifically dense in those areas of the flow field where some specific features need to be resolved. However, as we are not sure on the appearance of such phenomena conservative approach is employed with a high number of mesh points. But it is clear that this recipe does not solve the problem. Future solutions will hopefully provide means to automatically adapt the mesh and even the discretization accuracy to the local error information. Through the formulation of a so-called adjoint problem it is possible to compute gradient information by which the sensitivity of a quantity like lift, drag or moment against movement or placing of mesh points can be determined [10, 11, 12, 13].

Much progress has been achieved using modern iterative solution techniques. Effective preconditioning schemes are available in context with implicit and multi-grid iterative algorithms for the nonlinear equation system. Numerical dissipation is also more and more under control, thus minimizing the artificial or numerical effects in the CFD flow solutions. A next step will deal with mixed meshes, i.e. an integrated combination of structured and unstructured mesh discretizations. The essential element of this so-called HyperFlex approach [14] is to preserve the typical structured discretization accuracy in most of the flow field while allowing for the flexibility provided by unstructured meshes. This will free the need of an overall structured multi-block mesh for which there is no chance of full automation.

Further steps to come will be on higher order and fully mesh/order adaptive methods. However, a robust industrialized production code for complex applications based on these new methods is not foreseen before 2020. Many of the above mentioned topics are still open for further development.

Flight Envelope Prediction

Figure 5 shows a diagram that depicts the speed/load area which is the typical extended area of aircraft operation, i.e., the flight envelope. 'Flaps up' stands for the so-called cruise configuration, 'Flaps down' are for take-off or landing. The success of aircraft is based on the fact that air flow behaves in a rather controllable manner throughout a wide range of flight conditions. However, towards the border of the envelope some major changes happen, which need careful consideration. Two main aspects can be observed: shocks appear if Mach number increases beyond a certain geometry-dependent threshold, which makes the flow behavior non-linear. Heavy loading of the aircraft, i.e. large differences between the upper and lower surface pressures, has a major effect on the boundary layer, which could separate from the aircraft surface and thus radically lower the generated lift. While the transonic non-linearities (e.g. shock waves) are quite well understood, in their physics and their predictability, this is still not the case for flow separation (Figure 6).

Finally, aircraft design and data work requires a large amount of flow simulation data. However, this does not just mean a repetition by simulation of what has been done with the wind tunnel in the past. On the contrary, it is expected that new ways of sorting and organizing simulation processes will save quite a number of those computations. Sophisticated technologies like Design of Experiments, Variable Fidelity Methods, Reduced Order Methods and other techniques will be used to provide a full map of data at a minimum number of high fidelity simulations. For these techniques, error estimators and error propagation control will enable provision of results at guaranteed accuracy.

Flow is Unsteady

Looking into nature of flow there is nothing steady. It is only the small scales or high frequencies that are not really recognized by an aircraft and its passengers. This is a lucky point for aircraft flight overall, however, the more we go into detail with our analysis the more we detect that the non-deterministic unsteadiness of flow plays an essential role (Figure 7). Small scales are becoming more and more relevant, specifically in context of simulation of turbulence. But also larger scale unsteadiness poses a problem on numerical simulation. The numerical effort to solve the unsteady flow equations with certain accuracy in space and time is at least one order of magnitude higher than for the steady case.

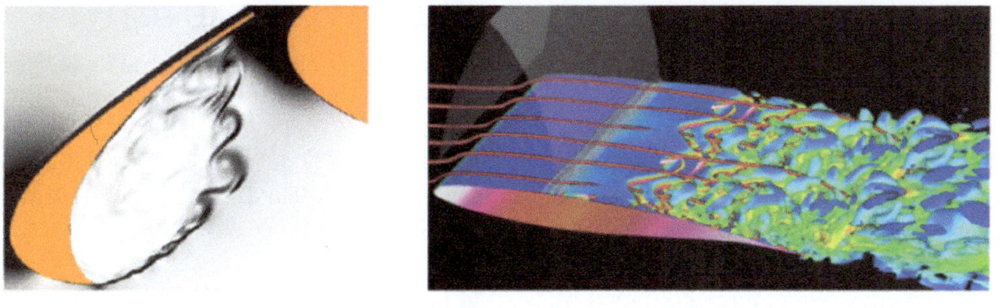

Figure 7: Slat cove and upper wing surface turbulent flow. Unsteady flow is present in many areas of the flow field. For aircraft development it is essential to know the scales of unsteady effects.

Seeking for higher accuracy of a flow solution via subsequent mesh refinement may lead us into the middle of the problem: Resolution of the flow down to very small scales in boundary layers with a steady flow solver probably provokes a non-converging iterative process, because the flow is inherently unsteady. Therefore new approaches have to be taken to allow automatic switching to an unsteady simulation if the steady solution does not converge. This is a topic for further investigation.

Multi-Disciplinary Interaction

Efficiency, reliability etc. of an aircraft is not only the result of a single discipline's work. Multiple interactions determine what the customer finally sees as the product performance. With increasing mono-disciplinary simulation accuracy it has become necessary now to also model and simulate all relevant interactions. A major link exists, for example, between the aircraft structure and aerodynamics. Structural deformation due to aerodynamic loads influences the aerodynamic efficiency. This circuit has to be converged until an equilibrium state is achieved. Numerically speaking we have to couple aerodynamic and structural simulation via a local feedback transmission scheme. This type of integrated process is more and more entering the routine simulation for static deformation. An example of static deformation on a complex configuration is depicted in Figure 8.

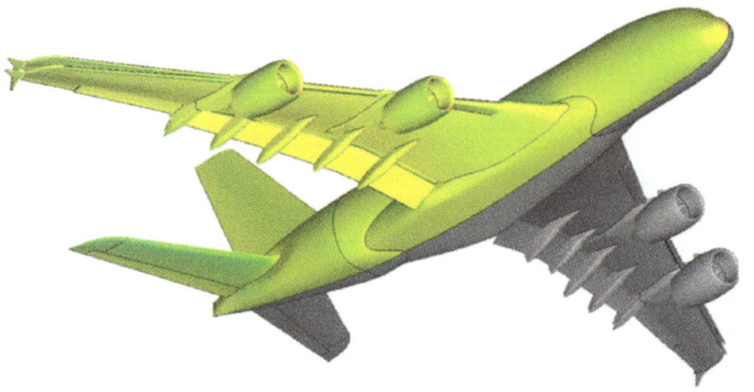

Figure 8: Static deformation on complex aircraft configuration. Demonstration of coupled aero-structures simulation capability on an A380 aircraft in high lift configuration. The picture shows the geometrical deformation.

More specific is the simulation of aero-elastic effects, like limit cycle oscillations, buffeting or flutter. Here people are interested in the time accurate behavior of the interacting mechanism which finally may lead to exceed the structural load limits of the aircraft which could be potentially catastrophic. This technology is still under development.

Management of Uncertainties

Even if we were able to do an absolutely exact numerical simulation of aircraft flight we will have to deal with problems: Weather conditions, air turbulence, payload distribution, fuel distribution, engine performance and other parameters may vary. In order to manage these type of uncertainties we need to know about the sensitivity of all of the aircraft coefficients to changing input parameters. This is quite a new mathematical challenge. Statistical and heuristic methods are being applied; however, the numerical effort can hardly be acceptable. Therefore more efficient approaches have to be developed that would allow a judgment on potential risks.

CONCLUSIONS

Flow simulation plays a major role in aerodynamic design and its predictive quality is crucially dependent upon both discretization techniques and the capabilities of turbulence modelling over a broad range of configurations and flow situations up to the borders of the flight envelope. Enhancing these capabilities, especially for critical regimes of unsteady and/or separated flows, is presently considered as one of the main objectives in the field. This will directly impact the quality of aircraft design and as a consequence in drag and weight reduction, which in turn lead to reduced fuel consumption and CO_2 emission. These are major objectives of the Green Aircraft Area.

With the clear tendency of the airframe industry to base their design cycles much more upon numerical simulation and to perform experiments with a significantly reduced frequency at a later point in the development cycle, it is of utmost importance to increase the reliability and the trust in numerical predictions.

It obvious that improved simulation capabilities will have a rather large impact on improving cost efficiency both with respect to aircraft development cost and aircraft operational cost. With advanced numerical simulation tools becoming less error-prone, this will not only improve the flow simulation alone, but also influence coupled computations, like design optimization, simulations of fluid-structure interaction or multi-disciplinary optimization. The quality of flow simulation has an even stronger impact in these fields where quantitative errors easily multiply. Thus the whole design chain will become not only more competitive, but also more productive, contributing to the reduction of the time-to-market of the products and to the reduction of aircraft development costs, leading in turn to stable or even reduced travel charges.

Airbus - together with major research partners and companies in the field - is working on the FuSim [15] initiative to develop Aerodynamics and Flight Physics towards a new paradigm of simulation. This treats all aspects of simulation (physics, turbulence modelling, mathematics, algorithms, hardware, software, computer science, information technology, man-machine interface, overall system, data handling, applications, etc.) which deliver essential contributions and provide their input and support to the superior cooperative effort. Enormous effort is needed to develop the simulation capabilities to the level required to be fully deployed for aircraft design. Major centres of expertise in numerical simulation in several countries are working together on this initiative with emphasis on specific aspects of simulation technology and application.

In this paper we have not tackled the extension to numerical optimization. This is another field of mathematical activities where we are looking for fast and comprehensive search algorithms for local and global optima of a variety of cost functions. This is a wider topic that will receive our attention in the coming decade.

ACKNOWLEDGEMENTS

The authors would like to thank all colleagues at Aerodynamics & Flight Physics who have helped in the preparation of this article, and a number of people from our major partners in the field of CFD development. Specific thanks go to Andreas Grimminger, Julien Delbove, Scott Shaw, Bernhard Eisfeld and Stefan Albensoeder.

AUTHORS' CONTRIBUTIONS

Adel Abbas is head of aerodynamic research and technology at Airbus.

Klaus Becker is head of Aerodynamic Strategies Sub-Domain at Airbus.

They have both worked in the field of numerical simulation methods for many years. They are now driving the strategy and future development of Airbus Aerodynamics numerical simulation capability. Both authors have participated in the paper preparation and drafted the manuscript.

REFERENCES

1. Kroll, N., Becker, K.: Numerical simulation of aircraft aerodynamics. In: Presentation Given at ISC07, Dresden, June 2007
2. Cambier, L., Veuillot, J.-P.: Status of the elsA CFD software for flow simulation and multidisciplinary

applications. AIAA Paper 2008-664, 46th AIAA Aerospace Science Meeting, Reno, USA (2008)

3. Schwamborn D, Gardner A, von Geyr H, Krumbein A, Lüdeke H, Stürmer A: Development of the TAU code for aerospace applications. 50th NAL International Conference on Aerospace Science and Technology 2008.
4. Gerhold T: Overview of the hybrid RANS code TAU. In Notes on Numerical Fluid Mechanics and Multi-Disciplinary Design. Edited by: Kroll N., Fassbender J.. Springer, Berlin; 2005:81–92.
5. Grimminger, A.: Airbus Internal Presentation PR0806223 - Issue 1, Bremen, April (2008)
6. Chalot F, Mallet M, Roge G: Review of recent developments and future challenges for the simulation-based design of aircraft. 27th Int. Congress of the Aeronautic Sciences (ICAS 2010) 2010.
7. www.3ds.com/catia
8. Baker T: Mesh generation: Art or science? Prog. Aerosp. Sci. 2005, 41: 29–63.
9. White FM: Viscous Fluid Flow. McGraw-Hill, New York; 1991.
10. Venditti, D.A.: Grid adaptation for functional outputs of compressible flow simulations. Dissertation, MIT, Boston, USA (2002)
11. Park, M.A.: Anisotropic output based adaptation with tetrahedral cut cells for compressible flows. Dissertation, MIT, Boston, USA (2008)
12. Dwight R: Heuristic a posteriori estimation of error due to dissipation in finite volume schemes and application to mesh adaptation. J. Comput. Phys. 2008, 227: 2845–2863.
13. Mani, K., Mavriplis, D.J.: Error estimation and adaptation for functional outputs in time-dependent flow problems. AIAA 2009-1496, USA (2009)
14. Becker, K.: HyperFlex CFD - Airbus approach to more accurate and flexible industrial CFD. Airbus internal presentation, Bremen (2009)
15. Klenner, J., Becker, K., Cross, M., Kroll, N.: Future simulation concept. Paper D07027256, CEAS Conference, Berlin (2007)

CHAPTER 3

DLR PROJECT DIGITAL-X: TOWARDS VIRTUAL AIRCRAFT DESIGN AND FLIGHT TESTING BASED ON HIGH-FIDELITY METHODS

N. Kroll[1], M. Abu-Zurayk[1], D. Dimitrov[4], T. Franz[1], T. Fu¨hrer[2], T. Gerhold[1], S. Go¨rtz[1], R. Heinrich[1], C. Ilic[1], J. Jepsen[3], J. Ja¨gersku¨pper[1], M. Kruse[1], A. Krumbein[1], S. Langer[1], D. Liu[1], R. Liepelt[4], L. Reimer[1], M. Ritter[4], A. Schwo¨ppe[1], J. Scherer[5], F. Spiering[1], R. Thormann[4], V. Togiti[1], D. Vollmer[1], J.-H. Wendisch[1]

[1] Institute of Aerodynamics and Flow Technology, German Aerospace Center (DLR), Lilienthalplatz 7, 38108 Braunschweig, Germany

[2] Institute of Composite Structures and Adaptive Systems, German Aerospace Center (DLR), Braunschweig, Germany

[3] Institute of Air Transportation Systems, German Aerospace Center (DLR), Hamburg, Germany

[4] Institute of Aeroelasticity, German Aerospace Center (DLR), Go¨ttingen, Germany

[5] Institute of Structures and Design, German Aerospace Center (DLR), Stuttgart, Germany

ABSTRACT

Numerical simulation is already an important cornerstone for aircraft design, although the application of highly accurate methods is mainly limited to the design point. To meet future technical, economic and social challenges in aviation, it is essential to simulate a real aircraft at an early stage, including all multidisciplinary interactions covering the entire flight envelope, and to have the ability to provide data with guaranteed accuracy required for development and certification. However, despite the considerable progress made there are still significant obstacles to be overcome in the development of numerical methods, physical modeling, and the integration of different aircraft disciplines for multidisciplinary analysis and optimization of realistic aircraft configurations. At DLR, these challenges are being addressed in the framework of the multidisciplinary project Digital-X (4/2012–12/2015). This paper provides an overview of the project objectives and presents first results on enhanced disciplinary methods in aerodynamics and structural analysis, the development of efficient reduced order methods for load analysis, the development of a multidisciplinary optimization process based on a multi-level/variable-fidelity approach, as well as the development and application of multidisciplinary methods for the analysis of maneuver loads.

INTRODUCTION

In recent years, the aeronautical industry has established numerical flow simulations as a key element in the aerodynamic design process, complementing wind tunnel and flight tests. The continuous development of physical models and numerical methods and the availability of increasingly powerful computers suggest using numerical simulations to a much greater extent than in the past, radically changing the way aircraft will be designed in the future. In addition to speeding up and improving the product design cycle, numerical simulation also provides the possibility to mathematically model all properties of the designed product with their interactions and to determine the behavior under realistic operating conditions. With suitable high-fidelity multidisciplinary simulation methods at hand, the flight characteristics of an aircraft can be determined through numerical computation and the flight envelope can be flown virtually before the real first flight is performed. The realization

Citation: Kroll, N., Abu-Zurayk, M., Dimitrov, D. et al. DLR Project Digital-X: Towards Virtual Aircraft Design and Flight Testing Based on High-Fidelity Methods CEAS Aeronaut J (2016) 7: 3. doi:10.1007/s13272-015-0179-7

of the vision of an aircraft performing its maiden flight in a virtual computer environment, denoted here by the synonym Digital-X, offers the reduction of development risks and in the medium and long-term significant cuts in development costs through stepwise certification.

In the context of the vast future challenges for the aircraft industry (Green aircraft, [1]), numerical simulation is considered to be a key technology for development of new or improvement of existing aircraft configurations. Thus, development and industrialization of advanced simulation methods and processes are being highly prioritized worldwide ([2–5]). At DLR, the multidisciplinary project Digital-X (04/2012–12/2015) represents a first significant building block for the progressive realization of the vision of digital aircraft design and virtual flight testing.

In this overview paper the main objectives of the project are explained in details and first results are presented from the various areas of work.

OBJECTIVES AND PROJECT SET-UP

The primary objective of the project Digital-X is the development and deployment of a flexible, parallel software platform for multidisciplinary analysis and optimization of aircraft and helicopters based on high-fidelity numerical methods for each discipline involved. This platform will provide a robust, integrated design process for aerodynamics and structural analysis. This will break up the predominantly sequential approach currently used in detail design and the full potential of multidisciplinary design will be made available. This new software platform will also make it possible to efficiently and reliably perform maneuver simulations throughout the entire flight envelope, and thus permit the determination of aerodynamic and aeroelastic data for evaluating the handling qualities based on high-fidelity numerical methods. The simulation capabilities of the platform will be demonstrated through application-oriented design tasks and maneuver scenarios. This multidisciplinary simulation and optimization system will enable DLR to evaluate innovative technologies for new aircraft configurations based on high-fidelity methods. Due to the multidisciplinary objectives of Digital-X several institutes of DLR are involved and they contribute their specific expertise in a wide range of disciplines. Participants are the Institute of Aerodynamics and Flow Technology, Institute of Aeroelasticity, Institute of Propulsion Technology, Institute of Structures and Design, Institute of Composite Structures and Adaptive Systems, Institute of Flight Systems, Institute of Air Transportation Systems, Institute of System Dynamics and Control and DLR Simulation and Software Technology. The overall coordination is carried out by the Institute of Aerodynamics and Flow Technology.

The project runs for 4 years and is scheduled to end in June 2016. Seven research fields are addressed as indicated in Fig. 1. Note that the methods and tools for structural analysis are further developed in the work package dealing with multidisciplinary optimization, while there is a dedicated work package for the development of aerodynamic methods.

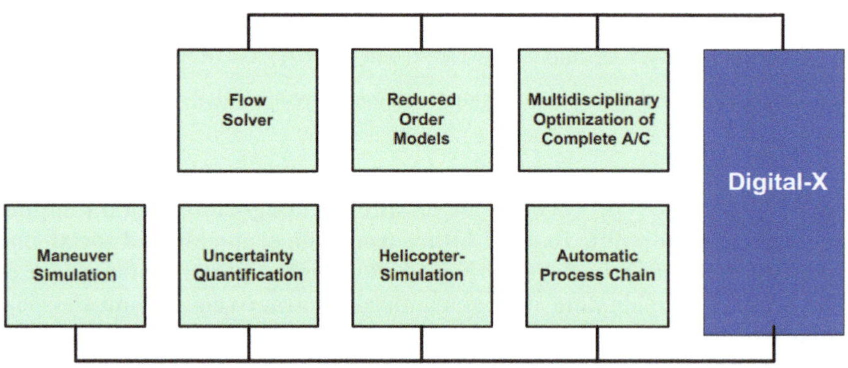

Figure 1: Main research fields of Digital-X.

In the frame of the Digital-X project DLR works in partnership with aircraft industry and selected universities. The associated partnership with Airbus allows on the one hand the consideration of operating conditions relevant to industrial applications of numerical simulation methods and processes at an early stage and, on the other hand, joint validation activities taking into account industrial experience and data sets. It was agreed that the Airbus large transport aircraft of the external Research Forum (XRF), the XRF-1, is used as a reference configuration for the developments in Digital-X. For this purpose global aircraft data, CAD geometry and structural models are being provided by Airbus. Specific knowledge of German universities in physical modeling as well as simulation and process chain developments are incorporated through a close link with collaborative research projects funded within the frame of the fourth aviation program of the German Federal Government.

CURRENT WORK AND RESULTS

Flow Solver

The unstructured DLR flow solver TAU routinely used in industry and research for aerodynamic applications is being further developed and improved in Digital-X for the required multidisciplinary flow simulations over the entire flight envelope. Furthermore, the development of the DLR 'next generation' CFD solver is being driven forward. In addition to consolidating existing methods and algorithms it will use innovative simulation approaches and provide the best possible utilization of future HPC systems.

Further Development of the DLR TAU Code

The key aspects of further developments in the hybrid TAU

code [6] are to improve physical modeling and to increase both robustness and efficiency of the solver algorithms.

Physical Modeling

Several lines of development in the area of modeling are being followed in Digital-X. One is to target improvements in RANS turbulence models, which in practice are the basis of most of the turbulent simulations. The focus is on the correct capture of moderate separated flows on curved surfaces which cause separation at wings and determine the size and angle of attack of maximum lift. Particular features available in the DLR TAU code are the Reynolds Stress Models (RSM) [7–9]. They represent the highest level of RANS modeling and describe essential physical phenomena, which cannot be predicted by classical turbulent viscosity models used in industry. In Fig. 2 computed polars for the wing/fuselage configuration of the NASA Common Research Model (CRM) are compared with experimental data [10]. Note, that lift is plotted in terms of idealized profile drag where the estimate of induced drag is subtracted from the drag coefficient, enabling the use of expanded drag scales. The figure shows that the Reynolds stress model results (SSG/LLR) clearly agree better with experimental data than results obtained with the single equation turbulence model of Spalart–Allmaras (SA). There is also a much lower dependency on mesh refinement. Efforts are currently being made to improve the numerical stability of the standard RSM in the TAU code for complex applications to ensure routine use under industrial conditions.

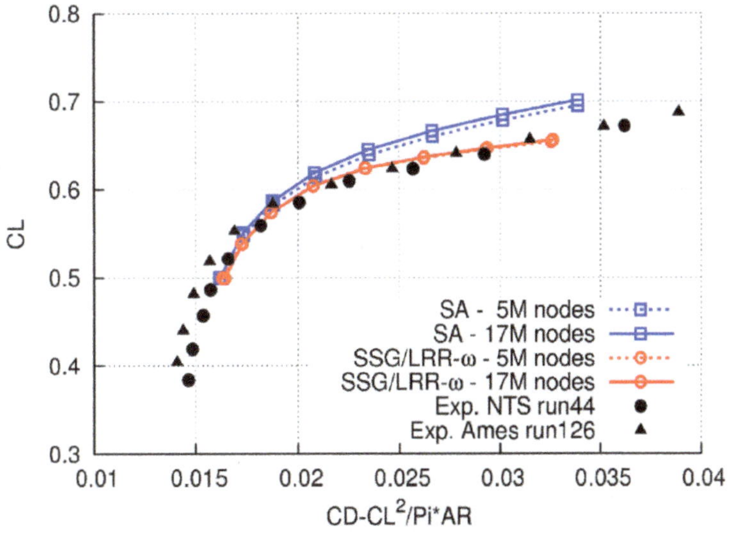

Figure 2: Comparison of polars for the CRM wing/fuselage configuration ($M = 0.85$, $Re = 5 \times 10^6$).

Techniques for scale resolving simulations (hybrid RANS/LES) have been expanded and improved as required to capture turbulent fluctuations in highly non-stationary processes, for example in the region downstream of components and attachments [11]. In particular for modeling weak flow separation on curved surfaces, a DES-based approach was developed using the thickness of the adjacent turbulent boundary layer and the separation point in the model to separate RANS and LES regions from each other as accurately as possible. Furthermore, the discretization scheme was adjusted so that LES results could be obtained with better quality.

For the simulation of transitional flows the range of applications for the automatic transition prediction was extended, which in its basic concept uses an e^N-method. As an alternative option a correlation-based transport equation transition model was implemented [12] and extended to the simulation of cross-flow instabilities [13]. Another line of development recently started was to couple this approach and the hybrid RANS/LES methods with the Reynolds stress models.

Improvements in Efficiency and Robustness

Implicit solution methods are superior to explicit algorithms through their increased robustness, which is accompanied by additional cost particularly in memory and CPU time. The implicit solver currently available in the DLR TAU code is based on the LU-SGS algorithm. This procedure is characterized by a comparatively low memory and computational effort, at the expense of significant simplifications in the model derivation and a simple iterative solution method to solve the associated linear system of equations. However, due to increased computing capacity a loss in both robustness and efficiency is frequently found with the LU-SGS method for current applications with increasing size and complexity.

Digital-X activities are focused on improvements to the LU-SGS method as a smoother for agglomerated multi-grid techniques and on the further development of implicit methods, which both improve the approximations in the derivatives and also integrate efficient solution methods for the resulting equations into the overall process [14]. The single equation turbulence model of Spalart and Allmaras was initially applied to improve the solver algorithms. Recommendations from [15] were incorporated in the TAU code and the LU-SGS adapted accordingly. For example the limit of the transport variable of the turbulence model was removed from the TAU code. In addition, within the multi-grid framework the smoothing properties of the LU-SGS in combination with various agglomeration techniques were investigated. It has been shown that the use of agglomeration techniques exploring specific features of structured grid regions in the boundary layer is advantageous. Furthermore, appropriate boundary conditions were implemented in the correction terms in the multi-grid and several smoothing steps were realized at each multi-grid level. These adjustments resulted in a significant increase in robustness and efficiency of the LU-SGS method in the TAU code.

As mentioned above improved preconditioning techniques were developed for implicit methods. These techniques are still based on the derivative computed by nearest neighbor information, but all derivative terms are consistently incorporated in the Jacobian matrix [16]. Furthermore, the symmetrical Gauss–Seidel algorithm is not terminated by a single iteration step, but several iterations are carried out to obtain better approximations to the solutions of the linear system of equations. To improve the convergence behavior for strongly anisotropic meshes, line iterative solution methods are applied.

The CRM wing/fuselage configuration from the 5th AIAA Drag Prediction Workshop was analyzed to demonstrate the algorithms developed [14]. A sequence of meshes was considered (Table 1) to assess the convergence behavior of the various methods. A reduction in residuals of density and of the turbulence equation was achieved up to machine accuracy for all meshes considered (Fig. 3). A CFL number of 1000 could be used for the newly implemented implicit method for all coarse meshes, although the achievable CFL number for the LU-SGS method varies between 2.5 and 25, depending on the mesh. On the finest mesh the CFL number of the newly implemented implicit method had to be reduced to 250.

Table 1: Mesh density for CRM wing/body-configuration

Level	No. of prisms	No. of tetrahedra	No. of points
L1	425,984	255,904	660,177
L2	1,437,696	8,626,176	2,204,089
L3	3,301,376	20,766,720	5,196,193
L4	11,261,952	69,728,256	17,441,905
L5	26,411,008	166,133,760	41,231,169

(a)

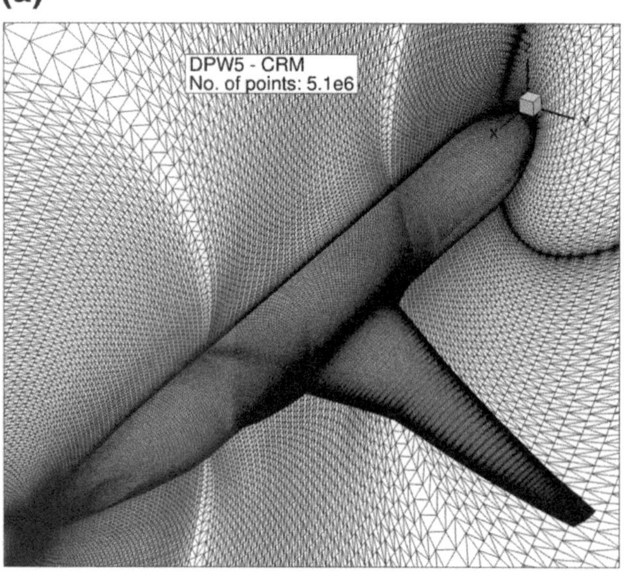

(b)

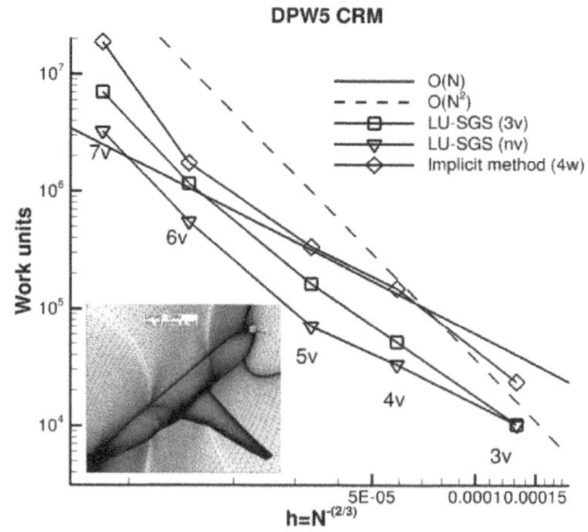

(c)

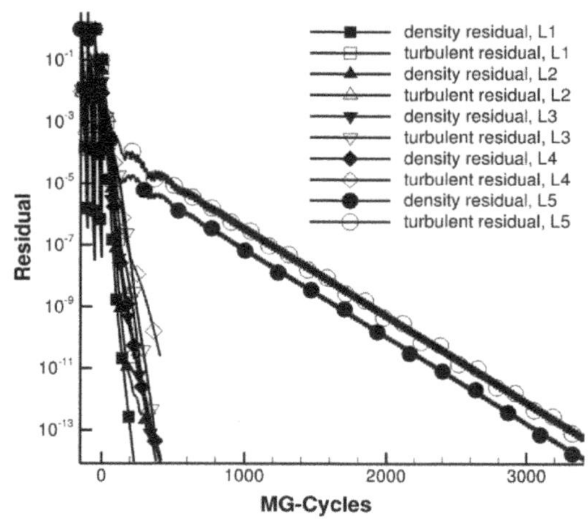

(d)

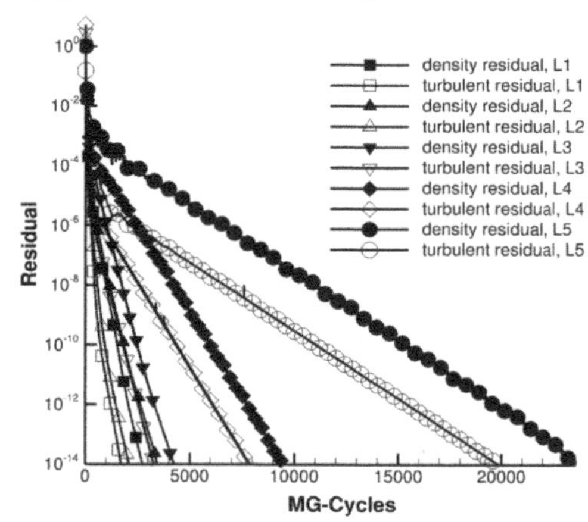

Figure 3: Convergence behavior of implicit methods for the CRM wing/body-configuration (M = 0.85, Re = 5 × 10^6): **a** mesh, **b** work units as function of mesh size, **c** convergence history for implicit methods, **d** convergence history for LU-SGS method.

The high CFL number demonstrates the increase in robustness for implicit methods. However, given the current state of development, a final conclusion could not be made regarding the efficiency of the newly integrated implicit method in the TAU code. This is the subject of further work.

Next Generation CFD Solver

This activity is concerned with the design and implementation of a flow solver of the next generation solver Flucs. The modular design of this software is constructed so that it can be used in highly parallel simulations for complex applications in the field of internal and external flows. The aim is to provide the basis for a consolidated flow solver using modern software techniques with high flexibility and high degree of innovation for a wide range of applications. This should make it possible to carry over the established models and methods of the DLR TAU code and combine them with new, innovative numerical approaches. An important aspect in developing the next generation solver Flucs is the efficient use of current and future parallel HPC systems.

Detailed specifications for Flucs were established based on an extensive survey of current and potential users of flow solvers in industrial and scientific fields. In addition to users, developers of current flow solvers were questioned to be able to make meaningful design decisions for the basic structure in many extension scenarios and thereby achieve high flexibility. After prioritization of the requirements, which implicitly defines a framework for the scope of work in Digital-X, a requirements specification was developed with concrete working points for implementation in a prototype. The focus here was on a common framework for two typical discretizations: second order finite volume and discontinuous Galerkin with variable order.

Not only future extensibility towards foreseeable developments, but also novel approaches are a core aspect of Flucs, but at the same time difficult to plan and control. Therefore, the development of a monolithic "all-embracing" flow solver was rejected, since the broad requirements cause potentially complex interactions and dependencies in various functions. Instead a "framework" (Fig. 4) was designed, whose data structures and functions serve as a basis for implementing lean modules, for example equations, discretizations and time integration procedures. The framework takes over many functions such as efficient implementation of loops, parallelization, support for algorithmic differentiation (AD) or the provision of required data.

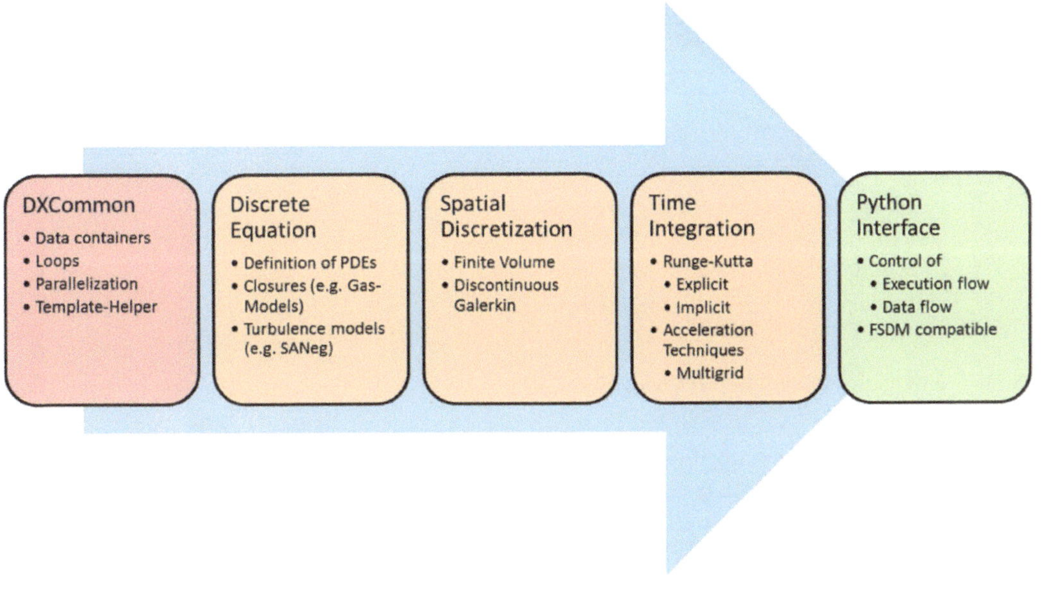

Figure 4: Modular structure of the next generation flow solver Flucs.

The highest control module is designed as compatible Python API for the simulation environment FlowSimulator (see Sect. 3.7.1) which can flexibly describe a wide variety of simulation scenarios, since flow solver applications take place increasingly in a multidisciplinary environment. C++11 was chosen as implementation language, which provides an object-oriented design (flexibility and modularity), together with a "close to the hardware" implementation of run-time critical program parts when required. An additional advantage is that certain features have only a small or no influence on the execution speed through the selective use of templates. Modern software development tools are used for the development, such as distributed version control (git), web-based code reviews (Gerrit) and continuous integration (Jenkins).

In the prototype to date a second order finite volume discretization and a higher order discontinuous Galerkin discretization were implemented on unstructured grids for the Euler equations with simple solution algorithms. The aim was to achieve the largest possible synergy between the two approaches. Next, the Flucs prototype will be extended to include additional functionality, for example RANS equations with one-equation turbulence model, coupling together several discretizations, SIMD and shared memory parallelization and improved implicit methods. These are not necessarily features that are the most important for users of flow solvers. On the other hand, they are aspects that potentially have a strong influence on existing software design.

Time for an evaluation phase is planned after completion of the prototype, which will assess the decisions made, the design and its implementation. A first release for selected target applications in internal and external aerodynamic flows should be available at the end of the project.

Parallelization

Modern HPC systems have various degrees of parallelization at different levels: a network-connected cluster consisting of multiple computing nodes, each of which comprises several CPUs, each of which again has available multiple processing cores. A flow solver must optimize the use of all possible parallelization levels for a high-performance simulation. Domain decomposition provides the basis for parallelization, by which the simulation is carried out on multiple computing nodes. Computation on these domains is synchronized through exchange of data at the domain boundaries over the network. For best possible performance it is important to have a good balance both in computational load and in the necessary communication between domains. In the DLR TAU code significant improvements have been achieved in domain partitioning through the integration of graph-based algorithms, compared with partitioning based on recursive bisection. The TAU code uses domain decomposition as the sole parallelization strategy, with one domain per computing core. Figure 5 shows that this one-level parallelization is not optimal even for current HPC systems.

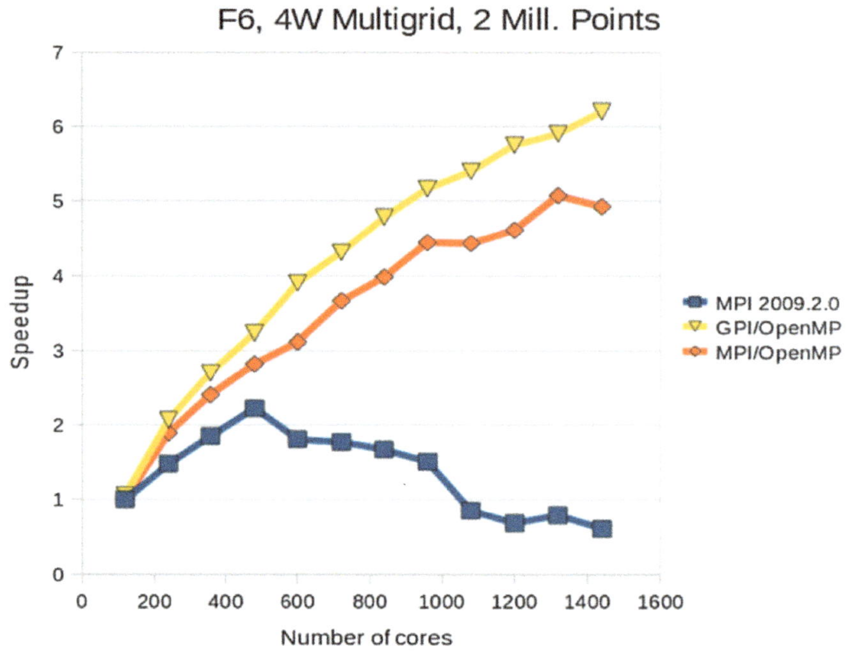

Figure 5: Comparison of different parallelization strategies.

For a very small mesh (2×10^6 discretization points) this one-level parallelization scales up to 480 processing cores before scaling breaks down at about 4000 discretization points per core due to increasing communication overhead as well as load imbalances (see Fig. 5, blue curve). A significantly improved scalability (red curve) is possible as a TAU code prototype [17] demonstrates which uses one domain per multicore CPU. Here a two-level parallelization is used, where the domains are further subdivided so that a domain is now processed in parallel by all cores of a CPU, without (explicit) communication. This considerably reduces the comparatively slow communication. The scaling can be further improved by overlapping communication with computation (yellow curve). These findings form the basis for design of multi-level parallelization in the next generation solver Flucs. In addition to two-level domain decomposition there is a 3rd level of parallelization, namely the use of single instruction multiple data (SIMD) for unstructured meshes, which is a particular challenge addressed in Digital-X to use current and future HPC systems efficiently for CFD simulations.

Reduced Order Aerodynamic Models for Loads Analyses

The loads analysis of an aircraft requires computation of thousands of parameter combinations. This includes variation of parameters describing the steady flow conditions in terms of Mach number, altitude and load factor, as well as parameters describing the flight dynamics, such as roll and pitch rate, and parameters defining the shape of the gust. Further, highly accurate predictions of the CFD code are required, in particular for transonic flow conditions. Their direct application for the entire flight envelope is, however, not practical due to the high computation times per calculation. Thus, reduced order models (ROM) are developed within the DLR Digital-X project to efficiently provide CFD-based aerodynamic data for loads analyses.

Parametric ROM Based on the Isomap Method

A parametric ROM for steady CFD data was developed based on isometric mapping (Isomap) [18], a nonlinear "manifold learning" (ML) method. It is assumed that the space of all CFD solutions forms a nonlinear manifold, which in turn constitutes a sub-manifold of RnRn of lower intrinsic dimensionality. This is a more elaborate approach compared to the "proper orthogonal decomposition" (POD) method [19], in which a linear subspace of RnRn is assumed. ML methods apply various approaches to identify the manifold geometry and to represent it in a low-dimensional Euclidean space. For this purpose Isomap uses the pairwise geodesic distances between previously generated CFD solutions at selected parameter combinations. After approximating these distances Isomap computes a data set of low-dimensional vectors in the so-called embedding space, whose pairwise Euclidean distances correspond to the approximated geodesic distances. A mapping from the embedding space back onto the manifold in the high-dimensional CFD solution space was developed to make predictions of CFD solutions at any point in the parameter space. When combined with an interpolation model between the parameter space and the embedding space, similar to POD with interpolation (POD + I, [20]), an Isomap-based ROM, called Isomap + I [21], is obtained.

Figure 6 provides a comparison of the predictions of Isomap + I and POD + I with the computed TAU reference solution for the pressure coefficient at three wing sections for the LANN wing [22] in inviscid flow. Isomap + I yields better predictions than a POD-based interpolation method, in particular for the location and magnitude of the shock wave at transonic flow conditions. The parameter space, defined by variations of the angle of attack, α and the Mach number, M, was previously sampled in the range [α × M] = [1°, 5°] × [0.76, 0.82] using a "Design of Experiment" (DoE) based on a Latin/Hypercube approach with 25 different α-Mach combinations. At these parameter combinations the corresponding CFD solutions were computed and used as input data for both ROMs. The required parameters for Isomap + I were determined automatically.

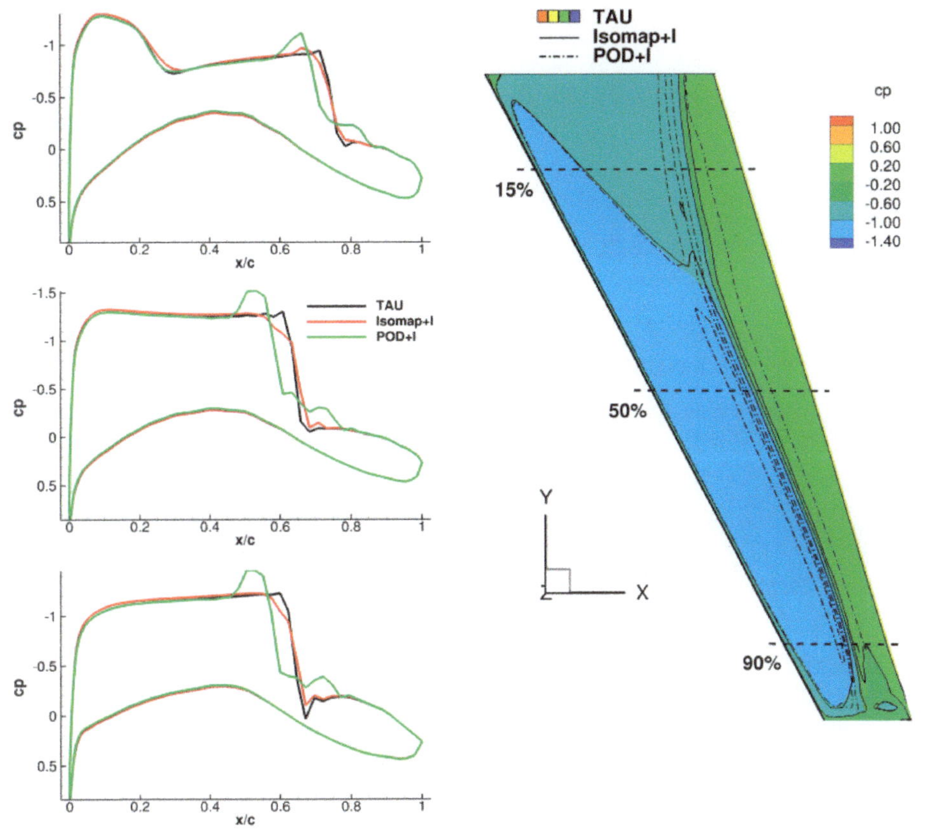

Figure 6: Steady Euler computation for the LANN wing compared to ROM predictions.

Correction to the Doublet Lattice Method

CFD solutions for selected parameter combinations have been used to correct the doublet lattice method (DLM), the standard method for calculating aerodynamic loads in aeroelasticity. The DLM solves the linear potential equations under the assumption of isentropic, inviscid flow, neglecting thickness effects. Thus, the method is valid only for subsonic flow conditions. Shocks and flow separation cannot be represented by the method and a correction is required in the transonic velocity range.

The DLM correction method Correcting Aerodynamic Matrices (CREAM) [23] was developed to improve predictions for unsteady aerodynamic loads. In addition to a quasi-steady correction (CREAM-0) an unsteady support point can be considered (CREAM-1). The more accurate solution (RANS method) and the faster approximate solution (DLM) are expanded as Taylor series with respect to the reduced frequency. The Taylor coefficients of the low-fidelity series are successively replaced by coefficients of the more accurate series. By further transposition the explicit Taylor series can be circumvented. In addition, it is assumed that the solution of the more exact RANS method can be represented by the AIC matrix (Aerodynamic Influence Coefficient) through multiplication by a correction matrix.

The additional correction term in CREAM-1 improves the phase component of the frequency response, thus enlarging the region of validity of the correction along the frequency axis. This is demonstrated in Fig. 7, using the example of a forced pitching motion. The calculated frequency response with DLM significantly underestimates the magnitude, with the phase response also shown incorrectly. CREAM-0 significantly improves the predictions of the magnitude. The correction also improves the phase, although it is not sufficient. The phase curve is corrected in CREAM-1 due to the additional frequency reference point, although the effect on the magnitude is fairly small. With increasing frequency, the deviation increases between the CREAM solutions and the nonlinear reference CFD solution. This can be explained because the CFD support points used in the corrections are at zero frequency, or close to zero, thus extrapolation to higher frequencies is represented by the DLM model only.

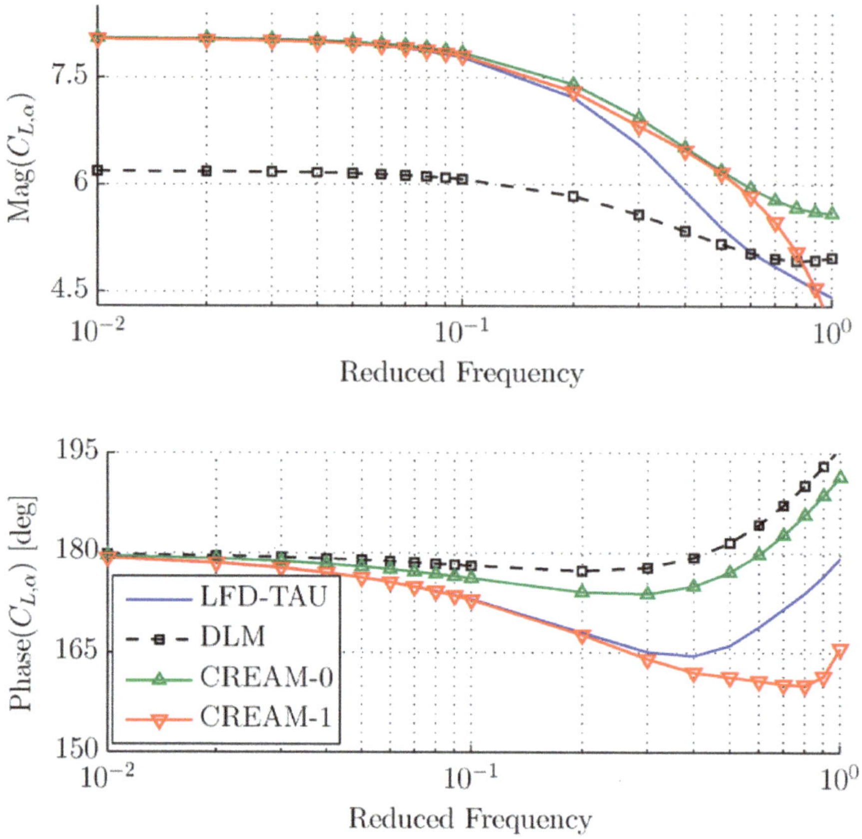

Figure 7: Frequency response dCL/dα for a pitching motion on the LANN wing, M = 0.82, α = 0.6°.

The method CREAM-0 was also used to calculate the aerodynamic response of "1-cos" gust loads. As a reference a gust load was simulated using the Field Velocity Method (FVM) [24] with the nonlinear CFD solver TAU. The gust load was calculated with both DLM and CREAM-0. The correction data were computed with the linearized TAU code in the frequency domain (LFD code, [25]) applying a rigid pitching motion at zero frequency. The effect of the correction can be clearly seen in Fig. 8. Compared to the nonlinear CFD solution, the DLM predicted amplitude is too low at these transonic flow conditions. In contrast, the result obtained by CREAM-0 agrees very well with the reference solution.

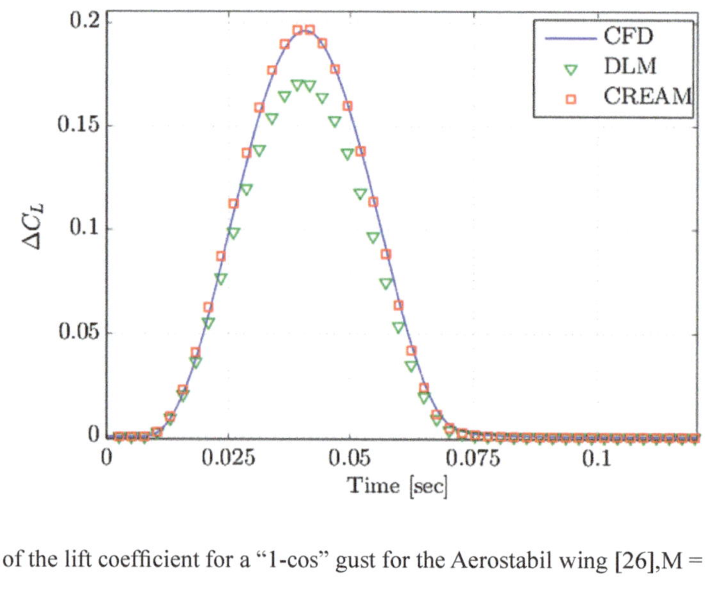

Figure 8: Aerodynamic response of the lift coefficient for a "1-cos" gust for the Aerostabil wing [26],M = 0.8, α = 0.0°.

Software Platform for Multidisciplinary Optimization of the Complete Aircraft

The main objective of this activity is the development, implementation and testing of suitable strategies for multidisciplinary optimization (MDO) of the complete aircraft based on highly accurate numerical methods. One major activity is therefore extending the MDO capabilities that were initially developed for preliminary design in DLR internal projects TIVA I/II [27] and VAMP [28] to high-fidelity methods. Initial work took place in the DLR project MDOrmec [29]. In Digital-X, a flexible software platform based on the parallel simulation environment FlowSimulator [30] (see Sect. 3.7.1) is being developed by considering tools and processes from all relevant disciplines. This platform will provide a robust, integrated design process for aerodynamics and structures, while taking the engine into account. To use the full potential of multidisciplinary design the idea is to replace the predominantly sequential approach to detailed design by pursuing a "multi-level" approach, which combines highly accurate multidisciplinary analysis (MDA) and MDO processes for aerodynamics and structures with simplified techniques for complete aircraft design and rapid methods for the identification of critical load cases. Rule-based design procedures are implemented for preliminary design. Coupling of the individual tools and components and execution of the process chain is controlled by DLR's workflow management tool RCE (Remote Component Environment) [31]. The description of aircraft geometry and exchange of data between components are achieved using DLR's XML-based Common Parametric Aircraft Configuration Scheme (CPACS) [32]. The aircraft geometry is modeled using DLR's geometry library TIGL and various tool-specific model generators which are able to read and write the CPACS data format. In the frame of Digital-X project focus is put on the optimization of conventional aircraft configurations.

MDO Architecture and Process Chain

Figure 9 shows schematically the sequential MDO process chain that has been developed. The formulation is considered a per-cycle "multidisciplinary feasible" approach, coupling the involved disciplines at different stages of fidelity. The process is driven by a single, gradient-free optimization algorithm, where all objective-relevant data are provided by the detailed level (i.e., coupled high-fidelity methods).

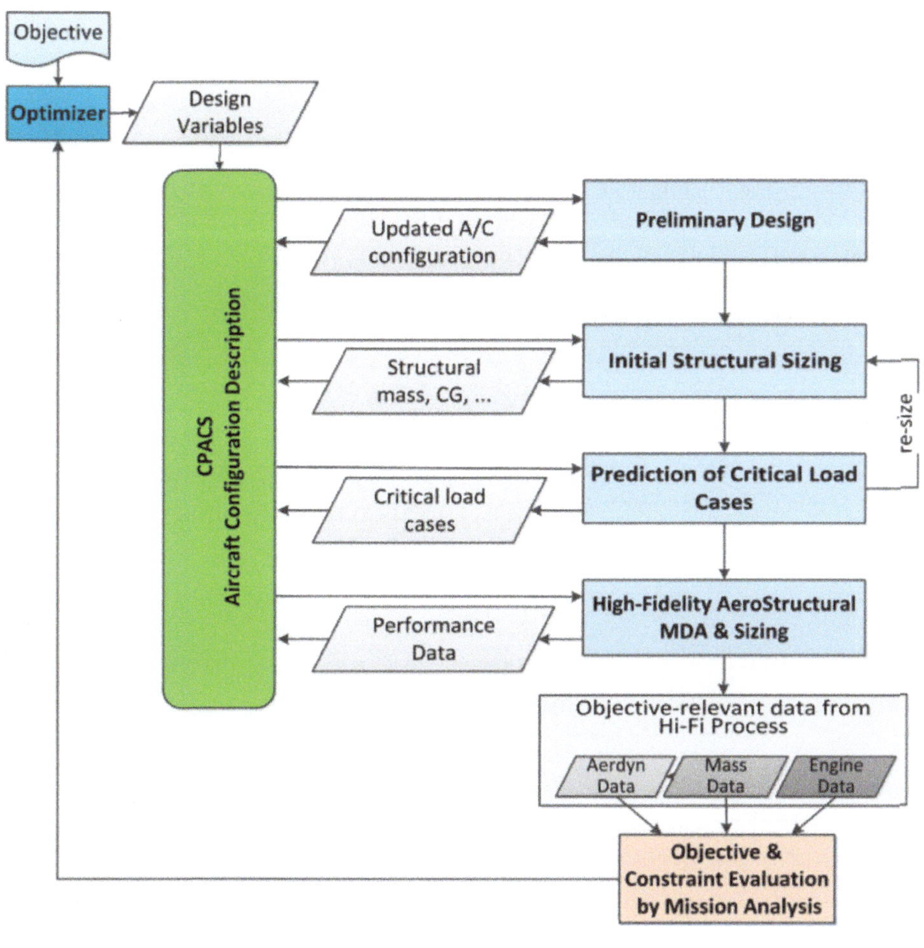

Figure 9: Digital-X MDO process chain (schematic).

The chosen "multi-level/variable fidelity" MDO approach allows for an efficient treatment of configurational design requirements by means of preliminary design methods. Assessment of configuration-specific critical load cases and initial structural sizing is done in the second stage. Employing high-fidelity coupled CFD/CSM (computational structural mechanics) simulations and refined structural sizing, the detailed level provides the required performance data for subsequent objective function evaluation. This automated process chain is currently being implemented as a prototype in DLR's distributed, simulation integration environment RCE.

Reference Configuration

The Airbus XRF-1 configuration is used as the reference geometry to demonstrate the capabilities of the MDO environment in a realistic application. The XRF-1 is a research configuration similar to an existing Airbus wide-body aircraft. Figure 10 shows the XRF1 geometry as a wing/fuselage/tail configuration. It was specified consistently in CPACS format and recalculated using improved preliminary design tools.

Figure 10: XRF-1 reference configuration.

Figure 11 shows results for the recalculation of the reference mission. This is a simplified 8000 nm mission consisting of climb, cruise, descent and landing as well as a flight to an alternate airport (200 nm). The results obtained with DLR's preliminary design tools perfectly match the reference results by Airbus, demonstrating that a consistent definition of the geometry and top-level aircraft requirements was specified.

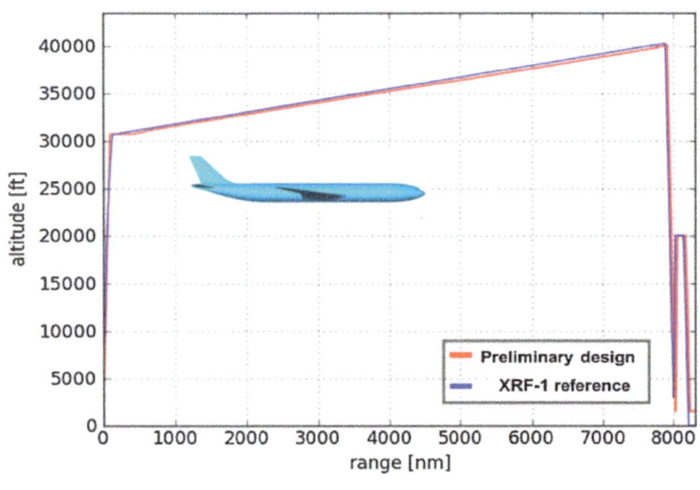

Figure 11: Recalculation of long-range mission with preliminary design tools.

Loads and Structural Sizing

For efficient calculation of critical structural design loads and load cases the Dynamic Master Model (DMM) of the XRF-1 configuration was generated [33]. Figure 12 shows the applied finite element (FE) model and the aerodynamic model which were created automatically from data in the corresponding CPACS file. The detailed Static Structural Model (SMM) for the XRF-1 wing/fuselage configuration for structural analysis in metallic design was also created with advanced, automated model generators [34] from the CPACS file and is shown in Fig. 13. The model generators were developed in previous projects and have been enhanced within Digital-X to match the requirements of the MDO process. The structural sizing of the SMM is based on critical loads calculated with the DMM, as well as those recalculated for selected critical load cases using highly accurate methods. Work is also under way to extend the static and dynamic structural analysis tools and model generators to deal with composite structures. Results for sizing the composite XRF-1 wing, fuselage and horizontal tail plane based on 17 generic DMM load cases are available.

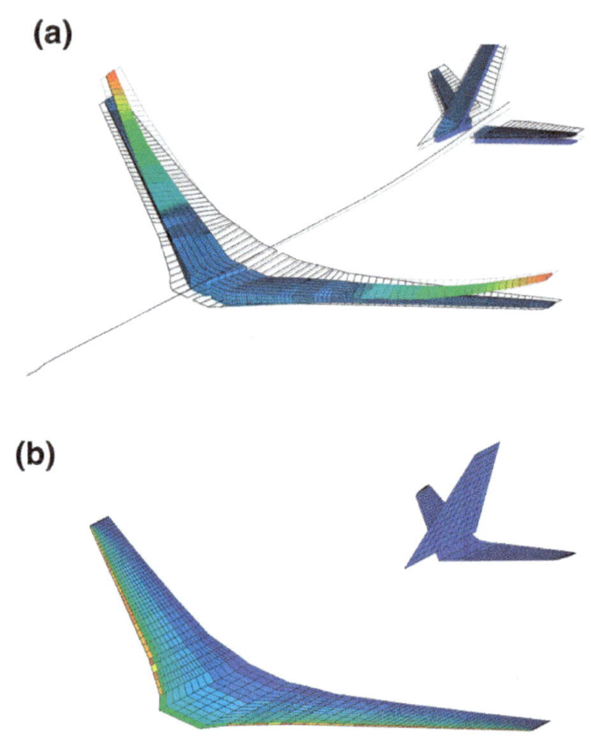

Figure 12: Dynamic model of XRF-1: **a** global FE model for modal analysis (first wing bending), **b**doublet-lattice model with pressure distribution.

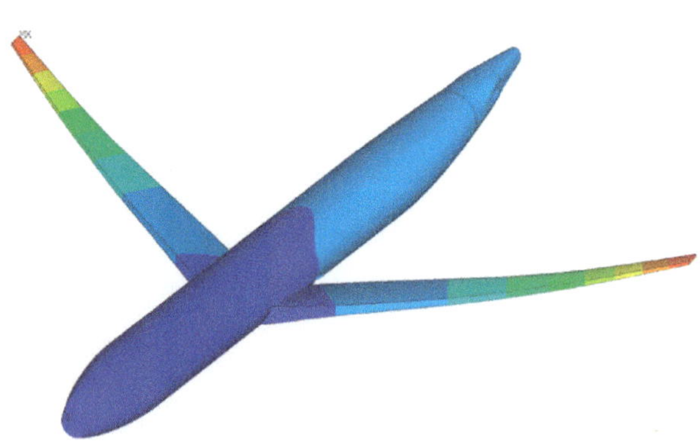

Figure 13: Detailed coupled structural model of XRF-1 configuration in metallic design.

Validation and Demonstration

After specifying the optimization tasks, the capabilities of the MDO process were first demonstrated using a simplified XRF-

1, consisting of wing and fuselage. Figures 14 and 15 show results of a multidisciplinary optimization of the block fuel mass.

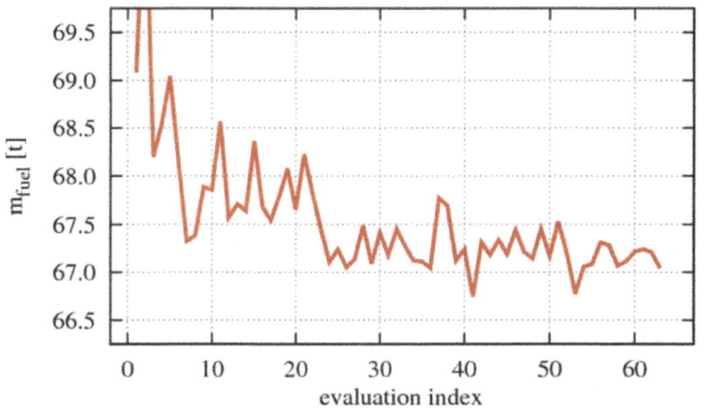

Figure 14: Optimization convergence history relative to baseline block fuel mass.

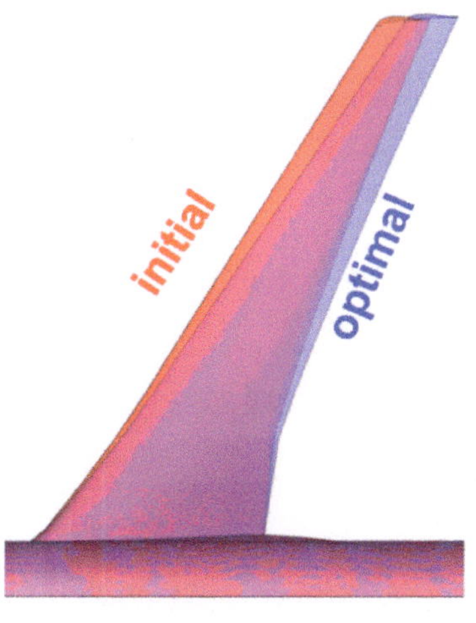

Figure 15: Result of multidisciplinary optimization of XRF-1 wing/fuselage configuration.

The fuel consumption was reduced by 3.6 %. This result was achieved with the MDO chain on the detailed level, using a Subplex algorithm. High-fidelity CFD calculations were used for the wing/fuselage aerodynamics, coupled with a finite element analysis of the wing structure to determine the static aeroelastic equilibrium. In addition, the wing structure was sized using two predefined load cases for each optimization step. The wing was parameterized using five geometry parameters (twist, aspect ratio and sweep) with planform area kept constant. Many other constraints that are necessary to achieve a realistic aircraft design were not yet taken into account at this time. For the evaluation of the objective function a mission analysis based on backwards-integration of an ordinary-differential equation (ODE) for a simplified mission over 5600 nm with a three-segment cruise climb from 11000 to 12000 m was carried out.

Gradient-Based Optimization

To improve efficiency, gradient-based optimization methods are used which make possible a large number of design parameters. Figure 16 shows the first results of a gradient-based structural optimization of a wing with 245 regions optimized and 5880 constraints. In this case the structural mass was optimized considering four different failure criteria and 12 load cases.

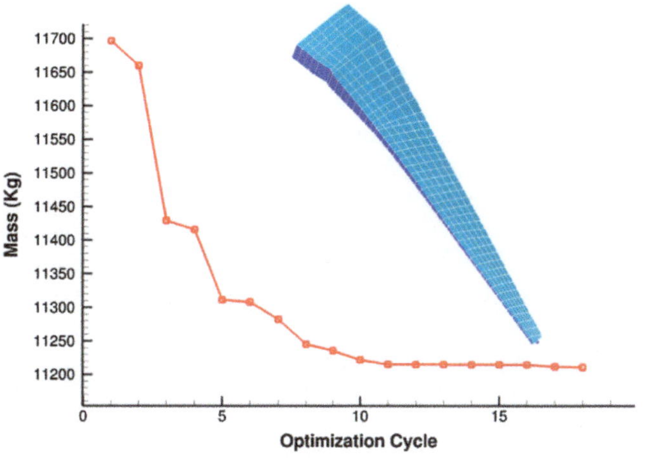

Figure 16: Gradient based optimization of the wing structural mass.

For this purpose, the suitability of various optimization algorithms was tested with a single objective function and many constraints. A sequential quadratic programming (SQP) method was applied here with gradients calculated using finite differences. As a next step the entire MDO process chain is to be converted to a gradient-based optimization.

In future developments separation methods are being followed in addition to gradient-based methods, which decouple sub-processes so that multiple optimizers can be used. The full possibilities of multidisciplinary optimization based on the "multi-level/variable fidelity" approach will be demonstrated with additional parameters and constraints, a more complex mission, and the trimmed complete XRF-1 aircraft configuration. Further design load cases are also to be considered in the context of the automated sizing of the complete aircraft with carbon fiber reinforced polymer composites (CFRP). Correction techniques and reduced order models should be applied to improve efficiency and accuracy in the loads process. Finally, the MDO processes developed are being evaluated and the optimization results are analyzed with regard to their physical improvements. At the end of the project a best practice guideline for multidisciplinary optimization of the complete aircraft will be delivered.

Simulation of Flight Maneuvers on the Borders of the Flight Envelope

The analysis of maneuver loads is intrinsic to the design and certification of aircraft. The aim of this activity is to provide methods and processes for the numerical simulation of free-flying elastic aircraft. For the demonstration of the capabilities of the methods developed, extensive and challenging simulation scenarios are planned, particularly on the borders of the flight envelope. Unlike the methods that are primarily in use in industry at present for flight dynamics analysis and load calculations, which in most of the cases are based on fast, yet very simplified aerodynamic methods, Digital-X puts emphasis on the development and coupling of high-fidelity simulation methods using CFD and CSM codes.

CFD/CSM Coupling

The flight dynamics simulations intended in Digital-X involve computations of the aircraft's aerodynamics using CFD, its structural dynamics using computational structural mechanics (CSM), and its flight mechanics (FM) and flight control. The computations are regarded as a multi-fields problem, that is, independent and highly specialized solvers for the single-field problems are suitably coupled together in space and time under consideration of the boundary conditions mutually existing between the single-field problems. The term spatial coupling refers to the operation that performs the required energy-conserving transfer of aerodynamic loads from the CFD surface mesh to the structural model, and, vice versa, the reverse transfer of structural deformations. The temporal coupling represents the appropriate synchronization of the individual single-field solver calls. In Digital-X, both tight and loose coupling schemes are foreseen. The former ensures correct energy transfer by performing several sub-iterations between the single-field solvers at each time step while the latter gives only an approximately correct energy transfer, but is less computationally intensive. In an ensuing step to the transfer of coupling data, the entire CFD volume mesh needs to be adapted to the deformation state of the CFD surface mesh. High requirements are set for this operation in terms of computational cost and preservation of mesh quality. Both requirements are difficult to fulfill, particularly in applications to complex geometries which involve control surface deflections and are subject to large deformations. Nevertheless, substantial progress has been made within Digital-X on further development and improvement of the mesh deformation using Radial Basis Functions (RBF).

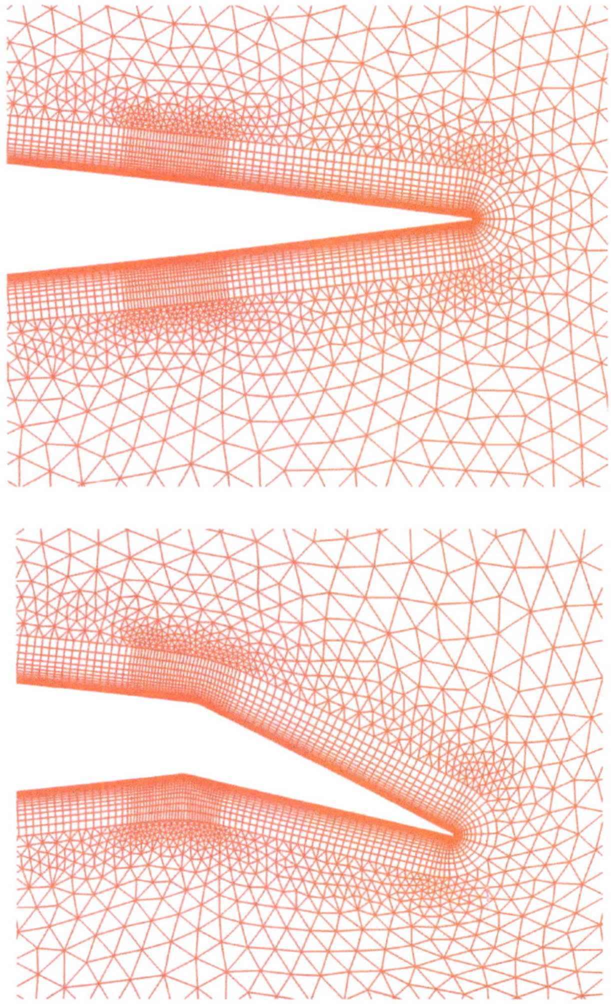

Figure 17: Deformation of the volume mesh around an airfoil-rudder configuration (without gap modeling) using the RBF technique.

Modeling Control Surfaces

In CFD-based simulations of aircraft maneuver, the capability is needed to carry out (time-dependent) motions of the primary control surfaces. Similarly in steady simulations of the aircraft trim state, (static) control surface deflections must be considered, at least of the horizontal tail plane. In recent years, considerable progress has been made on modeling movable control surfaces in previous projects [35] and in Digital-X. An established and robust method for the rotation of control surfaces is to use mesh deformation, which avoids the challenging, complex and computationally intensive application of the overset grid technique (also known as the Chimera method). The vortices emanating from gaps between the control surfaces and the main lifting surface, along with the circulation changes induced by them, are not considered in the procedure that is solely based on mesh deformation, since the gaps cannot be modeled. However, the main effect of the deflection may be captured very efficiently in this way and usually with sufficient accuracy. As an example, the deflection of a generic rudder using RBF-based mesh deformation is shown in Fig. 17. The quality of the original non deformed CFD volume mesh is maintained during mesh deformation, even at very high rudder deflections.

A further enhancement for modeling movable control surfaces in Digital-X is the combination of the so-called "patched grids" method with the RBF-based mesh deformation. In contrast to pure mesh deformation techniques, this method can conveniently take into account existing rudder gaps. The basic procedure of the technique is illustrated in Fig. 18 for a generic aileron. The control surface installation is based on four meshes. The wing mesh serves as a background mesh. The actual rudder and half of the gap to the wing on each side are embedded in an individual Chimera mesh. Two extra Chimera meshes represent the other halves of the rudder gaps, including the adjacent parts of the wing mesh. The meshes do only overlap along the positions indicated in Fig. 18. Without the patched grid method, additional Chimera meshes would be needed between the rudder and the neighboring gap meshes, see the planes marked by "sliding interface" in Fig. 18. In the initial state without rudder deflection, the mesh points on both sides of the sliding interface planes are matching. The new algorithm automatically generates sufficient mesh overlap in the gap region by mesh extrusion. In situations with flap deflection, interpolation of the flow quantities in the gap region is performed using the Chimera technique.

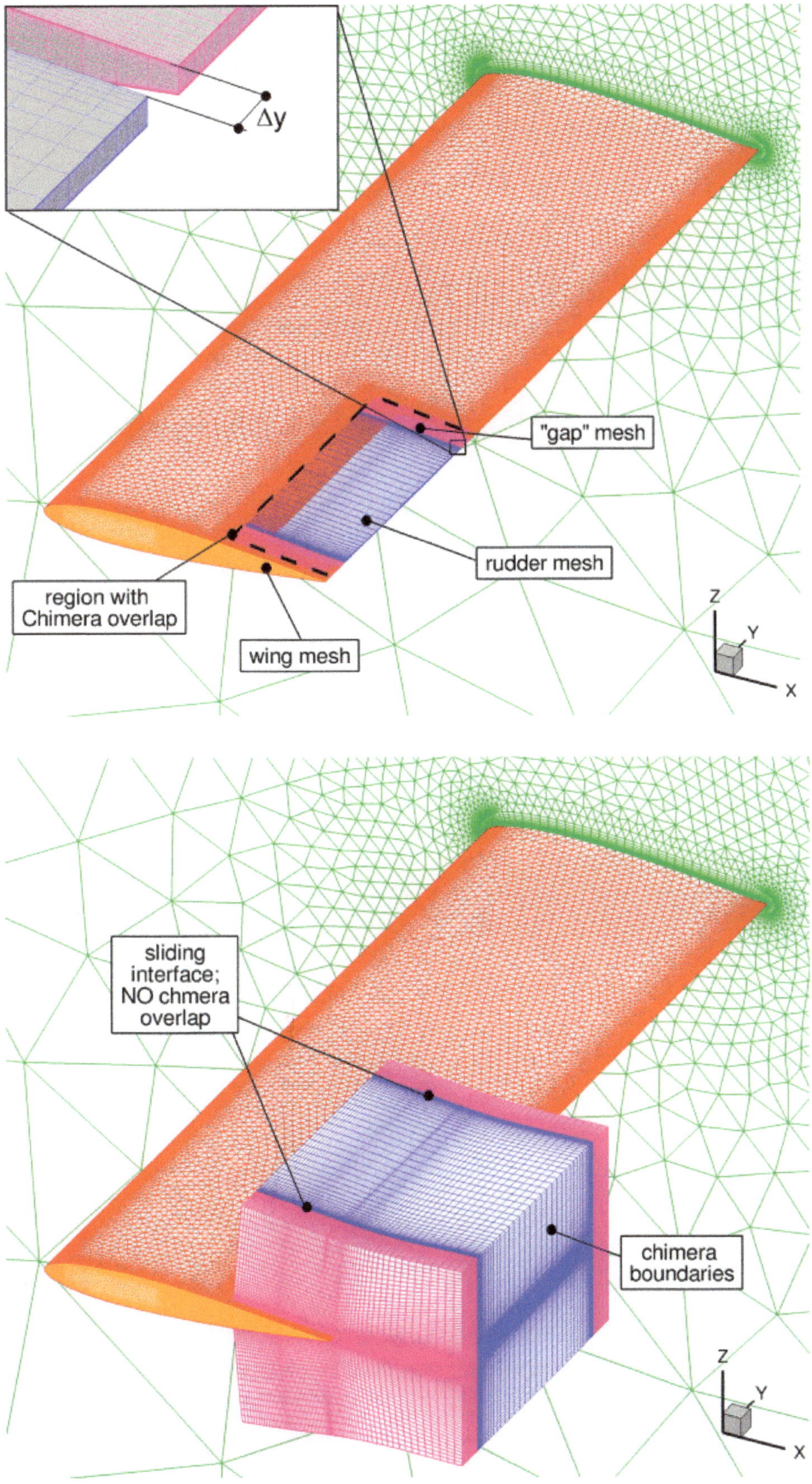

Figure 18: Application of the "sliding interface" technique for installation and deflection of a rudder.

For actually deflecting the flap, the RBF-based mesh deformation is applied to the mesh containing the flap only. In this operation, the outer boundaries of the Chimera meshes always remain fixed (see Fig. 19) so that the CPU-intensive Chimera search for donor cells only needs to take place once. The patched grids technique avoids the cumbersome and often very time-consuming manual generation of meshes having sufficient overlap in the rudder gap regions such that the Chimera mesh technique is applicable. Furthermore, many mesh points are saved by the technique, which would otherwise be required solely to ensure the Chimera overlap in the gap regions. This considerably reduces computing time compared with a pure non-automated Chimera-based control surface mesh setup. The new method is extremely flexible in its application and for example can also be used for modeling a segmented aileron (Fig. 20).

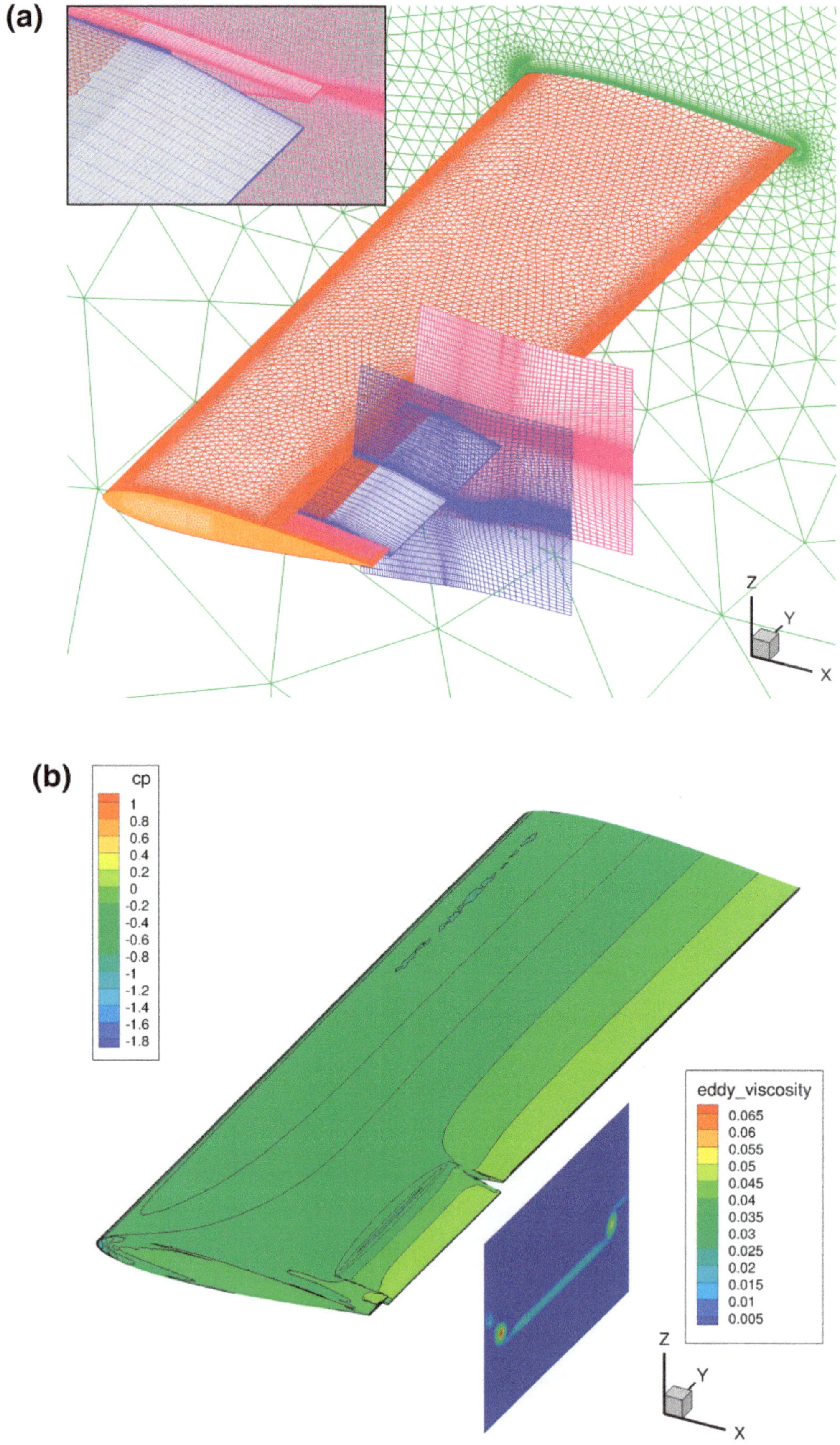

Figure 19: **a** Deflected generic aileron modeled by "sliding interface" technique, **b** resulting CFD solution (c_p and eddy viscosity).

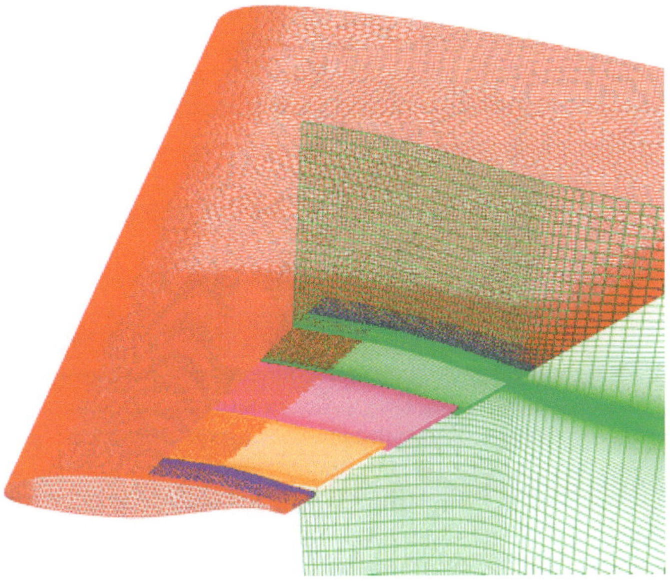

Figure 20: Three-part aileron modeled by "sliding interface" technique.

Flight Dynamics Module

A flight dynamics module has been developed for the unsteady simulation of an elastic aircraft in the time domain. This module provides the necessary spatial and temporal coupling of aerodynamics, structural dynamics, and flight mechanics. It is implemented in Python and uses the FlowSimulator framework for massively parallel computations. The translational and rotational motion of the center of gravity of the aircraft is calculated by Newton's second law and Euler's gyroscopic equations. For the decoupling of the rigid body and elastic degrees of freedom, the equations are formulated in a coordinate system which obeys the so-called "mean axes" conditions. The integrated structural dynamics solver operates under the assumption of a geometrically and physically linear structural behavior. The equations of motion of the free-flight elastic aircraft were derived from Hamilton's principle. They were spatially discretized with the finite element method and transformed into modal space using the elastic modes of the unconstrained aircraft. Time integration is carried out with the Newmark scheme. A "scattered data" interpolation with a thin plate spline RBF is provided by the flight dynamics module for the transformation of the aerodynamic loads onto the structural grid and for the interpolation of structural deformations onto the CFD grid. Figure 21 illustrates the loop implemented in the flight dynamics module for time integration of the multi-field problem described.

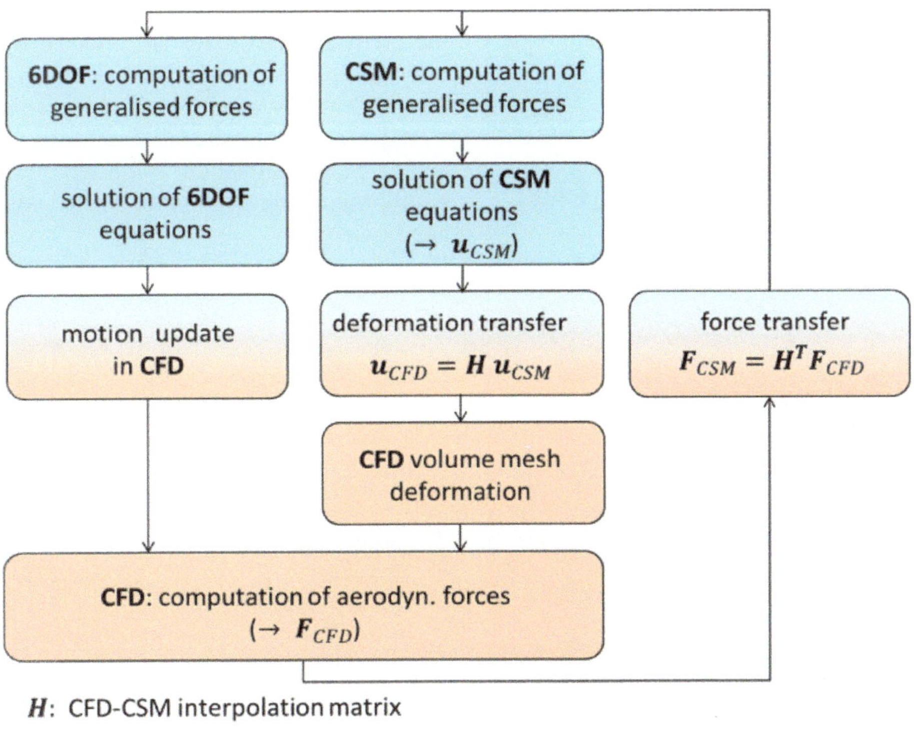

Figure 21: Schematic representation of the time integration loop in the flight dynamics module for spatial and temporal coupling of aerodynamics (CFD), structural dynamics (CSM) and flight mechanics (6DOF) codes.

As an example Fig. 22 shows the application of the flight dynamics module for simulating the interaction of an elastic aircraft with a "1-cos" gust. The plot shows the heave and pitch accelerations as function of time. The interaction of the elastic deformation with the rigid body motions of the aircraft becomes evident as higher frequency vibrations.

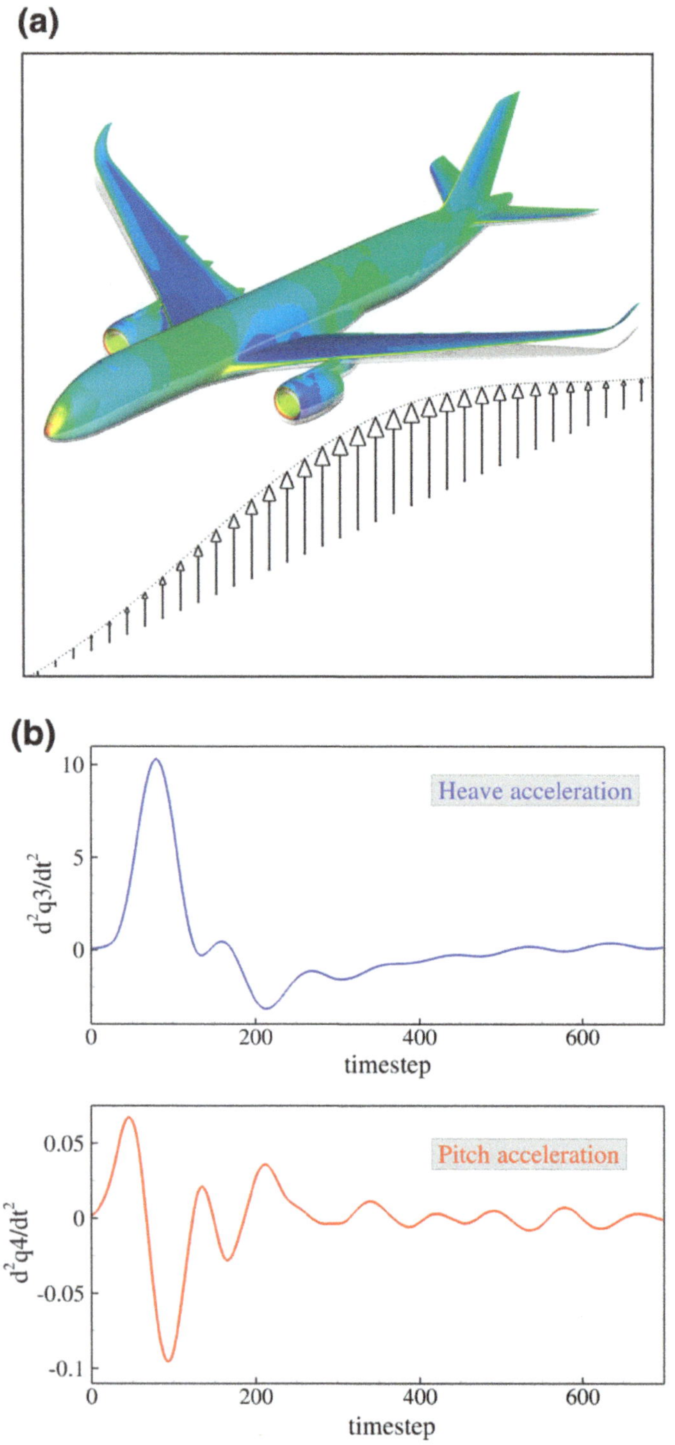

Figure 22: Interaction of a transport aircraft with a gust: **a** instantaneous elastic deformation and initial state (gray) corresponding trim solution, **b** heave and pitch accelerations at the center of gravity.

The initial condition for the simulation of response problems such as gust interaction is typically an horizontal flight with constant velocity. This initial (trim-) condition is computed by the flight dynamics module with a Newton algorithm which iteratively solves for static aeroelastic equilibrium.

Simulation Scenarios

In the context of Digital-X, aiming for more precise predictions of aircraft loads, a set of maneuver scenarios is addressed using highly accurate CFD-based simulation methods. During certification, proof must be provided that the designed aircraft structure withstands the loads occurring in these scenarios. As an example, the simulation of one scenario is discussed in detail below—the gust encounter of a modern elastic passenger aircraft.

To evaluate the potential of a CFD-based load analysis process, Airbus started the so-called CFD4Loads initiative in 2013. DLR was involved in this initiative, besides Universities and other European research organizations such as ARA and ONERA. Investigations were carried out on a realistic passenger aircraft test case (Fig. 23). The focus of DLR was on simulating gust interactions. Six gust load cases were considered. They differ in terms of Mach number, Reynolds number, flight altitude and aircraft weight, as well as gust wavelength, gust amplitude and direction (upwind or downwind). So far, only gusts which influence the aircraft longitudinal motion have been considered. In the following, exemplary results are presented for an investigated load case in transonic flow (M = 0.836, Re = 86 × 10^6). An upwind gust of about 12 m/s and a penetration depth of 350 ft (107 m) acts on the aircraft with low flight massm = 150 t. Compared to the load cases with subsonic flow, which also were investigated in CFD4Loads, the transonic case is of particular interest for CFD-based load analysis, since conventional process chains that are based on linear aerodynamics are likely to reveal their limitations in the transonic flow region. An important aspect of the CFD4Loads initiative was to examine whether the predicted gust loads are affected by different degrees of multidisciplinarity considered in the simulations and, if so, to what extent. For this purpose, the gust interaction simulations were carried out in three different ways. The starting point for each type of simulation is the aircraft's trimmed aeroelastic equilibrium configuration associated with the respective flight conditions.

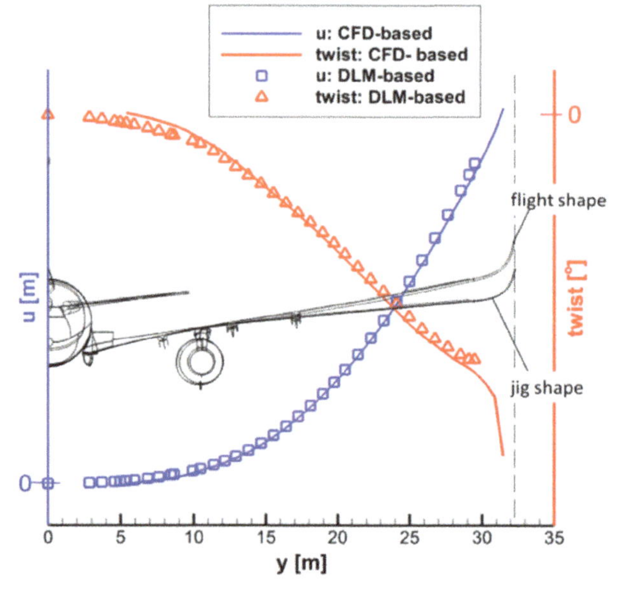

Figure 23: Spanwise bending and twist distribution of the trimmed aeroelastic equilibrium configuration, as predicted by an industrial reference process and the Digital-X process (subsonic test case).

- Pure CFD analysis: In this simulation, the aircraft is completely fixed in the initial position while being exposed to the gust. No flight dynamic or elastic degrees of freedom are considered.

- Coupled CFD-FD analysis: The aircraft has only flight dynamic degrees of freedom. During gust encounter, the aircraft in trim configuration is allowed to undergo rigid body motions only.
- Coupled CFD-FD-CSD analysis: The aircraft has flight dynamic and elastic degrees of freedom. The aircraft's reaction to the gust loads is a superposition of rigid body motions and elastic deformation.

DLR applied the Digital-X multidisciplinary process chain for the simulation of the CFD4Loads test cases, using TAU as CFD solver in fully turbulent RANS mode. For gust modeling, the existing TAU disturbance velocity approach was used [36], [37]. It considers the influence of the gust on the aircraft, but not the reaction of the aircraft on the gust. However, it has been shown that the disturbance velocity approach is sufficiently accurate if the ratio of the gust wavelength to the aircraft characteristic length exceeds the value of two [38]. This holds for all CFD4Loads gust cases investigated. "1-cos" gusts were simulated according to the definitions in CS 25.341 (a) [39].

Figure 23 shows the trimmed aeroelastic equilibrium configuration for a subsonic case (M = 0.45, Re = 70 × 10^6, m = 275 t). Here, the results of the CFD-based process and those of the conventional process which is based on linearized aerodynamics are in good agreement.

The gust interaction simulations were then carried out based on the trim configurations. Figure 24 shows the time histories of the predicted load factor n_z obtained for simulations with different degrees of multidisciplinarity (n_z is measured at an aircraft fixed reference point). The time t = 0 in the diagrams of Fig. 24 corresponds to the moment when the gust reaches the nose of the aircraft. In all simulations the time step was chosen so that the respective gust period was resolved with well over 100 time steps. Test simulations with close coupling did not show significantly different results compared with loose coupling. The latter was therefore used for all further simulations.

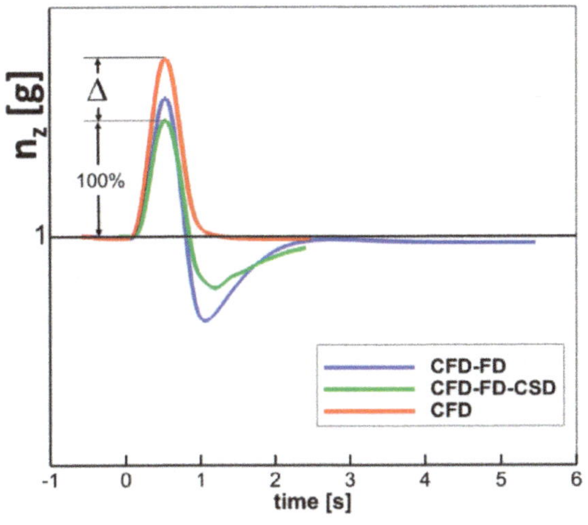

Figure 24: Effect of multidisciplinarity on the time histories of the load factor n_z during a gust interaction case in transonic flow.

The investigation of various levels of multidisciplinarity in the simulations shows the expected consistent behavior. The highest load factors occur in the single-discipline analysis (only CFD), the lowest when considering the full scope of multidisciplinarity (coupled CFD-FM-CSM). The differences in the maximum load factor between the two extremes is about Δ = 63 %.

Figure 25 shows the state of the aircraft in the transonic gust load case at the time of the peak load factor as predicted in Digital-X with CFD-FM-CSM coupling. The additional wing bending caused by the reaction to the gust load is clearly seen when compared to the shape of the trimmed aeroelastic equilibrium configuration.

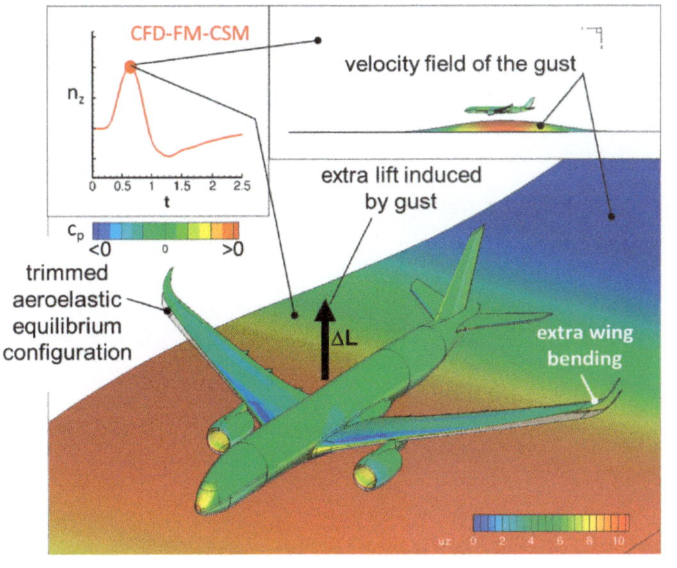

Figure 25: Visualization of the gust response of the analyzed aircraft as predicted by the Digital-X process (CFD-FM-CSM simulation of the transonic gust test case).

As a result of the findings in the CFD4Loads initiative regarding the prediction of aircraft-gust interactions using the Digital-X process, the selective use of CFD in the context of load analysis has now been intensified by the industry partner.

Quantifying Uncertainties Using Numerical Simulation

Aerodynamic and aeroelastic input parameters (flow conditions, geometry, material quantities, etc.) are often subject to significant uncertainties, which are not usually considered in the established deterministic simulation methods. However, quantitative information on the effects of such uncertainties is desirable for assessing simulation results and ultimately necessary for virtual certification. The objective of this work package is to provide efficient methods and tools for quantifying output uncertainty using numerical simulation.

Non-Intrusive Method

Based on input data considered as random fields, the random simulation results can be computed with intrusive or non-intrusive methods. The project Digital-X focuses on efficient non-intrusive methods which significantly reduce the computational cost compared to Monte Carlo methods, and make possible the treatment of distributed uncertainties such as those caused, for example, in geometry by manufacturing tolerances, or through wear, dirt or ice accumulation. Both gradient-assisted surrogate methods and reduced order models based on POD are employed.

Current research is concerned with parameterization of geometrical uncertainties for three-dimensional wings to model and analyze distributed uncertainties. One of the challenges was to implement the geometrical variations from a practical standpoint. To handle spatially correlated variation in the geometry an Eigen-decomposition of a very large, but rank deficient covariance matrix has to be made. In view of the computational complexity and memory required, methods have been developed for a coarse approximation to this matrix which maintains the geometric variance up to the machine precision. A parameterization in independent random variables is achieved by a subsequent Karhunen-Loève expansion of the random field based on the approximate matrix. After development the method was tested on a transport aircraft wing with 56,312 upper surface mesh points with uncertainties. In doing this particular assumptions were made about the correlation between any two mesh points such that the geometry variation is parameterized by only 600 instead of 56,312 random variables without any loss in the variance of the variation. The resulting eigenvalue problem could be solved very quickly and efficiently using a single core of a desktop computer. Figure 26 shows four examples of the random geometry parameterized with 600 variables.

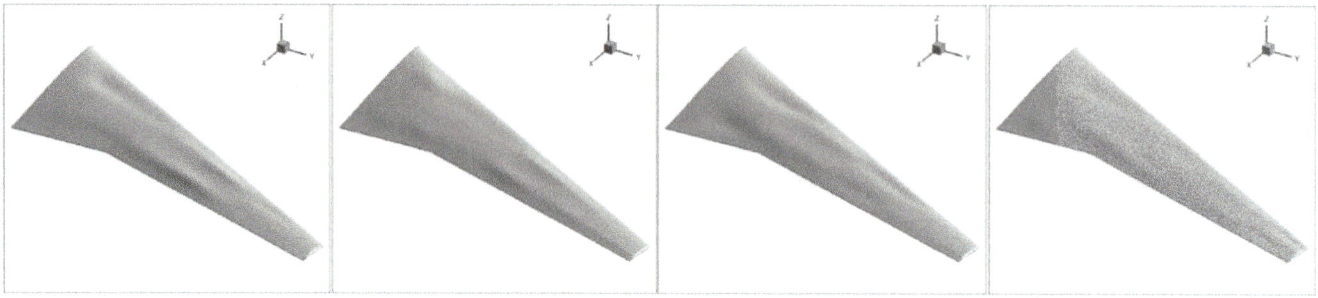

Figure 26: Examples of random geometry deformations of a wing (600 variables).

Methods for Robust Design

Building on the methods for efficient quantification of distributed uncertainties that were developed in the work package above, stochastic optimization techniques have been developed, which allow geometric variability in the design process to be considered and permit a robust design, i.e., a design that is less sensitive to small random perturbations.

The robust design formulation used is based on an expectation measure. The goal was to minimize the sum of the mean and standard deviation of the drag coefficient of the RAE 2822 airfoil for a given nominal lift coefficient. Here, both the flow conditions and the design parameters were considered uncertain. The nominal flow conditions were set to M = 0.734 andα = 2.79°. The standard deviations of these parameters were 0.005 and 0.1, respectively. The geometric uncertainty was parameterized with the KLE, resulting in ten normally distributed uncertain geometry variables with a maximum standard deviation of 0.00125. The airfoil itself was parameterized with ten deterministic design variables. For each optimization step 100 samples were computed with the TAU code to determine the stochastic variation of the drag coefficient. They were used to construct a Kriging surrogate model for the uncertain objective function, which in turn was used to efficiently perform a full Monte Carlo simulation with Gaussian distributed variables to evaluate the statistics. As a result, the mean of the drag coefficient could be reduced by 60 drag counts while its standard deviation was reduced from 12.4 counts to five counts, requiring a total of about 500 iterations of the subplex optimization algorithm.

Other measures of robustness that have been considered so far include the worst-case risk measure and the mean-risk approach, which are both reliability-based robust design formulations. In the mean-risk approach, the idea is to maximize the probability that the drag coefficient is within a certain range around the mean value, requiring both the mean value and the probability density function (PDF) of the drag coefficient to evaluate the objective function. The result of this optimization in terms of the PDF of the initial and the optimized airfoil is shown in Fig. 27. The probability could be increased from 60.5 to 81.4 % at the cost of a higher mean value. The influence of the different measures of robustness on the result of a 2D robust design optimization problem has also been investigated.

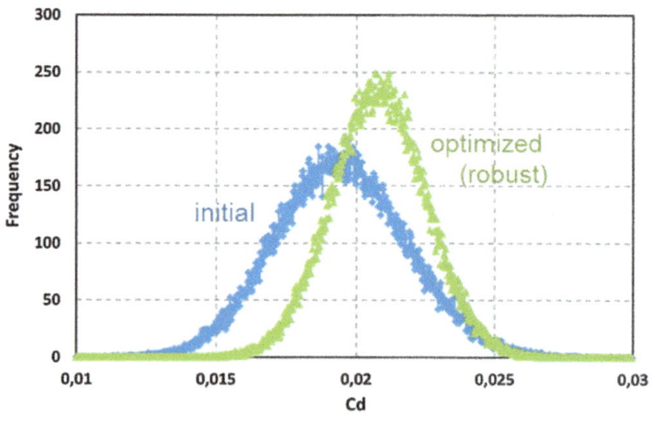

Figure 27: Comparison of output PDFs: RANS-based robust design optimization (mean-risk approach) of Rae 2822 airfoil with uncertain geometry.

Coupled CFD/CSM Simulation of the Trimmed Helicopter

The numerical simulation of the flow around helicopters is very expensive because it must be performed time-accurate and multidisciplinary. The calculations are complicated by a number of aerodynamic effects which occur during a revolution of the rotor, such as compression shocks, sheared boundary layers and flow separation. In addition, the blade tip vortex must be calculated very exactly to accurately predict blade/vortex interactions. Due to the strong aeroelastic deformation of the rotor blade, simulations must always be performed as a coupled fluid/structure computation. Furthermore, rotor trimming is required to set the required helicopter flight conditions.

In Digital-X, a multidisciplinary process chain for the complete trimmed helicopter including tail rotor under industrial conditions is developed, which builds on earlier work in the development and validation of simulation capabilities for isolated helicopter components. In contrast to previous work the unstructured TAU code is used because of the high geometric complexity and the requirement for automated mesh generation [40].

Coupling Methods

To take into account the elasticity of the rotor blades and rotor trim, the CFD solver TAU was coupled with the comprehensive rotor code HOST (helicopter overall simulation tool [41]) of Airbus Helicopters in a previous research project. Here the RANS solver TAU computes the aerodynamic loads based on the blade deformations and control angles provided by HOST. The aerodynamic loads are then used to correct HOST's simplified aerodynamic model. The iterative process is carried out in the form of a so-called weak coupling. Whereas previously coupling was limited to single main rotor blades, in Digital-X the coupling chain is improved and extended with focus on applications to the complete helicopter. An important aspect is the development of a suitable protocol for transferring aerodynamic loads, determined by the TAU code on unstructured grids, to the beam-structure model of HOST. The transfer method was implemented such that in future more accurate finite element structural solvers can be used instead of beam models.

The design of the extended coupling environment allows different simulation tools to be used. Thus, the TAU code can be replaced by an alternative flow solver, or instead of the HOST code the DLR internally developed rotor simulation code S4 [42] can be used. As a step towards industrialization the coupling chain can be connected as TAU Python "plug-in" to the parallel simulation environment FlowSimulator.

Chimera Extensions

For helicopter simulations the availability of the Chimera technique is essential because it allows a relative motion of the component meshes so that rotors, etc., can be simulated. In the Chimera technique the components of a configuration are first meshed independently. Subsequently the component meshes are combined into a mesh system, in which the meshes partially overlap each other. To ensure a valid solution, the data are transferred between the overlapping mesh blocks by interpolation. Additionally the nodes of a component mesh which are inside the body of another mesh must be excluded from the calculation.

For the hole cutting procedure so-called "hole definition geometries" are used, which consist of simple geometric shapes. In the case of a rotor simulation with large blade deformations it is possible that a deformed blade protrudes from the originally specified hole cutting geometry during the computation (Fig. 28). This leads to an irregular data transfer and termination of the computation.

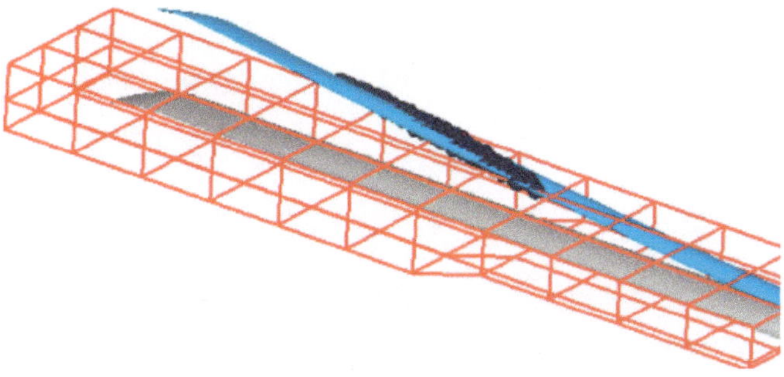

Figure 28: Position of the deformed rotor blade (blue) relative to the original position (gray) and the hole definition geometry (red), GOAHEAD rotor.

A possible solution would be the generation of a large overlap region for the blade mesh and background mesh as well as the definition of a sufficiently large hole geometry, so that the deformed blade does not leave the defined hole geometry during the simulation. However, this is not suitable in complex configurations as the proximity of the main and tail rotor to the fuselage does not allow large overlap regions.

An alternative approach is to consider the rotor blade deformation during the hole cutting procedure in determining the nodes to be eliminated. This can be achieved by deforming the hole definition geometries according to the blade deformation (Fig. 29). In this way narrow regions of grid overlap can be realized.

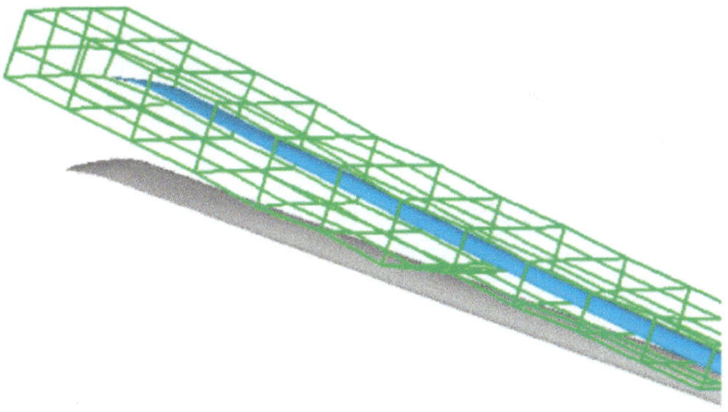

Figure 29: Position of the deformed rotor blade (blue) relative to the original position (gray) and the deformed hole definition geometry (green), GOAHEAD rotor.

First Results from the Coupled TAU/HOST Simulations

To validate the TAU/HOST coupling process chain, test data for cruise flight were selected from the GOAHEAD wind tunnel test campaign in the DNW/LLF [43]. The wind tunnel model (Fig. 30) consists of the ONERA 7AD main rotor, a BO105 tail rotor and a generic transport helicopter fuselage. In the first step the TAU/HOST simulations were performed for the main rotor [44]. In addition, numerical reference results are available from the validated Flower/HOST process chain. Flower is DLR's block-structured RANS solver. The TAU calculations were performed on the structured Flower grids, to exclude grid effects in the validation.

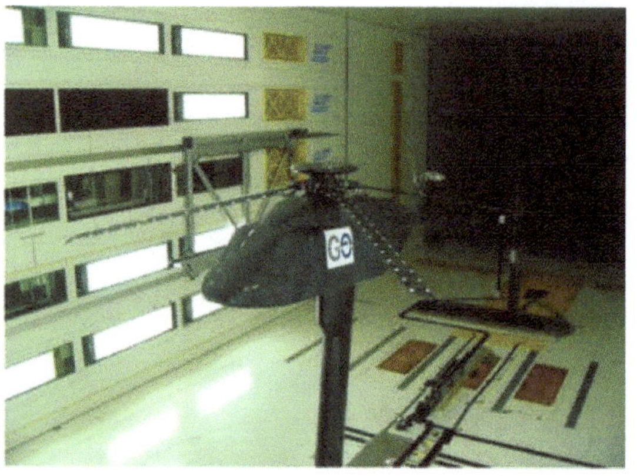

Figure 30: Wind tunnel set-up in GOAHEAD.

Figure 31 shows the convergence of the control angles. The first trim cycle was carried out after three rotor revolutions, with the following trim cycles performed after each further rotor rotation. The simulation was terminated after the incremental changes in all control angles were less than 0.001°.

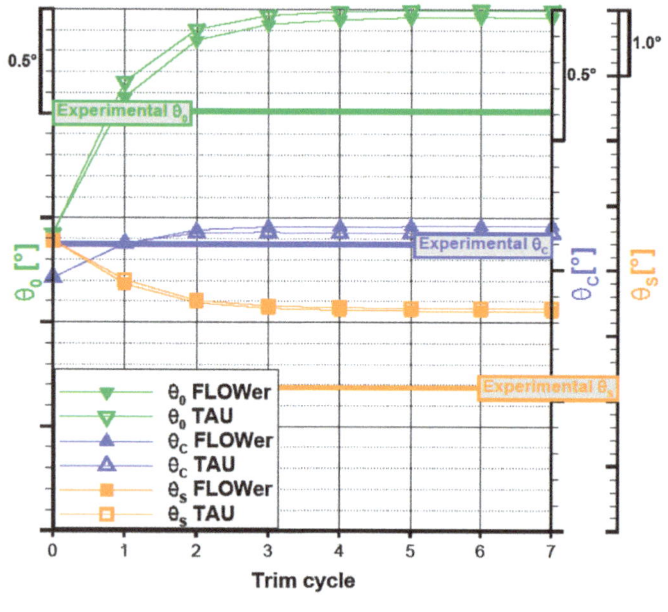

Figure 31: Convergence behavior of the control angles, comparison between TAU/HOST, FLOWer/HOST and experimental data for the GOAHEAD main rotor.

The convergence behavior of the two simulations is very similar. Convergence was reached after seven trim iterations. The numerical results agree very well with each other. The deviations from the experimental results in the collective (θ_0) and lateral (θ_C) control angles are relatively small under 0.5°. A larger deviation is observed only in the longitudinal control angle (θ_S). This is mainly caused by the lack of interaction with the fuselage. It is concluded from the detailed comparison of the results that the TAU/HOST coupling was successfully validated for the isolated 7AD main rotor. Future activities will focus on the simulation of the complete GOAHEAD configuration.

Improvement in Accuracy of Blade/Vortex Interactions

The accurate prediction of the blade/vortex interaction, which has a significant influence on aerodynamic and aeroacoustic loads, requires the correct transport of vortices through the flow field. The DLR TAU code uses a second-order discretization of the RANS equations, so that simulations of the blade tip vortices and blade-vortex interactions require extremely fine computational grids due to the relatively high inherent numerical dissipation.

This restriction should be lifted in Digital-X through coupling the TAU code with a RANS higher order method. The fourth order Flower version (FLOWer-4, [45]) was chosen for this. A coupling method for two CFD codes was already developed within the DFG Research Unit FOR1066, [46, 47]. This method is based on the Chimera technique and is implemented as a separate coupling module. The flow solvers require only an appropriate communication interface for data transfer and a suitable boundary condition for defining flow data on the coupling surface. This approach allows a flexible coupling between different flow solvers. In the context of Digital-X the necessary improvements and extensions to TAU and FLOWer-4 are currently being carried out so that specific helicopter applications can be performed with the coupled procedure. The communication interface in FLOWer-4 has already been implemented and successfully tested with a two-dimensional test case. Currently the coupling is being extended to allow for relative motion of component grids and the predictive accuracy of the coupled approach is tested on an isolated rotor.

Automatic Process Chain

Complex multidisciplinary simulations and optimizations require automated process chains. The main points here are the flexible yet efficient connection of different software components, the application of massively parallel supercomputers, as well as support in the use of automated processes. Thus, in Digital-X specific developments are being undertaken in the fields of a parallel multidisciplinary simulation environment and a workflow management system.

FlowSimulator Environment

The simulation environment in Digital-X is based on the FlowSimulator software which was developed by Airbus and DLR in cooperation with other European research partners [30]. This software manages in parallel CFD data (meshes, solvers, CAD and meta-data), so that various programs (e.g., CFD and CSM software at different levels, and pre- and post-processing) can be coupled without file-based data exchange (file-IO).

First extensive developments for multidisciplinary simulation were carried out in the framework of the $C^2A^2S^2E$ project [35]. FlowSimulator allows coupling of different programs within a so-called MPI communicator, in which the parallel distributed programs control data exchange via the shared memory of the individual MPI processes. The process chain is defined by a Python script which controls the sequence of programs per MPI process. The aim of the work in Digital-X is to integrate the newly developed capabilities and to provide necessary scripts for the planned multidisciplinary analysis and optimization scenarios. The high parallel efficiency of the computation chains plays a crucial role. Figure 32 gives a schematic overview of the essential applications of FlowSimulator.

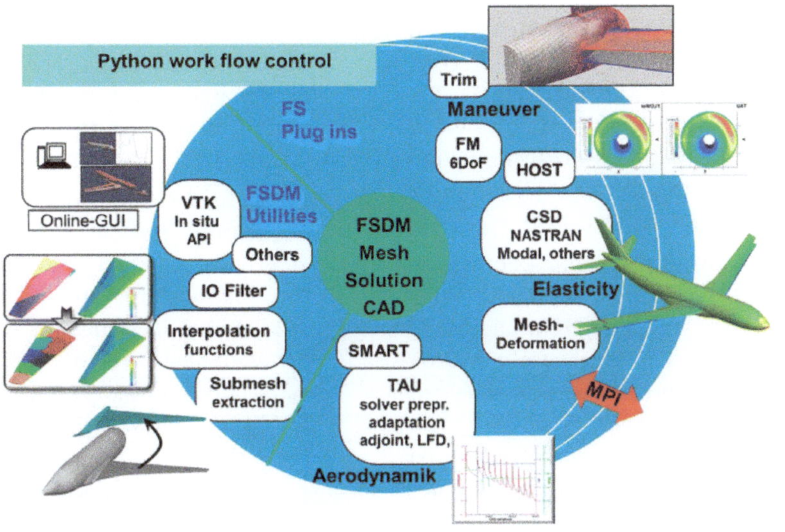

Figure 32: Application scenarios for the FlowSimulator.

RCE Workflow Management System

For setting up and controlling the computational workflow the management system RCE developed within DLR [31] is used. The main objective here is to further develop RCE for complex multidisciplinary simulations based on highly accurate computations (see also Sect. 3.3). The management task includes control of one or more completed remote simulations that take place simultaneously on one or more computer clusters. This involves job start and run-time control with data transfer from input and output files over the network between the workstation and target computer, taking into account that queuing systems must usually be addressed there. Furthermore, the optimizer workflow component should be extended to Digital-X requirements.

Figure 33 shows the RCE workflow of the detailed level process of Fig. 9, which was used to produce the results shown in section Fig. 14. It contains the components which represent the disciplinary tools (CFD, structural sizing, etc.) as well as service components, which adapt the inputs and outputs between different tools and establish iterative loops. To run all the tools in this particular workflow, RCE orchestrated the use of an HPC cluster, two Linux-based workstations, and two Windows-based workstations, spread over two different DLR institutes.

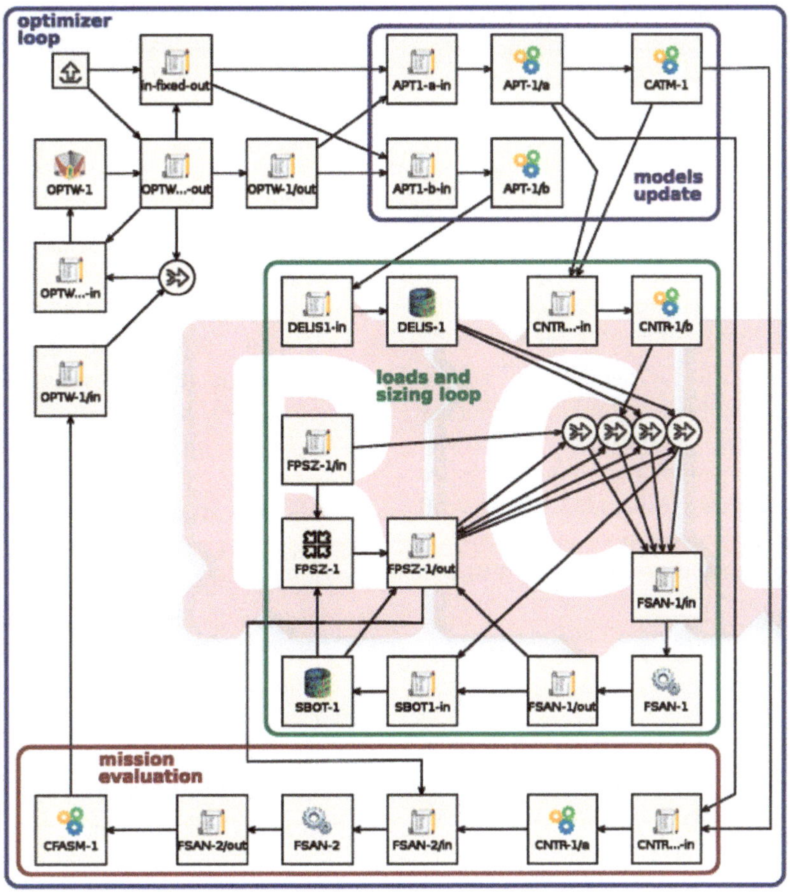

Figure 33: RCE workflow implementing a high-fidelity based aero-structure optimization process

SUMMARY AND NEXT STEPS

At DLR, a multidisciplinary research project has been set-up. It represents a first significant step towards realizing DLR's vision of the digital aircraft and virtual flight testing. Dedicated developments in disciplinary and multidisciplinary simulation methods are being addressed with a focus on multidisciplinary design and analysis of aircraft and helicopters.

The challenges in terms of physical modeling across the flight envelope require further improvements and enhancements of the DLR's well-established flow solver TAU. In Digital-X, the efficiency, robustness, reliability and level of automation of the TAU code is significantly improved and its range of applications is expanded. Given the technology development of high-performance computers, most CFD solvers used today have reached the limits of scalability when it comes to parallelization. Therefore, the design and implementation of a next generation flow solver are a key objective of the project. Besides providing the best possible utilization of future HPC systems, the new CFD code consolidates existing algorithms and incorporates innovative simulation approaches. The development of the next generation CFD solver is to be regarded as a future investment of DLR needed to address challenges in numerical simulation beyond the current project. The software specification has been completed and a prototype has already been implemented and evaluated. A first release for selected target applications is scheduled for the end of the project.

A multi-level/multi-fidelity MDO concept and process architecture have been defined and implemented based on improved disciplinary tools and disciplinary models that are generated automatically based on a common parametric description of the XRF-1 reference long-range transport aircraft configuration. An aero-structural optimization of a simplified XRF-1 with a metallic structure has been performed based on high-fidelity models and tools for a simplified mission and selected load cases, reducing the block fuel mass by 3.6 %. Next, the full multi-level MDO chain will be exercised on the full XRF-1 with additional design parameters and constraints and for a more complex mission. Further design load cases will also to be considered. Correction techniques and the developed reduced order models will be applied to improve efficiency and accuracy in the loads process. Finally, the entire MDO process chain is to be converted to a gradient-based optimization chain.

The results of the high-fidelity based multidisciplinary simulations obtained so far have demonstrated the general feasibility of such an advanced and ambitious venture. Planned simulations of the free-flying aircraft will include a flight control system to predict structural loads even more realistically. A series of steady and unsteady maneuvers including gust and wake vortex encounter scenarios will be simulated to further demonstrate the power of CFD-based aircraft analysis.

ACKNOWLEDGMENTS

The Digital-X project is a large, multidisciplinary project with many dedicated colleagues and external partners. The authors would like to thank all who were involved in the discussions and research activities, particularly those colleagues whose contributions were not explicitly mentioned in this overview article because of the size limitations.

REFERENCES

1. Flightpath 2050—Europe's vision for aviation, report of the high level group on aviation research. http://acare4europe.com, ISBN 978-92-79-19724-6, European union (2011)
2. Slotnick, J., Khodadoust, A., Alonso, J., Damofal, D., Gropp, W., Lurie, E., Mavriplis, D.: CFD vision 2030 study: a path to revolutionary computational aerosciences.http://ntrs.nasa.gov, NASA/CR-2014-218178, NASA Center for AeroSpace Information, 7115 Standard Drive, Hanover, MD 21076-1320, US (2014)
3. Dean, J.P., Clifton, J.D., Bodkin, D.J., Ratcliff, J.: High resolution CFD simulations of maneuvering aircraft using the CREATE-AV/Kestrel solver. In: AIAA 2011-1109, 49th AIAA Aerospace Sciences Meeting, Orlando, Florida, 4–7 Jan 2011
4. Wissink, A. et al.: Capability enhancements in version 3 of the helios high-fidelity rotorcraft simulation code. In: 50th AIAA Aerospace Science Meeting, Nashville, Tennessee, 9–12 Jan 2012
5. Kenway, G.K.W., Martins, J.R.R.A.: Multipoint high-fidelity aero-structural optimization of a transport aircraft configuration. J. Aircr. 51, 1 (2014)
6. Schwamborn, D., Gerhold, T. Heinrich, R.: The DLR TAU code: recent applications in research and industry. In: Proceedings of European Conference on Computational Fluid Dynamics, ECCOMAS CDF 2006, Delft, The Netherland (2006)
7. Eisfeld, B.: Numerical simulation of aerodynamic problems with the SSG/LRR-ω Reynolds stress turbulence model using the unstructured TAU code, In: Tropea, C., Jakirlic, S., Heinemann, H.-J., Henke, R., Hönlinger, H. (eds.), Contributions to the 15th STAB/DGLR Symposium, Darmstadt, Germany 2006, Notes on Numerical Fluid Mechanics and Multidisciplinary Design, vol. 96, pp. 356–363, Springer, Berlin Heidelberg (2008)
8. Cecora, R.-D., Eisfeld, B., Probst, A., Crippa, S., Radespiel, R.: Differential reynolds stress modeling for aeronautics. In: AIAA-Paper 2012-0465 (2012)
9. Eisfeld, B., Probst, A.: Industrial application of differential reynolds stress models, DLR-IB 2012, ISSN 16147790, 2012
10. Togiti, V., Eisfeld, B., Brodersen, O.: Turbulence model study for the flow around the NASA common research model. J. Aircr. 51(4), 1331–1343 (2014)
11. Probst, A., Reuß, S.: Scale-resolving simulations of wall-bounded flows with an unstructured compressible flow solver. In: 5th Symposium on Hybrid RANS-LES Methods, TEXAS A&M University, College Station, Houston, USA, 19–21 Mar 2014
12. Grabe, C., Krumbein, A.: Correlation-based transition transport modeling for three-dimensional aerodynamic configurations. J. Aircr. 2013(50), 1533–1539 (2013)
13. Grabe, C., Krumbein, A.: Extension of the γ-Reθt model for prediction of crossflow transition. In: AIAA 2014-1269, 52nd Aerospace Sciences Meeting (2014)

14. Langer, S., Schöppe, A. Kroll, N.: Investigation and comparison of implicit smoothers applied in agglomeration multigrid in the framework of the DLR TAU code. AIAA J. 53(8), 2080–2096 (2015)
15. Allmaras, et al.: Modifications and clarifications for the implementation of the Spalart-Allmaras turbulence MODEL, ICCFD7-1992. Big Island, Hawaai (2012)
16. Langer, S.: Agglomeration multigrid methods with implicit Runge-Kutta smoothers applied to aerodynamic simulations on unstructured grids. J. Comput. Phys. 277, 72–100 (2014)
17. Jägersküpper, J., Simmendinger, CH.: A novel shared-memory thread-pool implementation for hybrid parallel CFD solvers. In: Proceedings Euro-Par 2011, Lecture Notes in Computer Science, Band 6853, Springer, S. 182–193 (2011)
18. Tenenbaum, J.B., De Silva, V., Langford, J.C.: A global geometric framework for nonlinear dimensionality reduction. Science 290(5500), 2319–2323 (2000)
19. Holmes, P., Lumley, J., Berkooz, G.: Turbulence, Coherent Structures, Dynamical Systems and Symmetry. Cambridge University Press, New York (1996)
20. Bui-Thanh, T., Damodaran, M., Willcox, K.: Proper orthogonal decomposition extensions for parametric applications in compressible aerodynamics. In: AIAA 2003-4213, 21st AIAA Applied Aerodynamics Conference, Orlando, Florida, June 2003
21. Franz, T., Zimmermann, R., Görtz, S., Karcher, N.: Interpolation-based reduced order modelling for steady transonic flows via manifold learning, in special issue: reduced order modelling: the road towards real-time simulation of complex physics. Int J Comput Fluid Dynam 28(3–4), 106–121 (2014)
22. Zwaan, R.J.: LANN Wing. Pitching oscillation, compendium of unsteady aerodynamic measurements, Addendum No. 1, AGARD-R-702 (1985)
23. Thormann, R., Dimitrov, D.: Correction of aerodynamic influence matrices for transonic flow. CEAS Aeronaut J 5, 435–446 (2014)
24. Parameswaran, V., Baeder, J.D.: Indicial aerodynamics in compressible flow—direct computational fluid dynamic calculations. J. Aircr. 34(1), 131–133 (1997)C
25. Thormann, R., Widhalm, M.: Linear-frequency-domain predictions of dynamic-response data for viscous transonic flows. AIAA J. 51(11), 2540–2557 (2013)C
26. Dietz, G., Schewe, G., Kiessling, F., Sinapius, M.: Limit-cycle-oscillation experiments at a transport aircraft wing model. International Forum of Aeroelasticity and Structure Dynamics (IFASD), Amsterdam, The Netherlands (2003)
27. Liersch, C.M., Hepperle, M.: A distributed toolbox for multidisciplinary preliminary aircraft design. CEAS Aeronaut. J. 2(1–4), 57–68 (2011)
28. Nagel, B., Zill, T., Moerland, E., Böhnke, D.: Virtual aircraft multidisciplinary analysis and design processes—lessons learned from the collaborative design project VAMP. In: Proceedings 4th CEAS Air and Space Conference, Linköping, Schweden (2013)
29. Ronzheimer, A., Natterer, F.J., Brezillon, J.: Aircraft wing optimization using high fidelity closely coupled CFD and CSM methods. In: AIAA Paper 2010-9078, 13th AIAA/ISSMO Multidisciplinary Analysis Optimization Conference, Fort Worth, Texas, Fort Worth, Texas (2010)
30. Meinel, M., Einarsson, G.O.: The FlowSimulator framework for massively parallel CFD applications, PARA 2010. PARA2010, 6–9 June 2010, Reykjavik, Island (2010)
31. Seider, D., Zur, S., Flink, J., Mischke, R., Seebach, O.: RCE—Distributed, workflow-driven integration environment, EclipseCon Europe 2013, 29–31 Oct 2013, Ludwigshafen (2013)
32. Rizzi, A., Zhang, M., Nagel, B., Böhnke, D., Saquet P.: Towards a unified framework using CPACS for geometry management in aircraft design. In: AIAA Paper 2012-0549, AIAA Aerospace Sciences Meeting, Nashville, USA (2012)
33. Liepelt, R.; Chiozzotto, G.P., Schmidt, H.: Variable fidelity loads process in a multidisciplinary aircraft design environment. In: 4th CEAS Air and Space Conference (2013)
34. Freund, S., Heinecke, F., Führer, T., Willberg, C.: Parametric model generation and sizing of lightweight structures for a multidisciplinary design process. In: NAFEMS DACH-Tagung, 20–21 May 2014, Bamberg Germany (2014)
35. Kroll, N., Heinrich, R., Görtz, S., Krumbein, A., Gerhold, T.H.: Abschlussbericht des Projektes $C^2A^2S^2E$, DLR-IB 124-2014/3, ISSN 1614-7790 (2014)
36. Heinrich, R., Reimer, L., Michler, A.: Multidisciplinary simulation of maneuvering aircraft interacting with atmospheric effects using the DLR TAU code, RTO AVT-189 Specialists Meeting on Assessment of Stability and Control Prediction Methods for Air and Sea Vehicles, 12–14 Oct, Portsdown West, UK (2011)
37. Heinrich, R.: Simulation of Interaction of Aircraft and Gust Using the TAU code. In: Dillmann, A. et al. (eds.), New Results in Numerical and Experimental Fluid Mechanics IX, Notes on Numerical Fluid Mechanics and Multidisciplinary Design, Springer, Band 124, S. 503–511 (2014)
38. Heinrich, R., Reimer, L.: Comparison of different approaches for gust modeling in the CFD Code TAU, international forum on aeroelasticity and structural dynamics, 24–27 June, Bristol, UK (2013)
39. European Aviation Safety Agency (EASA), Certification Specifications for Large Aeroplanes CS-25. Volume Subpart C—Structure (2010)
40. Khier, W., Dietz, M., Schwarz, T., Wagner, S.: Trimmed CFD Simulation of a Complete Helicopter Configuration, 33rd European Rotorcraft Forum. Kazan, Russia (2007)
41. Benoit, B., Dequin, A.-M., Kampa, K. W., Gruenhagen, P.-M., Basset, P, B.: HOST, A General Helicopter Simulation Tool for Germany and France. In: 56th Annual Forum of the American Helicopter Society Virginia Beach, USA (2000)

42. Van der Wall, B.G., Yin, J.: DLR's S4 Rotor code validation with HART II data: the baseline case, 1st international forum on rotorcraft multidiscipinary technology, Seoul, Korea, 15–17 Oct 2007
43. Pahlke, K.: The GOAHEAD project. In: Proceedings of the 33rd European Rotorcraft Forum, Kazaan, Russia (2007)
44. Wendisch, J.-H., Raddatz, J.: Validation of an unstructured CFD solver for complete helicopter configurations with loose CSD-trim coupling, 40th European Rotorcraft Forum, Southampton, UK, 2.-5. September 2014
45. Enk, S.: Ein Verfahren höherer Ordnung in FLOWer für LES, DLR-IB-124-2007/8 (2008)
46. Spiering, F.: Coupling of TAU and TRACE for parallel accurate flow simulations. In: International Symposium on Simulation of Wing and Nacelle Stall, Braunschweig, Germany (2012)
47. Spiering, F., Kelleners, P.: Coupling of flow solvers with variable accuracy of spatial discretization, new results in numerical and experimental fluid mechanics IX. In: Contributions to the 18th STAB/DGLR Symposium, Stuttgart, Germany, pp. 415–423, Springer, ISBN 978-3-319-03157-6 (2012)

CHAPTER 4

CFD STUDY OF AN ANNULAR-DUCTED FAN LIFT SYSTEM FOR VTOL AIRCRAFT

Yun Jiang, Bo Zhang and Tao Huang

UAV Research Institute, Northwestern Polytechnical University, 127 West Youyi Road, Xi'an 710072, China

ABSTRACT

The present study aimed at assessing a novel annular-ducted fan lift system for VTOL aircraft through computational fluid dynamics (CFD) simulations. The power and lift efficiency of the lift fan system in hover mode, the lift and drag in transition mode, the drag and flight speed of the aircraft in cruise mode and the pneumatic coupling of the tip turbine and jet exhaust were studied. The results show that the annular-ducted fan lift system can have higher lift efficiency compared to the rotor of the Apache helicopter; the smooth transition from vertical takeoff to cruise flight needs some extra forward thrust to overcome a low peak of drag; the aircraft with the lift fan system enclosed during cruise flight theoretically may fly faster than helicopters and tiltrotors based on aerodynamic drag prediction, due to the elimination of rotor drag and compressibility effects on the rotor blade tips; and pneumatic coupling of the tip turbine and jet exhaust of a 300 m/s velocity can provide enough moment to spin the lift fan. The CFD results provide insight for future experimental study of the annular-ducted lift fan VTOL aircraft.

INTRODUCTION

The primary drawback to conventional fixed-wing aircraft is that they must have a runway to create sufficient airflow over the wings such that they may take off and land. Much effort has been directed towards the development of aircraft capable of vertical takeoff or landing (VTOL), which are not restricted to airport runways, but can land and takeoff from any relatively small open area. There are four types of successful and practical VTOL aircraft so far, which are helicopters, vectored jet aircraft, tiltrotors and ducted lift fan aircraft. These aircraft provide solutions to this problem, but also have some disadvantages. Helicopters have complex and costly rotor control system and usually have a limited forward speed of less than 200 knots due to compressibility effects on the rotor blade tips. Vectored jet aircraft, such as the AV-8A Harrier, utilize a turbofan engine for both hover and cruise propulsion. The nozzles are designed for efficient high speed forward thrust,

Citation: Jiang, Y.; Zhang, B.; Huang, T. CFD Study of an Annular-Ducted Fan Lift System for VTOL Aircraft. Aerospace 2015, 2, 555-580. doi:10.3390/aerospace2040555.

but are very inefficient in vertical lift mode, and the high speed and force of exhausting air requires takeoff and landing pads, which must be specially prepared so as not to be damaged. The tilt rotor concept, found in the V-22 Osprey aircraft, uses large diameter propellers powered by two cross-shafted turboshaft engines. Its disc loading is higher than a helicopter, but lower than a turbofan, thus being efficient in the vertical flight modes. However, the large propellers limit the top speed to about 300 knots. Another problem with tiltrotors involves stability control difficulties. Turbulent rotational flow on the propeller blades may occur in descent and cause a vortex-ring state, which causes unsteady shifting of the flow along the blade span and may lead to roughness and loss of aircraft control.

The last approach to VTOL aircraft is the use of a ducted lift fan or fans mounted in the airframe for developing vertical thrust aligned with the aircraft center of mass. The duct is usually closed off during level flight, while forward thrust is provided by horizontal jet engines. Compared to the conventional VTOL aircrafts, lift fan aircraft fly faster and require minimum ground preparation, without high discharge velocity and high surface temperatures to cause ground erosion and reingestion of hot gases to cause the lift engines to stall and lose lift [1,2]. Ducted fans have no exposed rotor, thus being safer than helicopters and tiltrotors. Its compact structure, low pneumatic noise and high safety make it a highly efficient, safe and quiet vertical propulsion method [3,4,5]. The latest development of ducted lift fan aircraft is the Lockheed Martin F-35B joint strike fighter [1].

However, the ducted fans so far are all a circular duct or its variants with a central inlet. The fans are submerged in the fuselage or wings. This design not only limits the size of the fans due to the constraints of the fuselage and wing size, but also increases drag because the wings containing the fans have to be made relatively thick to maintain the depth of the fan ducts. The thick wings create unacceptable drag during forward flight. If the thickness of the wings is made much smaller than the diameter of the fan ducts, the benefits of the duct will be reduced, and the vertical thrust produced by the fan system will be limited. Because of the relatively small size, the circular ducted lift fans thus far all have high disc loading in order to provide sufficient vertical thrust to raise the aircraft off the ground.

According to the momentum theory of ducted fans, high disc loading means higher power required to lift the aircraft, thus leading to lower lift efficiency (see below). To increase lift efficiency while maintaining the lift, the fan area has to be increased. However, the space in the conventional aircraft for circular ducted fans is limited. Like the Ryan XV-5A, the wing size is not big enough to contain larger low disc loading circular fans. Other designs, such as a huge circular lift fan in the center of the aircraft or a plurality of small circular ducted fans around the aircraft, suffer from other problems, such as difficult layout for the fuselage, thick wings and increased complexity, which have made them impractical so far.

It is therefore desirable to provide a design that increases the lift fan area and lift efficiency while keeping all of the benefits of a ducted fan and also avoiding the drawbacks of the conventional VTOL aircraft, such as slow cruise speed, dangerous exposed rotor blades, hot downward high speed and pressure exhausting air, poor stability, etc. We present here an annular-ducted fan lift system for VTOL aircraft with a central fuselage to solve this problem. The annular-ducted fan lift system is mounted between the central fuselage and the peripheral wing (see Section 3.1). With the fuselage in the center, the annular duct can be made much larger around the fuselage; thus, the fan area can be greatly increased to realize low disc loading and high lift efficiency that is comparable to a helicopter rotor. Meanwhile, the larger diameter of the annular duct does not require a very thick wing. A relatively thin wing containing the annular duct can provide enough duct effect. In cruise mode, the annular duct is closed off by upper shutters or aperture and lower louvers to reduce drag and provide aerodynamic lift; thus, the forward speed will not be limited by compressibility effects on the fan blade tips, as in the case of helicopters and tiltrotors.

In the last two decades, as a result of continuing efforts spent in computational fluid dynamics (CFD) calculations and more diffuse access to adequate computational resources, many aerodynamic researches have moved into the branch of CFD, which is more affordable, practical, systematic and reliable [6]. A numerical investigation of the novel annular-ducted fan system before expensive experiments is necessary to provide insights for future studies. The present work describes a CFD study of the annular-ducted fan lift system. Several CFD analyses were performed. First, the power and lift efficiency of the annular-ducted fan system in hover mode were studied and compared to the rotor of a helicopter. Next, the lift and drag of the aircraft with the annular duct system opened during transition from VTOL to horizontal flight were studied. Then, the drag and lift of the aircraft with the annular duct enclosed in cruise mode were predicted to estimate the theoretical maximum flight speed of the aircraft. Finally, pneumatic coupling of the tip turbine and jet exhaust gases for a gas-driven lift fan during VTOL was simulated.

MOMENTUM THEORY OF THE DUCTED LIFT FAN IN HOVER MODE

As shown in Figure 1, with zero velocity ($V_0 = 0$) of incoming flow to the ducted lift fan, the thrust and power can be calculated:

Total thrust of the lift fan system:

$$T = T_{rotor} + T_{duct} = \rho A V_1 V_2 = \rho \sigma A V_2^2 \quad (1)$$

$$V_1 = \sigma V_2 \quad (2)$$

where ρ is the density of air; ρ = 1.225 kg/m^3; A is area of lift fan; σ is the duct diffusion ratio, σ = 0.5–1 and for an unducted rotor σ = 0.5; V_1 is the velocity under the rotor; and V_2 is the far wake velocity.

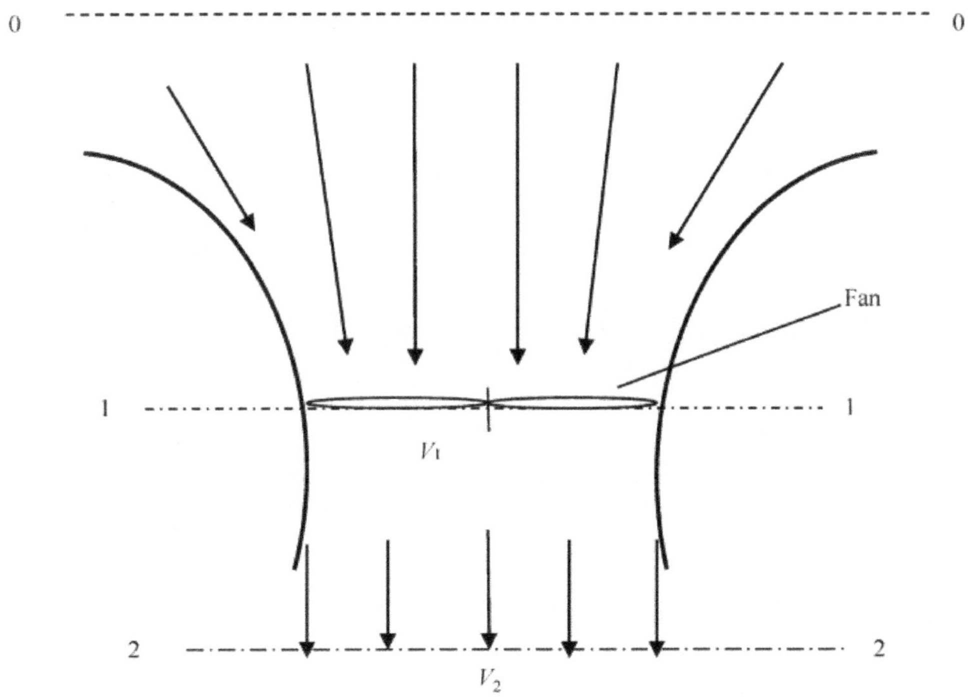

Figure 1. The momentum model of the ducted lift fan system.

Lift fan thrust:

$$T_{rotor} = A(\Delta p) = \frac{1}{2}\rho A V_2^2 \quad (3)$$

$$\frac{T_{rotor}}{T} = \frac{1}{2\sigma} \quad (4)$$

Power of the lift fan:

$$P\eta = T_{rotor}V_1 = \sigma T_{rotor}V_2 = \frac{T^{3/2}}{\sqrt{4\sigma\rho A}} \quad (5)$$

where η is the rotor efficiency.

Lift efficiency (power loading):

$$\frac{T}{P} = 2\frac{\eta}{V_2} = 2\eta\sqrt{\frac{\sigma\rho A}{T}} \quad (6)$$

The power required to provide the same thrust for the ducted fan and an unducted rotor:

$$\frac{P_{ducted}}{P_{unducted}} = \frac{\dfrac{T^{3/2}}{\sqrt{4\sigma\rho A_{ducted}}}}{\dfrac{T^{3/2}}{\sqrt{2\rho A_{unducted}}}} = \sqrt{\frac{A_{unducted}}{2\sigma A_{ducted}}} \quad (7)$$

Equations (1)–(7) describe the momentum model of the ducted lift fan in the VTOL design [5]. It can be seen from Equations (5) and (6) that high disc loading T/A results in higher power, thus lower lift efficiency.

GEOMETRY AND COMPUTATIONAL MESH

Geometry Definition

The annular-ducted lift fan system and the aircraft with the system considered in the paper are shown in Figure 2a. The central fuselage is connected to the peripheral wing through bridges (not shown). The annular-ducted lift fan system is mounted in the annular duct located between the central fuselage and the peripheral wing. The lift fan system has two counter-rotating fans mechanically or pneumatically coupled with two forward turbofan jet engines incorporated in the peripheral wing during VTOL. For this study, the diameter of the aircraft is set to 20 m (R = 10 m). The height of the central fuselage is 2 m with a diameter of 8 or 10 m. The annular duct system is in an area of a radius from 4 or 5 to 7 m, containing two fans of 36 blades (Figure 2b). The width of the annular duct and the length of the fan blades (angle of incidence 27°, chord length 0.53 m, thin blade) are 2 m (7–5 m) or 3 m (7–4 m). The depth of the annular duct is 1 m. The annular duct system is closed off in cruise mode by shutters, a diaphragm or aperture at the upper surface and louvers at the lower surface of the aircraft (Figure 2c). The weight of the aircraft is 10.433 tons.

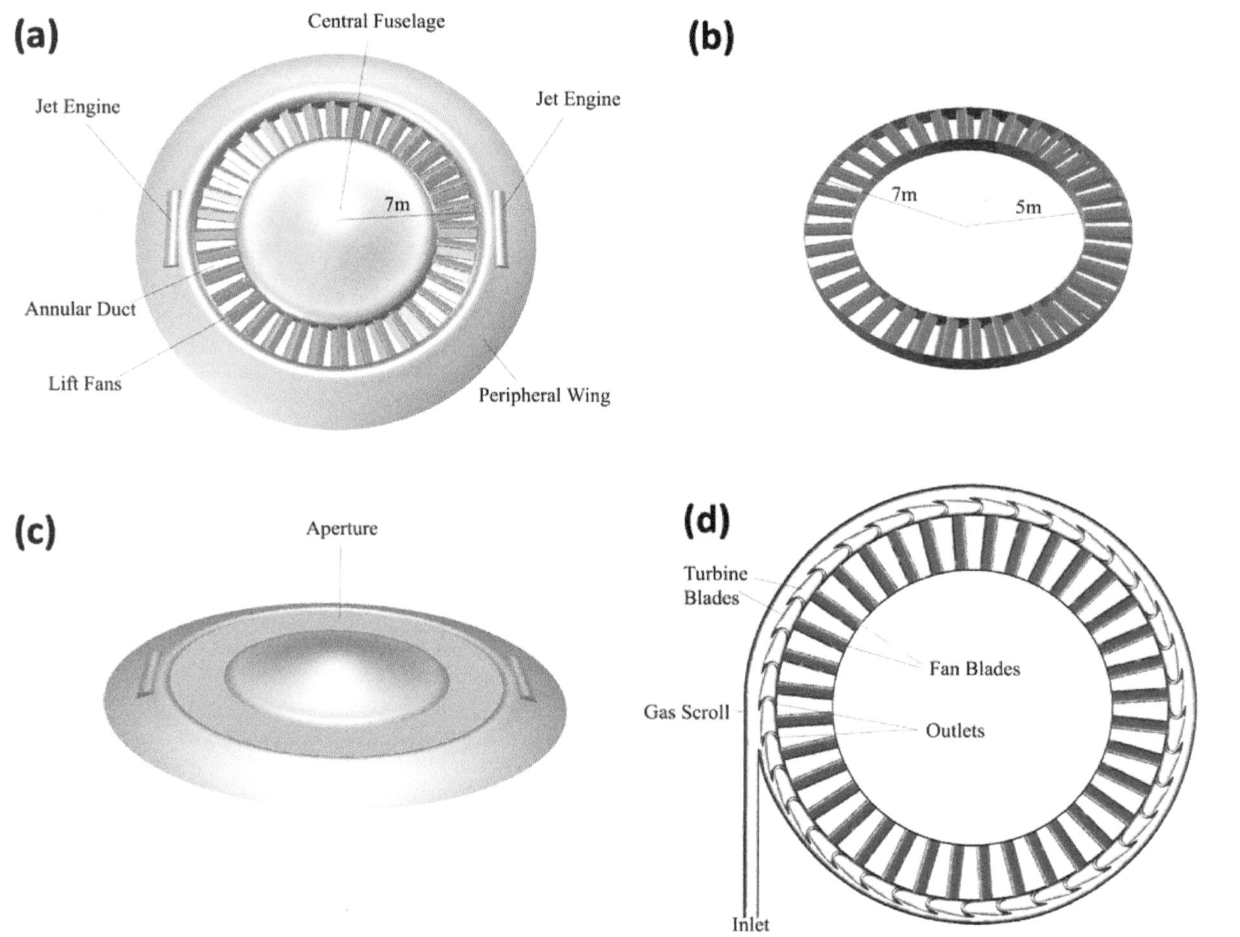

Figure 2. (**a**) The annular-ducted lift fan system and the aircraft with the annular duct opened; (**b**) the lift fans; (**c**) the top front view of the aircraft with the annular duct system enclosed by upper aperture and lower louvers; (**d**) the lift fan and its tip turbine in a gas chamber. The upper wall of the chamber has been removed to expose the tip turbine inside.

The size of the annular-ducted fan lift system and the weight of the aircraft were so determined in order to compare to the Boeing AH-64E Apache helicopter, because the length of the rotor blades of the Apache is 7.32 m. The rotor of the Apache (four rigid blades, length 7.32 m, chord length 0.53 m, angle of incidence 10°) with a simplified fuselage was also numerically simulated to validate the CFD simulation and to compare to the annular-ducted lift fan system.

One structural problem with the lift fan aircraft may be the weak bridges connecting the central fuselage and the peripheral wing. The bridges cannot be too coarse and too great that they may block the inlet and outlet of the annular duct. This is a challenge, but may be solved by balanced consideration of the aerodynamics of the duct and the mechanical strength of the aircraft.

The control of the aircraft can be performed not only by aerodynamic control surfaces but also by thrust vector control of the two turbofan engines respectively. Especially in transition mode, attitude control such as roll, pitch, yaw, can be manipulated by thrust vectoring together with the annular-ducted fan lift system.

The lift fans in the annular duct can be shaft driven or gas driven by forward turbofan engines in VTOL mode. For gas-driven mode, the lift fan can be pneumatically coupled with jet exhaust gases by the tip turbine in a gas chamber. Figure 2d shows the lift fan with its tip turbine in a gas chamber. The tip turbine is fixed to the fan blades. The exhaust gases from the jet engine flow along the chamber and exit to the annular duct from the outlets located between the turbine blades on the inner wall of the chamber. The circulating gases push the turbine and fan to rotate.

Computational Mesh

The unstructured mesh technique has been successfully applied to solve CFD problems [6,7,8,9]. CFD methods based on unstructured grids have an advantage of easily handling complex geometries and can improve the solution accuracy by refining cells locally as required [7]. In this study, unstructured volume meshes were created using ANSYS 14.5 Meshing.

The tetrahedral volume mesh of the annular-ducted lift fan system and the aircraft in a 20R × 10R × 10R flow field in hover and transition mode is shown in Figure 3a. A dimension sensitivity study was carried out to make sure that the dimension was big enough, so that the simulation results would not change with the computational fluid domain dimension (data not shown). The flow field was divided into four volumes in which the two small blocks were dedicated to the fans. The mesh contained 3.0 M cells with three prismatic boundary layers. The first layer height of the boundary layer was fixed to 1–3 mm to make sure that the y^+ value was within the range of 30–300 on most of the surface area.

In cruise mode for aerodynamic surface drag simulation, the annular duct was enclosed. The whole aircraft was like a blended wing body. Due to the symmetry of the aircraft, only half of the aircraft in a 20R × 5R × 10R flow field was simulated (Figure 3b). The mesh contained 650 K cells with 12 prismatic boundary layers. Very small prismatic cells were used in the boundary layer region, with the first layer height only 2 μm. This high resolution was needed to fully resolve the thin boundary layer down to the viscous sublayer. Because the thickness of the viscous sublayer decreases with increasing flow Re number and the Re numbers for airflow around the aircraft are very large (10^8), the viscous sublayer at the surface is very thin. As a result, a very high grid resolution is required close to the walls. This is important because boundary layer separation from the surface determines to a large extent the aerodynamic drag [10]. A value of $y^+ < 1$ was achieved for the attached flow on the surface of the aircraft.

The volume mesh with 1.7 M cells inside and outside the tip turbine chamber is shown in Figure 3c. The computational domain included the gas chamber and part of the annular duct. Hexahedral grids were created inside the chamber to reduce the number of cells and to save computing time.

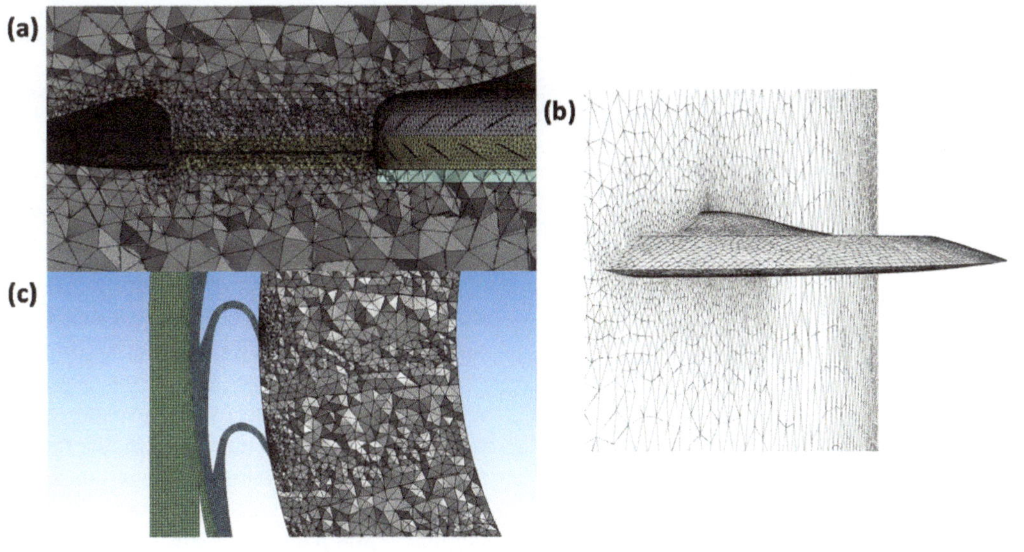

Figure 3. (**a**) Volume mesh of the annular-ducted lift fan system in the central plane for the hover and transition mode study. Different colors show different blocks; (**b**) surface mesh on the surface of the aircraft and the symmetry plane for the aerodynamic drag study; (**c**) volume mesh in the central plane of the gas chamber for the tip turbine study.

Mesh sensitivity studies were carried out in all simulations (Table 1, Table 2 and Table 3). Three cases were considered: coarse grid, medium grid and fine grid. The results show that the differences between medium and fine grid are small. Therefore, all simulations in the study were performed in the range of a medium–fine grid. The grid independency study was also performed in the validation studies to compare the simulation results with known experimental data. According toTable 4 and Table 5, the comparison of the results shows that the medium and fine grid are close and in agreement with the experimental data.

Table 1. Grid sensitivity analysis in the power prediction study

Parameters	Coarse Grid	Medium Grid	Fine Grid
Element number	1.5M	2.3M	3.0M
Power (kW)	2117	2097	2085

Table 2. Grid sensitivity analysis in the drag prediction study

Parameters	Coarse Grid	Medium Grid	Fine Grid
Element number	323K	650K	1.26M
C_D	0.00842	0.00680	0.00676

Table 3. Grid sensitivity analysis in the tip turbine study

Parameters	Coarse Grid	Medium Grid	Fine Grid
Element number	323K	670K	1.7M
Time-averaged moment (kN m)	162	158	156

Table 4. Grid independency study of the helicopter power

Parameters	Coarse Grid	Medium Grid	Fine Grid	Actual
Element number	1.12M	2.43M	4.68M	
Power (kW)	2845	2688	2685	2676

Table 5. Airfoil grid resolution study

Parameters	Coarse Grid	Medium Grid	Fine Grid	Experimental
Cell number	355K	638K	1.55M	
C_L	0.266	0.260	0.258	0.241
C_D	0.0110	0.0088	0.0075	0.0079

BOUNDARY CONDITION AND SIMULATION SETUP

Fluid dynamic simulations were performed using ANSYS Fluent 14.5, which has been used in various studies [6,9,10,11,12]. The 3D Reynolds-averaged Navier–Stokes (RANS) equations for the mean flow quantities were solved with different turbulence models. The RANS method uses the Boussinesq hypothesis to relate the Reynolds stresses to the mean velocity gradients to facilitate closure of the governing equations.

For the lift fan system in hover and transition mode, a pressure-based solver type with absolute velocity formulation, the Green–Gauss node-based gradient option and steady approach were used in the analyses. The shear stress transport (SST)k–ω turbulence model developed by Menter [13,14] was selected for the turbulence model. This model incorporates a blending function designed to activate the standard k–ω turbulence model in the near-wall region and to activate the k–ε turbulence model away from the surface. The SST k–ω turbulence model also incorporates a damped cross-diffusion derivative term in the ω equation, along with a modified definition of turbulent viscosity to account for the transport of the turbulent shear stress [15]. These features make the SST k–ω model more accurate and reliable for adverse pressure gradient flows, airfoils and transonic shock waves. The Semi Implicit Method for Pressure Linked Equations (SIMPLE) scheme was used to resolve the pressure–velocity coupling with the second order discretization for the pressure equation, the third order Monotonic Upstream-Centered Scheme for Conservation Laws (MUSCL) for the momentum, the turbulent kinetic energy and turbulent dissipation rate [6]. Multiple reference frames were used for the rotor fluid zone and the stationary fluid zone. Velocity-inlet, pressure-inlet and pressure-outlet boundary conditions were applied over the inlet and outlet sections. In transition mode, unsteady calculations were performed. The areas surrounding the fans were designated as a sliding mesh. Sliding interfaces separated the rotating domain from the stationary domain. The initial data for the unsteady run were obtained from steady calculations.

For aerodynamic surface drag prediction, a density-based solver type with the SST k–ω turbulence model was used. The second order upwind discretization was used for the flow, turbulent kinetic energy and specific dissipation rate. The air was treated as an ideal gas having constant specific heat, whereas viscosity was bound to the three coefficients the Sutherland law, which automatically enabled the energy equation resolution, because it is dependent on the actual temperature [6]. Velocity-inlet and pressure-outlet conditions were imposed over the inlet and outlet sections.

For the tip turbine study, the k-epsilon realizable model was selected for turbulence model [16]. This model is more precise for flows involving rotation, separation and recirculation. Pressure Staggering Option (PRESTO!) was used for discretization of the pressure equation and the third order MUSCL for the momentum, the turbulent kinetic energy and turbulent dissipation rate. The boundary condition used to simulate the jet exhaust at the inlet of the gas chamber was velocity-inlet. Unsteady calculations were performed using a sliding mesh to simulate the tip turbine blades rotating in the gas chamber. The initial data for the unsteady run were obtained from steady calculations.

All simulations were performed with double precision.

NUMERICAL MODEL VALIDATION

Because this is a concept study of a novel annular-ducted fan lift system, no experimental data are available now for the exact geometry. The validations were performed on similar models using the same settings and meshes to compare the predicted results with the known experimental data. The power of the AH-64E Apache helicopter, the lift and drag coefficients of a generic fan-in-wing configuration [1] and the lift and drag coefficients of a rectangular wing of the NACA 0012 airfoil were simulated and compared.

Validation of Power Prediction

A helicopter rotor of the VR-12 airfoil with a simplified fuselage in hover mode was created. The settings for the power prediction of the annular-ducted fan lift system were applied to study the power of the AH-64E Apache. A tetrahedral mesh of 2.43 M elements was created. The computational domain was a 20R × 10R × 10R flow field, which was divided into two volumes in which the small block was dedicated to the rotor. $y^+ = 30–300$.

The results show that the rotor generates a long wake in the flow field. Three-dimensional streamline analysis demonstrates that the wake is swirling around the central axis (Figure 4a). There is energy wasted on swirling and also on vortex at the rotor tips (Figure 4b).

When the rotor rotates at 344 rpm, the net lift of the helicopter is obtained as 102,243 N (10.433 tons, the maximal weight of the Apache).

According to Equation (5), the induced power of the rotor:

$$P_{induced} = \frac{T^{3/2}}{\sqrt{4\sigma\rho A}} = 1615\text{kW} \tag{8}$$

where $\sigma = 0.5$ for the unducted rotor, and rotor area $A = 167.33$ m^2.

The mechanical shaft power of the rotor can be calculated based on the moment of the rotor and the rotational speed:

$$P_{shaft} = \omega rF = 2\pi\frac{n}{60}rF = 2685\text{kW} \tag{9}$$

where moment rF = 74,548 N m, and rotational speed n = 344 rpm.

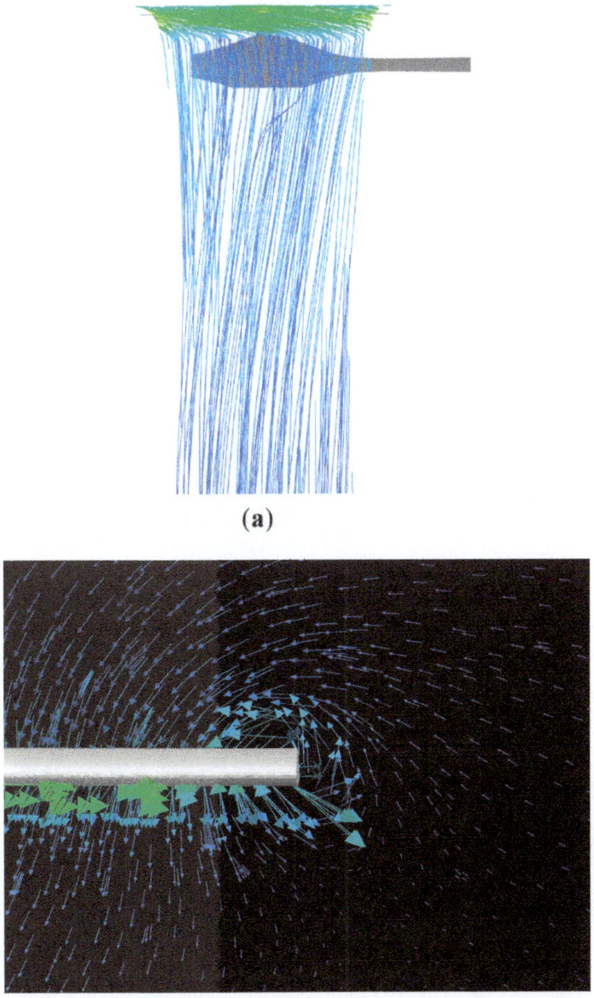

Figure 4. (a) Three-dimensional swirling streamlines in front of the central plane of the helicopter in hover mode; (b) the rotor tip vortex in the central plane of the helicopter.

Therefore, the rotor efficiency of the model is:

$$\eta = \frac{T^{3/2}}{P_{shaft}\sqrt{4\sigma\rho A}} = 0.601 \tag{10}$$

Lift efficiency $T/P_{shaft} = 10{,}433/2685 = 3.88$ kg/kW.

The simulated power 2685 kW is very close to the actual value of the Apache power $2 \times 1487 \times 0.9 = 2676$ kW, supposing that the mechanical transmission efficiency is 0.9. The actual rotor efficiency $\eta = P_{induced}/P = 0.603$, and lift efficiency $T/P = 3.89$ kg/kW. Therefore, the prediction of the power is quite accurate.

Validation of Transition Mode

The simulation of a three-dimensional configuration with a fan rotating in the plane of a wing [3] was repeated with an unstructured tetrahedral grid to validate the transition study. The model airfoil is of the NACA 16-020 type. One single circular fan is submerged at the rear part of the wing as in the case of the annular-ducted lift fan (Figure 5). In [3], a structured mesh was used, and three different freestream velocities and three different fan rotation speeds were investigated; however, here, only freestream velocity U = 30 m/s, corresponding to a Reynolds number $Re = 1.5 \times 10^6$, and rotation speed N = 21,000 rpm at different angles of attack were repeated to compare to the experimental data in [3]. The settings for the transition study were used. Ten chords were used in each direction from the wing to create the computational domain. The mesh contained 1.2 M tetrahedral cells. $y^+ = 30$–300 was achieved on the surface of the wing. However, for an angle of attack of 20°, flow separation happened, which caused significantly drop of lift, and the iterations could not converge; therefore, $y^+ = \sim 1$ on the surface of the wing was used.

The CFD results show a good agreement with the experimental data [3] (Figure 6), although the drag coefficients may be a little under-predicted. As shown in Figure 5, the main features encountered in a jet in crossflow are predicted by the simulation as in [3]. Even at the high angle of attack of 20°, the flow remains attached.

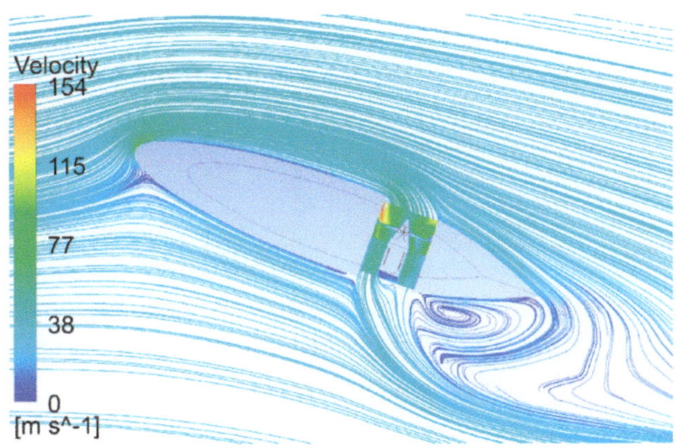

Figure 5. Streamlines in the central plane of the generic fan-in-wing configuration at an angle of attack of 20°. Freestream velocity U = 30 m/s, and fan speed N = 21,000 rpm.

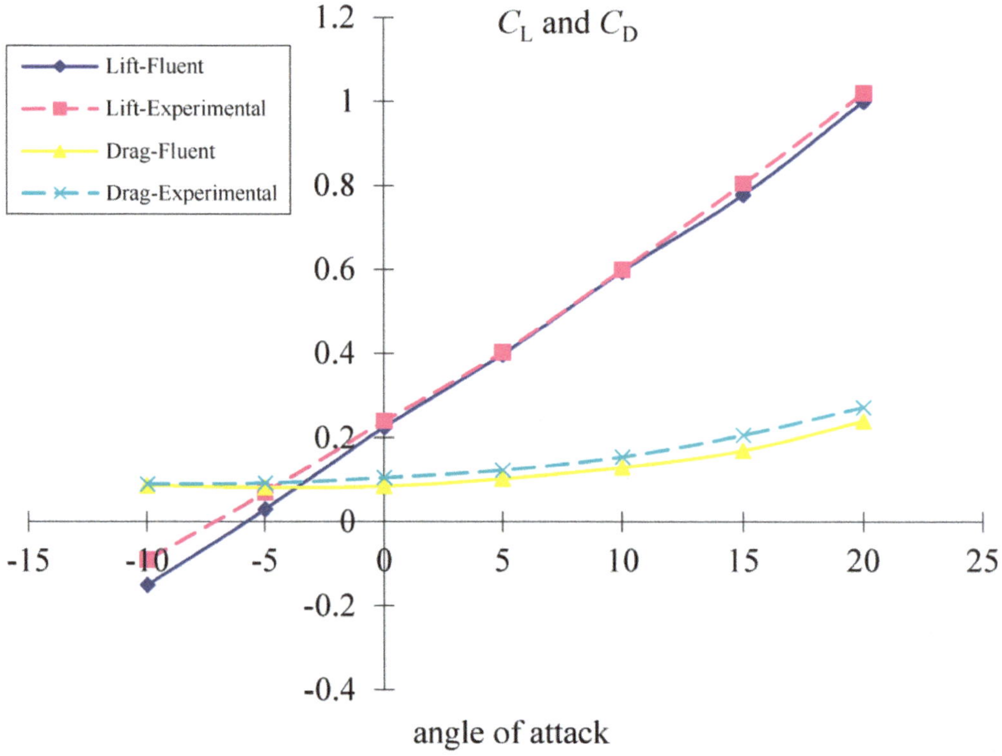

Figure 6. Lift and drag coefficients as a function of the angle of attack for U = 30 m/s and N = 21,000 rpm. The experimental data come from Thouault [3].

Validation of Aerodynamic Drag Prediction

The aerodynamic drag prediction was validated with a three-dimensional rectangular wing of the NACA 0012 airfoil (c = 0.635 m, b = 2.438 m) in an eight-foot wind tunnel at M = 0.7 and Re = 9×10^6 with the transition point fixed at 5% chord on both the upper and lower surfaces of the airfoil [17]. The unstructured grid and the settings for drag prediction were validated. Ten chord were used from the wing to create the computational domain with 1.55 M tetrahedral cells. $y^+ = \sim 1$ was achieved on the surface of the airfoil.

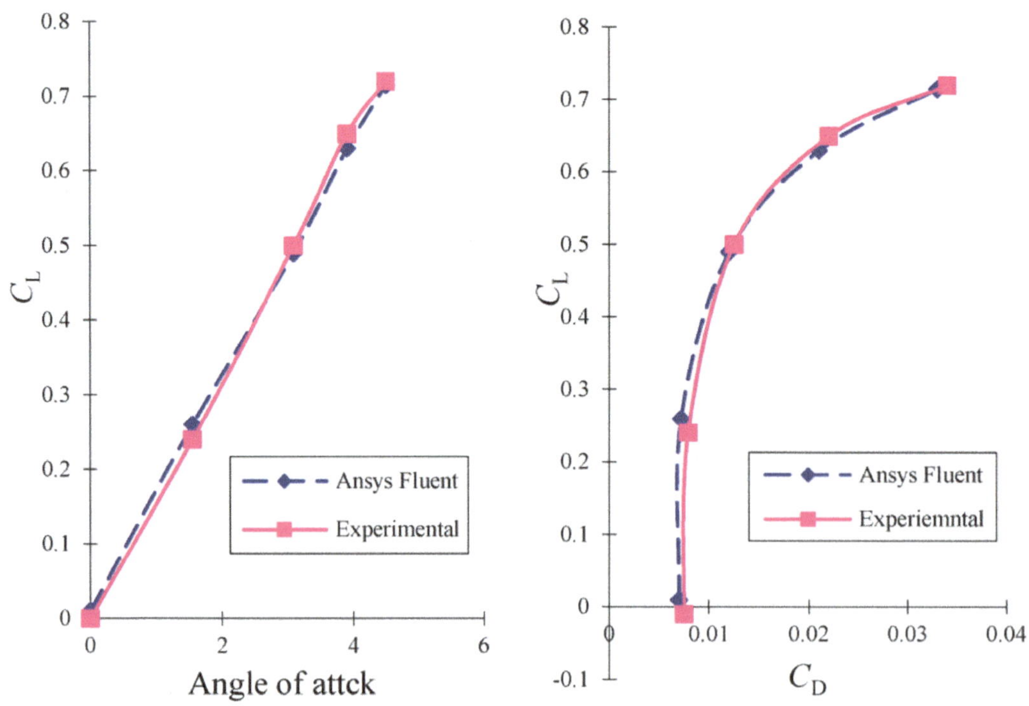

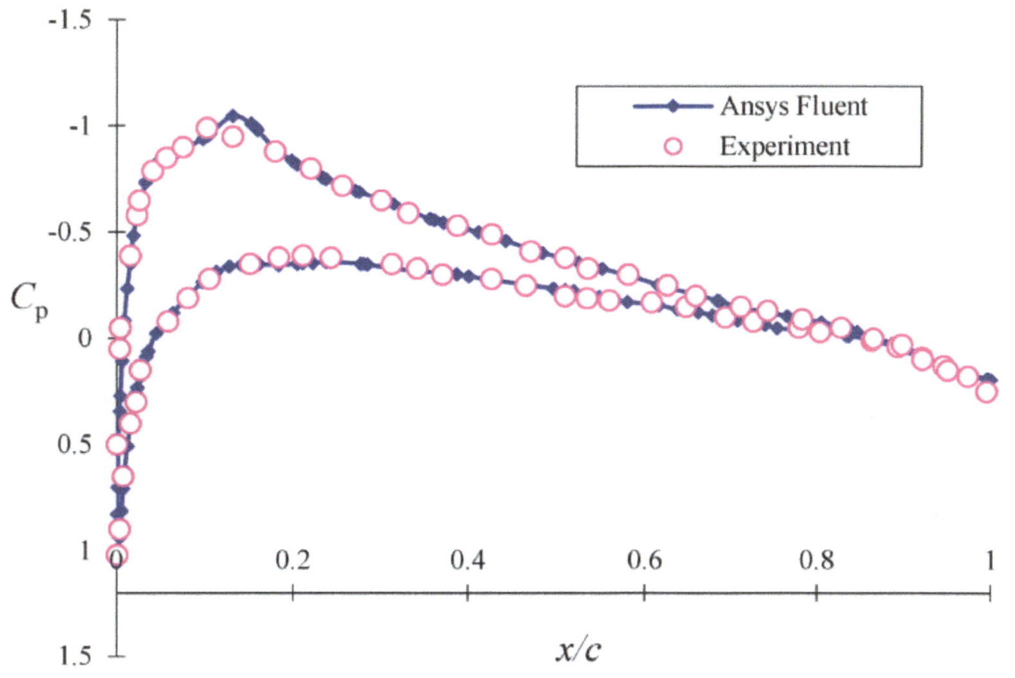

Figure 7. Comparison of computed and experimental data for the NACA-0012 airfoil at M = 0.7 and Re = 9 × 10[6]. (**a**) Lift vs. angle of attack; (**b**) lift vs. drag polar; (**c**) pressure coefficient distribution at z = 1 and α = 1.55°. The experimental data come from Harris [17].

As shown in Figure 7, good agreement at the given angle of attack α is obtained by correction of α and C_L for wall interference according to Harris's estimates [17,18,19]. The results indicate that the simulation may slightly under-predict lift and drag.

NUMERICAL SIMULATION RESULTS

Annular-Ducted Fan Lift System in Hover Mode

The annular-ducted fan lift system in hover mode was simulated. Figure 8 shows the streamlines and static pressure contour on the surface of the aircraft and inside the lift fan system in hover mode. The 3D streamlines through the annular duct show that the lift fan system generates a long wake in the flow field (Figure 8a). The wake is not swirling due to the counter rotation of the two fans; thus, there is no energy wasted on swirling air. The vortexes at the fan tips are also eliminated by the duct walls. The annular duct has a diffused outlet, which prevents the downstream flow from contraction and increases the duct lift. The wake streamlines still look a little contracted in Figure 8a because of the low pressure beneath the central fuselage, but the cross-sectional area of the flow is not decreased according to the surface streamlines in the central plane (Figure 8b). The pressures on the upper surfaces of the duct walls and fan blades are low (Figure 8c,d). The low pressure caused by the induced flow on the inlet lips of the duct and upper surface of the peripheral wing generates duct lift (Figure 8b,d).

For the 3-m annular duct, when the two fans contra rotate at +103 and −99 rpm producing the same torque, the system lift is T = 102,408 N (10.433 tons, the maximum weight of the Apache helicopter), among which fan thrust is T_{rotor} = 58,361 N; thus, according to Equation (4) of the momentum theory:

$$\sigma = \frac{T}{2T_{rotor}} = 0.88 \qquad (11)$$

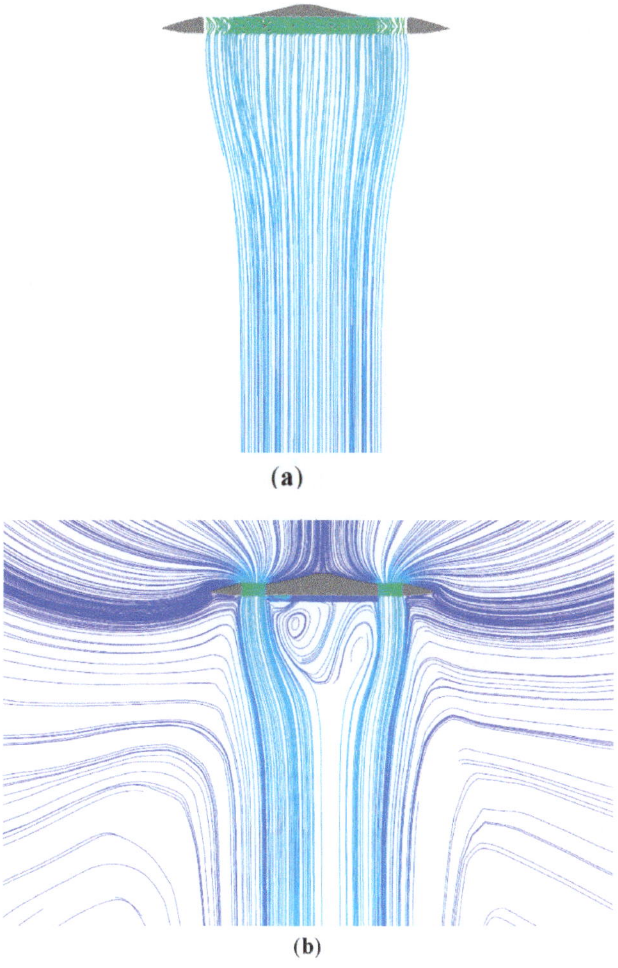

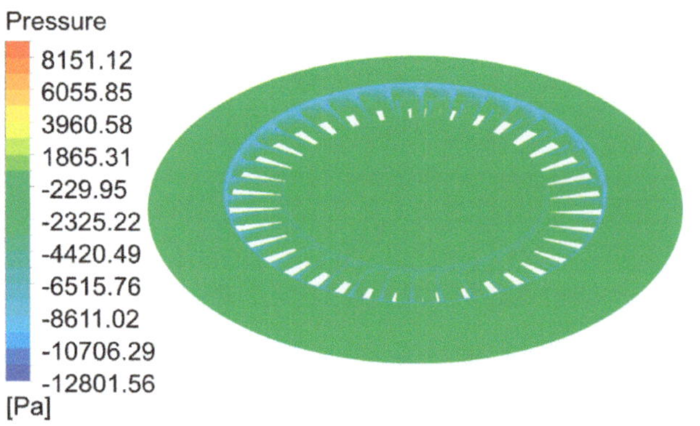

(c)

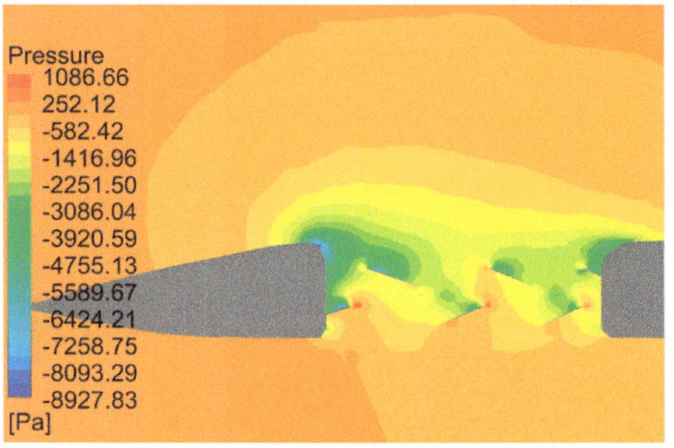

(d)

Figure 8. (**a**) Three-dimensional streamlines in front of the central plane in hover mode; (**b**) surface streamlines in the central plane of the aircraft; (**c**) static pressure on the upper surface of the fans, duct, fuselage and peripheral wing; (**d**) pressure contour inside the annular-ducted lift fan system.

The induced power according to Equation (5),

$$P = \frac{T^{3/2}}{\sqrt{4\sigma\rho A}} = 1548\text{kW} \tag{12}$$

where fan area A = 103.62 m^2.

Compared to the Apache rotor, according to Equation (7):

$$\frac{P_{ducted}}{P_{unducted}} == \sqrt{\frac{A_{unducted}}{2\sigma A_{ducted}}} = 0.95 \tag{13}$$

where $A_{unducted}$ = 167.33 m^2 for the rotor of the Apache helicopter.

It can be seen that, although the disk area of the Apache rotor (167.33 m^2) is much larger than that of the lift fan system (103.62 m^2), the induced power for the annular-ducted lift fan system to provide the same lift is even lower.

The moment of the upper fan is obtained as 98,925 N m, and the moment of the lower fan is 98,292 N m; thus, the mechanical shaft power of the lift fan system:

$$P_{shaft} = 2\pi\frac{n}{60}rF = 2085\text{kW} \tag{14}$$

Therefore, the fan efficiency is η = 1548/2085 = 0.74. The lift efficiency T/P_{shaft} = 10,433/2085 = 5.00 kg/kW.

For the AH-64E Apache helicopter, it is known that the maximum weight is T = 10.433 ton (102,243 N), power P = 2676 kW, A = 168.11 m^2, so the rotor efficiency $\eta = P_{induced}/P$ = 0.603, and the lift efficiency T/P = 3.89 kg/kW. Therefore, compared to the Apache rotor, the 3-m lift fan system is more efficient and can save 22% of the power.

The reasons why the annular-ducted fan can save energy are two-fold [20]: (1) elimination of rotor tip vortex loss, wake swirling loss and wake coning loss: the duct not only eliminates fan tip vortex, but also prevents the downstream flow from contraction; and (2) additional duct lift. Beside the fan thrust, there is an additional duct lift caused by the low pressure on the duct inlet lips and the upper surfaces of fuselage and peripheral wing, which can be almost as much as the fan thrust. The additional duct lift helps reduce the required fan thrust, thus reducing the drag on fan blades and the corresponding power required to run the fan. According to the simulation results, about 1/4 of the lift comes from the upper fan, 1/4 from the lower fan, 1/4 from the annular duct lips and 1/4 from the peripheral wing due to the induced flow over the wing (Figure 8b,d).

It is known that a conventional circular ducted fan is more efficient than an unducted fan or propeller, so this result that the annular-ducted fan system is more efficient than a rotor is not surprising.

For a 2 m-wide annular duct, when the fans contra rotate at +118.5 and −116 rpm, the system lift is T = 102,270 N, among which fan thrust is T_{rotor} = 56,698 N and fan area A = 75.39; thus, σ = 0.902; the induced power P = 1791 kW; and $P_{ducted}/P_{unducted}$ = 1.11.

The moment of the upper fan is 99,455 N m, and the moment of the lower fan is 100,603 N m; thus, the mechanical shaft power P_{shaft} = 2455 kW; the fan efficiency η = 0.73; and the lift efficiency T/P_{shaft} = 4.25 kg/kW. Therefore the 2-m annular-ducted lift fan system also needs less power and has higher lift efficiency than the Apache rotor to provide the same lift. Furthermore, if the ground effect is considered and the distance from the aircraft to the ground is 10 m, the shaft power will be only 1880 kW, much lower than the Apache's power. The lift efficiency T/P = 5.55, which is much higher than the Apache rotor.

Although the 2-m annular duct system requires a little more power than the 3-m duct due to the reduced ducted fan area, the compact structure may reduce the dead space of the lift fan system during horizontal flight. After all, the VTOL process only occupies a short time of the flight.

Transition from VTOL to Horizontal Flight

The low disc loading ducted lift fan system can introduce tremendous momentum drag during transition from vertical takeoff mode to cruise mode. Momentum drag is generally caused by a directional change of the airflow going through the lift fan system. To achieve successful transition, the aircraft with the lift fan system must overcome the momentum drag and increase its speed to a certain level to gain sufficient aerodynamic lift. As shown in Figure 9, the minimum speed for the aircraft to fly with the duct enclosed on aerodynamic lift is 30 m/s at an angle of attack of 15° (Spot A).

To approach this speed at an angle of attack of zero degrees (Figure 10a), the momentum drag forms a high peak at Spot B (Figure 9), which needs a forward jet thrust of 0.63 of the aircraft weight to be overcome. The peak exists because the momentum drag increases with the forward speed and the rotational speeds of the fans, while the rotational speeds of the fans needed to maintain the lift decrease with the forward speed. In the CFD simulation, the rotational speeds of fans were adjusted to keep the lift equal to the weight of the aircraft when the forward speed increased.

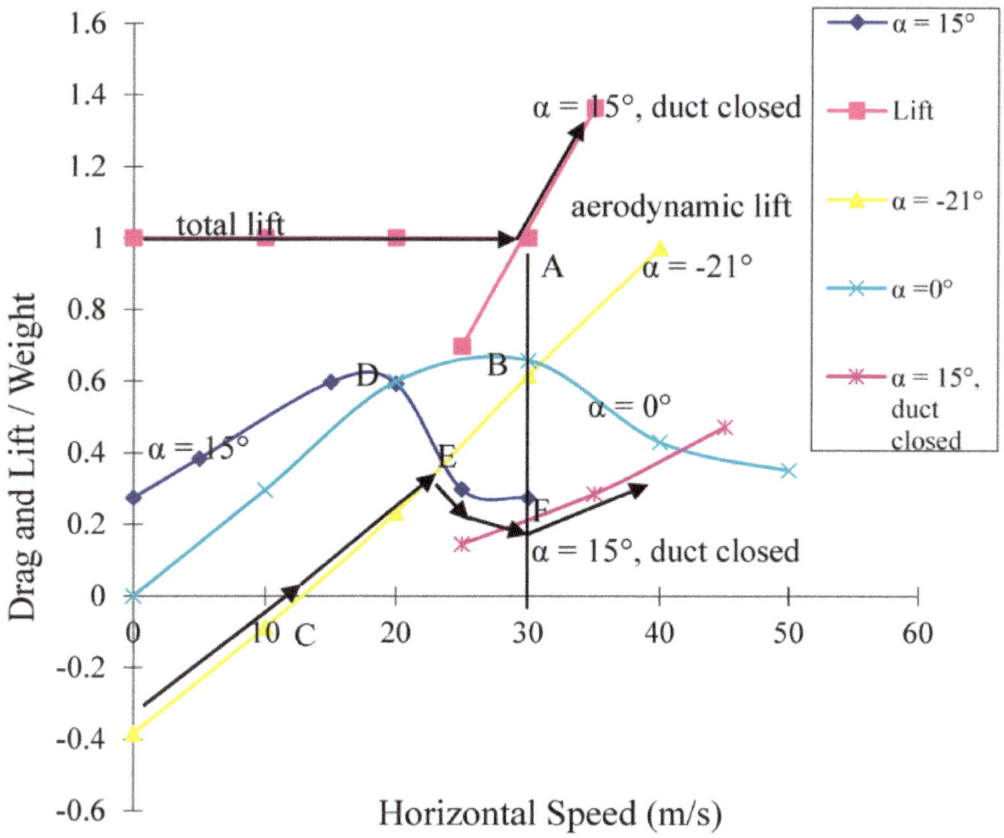

Figure 9. Computed time-averaged net drag and lift-to-aircraft weight ratio at different angles of attack in transition mode. The lift was maintained equal to the weight of the aircraft through the change of the rotational speed of the fans.

In Figure 9, the drag is not only aerodynamic drag, but net drag, including the horizontal force generated by the lift fan system. If the aircraft flies at a negative angle of attack, as shown in Figure 10b, there is a forward force produced by the lift fan system to offset aerodynamic drag, so the net drag can be negative. When the aircraft starts to fly at a −21° angle of attack (Figure 10b), the lift fan system generates a forward force to push the aircraft to fly forward without additional forward jet thrust. The aircraft can easily reach the speed of 13 m/s (Spot C) where the net drag is zero, which means that the forward force produced by the lift fan system equals the aerodynamic drag. After that point, the net drag increases rapidly with the increase of speed. The aircraft still needs additional forward thrust of 0.61 of the aircraft weight to reach the speed of 30 m/s.

If the aircraft starts at an angle of attack of 15° (Figure 10c), the lift fan system will produce a backward force, which is a positive net drag at a speed of 0 m/s. With the increase of the speed, there is also a high peak of drag at the speed of 18 m/s (Spot D), but after that point, the drag decreases rapidly.

Therefore, the best way to achieve efficient transition seems: (1) to rise at an angle of attack of zero degrees; (2) to change the angle of attack to −21° and fly forward without additional forward jet thrust (Figure 10b); (3) when the speed reaches about 13 m/s (Spot C), continue to fly with additional forward jet thrust; (4) when the speed reaches about 23 m/s (Spot E), change the angle of attack to 15° (Figure 10c), slow down the rotational speeds of the fans and continue to fly with additional forward jet thrust until the speed reaches 30 m/s (Spot F); (5) to stop the lift fan system and close off the duct; and (6) to continue to fly with aerodynamic lift and forward jet thrust. In this way, the peak of momentum drag is much lower (near Spot E), which only needs forward thrust of about 0.32 of the aircraft weight to overcome. The lift and net drag increase along the arrows in Figure 9.

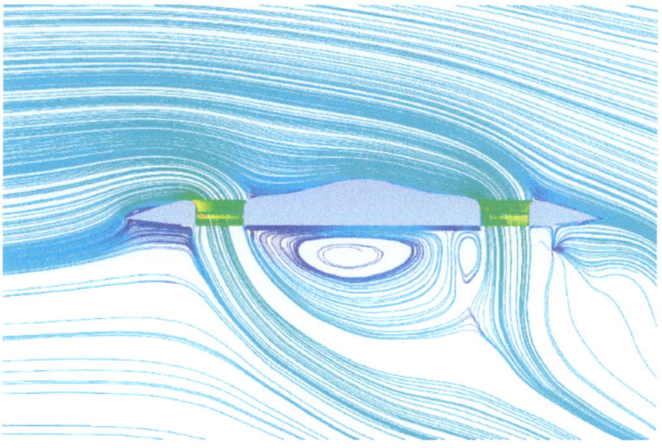

(a)

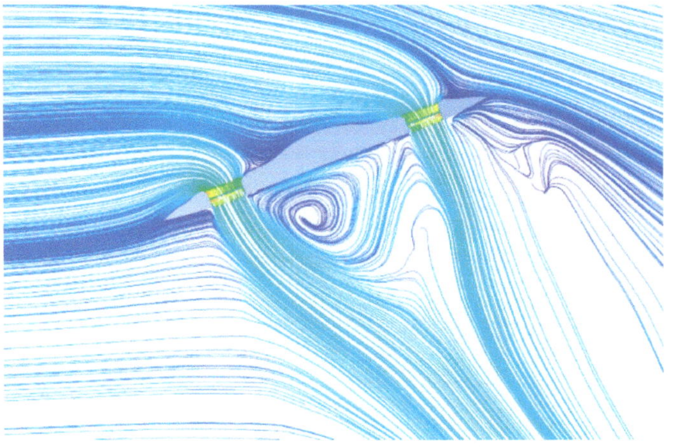

(b)

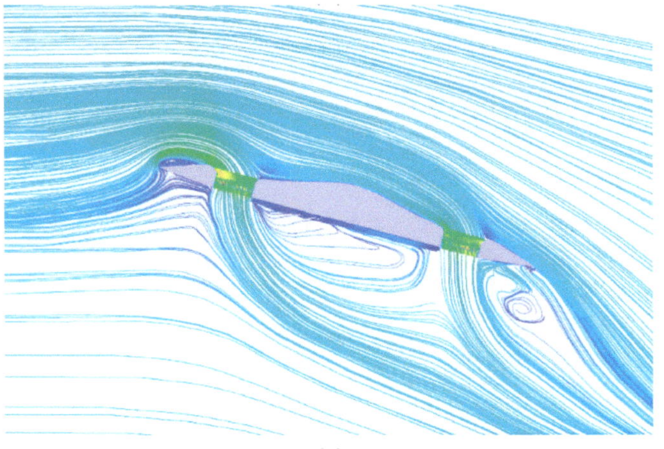

(c)

Figure 10. The aircraft attitudes and streamlines in the central plane of the aircraft in transition mode. The lift was maintained equal to the weight of the aircraft by adjusting the fan rotational speeds. (**a**) Angle of attack α = 0°, freestream speed U = 20 m/s, fan speeds N = +108, −117 rpm; (**b**) angle of attack α = −21°, U = 10 m/s, fan speeds N = +122, −121 rpm; (**c**) angle of attack α = 15°, U = 20 m/s, fan speeds N = +91, −97 rpm.

For vertical landing, the aircraft can start to reduce the forward speed at an angle of attack of 15°, then open the duct and start the lift fan system (Figure 10c). When the forward speed slows down, then change the angle of attack to zero degrees to land (Figure 10a).

The aircraft attitudes and the streamlines in the central plane in transition mode are presented in Figure 10a–c. The incoming flow on the aircraft's upper side is strongly entrained by the lift fan system, maintaining attached flow at the aircraft's trailing edge.

The attitude control can be performed through changing the direction of the thrust from the two jet engines respectively. Thrust vectoring can also be used to offset the nose up pitching moment and rolling moment during the transition.

Aerodynamic Drag and Horizontal Cruise Speed Predictions

During cruise flight, the annular-ducted fan lift system will be shut off to reduce drag and provide aerodynamic lift, so the compressibility effects on the fan blade tips and fan drag will be eliminated and replaced by aerodynamic drag. With the lift fan system enclosed during horizontal flight, the aerodynamic drag of a specific size may be smaller than the rotor drag, so the aircraft may possibly fly faster than helicopters and tiltrotors.

The drag and lift of the aircraft with the annular duct system enclosed during level flight were simulated. As shown in Figure 11, the airflow speed on the cambered upper surface of the aircraft increases (Figure 11a) and causes low pressure to generate lift (Figure 11b). The aerodynamic drag increases with flight speed (Figure 12).

Based on the balance of drag and jet thrust, when the drag equals the engine thrust, the maximum speed of the aircraft can be predicted. Therefore, when the flight speed increases to 0.35 Ma (428 km/h) at a zero degree angle of attack (Figure 12, Spot a), the drag is 15.7 kN. If the maximum jet thrust is just 15.7 kN, the maximum speed will be 428 km/h. Supposing the propulsive efficiency is 0.7, the jet power at this point is 2676 kW, which is the power of the Apache. While the maximum speed of the Apache is 293 km/h, the speed of 428 km/h is 46% faster than the Apache. At this point, the lift is 131 kN, enough to carry a weight of 102 kN. The configuration of the aircraft can be slightly modified (not so cambered) to reduce the lift to equal the weight without increasing the drag, so that the aircraft can fly in the minimum drag mode. The maximum speed is predicted here only based on drag prediction; some real-world effects, including engines, were not considered. The prediction of drag just gives some ideas about how fast the aircraft may theoretically fly without rotor drag.

If the jet thrust increases to 36.3 kN, the maximum speed will reach 0.52 Ma (625 km/h; Figure 12, Spot b). Supposing that the propulsive efficiency is 0.7, the jet engine power at this point is 9180 kW, which equals the power of the V-22 Osprey. The maximum speed of the Osprey is 509 km/h.

Without the constraints of rotor drag and the compressibility effects on the rotor blade tips, the speed of the aircraft can increase further if higher thrust turbofan jet engines are deployed. To reach the speed of 0.7 Ma (857 km/h), the aircraft will need jet thrust of 72.3 kN (Figure 12, Spot c). The lift at this point is 448 kN, much higher than the weight.

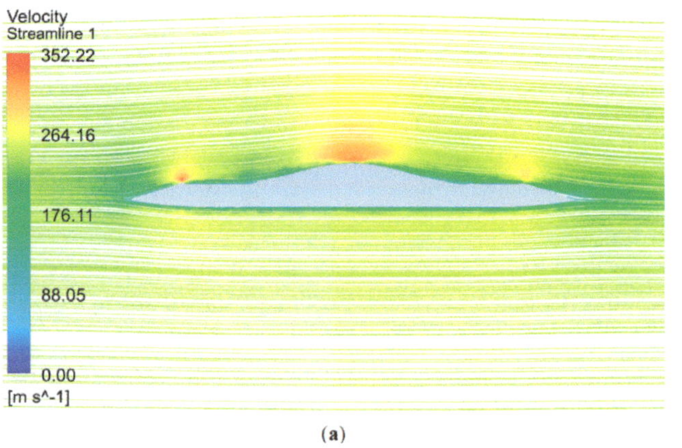

(a)

(b)

Figure 11. (**a**) Streamlines in the central plane of the aircraft in cruise mode; (**b**) pressure coefficient in the central plane of the aircraft at speed of 0.7 Ma.

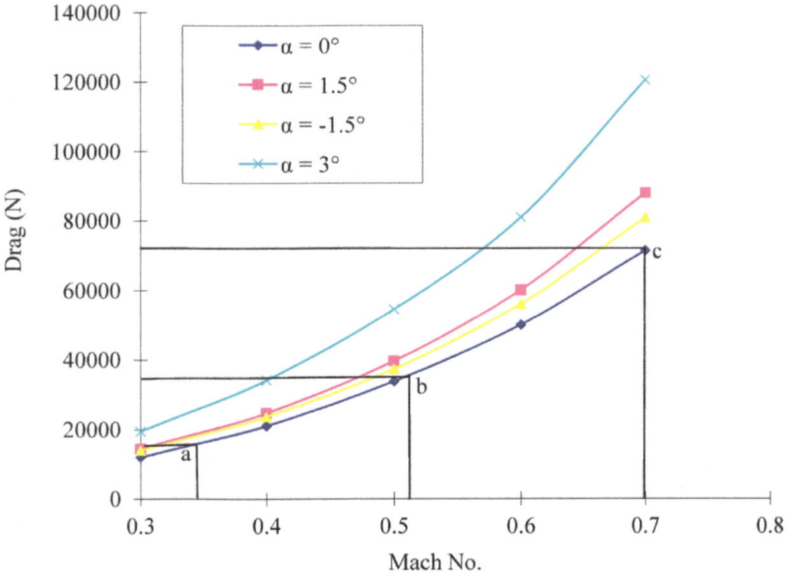

Figure 12. The computed drag (the engines were not considered) of the aircraft increases with cruise speed at different angles of attack.

The annular-ducted lift fan aircraft can also be in other configurations, rather than the above circular flying saucer shape, to reduce drag. A triangular or rhombic configuration (Figure 13), like a blended wing body [21,22], may fly faster and is easier to deploy conventional aerodynamic control surfaces. Aerodynamic drag is determined by the shape and size of the aircraft. A thinner and streamlined shape has less drag and thus may fly faster.

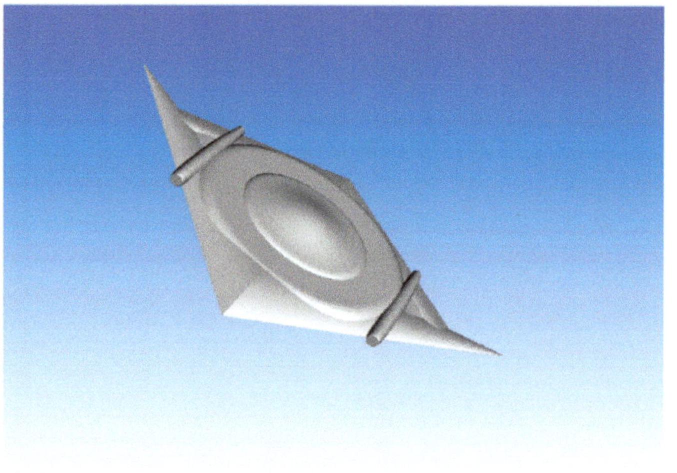

Figure 13. Rhombic-shaped annular-ducted lift fan aircraft.

Pneumatic Coupling of Tip Turbine and Engine Exhaust Gases

Shaft-driven and gas-driven annular-ducted lift fan systems are both feasible. The F-35 joint strike fighter finally adopted a shaft-driven lift fan [1], but a gas-driven annular-ducted lift fan system may be less expensive to develop than the shaft-driven system, because it does not require modifying the cruise engine. There are no gears or high power shafting required, and pneumatic coupling also provides balanced engine operation and engine out safety [2]. A preliminary gas-driven tip turbine CFD study was performed without validation due to a lack of experimental data.

As shown in Figure 14, the engine exhaust gases are ducted to the inlet of the gas chamber to spin the tip turbine. With a velocity inlet speed 300 m/s, a significant flow forms in the chamber (Figure 14a,b). The gas flow starts from the inlet, circulates in the chamber, passes through the turbine blades, exits from the outlets on the chamber and discharges itself to the annular duct. When the turbine and lift fan begin to rotate, the exiting gases mix with the rotating air (Figure 14c). The gas flows passing through the tip turbine blades cause a pressure difference on the two surfaces of a turbine blade (Figure 14d,e), which pushes the turbine to spin. A steady solution could not be achieved due to the strong unsteadiness on the turbine blade loading. Unsteady analysis showed that the forces on the blades slightly fluctuated with time, and time-averaged results could be obtained. With an inlet speed 300 m/s, when the turbine is still, the average moment of the gas flows on the turbine blades is about 156 kN m. When the turbine rotates at 120 rpm, the moment on the turbine blades is about 108 kN m, greater than the contra-moment on the lift fan of 100.6 kN m (see Section 6.1), thus enough to maintain the rotation of the lift fan. The gauge pressure inside the chamber is as low as about 0.1 atm (Figure 14d), making it easy to couple with the jet exhaust.

According to the lift fan experimental results with a fan diameter of 1.5 m, the fan augments the turbojet thrust by a factor of three to one for VTOL [2]. With the fan size increased to a diameter of 14 m, the fan may possibly augment jet thrust by more than five-fold.

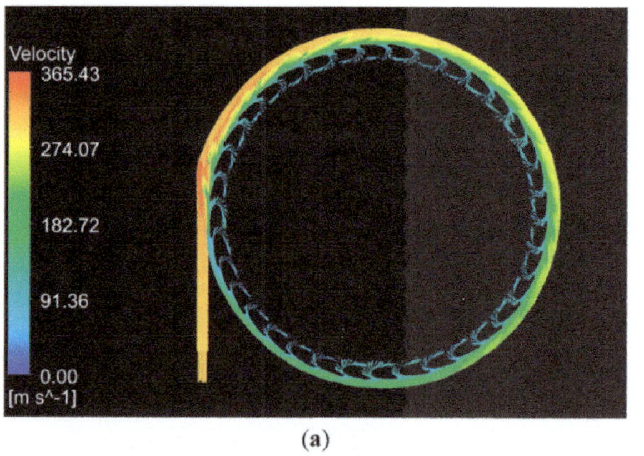

(a)

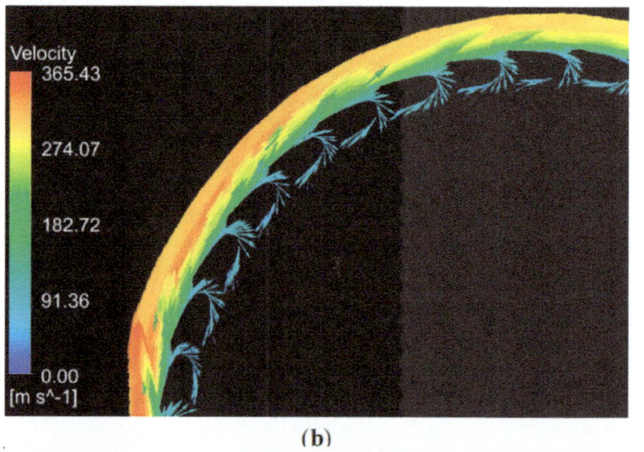

(b)

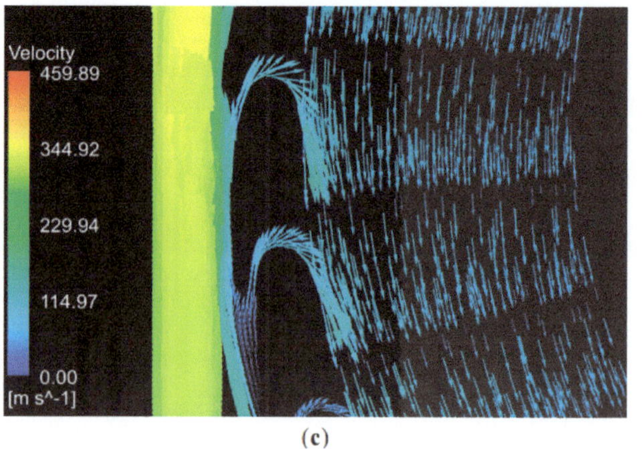

(c)

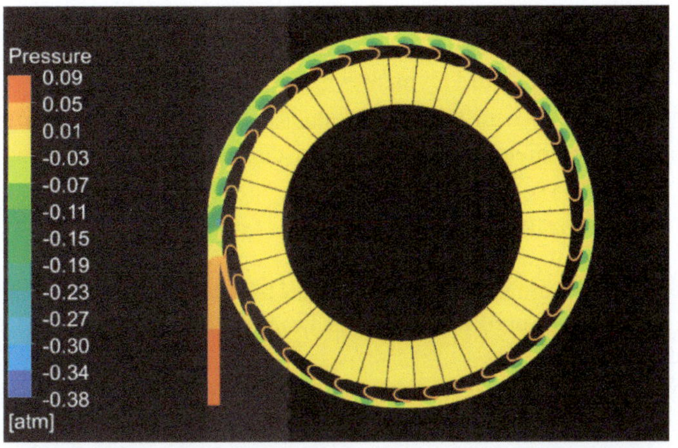

(d)

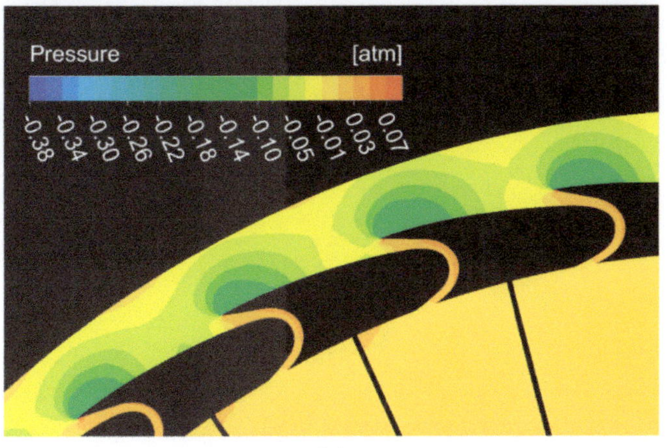

(e)

Figure 14. (**a**) Velocity vector in the central plane of the gas chamber when the tip turbine was still; (**b**) enlarged view of velocity vectors; (**c**) enlarged view of velocity vectors when the tip turbine rotated at a speed n = 120 rpm; (**d**) pressure contour in the central plane of the chamber; (**e**) the enlarged view of pressure on the turbine blades to show the pressure difference on two sides of the blades.

CONCLUSIONS

The CFD study allows assessing the power and lift efficiency of the annular-ducted lift fan system in hover mode, the lift and drag of the system in transition mode, the drag and flight speed of the aircraft in cruise mode and the pneumatic coupling of the tip turbine and exhaust gases of the jet engine. The predicted data provide insight for future experimental study of annular-ducted lift fan aircraft.

The CFD simulation results show that the annular-ducted lift fan system has higher lift efficiency, compared to the rotor of a helicopter, due to the additional duct lift and the elimination of energy losses. Transition from VTOL to cruise flight needs some extra forward thrust to overcome a low peak of drag, primarily caused by the momentum drag of the low disc loading fan system. It may fly faster than helicopters and tiltrotors in cruise mode based on aerodynamic drag prediction, owing to the replacement of rotor drag by aerodynamic drag as the annular duct system is closed off. Pneumatic coupling of the tip turbine and jet exhaust of 300 m/s can provide enough moment to spin the lift fan system.

ACKNOWLEDGMENTS

The authors would like to thank Hong-Wei Wang for the support of the investigations. The comments of the reviewers to help improve the paper are also gratefully acknowledged.

AUTHOR CONTRIBUTIONS

Yun Jiang proposed the ideas, performed the CFD simulation and wrote the manuscript. Bo Zhang and Tao Huang analyzed the data and provided software support.

REFERENCES

1. Bevilaqua, P.M. Genesis of the F-35 joint strike fighter. J. Aircr. 2009, 46, 1825–1836.
2. Asmus, F.J. Design and development of the tip turbine lift fan. Ann. N. Y. Acad. Sci. 1963, 107, 147–176.
3. Thouault, N.; Breitsamter, C.; Adams, N.A. Numerical and experimental analysis of a generic Fan-in-wing configuration. J. Aircr. 2009, 46, 656–666.
4. Zheng, Z.; Zhu, X.; Chang, M.; Zhou, Z. Analysis on the power characteristics of lift-fan VTOL. Appl. Mech. Mater.2011, 66–68, 1649–1656.
5. Dyer, K.G. Aerodynamic Study of a Small, Ducted VTOL Aerial Vehicle. Master's Thesis, Massachusetts Institute of Technology, Department of Aeronautics and Astronautics, Cambridge, MA, USA, 2002; pp. 31–49.
6. Simioni, N.; Ponza, R.; Benini, E. Numerical assessment of pneumatic devices on the wing/fuselage junction of a tiltrotor. J. Aircr. 2013, 50, 752–763.
7. Xu, H.Y.; Ye, Z.Y.; Shi, A.M. Numerical study of propeller slipstream based on unstructured overset grids. J. Aircr.2012, 49, 384–389.
8. Nam, H.J.; Park, Y.M.; Kwon, O.J. Simulation of unsteady rotor-fuselage aerodynamic interaction using unstructured adaptive meshes. J. Am. Helicopter Soc. 2006, 51, 141–149.
9. Kyrkos, A.; Ekaterinaris, J.A. Assessment of an unstructured mesh approach for CFD predictions of the NH90 fuselage rotor. Aerosp. Sci. Technol. 2012, 19, 77–85.
10. Blocken, B.; Defraeye, T.; Koninckx, E.; Carmekiet, J.; Hespel, P. CFD simulations of the aerodynamic drag of two drafting cyclists. Comput. Fluids 2013, 71, 435–445.
11. Malipeddi, A.K.; Mahmoudnejad, N.; Hoffmann, K.A. Numerical analysis of effects of leading-edge protuberances on aircraft wing performance. J. Aircr. 2012, 49, 1336–1344.
12. Qu, Q.; Lu, Z.; Liu, P.; Agarwal, R.K. Numerical study of aerodynamics of a Wing-in-Ground-Effect craft. J. Aircr.2014, 51, 913–924.
13. Menter, F.R. Zonal two equation k–ω turbulence models for aerodynamic flows. AIAA Pap. 1993, 93, 2906.
14. Menter, F.R. Two-equation eddy viscosity turbulence models for engineering applications. AIAA J. 1994, 32, 1598–1605.
15. Al-Garni, A.Z.; Saeed, F.; Al-Garni, A.M. Experimental and numerical investigation of 65 degree Delta and 65/40

degree double-delta wings. J. Aircr. 2008, 45, 71–75.

16. Shih, T.-H.; Liou, W.W.; Shabbir, A.; Yang, Z.; Zhu, J. A new k–ε eddy-viscosity model for high Reynolds number turbulent flows—Model development and validation. Comput. Fluids 1995, 24, 227–238.
17. Harris, C.D. Two-Dimensional Aerodynamic Characteristics of the NACA 0012 Airfoil in the Langley 8-Foot Transonic Pressure Tunnel; NASA: Washington, DC, USA, 1981; p. 81927.
18. Coakley, T.J. Numerical Simulation of Viscous Transonic Airfoil Flows; NASA: Washington, DC, USA, 1987; Volume 87, p. 4016.
19. Velichko, S.A.; Lifshitz, Y.B. Numerical simulation of viscous transonic flows over an airfoil. Theroret. Comput. Fluid Dynamics 1995, 7, 189–206.
20. Ohanian, O.J., III; Gelhausen, P.A.; Inman, D.J. Nondimentional modeling of ducted-fan aerodynamics. J. Aircr. 2012,49, 126–140.
21. Kanazaki, M.; Hanida, R.; Nara, T.; Shibata, M.; Nomura, T.; Murayama, M.; Yamamoto, K. Challenge of design exploration for small blended wing body using unstructured flow solver. Comput. Fluids 2013, 85, 71–77.
22. Scholz, O.; Gariepy, M.; Trepanier, J.Y. CFD optimization of an S-shaped engine diffuser with a boundary layer ingestion configuration. Can. Aeronaut. Space J. 2013, 59, 93–105.

CHAPTER 5

FLUID DYNAMICS ANALYSIS OF A SPACE VEHICLE ENTERING THE MARS ATMOSPHERE

Antonio Viviani[1] and Giuseppe Pezzella[2]

[1]Seconda Università degli Studi di Napoli, Aversa, Italy

[2]Centro Italiano Ricerche Aerospaziali, Capua, Italy

INTRODUCTION

This paper reports the results of design analyses of two Manned Braking Systems (MBS) entering the Mars atmosphere, with the aim of supporting design studies of a planetary entry system. Two lifting body configurations with rounded edge delta-like cross section have been analyzed. The preliminary aerodynamic and aerothermodynamic analyses have considered flight conditions compatible with a manned mission entering the Mars atmosphere. However, neither the mission architecture needed to reach Mars from Earth or neighbor Earth space, nor surface exploration have been addressed.

All the design analyses have been performed at several levels. Indeed, vehicle aerodynamic assessment has been extensively addressed through simplified design approach as, e.g., hypersonic panel methods (HPM); then, a number of fully three-dimensional computational fluid dynamics (CFD) simulations, both with Euler and Navier-Stokes approximations, of the hypersonic flowfield past the entry vehicle have been performed.

The results herein provided have been obtained for a Mars entry scenario compliant with an approach to the red planet both by direct planetary entry and entry after aerobraking (Polishchuk et al., 2006) (Viviani and Pezzella, 2009). These results may be used to provide numerical data for understanding requirements for the human exploration of Mars.

COMPUTATIONAL FLOWFIELD ANALYSIS

CFD analyses have been performed to assess the aerothermal environment that the MBS experiments during descent, thus evaluating several surface loading distributions (e.g., pressure and heat flux). To this aim, several fully three-dimensional numerical computations, both for perfect and chemically reacting gas approximation, have been performed.

The vehicle configurations, under investigation in this work, are shown in Fig. 1 These lifting bodies feature an aerodynamic configuration with a compact body about 8 [m] long with a rounded edge delta-like cross section (Hanley et al., 1964). A very preliminary internal layout for a crew of four astronauts is also reported in Fig. 1.

Citation: Antonio Viviani and Giuseppe Pezzella (2012). Fluid Dynamics Analysis of a Space Vehicle Entering the Mars Atmosphere, Applied Computational Fluid Dynamics, Prof. Hyoung Woo Oh (Ed.), InTech, DOI: 10.5772/25790.

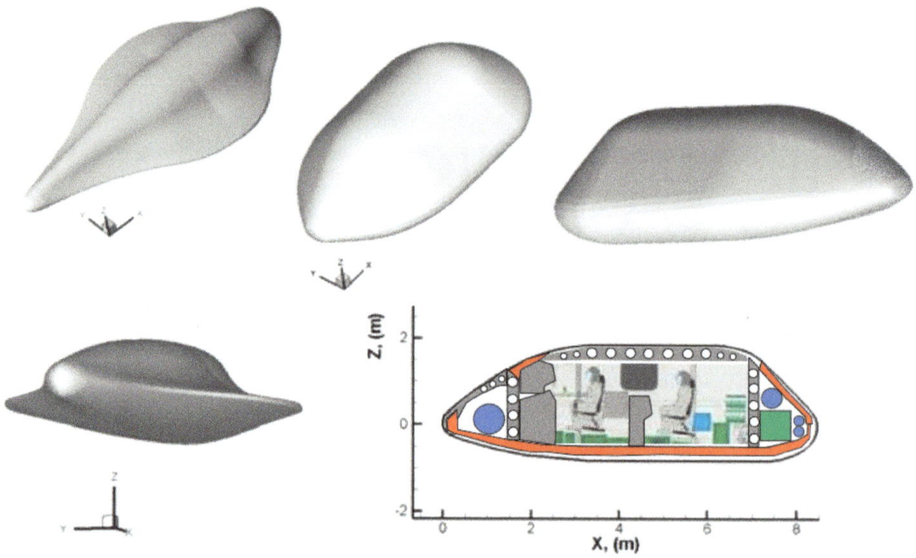

Figure 1: Vehicle configurations with quotes.

Freestream Conditions

The flight scenario considered so far is summarized in Table 1. They refer to entry conditions compatible with a vehicle entering the Mars atmosphere both from a hyperbolic orbit (HO), e.g., direct planetary entry, and an elliptic orbit (EO) e.g., planetary entry after aerobraking (Gupta et al., 1996).

Table 1: CFD freestream conditions

		Mach [-]	AoA [deg]	Altitude [km]
Perfect Gas		10	10	10.0
		15	40	60.0
		22	40	60.0
	EO PH	22	40	44.2
	HO PH	26	40	52.1
Reacting Gas		10	10	10.0
		10	20	10.0
		10	30	10.0
		10	40	10.0
EO PH	NCW	22	40	44.2
	FCW	22	40	44.2
HO PH	NCW	26	40	52.1
	FCW	26	40	52.1

EO PH (Elliptic Orbit Peak Heating)
HO PH (Hyperbolic Orbit Peak Heating)
NCW (Non-Catalytic Wall)
FCW (Fully Catalytic Wall)

Therefore, thirteen CFD numerical simulations have been performed. As one can see, CFD computations (both Euler and Navier-Stokes) have been performed, both in trajectory-based and space-based design approaches (Hanley et al., 1964). Several Mach numbers and different angles of attack (a) have been investigated and compared each other. The Fluent code together with user defined functions (UDFs), developed in order to simulate mixtures of gas in thermo-chemical non-equilibrium, have been used for CFD computations with a non-equilibrium chemical model suitable for Martian atmosphere (Gupta et al., 1996) (Mack et al., 2008) (Kustova et al., 2009).

For the reacting gas computations, the Martian atmosphere has been considered as a mixture of 95.7% carbon-dioxide, 1.6% Argon and 2.7% nitrogen. The flow has been modelled as a reacting gas mixture of 9 species (Ar, CO_2, N_2, O_2, CO, NO, N, O, C) involved in the chemical reactions of Table 2 (Park et al., 1994) (Anderson, 1989). The reaction mechanism and the related chemical kinetics, taken into account in the present work, are summarized in Table 2, where M is the reacting partner (third body) that can be any of the nine reacting species of the gas mixture.

Non-equilibrium computations have been performed since one of the most challenging problem facing the design of atmospheric entry vehicle is the phenomenon of "real gas behavior". At hypersonic speeds, the shock wave produced ahead of the vehicle suddenly elevates the gas temperature in the shock layer. So the gas thermal energy may be comparable with the energy associated with a whole range of gas chemical processes such as: molecular vibrational excitation; dissociation of atmospheric molecules into their atomic forms; formation of other chemical species through recombination reactions; and ionization of both molecular and atomic species (Park et al., 1994).

Table 2: Reactions mechanism and rate parameters (Park et al., 1994)

Reaction	Third Body M	A_r $[cm^3 mol^{-1} s^{-1}]$	β_r	T_a [K]
$CO_2 + M \rightarrow CO + O + M$	CO_2,CO,N_2,O_2,NO	$6.9x10^{21}$	-1.5	63275
	Ar	$6.9x10^{20}$		
	C,N,O	$1.4x10^{22}$		
$CO + M \rightarrow C + O + M$	CO_2,CO,N_2,O_2,NO	$2.3x10^{20}$	-1.0	129000
	Ar	$2.3x10^{19}$		
	C,N,O	$3.4x10^{20}$		
$N_2 + M \rightarrow N + N + M$	CO_2,CO,N_2,O_2,NO	$7.0x10^{21}$	-1.6	113200
	Ar	$7.0x10^{21}$		
	C,N,O	$3.0x10^{22}$		
$O_2 + M \rightarrow O + O + M$	CO_2,CO,N_2,O_2,NO	$2.0x10^{21}$	-1.5	59750
	Ar	$3.0x10^{21}$		
	C,N,O	$3.0x10^{22}$		
$NO + M \rightarrow N + O + M$	CO_2,C,N,O,NO	$1.1x10^{17}$	0.0	75500
	Ar	$5.0x10^{15}$		
	CO,N_2,O_2	$5.0x10^{15}$		
$C_2 + M \rightarrow C + C + M$	All	$2.0x10^{21}$	-1.5	59750
$NCO + M \rightarrow CO + N + M$	All	$6.3x10^{16}$	-0.5	24000
$NO + O \rightarrow N + O_2$		$8.4x10^{12}$	0.0	19450
$N_2 + O \rightarrow NO + N$		$6.4x10^{17}$	-1.0	38370
$CO + O \rightarrow C + O_2$		$3.9x10^{13}$	-0.18	69200
$CO_2 + O \rightarrow CO + O_2$		$2.1x10^{13}$	0.00	27800

Therefore, the gas mixture has to be considered in thermal and chemical non-equilibrium. Finally, the CFD analysis of the MBS have been preceded by a code validation phase performed considering the available numerical and experimental data for the Mars Pathfinder probe at entry peak heating conditions, as summarized in (Viviani et al., 2010) (Gnoffo et al., 1998) (Gnoffo et al., 1996) (Mitcheltree et al., 1995).

Numerical Results

The aerodynamic analysis of MBS is shown in term of lift (C_L), drag (C_D) and pitching moment (C_{My}) coefficients which are calculated according to Eq. (1) and Eq. (2), respectively.

$$C_i = \frac{F_i}{\frac{1}{2}\rho_\infty v_\infty^2 S_{ref}} \qquad i = L,D \tag{1}$$

$$C_{Mj} = \frac{M_j}{\frac{1}{2}\rho_\infty v_\infty^2 L_{ref} S_{ref}} \qquad j = Y \tag{2}$$

The reference parameters L_{ref} (e.g., longitudinal reference length) and S_{ref} (e.g., reference surface) are the vehicle length (e.g., 8 m) and planform area (e.g., 32 m²). The pitching moment is computed from the vehicle nose (i.e. 0, 0, 0). Engineering based aerodynamic analysis has been extensively performed by using a 3D Panel Methods code developed by CIRA, namely HPM (Viviani and Pezzella, 2009). This tool, at high supersonic and hypersonic speeds, is able to accomplish the aerodynamic and aerothermodynamic analyses of a complex re-entry vehicle configuration by using simplified approaches as local surface inclination methods and approximate boundary-layer methods, respectively. The SIM typical of hypersonics are based on Newtonian, Modified Newtonian, and Prandtl-Mayer theories (Anderson, 1989). Typical surface meshes of the MBS, used for the engineering level computations, are shown in Fig. 2.

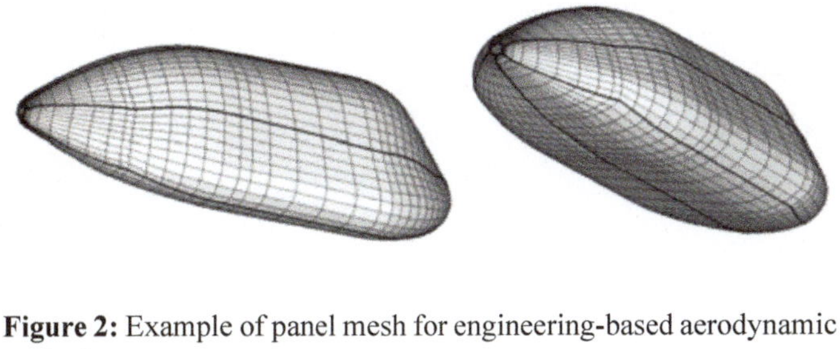

Figure 2: Example of panel mesh for engineering-based aerodynamic analysis.

MBS aerodynamic results, provided by engineering-based analysis, cover a ranging from 0 to 50 deg. Present CFD computations for the MBS have been carried out on a 3-D multiblock structured grid close to that shown in Fig. 3. The grid consists of about 20 blocks and 900.000 cells (half body). Both computational domains are tailored for the free-stream conditions of Table 1. The distribution of surface grid points has been dictated by the level of resolution desired in various areas of the vehicle such as the stagnation region and the base fillet one, according to the computational scopes. Fig. 3 shows also a close-up view of the 3-D mesh on the vehicle surface and pitch plane. Grid refinement in strong gradient regions of flowfield has been made through a solution adaptive approach.

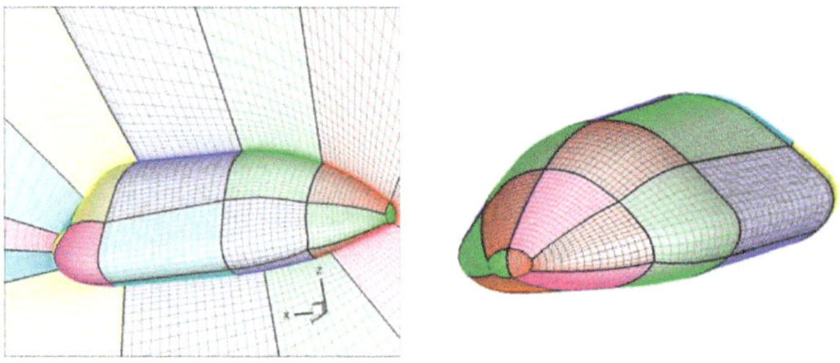

Figure 3: Example of computational mesh domains for Euler CFD simulations.

The preliminary results of CFD simulations performed so far are summarized hereinafter. For example, Fig. 6 shows the static temperature contours on the vehicle symmetry plane and static pressure contours on vehicle surface at $M_¥$=20 and a=20 deg, considering the Mars atmosphere as a reacting gas mixture. As shown, the MBS bow shock structure around the descent vehicle can be appreciated as well.

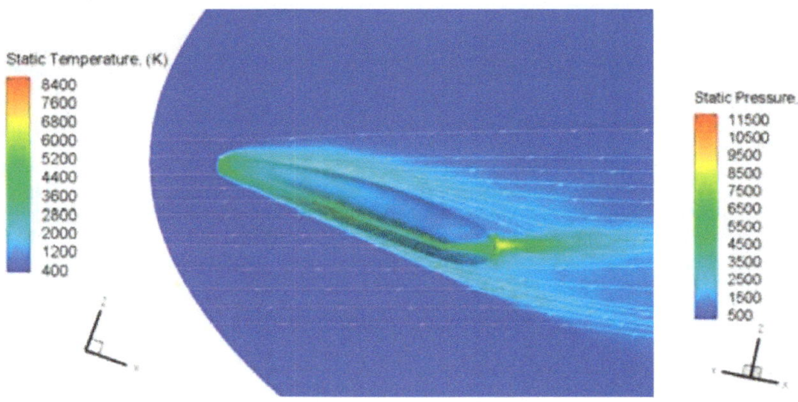

Figure 4: Results for M¥=20 and a=20 deg. (Left) static temperature field on vehicle symmetry plane; (right) static pressure contour on vehicle surface for non-equilibrium reacting gas.

At the same flight conditions, Fig. 5 reports on chemical dissociation of the flow in the shock layer considering the contours of CO2 mass fraction on MBS pitch plane. As a consequence, flow dissociation determines a large density ratio e across the strong bow shock compared with a flow of the same gas where no dissociation takes place (Viviani and Pezzella, 2009) (Anderson, 1989). This results in a thinner shock layer around the entry vehicle (e.g., lower stand-off distance).

Under conditions where dissociation exists, the aerodynamics of vehicle depends primarily on shock density ratio. In fact, the change of aerodynamic characteristics is the result of change in surface pressure acting on the vehicle forebody (Gnoffo et al., 1998) (Viviani et al., 2010).

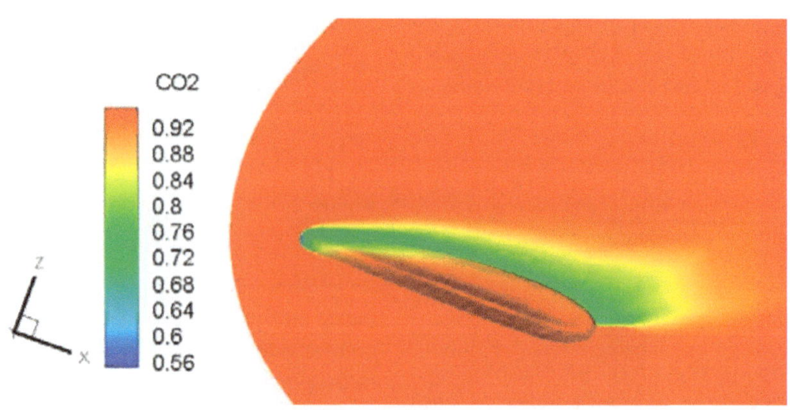

Figure 5: Results for $M_¥$=20 and a=20 deg. Contours of CO_2 mass fraction.

Moreover, Fig. 6 shows CFD results for $M_¥$=21 and a=40 deg. The left side reports pressure coefficient contours (C_p) on vehicle surface and on two cross sections; whereas on the right C_p contours on vehicle surface and Mach number contours on three cross sections have been shown.

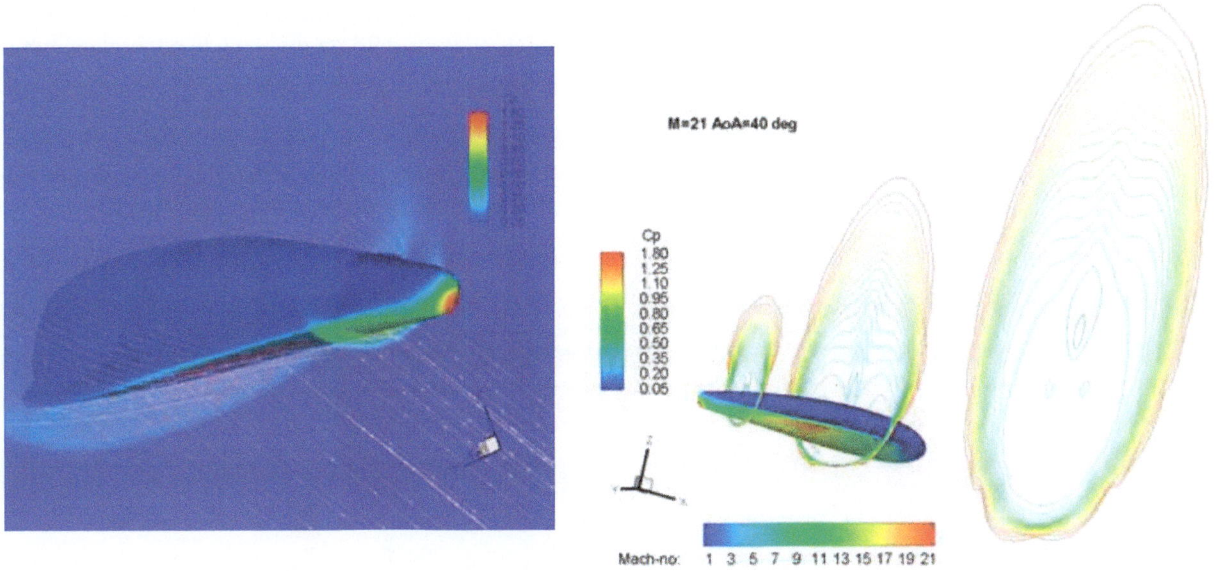

Figure 6: Results for M_{∞}=21 and a=40 deg. (Left) pressure coefficient contours (C_p) on vehicle surface and on two cross sections; (right) C_p contours on vehicle surface and Mach number contours on three cross sections.

The curves of lift and drag coefficients are shown in Fig. 7. Those curves collect MBS aerodynamic coefficients compared with available numerical data both for perfect gas and reacting gas approximations, reported in order to highlight accuracy of both numerical and engineering-based results (Viviani et al., 2010).

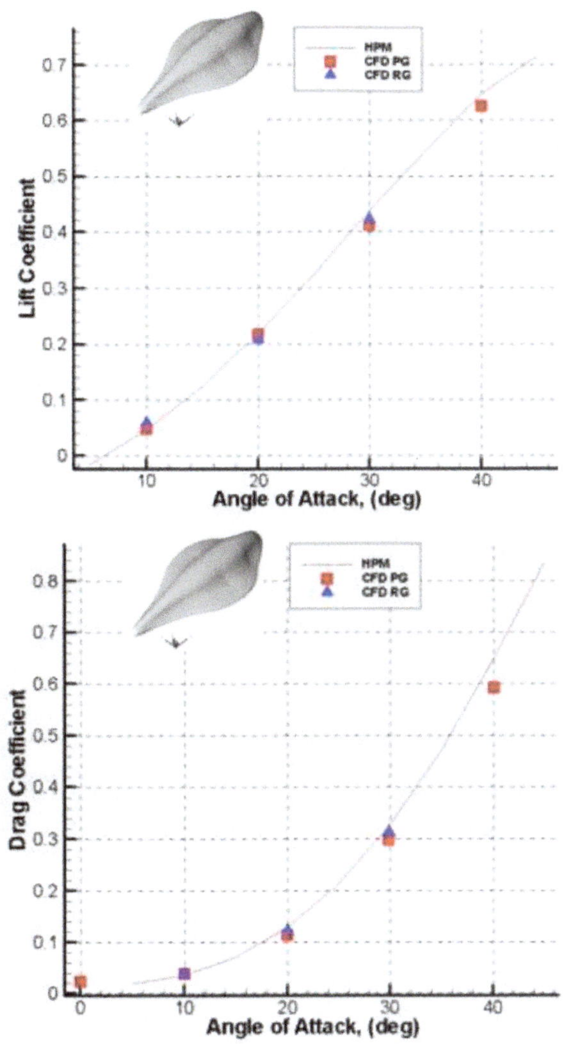

Figure 7: C_L and C_D versus a. Comparison between panel methods and CFD results for perfect and reacting gas approximations.

As one can see, engineering and numerical data compare very well, thus confirming that engineering-based estimations represent reliable preliminary aerodynamics of a Mars entry vehicle. Moreover, real gas effects increase the aerodynamic drag coefficient whereas the lift is only slightly influenced.

As far as CFD results for the second configuration are concerned, Fig. 8 shows the Mach number contour field that takes place around the vehicle when it is flying at the peak heating conditions of entry by EO (e.g., $M_{¥}$=22, a=40 deg, and H=44.20 km). In particular, the left side of Fig. 8 shows the Mach contour field on the vehicle pitch plane while at the right side of Fig. 8 gives an idea of the bow shock shape that envelopes the vehicle since the Mach field is reported on three different flowfield cross sections. As shown, a thin shock layer envelopes the entry vehicle windside with a strong expansion that characterizes the flow at the end of the vehicle.

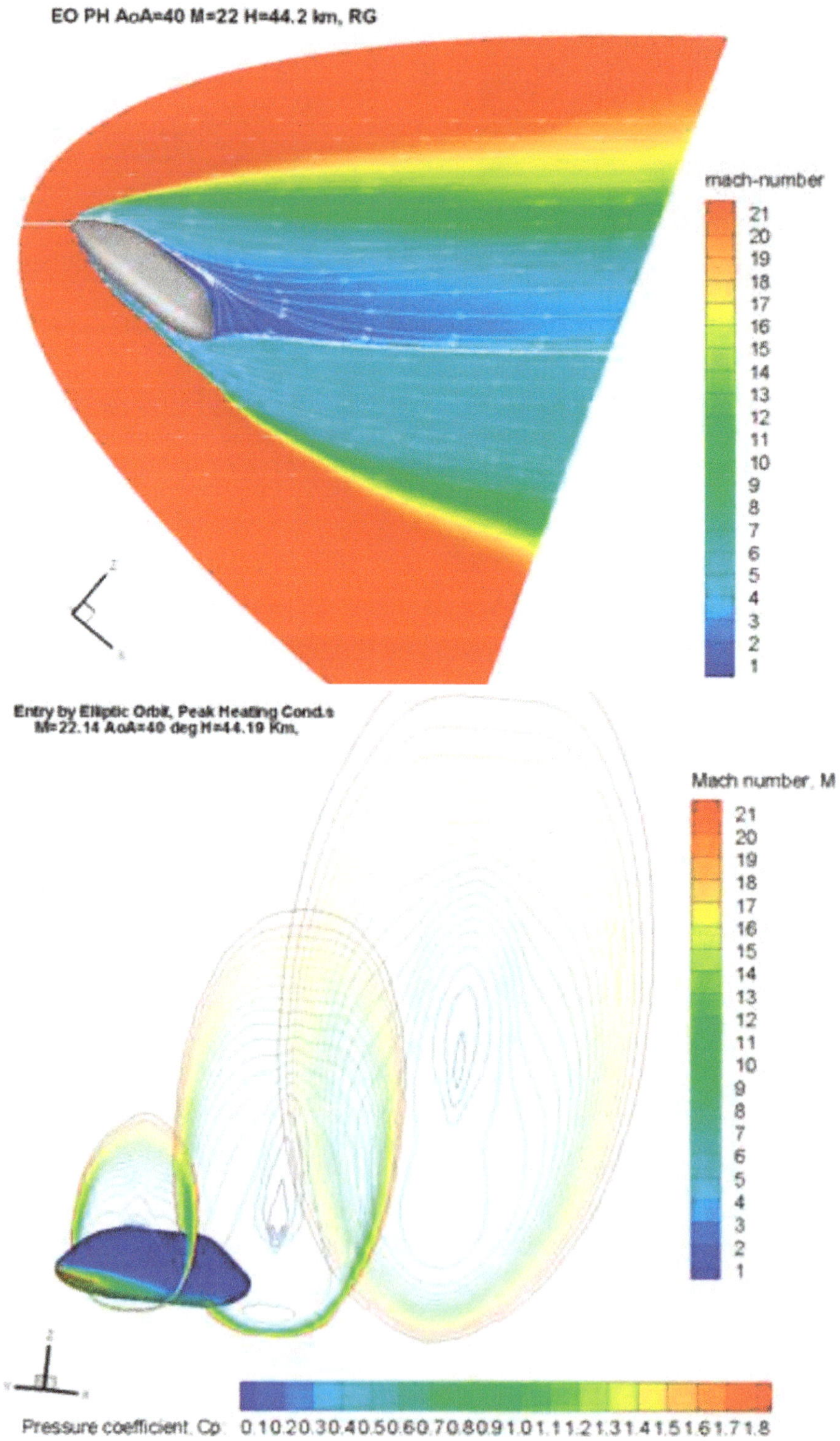

Figure 8: Mach contours on the vehicle pitch plane and three flowfield cross sections at the EOPH conditions.

The CO mass fraction field around the vehicle for the same freestream conditions is shown in Fig. 9 where some streamtraces colored by Mach number are also reported.

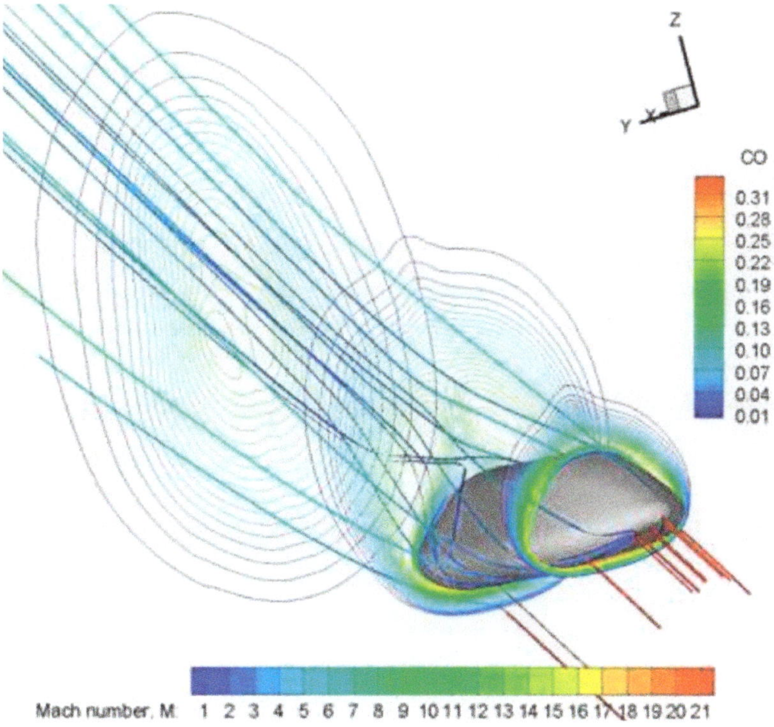

Figure 9: CO mass fraction at the EOPH conditions on three cross sections with streamtraces colored by Mach number.

As shown the CO concentration reaches its maximum value close to the body. Fig. 10 shows the temperature comparison among non-equilibrium flow (right side of pilot) and perfect gas computation, evaluated at three flowfield cross sections (x=1.5 [m], 5.5 [m] and 9.5 [m]). It is clearly evident how real gas phenomena affect the vehicle shock layer, thus confirming all the conclusions highlighted before.

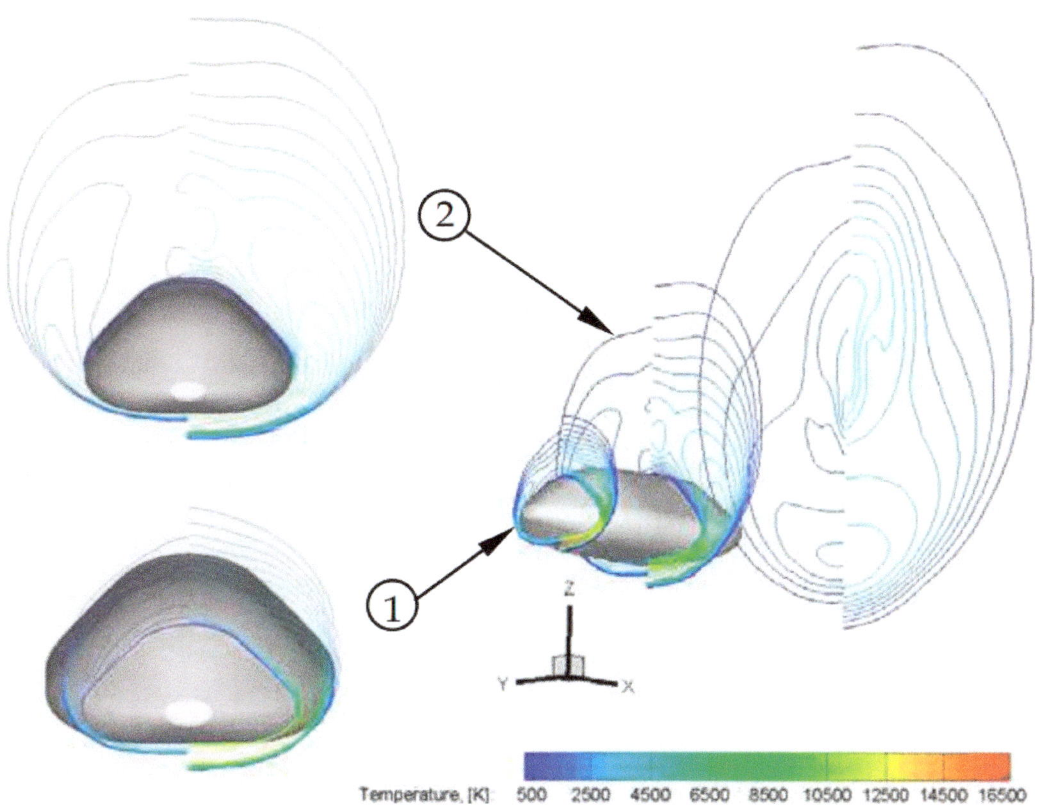

Figure 10: Temperature comparison between non-equilibrium flow (right side of pilot) and perfect gas computation at x=1.5 [m], 5.5 [m] and 9.5 [m] flowfield cross sections.

Finally, the curves of lift, drag, and pitching moment coefficients are shown in Fig. 11.

Real gas effects increase both drag and pitching moment coefficients, whereas the lift is only slightly influenced. Vehicle aerodynamics is also summarized in the table of Fig. 11.

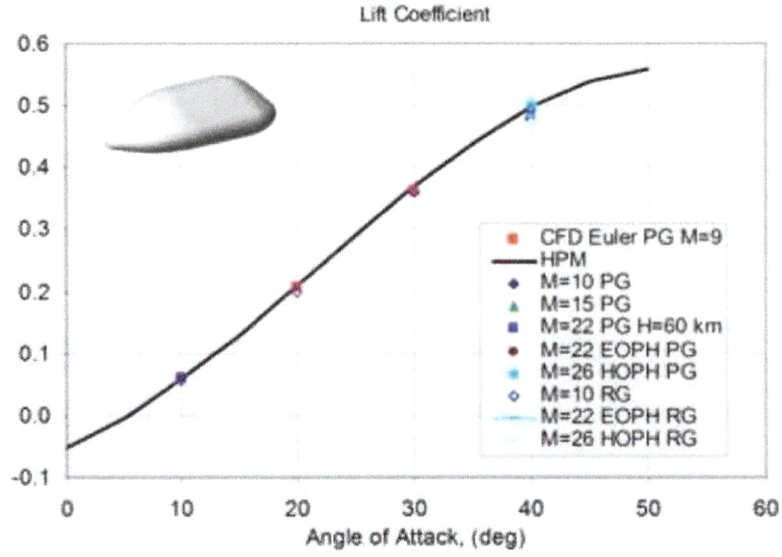

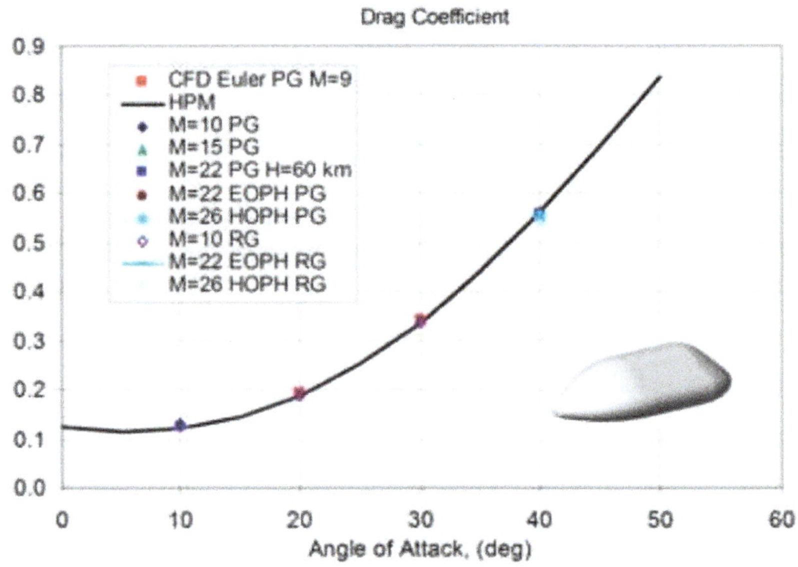

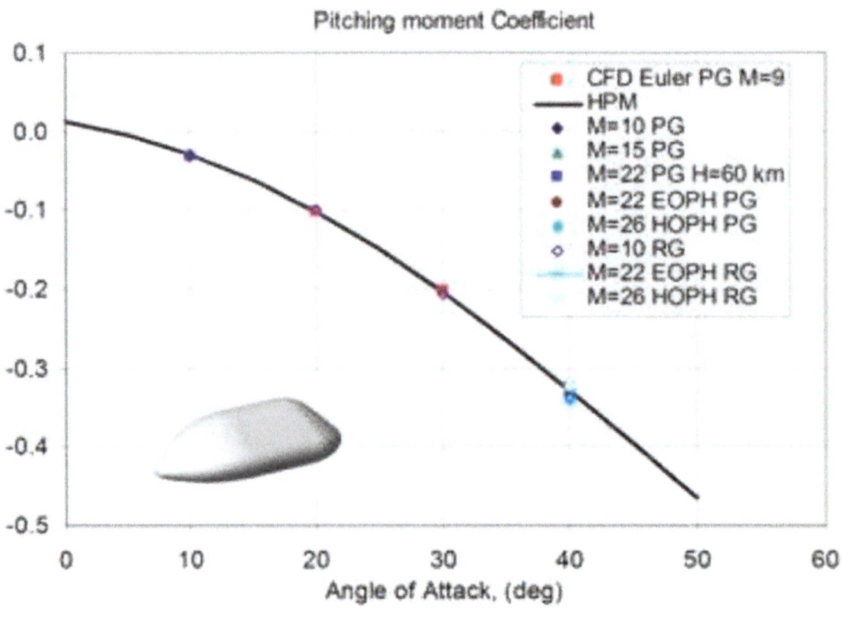

		Mach [-]	AoA [deg]	Altitude [km]	CL [-]	CD [-]	CM* [-]
Perfect Gas		10	10	10.0	0.06150	0.13060	-0.03170
		15	40	60.0	0.47830	0.55360	-0.32940
		22	40	60.0	0.47930	0.55540	-0.33010
	EO PH	22	40	44.2	0.48300	0.55500	-0.33200
	HO PH	26	40	52.1	0.49920	0.55930	-0.34330
Reacting Gas		10	10	10.0	0.05710	0.12450	-0.02950
		10	20	10.0	0.19940	0.18730	-0.09930
		10	30	10.0	0.35970	0.33420	-0.20650
		10	40	10.0	0.49340	0.56100	-0.33770
EO PH	NCW	22	40	44.2	0.48368	0.55014	-0.32980
	FCW	22	40	44.2	0.48369	0.55018	-0.32990
HO PH	NCW	26	40	52.1	0.47122	0.53627	-0.32220
	FCW	26	40	52.1	0.47875	0.54536	-0.32560

EO PH (Elliptic Orbit Peak Heating)
HO PH (Hyperbolic Orbit Peak Heating)
NCW (Non-Catalytic Wall)
FCW (Fully Catalytic Wall)
* pole @ vehicle nose

Figure 11: Lift, Drag and pitching moment coefficients versus a. Comparison between panel methods and CFD results for perfect and non-equilibrium gas computations.

CONCLUSION

The paper deals with the flowfield analysis of two braking systems for manned exploration mission to Mars.

A number of fully 3D Navier-Stokes and Euler CFD computations of the hypersonic flowfield past two lifting body vehicles have been performed for several freestream conditions of a proposed Mars entry loading environment. These evaluations have been aimed at carrying out only a preliminary design of the MBS configuration, in compliance with the Phase-A design level.

The range between Mach 2 and Mach 26 has been analyzed, to provide both aerodynamic databases according to both the space-based and trajectory-based design approaches. Numerical results show that real gas effects increase both the aerodynamic drag and pitching moment coefficient, whereas the lift is only slightly influenced.

REFERENCES

1. Anderson, J. D., Hypersonic and High Temperature Gas Dynamics, McGraw-Hill Book Company, New York, 1989.
2. Gnoffo, P., Weilmuenster, K., Braun, R., Cruz, C., Influence of Sonic-Line Location on Mars Pathfinder Probe Aerothermodynamics, Journal of Spacecraft and Rockets, vol. 33 No 2, March-April 1996.
3. Gnoffo, P., Braun, R., Weilmuenster, K., Mitcheltree, R., Engelung, W., Powell, R., Prediction and Validation of Mars Pathfinder Hypersonic Aerodynamic Data Base, 7th AIAA/ASME Joint Thermophysics and Heat Transfer Conference. June 15-18, Albuquerque, NM (USA). 1998.
4. Gupta, R., Lee, K., Scott, C., Aerothermal Study of Mars Pathfinder Aeroshell, Journal of Spacecraft and Rockets, vol.33 No 1, Jan.-Feb. 1996.
5. Hanley, G. M., Lyon, F. J., The Feasibility of Spacecraft Deceleration by Aerodynamic Braking at the Planet Mars, Proc. of the 1th AIAA Annual Meeting. Washington, D.C. June 29- July 2, 1964. AIAA-64-479.
6. Hannemann, V., Mack, A., Chemical Non Equilibrium Model of the Martian Atmosphere, Proc. of the 6th European Symposium on Aerothermodynamics for Space Vehicles. Versailles, France. 3-6 Nov. 2008 – ESA SP-659, January 2009.
7. Kustova, E.V., Nagnibeda, E.A., Shevelev, Y. D., Syzranova, N.G., Comparison of NonEquilibrium Supersonic CO2 Flows with Real Gas Effects near a Blunt Body, Proc. of the 6th European Symposium on Aerothermodynamics for Space Vehicles. Versailles, France. 3-6 Nov. 2008 – ESA SP-659, January 2009.
8. Mack, A., CFD Validation for CO2 Reentry Applications, 2nd Inter. ARA Days. Arcachon, France, 21-23 October 2008. AA-3-2008-37.
9. Mitcheltree, R. A., Gnoffo, P. A., Wake Flow about the Mars Pathfinder Entry Vehicle, Journal of Spacecraft and Rockets, Vol. 32 No 5, Sept.- Oct. 1995.
10. Polishchuk, G., Pichkhadze, K., Vorontsov, K., Pavela, V., Proposal on application of Russian technical facilities for International Mars Research Program for 2009–2015, Acta Astronautica, Vol. 59, pp.113-118. 2006.
11. Park, C., Howe, J. T., Jaffe, R. L., Candler, G. V., Review of Chemical-kinetic Problems of Future NASA Missions, II: Mars Entries, Journal of Thermophysics and Heat Transfer, vol. 8, No 1, pp. 9-23, Jan-Mar 1994.
12. Viviani, A., Pezzella, G., Aerodynamic Analysis of a Capsule Vehicle for a Manned Exploration Mission to Mars. 16th AIAA/DLR/DGLR International Space Planes and Hypersonic Systems and Technologies Conference. Bremen Oct. 2009. AIAA-2009- 7386.
13. Viviani, A., Pezzella, G., Golia, C., Aerodynamic and thermal design of a space vehicle entering the Mars atmosphere. The Fifth International Conference on Thermal Engineering Theory and Applications. May 10-14, 2010

14. Marrakesh. Morocco. Viviani, A., Pezzella, G., Golia, C., Aerothermodynamic analysis of a space vehicle for manned exploration missions to Mars. 27th International Congress of the Aeronautical Sciences. ICAS 2010. Nice, France, 19-24 September 2010.

CHAPTER 6 ADJOINT METHODS FOR CAR AERODYNAMICS

Carsten Othmer

Vehicle Technology, Group Research, Volkswagen AG, 38436 Wolfsburg, Germany

ABSTRACT

The adjoint method has long been considered as the tool of choice for gradient-based optimization in computational fluid dynamics (CFD). It is the independence of the computational cost from the number of design variables that makes it particularly attractive for problems with large design spaces. Originally developed by Lions and Pironneau in the 70's, the adjoint method has evolved towards a standard tool within the development processes of the aeronautical industries. Its uptake in the automotive industry, however, lags behind. The first systematic applications of adjoint methods in automotive CFD have interestingly not taken place in the classical shape design arena, but in a relatively young discipline of sensitivity-based optimization: fluid dynamic topology optimization. While being an established concept in structure mechanics for decades already, its transfer to fluid dynamics took place just ten years ago. We demonstrate that specifically for ducted flow applications, like airducts for cabin ventilation or engine intake ports, it constitutes a very powerful tool and has matured over the last years to a level that allows its systematic usage for various automotive applications. To drive adjoint-based shape optimization to the same degree of maturity and robustness for car applications is the subject of ongoing research collaborations between academia and the car industry. Achievements and challenges encountered during these efforts are presented.

BACKGROUND

Computational Fluid Dynamics (CFD) is a central element of the automotive development process. Besides the classical external aerodynamics for the prediction of drag and lift coefficients, there is a whole plethora of applications for ducted flows: airducts for cabin ventilation, engine intake ports, exhaust systems including catalytic converters, air intakes, water jackets of cylinder heads, cooling plates for electric vehicle battery packs and many more (Figure 1). The standard procedure of developing these parts still consists of manual iteration loops between designer and computational engineer, and automatic optimization methods are being used systematically only for some selected CFD applications (e.g. for airducts, Figure 2). Via an automatic process chain consisting of a CAD (Computer-Aided Design) system, meshing software, CFD solver and post-processor a black-box optimization algorithm, typically of the evolutionary strategy type, is employed to drive a parameterized CAD-model into an optimal state.

Citation: Carsten Othmer, "Adjoint methods for car aerodynamics," Journal of Mathematics in Industry20144:6 DOI: 10.1186/2190-5983-4-6.

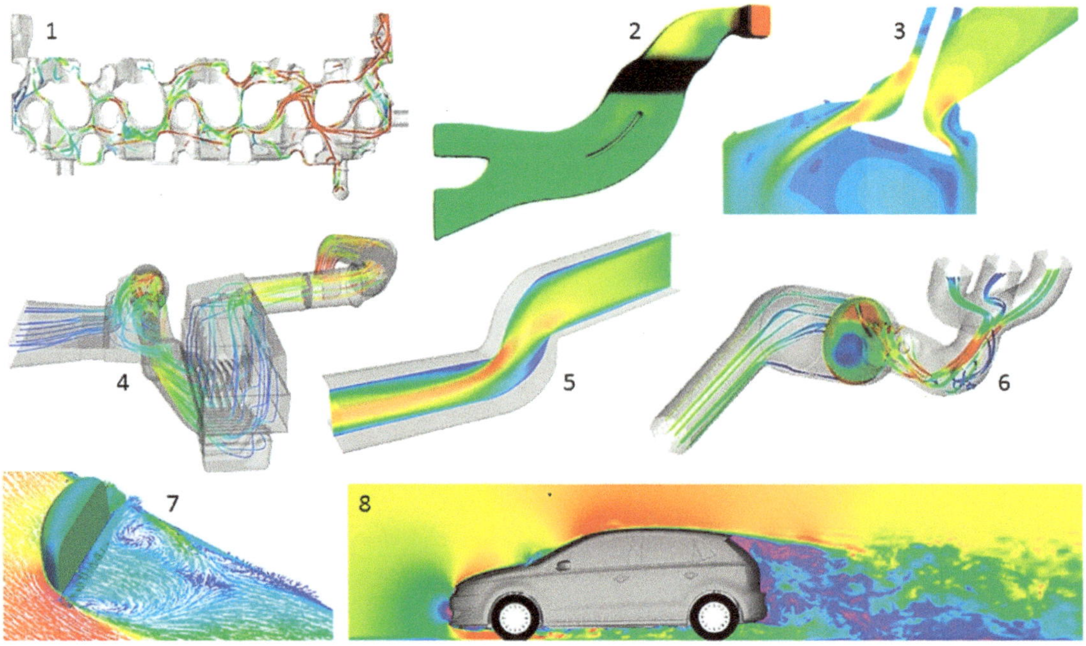

Figure 1: The ubiquity of CFD in automotive development. External aerodynamics (7 and 8) is only one aspect of automotive CFD. The many ducted flow applications include, among others, water jackets for cylinder heads (1), airducts for cabin ventilation (**2** and **5**), engine intake ports (**3**), raw air intakes (**4**) and exhaust systems (**6**).

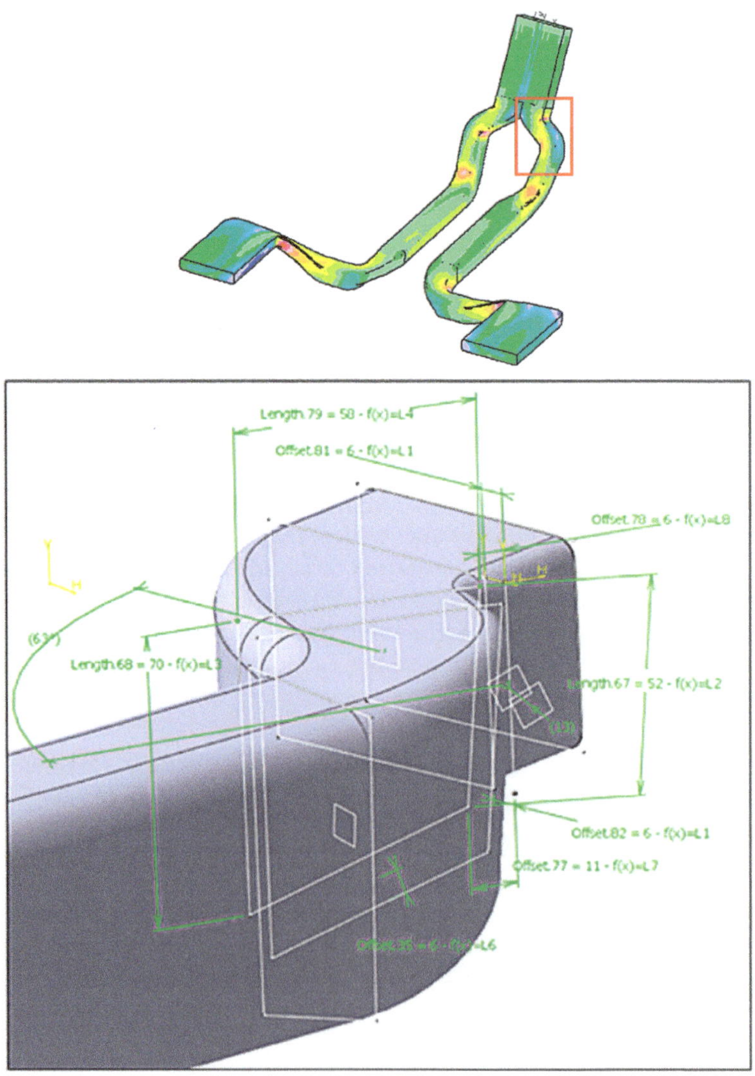

Figure 2: Automotive CFD optimization today. Example of a CAD-parameterization of an airduct segment (provided by S Brück and C Ehlers, Volkswagen AG) as input to a black-box optimization process chain.

Due to the high computational effort associated with black-box optimization, where the number of CFD evaluations scales roughly linearly with the number of design variables, the explorable design space is very limited: In our current practice, these methods cannot afford more than ten design parameters. An additional obstacle that inhibits the further adoption of black-box optimization within the regular automotive development process is the necessity of a versatile but stable CAD parameterization. For e.g. an engine intake, the setup of the parameterized model might take an experienced CAD engineer several weeks before it comes up to the stability requirements of an automatic optimization process chain. It is these limitations that call for a radically different approach to automotive CFD optimization - especially in view of applications like entire vehicle aerodynamics that entail shape optimization of highly complex free-form geometries with a practically unlimited design space.

Adjoint-based optimization [1, 2, 3, 4] has long been recognized to be efficient for tasks with large numbers of design variables. Its underlying principle, the adjoint method, allows for a very elegant computation of sensitivities, i.e. of the derivative of the objective function w.r.t. the design variables. The computational effort is thereby independent of the number of design variables and just involves one solve of the adjoint counterparts of the governing equation system, in our case the Navier-Stokes equations. When applied to a surface mesh representation or a volume mesh of the part to be optimized, information like the one depicted in Figure 3, for which we coined the term 'sensitivity maps', can be generated: A surface sensitivity map tells us for each and every surface node how the objective function changes w.r.t. an infinitesimally small normal displacement of this surface node. On the other hand, a volume sensitivity map represents for each volume cell the sensitivity of the objective function w.r.t. a change of permeability of this volume cell. Both sensitivity maps give precise indications on where and how to change the geometry - perturb the surface inwards or outwards, and remove counterproductive cells from the flow domain, respectively - in order to improve the objective function. Obviously, this wealth of information would be inaccessible without the adjoint method, and, as will be shown in the following, allows to develop very efficient optimization methods for automotive part design.

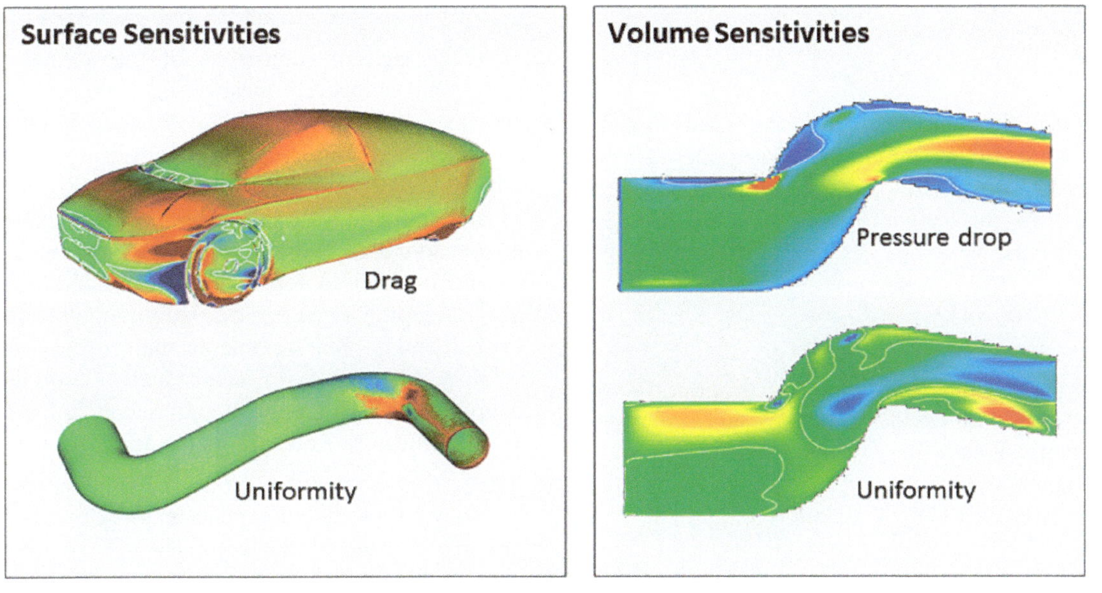

Figure 3: Sensitivity maps. Surface sensitivity maps (left) display the gradient of the cost function w.r.t. normal displacements of the surface. In red areas, a movement away from the fluid (i.e. inwards for the car, outwards for the pipe) would result in an improvement of the cost function. Contrarily, blue areas indicate the regions where a surface perturbation towards the fluid improves performance, while modifications in greenish surface sections have little effect on the cost function. The white lines on the car body are the isolines of zero sensitivity, i.e. the borderline between favorable inward and outward movement. Topological or volume sensitivities (shown on the right-hand side for a cut through the S-bend of Figure 2) represent the cost function gradient w.r.t. changes of the individual cell permeability. Blue volume sections are those where decreasing the cell permeability (by adding a porosity-based friction term) would improve the cost function. Those areas are thus counterproductive to the component's fluid dynamic performance and should be removed - the basis of CFD topology optimization.

Despite being already a long-established tool in the aerospace sector, the adjoint method has only recently started to also enter the automotive development processes. This backlog is mainly due to two obstacles specific to the automotive industry.

Obstacle #1 Rather than on in-house codes, the car industry relies almost exclusively on commercial CFD software.

The implementation of the adjoint method requires access to the source code - either to implement the analytically derived adjoint equations into the code (so-called continuous adjoint) or to perform a manual or automatic differentiation of the primal code (discrete adjoint, see [5] for a comparison). The sources of commercial CFD codes are, however, inaccessible for the user, and by the time we started our project to industrialise the adjoint method, the software vendors of the CFD solvers that are operational at Volkswagen declined to implement it into their codes. As a remedy, we selected the well-validated and versatile CFD toolbox OpenFOAM® [6] as our platform to develop adjoint-based optimization methods for car applications. In addition to its open source character, OpenFOAM comes with a high-level symbolic programming style, which makes it ideally suited to implement additional partial differential equations - like the adjoint Navier-Stokes equations.

The first continuous adjoint implementation in OpenFOAM [7] is based on the derivation outlined in [8]. It focused on steady-state, incompressible ducted flows as described by the RANS (Reynolds-averaged Navier-Stokes) equations:

$$(v\cdot\nabla)v=-\nabla p+\nabla\cdot(2\nu D(v))-\alpha v, \quad (1)$$

$$\nabla\cdot v=0, \quad (2)$$

where **v** and p stand for velocity and pressure, respectively, and α for the porosity distribution. The effective kinematic viscosity ν is the sum of molecular and turbulent viscosity, and the rate of strain tensor D(v) is defined as D(v)=12(∇v+(∇v) T). For a given cost function J composed of contributions from the design domain boundary Γ=∂Ω and from the interior of Ω,

$$J = \int_\Gamma J_\Gamma \, d\Gamma + \int_\Omega J_\Omega \, d\Omega, \quad (3)$$

the implemented adjoint field equations are

$$-(\mathbf{v}\cdot\nabla)\mathbf{u} - \nabla\mathbf{u}\cdot\mathbf{v} = -\nabla q + \nabla\cdot\big(2\nu \mathrm{D}(\mathbf{u})\big) - \alpha\mathbf{u} - \frac{\partial J_\Omega}{\partial \mathbf{v}}, \quad (4)$$

$$\nabla\cdot\mathbf{u} = \frac{\partial J_\Omega}{\partial p}, \quad (5)$$

with **u** and q denoting adjoint velocity and adjoint pressure, respectively. While the cost function contribution from the domain interior J_Ω gives rise to source terms in the adjoint field equations, the surface contribution J_Γ enters the adjoint boundary conditions:

$$\mathbf{n}\cdot\nabla q = 0, \qquad \mathbf{u}_t = 0, \qquad u_n = -\frac{\partial J_\Gamma}{\partial p} \quad (6)$$

for boundaries with Neumann conditions for the primal velocity and a Dirichlet condition for the primal pressure (usually inlet and wall boundaries), and

$$q = \mathbf{u}\cdot\mathbf{v} + u_n v_n + \nu(\mathbf{n}\cdot\nabla)u_n + \frac{\partial J_\Gamma}{\partial v_n}, \quad (7)$$

$$0 = v_n\mathbf{u}_t + \nu(\mathbf{n}\cdot\nabla)\mathbf{u}_t + \frac{\partial J_\Gamma}{\partial \mathbf{v}_t} \quad (8)$$

for boundaries with Dirichlet conditions for the primal velocity and a Neumann condition for the primal pressure (e.g. outlet boundaries, for details we refer again to [8]). By solving both primal and adjoint equation system, the desired (volume-normalized) topological and (surface-normalized) shape sensitivities can be computed as

$$\frac{\partial J}{\partial \alpha} = \mathbf{u}\cdot\mathbf{v} \quad \text{and} \quad \frac{\partial J}{\partial \beta} = -\nu\partial_n\mathbf{u}\cdot\partial_n\mathbf{v}, \quad (9)$$

where β denotes the normal node displacement.

Within the still ongoing collaboration with Prof. Giannakoglou's group at the National Technical University of Athens (NTUA) and Dr. de Villiers at Engys®, this basic implementation was subsequently enhanced towards an inclusion of an adjoint turbulence model [9], optimization of external flows [10], sensitivities for flow control [11], adjoint wall functions [12], the inclusion of heat conduction and constraints [13] and mildly compressible flows [14]. In addition, the uptake and further development of the basic adjoint code by other researchers lead to remarkable advances [15,16, 17, 18], especially in the area of topology optimization for exhaust systems [19, 20]. As a result of these development efforts, a versatile continuous adjoint code in OpenFOAM for steady-state RANS is now available.

Obstacle #2 The geometries of entire vehicles and even of single automotive parts are highly complex.

The complexity of typical automotive geometries, especially of entire vehicles, which consist of an assembly of several hundred parts, necessitates the use of an automatic process for the generation of the computational mesh. Even though the sophistication of automatic meshing software is continuously increasing, the achievable mesh quality is still inferior to a hand-made mesh, especially concerning the creation of layers along the surface of the flow domain. While this quality suffices for obtaining reasonable results for primal computations, it has been found that the adjoint equation system reacts much more delicately to imperfections of the computational mesh. This is particularly the case in the presence of flow separation, which is commonly encountered in flows in and around automotive geometries.

This lack of robustness of the adjoint when running on typical industrial meshes is a problem that has been accompanying the adjoint development in OpenFOAM from the very beginning - and for certain applications it is still inhibiting the systematic use of the adjoint method in the regular computational processes. The source of this instability is the so-called 'adjoint transpose convection' ∇u·v - the second convection term on the left-hand side of Eqn. (4). By developing several formulations for this term, as well as explicit and implicit treatments, major steps towards increased stability were made. As it introduces a high degree of cross-coupling between the three components of the velocity, recent efforts concentrate on solving the adjoint equation system as a single block rather than in a segregated manner [21].

While those developments are very promising, the current practice of running the adjoint - at least on meshes of the typically encountered quality in the car industry environment - still requires to trade robustness against accuracy: by using limiters, relaxed adjoint boundary conditions or by damping the adjoint transpose convection in an intelligent way that does not harm the qualitative correctness of the sensitivities. To report the achievements as well as the challenges of this pragmatic approach to solving adjoints for car applications is the intention of this article. It is organized as follows: The concepts of topology optimization and shape optimization for ducted flows are introduced in the next section, along with an application of these two complementary methodologies to engine port flows. Section 3 focuses on external aerodynamics, comprising a full-vehicle validation study, an example of a component optimization and a comparison of steady-state RANS sensitivities with those based on a time-averaged transient primal flow. The article finally closes with a comparatively young adjoint application area - car drag reduction by flow control.

DUCTED FLOWS

Geometric optimization of ducted flows in cars, like airducts for cabin climatization, engine air intakes or exhaust systems, are commonly subject to severe packaging constraints. This gave rise to the development and adoption of topology optimization methods for automotive applications. After a concise retrospective of these development efforts, it will be shown that especially for ducted flows, topology optimization is a perfect complement to the classical shape optimization.

Topology Optimization

Topology optimization is a well-established tool in computational structure mechanics [22] with widespread industrial use. In its simplest realization, a topological optimization starts from the available design domain filled up entirely with solid material of a certain density. In an iterative fashion, the given loads are applied, the stresses are computed all over the domain and the areas with low stresses are weakened by assigning a lower density to them. After several iterations, this method retains high-density material only in regions that are critical to fulfill the structural task, and in this way generates optimal lightweight structural designs. It delivers an un-biased design from scratch that automatically fulfills the installation space constraints.

It almost comes as a surprise that it took until 2003 until this elegant concept was transferred to fluid dynamics: Independently of each other Borrvall and Petersson [23] and Klimetzek [24] presented the first topological optimization methods for ducted flows. Analogously to its structural mechanic archetype, their fluid dynamic topology optimization starts from a completely flooded computational domain (Figure 4), uses some local criterion to identify those areas that are counterproductive w.r.t. the chosen objective function and iteratively 'punishes' or removes them from the fluid domain. As a result, the remaining areas constitute the optimal duct between inlet(s) and outlet(s) of the given installation space.

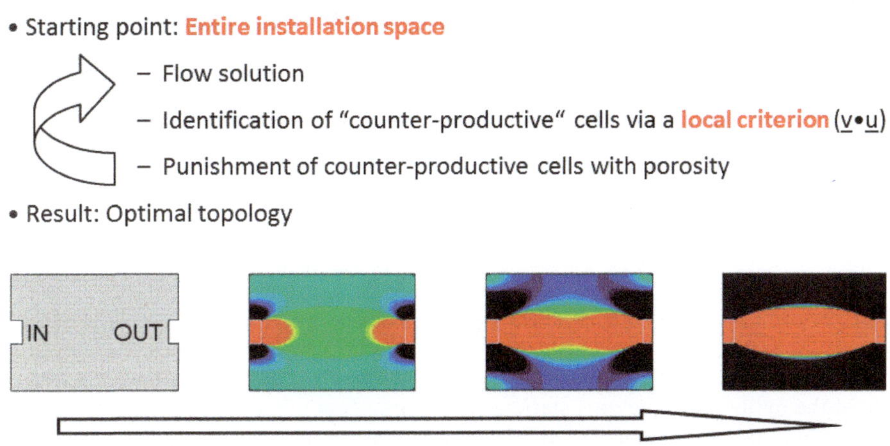

Figure 4: The principle of CFD topology optimization. Identified counter-productive cells are iteratively punished (black cells).

Despite being based on the same concept of flow optimization, these two approaches exhibit significant differences. Klimetzek's method is a rather pragmatic solution: The local criterion he employs to identify counterproductive cells is the deviation between the actual velocity direction and a desired direction computed as polygonal traces between inlet and outlet. If this deviation exceeds a certain angle, the corresponding cells are blocked for the fluid. His criterion allows to efficiently remove unwanted recirculations from the flow and - albeit being used for other cost function as well - is therefore ideally suited to reduce the pressure drop. From the date of its original publication, Klimetzek's method has already been applicable to three-dimensional turbulent Navier-Stokes flows and is seeing a continuously increasing number of successful applications in industrial flow optimization.

In contrast, the method proposed by Borrvall and Petersson was restricted to two-dimensional Stokes flows between parallel plates, and the punishment of counterproductive areas is performed by locally decreasing the distance between the plates - until zero upon convergence. What inspired a series of subsequent research works is, however, the versatility of their approach that lies in the way how they identify counterproductive fluid cells: by computing actual sensitivities via the adjoint method.

In order to extend the Borrvall and Petersson method towards three-dimensional flows, the punishment via plate distance was replaced by introducing a Darcy porosity term $-\alpha v$ into the governing Eqn. (1) (see [25] and [26] for a finite-element and a finite-volume implementation, respectively). The individual cell porosities allow a continuous transition between fluid (α=0) and solid (α=αmax) and act as the actual design variables of the optimization problem. With this setting, topological sensitivity maps like those shown in Figure 3 are nothing but the derivative of the cost function w.r.t. an increase in the cell porosity α.

Another requirement for the industrial applicability of adjoint-based topology optimization was its generalization from pure Stokes flows to turbulent Navier-Stokes flows. After the principal feasibility of laminar Navier-Stokes topology optimization was first demonstrated via Automatic Differentiation of an academic CFD code [27], the concept was implemented via a continuous adjoint also for turbulent flows [8] - albeit under the assumption of 'frozen turbulence', i.e. fully neglecting the variation of the turbulent flow quantities. Finally, the development of a continuous adjoint turbulence model [9] now allows to run topology optimization for turbulent Navier-Stokes flows under full consideration of variations in the turbulence fields [13].

As topology optimization is not only very fast but also mimics the classical way of part design when driven by installation space constraints, the uptake of this method in the regular automotive design process was quite straightforward. A number of ducted flow applications are today routinely optimized via adjoint-based topology optimization (see Figure 5 for examples from the Volkswagen Group and [19, 20] for others). Exemplarily, one of them - the optimization of an engine intake port - will be presented in more detail in the following.

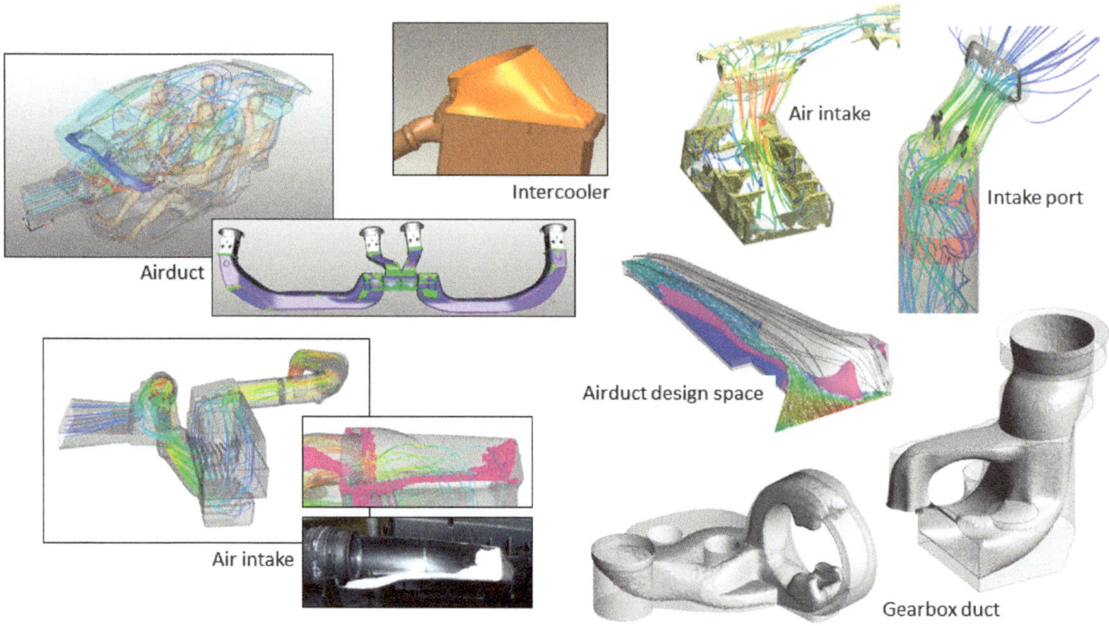

Figure 5: Sample applications of CFD topology optimization. Kindly provided by U Giffhorn, W Py, M Tomecki, C Ehlers, M Towara from Volkswagen AG and N. Peller from Audi AG.

The main components of the cylinder intake port static test configuration geometry that was the optimization target for this study can be seen in Figure 6. Only the volume enclosed by the two intake port arms was subjected to modification. The barrel, valves, valve seats and upstream inlet geometry (not shown) had to be kept clear of blockage via geometric and proximity based constraints. The working fluid for all the simulations was air modeled as a compressible perfect gas. Pressure at the outlet of the extended barrel was fixed, and at the inlet a fixed mass flow at ambient temperature was specified. For the adjoint system, density and turbulence was assumed 'frozen', thus no governing equations were solved for the adjoint systems beyond the momentum and pressure analogues. For the port flow the objectives were (1) to minimize the pressure loss between the inlet and the outlet and (2) to maximize the swirl, i.e. the angular momentum $r \times \rho v$ around the barrel axis in the cylindrical sub-zone indicated in Figure 6.

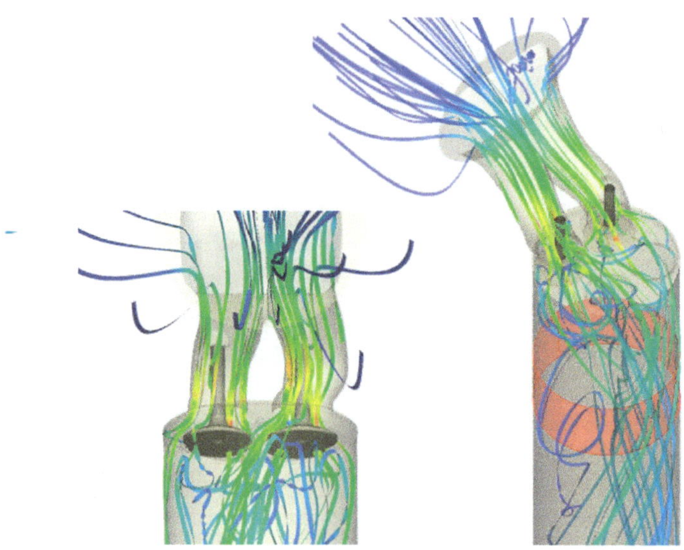

Figure 6: Intake port geometry and streamlines of the original flow. Only the two arms connecting the inlet plenum and the barrel were subject to geometrical modifications upon optimizing for (1) pressure drop between inlet and outlet and (2) swirl within the cylindrical slice indicated by the red mantle area.

One of the main issues with multi-objective adjoint optimization is the choice of a sensible relative weighting when combining the sensitivity fields from the two objectives for input into the topology update. As a first step, we therefore performed an exploration by varying the weighting factors dynamically within a wide range. After choosing three design points with different weightings (low, mid- and high swirl) that represented a likely range for desirable fluid dynamic performances of the intake port, we actually ran the topology optimization for these three trade-offs between swirl and pressure loss until convergence and thus obtained three Pareto-optimal states (Figure 7a-c, details see [14]).

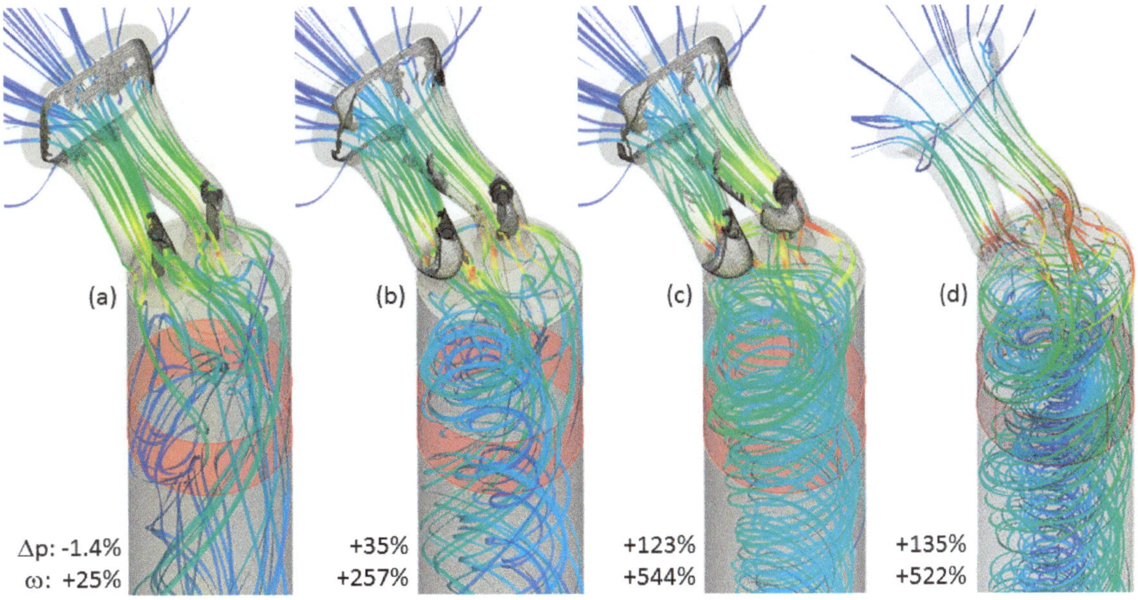

Figure 7: Intake port topology optimization results. The three figures (a)-(c) show the Pareto-optimal states for low, mid- and high swirl weighting, respectively. The grey areas within the intake port arms are the cells that were blocked during the topology optimization. Note the obvious improvement of swirl (ω, bottom numbers) from (a) to (c) at the cost of increasing pressure drop (Δp, top numbers), as compared to the original flow of Figure 6. The topology of the high swirl case (c) was recasted into a new CAD description and recomputed (**d**). Flow field, swirl and pressure drop confirm the results of the topology optimization (details see [14]).

For the low swirl case, it can be assumed that the geometry is only driven by the swirl sensitivities where the modifications have a very small detrimental effect on the pressure loss. As a result, we see a small reduction in pressure loss along with a modest increase in swirl. Predictably, the pressure loss starts to go up as the swirl weighting is increased and the fluid loses additional energy due to the highly swirling flow and the blockages required to produce it. What is encouraging is the extent to which the swirl can be reliably increased - by more than 500% - compared to the relatively modest relative increase in pressure loss (high swirl case of Figure 7c).

For topology optimization, it is critical that the design extracted from the optimization process be verified with a high accuracy simulation. To this end, the blocked element description from the high swirl case was extracted and incorporated into a new CAD description of the port flow geometry (Figure 7d). It is immediately apparent that the modified design is rather complex in its detail and not something a designer would likely create through intuition or experience. Neither could these geometries easily be arrived at via a process of parameter-based stochastic optimization. The most important point to note is that the accurately rebuilt geometry very nearly reproduces the results from the topology optimization run. While we do not contend that this will consistently be the case, it certainly proves the efficacy of the process in this instance.

Shape Optimization

The topology optimization method as introduced above, acts on the fluid domain like a sculptor on a piece of rock: It removes the unwanted portions and reduces the original geometry to its optimal shape. Since the original volume mesh is not changed during the optimization process, the final geometry consequently consists of a subset of those cells that made up the original shape. Thus, the inherent geometrical resolution of topology optimization corresponds to the cell size, and the final shape will inevitably have a ragged surface. Topology optimization is therefore to be regarded as a drafting method. For fine-tuning geometries that are already close to their optimal shape, it is ideally complemented by so-called shape optimization. Rather than on topological or volume sensitivities, shape optimization is based on the detailed information contained in surface sensitivity maps in order to morph the shape towards a further improvement of the cost function.

An example of a surface sensitivity map of an exhaust port is shown in Figure 8. Red areas are those that have to be moved outwards (away from the air) in order to reduce the pressure drop while blue regions need to be pushed inwards. It seems straightforward to translate this information into a shape modification that improves the performance of the exhaust port: Simply move every surface node in a steepest descent manner inwards or outwards by a distance corresponding to its sensitivity magnitude. Due to the inherent 'roughness' of the sensitivity distribution this would, however, quickly result in a collapse of the surface mesh. Therefore, a clever smoothing process has to be applied to the surface sensitivities before they are actually translated into a shape update. This smoothing process does not only have to generate a good compromise between ironing out the sharpest discontinuities of the sensitivity map and at the same time conserving the essential sensitivity information, but also has to be capable of keeping certain feature lines of the geometry.

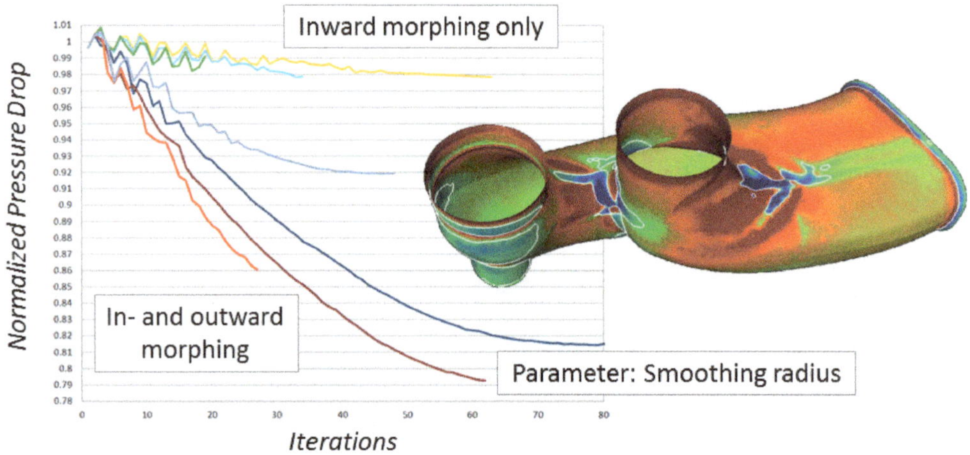

Figure 8: Exhaust port shape optimization. Based on a pressure drop surface sensitivity map (colors like in Figure 3 left), a one-shot shape optimization was conducted with the morphing routines from TUM <abbrgrp>2829</abbrgrp>. Expectedly, the achievable improvement is much bigger when shape deformations in both inward and outward direction are allowed (four bottom curves) as compared to enforcing only inward morphing (top three curves). Due to the small spatial structures of the sensitivity map, also the amount of smoothing applied to the shape morphing had a significant effect on the optimization. The meshed geometry and the morphing results were kindly provided by F Kunze from SEMCON® and M Hojjat and E Stavropoulou from TUM, respectively.

A surface regularization method that comes up to these requirements has been developed by Prof. Bletzinger and his team at the Technical University of Munich (TUM) [28, 29]. The results of applying their method to the exhaust port above are shown in Figure 8. Depending on the size of the smoothing radius and on whether morphing is allowed only in the inward direction or both inwards and outwards, different pressure drop reductions between 2% and 20% are achieved. Since this shape optimization process is set up such that primal and adjoint calculation are closely coupled to the shape update itself ('one-shot optimization'), the overall cost of the optimization amounts to not more than four primal computations.

As demonstrated above, topology optimization and shape optimization are complementary methods: the former for drafting, the latter for fine-tuning. Especially for the optimization of car components that are subject to design space constraints - and in the increasingly tighter getting installation spaces within the engine compartment and the passenger cabin this is the case for the majority of components - the combination of both methods suits the requirements of automotive component design very well. Starting from the feasible installation space, they are capable of delivering a fine-tuned optimal geometry, thereby each of them making use of the full potential of the available design space.

EXTERNAL AERODYNAMICS

The adjoint-based topology and shape optimization methods described above are obviously not restricted to ducted flows but can equally well be applied to external flows. The external aerodynamics of entire vehicles is, however, a peculiar application for optimization methods: Except for dedicated low emission cars, vehicle shapes are to a high degree driven by aesthetic considerations rather than aerodynamic performance. This has two implications: (1) Given the rather limited design freedom, external aerodynamic optimization is more about fine-tuning. Topology optimization is therefore not an adequate tool. (2) Since aesthetic requirements can impossibly be casted into mathematical constraints, the automatic optimization of vehicle shapes forbids itself - unless restricted to small portions like spoilers, mirrors or, of course, to the underbody.

Under these circumstances, the mere computation of surface sensitivity maps - with neither a topological nor a shape update - turned out to be a tool that fits very nicely into the specific requirements of the development of external vehicle shapes: The information contained in these maps provides the designer with concrete suggestions for aerodynamic improvements, and it is in his/her hand to translate this information into an aerodynamically improved shape without compromising the aesthetics.

Validation of RANS Shape Sensitivities for External Aerodynamics

In order to validate sensitivity maps and to explore their potential for external aerodynamics, the adjoint method was first applied to the Volkswagen XL1 - a dedicated low emission vehicle developed by Volkswagen in 2011 (Figure 9). Based on a steady-state RANS solution on a low-Reynolds mesh of a half-model of the XL1, our adjoint OpenFOAM code with adjoint turbulence model according to Zymaris et al. [9] was employed to compute a sensitivity map for drag (Figure 10).

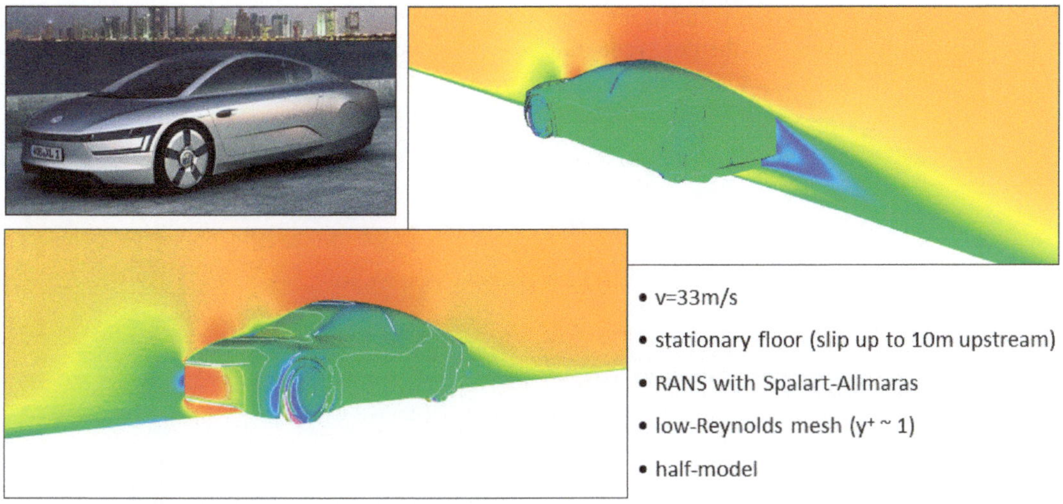

Figure 9: The primal simulation for the Volkswagen XL1. The symmetry plane is colored according to the velocity magnitude, while the car surface shows the pressure distribution.

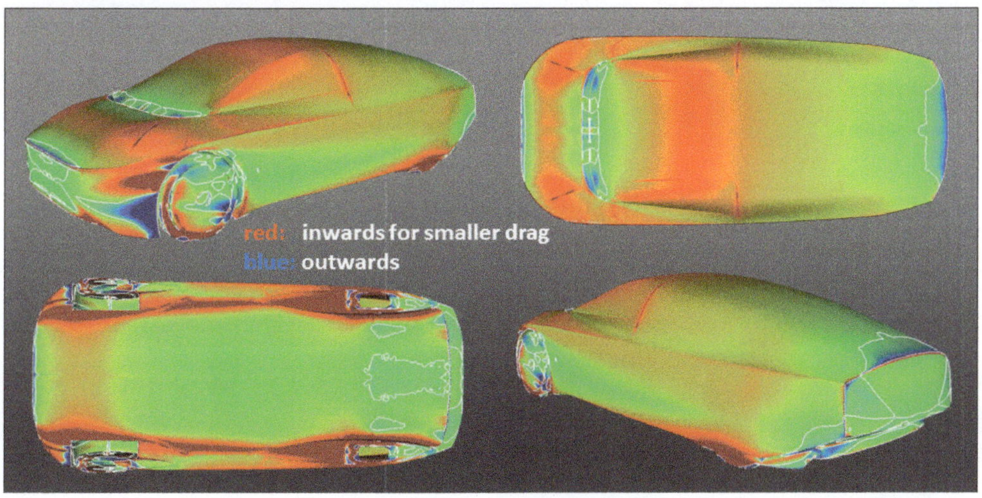

Figure 10: Drag sensitivity maps for the XL1. Isometric front and back view, bottom view (bottom left) and top view (top right).

Various areas are identified as being very sensitive (i.e. dark blue or dark red). In relation to the experience of a well-versed aerodynamicist, these areas are of different qualities: First of all, the 'usual suspects' of external car aerodynamics (the roof, the high-curvature regions of the bonnet, the areas around the wheels and the rear sillboard) are confirmed. For other sensitive areas like the spanwise blue stripe at the transition between the frontend and the bonnet (region 2 in Figure 11), it can be seen in the primal solution already that the flow detaches and requires the curvature to be reduced locally - an adjoint is not needed to get this piece of information. The added-value of the adjoint, however, becomes obvious for regions 1, 3 and 4: the rear end of the car where the adjoint 'suggests' a spoiler, the blue area of the front wing which should be further exposed, and the wheel spoiler area in front of the front tyres. This information cannot be deduced by looking only at the primal. The effect of changing the car shape here is non-local and happens further downstream of the sensitive regions themselves. It can therefore only be identified by the adjoint.

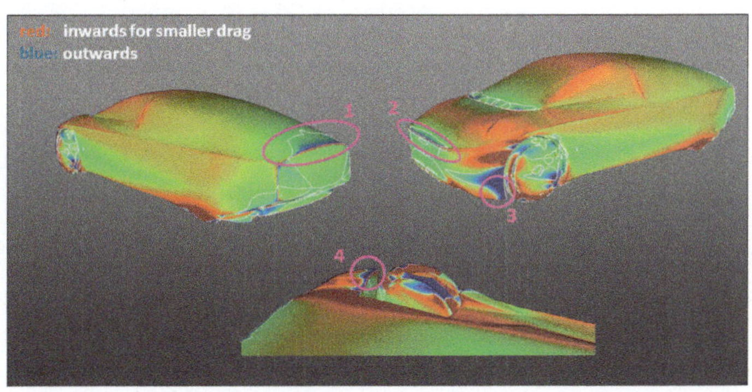

Figure 11: Selected areas for the validation of RANS surface sensitivity maps. Rear spoiler (1), front end of bonnet (2), front wing (**3**) and wheel spoiler area (**4**).

For the validation study, the four regions already mentioned and shown in Figure 11 were parameterized with free-form deformation boxes (Figure 12), and the car shape was morphed by 'macroscopic' distances (5 or 10 mm, respectively) both according to and opposite to the computed sensitivity map. Recomputations of the morphed car shapes confirmed the correctness of the sensitivity signs in all cases:

Morphing in the direction of the sensitivities resulted in a decrease of drag, while drag increased upon morphing opposite to the local sensitivities (Figure 13). Additionally, the drag reduction resulting from a rear spoiler (region 1) and from a wheel spoiler (region 4) were confirmed in wind tunnel experiments.

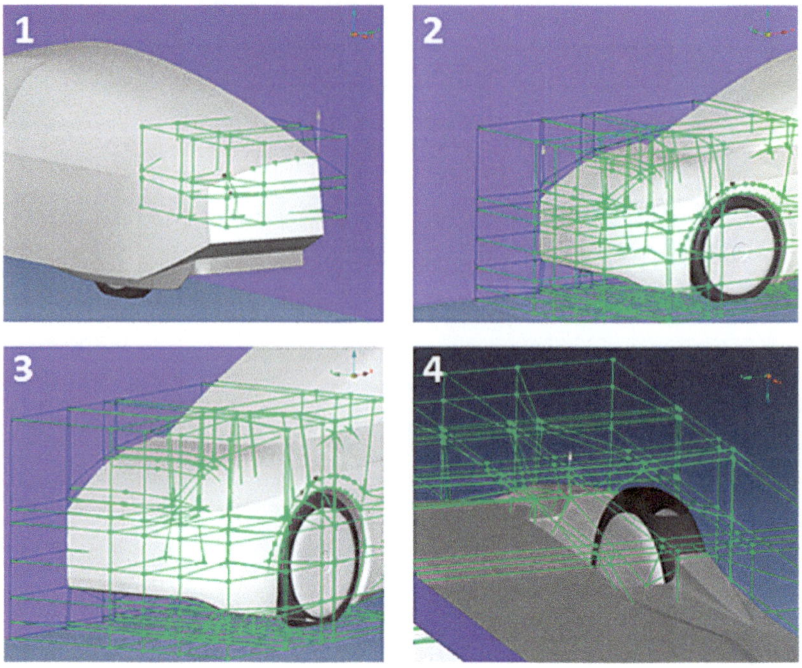

Figure 12: Definition of morphing boxes. The numbers refer to the region numbering of Figure 11 (kindly set up by BETA CAE Systems within their pre-processor ANSA®).

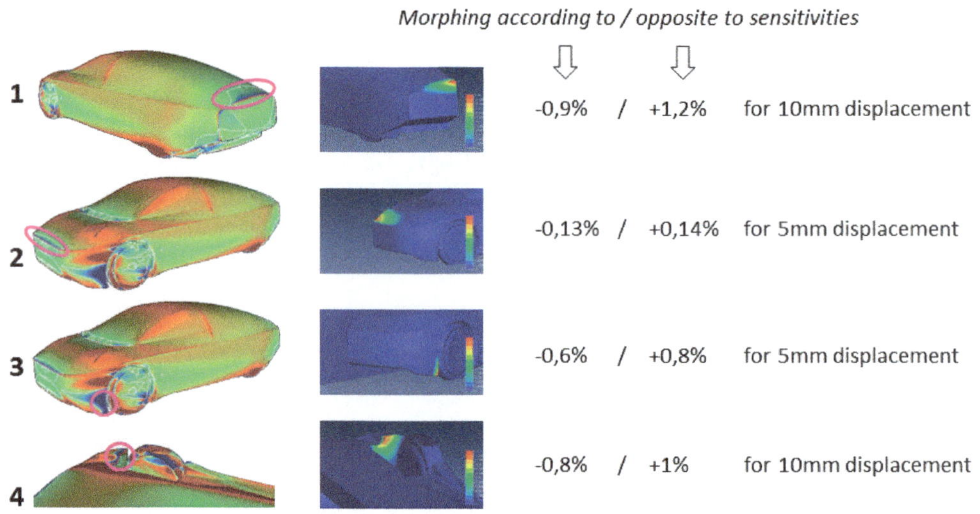

Figure 13: Recomputation of morphed geometries. Given are the drag variations upon morphing in the direction suggested by the sensitivity map (left column) and in the opposite direction (right column) for a deformation by 5 or 10 mm (assumed as sensible macroscopic distances for the respective region of the car).

The automation of vehicle shape morphing - at least for limited portions of the entire vehicle shape - can be accomplished by coupling the adjoint OpenFOAM solver with the free-form deformation tool of ANSA® [30]: Via the chain rule, the node sensitivities as computed by the adjoint solver are mapped onto the morphing control points of ANSA (see [10]). Their positions can then be optimized by any gradient-based algorithm. Exemplarily, this method was applied to the rear spoiler area of the XL1 (region 1). The rear edge of the half-model was parameterized by 5 morphing control points whose movements were restricted to the z-direction (i.e. up and down, Figure 14). Within a one-shot optimization driven by a simple steepest descent algorithm the z-coordinates of the morphing control points were moved towards their optimal position after roughly 20 iterations (Figure 14). The overall cost of the optimization corresponded to less than five primal flow solutions and resulted in a drag decrease of 2%. Given the fact that the XL1 with a drag coefficient of less than 0.2 was aerodynamically nearly perfect already, this was a significant reduction. As a beneficial side-effect, lift decreased by 30%.

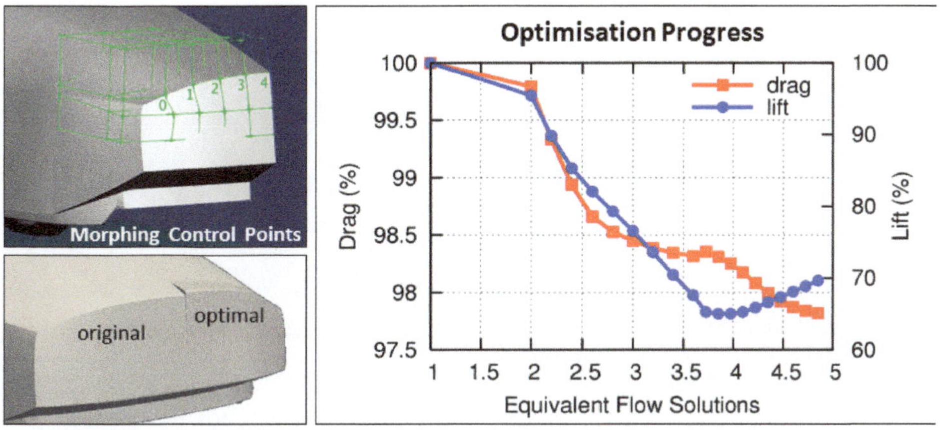

Figure 14: One-shot optimization of the XL1 spoiler shape. The rear edge of the XL1 half-model is parameterized with five morphing control points (top left). After twenty steepest descent driven shape updates in one-shot fashion, drag reduced by 2% and lift by 30% at a total cost of five equivalent flow solutions (right). The bottom left figure compares the original and the optimal shape of the rear car edge.

Rather than with free-form deformation boxes, which allow to modify shapes in a very controlled manner but are quite tedious to set up, another option of translating surface sensitivities into an improved shape are of course the node-based morphing algorithms from TUM as introduced in the previous section. Since they are capable of moving each and every node into its optimal position, they are ideally suited for the optimization of vehicle details - like external mirrors.

Figure 15 depicts an extract of the drag sensitivity map of the previous Volkswagen Passat. It was used in combination with the TUM morphing routines in order to optimize the shape of the external mirrors. The evolution of drag against the iterative progress of the one-shot optimization is shown in Figure 16. A remarkable 1% drag reduction of the entire car was achieved by only optimizing the mirror shapes and leaving the rest of the car unchanged. The bulgier upstream part and the more horizontal top edge reduce the size of the mirror wake and thus the drag (Figure 16).

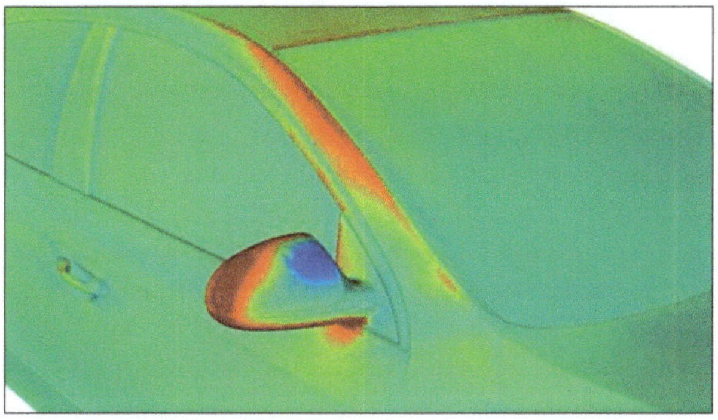

Figure 15: Shape optimization of an external mirror. Displayed are RANS-based drag sensitivities in the mirror region of the previous Volkswagen Passat. The coloring is as usual - red regions have to be pushed inwards and blue to be pulled out in order to reduce drag.

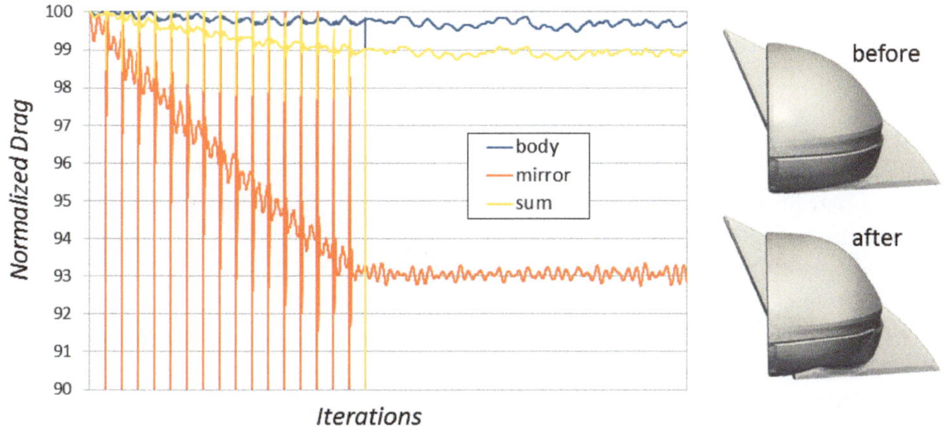

Figure 16: Shape optimization of an external mirror. Based on the drag sensitivities of Figure 15, the TUM morphing routines were employed to conduct a one-shot shape optimization. Only the mirror and the triangle mounted to the car were subject to deformation. The evolution of drag on the mirrors themselves (orange curve), the remaining car body (blue) and their sum effect on the entire vehicle (yellow) are shown. Due to a missing constraint on the mirror glass area, the optimization would ultimately generate inapplicably small mirror glasses and was therefore aborted before that happened. A constrained version of this optimization will be published in [29]. Note how the morphing affected the overall shape of the mirror (e.g. the bulgier upstream shape and the more horizontal upper egde), but was capable of keeping the essential feature lines of the original design. An illustrative movie of the shape evolution during optimization is provided.

RANS Sensitivities vs. Approximate Time-Averaged DES Sensitivities

The preceding section summarized the achievements in the development of adjoint-based shape optimization methods for external aerodynamics of vehicles. These methods rest upon steady-state RANS simulations. Meanwhile, however, the common practice of external aerodynamics computations has moved - at least in the largest part of the Volkswagen Group - from steady-state RANS to unsteady Detached-Eddy Simulations (DES, [31]). To create an adjoint to a time-varying primal like DES entails particular difficulties: Since the adjoint runs backwards in time, all primal states for the time period of interest have either to be stored or recomputed when running the adjoint. The huge amount of storage space and computational time can, in principle, be reduced to a manageable amount by clever check pointing algorithms [32, 33], but for the typical model sizes of car external aerodynamics an adjoint to a DES run is still not feasible yet.

Our provisional solution to compute sensitivity maps for primal DES computations therefore looks as follows: We plug the time-averaged velocity **v** and pressure p from the primal DES computation into a RANS turbulence model in order to solve for a turbulent viscosity vt. With these quantities - time-averaged **v** and p as well as RANS-vt - we run the existing RANS adjoint code to obtain the sensitivities. A validation study similar to the one above for the XL1 revealed that we can only expect qualitative accuracy from this procedure. Quantitatively, the sensitivities are not correct. Still, the following comparison will demonstrate that these approximate DES sensitivities are useful and actually have an added value against pure RANS sensitivities.

Drag sensitivities for the current Audi A7 (Figure 17) as computed by two different procedures are compared in Figures 18 to 20: RANS adjoint based on a RANS primal on the one hand, and the same RANS adjoint but run on a DES primal as described above on the other hand. For both RANS and DES primal the Spalart-Allmaras turbulence model was employed, while the adjoint made use of the frozen turbulence assumption. On the largest part of the car body, the sensitivities can be seen to be very similar. This does not come as a surprise, because here the flow is attached to the surface and RANS and DES flows do not differ dramatically for attached flows. Significant differences occur in the rear part of the car where the flow is close to detachment: While the RANS-based adjoint fails to predict the experimentally well-confirmed beneficial effect of a rear spoiler, it is clearly indicated by the DES-based adjoint. Moreover, the productive effect of a further 'boat-tailing', i.e. tapering of the car back, is almost missed by the RANS-based adjoint while clearly present in the DES counterpart. It is this kind of results that - despite the lack of their quantitative accuracy - gave a lot of credibility to our provisional solution of computing sensitivities based on DES primals and supported their integration into the regular computational processes for external aerodynamics (see Figure 21).

Figure 17: The current Audi A7. As simplifications for the primal flow model, the engine bay was completely sealed (i.e. no underhood flow) and the floor was assumed stationary.

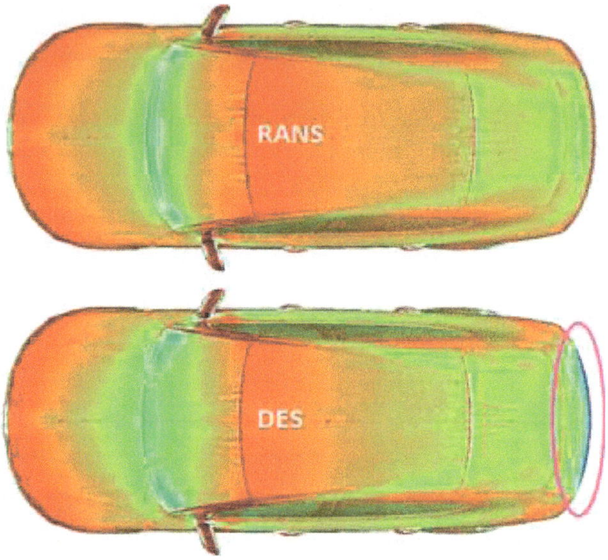

Figure 18: RANS vs. approximate DES sensitivities. Note the similarity between RANS- and DES-based drag sensitivities for the largest part of the car hat - and the dramatic difference at the rear part. While the RANS computation misses the favorable effect of a rear spoiler, it is clearly present in the DES results.

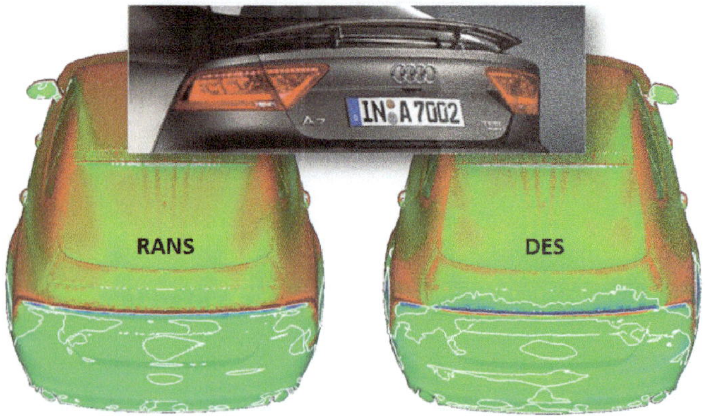

Figure 19: RANS vs. approximate DES sensitivities. A rear view of the sensitivity differences in the spoiler area, and the actual shape of the active spoiler in the current Audi A7.

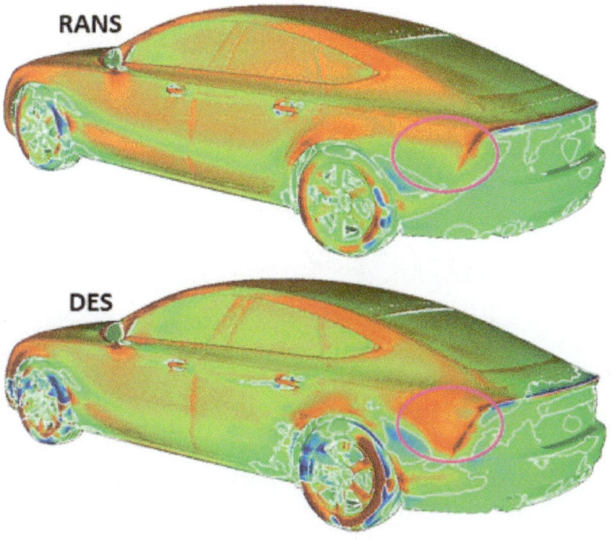

Figure 20: RANS vs. approximate DES sensitivities. The RANS-based sensitivities also fail to predict the beneficial effect of further lateral tapering of the car's back ('boat-tailing').

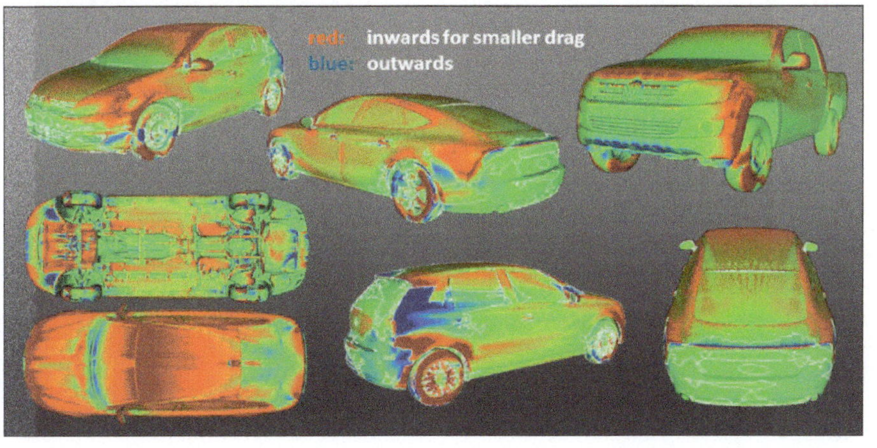

Figure 21: Drag sensitivity maps based on DES. Further examples of drag sensitivity maps for various car models of Volkswagen and Audi as computed by the outlined DES-based method.

Adjoint Methods for Flow Control

Geometrical modifications, be it via shape or via topology optimization, are not the only option of improving the fluid dynamic performance of car components or the entire vehicle. Recently there is a growing interest in qualifying active flow control devices for vehicle aerodynamics: Brunn et al. [34] demonstrated how steady blowing and oscillating jets can be successfully applied to reduce drag of the well-known Ahmed body [35]. Bideaux et al. [36] used the same bluff body shape to investigate the effect of a line of pulsed jet actuators on drag and lift, while Heinemann et al. [37] experimentally investigated continuous jets on a 1:4 model of a passenger production car.

Since flow control of entire vehicles is a rather young subject, experience for the optimal layout of jet configurations is still very limited. The questions to be answered when designing a flow control concept are: (1) What kind of jets should be applied - blowing or suction? (2) Where should the jets be positioned to have a maximum effect? (3) What is the achievable aerodynamic improvement?

To answer these questions, the adjoint method is again the tool of choice, as it allows to compute a sensitivity map of the whole car for the positioning of jets. While for shape sensitivity maps, the design variables were the normal displacements of all surface nodes, it is now the normal flow velocity v_n on each surface element that constitutes the design space. In other words, the car is regarded as having a perforated surface with initially zero normal mass flux through it, and the adjoint method is then applied to compute the sensitivity of drag F_x or lift F_z w.r.t. changing the mass flux through each of the 'holes' [11].

Figure 22 shows such a sensitivity map for the Ahmed body, i.e. the distribution of $\partial F_x/\partial v_n$. Blue areas are those where blowing ($v_n>0$) would result in drag reduction, while in red areas a suction jet would help to reduce drag. The colour intensity indicates the magnitude of the effect on the drag F_x. The computation was carried out on a low-Reynolds mesh ($y+\approx1$) using a RANS primal with the Spalart-Allmaras turbulence model and its adjoint counterpart.

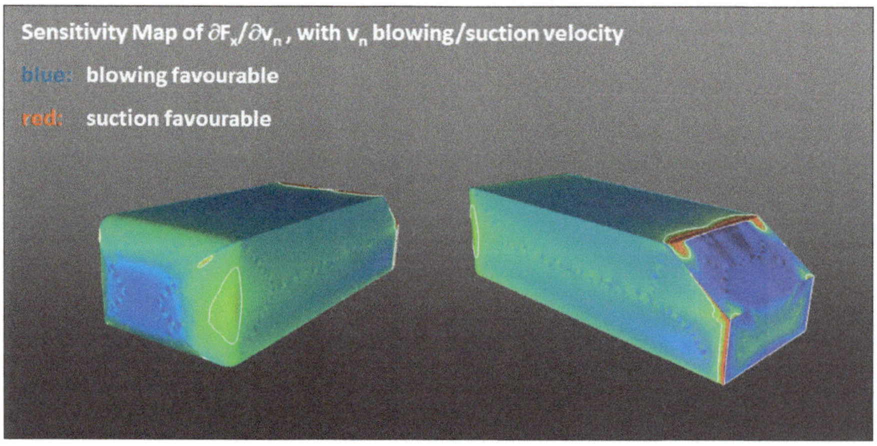

Figure 22: Flow control sensitivities for the Ahmed body. Based on a RANS computation, a flow control sensitivity map of dragFx w.r.t. normal jet velocity v_n was computed for the 35° case of the Ahmed body. The highest sensitivities are expectedly encountered at the rear part. Suction is favorable at the rear separation edge and at the origin of the longitudinal vortices, whereas on the slant surface blowing would reduce the drag by filling the wake.

The validation of these sensitivities is shown in Figure 23. At two different locations, one in the blue and another in the red region, various normal velocities ranging from −40 m/s (suction) to +40 m/s (blowing) were applied in order to compute their effect on drag. As expected, for small jet velocities of up to 5 m/s, the sensitivity captures the gradient $\Delta F_x/\Delta v_n$ very well. But surprisingly, still for higher jet velocities the sensitivity constitutes a good approximation to the actual effect of the jet. In conclusion, flow control sensitivity maps do not only indicate whether a suction jet or a blowing jet is favorable and where to place it, but also give a useful estimate of the achievable effect.

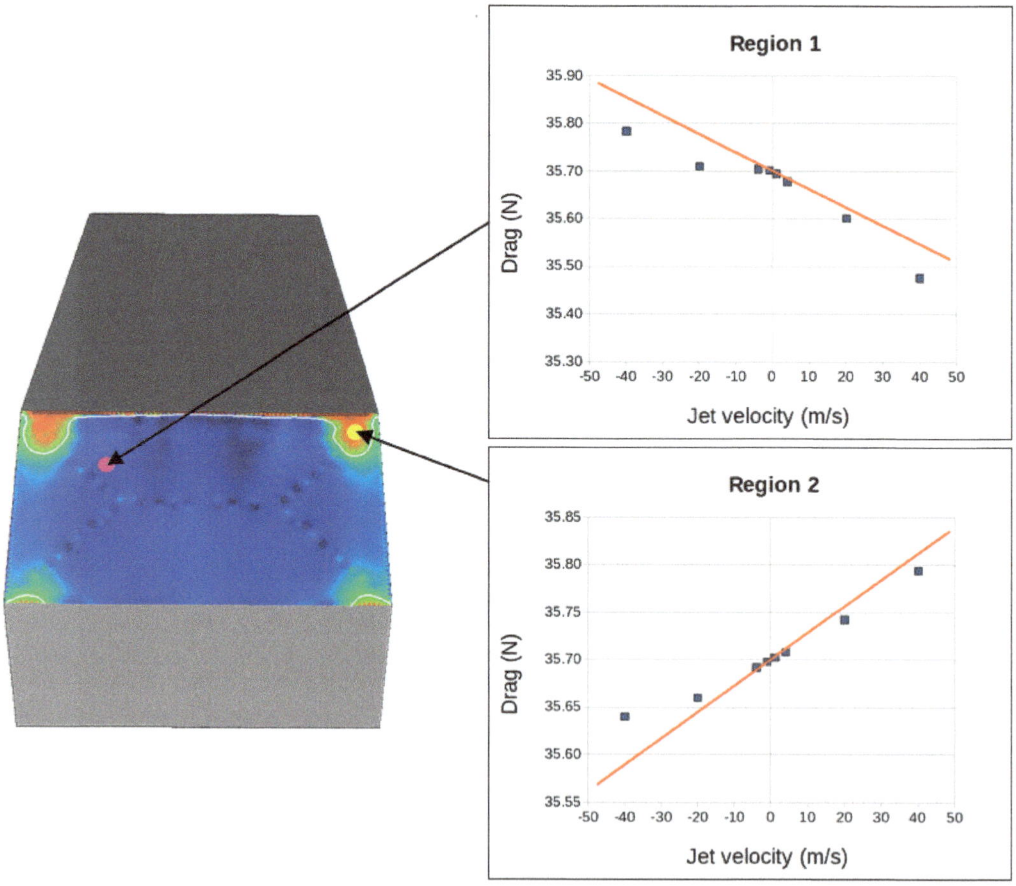

Figure 23: Flow control sensitivities for the Ahmed body. For validation purposes, normal jet velocities between −40 m/s (suction) and +40 m/s (blowing) were applied on surface elements at two different locations of opposite sensitivity signs (pink and yellow spot, respectively). The blue squares indicate the computed drag values, and the red lines correspond to the sensitivity of the respective locations. Note the qualitative and quantitative agreement even for high jet velocities.

For the Volkswagen XL1, apart from pure shape optimization as shown above, also the possibility of further drag reduction by blowing jets was investigated. A flow control sensitivity map revealed several promising areas for the application of blowing jets (blue areas in Figure 24). On the car hat, they correspond to a large extent to the blue areas of Figure 10, i.e. where an outward perturbation of the surface was found beneficial. The actual application of flow control, however, concentrated on the underbody - more precisely in a region behind the rear axis (pink circle in Figure 24), where we installed a number of blowing jets. Force measurements, oil film visualizations as well as Particle-Image-Velocimetry (PIV) measurements on a 1:4 model (Figure 25, details see [38]) revealed the effect of these jets on the flow around the XL1: The recirculation area behind the car became smaller, and the flow in the symmetry plane was directed in the upward direction (Figure 26). This was found to be due to a weakening of the longitudinal vortices that emanate from the rear corners of the car (Figure 27). As a result, lift dropped by 20%, and drag - in accordance with the computed sensitivities - decreased by about 2%. Even though this additional drag reduction of 2% is significant for a car like the XL1, it is unfortunately not enough to incorporate a jet system into a car and run it productively.

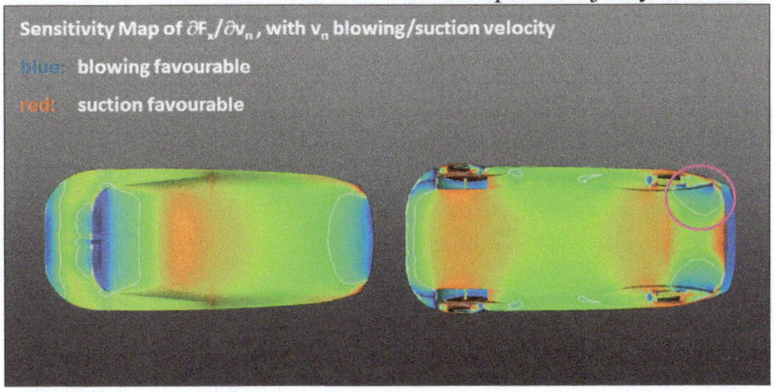

Figure 24: Flow control sensitivities for the Volkswagen XL1. Note the correspondence between 'blue' flow control sensitivities and shape sensitivities (Figure 10) on large parts of the car hat. In the hardware 1:4 model, blowing jets were applied symmetrically in the region within the pink circle.

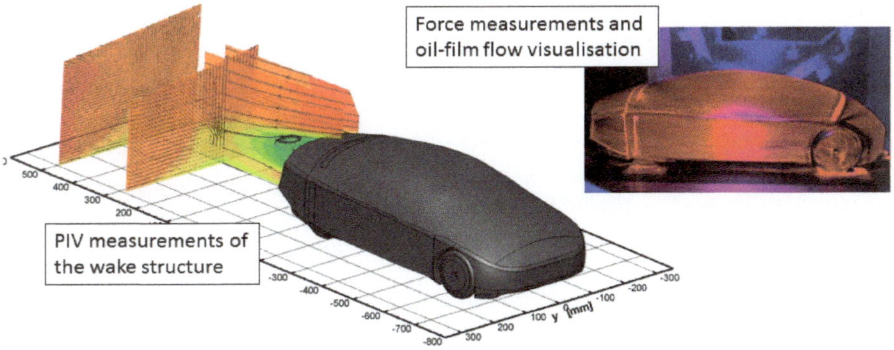

Figure 25: Wind tunnel measurements of the XL1. In addition to force measurements and oil film visualization, PIV measurements were conducted in four different planes in the car's wake, including the symmetry plane (data kindly provided by R Petzold and P Scholz from the Technical University of Braunschweig [38]).

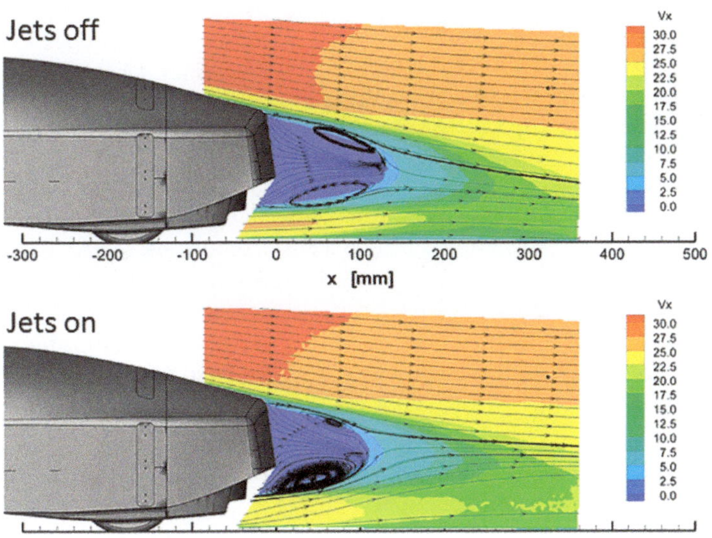

Figure 26: PIV measurements in the symmetry plane of the XL1 wake. Note the slightly smaller recirculation area and the upwardly directed flow when the jets are switched on (see [38] for details).

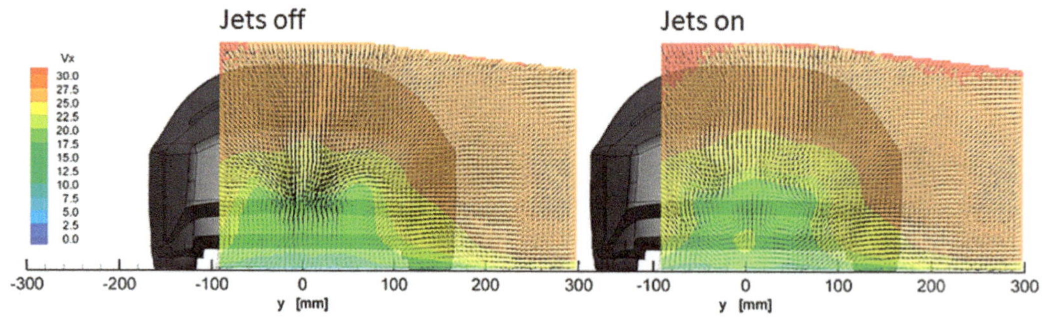

Figure 27: PIV measurements in a perpendicular plane behind the XL1. Upon switching on the jets, the two counter-rotating longitudinal vortices are effectively weakened. This goes along with less downwash and hence decreased lift (see [38] for details).

SUMMARY AND OUTLOOK

Starting from the obstacles of introducing the adjoint method into automotive CFD, we have reported on the efforts undertaken with various partners to develop an industrially viable adjoint solver. Based on OpenFOAM, powerful optimization methods - topology optimization for ducted flows, and shape optimization for ducted as well as external flows - have been developed. While topology optimization is being used systematically already, the implementation of shape optimization into the regular automotive development process has just started with some initial promising results. Both methods will most surely have an impact on overall vehicle performance and consumption in the future.

The adjoint development efforts have so far concentrated on steady-state phenomena. For an extension towards inherently transient flow applications, like aeroacoustics, flow control with pulsating jets and transient aerodynamics with DES, the adjoint has to become transient as well. This is a subject that is under investigation by several research groups already (e.g. [39, 40, 41]) and is currently addressed in the Marie-Curie International Training Network 'aboutFlow' [42]. Its industrialization can be regarded as the next challenge in developing adjoint methods for the automotive industry.

ACKNOWLEDGEMENTS

The author is truly grateful to E de Villiers and his colleagues at Engys, KC Giannakoglou and his team at NTUA (above all EM Papoutsis-Kiachagias and AS Zymaris), E Stavropoulou, M Hojjat and KU Bletzinger from TUM, JD Müller (Queen Mary University London) and the FlowHead team, as well as E Skaperdas and K Haliskos from BETA CAE Systems. The support from many CFD colleagues of the Volkswagen Group in the form of continuous interest in the adjoint method, provision of resources and of valuable feedback is also gratefully acknowledged.

REFERENCES

1. Lions JL: Optimal Control of Systems Governed by Partial Differential Equations. Springer, New York; 1971.
2. Pironneau O: On optimum design in fluid mechanics. J Fluid Mech 1974, 64: 97–110. 10.1017/S0022112074002023
3. Jameson A: Aerodynamic design via control theory. J Sci Comput 1988, 3: 233–260. 10.1007/BF01061285
4. Löhner R, Soto O, Yang C: An adjoint-based design methodology for CFD optimization problems. AIAA-03–02992003.
5. Nadarajah S, Jameson A: A comparison of the continuous and discrete adjoint approach to automatic aerodynamic optimization. AIAA-00–0667 2000.
6. OpenFOAM® - The Open Source CFD Toolbox [http://www.openfoam.com]
7. Othmer C, de Villiers E, Weller HG: Implementation of a continuous adjoint for topology optimization of ducted flows. AIAA-2007–3947 2007.
8. Othmer C: A continuous adjoint formulation for the computation of topological and surface sensitivities of ducted flows. Int J Numer Methods Fluids 2008, 58: 861–877. 10.1002/fld.1770
9. Zymaris AS, Papadimitriou DI, Giannakoglou KC, Othmer C: Continuous adjoint approach to the Spalart-Allmaras turbulence model for incompressible flows. Comput Fluids 2009,38(8):1528–1538. 10.1016/j.compfluid.2008.12.006
10. Othmer C, Papoutsis-Kiachagias EM, Haliskos K: CFD optimization via sensitivity-based shape morphing. Proceedings of the 4th ANSA & μETA Internat. Conf., Thessaloniki, Greece, BETA CAE Systems S.A. 2011.
11. Zymaris AS, Papadimitriou DI, Giannakoglou KC, Othmer C: Optimal location of suction or blowing jets using the continuous adjoint approach. Proceedings of the 5th European Conf. on Computational Fluid Dynamics, Lisbon, Portugal, ECCOMAS CFD 2010.
12. Zymaris AS, Papadimitriou DI, Giannakoglou KC, Othmer C: Adjoint wall functions: a new concept for use in aerodynamic shape optimization. J Comput Phys 2010,229(13):5228–5245. 10.1016/j.jcp.2010.03.037
13. Kontoleontos EA, Papoutsis-Kiachagias EM, Zymaris AS, Papadimitriou DI, Giannakoglou KC: Adjoint-based constrained topology optimization for viscous flows, including heat transfer. Eng Optim 2013,45(8):941–961. 10.1080/0305215X.2012.717074
14. de Villiers E, Othmer C: Multi-objective adjoint optimization of intake port geometry. SAE Technical Paper 2012–01–0905 2012.
15. Helgason E, Krajnovic S: Aerodynamic shape optimization of a pipe using the adjoint method. Proceedings of the ASME 2012 International Mechanical Engineering Congress & Exposition, Houston, Texas,

USA, IMECE 2012.

16. Lincke A, Rung T: Adjoint-based sensitivity analysis for Buoyancy-driven incompressible Navier-Stokes equations with heat transfer. Proceedings of the Eighth Internat. Conf. on Engineering Computational Technology, Dubrovnik, Croatia 2012.
17. Jakubek D, Wagner C: Shape optimization of train head cars using adjoint-based computational fluid dynamics. Proceedings of the First International Conference on Railway Technology, Las Palmas de Gran Canaria, Spain 2012.
18. Towara M: Numerical optimization of an oil intake duct with adjoint topological methods. Diploma thesis. RWTH Aachen University, Dept. of Software and Tools for Computational Engineering; 2011.
19. Hinterberger C, Olesen M: Automatic geometry optimization of exhaust systems based on sensitivities computed by a continuous adjoint CFD method in OpenFOAM. SAE Technical Paper 2010–01–1278 2010.
20. Hinterberger C, Olesen M: Industrial application of continuous adjoint flow solvers for the optimization of automotive exhaust systems. Proceedings of the ECCOMAS Thematic Conf. - CFD & Optimization, Antalya, Turkey, ECCOMAS 2011.
21. Giannakoglou KC: Continuous adjoint methods in shape, topology, flow-control and robust optimization. Open Source CFD International Conference, London, UK, ICON-CFD 2012.
22. Bendsoe MP, Sigmund O: Topology Optimization: Theory, Methods and Applications. Springer, Berlin; 2004.
23. Borrvall T, Petersson J: Topology optimization of fluids in Stokes flow. Int J Numer Methods Fluids 2003, 41: 77–107. 10.1002/fld.426
24. Moos O, Klimetzek F, Rossmann R: Bionic optimization of air-guiding systems. SAE Technical Paper 2004–01–13772004. Moos O, Klimetzek F, Rossmann R: Bionic optimization of air-guiding systems. SAE Technical Paper 2004-01-1377 2004.
25. Gersborg-Hansen A, Sigmund O, Haber R: Topology optimization of channel flow problems. Struct Multidiscip Optim2005, 30: 181–192. 10.1007/s00158-004-0508-7
26. Othmer C, Grahs T: Approaches to fluid dynamic optimization in the car development process. Proceedings of the EUROGEN Conference, Munich, Germany 2005.
27. Othmer C, Kaminski T, Giering R: Computation of topological sensitivities in fluid dynamics: cost function versatility. Proceedings of the ECCOMAS CFD Conference, Delft, The Netherlands 2006.
28. Stavropoulou E, Hojjat M, Bletzinger KU: In-plane mesh regularization for node-based shape optimization problems.Comput Methods Appl Mech Eng, in press [http://www.sciencedirect.com/science/article/pii/S004578251400070X]
29. Hojjat M, Stavropoulou E, Bletzinger KU: The vertex morphing method for node-based shape optimization. Comput Methods Appl Mech Eng 2014, 268: 494–513.
30. ANSA® pre-processor – BETA CAE Systems S.A. [http://www.beta-cae.gr/ansa.htm] ANSA® pre-processor – BETA CAE Systems S.A. [http://www.beta-cae.gr/ansa.htm]
31. Islam M, Decker F, de Villiers E, Jackson A, Gines J, Grahs T, Gitt-Gehrke A, Comas i Font J: Application of detached-Eddy simulation for automotive aerodynamics development. SAE Technical Paper 2009–01–0333 2009.
32. Griewank A: Achieving logarithmic growth of temporal and spatial complexity in reverse automatic differentiation.Optim Methods Softw 1992, 1: 33–54.
33. Wang Q, Moin P, Iaccarino G: Minimal repetition dynamic checkpointing algorithm for unsteady adjoint calculation.SIAM J Sci Comput 2009,31(4):2549–2567. 10.1137/080727890
34. Brunn A, Wassen E, Sperber D, Nitsche W, Thiele F: Active drag control for a generic car model. Notes Numer Fluid Mech Multidiscipl Des 2007, 95: 247–259. 10.1007/978-3-540-71439-2_15
35. Ahmed SR, Ramm R, Falting G: Some salient features of the time-averaged ground vehicle wake. SAE Technical Paper 84–0300 1984.
36. Bideaux E, Bobillier P, Fournier E, Gillieron P, El Hajem M, Champagne JY, Gilotte P, Kourta A: Drag reduction by pulsed jets on strongly unstructured wake: towards the square back control. Int J Aerodyn 2011,1(3/4):282–298. 10.1504/IJAD.2011.038846
37. Heinemann T, Springer M, Lienhart H, Kniesburges S, Becker S: Active flow control on a 1:4 car model. Proceedings of the 16th Int. Symp. on Applications of Laser Techniques to Fluid Mechanics, Lisbon, Portugal 2012.
38. Petzold R: Experimentelle Untersuchungen zur Strömungsbeeinflussung an einem 1:4-Fahrzeugmodell. Diploma thesis. TU Braunschweig, Inst. für Strömungsmechanik; 2011.
39. Torbert S, Löhner R: Checkpointing schemes for adjoint methods and strongly unsteady flow. AIAA-2012–05722012.
40. Economon TD, Palacios F, Alonso JJ: A coupled-adjoint method for aerodynamic and aeroacoustic optimization. AIAA-2012–5598 2012.
41. Carnarius A, Thiele F, Özkaya E, Nemili A, Gauger N: Optimal control of unsteady flows using a discrete and a continuous adjoint approach. Proceedings of the 25th Conference on System Modeling and Optimization, Berlin, Germany 2011, 318–327

CHAPTER 7

NUMERICAL INVESTIGATION ON THE AERODYNAMIC CHARACTERISTICS OF HIGH-SPEED TRAIN UNDER TURBULENT CROSSWIND

Mulugeta Biadgo Asress and Jelena Svorcan

Department of Aeronautics, Faculty of Mechanical Engineering, University of Belgrade, Kraljice Marije 16, Belgrade, Serbia

ABSTRACT

Increasing velocity combined with decreasing mass of modern high-speed trains poses a question about the influence of strong crosswinds on its aerodynamics. Strong crosswinds may affect the running stability of high-speed trains via the amplified aerodynamic forces and moments. In this study, a simulation of turbulent crosswind flows over the leading and end cars of ICE-2 high-speed train was performed at different yaw angles in static and moving ground case scenarios. Since the train aerodynamic problems are closely associated with the flows occurring around train, the flow around the train was considered as incompressible and was obtained by solving the incompressible form of the unsteady Reynolds-averaged Navier–Stokes (RANS) equations combined with the realizable k-epsilon turbulence model. Important aerodynamic coefficients such as the side force and rolling moment coefficients were calculated for yaw angles ranging from −30° to 60° and compared with the results obtained from wind tunnel test. The dependence of the flow structure on yaw angle was also presented. The nature of the flow field and its structure depicted by contours of velocity magnitude and streamline patterns along the train's cross-section were presented for different yaw angles. In addition, the pressure coefficient around the circumference of the train at different locations along its length was computed for yaw angles of 30° and 60°. The computed aerodynamic coefficient outcomes using the realizable k-epsilon turbulence model were in good agreement with the wind tunnel data. Both the side force coefficient and rolling moment coefficients increase steadily with yaw angle till about 50° before starting to exhibit an asymptotic behavior. Contours of velocity magnitude were also computed at different cross-sections of the train along its length for different yaw angles. The result showed that magnitude of rotating vortex in the lee ward side increased with increasing yaw angle, which leads to the creation of a low-pressure region in the lee ward side of the train causing high side force and roll moment. Generally, this study shows that unsteady CFD-RANS methods combined with an appropriate turbulence model can present an important means of assessing the crucial aerodynamic forces and moments of a high-speed train under strong crosswind conditions.

INTRODUCTION

Rail transport system brings enormous benefits to society by providing access and mobility that are essential for modern societies and economic growth and hence is a major form of passenger and freight transport in many countries [1]. In October 1964, the first high-speed rail in the world was put

Citation: Asress, M. B., & Svorcan, J. (2014). Numerical investigation on the aerodynamic characteristics of high-speed train under turbulent crosswind. Journal of Modern Transportation, 22(4), 225-234.

into operation with the highest speed of 210 km/h in Japan [2]. Since then, the last decades has witnessed the rapid development of high-speed rail system in many countries such as Germany (Fig. 1 shows a ICE-2 high-speed train made by Germany), France, Italy, Spain, China, and South Korea. Other emerging countries like Turkey and Brazil are also constructing high-speed rail networks to connect their major cities; some African countries such as Morocco, Algeria, and South Africa also proposed to build high-speed rail corridors in recent years.

Many of the current generation high-speed trains such as the Spanish AVE class 103, the German ICE-3, the French TGV Duplex, the South Korean KTX-II, the Chinese CHR C, and the Japanese Shinkansen E6 reach speeds of 300 km/h in regular operation. At these speeds, aerodynamic forces and moments are becoming more and more important for the running performance of the train. Strong crosswind may affect the running stability and riding comfort of the vehicles.

Figure 1: ICE-2 high-speed train.

The increases of the aerodynamic forces and moments due to crosswinds may deteriorate the train operating safety and cause the train to overturn. The stability of trains in crosswinds is of concern to a number of countries with high-speed rail networks [3]. Crosswind stability of rail vehicles has been a research topic during the last decades, mainly motivated by overturning accidents. Some crosswind-related accidents are shown in Fig. 2 [4, 5]. There have been 29 wind-induced accidents of vehicles since transport service was started in 1872 in Japan. Most of these accidents happened on narrow gage (1,067 mm) lines [6]. Therefore, understanding of crosswind stability for rail vehicles has to be a topic of recognized safety issues in the railway community of every country. Recently, the aerodynamics of a train under the influence of crosswinds has been taken as a safety relevant topic and covered in national standards of UK [7], Italy [8], and Germany [9], as well as in the European Community legislation and norm [10, 11] (Fig. 3).

Figure 2: Crosswind related accidents in Austria in 2002 (**a**) and Switzerland in 2007 (**b**) [3, 4].

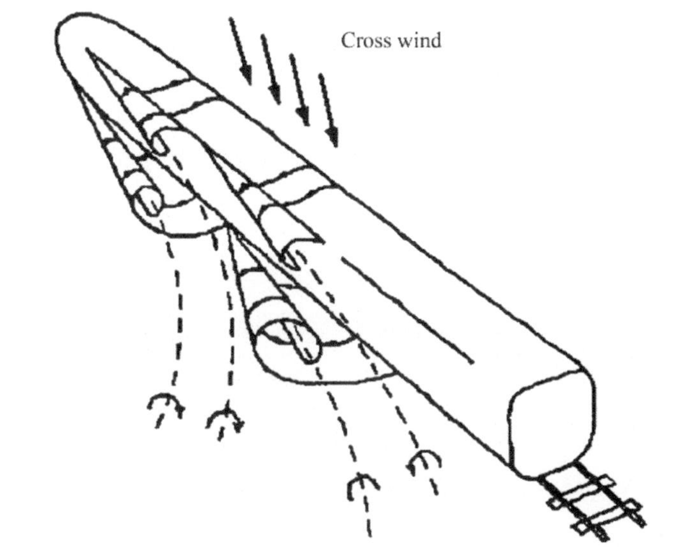

Figure 3: Flow behind a train in a crosswind.

The risk of crosswind-induced overturning depends on both the line infrastructure and vehicles' aerodynamic characteristics [12]. On the infrastructure side, sites with tall viaducts and high embankments call for attention. The combination of modern light weight and high speed leads to an increased concern regarding the stability of high-speed trains, especially when traveling on high embankments exposed to crosswinds and sudden, powerful wind gusts. Therefore, acquiring detailed and correct data on these scenarios is quite important due to the involved risks of accidents such as a train overturning.

On the vehicles side, the topic of train overturning due to crosswind exposure is closely linked to the crosswind sensitivity of the leading car of the train set, which is often the most sensitive part. This is because the front end of a railway car is usually subjected to the largest aerodynamic loads per unit length [13, 14].

The crosswind stability against overturning is a major design criterion for high speed railway vehicles and has been an experimental and/or numerical research topic for a number of scholars [15–22]. The experimental study allows to have a higher confidence in the absolute values of the measured aerodynamic forces where the numerical calculations allow to obtain a more detailed information of the flow field around the vehicle.

Among the experimental investigators, Orellano and Schober [18] have conducted a wind tunnel experiments on the aerodynamic performance of a generic high-speed train. The wind tunnel model used was a simplified 1:10 scaled ICE-2 train with and without simplified bogies. The model is known as aerodynamic train model (ATM). The study was confined to the aerodynamic loads on the stationary first car of the ATM for flat ground scenario, when exposed to yaw angles ranging from −30° to 60°. The flow speed was 70 m/s, which corresponds to Reynolds numbers of 1.4×10^6 based on the approximate model width of the train (0.3 m). In this experiment, the results have been presented through aerodynamic coefficients.

The objective of this study is to conduct a numerical investigation using unsteady Reynolds-averaged Navier–Stokes (RANS) method combined with the k-epsilon turbulence model on the aerodynamic characteristics of the leading and end cars of ICE-2 high-speed train subjected to a crosswind in static and moving ground case scenarios. The width, length, and height of the modeled train are 3.0, 29.3, and 3.9 m, respectively. For the static ground case, the numerical simulation scenario consists of a stationary train model exposed to a constant crosswind of 70 m/s at different yaw angles ranging from −30° to 60° in a similar way to the wind tunnel test performed by Orellano and Schober [18]. For the moving ground case, the numerical simulation scenario consists of a moving train exposed to effective crosswind (relative wind speed) of 70 m/s at different yaw angles ranging from −30° to 60°. The results were compared to the wind tunnel experimental data.

At present, feasible modeling technologies for turbulent flows are steady and unsteady RANS methods, large eddy simulation (LES), and detached eddy simulation (DES). Because of its relatively low computational cost, the unsteady RANS method was used in the simulations of this study. The aim is to assess the predicting capability of the unsteady RANS method by examining the behavior of the vehicle's aerodynamic coefficients numerically and comparing to the wind tunnel results.

GOVERNING EQUATIONS

The equations which govern the flow over the train are the continuity and Navier–Stokes equations [23–25]. The flow around the train in our particular problem is assumed to be incompressible. Hence, for turbulent flow, the incompressible unsteady RANS equations can be written as

$$\frac{\partial \overline{u}_i}{\partial x_i} = 0, \tag{1}$$

$$\frac{\partial \overline{u}_i}{\partial t} + \overline{u}_j \frac{\partial \overline{u}_i}{\partial x_j} = -\frac{1}{\rho}\frac{\partial \overline{p}_i}{\partial x_i} + \frac{\partial}{\partial x_j}\left(\mu \frac{\partial \overline{u}_i}{\partial x_j} - \rho\overline{u_i' u_j'}\right), \tag{2}$$

where $\overline{u}$ and $\overline{p}$ represent the mean (time averaged) velocity and pressure, respectively; ρ is the density of air, μ is the molecular viscosity, and the last nonlinear term $\overline{u_i' u_j'}$ is the turbulent stress tensor.

Turbulence Model

To model the turbulent stress tensor, the last nonlinear term in Eq. (2) and hence provide closure of the above open set of governing equations, the realizable k-epsilon turbulence model [26, 27] is used in our particular problem.

The k-epsilon model takes mainly into consideration how the turbulent kinetic energy is affected. In this model, turbulent viscosity is modeled as $\mu_t = \rho C_\mu k^2/\varepsilon$, where C_μ is the eddy viscosity coefficient, k is the turbulent kinetic energy, and □ is the rate of dissipation of turbulent kinetic energy. The realizable k-epsilon model has been widely used in various types of flow simulation. The transport equations for realizable k-epsilon model can be expressed as follows:

$$\frac{\partial k}{\partial t} + \rho\overline{u}_j \frac{\partial k}{\partial x_j} = \frac{\partial}{\partial x_j}\left(\left(\mu + \frac{\mu_t}{\sigma_k}\right)\frac{\partial k}{\partial x_j}\right) + P_k - \rho\varepsilon, \tag{3}$$

$$\rho\frac{\partial\varepsilon}{\partial t} + \rho\overline{u}_j\frac{\partial\varepsilon}{\partial x_j} = \frac{\partial}{\partial x_j}\left(\left(\mu + \frac{\mu_t}{\sigma_\varepsilon}\right)\frac{\partial\varepsilon}{\partial x_j}\right) + \rho C_1 S\varepsilon - \rho C_2\frac{\varepsilon^2}{k + \sqrt{v\varepsilon}}, \tag{4}$$

$$P_k = -\rho\overline{u_i' u_j'}\frac{\partial\overline{u}_i}{\partial x_j},\ C_1 = \max\left(0.43, \frac{\eta}{\eta + 5}\right),\ \eta = S\frac{k}{\varepsilon}, \tag{5}$$

$$S = \sqrt{2S_{i,j}S_{i,j}},\ S_{i,j} = \frac{1}{2}\left(\frac{\partial\overline{u}_j}{\partial x_i} + \frac{\partial\overline{u}_i}{\partial x_j}\right). \tag{6}$$

In these equations, P_k represents the generation of turbulence kinetic energy due to the mean velocity gradients; μ and μ_t represent the molecular viscosity and eddy (turbulent) viscosity, respectively; S is the modulus of the mean rate-of-strain tensor; v denotes the kinematic viscosity; σ_k, $\sigma_□$, and C_2 are model constants with default value of 1.0, 1.2, and 1.9, respectively.

The terms on left hand side of Eqs. (3) and (4) present the local rate of change of k and □ and transport of k and □ by convection, respectively. Whereas, the terms on the right hand side present the transport of k and □ by diffusion, rate of production of k and □, and rate of destruction of k and □, respectively.

Atmospheric Boundary Layer

According to Robinson and Baker [28], when a train moves, the inclusion of an atmospheric boundary layer (ABL) simulation is necessary for producing accurate flow physics. That is when the train moves, simulation of ABL is required. In particular, the train motion induces a skewed oncoming crosswind velocity profile (see Fig. 4). For the k-epsilon model, the vertical profiles for the mean wind speed $\overline{u}$, turbulent kinetic energy k, and turbulence dissipation rate ε in the ABL can be expressed as follows:

$$\overline{u}(z) = \frac{\boldsymbol{u}^*}{\boldsymbol{k}}\ln\left(\frac{z + z_0}{z_0}\right),\ k(z) = \frac{u^*}{\sqrt{C_\mu}},\ \varepsilon(z) = \frac{u^{*3}}{(z + z_0)}, \tag{7}$$

$$u^* = \boldsymbol{k}\boldsymbol{u}_{\boldsymbol{ref}}\left(\frac{z_{\text{ref}} + z_0}{z_0}\right), \tag{8}$$

where z is the height above the ground, z_0 is aerodynamic roughness length (ground roughness), u^* is the ABL friction velocity, κ is the von Karman constant, C_μ is a model constant of the realizable k-epsilon model, and u_{ref} is the reference velocity measured at the reference height z_{ref}. In the implementation of the wind alarm system, u_{ref} would be the wind speed measured at the nearest weather station at a railway line. These profiles given in Eq. (7) are commonly used as inlet profiles for computational fluid dynamics (CFD) simulations, when measured profiles are not available.

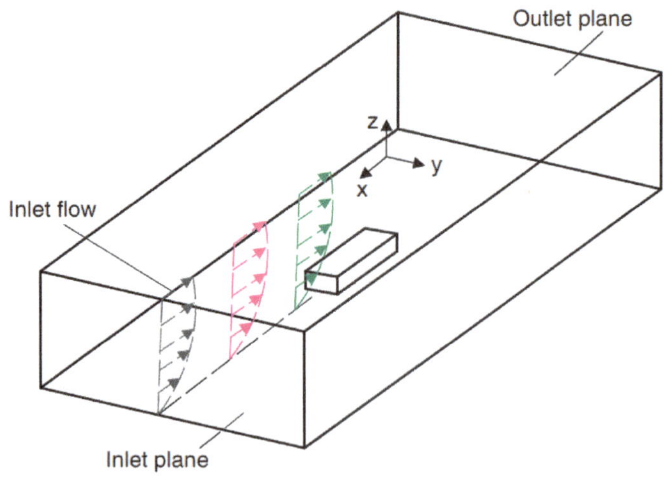

Figure 4: Computational domain with a train model for CFD simulation.

NUMERICAL SIMULATION METHOD

The experimental investigation by Orellano and Schober [18] was done only for a stationary train model on the flat ground influenced by a constant crosswind of 70 m/s at different yaw angles. However, in this paper, the numerical simulation scenario consists of static and moving trains exposed to a crosswind at different yaw angles. In a similar fashion to the experimental set up, the numerical simulation for static ground case scenario consists of a stationary train exposed to a constant crosswind of 70 m/s at different yaw angles ranging from −30° to 60°. When a crosswind of speed V_w impinges on a train of speed V_{tr}, the yaw angle β, the prevailing wind angle α, and the resultant relative wind speed V_{rel} are as shown in Fig. 5.

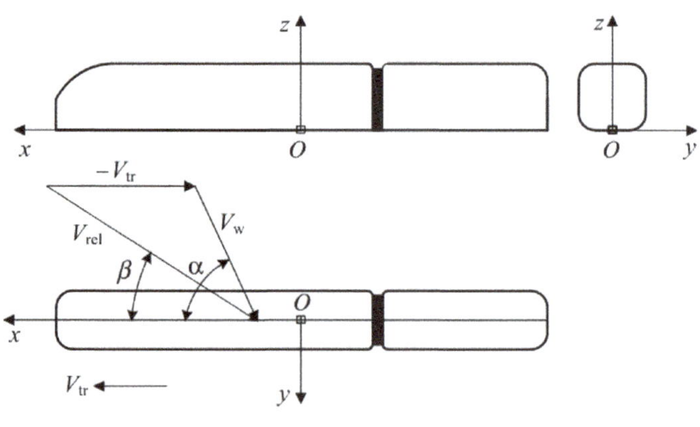

Figure 5: Definition of yaw angle (β).

On the other hand, to simulate the moving train, it is possible to consider the train static and move the ground with a speed of the train in opposite direction ($-V_{tr}$). In the moving case, a relative wind speed (effective crosswind speed) was set to be 70 m/s for all yaw angles considered. Then, the speed of the vehicle was determined for each yaw angle using the relative wind speed. Once the speed of the vehicle was known, the motion of the ground was simulated by presetting the longitudinal velocity component to the speed of travel. The Reynolds number based on the effective crosswind speed, and the width of train model is 1.4 × 10^7. The commercial CFD software FLUENT was used for the numerical simulations.

According to the coordinate system given in Fig. 5, the non-dimensionalized aerodynamic side force coefficient (C_s) and rolling moment coefficient (C_{mx}) can be calculated as follows:

$$C_s = \frac{F_y}{0.5\rho u_{\mathrm{rel}}^2 A}, \tag{9}$$

$$C_{mx} = \frac{M_x}{0.5\rho u_{\mathrm{rel}}^2 AL}, \tag{10}$$

where F_y is the force in the y direction, M_x is the moment about x-axis, ρ is the air density, u_{rel} denotes the approaching air speed, A represents a fixed reference area, and L represents a fixed reference length.

Description of Model Geometry

Real trains are often not used for aerodynamic studies owing to their geometrical complexities; instead, simplified, shortened models are used. Performing numerical simulation for a complete train with a length of about 205 m requires more advanced computational resources than those available. In addition, since the flow structure downstream of a certain distance from the nose of the train (less than one coach length) is more or less constant, a decrease in length does not alter the essential physical features of the flow [29].

The model studied in this work is a more realistic version of the ICE-2 high-speed train which consists of the leading car, an end car, and inter car gap. The model geometry has total length of 29.3 m, width of 3 m, and height of 3.9 m. The model has been created without bogies as shown in Fig. 6. The moment reference point is set to be located at ground level in the midway of the train length. The coefficients for the aerodynamic forces and moments have been obtained using a fixed reference area of 11.6 m² which corresponds to the cross-sectional area of the train model, and reference length of 3 m which presents the width of the train model.

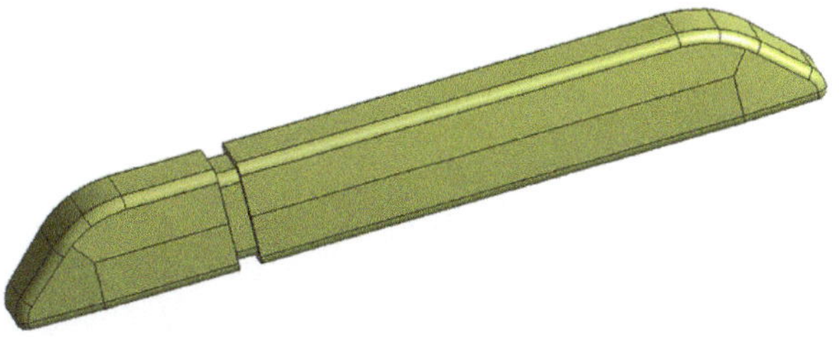

Figure 6: Leading and end car model without bogies used in the numerical simulation.

Computational Domain and Mesh

After the basic shape of the train has been created, a parallelepiped computational domain (see Fig. 4) with height of 50 m, width of 100 m, and length of 150 m is created for the numerical wind tunnel. In the computational domain, the model can be rotated about the z-axis by the required yaw angle for simulation. The distance between nose of vehicle and the inlet boundary is around 50 m, which is large enough to ensure that the velocity and pressure fields are uniform at the inlet and to allow the flow to develop by the time it reaches the train. The model is also sufficiently far from the top and side walls to minimize near wall effects. The clearance between the train and the flat ground (the computational domain floor) is set to be 0.15 times the height of the train.

The mesh of the computational domain was generated using a tetrahedron patch conforming method. Mesh refinement has been done on the train surfaces and surrounding areas. The generated mesh consists of about 3 million elements. The mesh resolution at the wall is very important and needs to be quantified. For standard or non-equilibrium wall functions, each wall-adjacent cell's centroid should be located within the log-law layer, $30 < y^+ < 300$. In the generated meshes, five prismatic cell layers of constant thickness were made on the train walls, and the first cell adjacent to the walls of the train was adjusted to meet the requirements of y^+. The cross-section of the meshes with refinement on the modeled train surfaces and surrounding areas are shown in Fig. 7.

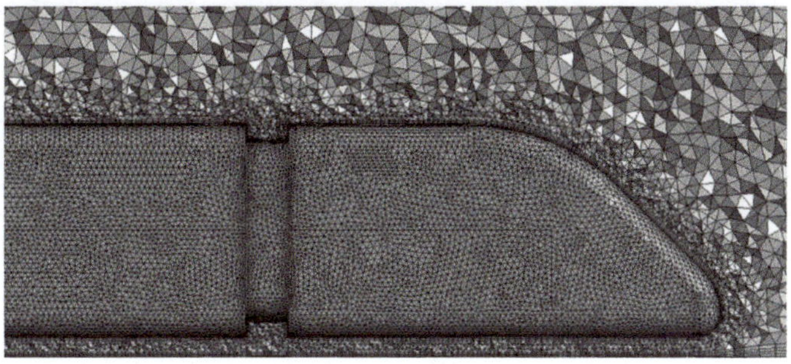

Figure 7: Cross-section of the mesh showing elements.

Boundary Condition

For stationary case, the flow enters the domain with a uniform velocity of 70 m/s. No-slip boundary conditions were used on the train surface and the ground floor. For moving case, a relative wind speed (effective crosswind speed) of 70 m/s was used as velocity inlet. The Reynolds number based on the effective crosswind speed and the width of train model is 1.4×10^7. Symmetry boundary conditions were used on the top and side walls. On the outlet, a homogeneous Neumann boundary condition is applied, meaning that the pressure gradient equal to zero. This will let the flow pass through the outlet without affecting the upstream flow, provided that the upstream distance to the aerodynamic body is large enough. No-slip boundary conditions were used on the train surface and the ground floor. The realizable k-epsilon model was adapted for the turbulence closure. The inflow turbulence intensity and length scale were set to be 3 % and 0.3 m, respectively. On the ground and solid surfaces, the non-equilibrium wall functions were used to determine the boundary turbulence quantities. All runs were performed in a transient mode with a time step of 0.08 s. The conventional SIMPLE algorithm was used to solve the coupled equations, where several iterations are performed in each time step to ensure convergence.

RESULTS AND DISCUSSIONS

The computed mean force and rolling moment coefficients were compared with experimental data and shown in Figs. 8 and 9, respectively. As can be seen in the graphs, the computed side force and rolling moment coefficients are in good agreement with the experimental data. However, at yaw angles of 50° and 60°, CFD slightly over-predicts the rolling moment. This may be due to the effect of end car and inter car gap that was included in the CFD model.

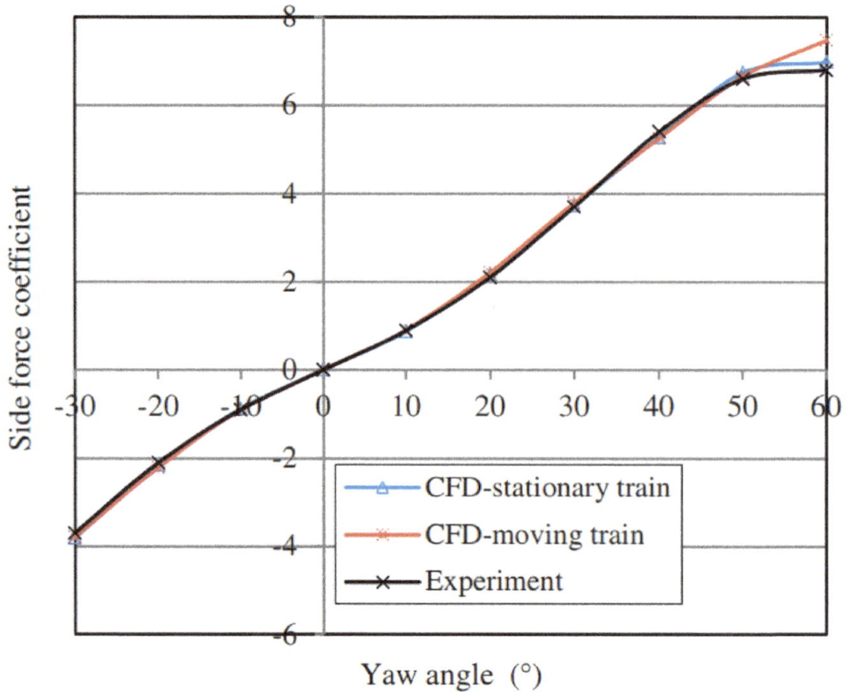

Figure 8: Side force coefficient versus yaw angle.

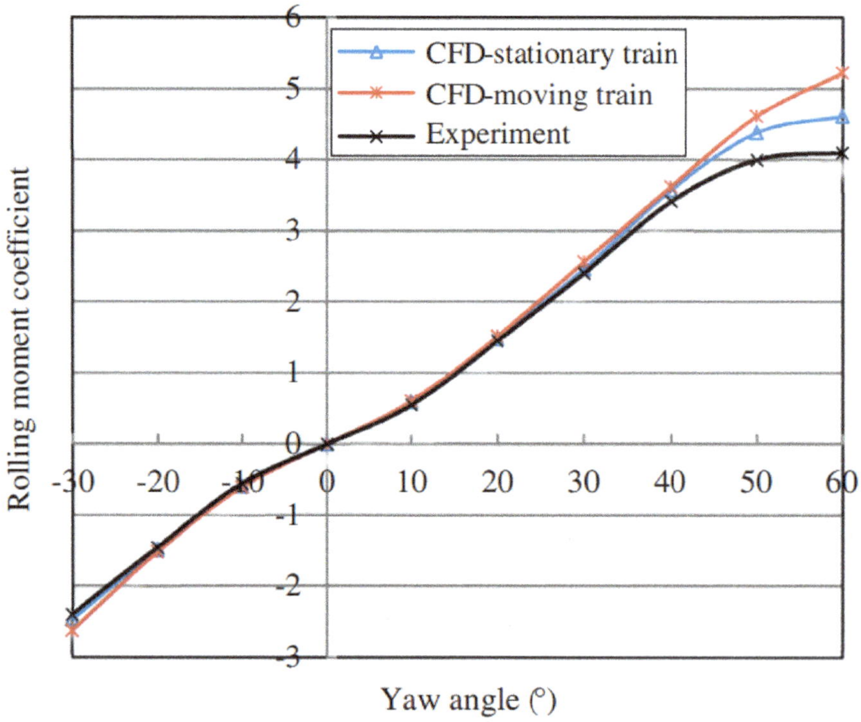

Figure 9: Rolling moment coefficient versus yaw angle.

The nature of the flow field and its structure are depicted by contours of velocity magnitude and streamline patterns along the train's cross-section are presented in Figs. 10, 11, 12, and 13. As expected, for large yaw angles, large flow separation zone exists on the leeward side of the train. The pressure coefficient (C_p) around the circumference of the train at different locations along its length is plotted in Figs. 14 and 16.

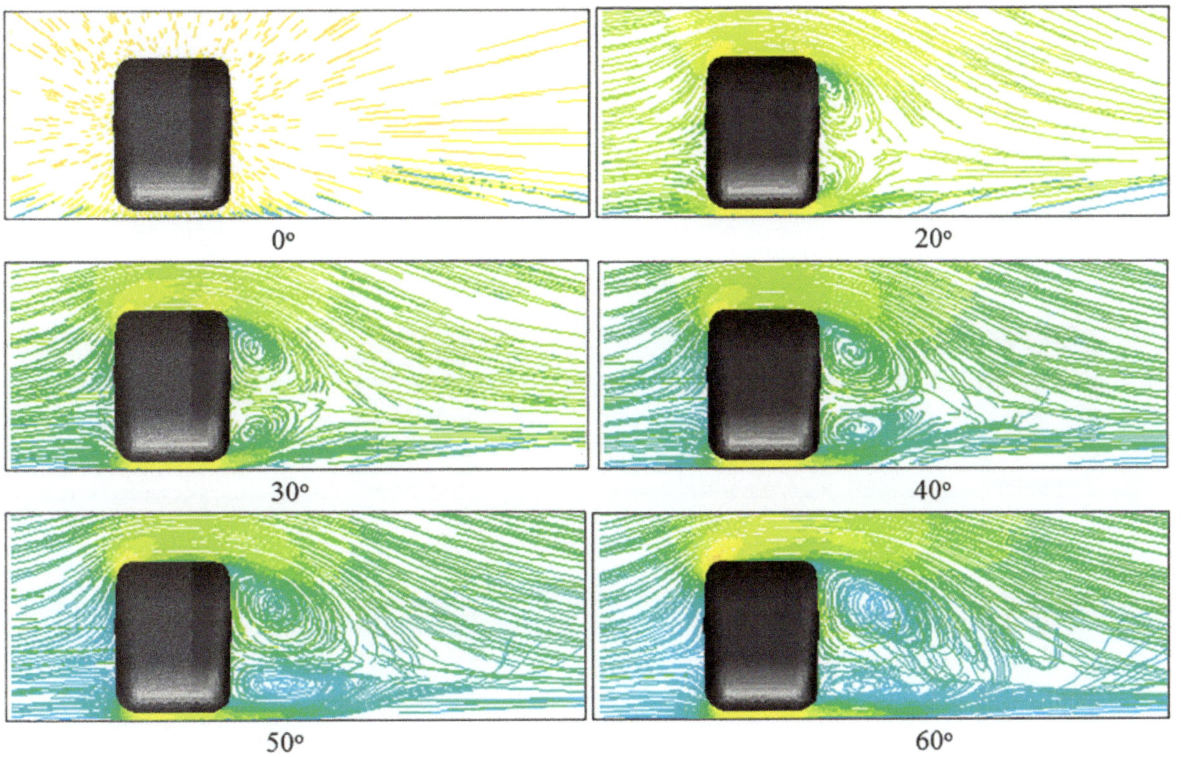

Figure 10: Streamlines along the train's cross-section at 6 m from the nose of the train.

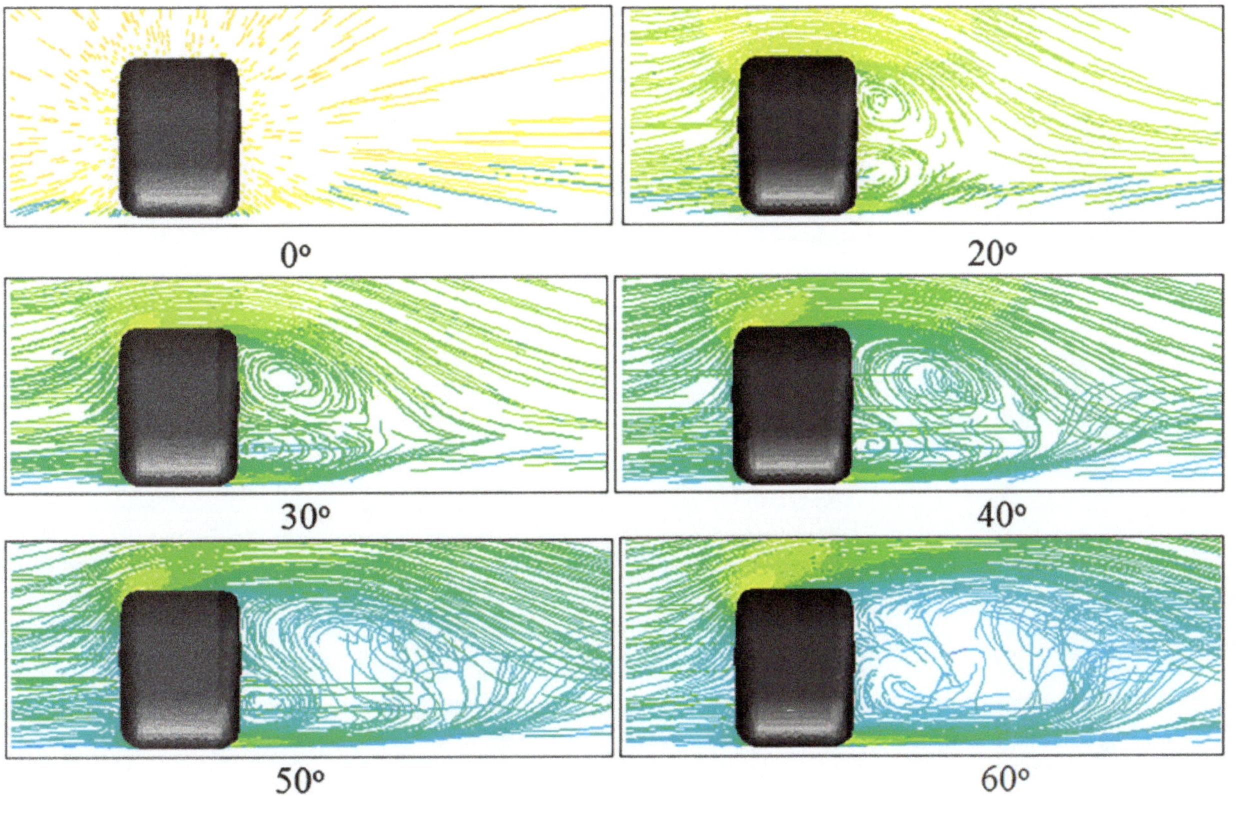

Figure 11: Streamlines along the train's cross-section at 14 m from the nose of the train.

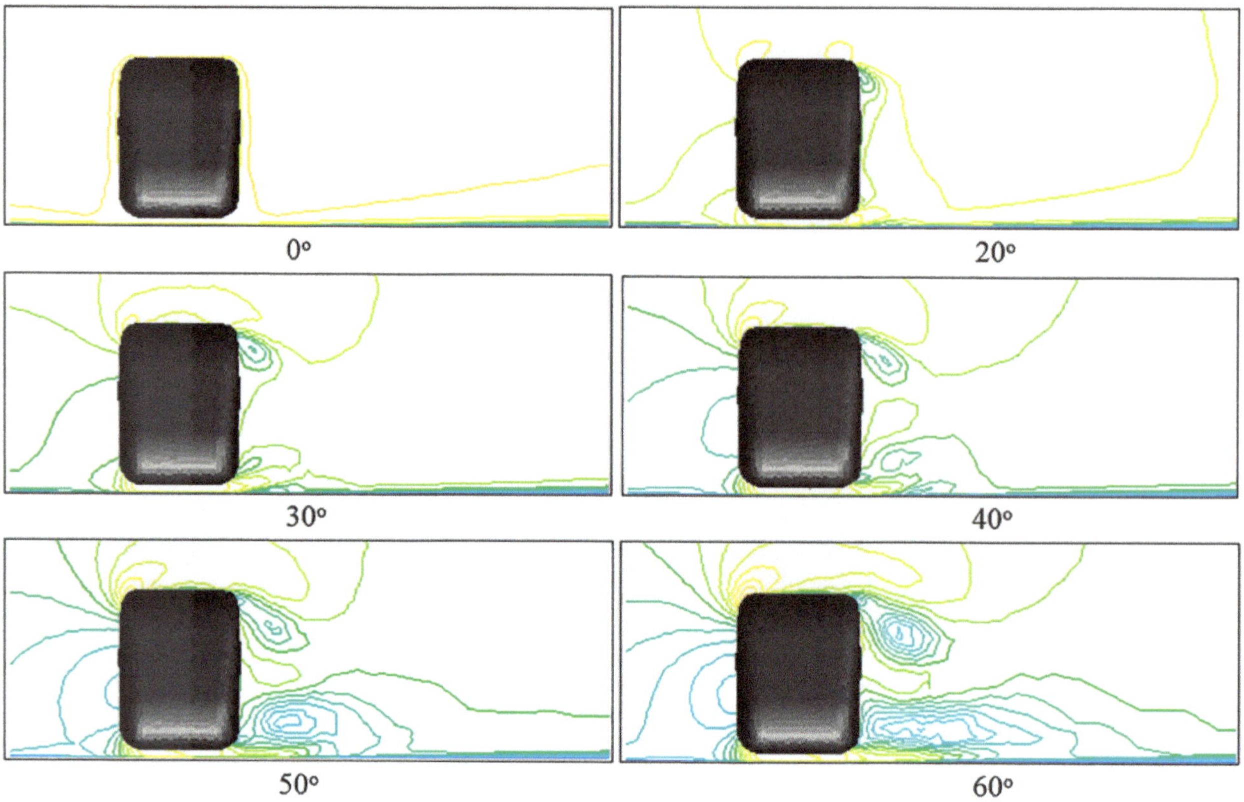

Figure 12: Contours of velocity magnitude along the train's cross-section at 6 m from the nose of the train.

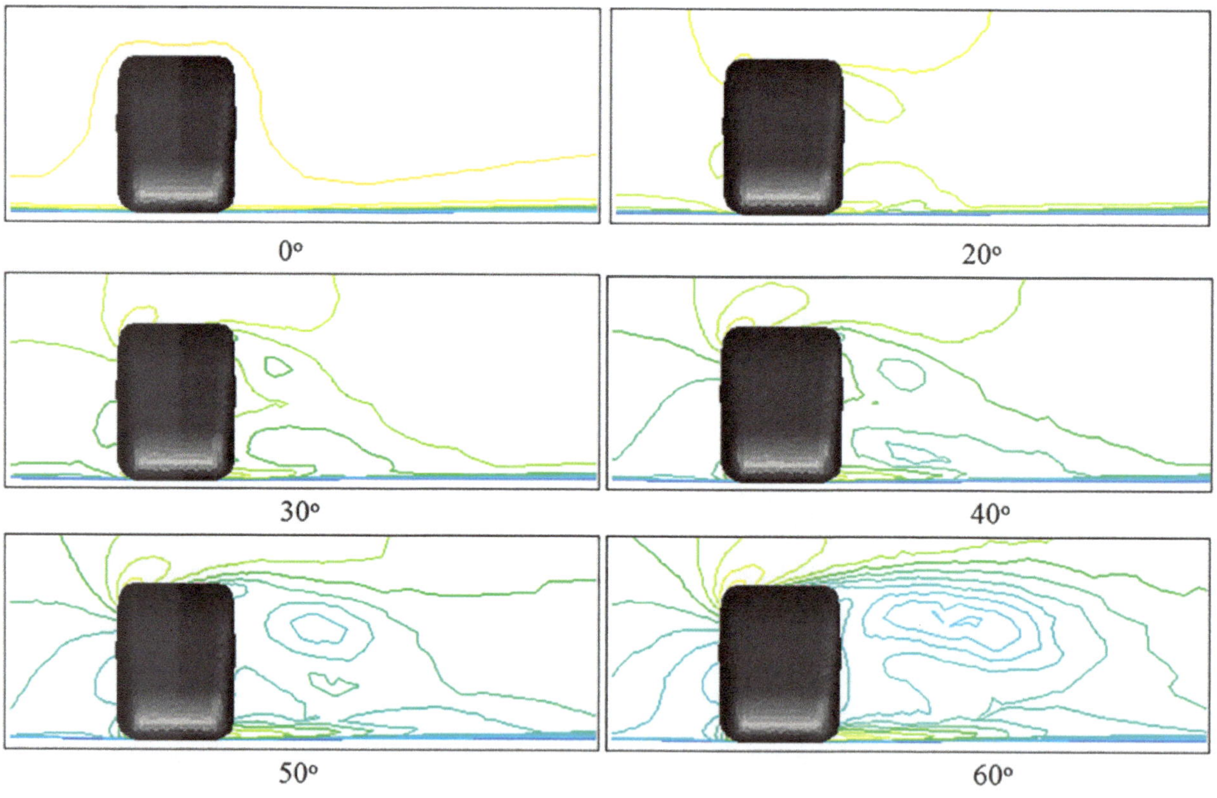

Figure 13: Contours of velocity magnitude along the train's cross-section at 14 m from the nose of the train.

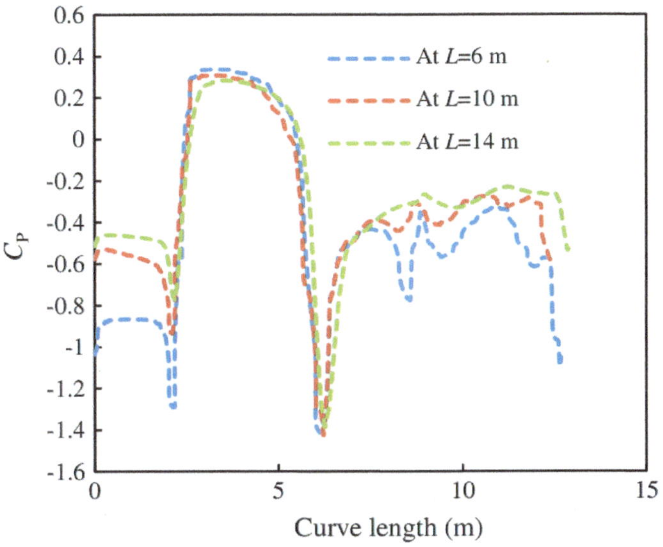

Figure 14: Pressure coefficient along the train's cross-section at different distance (L) from the nose of the train for yaw angle of 30°.

Side Force Coefficient

As can be seen from Fig. 8, the side force coefficient increases steadily with yaw angle till about 50° before it starts to exhibit an asymptotic behavior. For both cases, the computed side force coefficients are in a good agreement with the experiment. Side force is mainly caused by the pressure difference on the two sides of the train. The side force increases the wheel-track load on the leeward side and the wheel-rail contact force. Large side forces worsen the wear of the wheel and rail, and may cause train derailment, or even overturning.

Rolling Moment Coefficient

As can be seen from Fig. 9, the rolling moment coefficient varies in a similar fashion to the side force, and the results are in a good agreement with the experiment for both cases for lower yaw angles. However, at yaw angles of 50° and 60° CFD slightly over-predicts the rolling moment. This may be due to the effect of lift force and inter car gap that was included in the CFD model. The rolling moment is the result of both the lift and side forces with the side force being the main contributor. The rolling moment is responsible for the overloading of wheel-track on the leeward side and is found to be one of the most important aerodynamic coefficients regarding crosswind stability.

Flow Structure

The flow structure for different yaw angles is shown in detail by the two-dimensional streamlines at different locations along the train length (see Figs. 10, 11). As can be seen from Figs. 10 and11, on the lower and upper leeward edges of the train, a vortex is generated and grows steadily in the axial direction. This is in agreement with Fig. 3. The vortex distribution depends on the yaw angle. An increase in the yaw angle results in an advance of the formation and breakdown of vortex. Generally, the recirculation region caused by the vortex flow starts being adjacent to the walls of the train, and then it slowly drifts away from the surface as the flow develops further toward the wake. The flow separation takes place on both the lower and upper leeward edges. The pressure coefficient around the circumference of the train at different locations along its length has been computed for yaw angles of 30° and 60° and is shown in Figs. 14, 15, and 16. Obviously, the pressure distribution on the surface depends on the yaw angle. However, it does not change much along the train length except in a small region close to the nose as can be seen from Figs. 14 and 15. This shows that the pressure distribution around a high-speed train at higher yaw angles is almost independent on the axial position.

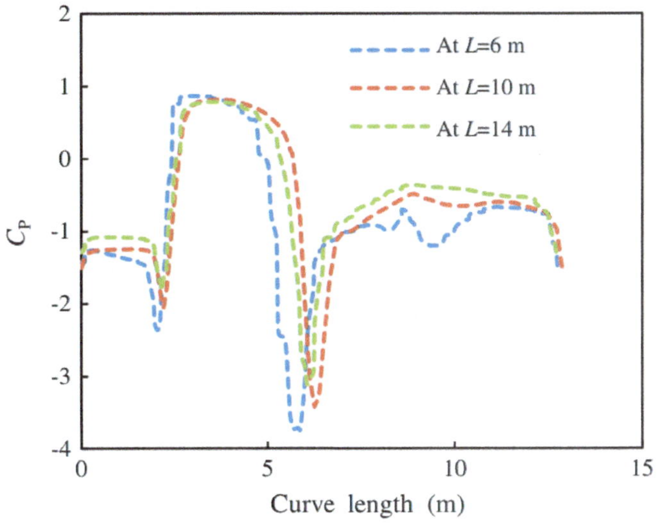

Figure 15: Pressure coefficient along the train's cross-section at different distance (L) from the nose of the train for yaw angle of 60°.

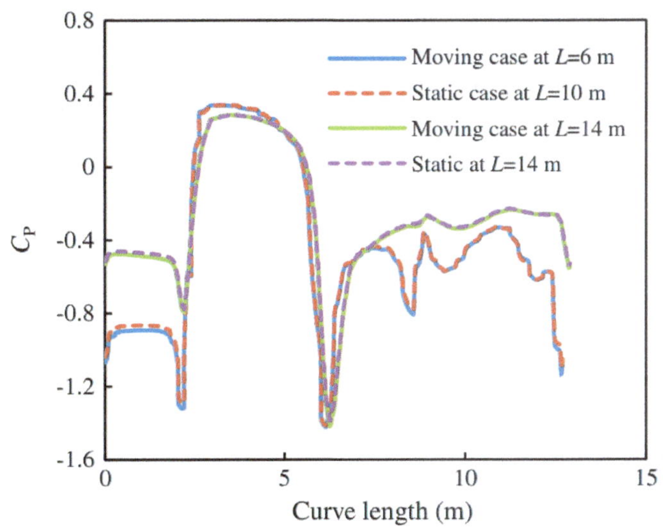

Figure 16: Pressure coefficient along the train's cross-section at different distance (L) from the nose of the train for yaw angle of 30°.

Contours of velocity magnitude have been computed at a cross-section of 6 m and 14 m from the nose of the train along its length for different yaw angles (see Figs. 12, 13). Wind direction has greater influence on flow structure in the rear wake. As can be seen from Figs. 12 and 13, in the absence of side wind (yaw angle 0°), symmetric condition existed. For yaw angles greater than 0 degree, more vortexes evolved on leeward side. The magnitude of rotating vortex in the lee ward side increased with increasing yaw angle. This leads to the creation of a low-pressure region in the lee ward side of the train causing high side force and rolling moment.

The pressure coefficient around the circumference of the train at different locations along its length is plotted in Figs. 14 and 15 for yaw angle of 30° and 60° for static ground case scenario. Obviously, the pressure distribution on the surface depends on the yaw angle. However, it does not change much along the train length except in a small region close to the nose (at L = 6 m from the nose). This shows that the pressure distribution around a high-speed train at higher yaw angles is almost independent on the axial position. Similar observations were reported in the experimental works of Robinson and Baker [28]. Figure 16 shows the pressure coefficient around the circumference of the train at different locations for both static and moving cases. As can be seen from the figure, the moving ground creates a little bit more negative pressure, which implies a little bit more lift.

CONCLUSION

The flow of turbulent crosswind over a more realistic ICE-2 high-speed train model has been simulated numerically by solving the unsteady three-dimensional RANS equations. The simulation has been done in static and moving ground case scenarios for different yaw angles ranging from −30° to 60°. The computed aerodynamic coefficient outcomes using the realizable k-epsilon turbulence model were in good agreement with the wind tunnel data. Both the side force coefficient and rolling moment coefficients increase steadily with yaw angle till about 50° before starting to exhibit an asymptotic behavior.

The nature of the flow field and its structure depict by contours of velocity magnitude and streamline patterns along the train's cross-section has been also presented for different yaw angles. As can be seen from the stream line patterns along the train's cross-section, on the lower and upper leeward edges of the train a vortex is generated and grows steadily in the axial direction. An increase in the yaw angle results in an advance of the formation and breakdown of vortex on the leeward edges. Contours of velocity magnitude were also computed at different cross-sections of the train along its length for different yaw angles. The result showed that magnitude of rotating vortex in the lee ward side pronounced with increasing yaw angle which leads to the creation of a low-pressure region in the lee ward side of the train causing high side force and roll moment.

The pressure coefficient around the circumference of the train at different locations along its length has been computed for yaw angles of 30° and 60°. Obviously, the pressure distribution on the surface depends on the yaw angle. However, it does not change much along the train length except in a small region close to the nose. This shows that the pressure distribution around a high-speed train at higher yaw angles is almost independent on the axial position.

Generally, this study shows that unsteady CFD-RANS methods combined with an appropriate turbulence model can present an important means of assessing the crucial aerodynamic forces and moments of a high-speed train under strong crosswind conditions. Since the observed variations between some of the CFD and wind tunnel results may be due to the turbulence parameters such as turbulence intensity and length scale; study on the influence of those parameters using advanced modeling like LES is vital. The aerodynamic data obtained from this study can be used for comparison with future studies such as the influence of turbulent crosswinds on the aerodynamic coefficients of high-speed train moving in dangerous scenarios such as high embankments.

REFERENCES

1. EuDaly K, Schafer M, Boyd J, Jessup S, McBridge A, Glischinksi S (2009) The complete book of north American railroading. Voyageur Press, Stillwater, pp 1–352
2. Zhou L, Shen Z (2011) Progress in high-speed train technology around the world. J Mod Transp 19(1):1–6
3. Baker CJ (2004) Measurements of the cross wind forces on trains. J Wind Eng Ind Aerodyn 92:547–563
4. Peters J-L (1990) How to reduce the cross wind sensitivity of trains. Siemens Transportation Systems, Muenchen
5. Thomas D (2009) Lateral stability of high-speed trains at unsteady crosswind. licentiate Thesis in Railway Technology, Stockholm, Sweden
6. Suzuki M, Tanemoto K, Maeda T (2003) Aerodynamic characteristics of train/vehicles under cross winds. J Wind Eng Ind Aerodyn 91:209–218
7. Blakeney A (2000) Resistance of railway vehicles to roll-over in gales. Railway group standard, Safety & Standards Directorate, Railtrack PLC, London
8. Mancini G, Cheli R, Roberti R, Diana G, Cheli F, Tomasini G (2003) Cross-wind aerodynamic forces on rail vehicles wind tunnel experimental tests and numerical dynamic analysis. In: The proceedings of the World congress on railway research, Edinburgh
9. DB Netz AG (2006) Richtlinie 80704 Aerodynamik/ Seitenwind, Frankfurt
10. EC (2006) TSI-technical specification for interoperability of the trans-European high speed rail system. Eur Law, Off J Eur Commun
11. CEN (2010) EN 14067-6: railway applications Aerodynamics Part 6: requirements and test procedures for crosswind assessment. European Committee for Standardization, Brussels
12. Elisa M, Hoefener L (ed.) (2008) Aerodynamics in the open air (AOA). WP2 Cross Wind Issues, Final Report, DeuFraKo Project, Paris
13. Diedrichs B (2008) Aerodynamic calculations of cross wind stability of a high-speed train using control volumes of arbitrary polyhedral shape. In the VI international colloquium on: bluff bodies aerodynamics & applications (BBAA), Milan
14. Carrarini A (2007) Reliability based analysis of the crosswind stability of railway vehicles. J Wind Eng Ind Aerodyn 95:493–509
15. Kwon H, Park Y-W, Lee D, Kim M-S (2001) Wind tunnel experiments on Korean high-speed trains using various ground simulation techniques. J Wind Eng Ind Aerodyn 89:1179–1195
16. Baker CJ (2002) The wind tunnel determination of crosswind forces and moments on a high speed train. In: Notes on numerical fluid mechanics, Vol. 79, Springer-Verlag, Berlin, p 46–60

17. Bocciolone M, Cheli F, Corradi R, Diana G, Tomasini G (2003) Wind tunnel tests for the identification of the aerodynamic forces on rail vehicles. In: the XI ICWE-international conference on wind engineering, Lubbock
18. Orellano A, Schober M (2006) Aerodynamic performance of a typical high-speed train. In: The proceedings of the 4th WSEAS international conference on fluid mechanics and aerodynamics, Elounda
19. Cheli F, Corradi R, Rocchi D, Tomasini G, Maestrini E (2008) Wind tunnel test on train scaled models to investigate the effect of infrastructure scenario. In: The VI international colloquium on: bluff bodies aerodynamics & applications (BBAA), Milan
20. Cheli F, Ripamonti F, Rocchi D, Tomasini G (2010) Aerodynamic behavior investigation of the new EMUV250 train to cross wind. J Wind Eng Ind Aerodyn 98:189–201
21. Krajnovic S, Ringqvist P, Nakade K, Basara B (2012) Large eddy simulation of the flow around a simplified train moving through a crosswind flow. J Wind Eng Ind Aerodyn 110:86–99
22. Guilmineau E, Chikhaoui O, Deng G, Visonneau M (2013) Cross wind effects on a simplified car model by a DES approach. Comput Fluids 78:29–40
23. Wendt JF (2009) Computational Fluid Mechanics: an Introduction. Springer-Verlag, Berlin
24. Currie IG (1974) Fundamental mechanics of fluids. McGraw-Hill, New York
25. Blazek J (2001) Computational fluid dynamics: principles and applications. Elsevier Science Ltd, Oxford
26. Wilcox DC (2006) Turbulence modelling for CFD, 3rd edn. DCW Industries, Inc, la Canada
27. Schlichting H, Gersten K (1999) Boundary layer theory. Springer-Verlag, New York
28. Robinson CG, Baker CJ (1990) The effect of atmospheric turbulence on trains. J Wind Eng Ind Aerodyn 34:251–272
29. Khier W, Breuer M, Durst F (2000) Flow structure around trains under side wind conditions: a numerical study. Comput Fluids 29:179–195

CHAPTER 8

NUMERICAL SIMULATION ON THE CAVITATION OF WATERJET PROPULSION PUMP

C Z Xia, L Cheng, Y N Shang, J R Zhou, F Yang, Y Jin

School of Hydraulic, Energy and power Engineering Yangzhou University, Yangzhou, Jiangsu, P.R.China, 225009

ABSTRACT

Waterjet propulsion system is widely used in high speed vessels with advantages of simple transmission mechanism, low noise underwater and good manoeuvrability. Compared with the propeller, waterjet propulsion can be used flow stamping to increasing cavitation resistance at high speed. But under certain conditions, such as low ship speed or high ship speed, cavitation problem still exists. If water-jet propulsion pump is run in cavitation condition for a long time, then the cavitation will cause a great deal of noise CFD is applied to analysis and predict the process of production and development of cavitation in waterjet propulsion pump.

Based on the cavitation model of Zwart-Gerber-Belamri and a mixture of homogeneous flow model, commercial CFD software CFX was taken for characteristics of cavitation under the three operating conditions. Commercial software ANSYS 14.0 is used to build entity model, mesh and numerical simulation. The grid independence analysis determine the grid number of mixed flow pump model is about 1.6 million and the grid number of water-jet pump system unit is about 2.7 million.

The cavitation characteristics of waterjet pump under three operating conditions are studied. The results show that the cavitation development trend is similar design and small rate of flow condition .Under the design conditions Cavitation bubbles are mainly gathered in suction surface of blade near the inlet side of the hub under the primary stage, and gradually extended to the water side in the direction of the rim with the loss of the inlet total pressure. Cavitation appears in hub before the blade rim, but the maximum value of gas content in blade rim is bigger than that in hub. Under large flow conditions, bubble along the direction of wheel hub extends to the rim gradually. Cavitation is found in the pressure surface of blade near the hub region under the critical point of cavitation nearby. When NPSHa is lower than critical point, the area covering by bubbles is about 40% in the suction surface of blade. It means that the critical point of cavitation of pump system is not the accrue point of install cavitation but cavitation has been developed to a certain stage.

INTRODUCTION

Along with the high-speed development of mechanical and military technology, high-speed vessels started to use a new way of special ship propulsion waterjet propulsion. Compared with the traditional propeller, waterjet propulsion system not only has the same good performance in acceleration

Citation: C Z Xia, L Cheng, Y N Shang, J R Zhou, F Yang and Y Jin, "Numerical simulation on the cavitation of waterjet propulsion pump," IOP Conf. Series: Materials Science and Engineering 129 (2016) 012011 doi:10.1088/1757-899X/129/1/012011.

and braking performance, but also has excellent high speed mobility [1]. The noise and vibration in the cabin of water jet propulsion ship is small. The waterjet system's daily maintenance is also relatively easy. Waterjet propulsion device in high speed ship can use water to flow stamping improve the cavitation resistance device. But in some special conditions, such as turning, speeding, maneuvering in reverse and only part of the pumps work, the waterjet propulsion system will get into the cavitation workspaces[2]. Cavitation will not only lead to sharp decline in pump hydraulic performance, also can make water jet propulsion pump thrust be reduced, the efficiency decline and reduce the service life of the impeller and bring cavitation, vibration and noise. So the research on water jet propulsion pump system has strong engineering application value and academic significance [3-4].

GEOMETRIC MODEL AND NUMERICAL METHOD

Geometric Model

By changing the propulsion pump in paper [5-6], new propulsion pump is used in this paper which is about the cavitation on the blade of the impeller. In the model, the inlet is extended so that accuracy and convergence for the simulation is solved, the outlet in connected with optimized nozzle. Figure 1 is the computational domains of propulsion pump.

The computational domains contain impeller, vane, inlet and outlet. In which, impeller has six blades and its rotating speed is 700 r/min, vane has seven blades, diameters of inlet and outlet are D0 and 1.19D0 separately. The straight pipe section shares the same diameter with the inlet. Plane 1 and plane 2 are chosen as the inlet and outlet to calculate the pump head.

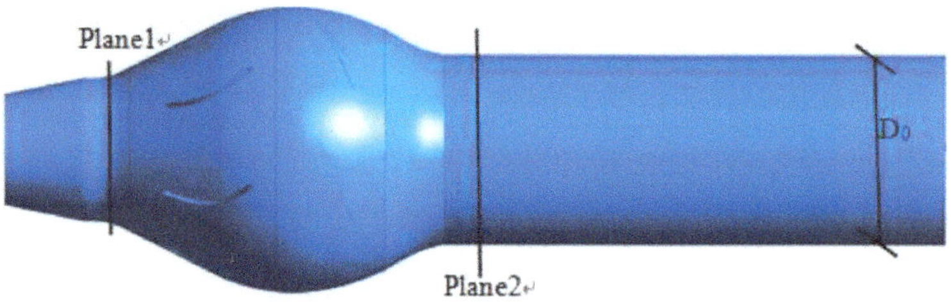

Figure 1: Computational domains.

Mesh Generation

Considering structure mesh accelerating CFD calculation speed, the mesh is generated by using ICEM CFD shown in Figure 2. The mesh quantity is 1.6 million.

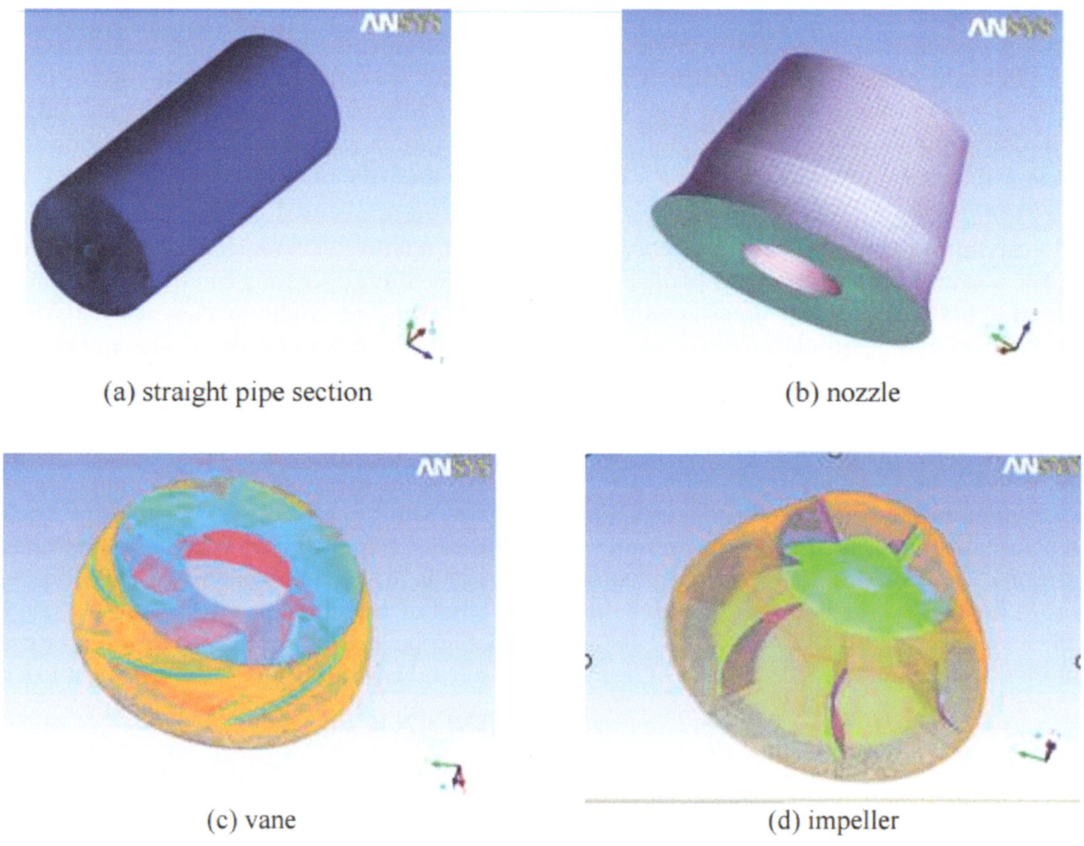

(a) straight pipe section

(b) nozzle

(c) vane

(d) impeller

Figure 2: Meshes.

In general, mesh quantity affects the accuracy and convergence speed. In theory, more mesh, higher accuracy. Actually, overmuch mesh will increase calculation time and slow convergence speed. The mesh independence discussion is necessary. Impeller is the central domain in the model and its mesh quantity is the key. The result shows that the head and efficiency stay steady when the mesh quantity rises up to 1.6 million. H_d, η_d are head and efficiency separately at design condition.

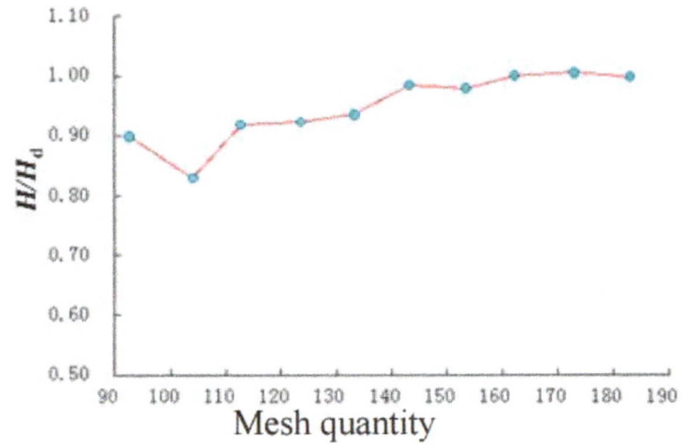

(a) head

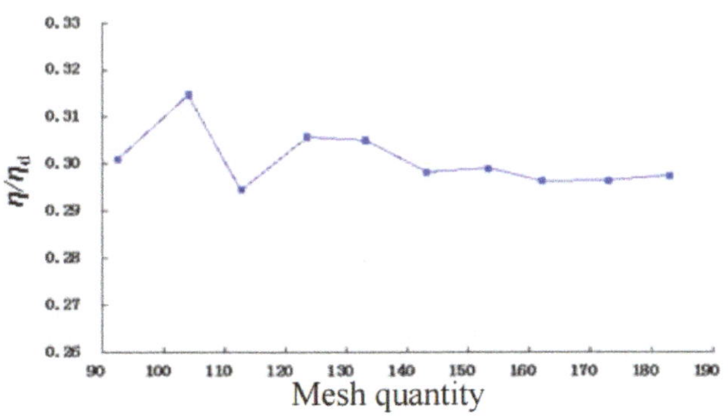

(b) efficiency

Figure 3: Mesh independence.

Boundary Condition

Inlet is set as total pressure stable, supposing no cavitation happens in the water. Outlet is set as mass flow rate. When cavitation model is applied, two physical parameters below are given: water's cavitation pressure PV equals to 3574Pa at normal atmospheric temperature, cavitation surface tension σ equals to 0.074N/M, Rayleigh−Plesset equation is used to describe bubble motion discipline. No slip wall condition is applied for the wall of hub, rim, blade surface and the other solid walls. The reference pressure is set as 0 atm. RNG k-ε model and mixture model are chosen as turbulent model and multiphase flow model separately [6-9].

Calculated result is more close to the test result at large flow rate coefficient by using ZwartGerber-Belamri cavitation model, so Zwart-Gerber-Belamri cavitation model [10-14] is chosen to simulate the cavitation in the impeller.

RESULTS AND ANALYSIS

Three flow rate conditions are simulated and compared which is design flow rate (Q_{bep}), small flow rate (0.7 Q_{bep}) and large flow rate (1.1 Q_{bep}).

Cavitation Performance at Design Flow Rate

Based on the pump performance, the best efficiency point is chosen as the design flow rate which can be dawn as Q_{bep}. Different total pressures on the inlet are applied which from 0.5atm to 0.1atm, pressure distributions on the pressure side and suction side of the impeller, cavitation area change is observed.

As shown in Figure 4 and Figure 5, there is pressure distribution on the pressure side and suction side of the impeller separately for different cavitation remainders. In the Figure, the blue area is low pressure zone which firstly appears at the rim of suction side inlet. As NPSHa reduces, low pressure zone increases along blade margin. When NPSHa is 0.76m, the entire suction side of the blade is low pressure zone. Meanwhile, there is also low pressure on the pressure side of blade whose shape is almost triangle, pressure at the hub is larger than it at the rim.

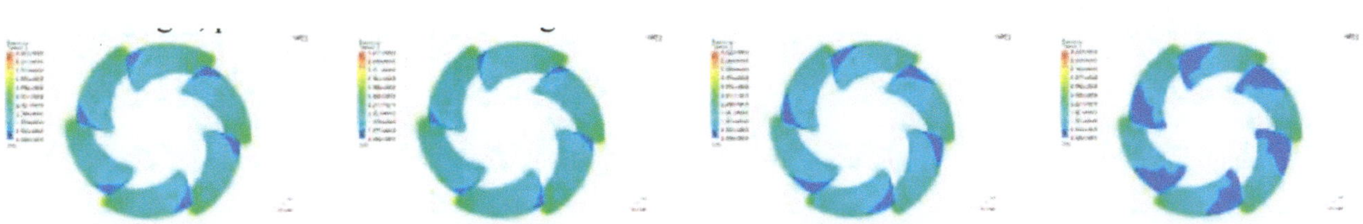

(a) NPSHa=2.41m (b) NPSHa=2.2m (c) NPSHa=1.87m (d) NPSHa=1.62m

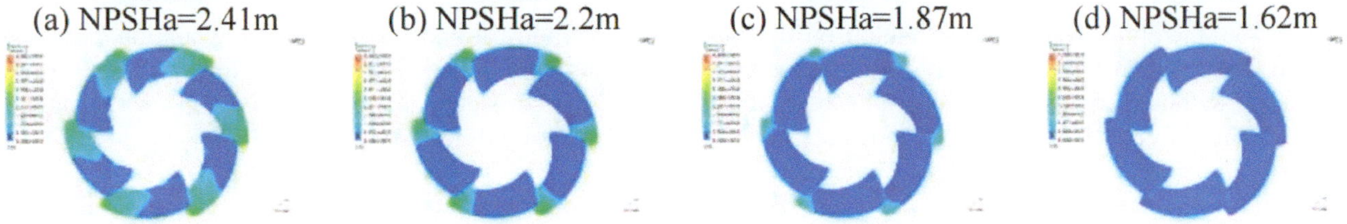

(e) NPSHa=1.38m (f) NPSHa=1.17m (g) NPSHa=0.96m (h) NPSHa=0.76m

Figure 4: Pressure distribution on the suction side of the blade at different cavitation remainders.

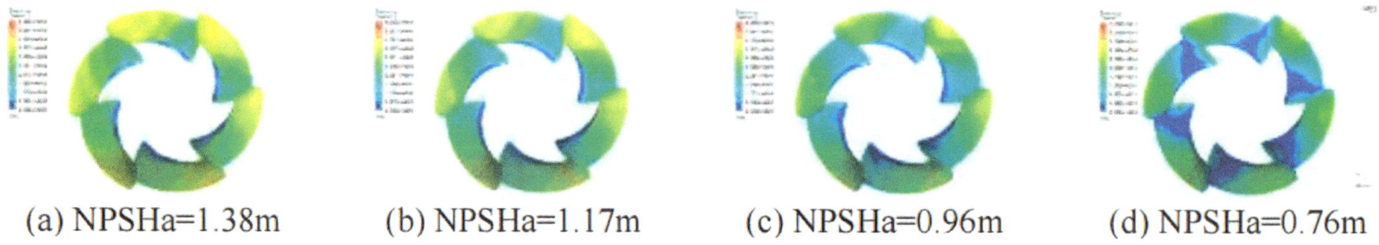

Figure 5: Pressure distribution on the pressure side of the blade at different cavitation remainders.

Figure 6 and Figure 7 are cavitation volume fraction distribution on the pressure side and suction side of the blade. In the Figure, red zone and blue zone represent large cavitation volume fraction and small cavitation volume fraction. By comparing the pressure distribution on the blade with different cavitation remainders, the same trend is shown. When NPSHa is 0.76m, cavitation on the whole suction pressure has been absolutely developed. The area around the outlet of the blade is affected worst, cavitation will block the entire impeller passage and even result in the cavitation happening on the pressure side of the blade.

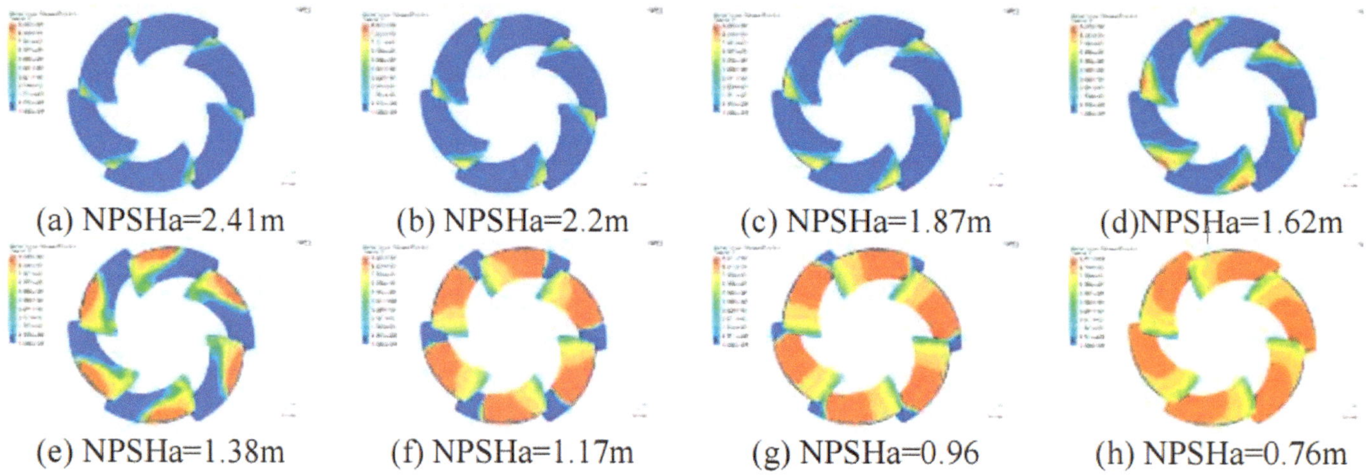

Figure 6: Cavitation volume fraction distribution on suction side of the blade.

Figure 7: Cavitation volume fraction distribution on the pressure side of the blade.

Figure 8 is the static pressure distribution on the blade at different spans. When NPSHa is 2.2m, low static pressure keeps steady. Cavitation happens at the rim of the blade at first. In Figure, on the above and below are the static pressure at the pressure side and suction side respectively. Cavitation will not happen until it happens at the whole suction side. Cavitation at suction side is caused by the low pressure. There is no obvious contact between the cavitation on the pressure side and pressure decrease.

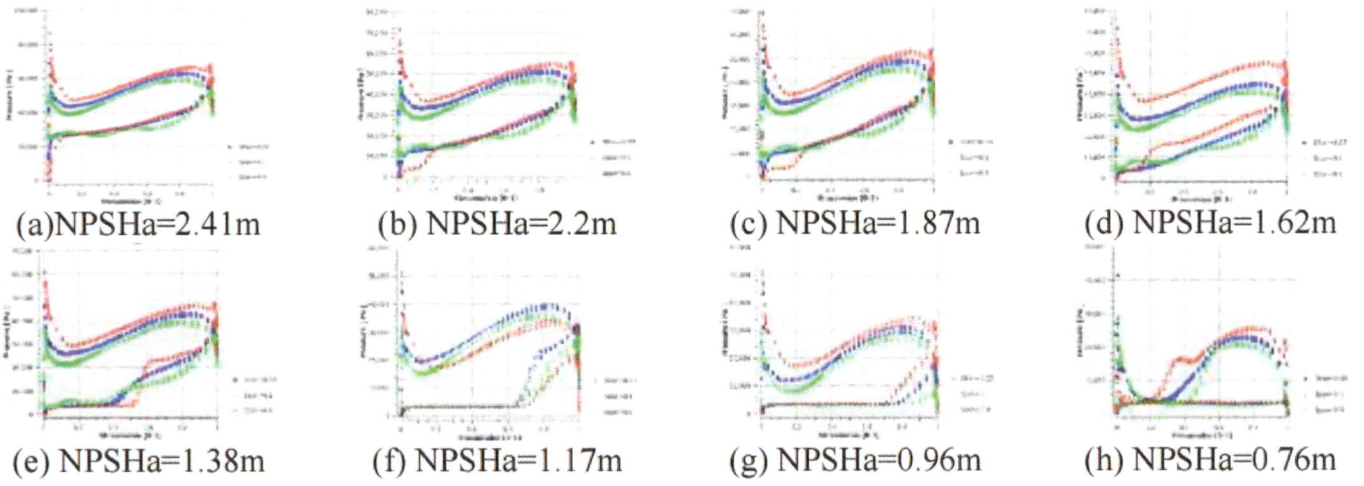

Figure 8: Static pressure distribution on the blade at different spans.

Partial Conditions of Cavitation Characteristics

Figure 9 shows the result of numerical simulation on cavitation in small flow rate. The relation between the distributions of water vapor volume fraction on the suction face of blades and the decreases of NPSHa is show in Figure 9. The result shows that the variations tend of the location of bubble volume fraction distribution are consistent with the design condition. The location of bubble on suction face of blade grandly extends to the outside from the inside of blades in same condition.

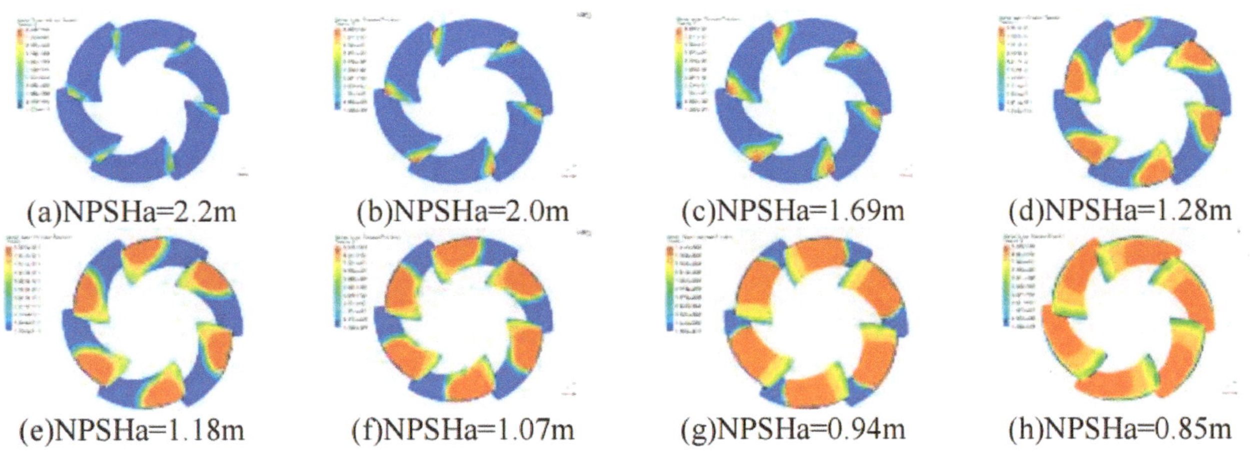

Figure 9: the distribution of bubble volume fraction.

Figure 10 and Figure 11 show the numerical simulation result of waterjet propulsion pump. The bubbles firstly appear on the edge of the impeller inlet in large flow rate condition, so as in small flow rate and design flow rate. But its variation tendency is greatly different with that in design condition and small flow rate with the decrease of NPSHa. When NPSHa reduces from 1.9m to 1.45m, the region covered by bubble firstly extends along the rim of blade and appears in the middle of blade near the hub. It develops from the hub to the rim of blade. The pressure is higher than the vapor pressure on pressure face of blade. So bubble cannot be found on pressure face of blade.

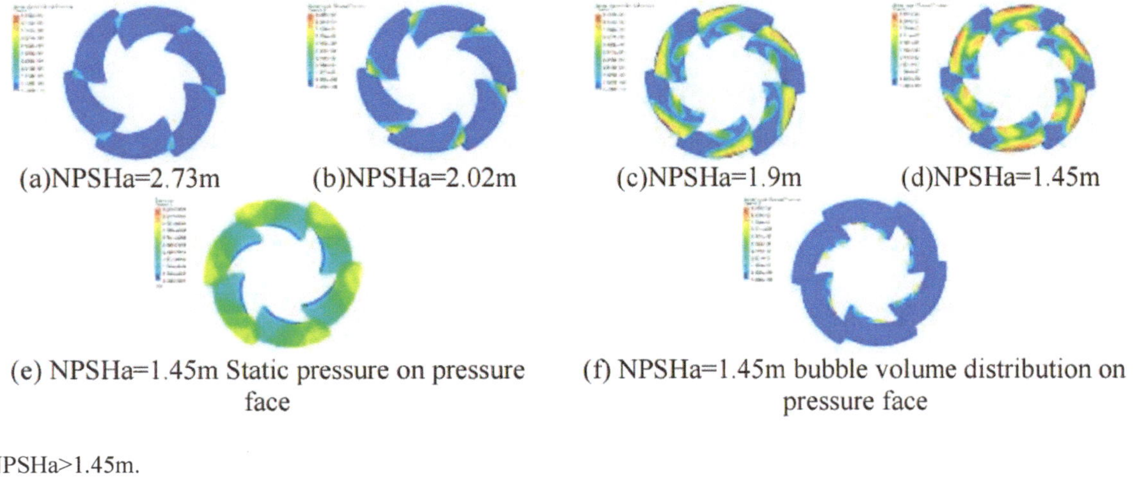

Figure 10: NPSHa>1.45m.

While NPSHa < 1.45m, cavitation is so hard that bubble blocks the passage of impeller. Then the efficiency of pump sharp drop and the pump cannot work. Flow separation is found at the outside edge of suction face of blades for large flow rate and negative angle. Bubble is found on the pressure face and Cavitation volume fraction extends to the rim from the hub at the outside of pressure face. The low pressure area that is blue area showed in Figure 11 is consistent with the distribution of bubble volume fraction. The bubble is found in the large flow rate condition while the development of cavitation is to some extent on the suction face of blade rather than the bubble cover suction face completely in design condition and small flow rate condition. Figure 10 and Figure 11 shows that the critical NPSHa and cavitation area are bigger than that in the other conditions.

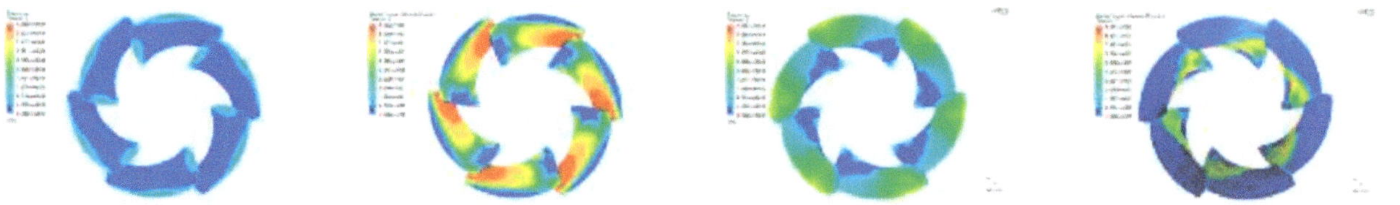

(a) Static pressure distribution on suction face

(b) Bubble volume distribution on suction face

(c) Static pressure distribution on pressure face

(d) Bubble volume distribution on pressure face

Figure 11: NPSHa=1.36m.

Figure 12 shows the pump cavitation characteristic curve in three working conditions. The flow rate and efficiency curves are showed little change with the decrease of NPSHa. The reason for that is that the flow is unstable in the pump at the begin of bubble produce. But the flow rate and efficiency curves will sharp drop after NPSHa is greater than critical value. The point of critical value is called critical cavitation point. The critical NPSHa decreases as the flow rate reducing and the dropping extent of pump cavitation characteristic curves is small while the flow rate is small. That means that the pump is hard to get cavitation while the flow rate is small.

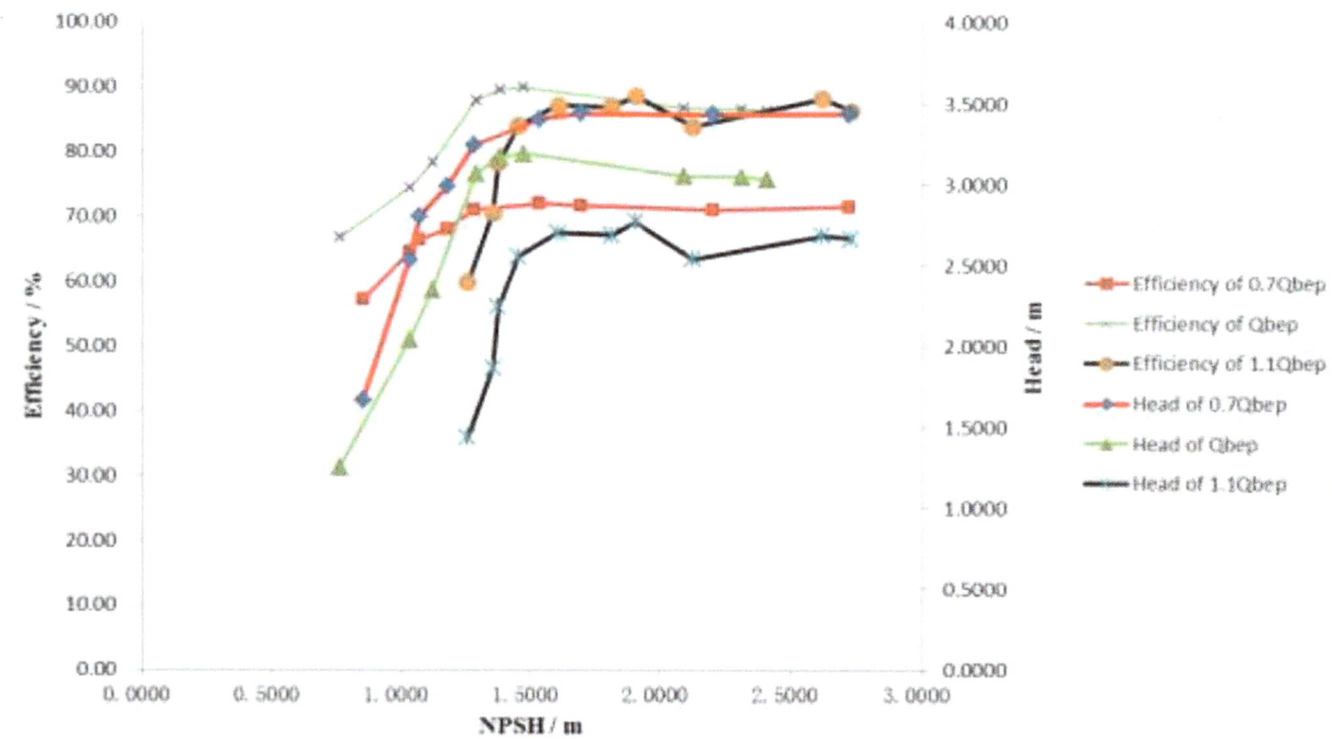

Figure 12: Pump cavitation characteristic curves.

CONCLUSIONS

The paper finds that the cavitation resistances of waterjet propulsion pump are grandly weaken as the critical NPSHa value grandly increase by the numerical simulation on the cavitation flow in waterjet propulsion pump. The harm of cavitation to blades is deepened while deviating from the design operation. The static pressure is bigger in the rim than that in the hub, and the distribution trends of static pressure are almost same in three working condition. Therefore, the bubble volume fraction in the rim is bigger than that in the hub. That means that the cavitation is harder in the rim. And that should be considered in actual design.

ACKNOWLEDGMENT

This work was financially supported by National science and technology plan project in rural areas (Grant No.2012BAD08B03), National Natural Science Foundation of China (Grant No. 51179167), Water conservancy science and technology project of Jiangsu Province (Grant No.2014046), The Young academic Leaders in Blue Project of Jiangsu, Prospective Joint Research Project of Jiangsu (Grant No. BY2015061-12) , A Project Funded by the Priority Academic Program Development of Jiangsu Higher Education Institutions and Major Natural Science Foundation of Universities in Jiangsu(Grant No.12KJA570001).

REFERENCES

1. Wang Y S, Ding J M 2004 Research on the relationship between waterjet power absorption and vessel speed[C]// In-ternational Conference on Waterjet Propulsion. London: RINA.
2. Verbeek R, Bulten N W H 1998 Recent development in waterjet design[C]//Proceedings of international conference on waterjet propulsion II. Amsterdam: RINA.
3. Harvey E N, Barnes D K, McElroy W D, et al. 1944 Bubble formation in animals. I. Physical factors[J]. Journal of Cellular and Comparative Physiology, 24(1): 1-22.
4. Knapp R T, Daily J W, Hammitt F G 1979 Cavitation[J].
5. Cheng L, Bart P M, Van E, Liu Chao et al. 2012 Radial forces of waterjet propulsion mixedflow pump[J]. Journal of Drainage and Irrigation Machinery Engineering, 30(6):636-640.
6. Bulten N W H, Verbeek R, Van E, Bart P M 2006 CFD simulation of the flow through a waterjet installation[C]// International Conference on Waterjet Propulsion. 4: 11-19.
7. Gopalan S, Katz J 2000 Flow structure and modeling issues in the closure region of attached cavitation[J]. Physics of Fluids(1994-present), 12(4): 895-911.
8. Pham T M, Larrarte F, Fruman D H 1999 Investigation of unsteady sheet cavitation and cloudcavitation mechanisms[J]. ASME Journal of Fluids Engineering, 121(2): 289-296.
9. Pellone C, Peallat J M 1995 Non-linear analysis of three-dimensional partially cavitating hydrofoil[C]// Proceedings of the International Symposium on Cavitation, Deauville, France. 1995: 63-67.
10. Yang F, Liu C, Tang F P 2012 Cavitation performance prediction of mixed-flow pump based on CFD[C]//26th IAHR Symposium on Hydraulic Machinery and System, Auguest 19-23, 2012, Beijing, China
11. Zwart P J, Gerber A G, Belamri T 2004 A two-phase flow model for predicting cavitation dynamics[C]// Proceedings of the Fifth International Conference on Multiphase Flow, Yokohama, Japan.
12. Orszag S A, Patterson Jr G S 1972 Numerical simulation of three-dimensional homogeneous isotropic turbulence[J]. Physical Review Letters, 28(2): 76-79.
13. Merkle C L, Feng J, Buelow P E O 1998 Computational modeling of the dynamics of sheet cavitation[C]// Proceedings of Third International Symposium on Cavitation , Grenoble,France, 1998:307－311.
14. Senocak I , Shyy W 2002 A pressure-based method for turbulent cavitating flow computations[J]. Journal of Computational Physics, 176(2):363-383.

SECTION II: CFD APPLICATIONS TO CARDIOVASCULAR HEMODYNAMICS AND BIOMIMETICS

CHAPTER 9

COMPUTATIONAL FLUID DYNAMICS SIMULATIONS OF BLOOD FLOW REGULARIZED BY 3D PHASE CONTRAST MRI

Vinicius C. Rispoli[1,2], Jon F. Nielsen[3], Krishna S. Nayak[4] and Joao L. A. Carvalho[1]

[1]Department of Electrical Engineering, University of Brasilia, Brasília, Brazil

[2] UnB Gama College, University of Brasilia, Brasília, Brazil

[3] fMRI Laboratory, Biomedical Engineering Department, University of Michigan, Ann Arbor, USA

[4] Magnetic Resonance Engineering Laboratory, Ming Hsieh Department of Electrical Engineering, University of Southern California, Los Angeles, USA.

ABSTRACT

Background

Phase contrast magnetic resonance imaging (PC-MRI) is used clinically for quantitative assessment of cardiovascular flow and function, as it is capable of providing directly-measured 3D velocity maps. Alternatively, vascular flow can be estimated from model-based computation fluid dynamics (CFD) calculations. CFD provides arbitrarily high resolution, but its accuracy hinges on model assumptions, while velocity fields measured with PC-MRI generally do not satisfy the equations of fluid dynamics, provide limited resolution, and suffer from partial volume effects. The purpose of this study is to develop a proof-of-concept numerical procedure for constructing a simulated flow field that is influenced by both direct PC-MRI measurements and a fluid physics model, thereby taking advantage of both the accuracy of PC-MRI and the high spatial resolution of CFD. The use of the proposed approach in regularizing 3D flow fields is evaluated.

Methods

The proposed algorithm incorporates both a Newtonian fluid physics model and a linear PC-MRI signal model. The model equations are solved numerically using a modified CFD algorithm. The numerical solution corresponds to the optimal solution of a generalized Tikhonov regularization, which provides a flow field that satisfies the flow physics equations, while being close enough to the measured PC-MRI velocity profile. The feasibility of the proposed approach is demonstrated on data from the carotid bifurcation of one healthy volunteer, and also from a pulsatile carotid flow phantom.

Results

The proposed solver produces flow fields that are in better agreement with direct PC-MRI measurements than CFD

Citation: Vinicius C. Rispoli, Jon F. Nielsen, Krishna S. Nayak and Joao L. A. Carvalho, "Computational fluid dynamics simulations of blood flow regularized by 3D phase contrast MRI," BioMedical Engineering OnLine201514:110. DOI: 10.1186/s12938-015-0104-7

alone, and converges faster, while closely satisfying the fluid dynamics equations. For the implementation that provided the best results, the signal-to-error ratio (with respect to the PC-MRI measurements) in the phantom experiment was 6.56 dB higher than that of conventional CFD; in the in vivo experiment, it was 2.15 dB higher.

Conclusions

The proposed approach allows partial or complete measurements to be incorporated into a modified CFD solver, for improving the accuracy of the resulting flow fields estimates. This can be used for reducing scan time, increasing the spatial resolution, and/or denoising the PC-MRI measurements.

BACKGROUND

Knowledge of blood flow patterns in the human body is a critical component in cardiovascular disease research and diagnosis. Two different approaches to 3D flow assessment are currently available to the researcher and clinician: direct, model-independent velocity mapping using phase contrast magnetic resonance imaging (PC-MRI) [1, 2, 3] or Doppler ultrasound, and model-based computational fluid dynamics (CFD) calculations [4, 5, 6, 7, 8, 9, 10, 11, 12, 13, 14, 15, 16]. Among the direct methods, PC-MRI has gained prominence in recent years due to its unrestricted 3D anatomical coverage and minimal operator dependence [1, 17, 18, 19]. The connection between MRI-based complex blood flow analysis (such as, turbulence [20] and helical blood flow [21]) and MRI-based biomarkers (such as, wall shear stress [22, 23, 24] and pressure gradients [25, 26, 27, 28, 29]) with disease progression and diagnosis are active and promising areas of research. However, PC-MRI provides limited spatial and temporal resolutions, which inevitably impacts the accuracy of MRI-based hemodynamic parameter estimates [30].

CFD is an alternative that has been used to predict flow patterns in various vascular geometries, including intracranial aneurysms [10], the thoracic aorta [11], and the carotid bifurcation, both in models [12, 13, 14, 15] and in vivo [16]. The equations describing Newtonian fluid flow are solved numerically for specified boundary and initial condition data. Such approach provides arbitrarily high spatial and temporal resolution, and is in principle capable of estimating flow fields for arbitrarily complex vessel geometries. Absolute hemodynamic parameter estimates can be obtained directly from the high-resolution flow fields produced by CFD, obviating the need for data smoothing or interpolation schemes.

The accuracy of conventional CFD routines hinges on many modeling assumptions that are not strictly true for in vivo vascular flow, including rigid vessel walls and uniform blood viscosity. Indeed, the proper choice of the underlying physics model is itself an open research question [10]. CFD predictions have so far shown variable agreement with PC-MRI measurements [10, 11, 13, 15], and the applicability of CFD to robust flow estimation is still being debated.

The use of fluid mechanics techniques for improving PC-MRI data is an active research field. Several algorithms from the literature use regularizations based on curl and divergence of the velocity field, which are associated with the irrotationality and incompressibility characteristics of the fluid flow, respectively. These include algorithms capable of improving streamlines [31, 32], and also algorithms for denoising the PC-MRI data [33, 34, 35, 36].

However, the solutions found by those methods do not necessarily satisfy the Navier-Stokes momentum equation. Other authors integrated point-based measurements within a CFD solver, by adding a "force" term to the momentum equations that is proportional to the difference between predicted and measured velocities for a given grid point [37, 38, 39]. They used such an approach to integrate, respectively, Doppler-ultrasound velocity measurements and cerebral aneurysm blood flow MRI data into a CFD solver.

More recently, a method to accelerate 4D cardiac flow MRI using CFD simulations was proposed. The image model was generated by integrating numerical blood flow simulations (calculated using openFOAM [40]) into the MRI image-reconstruction algorithm [41]. However, this approach cannot guarantee that the fluid physics model (momentum and continuity equations) is satisfied by the reconstructed velocity map.

Conventional CFD uses medical imaging data only to specify the vessel geometry and the flow at the inlet and outlet boundaries, or other previously known initial and boundary condition data, and uses the assumed fluid physics model to find the solution in the interior of the calculation domain [28]. The goal of this work is to develop a more general, flexible and easy to implement numerical framework for harnessing additional PC-MRI velocity measurements to construct a more robust and potentially more accurate CFD-based solution, considering PC-MRI data as ground truth. The proposed method is able to make use of full (or incomplete) PC-MRI measurements of one or more velocity components within the entire 3D volume. This is achieved through generalized Tikhonov regularization [42], obtaining a numerical solution that is close enough to the measured flow data; satisfies the fluid physics equations; reduces noise; and, in the clinical environment, can be used to reduce scan time.

Finally, this work is presented as a proof of concept of the CFD–MRI combined solver. All simulations herein were made using the finite volume method and SIMPLER algorithm in Cartesian grids, with unrealistic assumptions about the blood flow model, such as rigid walls and Newtonian viscosity. Nevertheless, the optimal numerical solution proposed in this work is general enough to be implemented: for any type of discretization method, such as finite differences, finite volume or finite element; for steady or unsteady flow; and, for any realistic physics model, such as non-Newtonian viscosity, elastic vessel walls, or slightly compressible flow. Even in the most realistic model, the discretization (finite differences, finite volume or finite elements) of the nonlinear set of differential equations produces a large and sparse system of linear equations that forms the basis of the proposed numerical solution. The feasibility of the proposed approach is demonstrated on data from the carotid bifurcation of one healthy volunteer, and from a pulsatile carotid flow phantom. Two implementations of the regularized computational solution were evaluated and compared: one using only the PC-MRI data corresponding to the main velocity component (*z* axis); and another, using PC-MRI data corresponding to all three velocity components.

THEORY

Blood Flow Model

The general model for fluid motion in 3D Euclidian space is given by the Navier–Stokes momentum equation [43]:

$$\rho\left(\frac{\partial \vec{v}}{\partial t} + \vec{v}\cdot\nabla\vec{v}\right) = -\nabla p - \nabla\cdot\hat{\tau} + \vec{b}, \tag{1}$$

where ρ is the fluid density, $\vec{v}$=(u,v,w) is the flow velocity vector (*u*, *v*, and *w* are the velocity components associated with spatial axes *x*, *y*, and *z*, respectively), *t* is time, ∇ is the gradient operator, *p* is the hydrodynamic pressure, $\hat{\tau}$ is the stress tensor, and $\vec{b}$ represents the body forces acting on the fluid during the flow. The stress tensor represents the momentum transferred in virtue of the molecular motions and interactions within the fluid. It is a function of the scalar invariants of the strain rate tensor $\hat{e} = (1/2)\left[\nabla\vec{v} + (\nabla\vec{v})^T\right]$, where TT denotes the transpose operation. For an incompressible fluid, it can be written as $\hat{\tau} = -\mu(\hat{e})\,\hat{e}$, where the scalar $\mu(\hat{e})$ is the generalized Newtonian viscosity for a given $\hat{e}$.

In this work, blood is modeled as a Newtonian, incompressible and isothermal fluid, with constant viscosity μ and constant density ρ. We are also assuming that there are no body forces acting on the blood flow. Then, the simplification of the general momentum equation, Eq. (1), provides our blood model equation [43]:

$$\rho\left(\frac{\partial \vec{v}}{\partial t} + \vec{v}\cdot\nabla\vec{v}\right) = -\nabla p + \mu\Delta\vec{v}, \tag{2}$$

where Δ is the Laplacian operator.

Since there are no sources of blood inside an artery, the flow field must also satisfy mass conservation [43], which is expressed by the continuity equation:

$$\nabla\cdot\vec{v} = 0. \tag{3}$$

SIMPLER Algorithm

Equations (2) and (3) must be solved for the unknown scalar field variables *u*, *v*, *w*, and *p*. Those equations are non-linear and coupled, and attempting to solve them directly in one step is a formidable, if not impossible, task.

The semi-implicit method for pressure-linked equations revised (SIMPLER) algorithm [44] is a well-known and established numerical routine for solving the momentum and continuity equations, subject to given boundary and initial conditions. It belongs to a class of algorithms capable of solving the non-linear coupled fluid dynamic equations, which also includes the SIMPLE, SIMPLEC, and PISO algorithms [45]. For our purposes, SIMPLER's major advantage is that it does not require an initial guess for the pressure field; instead an initial estimate of the velocity field is used.

The discretization of the momentum equation, Eq. (2), forms the basis of the iterative CFD routine, yielding three linear systems. Let *N* be the total number of grid points in the discrete 3D calculation domain, i.e., in a rectangular grid, $N=N_x\cdot N_y\cdot N_z$, where N_x, N_y, and N_z represent the number of grid points along the *x*, *y*, and *z* axes, respectively. Then, for the *n*-th iteration, we have:

$$\mathbf{S}_{u,n-1}\mathbf{u}_n = \mathbf{f}_{u,n-1} \tag{4}$$

$$\mathbf{S}_{v,n-1}\mathbf{v}_n = \mathbf{f}_{v,n-1} \tag{5}$$

$$\mathbf{S}_{w,n-1}\mathbf{w}_n = \mathbf{f}_{w,n-1}, \tag{6}$$

where $\mathbf{S}_{u,n-1}$, $\mathbf{S}_{v,n-1}$, and $\mathbf{S}_{w,n-1}$ are N×N square hepta-diagonal sparse matrices, each containing previous iteration information about all three velocity components, as well as the values of the density and viscosity constants; the three N×1 column vectors un, vn, and wn store the current iteration of the *u*, *v*, and *w* velocity component values, respectively, associated with all grid points in the 3D calculation domain (Fig. 1); and each of the constant N×1 column vectors $\mathbf{f}_{u,n-1}$, $\mathbf{f}_{v,n-1}$, and $\mathbf{f}_{w,n-1}$ contains previous iteration information about all three velocity components, current iteration pressure difference values, and the physical parameters ρ and μ.

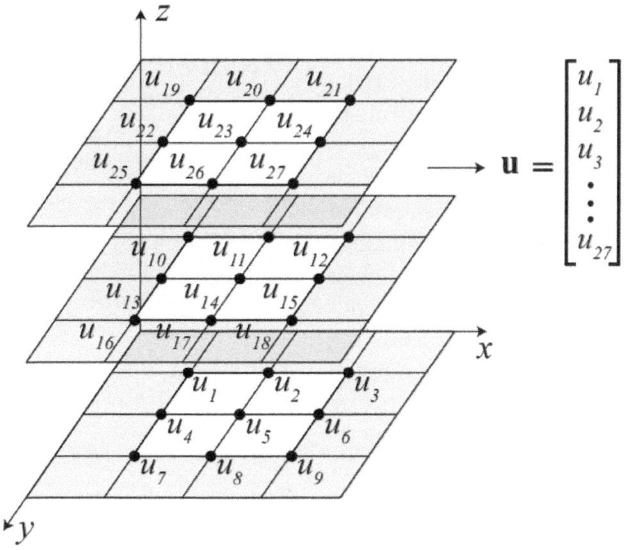

Figure 1: Discretization of the velocity component *u* on the computational grid, and representation as a stacked column vector.

The main difficulty when attempting to predict the flow of an incompressible fluid is that there is no equation for pressure. Hence, a discretized pressure equation is deduced applying Eqs. (4), (5), and (6) in the discretized continuity equation, giving rise to a discretization of the following Poisson equation:

$$\Delta p = \frac{1}{\delta t}\nabla \cdot \vec{v}, \tag{7}$$

where δt is the time step, i.e. the time increment between iterations (this is further discussed in the next section). For a given time instant $t=t_i$, convergence is achieved when $\nabla \cdot \vec{v} = 0$. The pressure field is updated at each iteration based on the current velocity field estimate, and hence does not appear explicitly in Eqs. (4), (5), and (6). It is important to note that *u*, *v*, and *w* values must be defined on regular grids, staggered by half a grid spacing (along the three directions) with respect to the grid on which pressure values are defined. This is to avoid non-physical wiggle solutions for the pressure and velocity fields [44].

The steps of the SIMPLER algorithm, for a given time instant $t=t_i$, are summarized in Algorithm 1. A detailed explanation of the discretization of the Navier–Stokes equations is provided in Refs. [44, 45].

Algorithm 1 Steps of the SIMPLER algorithm, for a given time instant $t = t_i$. [44]

1. Begin with a initial guess for the velocity components, $\mathbf{u}_0$, $\mathbf{v}_0$, and $\mathbf{w}_0$.
2. Using Eqs. (4), (5), and (6), the final velocity component estimates from the previous iteration ($\mathbf{u}_{n-1}$, $\mathbf{v}_{n-1}$, and $\mathbf{w}_{n-1}$) — or the initial guess, if this is the first iteration — and pressure field $\mathbf{p}_n = \mathbf{0}$ ($\mathbf{p}_n$ is a $N \times 1$ stacked column vector, as illustrated in Fig. 1; and $\mathbf{0}$ is the $N \times 1$ null vector), find a first estimate of $\mathbf{u}_n$, $\mathbf{v}_n$, and $\mathbf{w}_n$.
3. Using Eq. (7) and the velocity field found in step 2, find an estimate of the pressure field, $\mathbf{p}_n$.
4. Using Eqs. (4), (5), and (6), the final velocity component estimates from the previous iteration ($\mathbf{u}_{n-1}$, $\mathbf{v}_{n-1}$, and $\mathbf{w}_{n-1}$) — or the initial guess, if this is the first iteration — and the pressure field found in step 3, find an updated estimate of $\mathbf{u}_n$, $\mathbf{v}_n$, and $\mathbf{w}_n$.
5. Using Eq. (7) and the velocity field found in step 4, find an updated estimate of $\mathbf{p}_n$.
6. Using the pressure field found in step 5, find the final estimate of the velocity field, adding $\alpha\nabla\mathbf{p}_n$ — where α is a relaxation factor — to each of the velocity component estimates ($\mathbf{u}_n$, $\mathbf{v}_n$, and $\mathbf{w}_n$).
7. Repeat from step 2 until convergence, i.e., $\nabla \cdot \vec{v} = 0$, where $\vec{v}$ is the vector field represented by the velocity component estimates ($\mathbf{u}_n$, $\mathbf{v}_n$, and $\mathbf{w}_n$) found in step 6.

Note that correcting the pressure field after step 6 is pointless, as it will be reset to a null vector when step 2 is repeated.

METHODS

Algorithm Implementation

On constructing the numerical solution of the unsteady Navier–Stokes equation, we assume that a velocity field and the boundary conditions at a given time instant $t=t_0$ are known. For this initial set of data, the numerical solution for the next time step $t_1=t_0+\delta_t$ is constructed, and it converges toward the solution when the continuity equation is satisfied (step 7 in Algorithm 1). Starting with the solution for t_1, the same iterative procedure is repeated to obtain the solution for $t_2=t_1+\delta_t$, and so forth. In this manner, a time-dependent flow field is computed.

In unsteady flow simulations of the Navier–Stokes equations by implicit numerical routines, the nature of the transient SIMPLER iterative procedures is equivalent to steady-state SIMPLER calculations applied, until convergence, for each time instant [45]. In other words, solving transient problems using SIMPLER is equivalent to solving successive steady-state problems. Moreover, steady-state calculations may be interpreted as pseudo-transient solutions with spatially-varying time steps [45]. In other words, the steady-state solutions are, in practice, unsteady solutions, considering a virtual time step δt with fixed boundary conditions and initial data. This approach has been used by many recent authors [13, 16, 37, 38, 39]. While this was the numerical strategy adopted in this work, the approach proposed here can also be used for unsteady flow predictions.

The steady-state solution $\vec{v}_\infty$ corresponding to a given cardiac phase (a temporal frame within the cardiac cycle) is calculated using the MRI-measured inlet and outlet velocities for that cardiac phase. CFD calculations begin with an initial guess for $\vec{v}$, and simulations are carried forward in time until convergence, i.e.:

$$\frac{\|\vec{v}(t+\delta t) - \vec{v}(t)\|}{\delta t} < \varepsilon, \tag{8}$$

given a sufficiently small ε (‖·‖ denotes vector magnitude) and sufficiently small time step δt. Note that, here, time *t* is a simulation-only parameter, and is unrelated to time instants within the cardiac cycle. It is also not related with the iteration steps (*n*) of the SIMPLER algorithm (Algorithm 1), since the entire algorithm—with multiple iterations until convergence criterion $\nabla\cdot\vec{v}=0$ is satisfied—is performed at each time instant *t*, until the convergence criterion shown in Eq. 8 is satisfied. At this point, $\vec{v}\,\infty$ is obtained. If multiple cardiac phases were to be reconstructed, then $\vec{v}\,\infty$ would be independently calculated for each cardiac phase.

Our implementation of the SIMPLER algorithm was validated with the bidimensional lid-driven cavity flow problem, known in the literature as a benchmark for testing CFD algorithms [46, 47, 48]. All algorithms were implemented in Matlab (The MathWorks, Inc., Natick, MA, USA). Linear systems were solved using the biconjugate gradients stabilized method.

Proposed Numerical Solution

In this paper, we solve for a simulated velocity field, $\vec{v}_\infty = (u_\infty, v_\infty, w_\infty)$, that is close enough to the MRI-measured vector field $\vec{v}_{mri} = (u_{mri}, v_{mri}, w_{mri})$, and satisfies the fluid dynamics equations, Eqs. (2) and (3).

Let M be the total number of voxels in the reconstructed $\vec{v}_{mri}$ 3D velocity field, i.e., $M=M_x \cdot M_y \cdot M_z$, where M_x, M_y, and M_z represent the number of voxels along the x, y, and z axes, respectively. Consider u_{mri}, v_{mri}, and w_{mri} as the stacked M×1 column vectors with the PC-MRI measurements. Since the numerical solution of the Navier–Stokes continuity-equation system is based on the solution of linear systems, we propose that our numerical optimal solution is obtained by minimizing, for each velocity component, at iteration n, the following equations:

$$J_u(\mathbf{u}_n) = \frac{1}{2}||\mathbf{S}_u\mathbf{u}_n - \mathbf{f}_u||^2 + \frac{\lambda_u}{2}||\Gamma_u\mathbf{u}_n - \mathbf{u}_{mri}||^2 \tag{9}$$

$$J_v(\mathbf{v}_n) = \frac{1}{2}||\mathbf{S}_v\mathbf{v}_n - \mathbf{f}_v||^2 + \frac{\lambda_v}{2}||\Gamma_v\mathbf{v}_n - \mathbf{v}_{mri}||^2 \tag{10}$$

$$J_w(\mathbf{w}_n) = \frac{1}{2}||\mathbf{S}_w\mathbf{w}_n - \mathbf{f}_w||^2 + \frac{\lambda_w}{2}||\Gamma_w\mathbf{w}_n - \mathbf{w}_{mri}||^2. \tag{11}$$

The first term on the right hand side of Eqs. (9)–(11) is related to the numerical solution of the Navier–Stokes continuity equations, and the second term is related to the comparison between the numerical solution and the PC-MRI velocity field. The SS matrices and f vectors are defined in Eqs. (4)–(6), but note that we dropped the "n−1" subscripts for simplicity and clarity; these are updated by velocity and pressure values calculated in the previous iteration. Coefficients λ_u, λ_v, and λ_w are regularization factors, which weight consistence with PC-MRI data against conformance with the momentum equations. Matrices Γ_u, Γ_v, and Γ_w are of size M×N, and model the blurring effects due to finite k-space coverage in PC-MRI (this is further discussed below), while adjusting the number of points on the CFD grid in order to allow a comparison between $\vec{v}_n$ and $\vec{v}_{mri}$. In this approach, the number of grid points in the CFD and MRI grids are not necessarily the same; we can use a finer grid in CFD than in MRI, for example. The optimal solutions for Eqs. (9)–(11) are straightforward [42], and given by

$$\mathbf{u}_n = \left(\mathbf{S}_u^T\mathbf{S}_u + \lambda_u\Gamma_u^T\Gamma_u\right)^{-1}\left(\mathbf{S}_u^T\mathbf{f}_u + \lambda_u\Gamma_u^T\mathbf{u}_{mri}\right) \tag{12}$$

$$\mathbf{v}_n = \left(\mathbf{S}_v^T\mathbf{S}_v + \lambda_v\Gamma_v^T\Gamma_v\right)^{-1}\left(\mathbf{S}_v^T\mathbf{f}_v + \lambda_v\Gamma_v^T\mathbf{v}_{mri}\right) \tag{13}$$

$$\mathbf{w}_n = \left(\mathbf{S}_w^T\mathbf{S}_w + \lambda_w\Gamma_w^T\Gamma_w\right)^{-1}\left(\mathbf{S}_w^T\mathbf{f}_w + \lambda_w\Gamma_w^T\mathbf{w}_{mri}\right). \tag{14}$$

To understand the construction of the ΓΓ matrices, consider that in the absence of noise, artifacts, and distortions, the MRI-measured vector field $\vec{v}_{mri} = (u_{mri}, v_{mri}, w_{mri})$ is a blurred version of the true vector field $\vec{v} = (u, v, w)$. For the u component, for example, we can write:

$$u_{mri}(x,y,z) = u(x,y,z) * \psi_u(x,y,z), \tag{15}$$

where $*$ denotes convolution, and blurring kernel $\psi_u(x,y,z)$ is the point-spread function associated with the k-space coverage that was used when measuring u_{mri}. Similarly, we can write $v_{mri}=v*\psi v$ and $w_{mri}=w*\psi w$. If all three PC-MRI velocity components are measured using the same k-space coverage, then $\psi_u=\psi_v=\psi_w$. For a 3DFT acquisition, these spatial blurring kernels are equal to

$$\psi(x,y,z) = \text{sinc}\left(\frac{x}{\delta x}\right) \times \text{sinc}\left(\frac{y}{\delta y}\right) \times \text{sinc}\left(\frac{z}{\delta z}\right), \tag{16}$$

where δx, δy and δz are the spatial resolutions of $\vec{v}_{mri}$ along the x, y, and z axis, respectively.

We want the CFD-estimated vector field $\vec{v}_\infty=(u_\infty, v_\infty, w_\infty)$ to be an accurate representation of the true vector field $\vec{v}$. If this is so, then we should expect $u_{mri} \approx u_\infty * \psi_u$, $v_{mri} \approx v_\infty * \psi_v$, and $w_{mri} \approx w_\infty * \psi_w$. The discretization of these equations yields three linear systems. Then, using the same notation introduced earlier, for the nth iteration of the CFD algorithm, we can write:

$$\mathbf{u}_{mri} \approx \Gamma_u\mathbf{u}_n \tag{17}$$

$$\mathbf{v}_{mri} \approx \Gamma_v\mathbf{v}_n \tag{18}$$

$$\mathbf{w}_{mri} \approx \Gamma_w\mathbf{w}_n. \tag{19}$$

The coefficients of Γ_u, Γ_v, and Γ_w are calculated from ψ_u, ψ_v, and ψ_w, respectively. If all three PC-MRI velocity components are measured using the same k-space coverage, and reconstructed onto identical grids, then $\Gamma_u=\Gamma_v=\Gamma_w$.

The MRI-guided CFD estimate corresponding to one cardiac phase was calculated as a steady-state solution $\vec{v}_\infty$. All three components of the PC-MRI velocity field $\vec{v}_{mri}$ measured at the z positions at the boundaries of the calculation domain were used as inlet and outlet boundary conditions for that cardiac phase. Note that this steady state solution $\vec{v}_\infty$ is the closest fit in the least-squares sense to the direct PC-MRI measurements that satisfy both momentum equation (Eq. 2) and continuity equation (Eq. 3). This is guaranteed by the fact that the optimal solutions Eqs. (12)–(14) are solved in each iteration of the SIMPLER algorithm (steps 2 and 4, in Algorithm 1), and by the convergence criterion (step 7).

In each of our experiments, all three PC-MRI velocity components were measured using the same k-space coverage, and reconstructed onto identical grids. In the phantom experiments, we used the same grid size for both $\vec{v}_{mri}$ and $\vec{v}_\infty$, because the phantom data were measured with high spatial resolution. In these experiments, $\vec{v}_{mri}$ was reconstructed without zero-padding, i.e., onto $\delta_x \times \delta_y \times \delta_z$ voxels, and the CFD grid points were defined at the center of each of $\vec{v}_{mri}$'s voxels. Hence, $\Gamma_u=\Gamma_v=\Gamma_w$ was defined as an identity matrix. In the in vivo experiments, $\vec{v}_{mri}$ was reconstructed using 2-fold zero-padding along each of the spatial axes, since the data was acquired with low spatial resolution. Then, Γ was an N×N symmetric matrix, with coefficients calculated from the point spread function ψ(x,y,z), defined in Eq. (16). This infinite support point spread function was truncated by multiplication with the box function

$$B(x,y,z) = \text{rect}\left(\frac{x}{2\delta x}\right) \times \text{rect}\left(\frac{y}{2\delta y}\right) \times \text{rect}\left(\frac{z}{2\delta z}\right), \tag{20}$$

where rect (w)=1, if −1≤w≤1, and rect (w)=0, otherwise.

Experimental Setup: Phantom Demonstration

PC-MRI data of a pulsatile carotid flow phantom (Phantoms by Design, Inc., Bothell, WA) (Fig. 2) were obtained with high spatial resolution and high signal–to–noise ratio, from four time-resolved 3DFT FGRE image volumes (three acquired each with a velocity encoding bipolar gradient on one of the three axes, and one without a bipolar gradient). The scan parameters were: $0.5\times0.5\times1.0\text{m}^3$ spatial resolution; FOV $4.0\times3.5\times5.0$ cm^3; TR 11.4 ms; flip angle 8.5°; temporal resolution 91.2 ms; VENC 50 cm/s; 40 min per scan; 9 NEX. The data were acquired on a GE Discovery MR750 3T system (50 mT/m and 200 T/m/s max gradient amplitude and slew rate), with a 32-channel receive-only head coil array (Nova Medical, Inc., Wilmington, MA, USA). The through-slab (z) axis was oriented along the S/I direction. The phantom's pulse cycle was set to 60 bpm.

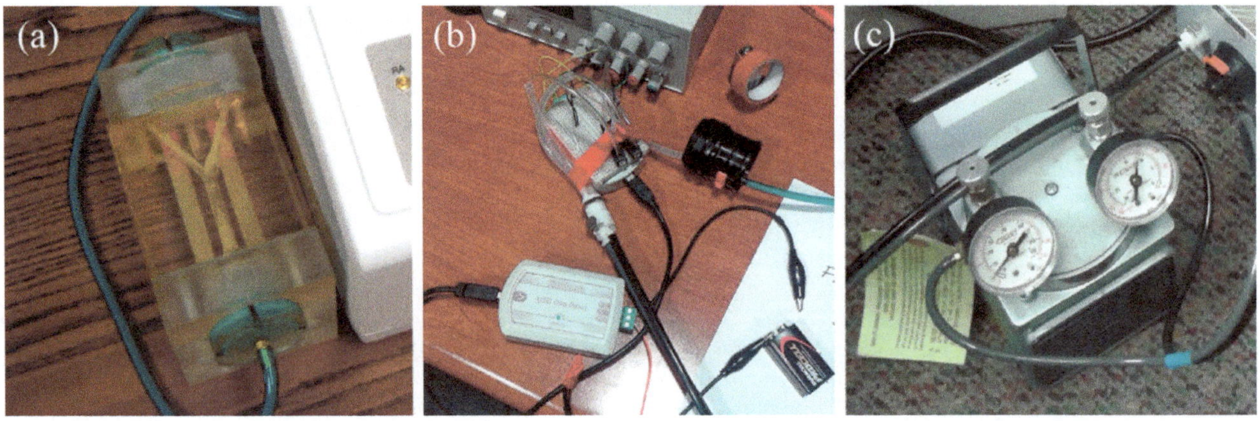

Figure 2: a Pulsatile carotid flow phantom (Phantoms by Design, Inc., Bothell, WA, USA) used to validate the proposed method; **b** pump controller that regulates flow frequency and gating signal; **c** air pump that generates the flow inside the phantom.

Only the temporal frame corresponding to peak flow was reconstructed. PC-MRI velocity component maps u_{mri}, v_{mri} and w_{mri} were calculated using data from all channels of the receive coil array. The lumen was segmented by manually outlining the vessel borders from a stack of 2D axial images, obtained from the reconstructed 3D volume. A few voxels presented phase-wrap artifacts; these voxels were manually identified and their velocities were corrected by adding 2π to their values.

The combined solver calculations assumed fluid viscosity of $\mu = 0.005$ Pa s and density of $\rho = 1100$ kg/m^3 (these values were provided by the phantom manufacturer). Calculations were performed with time step $\delta t = 0.1$ ms on a Cartesian grid of $0.5\times0.5\times1.0$ m^3 voxel size.

The CFD simulation domain was rectangular, of size $32.5\times9.0\times41.0\text{m}^3$. Each iteration required about 10 seconds of computation time on an Intel Core i7 processor running at 2.8 GHz.

Three simulated steady-state velocity fields $\vec{v}\infty$ were obtained: using the conventional SIMPLER algorithm, i.e., not using the PC-MRI data to constrain the CFD solution ($\vec{v}_{mri}$ was used only as inlet and outlet velocities for the geometry);

- using the velocity component associated with the main flow axis (z) measured with PC-MRI (w_{mri}) to constrain the CFD solution (u and v components were determined solely from the fluid physics model); and
- using all three velocity components measured with PC-MRI (u_{mri}, v_{mri}, and w_{mri}) to constrain the CFD solution.

The first approach is equivalent to making $\lambda_w=\lambda_u=\lambda_v=0$; in the second approach, we used $\lambda_w=1$and $\lambda_u=\lambda_v=0$; in the third approach, we used $\lambda_u=\lambda_v=\lambda_w=1$. All three approaches used all three components of the PC-MRI velocity field $\vec{v}_{mri}$ measured at the z positions at the boundaries of the calculation domain as inlet and outlet boundary conditions. The number of iterations until convergence for the above simulations was 89, 40 and 5 iterations, respectively.

Experimental Setup: In Vivo Demonstration

PC-MRI data of the carotid bifurcation of one healthy volunteer were obtained from four time-resolved 3DFT FGRE image volumes (three acquired each with a velocity encoding bipolar gradient on one of the three axes, and one without a bipolar gradient). The scan parameters were: $1.0\times1.0\times2.5$ m^3 spatial resolution; FOV $7.5\times12.0\times36.0$ cm^3; TR 7.0ms; flip angle 15°; temporal resolution 56ms; VENC 160 cm/s; 7 min per scan; 1 NEX. The data were acquired on a GE Signa 3T EXCITE HD system (40 mT/m and 150 T/m/s max gradient amplitude and slew rate), with a 4-channel neck receive coil array. The through-slab (z) axis was oriented along the S/I direction. The institutional review board of the University of Southern California approved the imaging protocols. The subject was screened for MRI risk factors and provided informed consent in accordance with institutional policy.

Only the cardiac phase corresponding to peak flow was reconstructed. PC-MRI velocity component maps u_{mri}, v_{mri} and w_{mri} were calculated using data from only one channel of the receive coil array. Residual linear velocity offsets in each velocity component map (e.g., due to eddy-currents) were removed by performing a linear fit to manually defined 3D regions containing only stationary tissue. The lumen was segmented by manually outlining the vessel borders from a stack of 2D axial images, obtained from the reconstructed 3D volume.

The combined solver calculations assumed blood viscosity $\mu = 0.0032$ Pa s and density of $\rho = 1060$ kg/m^3 [49]. Calculations were performed with time step $\delta t = 0.25$ ms on a Cartesian grid of $0.50\times0.50\times1.25\text{mm}^3$ voxel size. The CFD simulation domain was rectangular, of size $30\times37\times125\text{mm}^3$ (the PC-MRI data was cropped to match this grid size). Each iteration required about 180 s of computation time on an Intel Core i7 processor running at 2.8 GHz.

Three simulated steady-state velocity fields $\vec{v}\ \infty$ were obtained, using the same three approaches used in the phantom experiment. The number of iterations until convergence for the simulations was 1058, 190 and 6, respectively.

Quantitative Evaluation

The CFD-simulated velocity fields were quantitatively compared with the PC-MRI measurements by means of the signal-to-error ratio (SER). The SER measures the ratio between the energy of the signal and the energy of the estimation error. We considered the PC-MRI velocity field, $\vec{v}_{\mathrm{mri}} = (u_{\mathrm{mri}}, v_{\mathrm{mri}}, w_{\mathrm{mri}})$, as our ground-truth "signal"; consequently, the estimation error is the vector difference between the CFD-estimated velocity field, $\vec{v}_{\infty} = (u_{\infty}, v_{\infty}, w_{\infty})$, and the ground-truth field, $\vec{v}_{\mathrm{mri}}$. Thus, the SER is calculated (in decibels) as:

$$\mathrm{SER}_{\vec{v}} = 10\log_{10}\left(\frac{\sum_{i,j,k}\left\|\vec{v}_{\mathrm{mri}}(i,j,k)\right\|^2}{\sum_{i,j,k}\left\|\vec{v}_{\infty}(i,j,k) - \vec{v}_{\mathrm{mri}}(i,j,k)\right\|^2}\right), \quad (21)$$

where integers i, j, and k represent grid-point indexes along the x, y, and z axes, respectively. Similarly, the SER was also calculated individually for each of the velocity components, as:

$$\mathrm{SER}_u = 10\log_{10}\left(\frac{\sum_{i,j,k} u_{\mathrm{mri}}(i,j,k)^2}{\sum_{i,j,k}\left[u_{\infty}(i,j,k) - u_{\mathrm{mri}}(i,j,k)\right]^2}\right) \quad (22)$$

$$\mathrm{SER}_v = 10\log_{10}\left(\frac{\sum_{i,j,k} v_{\mathrm{mri}}(i,j,k)^2}{\sum_{i,j,k}\left[v_{\infty}(i,j,k) - v_{\mathrm{mri}}(i,j,k)\right]^2}\right) \quad (23)$$

$$\mathrm{SER}_w = 10\log_{10}\left(\frac{\sum_{i,j,k} w_{\mathrm{mri}}(i,j,k)^2}{\sum_{i,j,k}\left[w_{\infty}(i,j,k) - w_{\mathrm{mri}}(i,j,k)\right]^2}\right). \quad (24)$$

Using these SER values, the three CFD approaches—pure CFD, CFD driven by one PC-MRI velocity component, and CFD driven by all three PC-MRI velocity components—were quantitatively evaluated and compared.

Evaluation of Denoising Properties

Under our hypothesis, CFD simulations provide a smooth, noise-free flow field. Therefore, we expect that the proposed approach can be used as a denoising mechanism for PC-MRI flow assessment. In order to verify the denoising effects of the combined solver, we added zero-mean Gaussian noise with standard deviation 8 cm/s to the phantom's measured velocity field, $\vec{v}_{\mathrm{mri}}$.

This noisy flow field was used to constrain the CFD calculations, using the approach in which all three velocity components measured with PC-MRI are used. In this experiment, we used $\lambda_u=\lambda_v=\lambda_w=\lambda$; and four different values of λ were evaluated: 5×10^{-9}, 5×10^{-8}, 5×10^{-7}, and 5×10^{-6}. The SER between the proposed approach and the original PC-MRI measurements was calculated, and compared with the SER of the noisy flow field. The pure CFD approach, in which the noisy PC-MRI data are used only as inlet and outlet velocities for the geometry, was also evaluated (this is equivalent to making $\lambda=0$).

The phantom data was used in this denoising experiment, because it was acquired using 9 NEX—which results in high signal-to-noise ratio (SNR), while the in vivo data was acquired using only 1 NEX. The noise levels on the phantom's measured velocity components—estimated as the standard deviation in regions of uniform mean velocity—are lower than 3 cm/s; while the SNR of the magnitude images exceeds 26 dB. The velocity-to-noise ratio (VNR) [50, 51] for the u and w components reach 28 and 31 dB, respectively (the VNR for the v component was not calculated, because v is approximately null over the entire geometry).

Finally, in order to justify our denoising experiment, we analyze the noise distribution in PC-MRI images. We note that, from a maximum likelihood perspective, Eqs. (9)–(11) assume that the PC-MRI data is degraded by Gaussian noise. Under certain conditions, one can prove that velocity field noise in PC-MRI satisfies a zero-mean Gaussian distribution [52]. Therefore, the additive noise acting on the velocity fields can be assumed to be Gaussian distributed [27, 52]. Hence, the proposed minimization is well-suited in terms of the MR noise distribution.

RESULTS

Phantom Demonstration

Figure 3 shows a qualitative velocity-map comparison between the PC-MRI phantom measurements and the three simulated results. The PC-MRI velocity field does not satisfy the continuity equation, since its divergence is nonzero within the lumen (Fig. 3a). The pure CFD solution produced a velocity field that satisfies the physical model, but is considerably smoother than the PC-MRI measurements (Fig. 3b). Using one MRI-measured velocity component (w_{mri}) to guide the CFD simulation resulted in a solution that is qualitatively more similar to the MRI-measured field, while still satisfying the continuity and momentum equations (Fig. 3c). Even better agreement was achieved when all three MRI-measured velocity components (u_{mri}, v_{mri}, and w_{mri}) were used to guide the CFD simulation (Fig. 3d). These improvements can be also appreciated on a vector field visualization of the flow field over the entire tridimensional volume (Fig. 4).

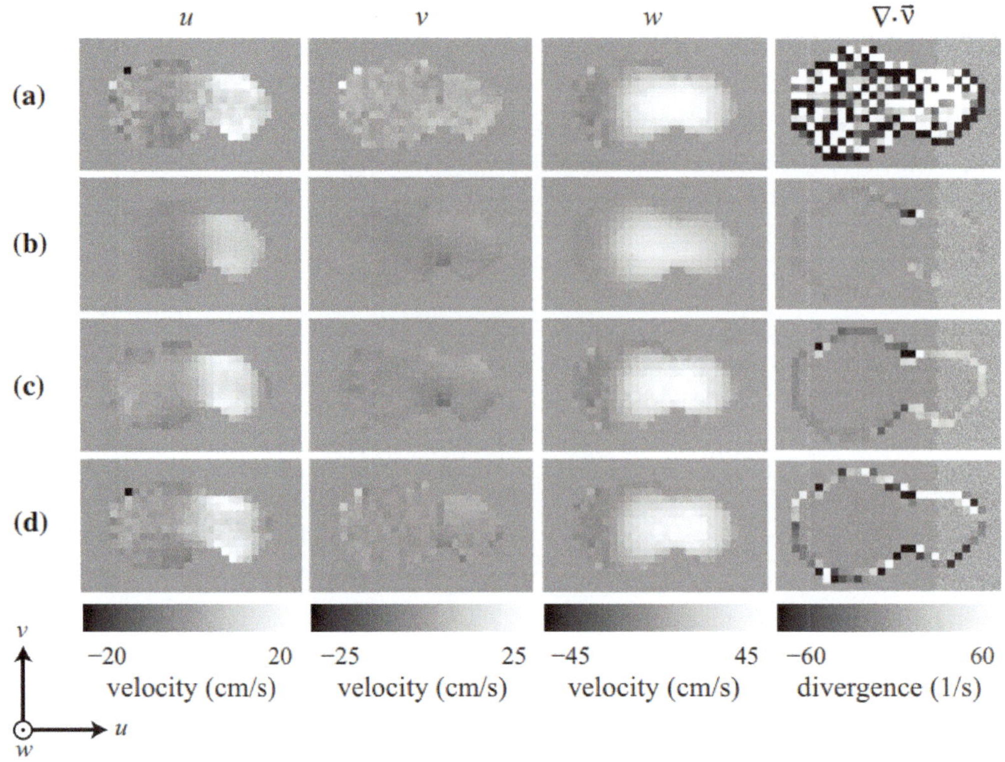

Figure 3: Velocity maps for the individual velocity components (u, v, and w) and divergence map of the velocity field ($\vec{v}$), for an axial slice at the bifurcation of the carotid flow phantom: **a** PC-MRI; **b** CFD; **c** CFD guided by PC-MRI data corresponding to the main velocity component (w_{mri}); and **d** CFD guided by PC-MRI data corresponding to all three velocity components (u_{mri}, v_{mri}, and w_{mri}).

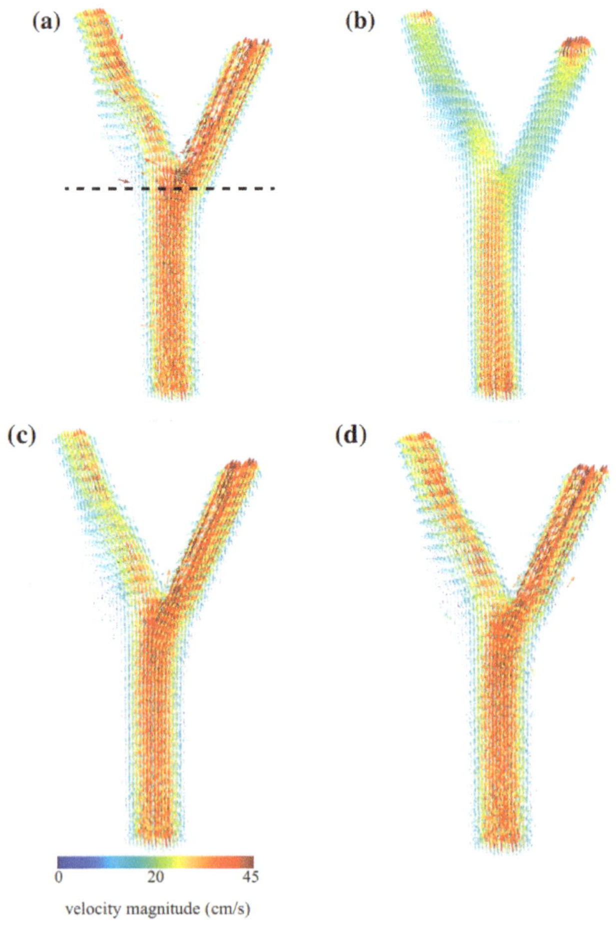

Figure 4: Vector field visualization of the velocity field ($\vec{v}$) over the entire tridimensional volume of the carotid flow phantom:**a** PC-MRI; **b** CFD; **c** CFD guided by PC-MRI data corresponding to the main velocity component (w_{mri}); and **d** CFD guided by PC-MRI data corresponding to all three velocity components (u_{mri}, v_{mri}, and w_{mri}). The *dotted line* in **a** indicates the position of the slice shown in Figs. 3 and 5.

Table 1: Signal-to-error ratio (in dB) between the phantom experiment results from each of the three CFD approaches—pure CFD (labeled "CFD"); CFD driven by one PC-MRI velocity component (labeled "CFD + 1D"); and CFD driven by all three PC-MRI velocity components (labeled "CFD + 3D"), relative to the MRI-measured velocity field

	CFD	CFD + 1D	CFD + 3D
SER_u	2.97	4.16 (↑)	6.74 (↑)
SER_v	−0.25	−0.30 (≈)	2.03 (↑)
SER_w	5.44	16.53 (↑↑)	13.46 (↑)
$SER_{\vec{v}}$	6.57	8.38 (↑)	13.13 (↑↑)

SER values were calculated according to Eqs. (21)–(24)

Table 1 shows the SER between the phantom experiment results from each of the three CFD approaches, relative to the MRI-measured velocity field. The approach using only one PC-MRI velocity component (w_{mri}) to drive the CFD calculations (labeled "CFD + 1D") provided a 11.09 dB improvement in SER relative to pure CFD (labeled simply "CFD") with respect to the *w* component; however, the improvement was only 1.81 dB when considering all three components. The approach using all three PC-MRI velocity components (u_{mri}, v_{mri}, and w_{mri}) to drive the CFD calculations (labeled "CFD + 3D") provided a 6.56 dB improvement in SER relative to pure CFD when considering all three components, while still providing a 8.02 dB improvement when considering only the *w* component.

Note that, for each of the CFD approaches, the SER was lower for the *u* and *v* components than it was for *w*(Table 1). This can be explained by the fact that the same VENC was used for measuring all three PC-MRI velocity components (which implies similar noise levels), but the velocities along the *z* axis are considerably higher than those along the *x* and *y* axes (the energy of w_{mri} was 11.42 dB higher than that of u_{mri}, and 15.72 dB higher than that of v_{mri}). As a consequence, the SNR of w_{mri} is substantially higher than that of u_{mri} and v_{mri}. Thus, even if the energy of the absolute error between the CFD-estimated velocities and the MRI-measured velocities was the same for the three components, SERw would be higher than SERu and SERv. Also, note that CFD approaches provide a smooth, (ideally) noise-free velocity field, but the SER was calculated with respect to a noisy PC-MRI field. This means that the denoising properties of the CFD approaches could actually hurt the SER, especially if noise levels are relatively high when compared to the velocity values (this is the case for the *u* and *v* components). However, denoising is a desirable feature, so this does not necessarily indicate an unwanted result.

Figure 5 and Table 2 show the results of the experiment in which the denoising properties of the proposed approach were evaluated. CFD calculations were constrained by all three PC-MRI velocity components, with added Gaussian noise. The combined solver improved the SER of each individual velocity component, for all different weight parameters we evaluated (Table 2). The CFD solution constrained by PC-MRI using $\lambda=5\times10^{-8}$ (Fig. 5e) provides a velocity field that is less noisy and visually more similar to the measured PC-MRI velocity field (Fig. 5a) than the pure CFD solution obtained using the noisy PC-MRI velocity field as boundary data (Fig. 5c). Using smaller values of λ (Fig. 5d) results in solutions that are closer to the pure CFD solution (Fig. 5c). Using larger values of λ (Figs. 5f, g) results in solutions that are closer to the noisy MRI data used to constrain the solution (Fig. 5b). These results can be appreciated quantitatively on Table 2. Relative to the noisy PC-MRI velocity field, the overall ($\vec{v}$) SER gain was 1.23 dB for $\lambda=5\times10^{-9}$; 2.47 dB for $\lambda=5\times10^{-8}$; 2.14 dB for $\lambda=5\times10^{-7}$; and 1.23 dB for $\lambda=5\times10^{-6}$. Moreover, all constrained CFD solutions presented better quantitative results than pure CFD. The overall ($\vec{v}$) SER gain, relative to pure CFD, was 0.55 dB for $\lambda=5\times10^{-9}$; 1.79 dB for $\lambda=5\times10^{-8}$; 1.46 dB for $\lambda=5\times10^{-7}$; and 0.55 dB for $\lambda=5\times10^{-6}$.

Table 2: Signal-to-error ratio (in dB) between noisy and original PC-MRI measurements; and between the MRI-guided CFD estimates and the original PC-MRI measurements

	Noisy PC-MRI	CFD $\lambda = 0$	CFD+3D $\lambda = 5 \times 10^{-9}$	CFD+3D $\lambda = 5 \times 10^{-8}$	CFD+3D $\lambda = 5 \times 10^{-7}$	CFD+3D $\lambda = 5 \times 10^{-6}$
SER_u	0.80	2.66	2.87	3.22	2.34	1.75
SER_v	−2.46	−0.50	−0.41	−0.39	−1.18	−1.67
SER_w	4.64	5.03	5.82	7.20	5.88	5.06
$SER_{\vec{v}}$	5.71	6.39	6.94	8.18	7.85	6.94

Additive zero-mean Gaussian noise with standard deviation of σ=8σ=8 cm/s was used. CFD estimates were obtained using the combined solver with different values of the weight parameter λ, constrained by the noisy PC-MRI measurements. All three PC-MRI velocity components were used to guide the CFD calculations. SER values were calculated according to Eqs. (21)–(24).

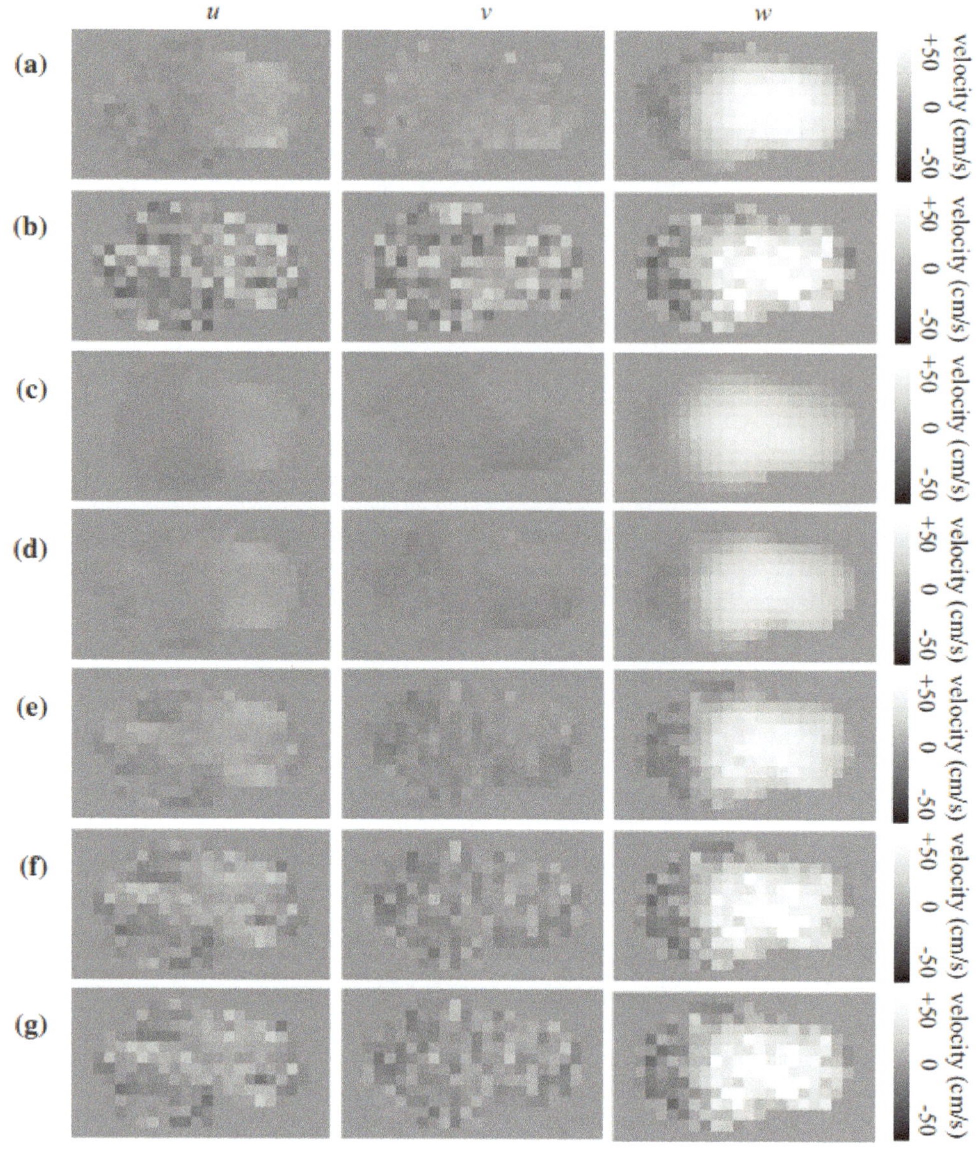

Figure 5: Velocity maps for the individual velocity components (*u*, *v*, and *w*), for an axial slice at the bifurcation of the carotid flow phantom: **a** PC-MRI; **b** PC-MRI with added Gaussian noise (σ=8 cm/s); **c** pure CFD solution using noisy PC-MRI data as inlet and oulet velocities; **d** CFD guided by the noisy PC-MRI data, with $\lambda=5\times10^{-9}$; **e** CFD guided by the noisy PC-MRI data, with $\lambda=5\times10^{-8}$; **f** CFD guided by the noisy PC-MRI data, with $\lambda=5\times10^{-7}$; **g** CFD guided by the noisy PC-MRI data, with $\lambda=5\times10^{-6}$.

These results illustrate the potential of the proposed numerical framework also as a denoising technique for PC-MRI.

In Vivo Demonstration

Figure 6 provides a qualitative velocity-map comparison between the PC-MRI in vivo measurements and the three simulated results. As in the phantom experiment, the PC-MRI velocity field does not satisfy the continuity equation, since its divergence is nonzero within the lumen (Fig. 6a). The pure CFD solution produced a velocity field that satisfies the physical model, but differs considerably from the PC-MRI measurements (Fig. 6b). Using one MRI-measured velocity component (w_{mri}) to guide the CFD simulation resulted in a solution that is qualitatively more similar to the MRI-measured field, while still satisfying the continuity and momentum equations (Fig. 6c). Even better agreement was achieved when all three MRI-measured velocity components (u_{mri}, v_{mri}, and w_{mri}) were used to guide the CFD simulation (Fig. 6d). These improvements can be also appreciated on a vector field visualization of the flow field over the entire tridimensional volume (Fig. 7).

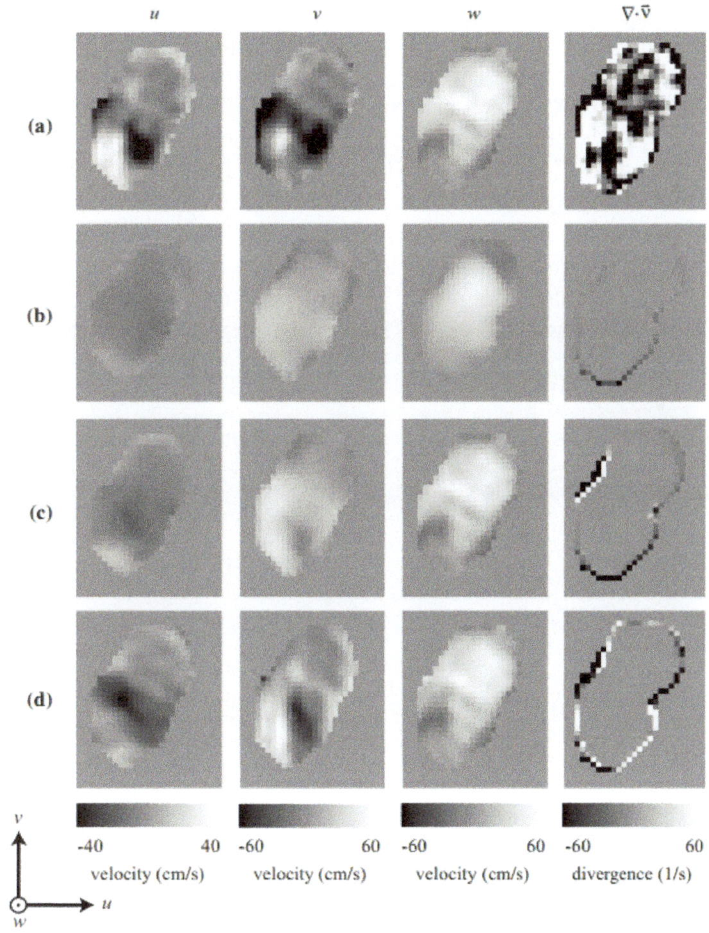

Figure 6: Velocity maps for the individual velocity components (*u*, *v*, and *w*) and divergence map of the velocity field (v^*), for a slice perpendicular to the carotid bifurcation of a healthy volunteer: **a** PC-MRI; **b** CFD; **c** CFD guided by PC-MRI data corresponding to the main velocity component (w_{mri}); and **d** CFD guided by PC-MRI data corresponding to all three velocity components (u_{mri}, v_{mri}, and w_{mri}).

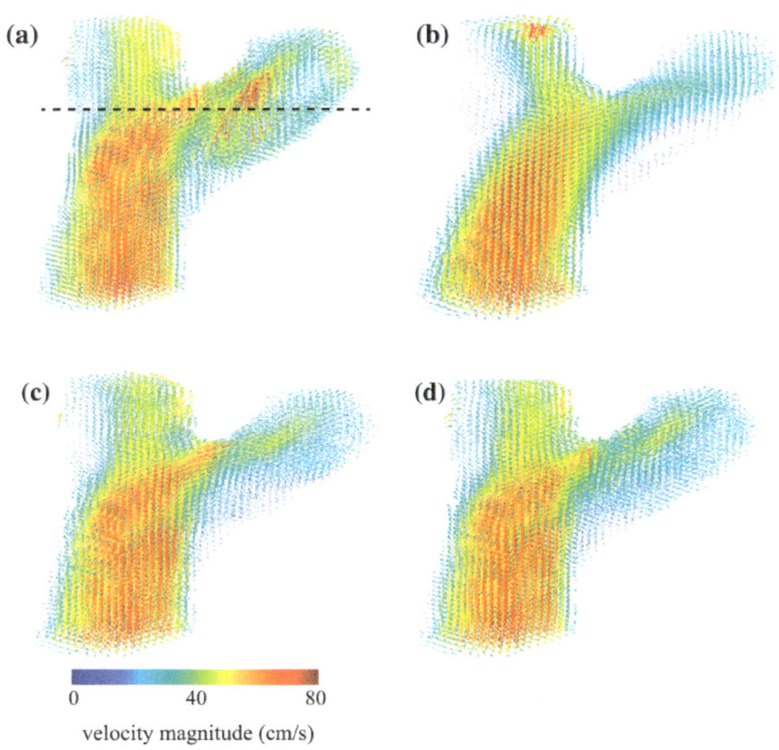

Figure 7: Vector field visualization of the velocity field (v^*) over the entire tridimensional volume of the carotid bifurcation of a healthy volunteer: **a** PC-MRI; **b** CFD; **c** CFD guided by PC-MRI data corresponding to the main velocity component (w_{mri}); and **d** CFD guided by PC-MRI data corresponding to all three velocity components (u_{mri}, v_{mri}, and w_{mri}). The *dotted line* in **a** indicates the position of the slice shown in Fig. 6.

Table 3 shows the SER between the in vivo experiment results from each of the three CFD approaches, relative to the MRI-measured velocity field. The approach using only one PC-MRI velocity component (w_{mri}) to drive the CFD calculations (labeled "CFD + 1D") provided a 9.79 dB improvement in SER relative to pure CFD (labeled simply "CFD") with respect to the *w* component; however, the SER was only 0.76 dB higher when considering all three components, since the agreement with PC-MRI for the *v* component was made worst. The approach using all three PC-MRI velocity components (u_{mri}, v_{mri}, and w_{mri}) to drive the CFD calculations (labeled "CFD + 3D") improved the SER for all three components, by 0.24, 0.59, and 6.00 dB, respectively; as a result, there was a 2.15 dB overall improvement in SER relative to pure CFD, and a 1.39 dB improvement relative to CFD guided only by w_{mri}. As in the phantom experiments, the SER was lower for the *u* and *v* components than it was for *w*, for all three CFD approaches. This can again be explained by the fact that the same VENC was used for all three velocity components, while the velocities along the *z* axis are considerably higher than those along the *x* and *y* axes (the energy of w_{mri} was 29.76 dB higher than that of u_{mri}, and 27.09 dB higher than that of v_{mri}), and is a direct (and desirable) consequence of the denoising properties of the proposed approach, as previously discussed.

Table 3: Signal-to-error ratio (in dB) between the in vivo experiment results from each of the three CFD approaches—pure CFD (labeled "CFD"); CFD driven by one PC-MRI velocity component (labeled "CFD + 1D"); and CFD driven by all three PC-MRI velocity components (labeled "CFD + 3D"), relative to the MRI-measured velocity field

	CFD	CFD + 1D	CFD + 3D
SER_u	0.51	0.50 (≈)	0.75 (↑)
SER_v	−1.51	−1.87 (↓)	−0.92 (↑)
SER_w	4.74	14.53 (↑↑)	10.74 (↑)
$SER_{\vec{v}}$	4.13	4.89 (↑)	6.28 (↑↑)

SER values were calculated according to Eqs. (21)–(24)

DISCUSSION

The proposed methodology uses the three-dimensional SIMPLER algorithm with Cartesian uniform meshes to perform blood flow simulations under the influence of three-dimensional MRI-measured velocity profiles. The combined MRI–CFD methodology attempts to correct the MRI-measured flow field, forcing it to satisfy the fluid mechanics equations. The choice of the SIMPER algorithm and Cartesian discretization was performed in order to facilitate implementation of the algorithm. We showed that the proposed technique provides better agreement with the PC-MRI measurements than pure CFD simulations. We also showed that this MRI-guided CFD approach can be used as a means of reducing noise in the PC-MRI measurements. It can also be used to reduce computation time: when all three MRI-measured velocity components are used to guide the CFD simulations, only a few iterations are required for convergence, i.e., for finding the flow field that is the most similar to the PC-MRI measurements (in the least squares sense) while satisfying the fluid mechanics equations.

We believe that the proposed method can also be used as a technique for reducing scan time. The proposed methodology allows the use of only part of the velocity measurements obtained with MRI to guide computational solutions by appropriately choosing the Γ matrices in Eqs. (12)–(14). In this paper, we used identical grids for MRI and CFD. Therefore, the Γ matrices were square with size N×N. In each of the following suggested approaches, at least one of the Γu, Γv or Γw matrices may be of size M×N, with M<N. Possible approaches for reducing scan time include:

- acquiring the MRI data with reduced spatial resolution, and using the MRI-guided CFD simulation to improve the spatial resolution;
- measuring portion(s) of the volume (e.g., carotid inlet and outlets) with full spatial resolution, while measuring the rest of the volume with reduced spatial resolution;
- acquiring only a few slices along the bifurcation, and using the MRI-guided CFD simulation to fill in the gaps;
- measuring one or two velocity components with full spatial resolution, while measuring the other component(s) with reduced spatial resolution;
- measuring one or two velocity components across the entire volume, while measuring the other component(s) for only a few slices; and
- any combination of these approaches. We plan to explore these ideas in future studies.

It is well known that blood is a non-Newtonian fluid, therefore its viscosity is not uniform. While there exists many constitutive models and studies in the literature regarding the non-Newtonian rheological properties of blood [43, 53, 54], no gold-standard constitutive model exists, and the assumption of constant whole blood viscosity is common practice [13, 16, 33, 37, 55, 56]. Therefore, we used the Newtonian blood flow model (constant whole blood viscosity), which greatly simplified the implementation.

It is often desirable to estimate the wall shear rate at the carotid bifurcation. Determining wall shear rates from CFD simulation results would require using highly refined meshes around the neighborhood of the vessel wall. Our implementation of the SIMPLER algorithm does not allow using spatially-varying grid spacing, and the mesh is uniform all over the integration domain. Using a very fine grid over the entire integration domain is possible, in principle. However, this would drastically increase the computational complexity, and could make the proposed methodology impractical in a clinical environment, for example.

Finally, the proposed approach does not take the effects of vessel wall elasticity into consideration. While the pulsatile carotid flow phantom we used has rigid tube walls, the human carotid vessel wall is generally elastic. While there exists blood flow CFD simulation methods that incorporate elastic wall effects [57, 58], the assumption of rigid vessel walls is another common practice [13, 16, 33, 37, 55]. The SIMPLER algorithm implemented in this work uses a Cartesian uniform mesh, and does not allow the use of the elastic wall models.

Both wall shear stress calculations and elastic vessel walls could be properly addressed with an implementation using finite elements—which would allow an easier adaptation of the mesh near the wall and also allow simulating the effects of fluid–structure interaction. Despite the fact that the implementation of a finite-element solver would be substantially more complex than our implementation, all the proposed methodology described in this proof of concept paper can still be applied in the same fashion, since the problem of solving the set of differential equations in a finite element discretization is replaced by a sparse system of linear equations similar to the ones obtained in this work.

In future works, this MRI-guided CFD methodology for unsteady flow will be implemented on FreeFem++, a partial differential equation solver capable of solving the Navier–Stokes equations using finite elements [59]. FreeFem++ allows access to the linear systems that can be modified in order to make use of the principles introduced here. With this approach, we expect more general, higher-quality solutions, allowing the calculation of biomarkers, such as wall shear stress, since FreeFem++ can handle different types of triangular finite elements, large varieties of linear system solvers, and automatic mesh adaptation. Note that there are other free software that could be used to reproduce this methodology, such as openFOAM, which uses finite volume discretization. Both FreeFem++ and openFOAM allow modifications of the linear systems, and have CFD solvers already implemented, which could facilitate the implementation of the methodology proposed in this study [40, 59].

All the assumptions and simplifications discussed above (Newtonian blood flow, imprecise geometry, non-compliant walls) contribute non-linearly to the differences observed between CFD solutions and MRI-measured velocity fields. Using the MRI-measured velocity field to constrain the CFD solution indirectly addresses these simplifications, and provides a more realistic CFD solution. However, we are still unable to clearly identify the main factors responsible for the disagreement between MRI-measured and CFD-computed velocity fields. Nevertheless, an implementation using a more robust solver, as proposed above, could improve on these limitations and further reduce the gap between MRI-measured and CFD-computed results.

CONCLUSION

We have proposed a framework for obtaining flow field estimates that are influenced by both PC-MRI measurements and a fluid physics model. The results showed that the proposed technique provides better agreement with the PC-MRI measurements than pure CFD simulations, and has reduced computation time (faster convergence). MRI-guided CFD can be used to correct the MRI-measured flow field, forcing it to satisfy the fluid mechanics equations. It can also be used as a means of reducing noise in the PC-MRI measurements, and has potential as a method for reducing scan time.

The proposed framework offers a general approach to in vivo blood flow assessment that is complementary to improvements in PC-MRI acquisition and reconstruction techniques, and can be applied to the study and diagnosis of a broad range of cardiovascular flow mapping applications.

AUTHORS' CONTRIBUTIONS

VCR: CFD–MRI code implementation, phantom data acquisition, analyzed and interpreted the data, drafted the manuscript. JFN: assisted CFD–MRI code implementation, in vivo and phantom data acquisition/reconstruction, drafted and critically revised the manuscript, supervised the research. KSN: in vivo data acquisition/reconstruction, critically revised the manuscript. JLC: design of the study, analyzed and interpreted the data, critically revised the manuscript, supervised the research. All authors read and approved the final manuscript.

ACKNOWLEDGEMENTS

This work was supported in part by the Brazilian Coordination for the Improvement of Higher Education Personnel (CAPES), scholarship Grant number 8195-13-7.

REFERENCES

1. Pelc NJ, Herfkens RJ, Shimakawa A, Enzmann DR. Phase-contrast cine magnetic resonance imaging. Magn Reson Q. 1991;7:229–54.
2. Gonzalez EG, Carvalho JLA. Does phase contrast MRI provide the mean velocity of the spins within a voxel? In: Proceedings of the International Society for Magnetic Resonance in Medicine, vol 22. Milan; 2014. p. 2480.
3. Sigfridsson A, Petersson S, Carlhäll C-J, Ebbers T. Four-dimensional flow MRI using spiral acquisition. Magn Reson Med. 2012;68:1065–73. doi:10.1002/mrm.23297.
4. Steinman DA, Taylor CA. Flow imaging and computing: large artery hemodynamics. Ann Biomed Eng. 2005;33:1704–9. doi:10.1007/s10439-005-8772-2.
5. Giddens DP, Zarins CK, Glagov S. The role of fluid mechanics in the localization and detection of atherosclerosis. J Biomech Eng. 1993;115:588–94. doi:10.1115/1.2895545.
6. Steinman DA. Image-based computational fluid dynamics: a new paradigm for monitoring hemodynamics and atherosclerosis. Curr Drug Targets Cardiovasc Hematol Disord. 2004;4:183–97. doi:10.2174/1568006043336302.
7. Lei M, Kleinstreuer C, Truskey GA. Numerical investigation and prediction of atherogenic sites in branching arteries. J Biomech Eng. 1995;117:350–7. doi:10.1115/1.2794191.
8. Hyun S, Kleinstreuer C, Archie JP. Hemodynamics analyses of arterial expansions with implications to thrombosis and restenosis. Med Eng Phys. 2000;22:13–27. doi:10.1016/S1350-4533(00)00006-0.

9. Longest PW, Kleinstreuer C. Comparison of blood particle deposition models for non-parallel flow domains. J Biomech. 2003;36:421–30. doi:10.1016/S0021-9290(02)00434-7.

10. Boussel L, Rayz V, Martin A, Acevedo-Bolton G, Lawton MT, Higashida R, Smith WS, Young WL, Saloner D. Phase-contrast magnetic resonance imaging measurements in intracranial aneurysms in vivo of flow patterns, velocity fields, and wall shear stress: comparison with computational fluid dynamics. Magn Reson Med. 2009;61(2):409–17. doi:10.1002/mrm.21861.

11. Canstein C, Cachot P, Faust A, Stalder AF, Bock J, Frydrychowicz A, Kuffer J, Hennig J, Markl M. 3D MR flow analysis in realistic rapid-prototyping model systems of the thoracic aorta: comparison with in vivo data and computational fluid dynamics in identical vessel geometries. Magn Reson Med. 2008;59(3):535–46. doi:10.1002/mrm.21331.

12. Milner JS, Moore JA, Rutt BK, Steinman DA. Hemodynamics of human carotid artery bifurcations: computational studies with models reconstructed from magnetic resonance imaging of normal subjects. J Vasc Surg. 1998;28:143–56. doi:10.1016/S0741-5214(98)70210-1.

13. Papathanasopoulou P, Zhao S, Köhler U, Robertson MB, Long Q, Hoskins P, Xu XY, Marshall I. MRI measurement of time-resolved wall shear stress vectors in a carotid bifurcation model, and comparison with CFD predictions. J Magn Reson Imaging. 2003;17:153–62. doi:10.1002/jmri.1220.

14. Thomas JB, Milner JS, Rutt BK, Steinman DA. Reproducibility of image-based computational fluid dynamics models of the human carotid bifurcation. Ann Biomed Eng. 2003;31:132–41. doi:10.1114/1.1540102.

15. Marshall I, Zhao S, Papathanasopoulou P, Hoskins P, Xu Y. MRI and CFD studies of pulsatile flow in healthy and stenosed carotid bifurcation models. J Biomech. 2004;37:679–87. doi:10.1016/j.jbiomech.2003.09.032.

16. Long Q, Xu XY, Ariff B, Thom SA, Hughes AD, Stanton AV. Reconstruction of blood flow patterns in a humancarotid bifurcation: a combined CFD and MRI study. J Magn Reson Imaging. 2000;11:299–311.

17. Pelc NJ, Sommer FG, Li KC, Brosnan TJ, Herfkens RJ, Enzmann DR. Quantitative magnetic resonance flow imaging. Magn Reson Q. 1994;10:125–47.

18. Markl M, Chan FP, Alley MT, Wedding KL, Draney MT, Elkins CJ, Parker DW, Wicker R, Taylor CA, Herfkens RJ, Pelc NJ. Time-resolved three-dimensional phase-contrast MRI. J Magn Reson Imaging. 2003;17:499–506. doi:10.1002/jmri.10272.

19. Harloff A, Albrecht F, Spreer J, Stalder AF, Bock J, Frydrychowicz A, Schllhorn J, Hetzel A, Schumacher M, Hennig J, Markl M. 3D blood flow characteristics in the carotid artery bifurcation assessed by flow-sensitive 4D MRI at 3T. Magn Reson Med. 2009;61:65–74. doi:10.1002/mrm.21774.

20. Dyverfeldt P, Gårdhagen R, Sigfridsson A, Karlsson M, Ebbers T. On MRI turbulence quantification. Magn Reson Imaging. 2009;27:913–22. doi:10.1016/j.mri.2009.05.004.

21. Morbiducci U, Ponzini R, Rizzo G, Cadioli M, Esposito A, Montevecchi FM, Redaelli A. Mechanistic insight into the physiological relevance of helical blood flow in the human aorta: an in vivo study. Biomech Model Mechanobiol. 2011;10:339–55. doi:10.1007/s10237-010-0238-2.

22. Tsuji T, Suzuki J, Shimamoto R, Yamazaki T, Nakajima T, Nagai R, Komatsu S, Ohtomo K, Toyo-Oka T, Omata M. Vector analysis of the wall shear rate at the human aortoiliac bifurcation using cine MR velocity mapping. Am J Roentgenol. 2002;178:995–9. doi:10.2214/ajr.178.4.1780995.

23. Stalder AF, Russe MF, Frydrychowicz A, Bock J, Markl JHM. Quantitative 2D and 3D phase contrast MRI: optimized analysis of blood flow and vessel wall parameters. Magn Reson Med. 2008;60:1218–31. doi:10.1002/mrm.21778.

24. Dyverfeldt P, Sigfridsson A, Knutsson H, Ebbers T. A novel MRI framework for the quantification of any moment of arbitrary velocity distributions. Magn Reson Med. 2011;65(3):725–31. doi:10.1002/mrm.22649.

25. Ebbers T, Wigstrm L, Bolger AF, Wranne B, Karlsson M. Non-invasive measurement of time-varying three-dimensional relative pressure fields within the human heart. J Biomech Eng. 2002;124:288–93. doi:10.1115/1.1468866.

26. Thompson RB, McVeigh ER. Fast measurement of intracardiac pressure differences with 2D breath-hold phase-contrast MRI. Magn Reson Med. 2003;49:1056–66. doi:10.1002/mrm.10486.

27. Wang Y, Amini AA. Integrable pressure gradients via harmonics-based orthogonal projection. In: Christensen GE, Sonka M, editors. Information processing in medical imaging: 19th international conference. Glenwood Springs; 2005. p. 431–42.

28. Nayak KS, Nielsen JF, Bernstein MA, Markl M, Botnar RM, Gatehouse PD, Saloner D, Lorenz C, Wen H, Hu BS, Epstein FH, Oshinski JN, Raman SV. Cardiovascular magnetic resonance phase contrast imaging. J Cardiovasc Magn Reson. 2015;17:71. doi:10.1186/s12968-015-0172-7.

29. Donati F, Figueroa CA, Smith NP, Lamata P, Nordslettena DA. Non-invasive pressure difference estimation from PC-MRI using the work-energy equation. Med Image Anal. 2015;26:159–72. doi:10.1016/j.media.2015.08.012.

30. Harloff A, Nussbaumer A, Bauer S, Stalder AF, Frydrychowicz A, Weiller C, Hennig J, Markl M. In vivo assessment of wall shear stress in the atherosclerotic aorta using flow-sensitive 4D MRI. Magn Reson Med. 2010;63(6):1529–36. doi:10.1002/mrm.22383.

31. Buonocore M. Algorithms for improving calculated streamlines in 3-D phase contrast angiography. Magn Reson Med. 1994;31(1):22–30. doi:10.1002/mrm.1910310104.

32. Fatouraee N, Amini AA. Regularization of flow streamlines in multislice phase-contrast MR imaging. IEEE Trans Med Imaging. 2003;22(6):699–709.

doi:10.1109/TMI.2003.814786.

33. Song SM, Napel S, Glover GH, Pelc NJ. Noise reduction in three-dimensional phase contrast MR velocity measurements. J Magn Reson Imaging. 1993;3:587–96. doi:10.1002/jmri.1880030407.

34. Verdugo AMP, Mura J, Uribe S. Enforcing divergence free to velocity data from 4D flow MR images. In: Proceedings of International Society for Magnetic Resonance in Medicine, vol 22. Milan;2014. p. 2485.

35. Ong F, Uecker M, Tariq U, Hsiao A, Alley M, Vasanawala S, Lustig M. Compressed sensing 4D flow reconstruction using divergence-free wavelet transform. In: Proceedings of international society for magnetic resonance in medicine, vol 22. Milan;2014. p. 0326.

36. Bostan E, Lefkimmiatis S, Vardoulis O, Stergiopulos N, Unser M. Improved variational denoising of flow fields with application to phase-contrast MRI data. IEEE Signal Process Lett. 2014;22(6):762–6. doi:10.1109/LSP.2014.2369212.

37. Liu L, Funamoto K, Hayase T. Numerical experiment for ultrasonic-measurement-integrated simulation of developed laminar pipe flow using axisymmetric model. J Biomech Sci Eng. 2008;3:101–15. doi:10.1007/s10439-008-9519-7.

38. Kato T, Funamoto K, Hayase T, Sone S, Kadowaki H, Shimazaki T, Jibiki T, Miyama K, Liu L. Development and feasibility study of a two-dimensional ultrasonic-measurement-integrated blood flow analysis system for hemodynamics in carotid arteries. Med Biol Eng Comput. 2014;52(11):933–43. doi:10.1007/s11517-014-1193-3.

39. Funamoto K, Suzuki Y, Hayase T, Kosugi T, Isoda H. Numerical validation of MR-measurement-integrated simulation of blood flow in a cerebral aneurysm. Ann Biomed Eng. 2009;37(6):1105–16. doi:10.1007/s10439-009-9689-y.

40. Weller HG, Tabor G, Jasak H, Fureby C. A tensorial approach to computational continuum mechanics using object-oriented techniques. Computers Phys. 1998;12(6):620–31. doi:10.1063/1.168744.

41. Christodoulou AG, Ramb R, Menza M, Hennig J, Liang ZP. 4d flow imaging incorporating a fluid dynamics model. In: Proceedings of the international society for magnetic resonance in medicine, Toronto, vol 23. 2015. p. 2735.

42. Seo JK, Woo EJ. Nonlinear inverse problems in imaging. 1st ed. Chichester: Willey; 2013.

43. Bird R, Armstrong R, Hassager O. Dynamics of polymeric liquids: fluid mechanics, vol. 1, 2nd ed. New York: Willey; 1987.

44. Patankar SV. Numerical heat transfer and fluid flow. 1st ed. New York: Hemisphere Publishing Corporation; 1980.

45. Versteeg H, Malalasekera W. An introduction to computational fluid dynamics: the finite. 2nd ed. Glasgow: Prentice-Hall; 2007.

46. Griebel M, Dornseifer T, Neunhoeffer T. Numerical simulation in fluid dynamics. 1st ed. Philadelphia: Society for Industrial and Applied Mathematics; 1998.

47. Ghia U, Ghia K, Shin C. High-resolutions for incompressible flow using the Navier–Stokes equations and a multigrid method. J Comput Phys. 1982;48:387–411. doi:10.1016/0021-9991(82)90058-4.

48. Gupta M, Kalita J. A new paradigm for solving Navier–Stokes equations: streamfunction-velocity formulation. J Comput Phys. 2005;207:52–68. doi:10.1016/j.jcp.2005.01.002.

49. Westerhof N, Stergiopulos N, Noble M. Snapshots of hemodynamics: an aid for clinical research and graduate education. 1st ed. New York: Springer; 2005.

50. Papaharilaou Y, Doorly DJ, Sherwin SJ. Assessing the accuracy of two-dimensional phase-contrast MRI measurements of complex unsteady flows. J Magn Reson Imaging. 2001;14:714–23. doi:10.1002/jmri.10008.

51. Ha H, Kim GB, Kweon J, Kim Y-H, Kim N, Yang DH, Lee SJ. Multi-VENC acquisition of four-dimensional phase-contrast MRI to improve precision of velocity field measurement. Magn Reson Med. 2015. doi:10.1002/mrm.25715.

52. Gudbjartsson H, Patz S. The rician distribution of noisy MRI data. Magn Reson Med. 1995;34(6):910–4. doi:10.1002/mrm.1910340618.

53. Kim S. A study of non-newtonian viscosity and yield stress of blood in a scanning capillary-tube rheometer. PhD thesis, Drexel University. 2002.

54. Kim S, Namgung B, Ong P, Cho Y, Chun K, Lim D. Determination of rheological properties of whole blood with a scanning capillary-tube rheometer using constitutive models. J Mech Sci Technol. 2009;23(6):1718–26. doi:10.1007/s12206-009-0420-6.

55. Steinman DA, Thomas JB, Ladak HM, Milner JS, Rutt BK, Spence JD. Reconstruction of carotid bifurcation hemodynamics and wall thickness using computational fluid dynamics and MRI. Magn Reson Med. 2002;47(1):149–59. doi:10.1002/mrm.10025.

56. Carvalho JLA, Nielsen JF, Nayak KS. Feasibility of in vivo measurement of carotid wall shear rate using spiral Fourier velocity encoded MRI. Magn Reson Med. 2010;63:1537–47. doi:10.1002/mrm.22325.

57. Perktold K, Rappitsch G. Computer simulation of local blood flow and vessel mechanics in a compliant carotid artery bifurcation model. J Biomech. 1995;28(7):845–56. doi:10.1016/0021-9290(95)95273-8.

58. Figueroa CA, Vignon-Clementela IE, Jansenc KE, Hughesd TJR, Taylorb CA. A coupled momentum method for modeling blood flow in three-dimensional deformable arteries. Computer Methods Appl Mech Eng. 2006;195:5685–706. doi:10.1016/j.cma.2005.11.011.

59. Hecht F. New development in freefem++. J Numer Math. 2012;20(3–4):251–65. doi:10.1515/jnum-2012-0013.

CHAPTER 10

AN ADVANCED COMPUTATIONAL BIOHEAT TRANSFER MODEL FOR A HUMAN BODY WITH AN EMBEDDED SYSTEMIC CIRCULATION

Alberto Coccarelli[1], Etienne Boileau[2], Dimitris Parthimos[1] and Perumal Nithiarasu[2]

[1] Wales Heart Research Institute, School of Medicine, Cardiff University, Cardiff, UK

[2] Zienkiewicz Centre for Computational Engineering, College of Engineering, Swansea University, Swansea, UK

ABSTRACT

In the present work, an elaborate one-dimensional thermofluid model for a human body is presented. By contrast to the existing pure conduction-/perfusion-based models, the proposed methodology couples the arterial fluid dynamics of a human body with a multi-segmental bioheat model of surrounding solid tissues. In the present configuration, arterial flow is included through a network of elastic vessels. More than a dozen solid segments are employed to represent the heat conduction in the surrounding tissues, and each segment is constituted by a multilayered circular cylinder. Such multi-layers allow flexible delineation of the geometry and incorporation of properties of different tissue types. The coupling of solid tissue and fluid models requires subdivision of the arterial circulation into large and small arteries. The heat exchange between tissues and arterial wall occurs by convection in large vessels and by perfusion in small arteries. The core region, including the heart, provides the inlet conditions for the fluid equations. In the proposed model, shivering, sweating, and perfusion changes constitute the basis of the thermoregulatory system. The equations governing flow and heat transfer in the circulatory system are solved using a locally conservative Galerkin approach, and the heat conduction in the surrounding tissues is solved using a standard implicit backward Euler method. To investigate the effectiveness of the proposed model, temperature field evolutions are monitored at different points of the arterial tree and in the surrounding tissue layers. To study the differences due to flow-induced convection effects on thermal balance, the results of the current model are compared against those of the widely used modelling methodologies. The results show that the convection significantly influences the temperature distribution of the solid tissues in the vicinity of the arteries. Thus, the inner convection has a more predominant role in the human body heat balance than previously thought. To demonstrate its capabilities, the proposed new model is used to study different scenarios, including thermoregulation inactivity and variation in surrounding atmospheric conditions.

INTRODUCTION

A fundamental understanding of heat transport in a human body is important for studying a broad range of situations, such as the effect of temperature controlled surgeries and the impact of dramatic change in atmospheric or surrounding temperature. The thermal manipulation of body temperature has been used for a long time in cryosurgery and cancer treatment (hyperthermia) (Wells et al. 2005; Zohdi 2014; Zhu et al. 2005). It is also known that dramatic changes in weather (extremely cold seasons or heat waves) can lead to adverse health conditions and in some cases death. In thermally

Citation: Coccarelli, A., Boileau, E., Parthimos, D. et al. An advanced computational bioheat transfer model for a human body with an embedded systemic circulation (2015). doi:10.1007/s10237-015-0751-4.

controlled surgeries, understanding the heat or cold dissipation mechanism in a human body can be critically important. In a more ordinary situation, seasonal changes in atmospheric temperature require the body to adapt fast to keep the stimuli under control. Despite all the regulatory mechanisms of the human body, whenever temperature reaches a threshold value, tissue damage and/or alterations to biological processes may occur. When subjected to varying temperature conditions, the thermoregulatory system enables the control mechanisms of the body to keep tissue temperature within the threshold limit. Modelling such a control mechanism and resulting temperature behavior in time and space within a human body is extremely complex. A number of modelling attempts have been made in the past using lumped models or models that have only accounted for conduction and perfusion to understand the heat transfer in a human body. While these models provide a good starting point, they are not comprehensive as they do not include the effect of pulsatility and flow in arteries. A brief overview of the existing models are given below.

The available bioheat transfer models for a human body may be conveniently divided into lumped models, segmented models, and multi-dimensional models. In addition, there are models that fall between these categories. The lumped models are the simplest as they treat the human body as a single point with temperature change allowed only in time. Since the thermoregulation effects and heat transfer processes within the body were not accounted in the lumped models, application of such models are extremely limited. By comparison, fully three-dimensional models represent thermal changes in a human body close to reality. Such representations are, however, complex, extremely expensive, and difficult to employ. Thus, the segmented models that carefully account for various biological and physical processes are probably the best models for understanding the human body bioheat transfer.

Alternatives to lumped models were developed in the late sixties, and the first of such models consists of two nodes, representing a human body using two concentric shells. The first central shell represents internal organs, bone, muscle, and subcutaneous tissue, and the outer second shell represents the skin layer. The model presented by Gagge et al. (1967) is one of the best-known two-node models. It calculates the thermal response by means of two energy balance equations, one for the core node and one for the skin node. Gagge et al.'s model accounts for the effects of heat accumulation, conductive and convective heat transfer via blood flow between the core and skin shells, and metabolic heat generated during exercise and shivering. The energy exchange with environment has been modelled by considering respiration, convection, radiation, and evaporation of moisture. Gagge et al.'s model is simple to use, but it can only be applied to situations with moderate levels of activity and uniform environmental conditions. Although this model is an improvement of lumped models, it does not allow for the computation of detailed body temperature distribution.

An obvious extension to the two-node model is to introduce multi-nodes to discretely represent different parts of the body. Stolwijk (1971, 1977), Stolwijk and Hardy (1966) divided the human body into five cylindrical parts to individually represent the trunk, arms, hands, legs and feet, and a spherical body part for the head. Each part was further divided into four concentric shells representing the core, muscle, fat, and skin layers. In this model, the blood circulation system is represented by a blood pool located in the trunk. The trunk is connected to all tissue nodes by a network of blood vessels. The heat transfer between tissue nodes occur by conduction, while between the central blood node and the adjacent tissue nodes by convection. The energy balance equation includes heat accumulation, blood convection, tissue conduction, metabolic activity, respiration, and heat transfer to the environment by convection, radiation and evaporation. A thermal control system is also included. A fundamental limitation of this methodology is that it is restricted to isothermal blood flow.

The model developed by Wissler (1964, 1985), consists of six (later 15) elements connected by a vascular system. The vascular network is composed of arteries, veins, and capillaries and the blood temperatures are assumed to be uniform. Arteries connect the heart to the arterial pool of each element and further into the capillaries. From the capillaries, the blood circulates to the venous pool of the element and back to the heart. Between large arteries and large veins countercurrent heat exchange is modelled. It also accounts for breathing losses and is able to simulate transient states. The 3D transient multi-element model developed by Smith (1991) is based on a more realistic representation of the entire human body than previous models. The body here is composed of 15 elements, which are connected by the central macrocirculation and superficial veins. For the blood network, 1D steady-state Newtonian flow is assumed. An accurate evaluation of breathing losses is proposed by considering the respiration cycle. In this thermoregulatory system proposed, variation of skin blood vessel radii during vasomotor response, the sweat rate and the shivering metabolic rate are functions of the core and mean skin temperatures. The 3D approach makes the model suitable for situations with high-temperature gradients or highly non-uniform thermal conditions. However in such model fat and skin layers are modelled as a single layer; this may affect the heat convection to the skin surface carried by blood flow and thus the thermal response of entire body. Moreover, blood perfusion occurring in capillary beds is not considered.

Another relevant work is that of Fiala (1999), Fiala et al. (2001), who divide the body into cylindrical and spherical elements. Such subdivision was enforced whenever a significant change of body tissue properties occurred. The heat produced within the body is dispersed to the environment by convection, radiation, and moisture evaporation at the skin, and in the lungs/respiratory tract. Such multi-layered model consists of annular concentric tissue layers and uses seven different tissue materials: brain, lung, bone, muscle, viscera, fat, and skin. For the solid conduction problem 1D Penne's equation was used. Body elements are supplied with warm blood from the central pool by the major arteries. Along the pathway, arterial blood exchanges energy with returning veins as in a countercurrent heat exchanger. This was introduced in order to obtain a more realistic distribution of the arterial blood temperature instead of assuming a constant arterial blood temperature for all body elements equal to the temperature of the central blood pool. Perfusion is used as the mechanism of exchange between blood and tissues. The effects of the thermoregulatory system are also accounted. Cropper et al. (2008) coupled the human model of Fiala (1999) with a CFD for external airflow and obtained a tool able to predict the response of body temperature to several detailed local environmental conditions. Tanabe et al. (2002) modelled human thermal system in a similar way.

In this case, each individual body elements consists of a core layer and a skin layer and in the center of each core layer there are artery and vein blood pools. Between the artery and the superficial vein blood pools, an additional vessel is introduced to account for changes in blood flow due to changes in the ambient environment.

The model developed by Huizenga et al. (2001) is based on the Stolwijk model Stolwijk (1971,1977), Stolwijk and Hardy (1966) as well as on work by Tanabe et al. (2002), but includes several significant improvements, as it can simulate an arbitrary number of human body segments. Each of these segments consists of four body layers (core, muscle, fat, and skin tissues) while a clothing node has been added to model heat and moisture capacitances. The improved blood flow model includes central artery/vein countercurrent heat exchange and blood perfusion model to estimate blood flow to local tissue. The model also calculates the heat transfer by conduction to surfaces in contact with the body. A better estimation of the convection and radiation transfer coefficients, an explicit radiation heat transfer calculation using angle factors and the addition of a radiation heat flux model are notable additions. Besides this, the model allows simulation of any sequential combination of environmental, clothing, and metabolic conditions. Although the latter models with arterial systems (Fiala1999; Fiala et al. 2001; Tanabe et al. 2002; Huizenga et al. 2001) represent a step forward, they do not explicitly include a systemic circulation and the resulting inner convection occurring between the arterial blood and wall.

Other notable models include the one developed by The National Renewable Energy Laboratory (NREL) (Rugh et al. 2004), which contains a detailed simulation of human internal thermal physiological systems and thermoregulatory responses. Another multi-segmented human thermal model developed by Salloum et al. (2007) for bare human body consists of a comprehensive blood network. Flow rates are based on exact physiological data, real dimensioning, and anatomic positions of the arteries in the body. Holopainen (2012) combined the human thermal modelling with a thermal sensation and comfort model inside a simulated building. The models by Karaki et al. (2013) and Rida et al. (2014) also incorporate dynamic thermal response associated with arterio-venous anastomoses (AVA) functions.

Several works including Daanen (1991), Koscheyev et al. (1998), and Vanggaard et al. report that AVAs in the distal parts of the extremities play a significant role in the heat exchange with the environment. For example, exposure to extremely cold environment causing cold induced vasodilation (CIVD) to protect hands or feet from cold injury and very high-temperature environments causing heat-induced vasoconstriction (HIVC) so that the warm blood cannot reach the human core easily. However, as reported in Daanen (1991), these two exceptions for the AVA function during CIVD and HIVC do not apply to the vasoconstriction or vasodilation of the arterial system triggered by decreased or increased body core temperature.

The ability to appropriately characterize the inner convection between tissue and vessels, introduced here, has been a weak point in many recent models. Following the work by Smith (1991), Sun (2012) derived a comprehensive 3D model that is able to highlight the heat transfer for walking conditions. Although a blood network was included within solid tissues, this was under the assumption that blood nodes exchange heat with surrounding tissue only via conduction. Moreover, vessels were considered inelastic and thus pulsatile velocity was not accounted for. Ferreira and Yanagihara (2009) proposed a 3D conduction model, where arterial and venous flows are considered as reservoirs and the tissue temperature does not account for any flow temperatures. In subsequent work Ferreira and Yanagihara (2012) modelled heat transfer at a steady state in the upper limbs. Although tissue matter was modelled via partial differential equations, the conditions considered were stationary. Furthermore only two different tissues were used, while a reduced, arterial network was adopted.

In the literature, there is no clear evidence on the role of the venous system on the global thermal balance of the body. Indeed the importance of the heat exchanged between arteries and veins is still a matter of debate, as highlighted by an analytical model by Mitchell and Myers (1968), demonstrating no significant countercurrent effect in the human arm. This can be mainly justified because the distance between large arterial and venous vessels is significant; the high-velocity blood flow and the too short length of the vessels may affect further the countercurrent heat exchange. Vanggaard (1975) confirmed that countercurrent heat exchange is of minor importance in total heat exchange. He concluded that countercurrent heat exchange either had to be always 100 % effective or negligible, and naturally opted for the latter.

Some studies have reported heat transfer in the blood but without the surrounding body tissues. For example, Craciunescu and Clegg (2001) analyzed the effect of a pulsating blood velocity field on temperature. In their studies they obtained some important results on the relationship between the pulsating axial velocity and temperature profile and the effect of the Womersley number variation. In the work proposed by Bommadevara and Zhu (2002) a sophisticated bioheat transfer model representing neck is presented. They evaluate temperatures along common and internal carotid arteries for various environmental conditions. However, blood vessels in this study are treated as rigid tubes, and thus the effects of area variations are not accounted for. Ying et al. (2004) proposed a thermofluid model valid for a circulation system of the upper limb which involves arteries, capillaries, and veins. Here, the temperature is evaluated along the network by considering the effects of blood flow rate, transmural pressure, cross-sectional area, and elasticity. However, this model is not comprehensive as reflections due to variations in vessel topology and properties are not accounted. To enhance the accuracy of heat transfer predictions in flexible tubes and tube networks, a robust method has been proposed in Coccarelli and Nithiarasu (2015).

It is obvious from the reported studies that a step change in modelling approach is needed to more effectively address the human body bioheat transfer. It is also necessary to produce a new generation model that forms the basis for future development. In this regard, the present study combines a state-of-the-art systemic circulation model with heat transfer to a segmentation model for body tissue. The present model has been conceived as a combination of a multi-segmental solid model, derived mostly from Fiala et al. (2001) and the arterial modelling methodology proposed by Mynard and Nithiarasu (2008). For the thermoregulatory equations, we refer to Smith (1991). Coupling these components in a single model represents a novel approach to bioheat transfer studies, as it allows for computing the temperature simultaneously in each system, following a methodology recently reported by Coccarelli and Nithiarasu (2015). The blood flow is considered laminar, and a nonlinear wall law is used for describing blood-wall interaction. With these assumptions, the final expressions of governing equations depend only on cross-sectional area, velocity, and temperature of blood. The physical model of the arterial system is adopted from Low et al. (2012), where the flow and pressure distributions in the arterial system are extensively compared to measurement data. The methods for dealing with ventricle, valve, bifurcations, coronary arteries, and peripheral boundaries are detailed in Mynard and Nithiarasu (2008), Boileau et al. (2015) and not discussed in detail in the present work.

As the present methodology is based on two robust models, it is able to respond to a wide spectrum of conditions without losing integrity of the solution. An example to this effect is the straightforward calculation of conduction in the solid system. The inherent robustness of the proposed model is one of its main advantages when compared to relevant recent works such as Salloum et al. (2007) and Karaki et al. (2013). It would be fair to say that these modelling works have demonstrated good performance in simulating various situations, such as the evaluation of local thermal comfort and human physiological responses to cold water immersion. By comparison, the main aim of the proposed work is not to provide an analysis of the thermal performance of a specific subsystem, but to characterize the heat exchanges occurring within the multilayer solid tissues. A further aim is to investigate how the two intrinsically coupled subsystems interact whenever the body is exposed to various external conditions (especially during non-thermal neutral settings). We were thus able to demonstrate that, depending on the conditions, flow may have either a rewarming or cooling effects on the surrounding tissues. These results have emphasized the modulatory role of arterial inner convection. With regard to important processes involving CIVD and AVA, we are aware they would increase the prediction quality of our model. However for the target we aim, accounting for these processes would involve to disproportionately complicate the model. Indeed the present methodology was not developed just to improve on the accuracy of existing models, but rather to provide a robust tool that can probe temperature distribution along tissues and thus provide a different perspective in the study of bioheat transfer within the human body.

In the work presented here, the systemic blood circulation is embedded into a human body model. The human body model consists of multiple segments of solid cylinders representing head, neck, shoulders, thorax, abdomen, thighs, legs, arms, and forearms. The interface between fluid and tissue systems is explained below. The large arteries exchange heat with tissues only by convection and are subdivided into central vessels and transversal arteries. Each artery belongs to one or more predetermined segments. The central arteries are placed along a cylindrical segment axis, while transversal ones pass through the cylinder transversely in multiple directions. The energy exchange in and around small arteries in tissues is accounted for via perfusion. Each tissue layer of a segment is characterized by specific metabolism and perfusion rates that are adjusted by the thermoregulation system.

The numerical scheme used for solving the set of flow and temperature equations is the explicit form of locally conservative Taylor–Galerkin method (LCTG) (Nithiarasu 2004; Thomas and Nithiarasu 2008; Thomas et al. 2008). The conduction problem in the solid part along the radial direction is solved using an implicit finite difference method. Simulations are carried out on a bare human body.

EQUATIONS, COMPUTATIONAL METHOD, AND PHYSICAL MODEL

Governing Equations

The variables considered in the system are cross-sectional area (A), the average values of velocity (u) and temperature (T) over the artery cross section. Fluid pressure (p) is linked to area via the following nonlinear relation (Formaggia et al. 1999; Olufsen et al. 2000).

$$p = p_{\text{ext}} + \beta(\sqrt{A} - \sqrt{A_0}) \tag{1}$$

where pextpext is the external pressure acting on the walls of the tube, A0A0 is the unstressed cross-sectional area, and ββ is a characteristic property of elastic material, given as

$$\beta = \frac{\sqrt{\pi} h E}{A_0(1-\sigma^2)} \tag{2}$$

here h is the artery wall thickness, E is the Young's modulus of arterial wall, and σ is the Poisson's ratio.

In the flow model employed, the density (ρ) and viscosity (μ) are assumed to be constant. The thermal properties, specific heat (c_p), and thermal conductivity (k) of the blood are also assumed to be constant. Following some relevant works (Mynard and Nithiarasu 2008; Ying et al. 2004; Coccarelli and Nithiarasu 2015; Sherwin et al. 2003), the conservation laws for a flow in elastic vessels can be written in the following compact form:

$$\frac{\partial \bar{\mathbf{U}}}{\partial t} + \mathbf{H}\frac{\partial \bar{\mathbf{U}}}{\partial x} + \frac{\partial \bar{\mathbf{G}}}{\partial x} = \bar{\mathbf{S}} \tag{3}$$

with:

$$\bar{\mathbf{U}} = \begin{bmatrix} A \\ u \\ T \end{bmatrix}; \quad \mathbf{H} = \begin{bmatrix} u & A & 0 \\ \frac{\beta}{2\rho\sqrt{A}} & u & 0 \\ 0 & 0 & u \end{bmatrix};$$

$$\bar{\mathbf{G}} = \begin{bmatrix} 0 \\ 0 \\ -\alpha\frac{\partial T}{\partial x} \end{bmatrix} \quad \text{and} \quad \bar{\mathbf{S}} = \begin{bmatrix} 0 \\ -\frac{8\pi\mu}{\rho}\frac{u}{A} \\ \left(\frac{2h_{\text{con,in}}}{\rho c_p\sqrt{A/\pi}}\right)(T_t - T) \end{bmatrix} \tag{4}$$

where $\bar{U}$, $\bar{G}$ and $\bar{S}$ are the vectors of primitive variables, the diffusive and source terms, respectively, and H is the Jacobian matrix. In order to assign boundary conditions and to apply the Taylor–Galerkin method, it is convenient to write the whole system in a linearized de-coupled form. If diffusion and sources are considered negligible ($\frac{\partial \bar{G}}{\partial x} = 0$ and $\bar{S} = 0$), the characteristic variables of Eq. (3) may be obtained as in Mynard and Nithiarasu (2008), Coccarelli and Nithiarasu (2015), Formaggia et al. (1999), Sherwin et al. (2003)

$$w_1 = u + 4\sqrt{\frac{\beta\sqrt{A}}{2\rho}}; \quad w_2 = u - 4\sqrt{\frac{\beta\sqrt{A}}{2\rho}}; \quad w_3 = T \tag{5}$$

Rearranging the expressions in Eq. (5), it is possible to express A and u by means of characteristic variables as

$$A = \frac{(w_1 - w_2)^4}{1024}\left(\frac{\rho}{\beta}\right)^2; \quad u = \frac{1}{2}(w_1 + w_2) \tag{6}$$

These relationships are employed at the boundaries to apply boundary conditions.

In the current model a local thermal equilibrium between the venous blood and the tissue temperatures is assumed. The heat transfer by perfusion is assumed to be proportional to the temperature difference between arterial blood entering the tissue and the tissue. In order to evaluate tissue temperature (T_t), the one-dimensional bioheat transfer equation in cylindrical coordinates is solved. For a single layer, the heat conduction is described with the following expression (Pennes 1948)

$$\rho_t c_t \frac{\partial T_t}{\partial t} - k_t \frac{1}{r}\frac{\partial}{\partial r}\left(r\frac{\partial T_t}{\partial r}\right) = q_m + \phi\rho c_p (T - T_t) \tag{7}$$

In the above equation, r is the radial coordinate, q_m is the volumetric heat generation associated with the metabolism, and ρ_t, c_t, k_t, ϕ are the density, specific heat, thermal conductivity, and perfusion coefficient of the tissue, respectively. The metabolism term q_m represents mainly the energy generation due to biological processes. If the body is subjected to work, an enhancement of metabolism in muscle tissue occurs.

Numerical Schemes

In this section, a brief overview on the numerical method is provided. Equation (3) requires a scheme with a stabilization term to obtain a stable solution. Thus, in this study the locally conservative Taylor Galerkin (LCTG) method is used (Nithiarasu 2004; Thomas et al. 2008). Applying the LCTG method to mass and momentum equations, it is possible to obtain (Mynard and Nithiarasu 2008)

$$[M_e]\{\Delta U\}^{n+1} = \Delta t\left([K_e]\{F\}^n + [L_e]\{S\}^n + f_{\Gamma_e}{}^n\right) \tag{8}$$

where $[M_e]$, $[K_e]$ and $[L_e]$ are the mass matrix, coefficient matrices for convection, Taylor–Galerkin, and source terms, respectively. Each of these are 2×2 matrices (for each equation), and this system of equations is solved on individual elements, independent of surrounding elements. Information is transmitted between elements via the numerical flux term that is imposed, for each element, on the boundary (Mynard and Nithiarasu 2008). The 2×2 matrices can be evaluated and inverted at the preprocessing stage, which removes the need for any matrix inversions.

Applying LCTG method to the energy equation gives (Coccarelli and Nithiarasu 2015)

$$[M_e]\{\Delta T\}^{n+1} = \Delta t\{([K_{eT}] + [D_{eT}])\{T\}^n + [L_{eT}]\{T - T_t\}^n + q^n_{\Gamma_e}\} \tag{9}$$

where the matrix $[D_{eT}]$ is the coefficient matrix for diffusion and q_{Γ_e} is the numerical conduction flux exchanged between two adjacent elements (Nithiarasu 2004).

The time step restrictions for the fluid solver may be computed using the condition (Mynard and Nithiarasu 2008):

$$\Delta t = 0.9\frac{\Delta x_{\min}}{c_{\max}}. \tag{10}$$

where Δx_{min} is the minimum element length and c_{max} is the maximum intrinsic wave speed. For the problem of heat conduction through the surrounding tissue, the forward Euler method is used. At the interface between two layers, continuity of flux is imposed. Since the matrix of the linear system is tridiagonal, Thomas algorithm is used to solve the system.

Physical Model Architecture and Boundary Conditions

The human body bioheat transfer may be modelled using a combination of "passive" and "active" systems. The passive part consists of transport in arteries and solid tissues, and the active system is the thermoregulatory part of the model that attempts to keep the body temperature within a predetermined threshold.

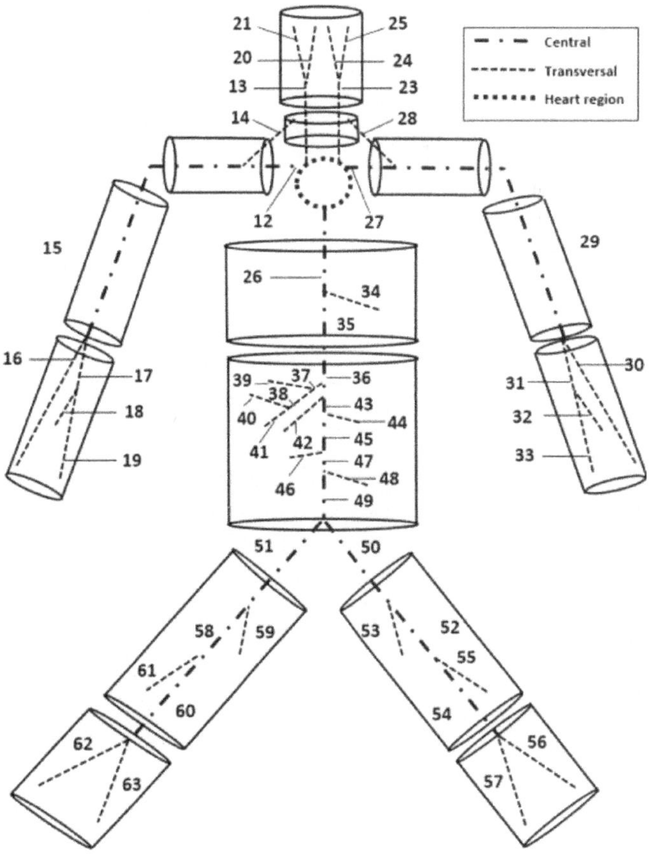

Figure 1: Arterial network considered.

Arterial Network

The systemic circulation is subdivided into large and small vessels. The large arteries are shown in Fig. 1 as proposed in Low et al. (2012). In the present study, only major arteries are included. The micro-circulation is represented by tapering vessels at the extremities of the network (Mynard and Nithiarasu 2008) and the energy exchange occur only through perfusion mechanism. The venous system is not included for the reasons mentioned previously. The whole network is composed by 91 segments (28 are tapering vessels), 6288 elements, and 6379 nodes. Full details about the parameters and dimensions of the network are reported in Low et al. (2012).

Table 1: Tissue distribution within body

Cylinder	Tissues	Layer radii (cm)	Length (cm)
Head	Brain, bone, fat, skin	6.6, 7.6, 7.8, 8.0	23.5
Neck	Bone, muscle, fat, skin	1.9, 5.4, 5.6, 5.8	7.9
Thorax	Lung, bone, muscle, fat, skin	7.7, 8.9, 12.3, 12.6 12.9	15.6
Abdomen	Viscera, bone, muscle, fat, skin	7.9, 8.3, 10.9, 12.4, 12.6	24.8
Shoulder	Bone, muscle, fat, skin	3.7, 3.9, 4.4, 4.6	13.4
Arm	Bone, muscle, fat, skin	1.5, 3.4, 4.0, 4.2	29.6
Forearm	Bone, muscle, fat, skin	1.5, 3.4, 4.0, 4.2	23.7
Thigh	Bone, muscle, fat, skin	2.2, 4.8, 5.3, 5.5	58.5
Leg	Bone, muscle, fat, skin	2.2, 4.8, 5.3, 5.5	34.3

Note that the thorax length is smaller than the real average size as heart region is not included

Inlet Flow Conditions

Modelling heart's pumping action is implemented using the method proposed in Mynard and Nithiarasu (2008), Low et al. (2012). The action of the heart allows one to set inlet boundary conditions and the system includes the left ventricle (LV) and aortic valve (AV) models. The LV is treated as a prescribed forward pressure source, which describes the cardiac cycle and the number of heart beats per unit time or heart rate (HR). The input of the model is a ventricular (forward) pressure prescribed in the ventricle's point just before the valve. Prescribing inlet and outlet variables is carried out by means of characteristic variables. Rearranging formulations in Eqs. (1) and (5) and by prescribing forward pressure (p_{in}), it is possible to evaluate the forward characteristic at the inlet (w_{1in}):

$$w_{1\text{in}}^{n+1} = w_2^0 + 4\sqrt{\frac{2}{\rho}}\sqrt{\left(p_{\text{in}}^{n+1} - p_{\text{ext}}\right) + \beta\sqrt{A_0}} \tag{11}$$

where w_2^0 is the initial value of w_2 and is also equal to the value of w_2 at any time if no backward-running waves reach the inlet. The backward characteristic variable (w_2) may be evaluated via linear extrapolation in the x–t plane, where for the next time step n+1,

$$w_2^{n+1}|_{x=x_0} = w_2^n|_{x=x_0-\lambda_2^n \Delta t} \tag{12}$$

Primitive variables A and u at the inlet node can be evaluated by using Eq. (6). The behavior of the AV is represented by a time-varying transmitter and reflector at the inlet. For each impedance of the network, a characteristic reflection coefficient (R_z) could be defined as

$$R_z = -\frac{\Delta w_2}{\Delta w_1} = \frac{w_2^{n+1} - w_2^0}{w_1^{n+1} - w_1^0} \tag{13}$$

where w_1^0 is the initial value (corresponding to no-pulse situation). Including the contribution from the AV, the total forward characteristic variable $(w_{1\text{in}}^*)$ can be written as

$$w_{1\text{in}}^* = w_{1p} + w_{1r} + w_1^0 \tag{14}$$

where w_{1p} is the change in the incoming characteristic associated with the ventricular pump and w_{1r} is the change associated with backward-travelling waves that are partially or completely reflected from the valve.

Using Eq. (13) to model the AV impedance, it is possible to write:

$$w_{1r} = R_{Vr}(t)\Delta w_2 \tag{15}$$

where $R_{Vr}(t)$ is a time-varying valve reflection coefficient for backward-travelling waves. It is assumed that $R_{Vr}=0$ when the valve is open, $R_{Vr}=1$ when it is closed, and that this value varies exponentially when the valve is opening or closing. Further details on the boundary conditions may be obtained from Mynard and Nithiarasu (2008), Low et al. (2012).

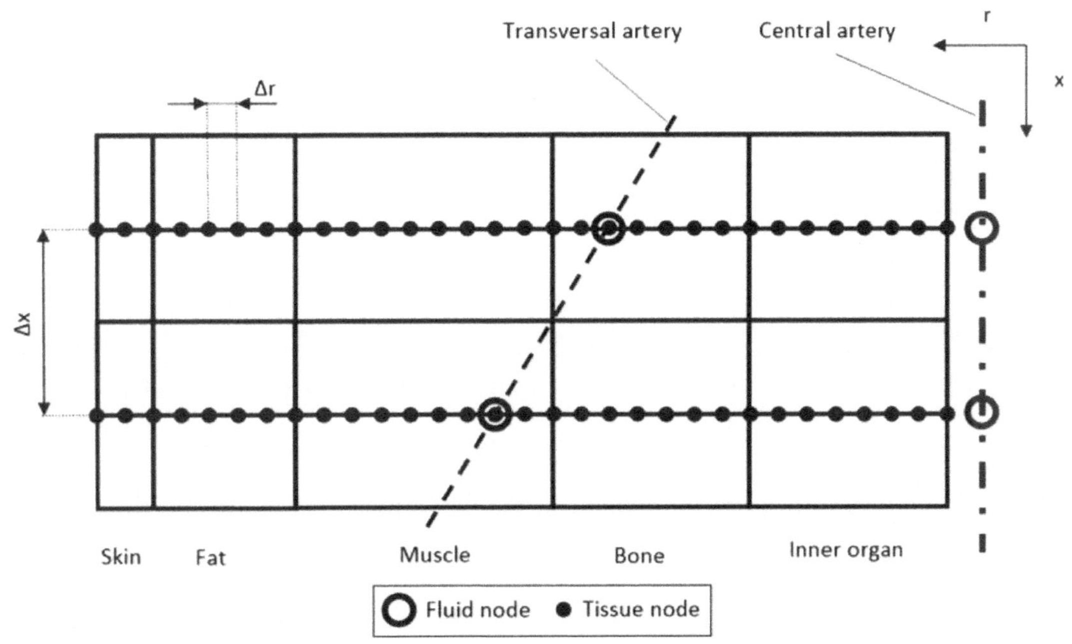

Figure 2: Longitudinal and radial discretizations for one cylinder.

Extremities and Branching Points

To model branch ending, tapering vessels are used. These terminal tubes present multiple step decreases in A_0 or step increases in β. Thus characteristic reflections of the downstream vasculature are accounted for. Calculation of the backward characteristic variable on the exit node may also be performed by prescribing reflections at the exit. The reflection coefficient of the terminal vessel (R_t) can be determined again by means of Eq. (13), while the value of w_1 for the next time step (t=n+1) is extrapolated. Thus the unknown (w_2^{n+1}) is

$$w_2^{n+1} = w_2^0 - R_t\left(w_1^{n+1} - w_1^0\right) \tag{16}$$

In the present work, R_t is set equal to 0 for each tapering vessel. Further details on the flow boundary conditions may be obtained from relevant published work (Mynard and Nithiarasu2008). At the extremities, we assume that the incoming flow is in thermal equilibrium with the surrounding tissue nodes. When the flow is outgoing from the system, the temperature at the downstream point is assigned by characteristic variable extrapolation. As tapering vessels used at the exit boundaries represent microcircualtion, adiabatic conditions are assumed within these vessels (no heat exchange through convection occurs). A robust modelling requires consideration of branching points such as bifurcations or discontinuities in geometrical and material properties. The works of Mynard and Nithiarasu (2008) and Coccarelli and Nithiarasu (2015) are adopted in the present work.

Solid Tissues

For the solid tissue representation, we follow the work of Fiala (1999). The body consists of 14 multilayered cylindrical elements representing head, neck, shoulders, thorax, abdomen, arms, forearms, thighs, and legs (details are reported in Table 1). The segments representing shoulders, legs, thighs, arms, and forearms are constituted by four layers of materials with different properties; from inside to outside the cylinder consists of bone, muscle, fat tissue, and skin layers. In the head, thorax and abdomen segments, inner organs brain, lung and viscera respectively are also included. We notice that some geometrical parameters differ because cylinders have been adapted to the arterial tree.

We note that, in contrast to Fiala (1999), we model head as a cylinder and not like a sphere; thus, the layer radii are resized keeping the head volume constant.

Coupling Blood with Solid Systems

The coupling between the blood vessels and surrounding tissue is critical to obtain sensible results. As mentioned previously a one-dimensional bioheat transfer model along the radial direction of body segment is used. The blood vessels are embedded into these segments as shown in Fig. 1. This approach is considerably more advanced than the common assumption of a single core node, which implies that in each cylinder all type of convection losses are depending only on a scalar value. Furthermore, such a model is a good compromise between computational cost and accuracy (Charny et al. 1990).

The artery locations within the solid body are estimated from Uflacker (2006). As reported previously, the heart represents the inlet to the fluid system and is not part of any cylindrical segment. The large arteries proposed in Low et al. (2012) are subdivided into three categories of heart region, central and transversal vessels. As arteries in the heart region are not included in the tissue discretization, no heat transfer with solid tissues occurs; the only exception is represented by the inlet flow node (which is isothermal with surrounding tissues). Each and every central artery is assumed to coincide with the axis of one or more cylindrical segments, while transversal vessels cross transversely one or more cylinders. The arrangement of these vessels within the cylindrical segment is shown in Fig. 1, where central arteries are depicted as chained lines while transversal ones are represented by dashed lines. Figure 2 shows a typical section of the cylindrical segment with embedded central and transversal arteries. It should also be noted that while the geometrical and mechanical properties of elastic vessel may be allowed to change along the longitudinal coordinate, the solid tissue properties along the axial direction of the cylindrical segment are fixed. In addition, cylindrical segments are not considered deformable. The geometrical, thermophysical, and basal physiological properties of tissue materials and the body features are adopted from Fiala (1999). The inner wall heat transfer coefficient is set up following (Shitzer and Eberhart 1985) (Nusselt number is assumed be equal to 4).

Figure 2 clearly shows the spatial discretization adapted in the present study. The body is assumed to be axisymmetric, and the nodes of the central arteries are linked to the first node of the surrounding tissue layers as shown via a convective boundary condition. For every node along the central artery, a matching radial set of nodes are introduced into the surrounding solid tissues. Each central vessel node is therefore identified by two coordinates, a longitudinal and radial coordinate. As shown, the transversal vessels are embedded into the cylindrical segment and at the intersecting point of tissue mesh and transversal artery, a flow node is introduced that coincides with the solid node (see Fig. 2).

The temperature calculation at fluid–solid interface nodes includes the following steps. The temperature transported through the systemic circulation network forms the basis for the boundary condition to Eq. (7). The fluid inlet node (first node of seg. 1) is assumed to be in thermal equilibrium with a tissue node located in the middle of the thorax, having radial coordinate equal to 8 cm. The nodal temperatures of the central arteries provide the wall temperature for the convective boundary condition between the blood and arterial wall (first tissue node). Where a transversal vessel node coincides with the tissue node, a volumetric source term is explicitly evaluated based on the expected convection contribution and added to the discrete heat conduction equation of the tissue (increasing the term q_m). In cylinders representing head, neck, legs, and forearms there are no central arteries. Thus, along the axis of these cylindrical segments, an adiabatic condition is adopted.

The perfusion in solid tissue segments is modelled through the perfusion coefficients (see Eq. 7). The temperature difference in the perfusion term is calculated as the difference between the section average blood (mean between all vessels crossing the section) and tissue temperatures. Equation (7) is applied to all tissue nodes by setting the appropriate material constants k_t, ρ_t, and c_t, q_m and ϕ. It should be noted that q_m and ϕ are variables regulated by the thermoregulatory system (for further details see Sect. 2.4). In the present study, the tissue temperatures are computed after the evaluation of blood temperatures every time step. All the components of q_m are evaluated before computing the tissue temperature at each time step.

The respiration losses are incorporated by considering a negative volumetric heat source q_{bre} at all lung nodes. To estimate such losses the following formulation has been used (Smith 1991)

$$q_{\text{bre}} = \frac{1}{V_{\text{lung}}}[0.0014\ Q_{m,\text{glob}}\ (34 - T_{\text{out}}) + 0.0173\ Q_{m,\text{glob}}\ (5.87 - P_{\text{out}})]\ (W/cm^3) \quad (17)$$

where $Q_{m,glob}$ is the global metabolic heat generation rate, V_{lung} is the lung volume (respectively 58.2 W/m² and 5631.41cm³) and Pout is the ambient water vapor pressure. Further details may be found in Smith (1991).

Heat Exchanged with the Environment

The body exchanges heat with the environment through the skin and breathing. The skin is represented by the outer most part of the cylindrical segment. The flux exchanged between the skin layer and outside environment q_{skin} is the sum of the convection ($q_{con,out}$), radiation (q_{rad}) and evaporation (q_{eva}) losses. The Neumann boundary condition used in the present study is

$$-k_t A_{\text{out}} \frac{\partial T_s}{\partial r}|_{r_{\text{out}}} = q_{\text{con,out}} + q_{\text{rad}} + q_{\text{eva}} \quad (18)$$

For the evaluation of $q_{con,out}$ and q_{rad} the methodology proposed by Fiala (1999) is followed. The convective heat transfer between skin node and the external environment may be evaluated with the following expression.

$$q_{\text{con,out}} = h_{\text{con,out}}(T_t(r_{\text{out}}) - T_{\text{out}}) \quad (19)$$

where $h_{out,con}$ is the convection heat transfer coefficient and it is a function of the node location in the body, the air velocity and the temperature difference between the outer surface and environment. For the radiative exchange the evaluation of the mean temperature of the surrounding surfaces ($T_{sur,m}$) is necessary before applying

$$q_{\text{rad}} = h_{\text{rad}}(T_t(r_{\text{out}}) - T_{\text{sur},m}) \quad (20)$$

where h_{rad} is the radiative heat transfer coefficient depending on the temperatures, the emission coefficients and the view factors of the surrounding surfaces considered. As sweating (evaporation) is part of the thermoregulatory system, it is discussed in Sect. 2.4 below.

Control system

Studies show that a state of thermoneutrality exists when the core and mean skin temperatures of the body are respectively 36.8 and 33.7∘C33.7∘C (Holopainen 2012). When an imbalance in energy exchange between the body and environment occurs, the thermoregulatory system is activated to maintain the body homeostasis. The core body temperature is controlled by the thermoregulatory system consisting of thermoreceptors and hypothalamus. Three control mechanisms, shivering (lower skin temperatures), sweating (higher skin temperatures), and vasomotion (flow control), are considered here. We define T_{core} as the mean temperature between the first layer inner nodes of head, neck, thorax and abdomen, while T_{skin} is the average value on the skin surface. With such integral variables, we can evaluate the shivering and vasomotion contributions (Smith 1991). The shivering heat per unit volume, q_{shiv}, may then be obtained by dividing the total segmental heat production by muscle volume. In the present study, the basal and vasomotor blood flows are taken from Smith (1991). The corresponding perfusion rate ϕ is evaluated by dividing the flow rate by the skin mass of the segment considered. The total evaporative heat loss q_{swe} is computed following (Refrigerating American Society of Heating and Air-Conditioning Engineers 1993). For such calculations we assume that the vapor pressure on skin is equal to that of saturated water vapor at skin temperature Fanger (1970), while the evaporative heat transfer coefficient (h_{swe}) is taken from Kerslake (1972). We include also the clothing model proposed in Holopainen (2012).

Table 2: Fluid parameters and properties used in the simulations

Density of fluid, ρ (g/cm^3)	1.060
Viscosity of fluid, μ (poise)	0.035
Thermal conductivity of fluid, k (W/cm°C)	0.050
Specific heat of fluid, c_p (J/g°C)	3.900

Table 3: Solid properties used in the simulations

Tissue	c_t (J/g K)	$q_{m,0}$ (W/cm^3)	ρ_t (g/cm^3)	k_t (W/cmK)	ϕ (1/s)
Brain	3.850	0.013400	1.080	0.0049	0.011320
Lung	3.718	0.000600	0.550	0.0028	0.004310
Viscera	3.697	0.004100	1.000	0.0053	0.000500
Bone	1.700	0.000000	1.375	0.0075	0.000000
Muscle	3.700	0.000727	1.085	0.0042	0.000538
Fat	2.300	0.000003	0.850	0.0016	0.000004
Skin	3.680	0.001096	1.085	0.0047	*Variable*

For the cutaneous perfusion, we adopted a specific coefficient for each cylinder [more details can be found in Holopainen (2012)]

RESULTS AND DISCUSSION

Although the model proposed in the present work is new and comprehensive, different components of the model have undergone extensive testing in the past. The systemic circulation model, for example, has been extensively used in different studies and compared against experimental flow and pressure measurements (Mynard and Nithiarasu 2008; Low et al. 2012). The temperature transport in the one-dimensional flexible pipes has been studied in detail recently (Coccarelli and Nithiarasu 2015). Thus, the focus of the results in the present study is the bioheat transfer within a human body for various governing parameters. We consider a bare body and thus $R_{swe,cl}$ and f_{cl} are set respectively equal to 0 and 1 (see "Appendix"). For the cases considered, we assume the same radiative parameters presented in Fiala (1999), while air velocity (v_{air}) is set equal to 4 m/min. The initial temperature at all nodes are set at 36.8°C36.8°C in order to reflect an initial thermo neutral condition. The fluid properties and outside conditions used in the study are listed in Tables 2 and 3.

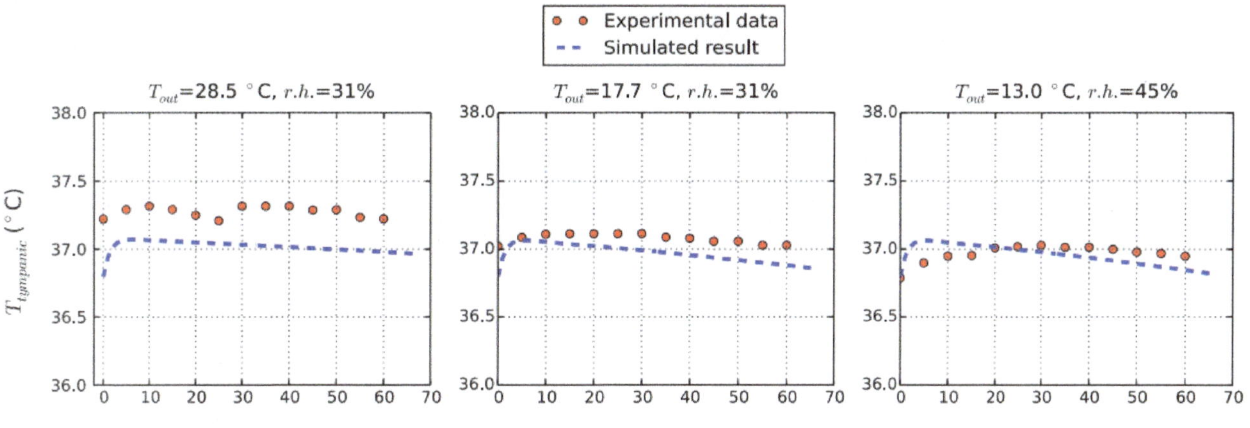

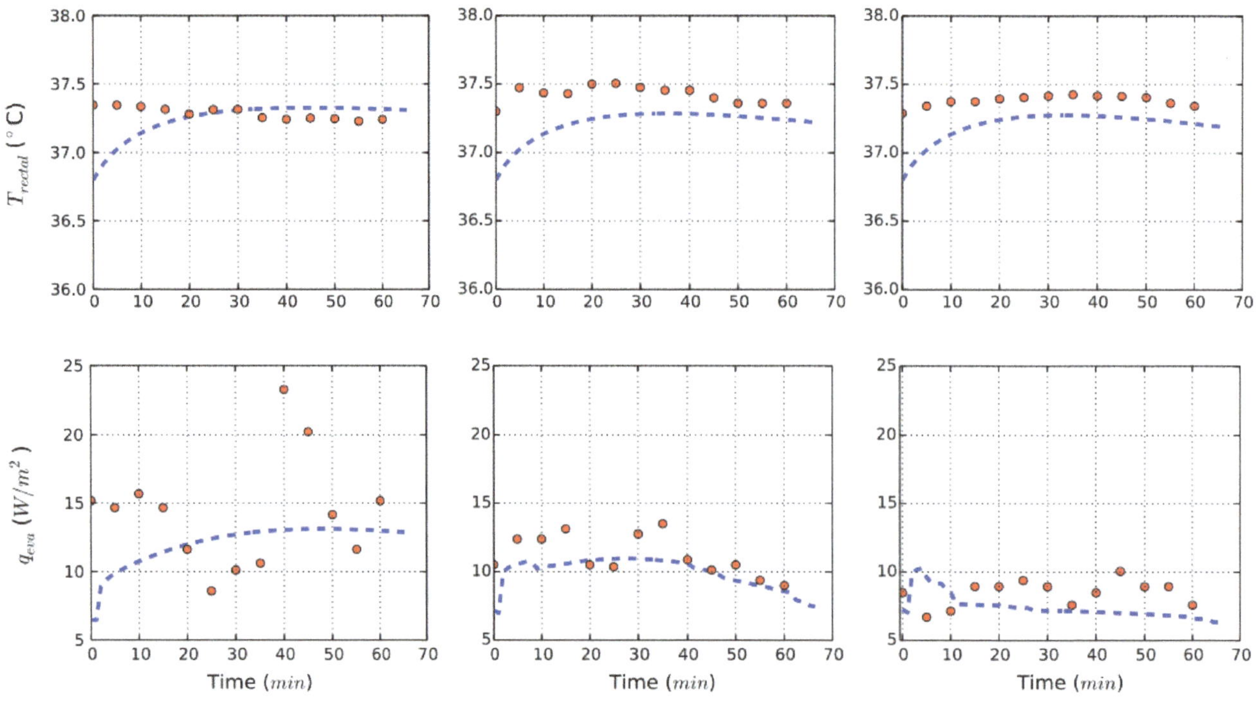

Figure 3: Tympanic and rectal temperatures for various external conditions.

In the following subsection, a comparison of the current model for various atmospheric conditions against measurement is provided. This is followed by an investigation on the contribution of inner convection to the body thermal balance and finally the thermoregulatory response of the body in a cold environment is quantified.

Comparison against Measurements

At first a validation of the model with experimental data is presented. For doing this the relevant works by Stolwijk and Hardy (1966), and Hardy and Stolwijk (1966) are used. In these studies volunteers have undergone to various environmental conditions. Temperatures were recorded for the tympanic and rectal regions and also the evaporative losses (sweating and breathing latent losses) were evaluated. In order to test systematically the current model, we simulate the body response for three different external exposures. The considered conditions are (T_{ext}=28.5°C–r.h.=31%), (T_{ext}=17.7°C–r.h.=31%) and (T_{ext}=13.0°C–r.h.=45%). For the rectal temperature calculation, we use the tissue node at an axial distance of 22 cm from the top of abdominal cylinder and at a radius of r=3.5 cm. The tympanic site is assumed to be at a distance of 12 cm from the bottom of the head cylinder and at r=5.0 cm. The evaporative losses are evaluated by summing the contributions of each cylinder section and then dividing by the total skin surface. We note that the initial temperature field imposed slightly differs from the one of a body under thermoneutral conditions. However, after a long transient all results have to converge to the same value range.

In Fig. 3 the time evolutions of tympanic, rectal temperatures and evaporative losses are reported. For all exposure conditions considered, the simulated results match very well the experimental data. It is possible to see that the temperature errors decrease significantly with the time. At quasi-steady state, the largest difference in temperature is less than 0.25°C0. The accuracy of the evaporative losses calculated is difficult to evaluate as the experimental data is widely scattered.

Next, we report the thermal body response under controlled external conditions providing comparisons with experimental measurements and other numerical models. Specifically, the model is tested under exposure to heat for 1h at (28.1°C,43% R.H.), 2 h at (47.8°C,27% R.H.) and 1h at (28.3°C,44% R.H.). Findings for these simulated conditions are compared with experimental data (Hardy and Stolwijk 1966) and solutions provided by "Smith" and "Karaki" models, respectively, presented in Smith (1991), Karaki et al. (2013). The core and mean skin temperature behaviors in time are shown in Fig. 4. Our results are inline with the expectations. As seen the mean skin temperature curve rises suddenly as the step change occurs, but then it remains within an acceptable range of temperatures.

The model is also tested when the naked body is exposed to cold conditions. The core temperature prediction is compared against the findings reported in one of the most recent work (Karaki et al. 2013). Here the body is exposed to (13°C, 45% R.H.) for 65 min. The results are reported in Fig. 5. As seen, a good agreement with experimental data is evident.

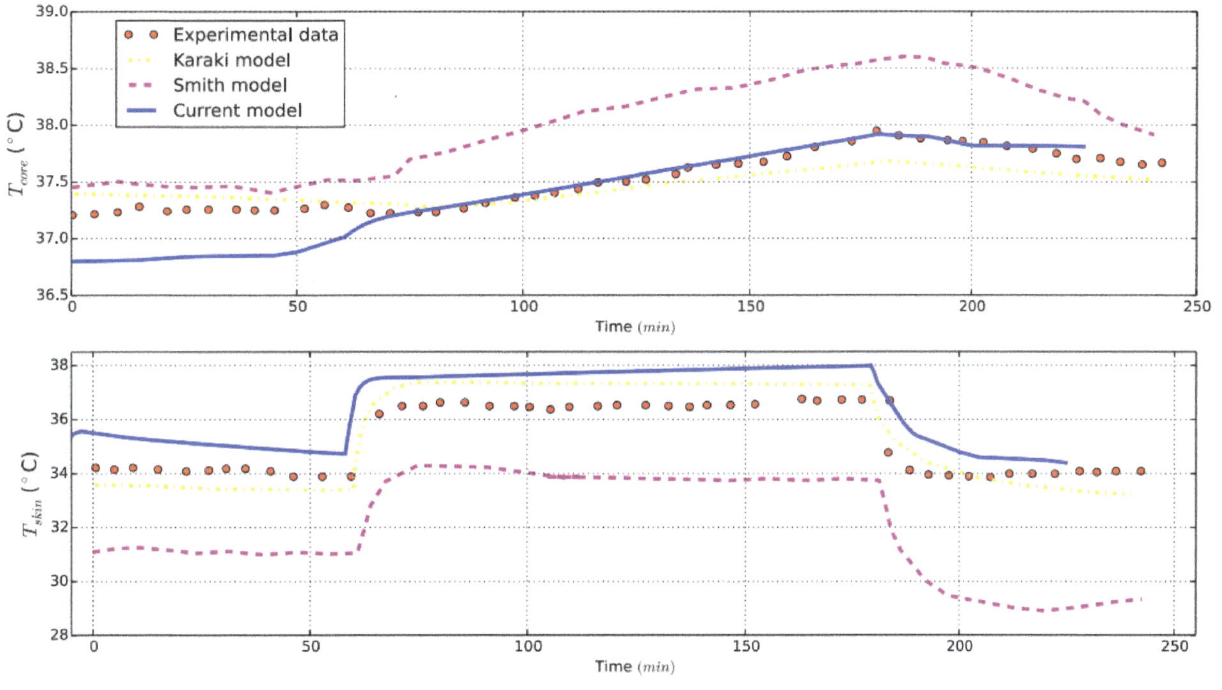

Figure 4: Benchmark case for naked body under hot exposure.

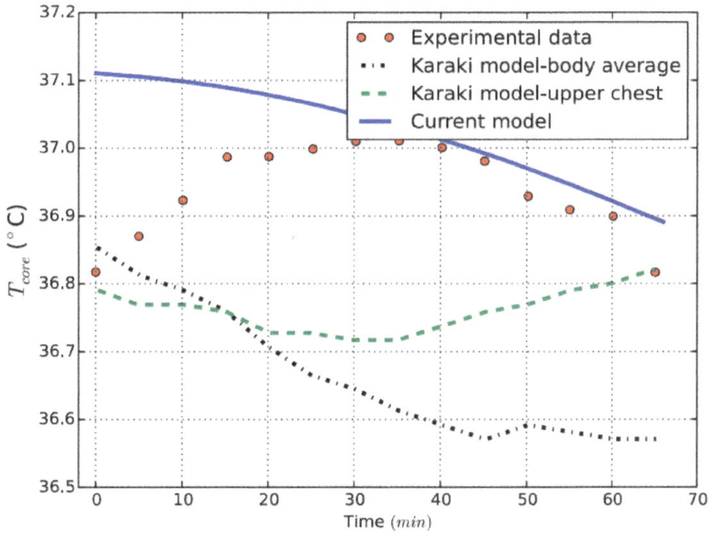

Figure 5: Benchmark case for naked body under cold exposure.

Role of Inner Convection

To understand the temperature changes in blood, four representative arteries, abdominal aorta II (seg. 43, abdomen), left external carotid (seg. 25, head), right external iliac (seg. 58, right thigh), and right radial (seg. 16, right arm), are selected (for more detail about artery labelling see Low et al. 2012). The temperatures at these locations are recorded once a quasi-steady state is reached. The tissue temperature distributions are recorded for the sections corresponding to the nodes selected in the mentioned arteries (abdomen, head, thigh, and arm).

Since the flow is pulsatile in nature (Low et al. 2012), pulsatility of temperature is also anticipated. In addition, the wave nature of the flow leads to reflected temperature waves. Although a number of different parameters such as elastic properties of the vessels can be tested using the proposed model, all the material properties, heart rate and flow boundary conditions at the extremities are fixed to produce an understanding of a normal human body behavior. Note that describing bioheat transfer in a body subjected to some disease states or extreme environmental conditions needs parameter changes.

Figure 6 shows the blood temperature at three selected monitoring points in the systemic circulation. As anticipated, the temperature follows a mild periodic pattern inline with the velocity changes. As seen the frequency and amplitude of oscillations differ for different environmental conditions. In general the amplitude of the temperature waves is low, and thus no dramatic local change in temperature is possible. The blood temperature is mildly influenced by the atmospheric temperature in the core part of the body. The pronounced effect in the radial artery is due to the smaller dimensions of the forearm and to the absence of any metabolic active tissue.

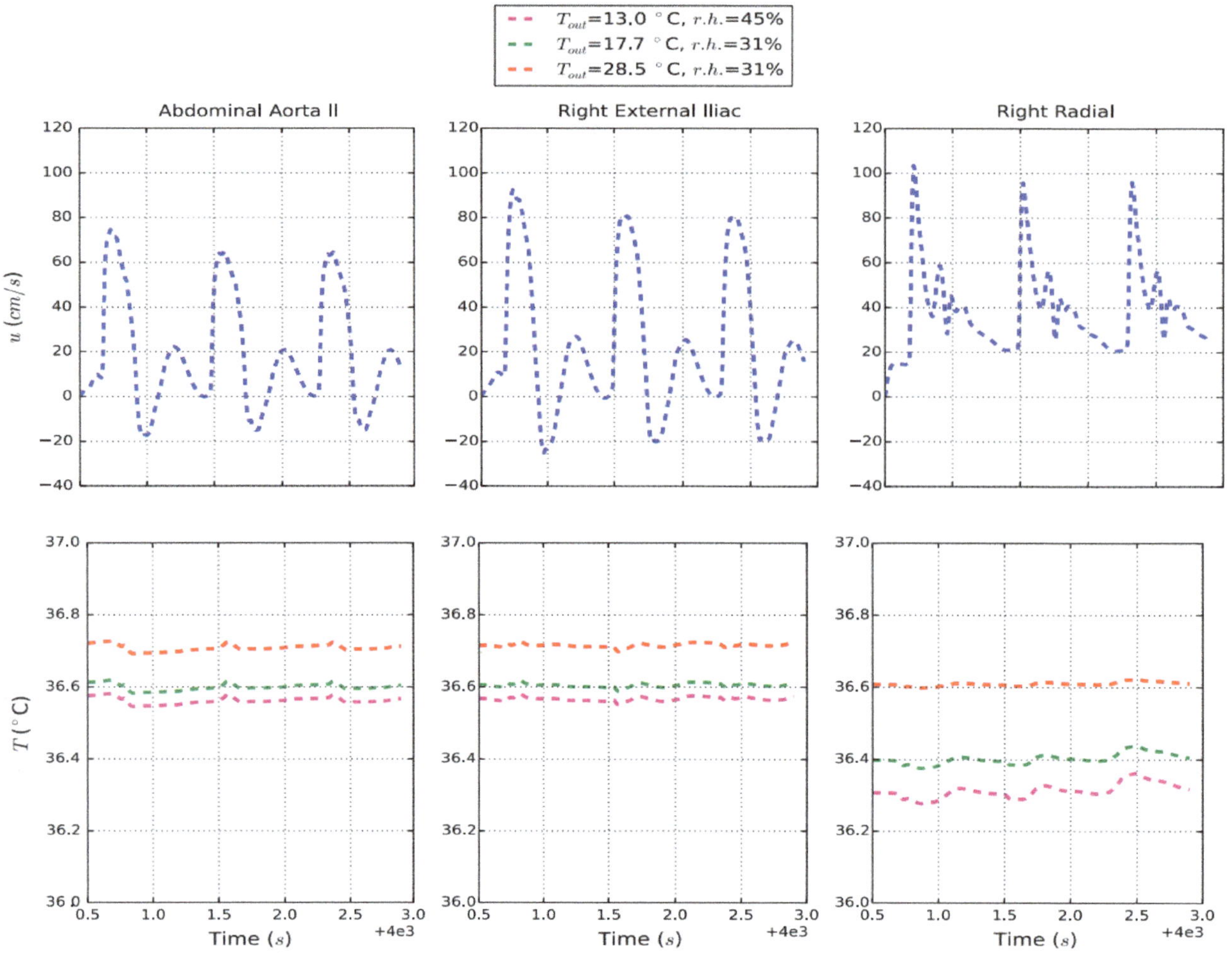

Figure 6: Blood velocity and temperature along the arterial tree for various external conditions.

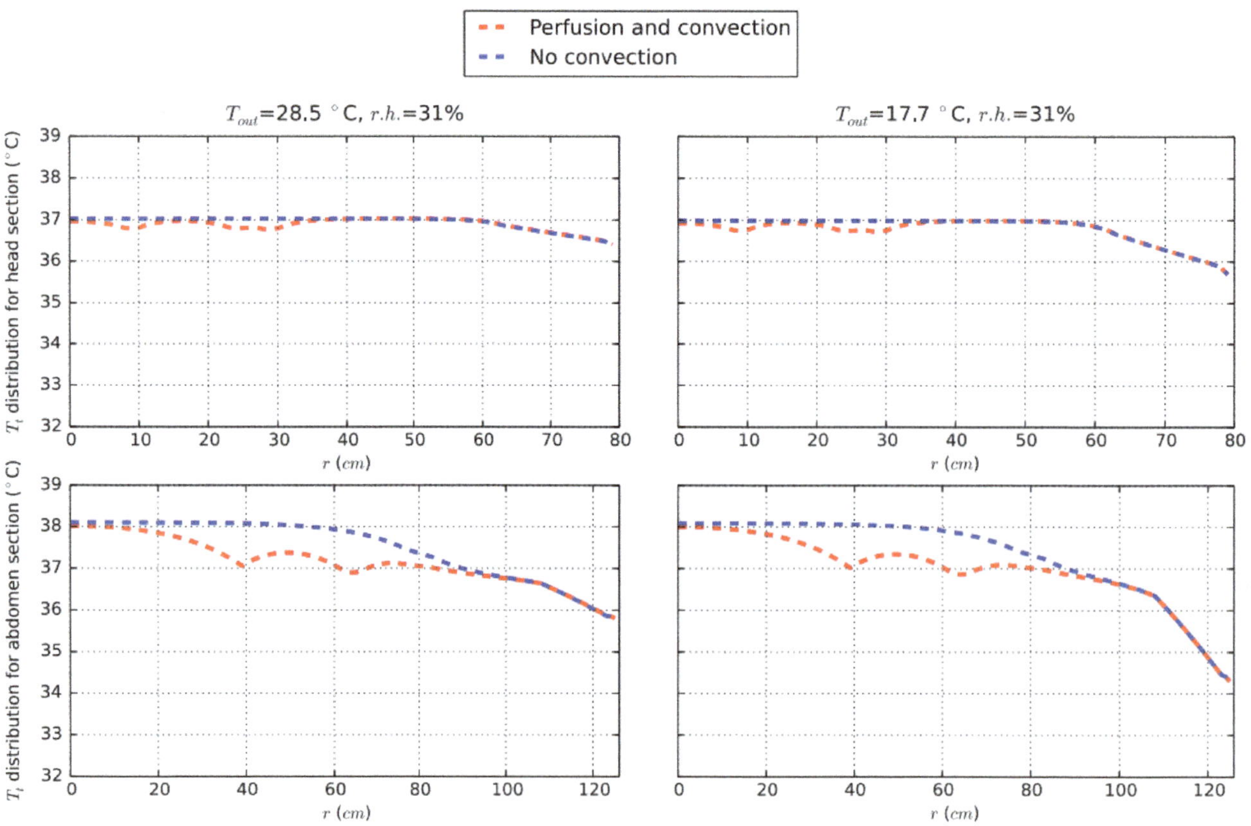

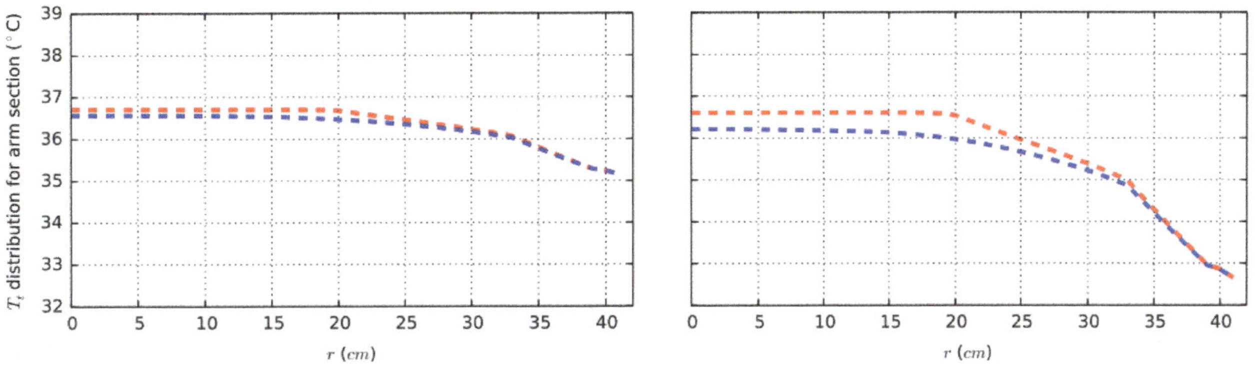

Figure 7: Tissue temperatures for two different modelling approaches at t = 33.0 min

In order to evaluate the effect of heat convection on tissues, the results obtained from the proposed model are compared to the approach used in other reference works (Fiala 1999; Tanabe et al. 2002) where heat conduction is exclusively used to model heat transfer occurring between blood and tissue system. Figure 7 shows the temperature distribution with and without heat convection and perfusion in arteries. As seen a local temperature variation of more than 1.0°C is observed in tissues in the abdominal area. Although this variation decays as we approach the skin layer, this finding is important for further investigation. In the abdomen and head, convection involves a smaller average tissue temperature compared to the case without convection. The situation in the arm instead is the opposite. This can suggest that, with convection, a more uniform energy redistribution is enforced. It can therefore be reasonably concluded that flow and convection heat transfer play an important regulatory role that may be further enhanced in abnormal conditions such as high blood pressure and stiffer arteries.

Influence of Thermoregulation

It is often difficult to evaluate the effect of thermoregulation as this is highly coupled with different external parameters. Thus, in this section an example is provided to demonstrate the effect of thermoregulation when the body is subjected to a cold exposure. For doing this we consider also a case in which all control mechanisms (shivering, cutaneous vasomotion and sweating) are shut down. All other parameters are assumed to be the same as the previous subsections.

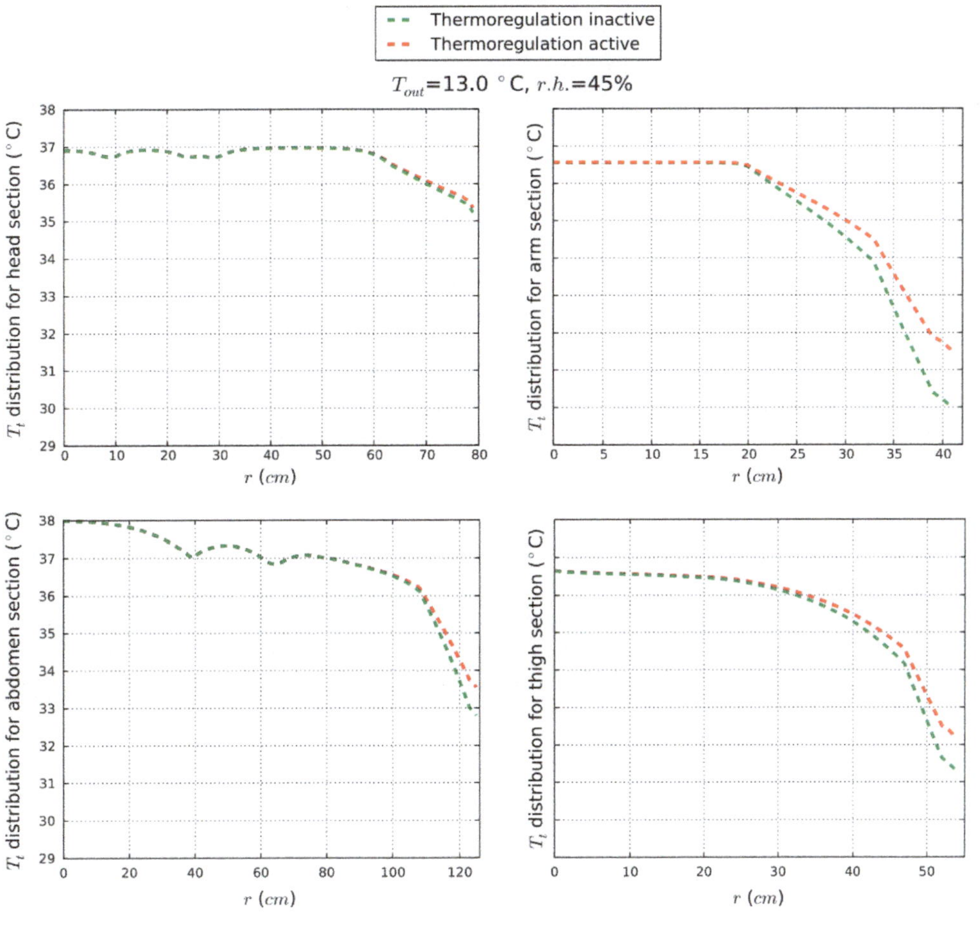

Figure 8: Thermoregulation effects on tissue temperatures at t = 33.0 min.

Figure 8 highlights the influence of the thermoregulatory system in all the four regions considered. While the core temperature remains approximately the same, the temperature at the periphery has dropped without thermoregulation. Most consistent temperature variations occur in the peripheral body cylinders (arm and thigh). The reduction after 33.0 min is as high as 1°C. The shivering effect is not included as the core temperature needed to trigger shivering has not been reached. For a longer time or more extreme external conditions, such profiles could change significantly.

CONCLUSIONS

A novel, next-generation bioheat model for the human body has been developed and tested. The systemic circulation embedded human body model is more comprehensive than existing models. Further improvements are nevertheless possible by including more arteries and veins. The proposed model in its present form can test various parameters including artery stiffness, blood pressure, various dimensions, tissue properties, surrounding conditions and many more. The results produced very clearly highlight the effect of arterial heat convection on the surrounding tissues. The heat convection and perfusion enhances the energy exchange between the blood and surrounding tissues. As expected surrounding temperature changes significantly affect the skin temperature; however, the control system limits the rapid variation of temperature whenever outside temperature is far from thermally neutral conditions.

There are numerous potential applications of the proposed model, such as better understanding of hyperthermia/hypothermia and the detailed study of resulting temperature transport and distribution. Besides these applications, the proposed model can study the influence of disease conditions such as hypertension and arterial dysfunction and also ageing on energy exchange. The model can also be used to study changing environments as a condition for enhanced quality of life.

ACKNOWLEDGMENTS

The authors acknowledge the financial support provided by the Ser Cymru National Research Network in Advanced Engineering and Materials.

REFERENCES

1. Boileau E, Nithiarasu P, Blanco JB, Muller LO, Fossans FEE, Helleviks LR, Doners WP, Huberts W, Willemet M, Alastruey J (2015) A benchmark study of 1-d numerical schemes for arterial blood flow modelling. Int J Numer Methods Biomed Eng 31. doi:10.1002/cnm.2732
2. Bommadevara M, Zhu L (2002) Temperature difference between the body core and arterial blood supplied to the brain during hyperthermia or hypothermia in humans. Biomech Model Mechanobiol 1:137–149
3. Charny CK, Weinbaum S, Levin RL (1990) An evaluation of the Weinbaum-Jiji bioheat equation for normal and hyperthermic conditions. ASME J Biomech Eng 112:80–87
4. Coccarelli A, Nithiarasu P (2015) A robust finite element modelling approach to conjugate heat transfer in flexible elastic tubes and tube networks. Numer Heat Transf Part A Appl 67:513–530
5. Craciunescu OI, Clegg ST (2001) Pulsatile blood flow effects on temperature distribution and heat transfer in rigid vessels. Trans ASME J Biomech Eng 123:500–505
6. Cropper PC, Yang T, Cook MJ, Fiala D, Yousaf R (2008) Exchange of simulation data between cfd programmes and a multisegmented human thermal comfort model. In: Proceedings of conference: air conditioning and the low carbon cooling challenge, Cumberland Lodge, Windsor, pp 27–29
7. Daanen HAM (1991) Arterio-venous anastomoses and thermoregulation. Report no. izf 1991 b-12, TNO Institute for Perception Group: thermophysiology, Soesterberg
8. Fanger PO (1970) Thermal comfort. McGraw-Hill, New York
9. Ferreira MS, Yanagihara JI (2009) A transient three-dimensional heat transfer model of the human body. Int Commun Heat Mass Transfer 36:718–724
10. Ferreira MS, Yanagihara JI (2012) A heat transfer model of the human upper limbs. Int Commun Heat Mass Transfer 39:196–203
11. Fiala D (1999) A computer model of human thermoregulation for a wide range of environmental conditions: The passive system. J Appl Physiol 87:1957–1972
12. Fiala D, Lomas KJ, Stohrer M (2001) Computer prediction of human thermoregulatory and temperature response to a wide range of environmental conditions. Int J Biometeorol 45:143–159
13. Formaggia L, Nobile F, Quarteroni A, Veneziani A (1999) Multiscale modelling of the circulatory system: a preliminary analysis. Comput Vis Sci 2:75–83
14. Gagge AP, Stolwijk JAJ, Hardy JD (1967) Comfort and thermal sensation and associated physiological responses at various ambient temperatures. Environ Res 1:1–20
15. Hardy JD, Stolwijk JAJ (1966) Partitional exposures calorimetric studies of man during to thermal transients. J Appl Physiol 21:1799–1806
16. Holopainen R (2012) A human thermal model for improved thermal comfort. PhD thesis, VTT, Technical Research Centre of Finland
17. Huizenga C, Hui Z, Arens E (2001) A model of human physiology and comfort for assessing complex thermal environments. Build Environ 36:691–699
18. Karaki W, Ghaddar N, Ghali K, Kalev K, Holmer I, Vanguard LL (2013) Human thermal response with improved AVA modeling of the digits. Int J Therm Sci 67:41–52
19. Kerslake DM (1972) The stress of hot environments. University Press, Cambridge
20. Koscheyev VS, Paul S, Leon GR, Tanchida D, Taylor TJ, Koscheyev IV (1998) Body surface temperature tuning as a comfort support system in space and other extreme environments. In: Proceedings of the 28th international conference on environmental systems, pp. 1e8, SAE Technical Paper Series 981723, Danvers, MA, pp 13–16

21. Low K, van Loon R, Sazonov I, Bevan RLT, Nithiarasu P (2012) An improved baseline model for a human arterial network to study the impact of aneurysms on pressure-flow waveforms. Int J Numer Methods Biomed Eng 28:1224–1246
22. Mitchell JW, Myers GE (1968) An analytical model of the counter-current heat exchange phenomena. Biophys J 8:897–911
23. Mynard JP, Nithiarasu P (2008) A 1D arterial blood flow model incorporating ventricular pressure, aortic valve and regional coronary flow using locally conservative Galerkin (LCG) method. Commun Numer Methods Eng 24:367–417
24. Nithiarasu P (2004) A simple locally conservative galerkin (LCG) finite-element method for transient conservation equations. Numer Heat Transf Part B Fundam 46:357–370
25. Olufsen MS, Peskin CS, Kim WY, Pedersen EM, Nadim A, Larsen J (2000) Numerical simulation and experimental validation of blood flow in arteries with structured-tree outflow conditions. Ann Biomed Eng 28:1281–1299
26. Pennes H (1948) Analysis of tissue and arterial blood temperatures in the resting human forearm. J Appl Physiol 1:91–122
27. Refrigerating American Society of Heating and Air-Conditioning Engineers (1993) Physiological principles and thermal comfort
28. Rida M, Karaki W, Ghaddar N, Ghali K, Hoballah J (2014) A new ematical model to simulate ava cold-induced vasodilation reaction to local cooling. Int J Biometeorol 58:1905–1918
29. Rugh JP, Farrington RB, Bharathan D, Vhalinos A, Burke R, Huizenga C, Zhang H (2004) Predicting human thermal comfort in a transient nonuniform thermal environment. Eur J Appl Physiol 92:721–727
30. Salloum M, Ghaddar N, Ghali K (2007) A new transient bioheat model of the human body and its integration to clothing model. Int J Therm Sci 46:371–384
31. Sherwin SJ, Franke V, Peiró J, Parker KH (2003) One-dimensional modelling of a vascular network in space-time variables. J Eng 47(3):217–250
32. Shitzer A, Eberhart RC (1985) Heat transfer in medicine and biology-analysis and applications. Plenum Press, New York
33. Smith C (1991) A transient, three dimensional model of the human thermal system. PhD thesis, Kansas State University
34. Stolwijk JAJ (1971) A ematical model of physiological temperature regulation in man. Technical Report NASA CR-1855. National Aeronautics and Space Administration
35. Stolwijk JAJ (1977) Control of body temperature. In: Handbook of physiology-reaction to environmental agents, pp 45–67
36. Stolwijk JAJ, Hardy JD (1966) Partitional calorimetric studies of responses of man to thermal transients. J Appl Physiol 21:967–977
37. Sun X (2012) Development of an improved thermal model of the human body and an experimental investigation of heat transfer from a moving cylinder. PhD thesis, Kansas State University
38. Tanabe S, Kobayashi K, Nakano J, Ozeki Y, Konishi M (2002) Evaluation of thermal comfort using combined multi-node thermoregulation (65mn) and radiation models and computational fluid dynamics (cfd). Energy Build 34:637–646
39. Thomas CG, Nithiarasu P, Bevan RLT (2008) The locally conservative galerkin (LCG) method for solving the incompressible Navier–Stokes equations. Int J Numer Meth Fluids 57:1771–1792
40. Thomas CG, Nithiarasu P (2008) An element-wise, locally conservative galerkin (LCG) method for solving diffusion and convection-diffusion problems. Int J Numer Meth Eng 73:642–664
41. Uflacker R (2006) Atlas of vascular anatomy: an angiographic approach. Lippincott Williams and Wilkins, Philadelphia
42. Vanggaard L. (1975) Physiological reactions to wet cold. Aviat. Space Environ Med 46:33–36
43. Vanggaard L, Kuklane K, Holmer I, Smolander J. (2012) Thermal responses to whole-body cooling in air with special reference to arteriovenous anastomoses in fingers. Clin Physiol Funct Imaging 32:463–469
44. Wells PB, Thomsen S, Jones MA, Baek S, Humphrey JD (2005) Histological evidence for the role of mechanical stress in modulating thermal denaturation of collagen. Biomech Model Mechanobiol 4:201–210
45. Wissler EH (1964) A ematical model of the human thermal system. Bull Biophys 62:66–78
46. Wissler EH (1985) ematical simulation of human thermal behavior using whole-body models. Plenum Press, New York
47. Ying H, Hao L, Ryutaro H (2004) A one-dimensional thermo-fluid model of blood circulation in the human upper limb. Int J Heat Mass Transf 47:2735–2745
48. Zhu L, Pang L, Xu LX (2005) Simultaneous measurements of local tissue temperature and blood perfusion rate in the canine prostate during radio frequency thermal therapy. Biomechan Model Mechanobiol 4:1–9
49. Zohdi TI (2014) Modeling electrical power absorption and thermally-induced biological tissue damage. Biomech Model Mechanobiol 13:115–121

CHAPTER 11

INVESTIGATION OF BLOOD FLOW IN THE EXTERNAL CAROTID ARTERY AND ITS BRANCHES WITH A NEW 0D PERIPHERAL MODEL

Yoshihito Ohhara[1], Marie Oshima[2,3], Toshinori Iwai[1], Hiroaki Kitajima[1], Yasuharu Yajima[1], Kenji Mitsudo[1], Absy Krdy[2,3] and Iwai Tohnai[1]

[1] Department of Oral and Maxillofacial Surgery, Yokohama City University Graduate School of Medicine, 3-9 Fukuura, Kanazawa-ku, Yokohama 236-0004, Japan

[2] Department of Interfaculty Initiative in Information Studies, The University of Tokyo, 7-3-1 Hongo, Bunkyo-ku, Tokyo 113-0033, Japan

[3] Institute of Industrial Science, The University of Tokyo, 4-6-1 Komaba, Meguro-ku, Tokyo 153-8505, Japan.

ABSTRACT

Background

Patient-specific modelling in clinical studies requires a realistic simulation to be performed within a reasonable computational time. The aim of this study was to develop simple but realistic outflow boundary conditions for patient-specific blood flow simulation which can be used to clarify the distribution of the anticancer agent in intra-arterial chemotherapy for oral cancer.

Methods

In this study, the boundary conditions are expressed as a zero dimension (0D) resistance model of the peripheral vessel network based on the fractal characteristics of branching arteries combined with knowledge of the circulatory system and the energy minimization principle. This resistance model was applied to four patient-specific blood flow simulations at the region where the common carotid artery bifurcates into the internal and external carotid arteries.

Results

Results of these simulations with the proposed boundary conditions were compared with the results of ultrasound measurements for the same patients. The pressure was found to be within the physiological range. The difference in velocity in the superficial temporal artery results in an error of 5.21 ± 0.78 % between the numerical results and the measurement data.

Citation: Yoshihito Ohhara, Marie Oshima, Toshinori Iwai, Hiroaki Kitajima, Yasuharu Yajima, Kenji Mitsudo, Absy Krdy and Iwai Tohnai, "Investigation of blood flow in the external carotid artery and its branches with a new 0D peripheral model", BioMedical Engineering OnLine201615:16. DOI: 10.1186/s12938-016-0133-x

Conclusions

The proposed outflow boundary conditions, therefore, constitute a simple resistance-based model and can be used for performing accurate simulations with commercial fluid dynamics software.

BACKGROUND

The standard treatment for oral cancer is surgery, but radical surgery for advanced oral cancer often causes severe oral dysfunction, including speech and swallowing disorders. However, organ and function preservation can be maximized by adopting a multidisciplinary approach that combines radiotherapy and chemotherapy [1]. Selective or superselective intra-arterial chemotherapy, in particular, plays an important role in obviating the need for radical surgery. However, the side effects can be severe and the distribution of the anticancer agent to tumour-feeding arteries is unclear. Therefore, it is necessary to clarify the optimal dose of anticancer agent for each patient, and it is critical to obtain accurate information on flow distribution in each vessel for optimal drug delivery. Even though the final target of this study is to understand the distribution of the anticancer agent in the carotid artery area, including the branches of the external carotid artery (ECA) which act as feeding arteries for oral cancer, the present study focuses on patient-specific blood flow simulations in the area relevant for intra-arterial chemotherapy since anticancer agent flows are correlated with the blood flow due to their small concentration. Despite the recent progress of measurement methods, such as ultrasound-based methods, it is still difficult to obtain accurate information about the flow distribution in vessels, such as branches of the ECA, in deep soft tissues.

Due to remarkable progress over the last two decades, computational methods have allowed us to conduct realistic haemodynamic simulations. Simulations have also recently been performed in the fields of neurosurgery and cardiovascular surgery for atherosclerosis and aneurysm [2, 3, 4, 5, 6, 7, 8, 9, 10, 11, 12, 13,14, 15, 16, 17, 18]. Simulation is also used as a preoperative examination tool [19, 20, 21, 22, 23]. More recent patient-specific 3D model studies have been oriented to the carotid haemodynamics in order to establish correlation between blood flow-induced wall shear stress and atherosclerosis using patient-specific carotid models [24, 25, 26, 27, 28, 29]. Using patient-specific internal carotid artery (ICA) model, mechanical stresses also has been analyzed by blood flow simulation [30]. In this regard, simple and versatile computational haemodynamic simulation using patient-specific imaging data before treatment is expected to improve treatment quality and allow surgeons to predict treatment results. Also, by using patient-specific imaging data, the postoperative outcome can be compared with the preoperative state. However, it is difficult for surgeons to conduct numerical simulations since they require a certain set of skills related to computers and numerical algorithms, which are outside the scope of clinical practice. Therefore, the use of simulation as a tool to assist treatment can be very challenging. Furthermore, the simulation must yield accurate results within a reasonable time in order for it to be applicable to individual patients.

To achieve a more realistic simulation, the boundary conditions should include the effects of the entire circulatory system. Although a number of numerical models for these boundary conditions have been developed, most models, such as the Windkessel model [31, 32, 33] and the lumped parameter model [34, 35,36], rely on physiological parameters, such as resistance, compliance and conductance, which can be determined only from measurement data. Another approach is to conduct a multi-scale simulation combined with a one dimensional–zero dimensional (1D–0D) simulation including the entire circulatory system developed by Liang et al. [37]. Although the multi-scale simulation with the entire circulatory system is more physiologically accurate, it is complicated and time consuming. Since our region of interest is limited to the head and neck area, it is not necessary to conduct a simulation of the entire circulatory system. However, it is necessary to consider the influence of the peripheral blood vessels. Therefore, in this study, we develop a systematic method to determine physiological parameters, such as the resistance of the peripheral network, based on anatomical and physiological knowledge and model the boundary conditions with a 0D model, in order to simulate the effects of the entire circulatory system. The 0D resistance model of vessels is created in such a way that the peripheral vessel network is constructed as a binary symmetric tree that is attached to the outlets of a three-dimensional (3D) geometric structure. As a result, the resistance model defines a relationship between pressure and flow at the outlets such as the outflow boundary conditions in the 3D blood flow simulation. Since the peripheral vessel network of the ECA consists of smaller blood vessels, they tend to be mechanically stiff so the model takes into account changes in the ratio of vessel length to diameter, obtained either from medical imaging data or from anatomical knowledge of smaller vessels. The proposed model also considers non-Newtonian effects in blood viscosity with respect to changes in vessel diameter by using the constitutive relation between the apparent viscosity of blood and the haematocrit.

Because it is more convenient to use commercial software rather than custom-developed software in clinical studies, we have developed a simple boundary condition which can be implemented in commercial software. The model was applied to two patients in order to investigate the flow rate in branching arteries over a large number of cardiac cycles. A comparison between the numerical simulations and the ultrasound measurement data of the flow rates in the superficial temporal artery (STA) of the two patients was performed.

METHODS

This study was reviewed and approved by the Ethics Committee of Yokohama City University (No. B110512003). Written informed consent to use image data was obtained from oral cancer patients.

Patient-Specific Geometric Model

Computed Tomography Angiography

Patients with oral cancer underwent computed tomography (CT) angiography before intra-arterial chemotherapy. We used a 64-slice CT scanner (Aquilion 64; Toshiba Medical Systems, Tokyo, Japan), and non-ionic contrast medium (100 ml) was injected at a rate of 4.0 ml/s through an antecubital vein with an automatic power injector. A bolus-tracking technique was used to select the individual scan delay of the arterial phase. Repetitive low-dose scans were performed with a delay of 8 s at a level inferior to the carotid bifurcation. To measure the bolus arrival time, the region of interest was chosen to lie in the common carotid artery (CCA). As soon as an enhancement level of 90 Hounsfield units was reached, the scanning procedure started automatically. For the arterial phase scan, the scanning volume included the inferior margin of the thyroid cartilage/ bottom of C6 and the superior margin of the orbit. The scanner settings were 120 kV, 250 mA, 64 × 0.5 mm slice collimation, table speed 20.5 mm/rotation (pitch 0.641), and rotation time 0.75 s. The resolution matrix was 512 × 512 pixels with a slice thickness of 1 mm.

Patient-Specific Model

Digital imaging and communication in medicine (DICOM) data from the CT angiography of the two patients (right lower gingival cancer and right tongue cancer) were input into Mimics software (Materialise, Leuven, Belgium). The bilateral carotid arteries of each patient were segmented, and thus four separate three-dimensional patient-specific analysis models were created (Fig. 1a, b). The models included the CCA, the ICA, the ECA and ECA branches, such as the superior thyroid artery (SThA), lingual artery (LA), facial artery (FA), occipital artery (OA), maxillary artery (MA), posterior auricular artery (PAA), STA, middle meningeal artery (MMA) and transverse facial artery (TFA). Due to limitations in CT resolution, models 1 and 2 included no PAA, models 2–4 had no TFA, and model 2 had no MMA. Therefore, models 1, 3 and 4 had nine branches, and model 2 had seven branches.

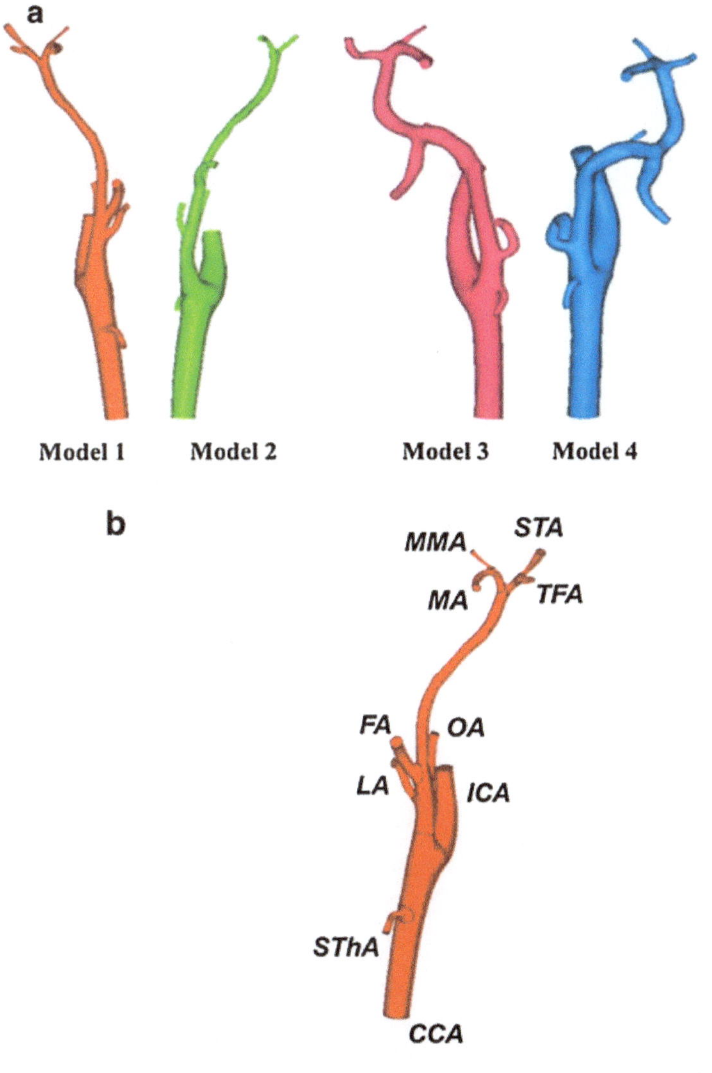

Figure 1: 3D models constructed from computed tomographic angiography images (**a**) and branches of the external carotid artery in the analysis area (**b**). The figure shows the common carotid artery (CCA), the internal carotid artery (ICA), the superior thyroid artery (SThA), the lingual artery (LA), the facial artery (FA), the occipital artery (OA), the maxillary artery (MA), the superficial temporal artery (STA), the middle meningeal artery (MMA) and the transverse facial artery (TFA).

Mesh Generation

Before mesh generation for a simulation, the diameters of each artery immediately after branching were measured. In addition, each artery was cut off at a length 5 times the diameter in order to minimize the influences of both inflow and outflow boundary conditions on the flow distributions (Fig. 2a). Numerical analyses with the finite volume method (FVM) were performed using the general-purpose fluid dynamics software FLUENT (Ansys Inc, Canonsburg, PA). The meshes for the FVM were generated using ICEM-CFD software (Ansys Inc., Canonsburg, PA). The mesh consists of tetrahedral cells in the artery core and prismatic cells in the region near the artery wall, as shown in Fig. 2a, and the total number of cells was approximately 3,000,000. Due to this mesh configuration, the cells in the core region are regular while orthogonality was maintained near the wall. In order to resolve the boundary layer, it is necessary to create a fine mesh near the wall, but the calculation time results in increase. Because we suppose to use as a future clinical tool, the mesh requires to minimize within a range that does not affect the present analysis. Therefore, we adopted a tetra prism type mesh. In this paper, the first grid point from the wall was located within $y^{+}<1$, as shown in Fig. 2b. There were four layers of prismatic cells in the viscous layer, and the mesh resolution was good enough for the analysis.

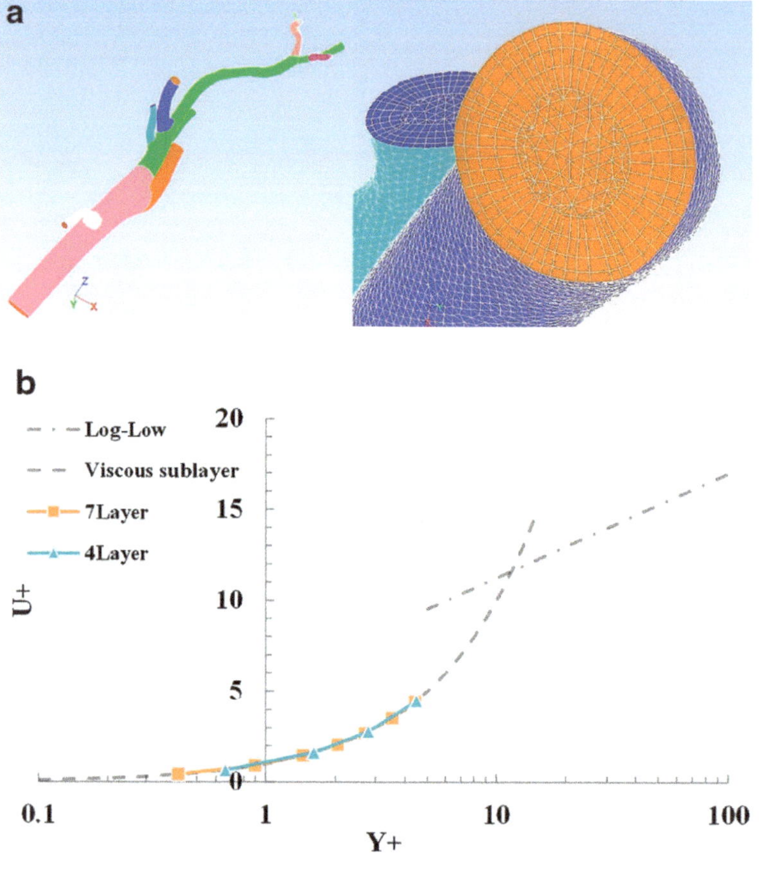

Figure 2: Tetrahedral mesh with a prismatic boundary layer (**a**) and mesh value of y+ less than one in each vessel (**b**)

Outflow Boundary Conditions

We performed 3D blood flow simulations using three types of outflow boundary conditions as the alternative boundary conditions for patient-specific vessels models to be described later in detail: (1) zero-pressure outflow boundary conditions, (2) the pressure boundary conditions with the conventional 0D resistance model and (3) the pressure boundary conditions with the present 0D resistance model. In both 0D resistance models, a capillary with a radius of 12μm or less was set to be the end of a branch. The conventional model used a constant λ (the ratio of the vessel length to its radius of 30) while λ in the present model varied depending on a size of a blood vessel diameter. Also, the terminal resistance in the present model was adjusted so that the terminal pressure became 30 mmHg, which is within the physiological range of the blood pressure. In two types of 0D resistance model, the flow rates acquired from the 3D analysis became the inflow boundary conditions for the 0D analysis while the pressure from the 0D analysis became the outflow boundary conditions for the 3D analysis. For the zero-pressure boundary condition, the pressures in all outlets were fixed at zero pressure in all outlets.

Numerical Methods

Since the 3D simulation was conducted by regarding blood as an incompressible Newtonian fluid (1050 kg/m^{3}density and 0.0046 Pa•s viscosity) in laminar flow with the average properties of blood, blood vessel walls can be regarded as rigid and the no-slip condition can be applied. Inflow boundary conditions of a constant blood flow velocity (Ultrasound system) were prescribed. Since the time scale of administration of drugs in intra-arterial chemotherapy is at least 1 h, a steady state blood simulation is performed instead of an unsteady pulsatile simulation. Three types of outflow boundary conditions for pressure on the outlets (0D resistance models) were prescribed.

The finite volume commercial software Ansys FLUENT was used to carry out steady states fluid dynamics simulations. A SIMPLE scheme was used to couple velocity and pressure and the discretization was accurate to the second order. The 0D resistance models are used as input data for user-defined functions in the code.

Simulations were performed on a PC running the Microsoft Windows 7 Professional operating system. The CPU was a dual-core Intel Xeon W5590 (clock frequency: 3.33 GHz) and there was 64 GB of RAM per core. The convergence of the simulations was verified by monitoring the mass flow rate of each outlet and the three velocity components. Convergence was achieved when each of these monitored values and the residuals showed oscillations of the order of 10^{-4} or less around a constant mean value. Most of the simulations required a total wall-clock time of less than 26 h.

Mathematical Modelling of Outflow Boundary Conditions

Physiological Background

In order to achieve a realistic blood flow simulation, it is important to model appropriate physiological conditions, particularly boundary conditions. Even though the systemic arteries are compliant, vessels become stiffer with a smaller radius. Blood vessels are organized in a bifurcating tree in which the total cross-sectional area of the vessels expands from approximately 5 cm^2 at the aortic root to approximately 400 cm^2 at the arterioles [38]. This expansion in the total cross-sectional area occurs despite the decrease in diameter of individual vessels because the number of daughter vessels doubles at each bifurcation, as shown in Fig. 3. As the number of bifurcations increases, the flow resistance becomes higher and the pressure falls as blood flows through the arterial tree. In this paper, we adopt outflow boundary conditions based on a 0D resistance model to reflect the effects of the peripheral network on the 3D blood flow simulation; this allows us to conduct simulations using commercial software within a reasonable computation time. Since the diameter of the outlet in the 3D model varies, our 0D resistance model is divided into six groups (aorta, large arteries, main artery branches, terminal artery branches, arterioles and capillaries) depending on vessel diameter, as summarized in Fig. 4.

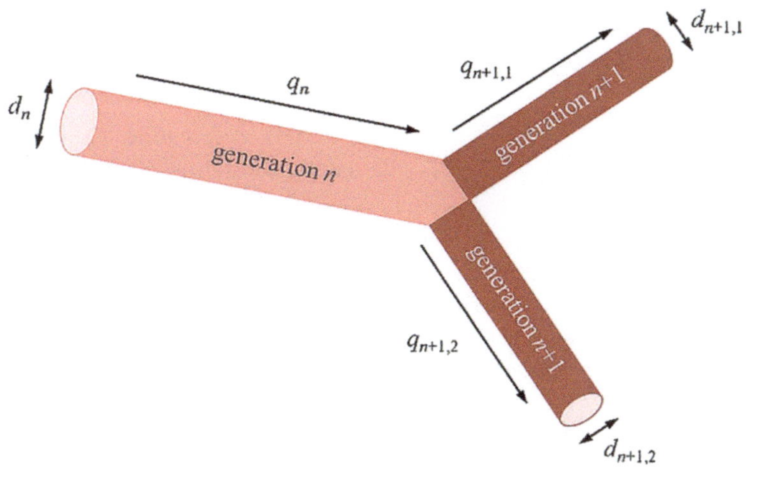

Figure 3: A parent vessel splitting into two daughter vessels.

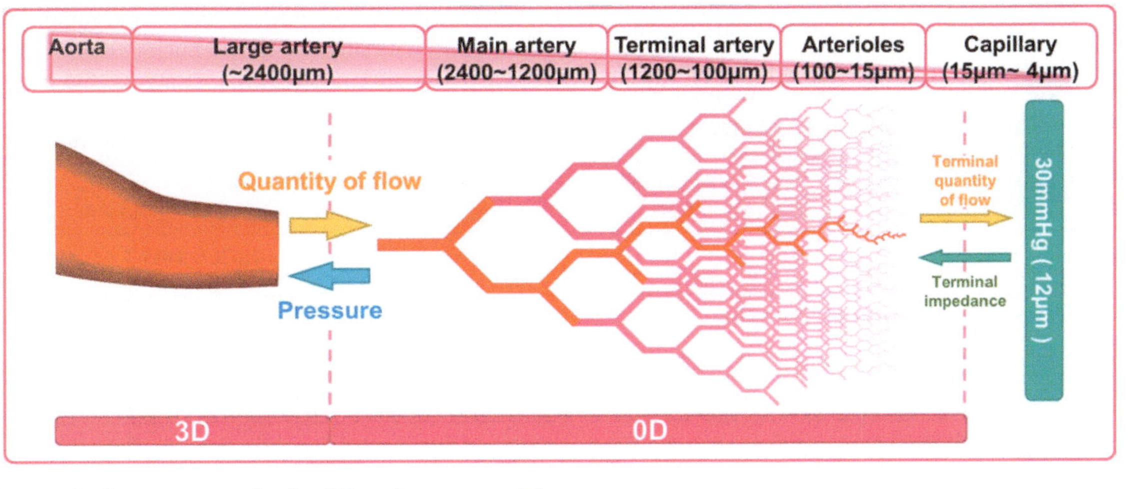

Figure 4: Anatomical components in the 0D resistance model.

Geometrical Modelling of the Peripheral Vessel Network

The resolution of CT or magnetic resonance (MR) imaging is not sufficiently high for accurate geometric modelling of a peripheral blood vessel network with small arteries, arterioles and capillaries. Instead, the peripheral network is constructed with a self-similar geometry. The parent blood vessel of generation i has a diameter d_i, and the flow rate q_i is divided into two branches in two daughter generations (i + 1, 1) and (i + 1, 2), as shown in Fig. 3. The geometry of peripheral blood vessels network of the 0D model is built on the basis of fractal due to anatomical knowledge [39]. In this study, we assume that the flow rate is distributed equally between the two symmetrical daughter vessels [40]:

$$q_i = q_{i+1,1} + q_{i+1,2} \quad (1)$$

$$q_{i+1,1} = q_{i+1,2} = 0.5q_i, \quad i = 1, \dots, n \quad (2)$$

where n denotes the total number of generations.

The diameters of the daughter generations are determined in accordance with a power law. This relation was suggested by Uylings [41] and was derived on the basis of the principle of minimum energy in the arterial system:

$$d_i^{\varepsilon} = d_{i+1,1}^{\varepsilon} + d_{i+1,2}^{\varepsilon} \quad (3)$$

$$d_{i+1,1}^{\varepsilon} = d_{i+1,2}^{\varepsilon} = 0.5d_i^{\varepsilon} \quad (4)$$

where ε = 3.0 for laminar flow and ε = 2.33 for turbulent flow. In this study, ε was set to 3.0 because the Reynolds number was smaller than 1000.

0D Resistance Model

The outflow boundary conditions were developed as a 0D model to represent only the resistance of smaller arteries in the peripheral vessel network because this is easy to implement in the software used for the flow simulation. Since the peripheral network consists of relatively small blood vessels, the influence of resistance can be assumed to be more pronounced than that of compliance [42]. However, the elasticity of the arterioles and capillaries was reduced to reflect their properties.

Since the length of small arteries cannot be determined by medical imaging, many authors have discussed methods for determining the length of vessels. For example, Olufsen et al. [43] considered the ratio of vessel length to vessel radius λ = l/r and set the ratio λ to 50 ± 10 [44, 45, 46]. In the 0D models developed by Olufsen et al. [43], it was possible to produce a physiological pressure drop, but it was not possible to calculate the physiological pressure. Therefore, in this research, to calculate the physiological pressure, a pressure of 30 mmHg was applied to the capillary vessel domain at the end of the conventional 0D models. In the basic experiment, we calculated the applied physiological pressure to the peripheral vessel with λ fixed at 20, 30, 40 or 50. As a result, the average outlet pressure in 3D became 75, 88, 100, or 122 mmHg. The average physiological pressure from arteries to arterioles would be too high at 100 mmHg, and excessively low at 75 mmHg. Therefore, in the conventional 0D resistance model used for comparison in this research, the value of λ was set to 30, and terminal resistance values were set to 0 at the end of the network of blood vessels.

We can write the relationship between pressure drop and flow rate in the time domain as follows:

$$\frac{P_n(L)}{Q_n(L)} = R_n(L) \quad (5)$$

where P, Q and R denote pressure, quantity of flow and resistance, respectively, and the suffix n denotes the generation of bifurcation. The present resistance model involves two steps, as described above.

The first step is to calculate the resistance $R_i(0)$ at the inlet of the artery at the i-th generation:

$$R_i(0) = \frac{8\,\mu\,\lambda}{\left(\pi r_i^3\right)} + R_i(L) \quad (6)$$

where μ is the viscosity of blood, λ is the ratio of the vessel length to its radius, and $R_i(L)$ is the resistance at outlet at the i-th generation, in which L is the distance from the inlet to the outlet at the i-th generation.

At a bifurcation point in the peripheral network, the resistance follows the same logic as resistance in electric circuits. If we assume that the parent vessel bifurcates into two daughter vessels symmetrically, the relation between the resistances of the parent and daughter vessels is given by

$$\frac{1}{R_i(L)} = \frac{1}{R_{i+1,1}(0)} + \frac{1}{R_{i+1,2}(0)} \quad (7)$$

The second step is to compute the resistance $R_i(L)$ at the outlet of the artery at the i-th generation:

$$R_i(L) = \frac{1}{2}R_{i+1,1}(0) \quad (8)$$

We categorized the vessels in the peripheral network into six groups based on their diameter and length, in accordance with a previous study [47], and the average value of λ for each group was determined. If the radius of the artery at the i-th generation indicates that the vessel is in the k-th group, λ_i is given by the value of λ for the k-th group shown in Table 1.

Table 1: Diameter, length and ratio of vessel length to vessel radius (λ) of each vessel in our peripheral network

Blood vessel	Ave. diameter (mm)	Min. diameter (mm)	Length (mm)	λ
ICA (all mean)	5.4834	4	117.3925	50.0154
Large arteries	6.5	2	200	61.5385
Main artery branches	2.4	1.2	100	83.3333

Terminal artery branches	1.2	0.1	10	16.6667
Arterioles	0.1	0.015	2	40
Capillaries	0.008	0.004	1	250

The values for the diameter and length of the internal carotid artery (ICA) are mean values over the four 3D models, and the λ value of each vessel is calculated from its length and half its averaged diameter

The advantage of the 0D resistance model is its flexibility in terms of setting the terminal resistance to a value in the physiological range, which is impossible with the structure impedance model [43, 48, 49]. The terminal resistance is simply given by

$$R_n(L)_{terminal} = \frac{p_{terminal}}{Q_n(L)_{terminal}} \quad (9)$$

Equation (9) allows us to update the terminal resistance at every step to match the physiological conditions of $P_{terminal}$. The terminal pressure is set in accordance with medical data [49, 50]. These modifications to the 0D resistance model allow for the tree model to be extended to the terminal radius (12 μm in our model), where the pressure was set to 30 mmHg (Fig. 4).

The special shape of the ICA, which does not branch until it enters the skull, was not included in this study as it lies outside of the region of interest. Therefore, in this study, the length and diameter of the ICA were determined on the basis of actual medical data.

Modelling of Non-Newtonian Viscosity

In large and medium arteries, blood behaves as a Newtonian fluid, and the viscosity. Blood has usually been treated as a Newtonian fluid in CFD studies of large and medium arteries. It does not exhibit a constant viscosity at all flow rates and is especially non-Newtonian in the microcirculatory system. Blood also exhibits non-Newtonian behavior in small branches and capillaries, where the cells squeeze through and a cell-free skimming layer reduces the effective viscosity through the tube. To take into account non-Newtonian effects, the apparent viscosity is given as a function of both the diameter and the haematocrit [47, 51, 52, 53]. First, the haematocrit is given by

$$Hct = \begin{cases} 0.45 & \text{if } d > 300\mu m \\ 0.45(0.196\ \log d - 0.117) & \text{if } d \le 300\mu m \end{cases} \quad (10)$$

where d is the vessel diameter.

When the vessel diameter becomes smaller than 10μm, the apparent viscosity decreases in accordance with the Fåhraeus–Lindqvist effect. By the inverse Fåhraeus–Lindqvist effect, the apparent viscosity increases when the vessel diameter is larger than a certain value. Equation (10) incorporates both effects.

The red line (Hct = f (d)) in Fig. 5 shows the behaviour of Hct in this study. The apparent viscosity is given in previous studies [54, 55]. According to Freitas [47], the haematocrit changes when the diameter d becomes smaller than 300μm. We used Eq. (11), which was proposed by Pries et al. [54, 55], to obtain the following relation for the changes in the viscosity μ_{app} associated with changes in vessel diameter d.

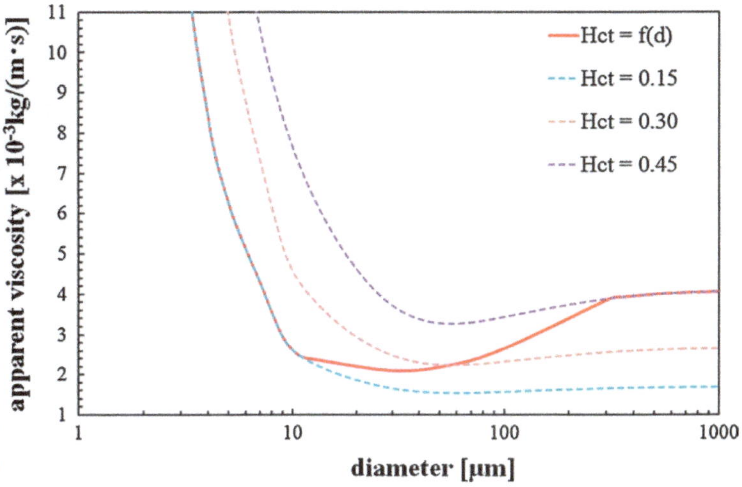

Figure 5: Apparent viscosity with variable haematocrit.

$$\mu_{app} = \left[1 + (\mu_{0.45} - 1)\frac{(1-Hct)^{Cd}}{(1-0.45)^{Cd}}\left(\frac{d}{d-1.1}\right)^2\right]\left(\frac{d}{d-1.1}\right)^2 \quad (11)$$

where

$$Cd = (0.8 + \exp[-0.075\,d])\left(-1 + \frac{1}{1+10^{-11}d^{12}}\right) + \frac{1}{1+10^{-11}d^{12}} \quad (12)$$

and

$$\mu_{0.45} = 6\exp[-0.085\,d] + \mu_{Newtonian} - 2.44\exp\left[-0.06\,d^{0.645}\right] \quad (13)$$

When μ (the viscosity of the haematocrit) is 0.45, $\mu_{Newtonian}$ is considered to be 4.06×10^{-3} (Pa·s).

Doppler Ultrasound Measurement

We defined the uniform flow velocity in a circumferential direction at the inlet to be the mean of the blood flow velocity for the cardiac cycle obtained by averaging four consecutive cardiac cycles. Blood flow measurement as a preoperative examination for intra-arterial chemotherapy was conducted using a commercial ultrasound scanner (HDI 5000; ATL-Philips Medical Systems, WA). Doppler velocity measurement accuracy is ±1 % or 1/2 pixel. Doppler velocity measurements are most accurate when the acoustic beam is aligned parallel with blood flow. All the data for this study were obtained by an experienced sonographer.

In each patient, B-mode ultrasound and color Doppler imaging showed that each artery had no severe stenosis, obstructive lesions or meandering. Using color Doppler imaging, velocities were determined at the centerlines of the CCA and STA. Pulsatile waveforms were obtained at the centerline of the distal CCA approximately 1–2 cm before the bifurcation without relevant flow disturbances, and at the centerline of the STA, velocities were determined at the front of the ear. The sample volume was set to be 2/3 of the vessel lumen; the center of the measurement site was in the long axis view; the angle correction was activated, and an angle of insonation of 60° was maintained whenever possible.

RESULTS

The simulations were conducted for 4 different types of vessel models (CCA, ICA, ECA and ECA branches). Each simulation took approximately 26 h (93,600 s) to complete.

Figure 6 shows the results for the pressure at each outlet in the 0D simulation. The pressures obtained with the present 0D resistance model were within the physiological range. The velocity magnitudes at the outlet of each vessel are summarized in Fig. 7. The velocity magnitude for the STA as given by both 0D resistance models is in agreement with the ultrasound data.

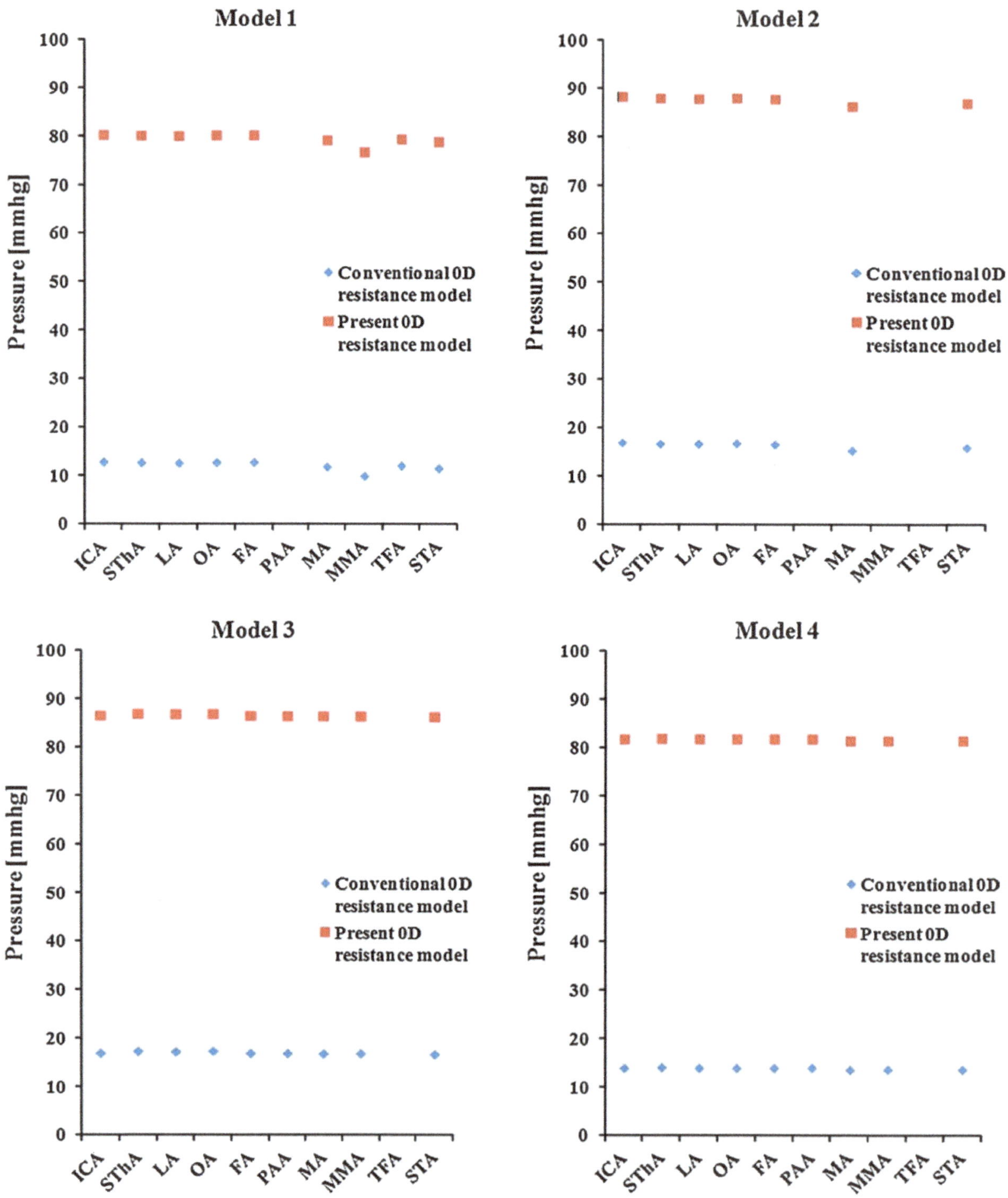

Figure 6: Pressure at the outlet of each blood vessel in the conventional 0D resistance model and in the present 0D resistance model. The pressure in the case of the present 0D resistance model was within the physiological range.

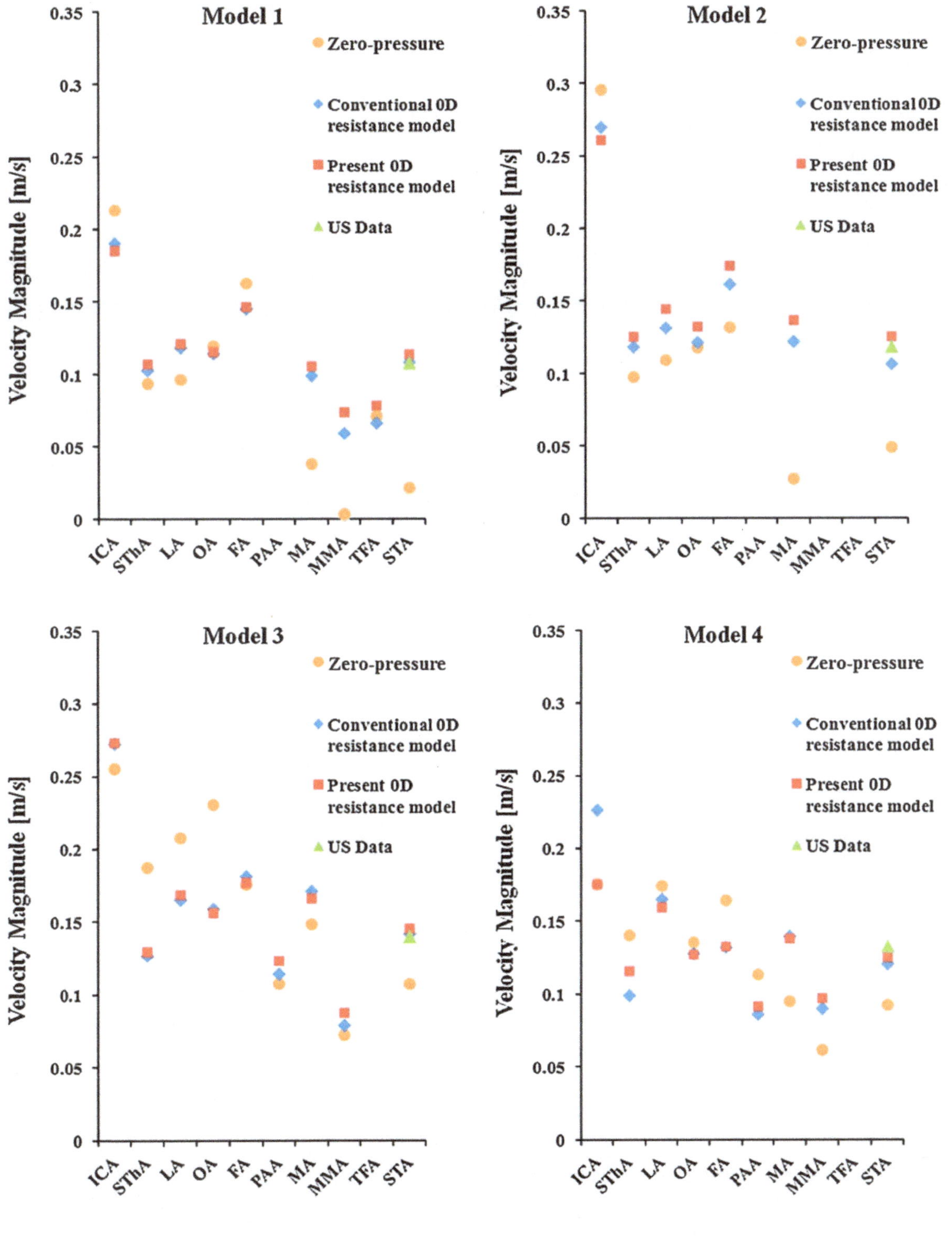

Figure 7: Velocity magnitude at the outlet of each blood vessel obtained with the zero-pressure outflow model, the conventional 0D model and the present 0D model. From the ultrasound data, only the velocity magnitude for the STA is shown.

To verify the simulation results, the blood velocities in the STA in each simulation were compared with those obtained from ultrasound measurements, as shown in Table 2. The mean error was defined as the average of the errors of the four models compared with the actual value obtained from ultrasound measurements, and the error value was calculated using the above-mentioned three types of boundary conditions. The mean error in the STA velocities in the simulation was 48.02 ± 22.66 %, 5.32 ± 4.52 % and 5.21 ± 0.78 % for the zero-pressure outflow model, the conventional 0D resistance model and the present 0D resistance model, respectively (Fig. 8). The simulation based on the zero-pressure outflow boundary condition yielded the largest mean error of the three models, whereas both 0D resistance models provided small mean errors. However, the standard deviation of the present 0D resistance model was smaller than that of the conventional 0D resistance model, which implies that simulation based on the present model is more precise.

Table 2: Velocities for the superficial temporal artery obtained for the four models with each outlet boundary condition together with ultrasound measurement data

	Model 1 (cm/s)	**Model 2 (cm/s)**	**Model 3 (cm/s)**	**Model 4 (cm/s)**
Zero-pressure outflow	2.167	4.917	10.760	9.260
Conventional 0D resistance model	10.834	10.643	14.196	12.049
Present 0D resistance model	11.368	12.538	14.565	12.525
Ultrasound data	10.796	11.854	14.016	13.304

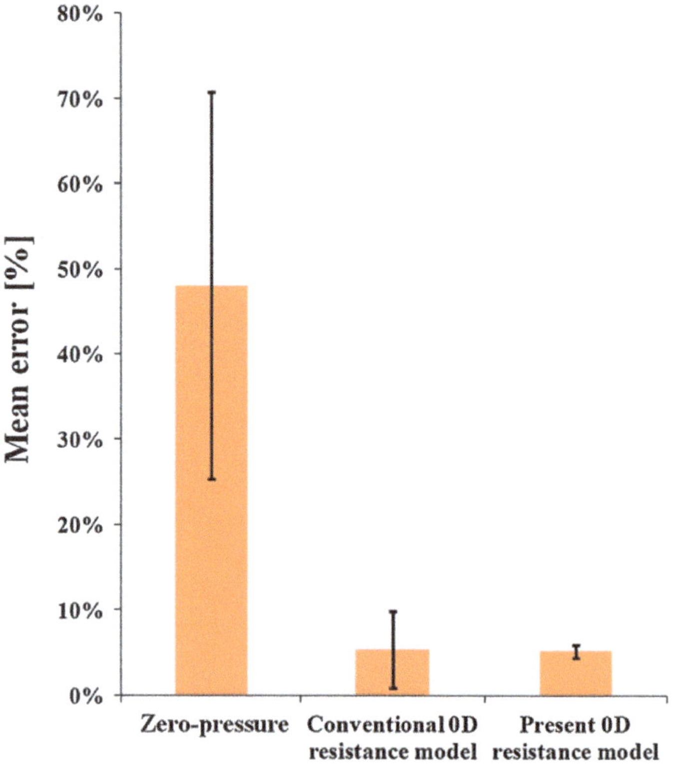

Figure 8: Mean error of the velocity for the STA obtained with the zero-pressure outflow model, the conventional 0D model and the present 0D model.

DISCUSSION

This study focused on head and neck haemodynamics for the treatment of oral cancer, especially in cases where intra-arterial chemotherapy is necessary. There have been few patient-specific blood flow simulations for oral cancer chemotherapy to date. Although Rhode et al. [56] previously reported the results of simulations of haemodynamic flow in head and neck cancer chemotherapy, where a patient-specific vessel model was created from CT images, branches of the ECA, such as the OA, MA and STA, were not modelled, and the peripheral vascular network was not considered. Our present 0D resistance model is clinically useful because it can provide an accurate estimation of the pressure and the flow distribution in vessels. The application of the present 0D resistance model as a boundary condition yields more realistic blood flow simulations the results of which can be used to achieve optimal tumour control with minimum accompanying side effects in oral cancer chemotherapy. In this paper, we consider a novel method for haemodynamic flow simulation applicable to intra-arterial chemotherapy for oral cancer.

The usefulness of our peripheral blood vessel model (0D resistance model) was verified using a patient-specific 3D model of the carotid arteries and branches together with ultrasound data of the CCA (the inlet of the blood flow) and the ECA (the outlet of the blood flow). The zero-pressure outflow boundary condition underestimated the STA velocity (compared with the results of ultrasound measurements) with a mean error of 48.02 ± 22.66 %. Since the zero-pressure boundary condition considers none of the physiological effects of the peripheral vascular system, the flow distribution is determined by only the geometry of the analysis model constructed from medical images. On the other hand, the conventional and proposed 0D resistance models consider the effects of the circulatory system. Thus, the mean error of the STA velocities was drastically reduced to 5.32 ± 4.52 % and 5.21 ± 0.78 %, respectively. The results obtained with both 0D resistance models were in closer agreement with the measurement data than the results obtained using the free outflow boundary condition. Although the peripheral resistance in the present 0D resistance model tended to be greater than that in the conventional 0D resistance model, there was no significant difference in flow rate between the two 0D models. However, the present model correlated well with the experiment, yielding a standard deviation of 0.78 % compared with 4.52 % in the case of the conventional model. This improved precision may be because the present 0D resistance model determines λ automatically and systematically based on anatomical knowledge, while the conventional 0D model uses a fixed value ($\lambda = 30$).

This study assumes a healthy blood vessel, but angiogenesis has been observed in the region of tumour progression. Therefore, it may be not reasonable to assume the parameters are the same for healthy blood vessels in the region with the tumour and further studies should be performed in the future.

CONCLUSIONS

We have presented some simple yet accurate outflow boundary conditions for conducting patient-specific blood simulations within a reasonable computation time for clinical applications. Since the region of interest in this study was the head and neck, the peripheral blood vessel network consists of mostly small blood vessels, which tend to be rigid. The outflow boundary conditions were designed based on the characteristics of the peripheral network by using the resistance of the blood vessel rather than the impedance, which combines compliance and resistance. To obtain an even more realistic simulation with the 0D model, the parameter λ was adjusted in accordance with anatomical knowledge.

The present 0D resistance model was used together with patient-specific 3D models of the carotid arteries to verify the simulation results, where ultrasound measurement results for the CCA and STA were used as reference data. The simulation based on the proposed 0D resistance model boundary conditions correlated well with the ultrasound measurement data and was more precise than those based on either the zero-pressure outflow condition or the conventional 0D resistance model. Furthermore, the pressure in the present 0D model was within the physiological range, in contrast to that of the conventional 0D model.

Our 0D resistance model is clinically useful because it can provide more realistic and accurate blood flow simulations, which may be helpful for achieving optimal tumour control with minimum accompanying side effects in oral cancer chemotherapy.

ABBREVIATIONS

0D: zero dimension

ECA: external carotid artery

ICA: internal carotid artery

1D–0D: one dimensional–zero dimensional

3D: three-dimensional

STA: superficial temporal artery

CT: computed tomography

CCA: common carotid artery

DICOM: digital imaging and communication in medicine

SThA: superior thyroid artery

LA: lingual artery

FA: facial artery

OA: occipital artery

MA: maxillary artery

PAA: posterior auricular artery

MMA: middle meningeal artery

TFA: transverse facial artery

FVM: finite volume method

MR: magnetic resonance

COE: Centers of Excellence

AUTHORS' CONTRIBUTIONS

YO conceived and designed the study, carried out the model parameterization and simulation studies, contributed in the data analysis and drafted the manuscript. MO participated in design of the study and coordinated the simulation activities and helped to draft the manuscript. TI carried out the clinical examination and participated in collecting the imaging data, helped in design of the study and participated in coordination and contributed in drafting the manuscript and revising it critically for important intellectual content. HK and YY participated in the work of simulation. AK participated in the work of model construction and provided technical supports. KM carried out the clinical examination and provided clinical supports. IT supervised the overall project, and conceived of the study. All authors read and approved the final manuscript.

ACKNOWLEDGEMENTS

This research was supported in part by the Global COE (Centers of Excellence) Program (Grant Number J07) and a Grant-in-Aid for Scientific Research (B) (JSPS KAKENHI Grant Number 21390542), JSPS, Japan.

REFERENCES

1. Mitsudo K, Shigetomi T, Fujimoto Y, Nishiguchi H, Yamamoto N, Furue H, Ueda M, Itoh Y, Fuwa N, Tohnai I. Organ preservation with daily concurrent chemoradiotherapy using superselective intra-arterial infusion via a superficial temporal artery for T3 and T4 head and neck cancer. Int J Radiat Oncol Biol Phys. 2011;79:1428–35.
2. Cebral JR, Yim PJ, Löhner R, Soto O, Choyke PL. Blood flow modeling in carotid arteries with computational fluid dynamics and MR imaging. Acad Radiol. 2002;9:1286–99.
3. Keshavarz-Motamed Z, Garcia J, Kadem L. Fluid dynamics of coarctation of the aorta and effect of bicuspid aortic valve. PLoS One. 2013;8:e72394.
4. Kono K, Shintani A, Terada T. Hemodynamic effects of stent struts versus straightening of vessels in stent-assisted coil embolization for sidewall cerebral aneurysms. PLoS One. 2014;9:e108033.
5. Leng X, Scalzo F, Ip HL, Johnson M, Fong AK, Fan FSY, Chen X, Soo YOY, Miao Z, Liu L, Feldmann E, Leung TWH, Liebeskind DS, Wong KS. Computational

fluid dynamics modeling of symptomatic intracranial atherosclerosis may predict risk of stroke recurrence. PLoS One. 2014;9:1–8.

6. Lou Z, Yang WJ. A computer simulation of the non-Newtonian blood flow at the aortic bifurcation. J Biomech. 1993;26:37–49.

7. Milner JS, Moore JA, Rutt BK, Steinman DA. Hemodynamics of human carotid artery bifurcations: computational studies with models reconstructed from magnetic resonance imaging of normal subjects. J Vasc Surg. 1998;28:143–56.

8. Molla MM, Paul MC, Roditi G. LES of additive and non-additive pulsatile flows in a model arterial stenosis. Comput Methods Biomech Biomed Engin. 2010;13:105–20.

9. Paul MC, Larman A. Investigation of spiral blood flow in a model of arterial stenosis. Med Eng Phys. 2009;31:1195–203.

10. Paul MC, Mamun Molla M, Roditi G. Large-Eddy simulation of pulsatile blood flow. Med Eng Phys. 2009;31:153–9.

11. Paul MC, Molla MM. Investigation of physiological pulsatile flow in a model arterial stenosis using large-eddy and direct numerical simulations. Appl Math Model. 2012;36:4393–413.

12. Perktold K, Peter RO, Resch M, Langs G. Pulsatile non-Newtonian blood flow in three-dimensional carotid bifurcation models: a numerical study of flow phenomena under different bifurcation angles. J Biomed Eng. 1991;13:507–15.

13. Steinman DA, Thomas JB, Ladak HM, Milner JS, Rutt BK, Spence JD. Reconstruction of carotid bifurcation hemodynamics and wall thickness using computational fluid dynamics and MRI. Magn Reson Med. 2002;47:149–59.

14. Steinman DA, Hoi Y, Fahy P, Morris L, Walsh MT, Aristokleous N, Anayiotos AS, Papaharilaou Y, Arzani A, Shadden SC, Berg P, Janiga G, Bols J, Segers P, Bressloff NW, Cibis M, Gijsen FH, Cito S, Pallarés J, Browne LD, Costelloe JA, Lynch AG, Degroote J, Vierendeels J, Fu W, Qiao A, Hodis S, Kallmes DF, Kalsi H, Long Q, et al. Variability of computational fluid dynamics solutions for pressure and flow in a giant aneurysm: the ASME 2012 Summer Bioengineering Conference CFD Challenge. J Biomech Eng. 2013;135:021016.

15. Taylor CA, Hughes TJ, Zarins CK. Effect of exercise on hemodynamic conditions in the abdominal aorta. J Vasc Surg. 1999;29:1077–89.

16. Taylor CA, Fonte TA, Min JK. Computational fluid dynamics applied to cardiac computed tomography for noninvasive quantification of fractional flow reserve: scientific basis. J Am Coll Cardiol. 2013;61:2233–41.

17. Zhao SZ, Ariff B, Long Q, Hughes AD, Thom SA, Stanton AV, Xu XY. Inter-individual variations in wall shear stress and mechanical stress distributions at the carotid artery bifurcation of healthy humans. J Biomech. 2002;35:1367–77.

18. Takizawa K, Schjodt K, Puntel A, Kostov N, Tezduyar TE. Patient-specific computer modeling of blood flow in cerebral arteries with aneurysm and stent. Comput Mech. 2012;50:675–86.

19. Fishman EK, Kuszyk BS, Heath DG, Cabral B. Surgical planning for liver resection. Computer (Long Beach Calif). 1996;29:64–72.

20. Gijsen FJ, van de Vosse FN, Janssen JD. The influence of the non-Newtonian properties of blood on the flow in large arteries: steady flow in a carotid bifurcation model. J Biomech. 1999;32:601–8.

21. Nakajima S, Atsumi H, Bhalerao AH, Jolesz FA, Kikinis R, Yoshimine T, Moriarty TM, Stieg PE. Computer-assisted surgical planning for cerebrovascular neurosurgery. Neurosurgery. 1997;41:403–9.

22. Pahlavian SH, Yiallourou T, Tubbs RS, Bunck AC, Loth F, Goodin M, Raisee M, Martin BA. The impact of spinal cord nerve roots and denticulate ligaments on cerebrospinal fluid dynamics in the cervical spine. PLoS One. 2014;9:1–14.

23. Taylor CA, Hughes TJR, Zarins CK. Finite element modeling of blood flow in arteries. Comput Methods Appl Mech Eng. 1998;158:155–96.

24. Lee C-J, Uemiya N, Ishihara S, Zhang Y, Qian Y. A comparison of estimation methods for computational fluid dynamics outflow boundary conditions using patient-specific carotid artery. Proc Inst Mech Eng H. 2013;227:663–71.

25. Mei Y, Müller-eschner M, Chen D: The Hemodynamic Comparison of Different Carotid Artery Bifurcation Angles Based on Patient-Specific Models. In: Ma S, Jia L, Li X, Wang L, Zhou H, Sun X, editors. Life system modeling and simulation. Springer, Berlin, Heidelberg; 2014. p. 170–7.

26. Malvè M, Chandra S, García A, Mena A, Martínez MA, Finol EA, Doblaré M. Impedance-based outflow boundary conditions for human carotid haemodynamics. Comput Methods Biomech Biomed Engin. 2014;17:1248–60.

27. Sousa LC, Castro CF, António CC, Santos A, Santos R, Castro P, Azevedo E, Tavares JMRS. Haemodynamic conditions of patient-specific carotid bifurcation based on ultrasound imaging. Comput Methods Biomech Biomed Eng Imaging Vis. 2014;2(3):157–66.

28. Seo T. Numerical simulations of blood flow in arterial bifurcation models. Korea Aust Rheol J. 2013;25:153–61.

29. Piskin S, Serdar Celebi M. Analysis of the effects of different pulsatile inlet profiles on the hemodynamical properties of blood flow in patient specific carotid artery with stenosis. Comput Biol Med. 2013;43:717–28.

30. Xu B, Zhong H, Duan S. Modeling of internal carotid artery aneurysm and blood flow simulation. Technol Heal Care. 2015;23:S43–8.

31. De Pater L, Van Den Berg JW. An electrical analogue of the entire human circulatory system. Med Electron Biol Eng. 1963;1964(2):161–6.

32. Westerhof N, Bosman F, De Vries CJ, Noordergraaf A. Analog studies of the human systemic arterial tree. J Biomech. 1969;2:121–43.

33. Noordergraaf A. Circulatory system dynamics. (Biophysics and Bioengineering Series, 1): V. 1. Academic Press Inc; 1978.
34. Mantero S, Pietrabissa R, Fumero R. The coronary bed and its role in the cardiovascular system: a review and an introductory single-branch model. J Biomed Eng. 1992;14:109–16.
35. Alastruey J, Parker K, Peiró J, Sherwin SJ. Lumped parameter outflow models for 1-D blood flow simulations : effect on pulse waves and parameter estimation. Commun Comput Phys. 2008;4:317–36.
36. Mirzaee MR, Ghasemalizadeh O, Firoozabadi B. Exact simulating of human arteries using lumped model and probing constriction in femoral and carotid arteries. Am J Appl Sci. 2009;6:834–42.
37. Liang FY, Takagi S, Himeno R, Liu H. Biomechanical characterization of ventricular-arterial coupling during aging: a multi-scale model study. J Biomech. 2009;42:692–704.
38. Caro CG, Pedley TJ, Schroter RC SW. The Mechanics of Circulation. J Fluid Mech. 90; 1979.
39. Zamir M. Nonsymmetrical bifurcations in arterial branching. J Gen Physiol. 1978;72:837–45.
40. Oshima M, Torii R, Tokuda S, Yamada S, Koizumi A. Patient-specific modeling and multi-scale blood simulation for computational hemodynamic study on the human cerebrovascular system. Curr Pharm Biotechnol. 2012;1–13:2153–65.
41. Uylings HBM. Optimization of diameters and bifurcation angles in lung and vascular tree structures. Bull Math Biol. 1977;39:509–20.
42. Rhodin JAG. Architecture of the vessel wall. In: Bohr DF, Somlyo AP, Sparks HV, editors. Handbook of physiology. Washington: American Physiological Soc; 1980. p. 1–31.
43. Olufsen MS, Peskin CS, Kim WY, Pedersen EM, Nadim A, Larsen J. Numerical simulation and experimental validation of blood flow in arteries with structured-tree outflow conditions. Ann Biomed Eng. 2000;28:1281–99.
44. Iberall AS. Anatomy and steady flow characteristics of the arterial system with an introduction to its pulsatile characteristics☆. Math Biosci. 1967;1:375–95.
45. Suwa N, Niwa T, Fukasawa H, Sasaki Y. Estimation of intravascular blood pressure gradient by mathematical analysis of arterial casts. Tohoku J Exp Med. 1963;79:168–98.
46. Bassingthwaighte JB, van Beek JHGM. Lightning and the heart: fractal behavior in cardiac function. Proc IEEE Inst Electr Electron Eng. 2002;76:693–9.
47. Freitas RA. Nanomedicine, Volume I: Basic Capabilities. Landes Bioscience. 1999.
48. Steele BN, Olufsen MS, Taylor CA. Fractal network model for simulating abdominal and lower extremity blood flow during resting and exercise conditions. Comput Methods Biomech Biomed Engin. 2007;10:39–51.
49. Mahy IR, Tooke JE, Shore AC. Capillary pressure during and after incremental venous pressure elevation in man. J Physiol. 1995;485:213–9.
50. Rowell LB. Human Cardiovascular Control. USA: Oxford University Press; 1993.
51. Pries AR, Secomb TW, Gaehtgens P, Gross JF. Blood flow in microvascular networks experiments and simulation. Circ Res. 1990;67:826–34.
52. Barbee JH, Cokelet GR. The Fahraeus effect. Microvasc Res. 1971;3:6–16.
53. Barbee JH, Cokelet GR. Prediction of blood flow in tubes with diameters as small as 29 microns. Microvasc Res. 1971;3:17–21.
54. Pries AR, Neuhaus D, Gaehtgens P. Blood viscosity in tube flow: dependence on diameter and hematocrit. Am J Physiol. 1992;263:H1770–8.
55. Pries AR, Secomb TW, Gessner T, Sperandio MB, Gross JF, Gaehtgens P. Resistance to blood flow in microvessels in vivo. Circ Res. 1994;75:904–15.
56. Rhode S, Paul MC, Martens E, Campbell DF. Simulation of haemodynamic flow in head and neck cancer chemotherapy. Biomed Eng Online. 2011;10:104.

CHAPTER 12

SIMULATION OF HAEMODYNAMIC FLOW IN HEAD AND NECK CANCER CHEMOTHERAPY

Stephan Rhode[1], Manosh C Paul[2], Eckhard Martens[3] and Duncan F Campbell[4]

[1] Department of Mechanical Engineering, Karlsruhe Institute of Technology, 76131 Karlsruhe, Germany

[2] School of Engineering, University of Glasgow, Glasgow G12 8QQ, UK

[3] Department of Mechanical Engineering, University of Applied Sciences, 76133 Karlsruhe, Germany

[4] Department of Oral & Maxillofacial Surgery, Queen Margaret's Hospital, Fife, UK.

ABSTRACT

Background

In recent years, intra-arterial chemotherapy has become an important component in head and neck cancer treatment. However, therapy success can vary significantly and consistent treatment guidelines are missing. The purpose of this study was to create a computer simulation of the chemical agent injection in the head and neck arteries to investigate the distribution and concentration of the chemical.

Methods

Realistic three dimensional patient specific geometry was created from image scan data. Pulsatile blood flow, turbulence, the chemical agent injection via a catheter, and the mixture between blood and the chemical were then simulated through the arterial network by computational fluid dynamics software.

Results

The results show a consistent chemical distribution throughout all the arteries and this is ineffective. In addition, due to high wall shear stress and turbulence at the inner bifurcation wall, serious complications during the treatment could occur, for instance haemolysis or thrombosis.

Conclusions

The modelled catheter position is insufficient to provide a high chemical agent concentration in the desired tumour feeding artery, which is vital for therapy success.

Citation: Stephan Rhode, Manosh C Paul, Eckhard Martens and Duncan F Campbell, "Simulation of haemodynamic flow in head and neck cancer chemotherapy," BioMedical Engineering OnLine201110:104. DOI: 10.1186/1475-925X-10-104

BACKGROUND

In head and neck cancer treatment and particularly in oral cancer treatment, surgery or radiotherapy alone or radiotherapy in combination with chemotherapy is generally administered. All modalities can lead to significant problems with swallowing and speech with consequent reduction in life quality. The general aim in cancer treatment is to approach maximum tumour control with minimum related side effects. In recent years, chemotherapy has become more relevant and is now an important treatment component in combined modality approaches, several studies in this subject are summarized in [[1], p. 127]. Studies, using concurrent chemoradiation in comparison with radiation alone, are commonly randomized trials, which vary in radiation dose, fractionation schedule, and administered chemotherapy. The dosage scheme, which is usually given in mass per body surface area (mg/m^2), as well as the administration method of the chemical is vital for the prognosis.

Using intra-arterial infusion improved the success of treatment. This is because of the arteries that supply specific tissue, detailed in [2, 3]. This technique allows the supply of much higher chemical agent concentration directly to the tumour than intra-venous infusion. The latter, distributes the chemical agent throughout the whole body and causes increased side effects. Because of the patient specific artery geometry and resultant haemodynamics, there is no method that always leads to success, and this is discussed in [4]. The knowledge about the chemical agent concentration in blood and the technique by which the tumours were infused is vital for therapy success [5]. The aim is to supply a high application rate of chemical in a specified target region inside the arterial network.

Advances in computational methods and three-dimensional imaging techniques allow improved understanding of haemodynamic flow in the cardiovascular network. Computational fluid dynamics (CFD) solves differential equations of fluid flows numerically and is an excellent technique for analyzing and displaying blood flow in the head and neck arterial network. In contrast to previous studies [6, 7, 8, 9], using two or three-dimensional ideal geometry, this project deals with real geometry obtained from image-scan segmentation of a patient [10, 11, 12, 13, 14].

Simulation of patient specific two-component flow, using various injection positions and parameters, for instance the mass flow rate of the chemical agent, should deepen the knowledge of chemical agent distribution in the artery network, which is the aim of this study. If an accurate computer simulation is achievable in a sufficiently short time, patient specific computation would provide the basis of comparing the effect of several treatment methods and parameters prior to therapy. A two-component transient CFD model can assist medical scientists to select the most appropriate treatment strategy.

METHODS

Considering the pulsatile character of blood flow, a transient CFD model is required. The model is reduced to the interesting region of the common carotid artery (CCA) bifurcation, the lingual artery (LA), facial artery (FA) and superior thyroid artery (STA) branch, including two-component flow. In particular the CFD model comprises the two liquids blood (continuous component or primary phase) and the chemical agent cisplatin (dispersed component or secondary phase). Cisplatin is injected via a catheter into the arterial blood flow. Due to this, the catheter has to be considered as thin pipe inside the CCA, because the catheter walls cause wall friction and influence the haemodynamic flow pattern.

A patient specific model was created with the image segmentation software Scanip™ [15] from CT-scans, presented in Figure 1. The catheter is an ideal pipe and in clinical practice has an inner diameter of 0.625 mm, an outer diameter of 0.924 mm and the common dose rate is approximately 50 ml/20 s. Four different injection positions are common:

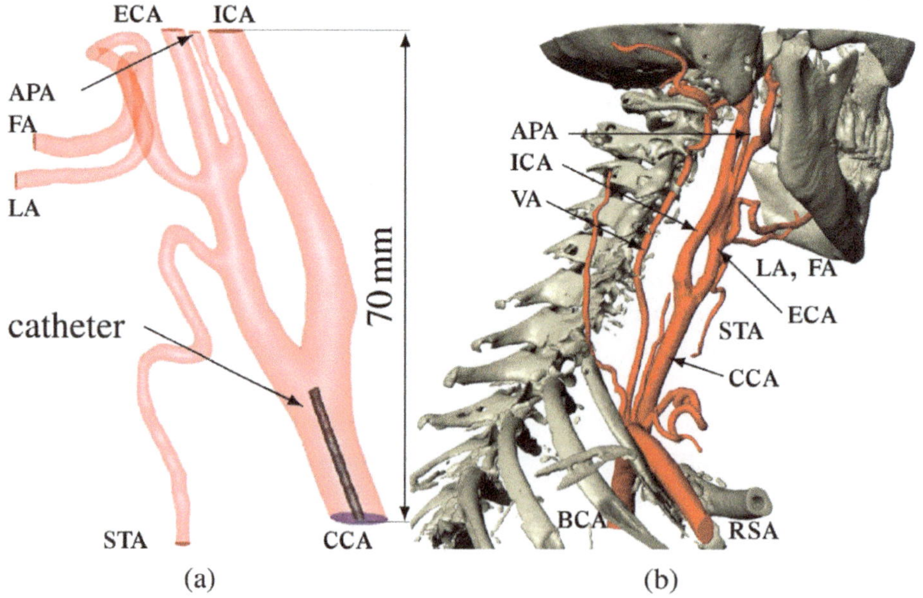

Figure 1: Two-component CFD model considering the cisplatin injection through a catheter, right before the CCA bifurcation in Figure (a). Figure (b) shows the right arterial network and skull source volume model as result of the CT image segmentation in Scanip™. APA-ascending

pharyngeal artery; BCA-brachiocephalic artery; CCA-common carotid artery; ECA-external carotid artery; FA-facial artery; ICA-internal carotid artery; LA-lingual artery; RSA-right subclavian artery; STA-superior thyroid artery; VA-vertebral artery.

- Short before the common carotid bifurcation.
- Internal carotid artery at the height of the superior thyroid branch.
- Lingual shunt.
- Inside the lingual artery.

It was decided to compute the first injection position, because the catheter can be modelled as straight ideal pipe, shown in Figure 1(a). Small pipe geometry creation using few voxels is complicated in Scanip™. Hence, it is impossible to model the catheter entirely including the inner wall in such small dimensions. A catheter outer wall diameter of approximately 1.2 mm (this is 30% larger than the real catheter) and tolerable wall smoothness was achieved with the CAD primitive generation tools in Scanip™. An additional volume at the top end of the catheter was created to obtain a catheter-blood interface in the following mesh import in Gambit™ [16]. Using the split face tool in Gambit™, this interface was divided by a circle of 0.6 mm diameter and an ideal circular face for the catheter outflow boundary condition was obtained. This face is slightly smaller than the real catheter inner diameter. Since the catheter inner wall is not considered in the model, it was decided to neglect the cisplatin flow inside the catheter entirely and the catheter volume remains unmeshed. Otherwise recirculation would occur inside the catheter near the catheter-blood interface. Additionally the steady ideal pipe flow of cisplatin inside the catheter is entirely solvable with analytical equations.

The mesh consists of hexahedron elements in the vessel core and tetrahedron elements adjacent to walls. A mesh independence test was done in a previous steady simulation on the highest blood flow rate of CCA using wall y^+ and pressure gradient adaption [17]. Due to just slightly monitor data changes after the mesh adaption, the initial mesh of counting 224,115 cells was taken.

A prior one-phase steady cause and effect study was carried out with various Reynolds numbers and viscous models for blood [17]. In this previous study it was discovered that a non-Newtonian material model was compulsory for blood. Comparing the velocity magnitude on diametrical line probes at several outlets, significant lower velocity values were observed for the non-Newtonian models in small arteries, shown in Figure 2. In particular, a maximum offset of 20% was computed at the FA outlet between the Newtonian and the power-law non-Newtonian model due to the very small FA diameter, approximately 2.9 mm. Moreover, the shear stress transport (SST) k-ω turbulence model was found to be appropriate for the whole velocity range and of superior robustness.

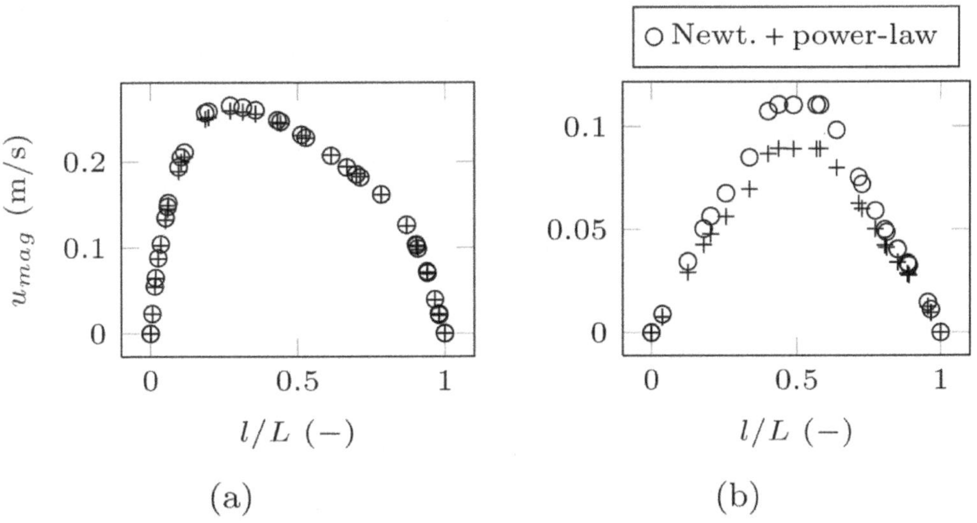

Figure 2: Velocity profiles over dimensionless vessel diameter with various blood models on a low Reynolds number, **taken** from our previous study [17]. Figure (a) shows the RSA with an approximately diameter of 9.3 mm, Figure (b) shows the velocity profile for the smaller FA with a vessel diameter of 2.9 mm. The non-Newtonian power-law approach results in lower velocity values, due to more viscosity on low shear rate in the FA in Figure (b). This effect was not observed in the larger RSA in Figure (a).

The boundary conditions comprise two mass flow inlets, CCA and catheter, and six outflow outlets with adjusted flow rate weighting. Several measured data dealing with CCA blood flow pattern is presented in [18,19, 20, 21]. It is assumed that the additional catheter inlet in the present model does not change the general flow division in the CCA bifurcation, due to the modelled catheter position still being inside the CCA. This is not advisable if the injection position moves downstream, for instance to the external carotid artery (ECA). Unfortunately it is impossible to consider a time dependent flow rate weighting of the CCA into the ECA and internal carotid artery (ICA) in the transient solver of Fluent™ as a boundary condition for each outlet. Hence, only a time averaged flow rate weighting could be used. The time averaged mass loss (ICA + ECA)/CCA is reported at roughly 5% in [21] and compared with two additional sources in Table 1. Regarding this 5% to the STA branch, it is assumed that each branch of the ICA will have the same proportion of approximately 13.5% of the ICA mass flow. The time averaged mass flow division was calculated for the ECA, STA, LA, FA and ascending pharyngeal artery (APA) outlet and assumed for the ICA outlet. Figure 3(a) provides an overview of the time continuous flow rate weighting for all outflow conditions.

Table 1: Measurements of the cycle averaged CCA rate of flow division in ml/min and the relative outflow/inflow mass loss

reference	V˙ ICA	V˙ ECA	V˙ CCA	loss
[21]	245	125	389	4.9%
[32]	248	95	370	7.0%
[33]	255	95	376	6.9%

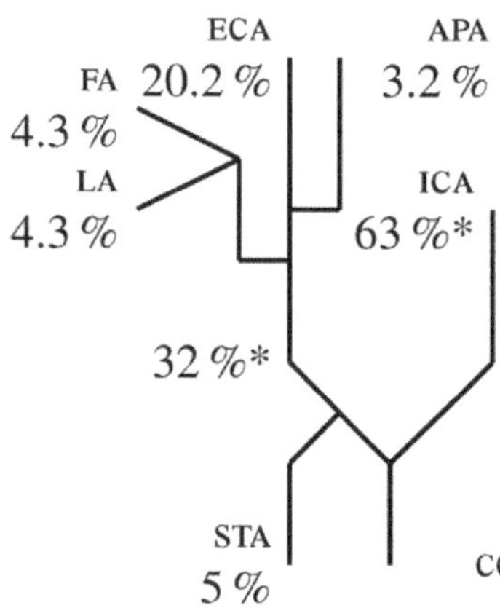

(a)

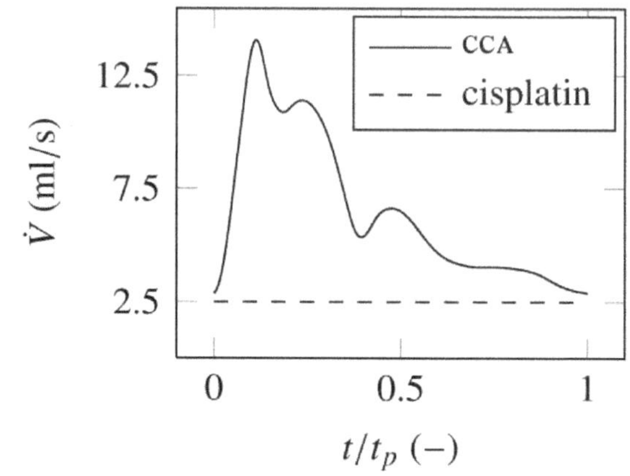

(b)

Figure 3: Time continuous flow rate weighting off all outlets using the outflow boundary condition in Figure (a). The asterisked values are the original time averaged data of [21]. The CCA flow rate inlet condition in Figure (b) is realized as a transient table with 200 equal distant timesteps (0.0047s timestep size). The cycle time t_p is 0.923s, which is equal to 65 heart beats per minute.

The most appropriate data for the CCA inlet is given in [21] obtained from older volunteers and shown with the solid line in Figure 3(b). However, it can be expected that because of the stress during chemotherapy, the heart rate should be higher than the reported 65 beats per minute in this source. Additionally, the study was made with healthy volunteers. Hence, it is not known how a catheter will affect the reported rate of flow range. These two inlet conditions are converted into a mass flux boundary in the Fluent™ simulation. In particular a transient table for the CCA and a constant value for cisplatin is used. The volume fraction of the dispersed phase cisplatin was set to 100% at the cisplatin inlet and 0% at the CCA inlet.

Blood is modelled as a non-Newtonian fluid using the Carreau model (1) with a constant density of ρ = 1056 kg/m³. The material constants for the haematocrit value of Ht = 45% and the temparature of T = 310.15 K are given in [[22], pp. 236-237]. The maximum Reynolds number of the pulsatile blood flow is 856, which indicates laminar blood flow through the whole cycle.

$$\mu(\dot{\gamma}) = \mu_\infty + (\mu_0 - \mu_\infty)[1 + (\lambda_t\dot{\gamma})^2]^{(n-1)/2},$$
$$\text{where} \quad \mu_0 = 0.056 \text{ Pas}, \quad \mu_\infty = 0.00345 \text{ Pa s},$$
$$\lambda_t = 3.313 \text{ s}, \quad n = 0.3568. \tag{1}$$

The Newtonian material model was used for cisplatin with properties of water, ρ = 998.2 kg/m³and μ = 0.001003 Pa s.

The Reynolds number of the steady cisplatin flow inside the catheter is 5331. This is far above the commonly known threshold for laminar pipe flow Re = 2300. Turbulence is considered through the SST k-ω model. At the laminar CCA inlet k and ω is set to 0. At the turbulent cisplatin inlet on top of the catheter k = 0.359 m²/s² and ω = 26165 s⁻¹, determined by using an approach for fully developed turbulent pipe flows (2) dealt with in [[16], ch. 7.2.2].

$$k = \frac{3}{2}(u_{mag} \cdot I)^2 \quad \text{with} \quad I = 0.16 \cdot Re^{-\frac{1}{8}},$$
$$\omega = \frac{k^{1/2}}{C_\mu^{1/4} \cdot l_T} \quad \text{with} \quad C_\mu = 0.09 \quad \text{and} \quad l_T = 0.07 \cdot d_h. \tag{2}$$

All walls are rigid and a non-slip condition is prescribed. The mixing model was used including slip velocity and implicit body force effects. Assuming a bubble size of 0.1 mm for the dispersed phase cisplatin, a hydraulic diameter of d_h = 6 mm for the CCA and a maximum blood velocity of 0.8 m/s, a Stokes number of 0.021 was calculated using the equations (3), dealt with in [[23], p. 194]. This is much smaller than one and means that all cisplatin bubbles will closely follow the blood flow. A strong coupling between blood and cisplatin is supposed.

$$St = \frac{\tau_d}{\tau_c} \quad \text{with} \quad \tau_d = \frac{\rho_d \cdot {d_d}^2}{18\mu_c} \quad \text{and} \quad \tau_c = \frac{d_h}{u_{mag}} \tag{3}$$

After testing one initial cycle using 100 time steps, the transient simulation proceeded in batch mode using 200 equal distant time steps for one additional cycle. Double precision, first order discretization and the pressure based segregated PISO solver including skewness and neighbor correction of one, were used with a convergence criteria of $1 \cdot 10^{-7}$ for the cisplatin volume fraction residual. All other settings remained at Fluent™ default values. Data for each 10 th time step was saved. The computing time required approximately four days, shared by three parallel processors.

RESULTS AND DISCUSSION

Figure 4(a) shows vectors of velocity magnitude in several cross sections colored by the turbulent kinetic energy k in the time step with the highest blood volume flow rate. The catheter inlet vectors are not exactly perpendicularly aligned with the catheter axis. The cisplatin flow is frontal orientated towards the inner wall of the CCA bifurcation with a small incline towards the ECA. Creation of the catheter geometry in CAD software would be useful to obtain a proper plain catheter front surface. This would be perpendicular, ideally cylindric and the catheter wall would be smooth. Because of the turbulent Reynolds number inside the catheter, turbulence is introduced in the blood flow through the cisplatin inlet. And it occurs at both inner walls after the CCA bifurcation in the ICA and ECA. Sutera and Mehrjardi [24] reported in their study that the red blood cells deform gradually toward a smooth ellipsoidal shape by the alternation of the magnitude in the shear stress from 10 Pa to 250 Pa. The region of fluctuating stresses associated with high turbulent kinetic energy, that indicates high turbulence and is shown in Figure 4(b), is therefore critical in matters of red blood cell damage (haemolysis), thrombosis (blood clotting) or even a stroke. Additionally, the CCA bifurcation is known as region with a high possibility of arteriosclerosis for elderly patients. The turbulent cisplatin flow could have an abrasive effect and release fatty particles.

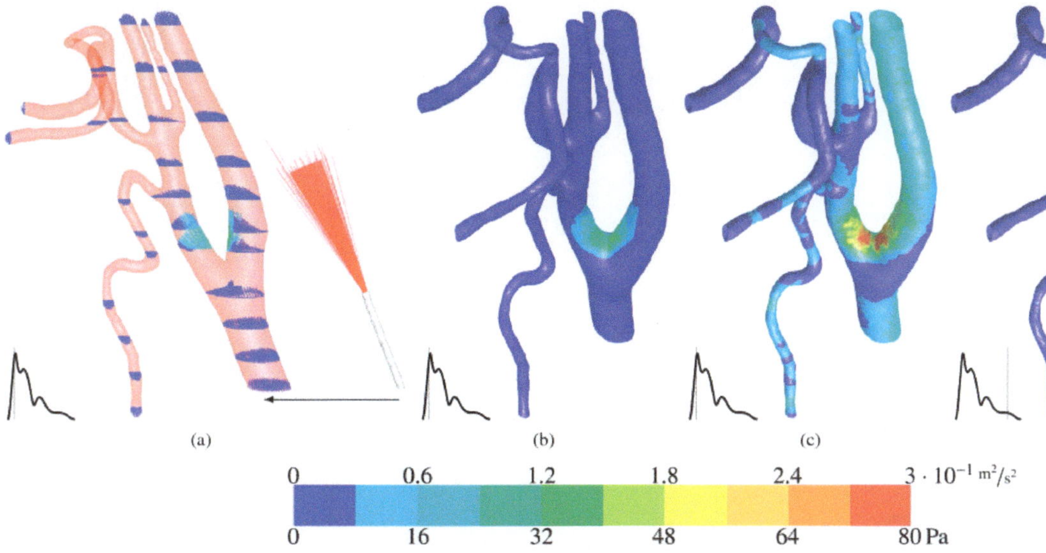

Figure 4: In (a) vectors of velocity magnitude colored by turbulent kinetic energy k are presented (range above the color bar). The catheter on the right is displaced from its origin in the CCA center. The mean velocity at CCA inlet is 0.38 ms^{-1} and at cisplatin inlet 7.66 ms^{-1}. Figure (b) shows the related contour plot of k. Figure (c) and (d) are plots of wall shear stress in systole and diastole (range below the color bar).

Furthermore, in consequence of the frontal orientated cisplatin inlet stream towards the CCA bifurcation wall, the wall shear stress shows a peak ($\approx$ 80 Pa) in this region, presented in Figure 4(c) and 4(d) for systole and diastole, respectively. This level of high wall shear stress can potentially cause acute vascular endothelial changes as it was found in [25] that the acute yield stress is approximately 37.9 Pa.

Due to the toxic properties of cisplatin, this catheter outflow position can cause several complications, such as local inflammation and neurological complications, for instance headache and facial paresis, [3].

The cisplatin distribution during one cardiac cycle can be described in detail by using Figure 5. In general, the chemical agent cisplatin distributes in all arteries, as to be expected for this catheter position. This is undesirable, because only a high cisplatin concentration in a specific region gives maximum tumour control with minimum related side effects. Prior systole, the path lines indicate recirculation upstream in the CCA. The cisplatin volume fraction in the ECA is larger than in the ICA. This is explained by the flow rate weighting conditions of Figure 3(a). With the start of systole and increasing blood flow, the cisplatin concentration decreases in the larger arteries ECA and ICA, Figure 5 (c)-(e), and the recirculation region disappears. There is a time shift in cisplatin volume fraction reduction to note between ECA and LA with FA. This is shown in Figure 5(d). This shift is retained until the beginning of the diastole in Figure 5(f) when the volume fraction in ECA, LAand FA are approximately equal. At the end of diastole in Figure 5(g) a similar condition is observed as in Figure 5(a). This suggests a periodic cisplatin distribution characteristic controlled by the cardiac cycle. It also shows the importance of considering a pulsatile blood flow boundary condition. If the simulation was steady, a delay in cisplatin concentration at various artery limits would be invisible in the simulation results.

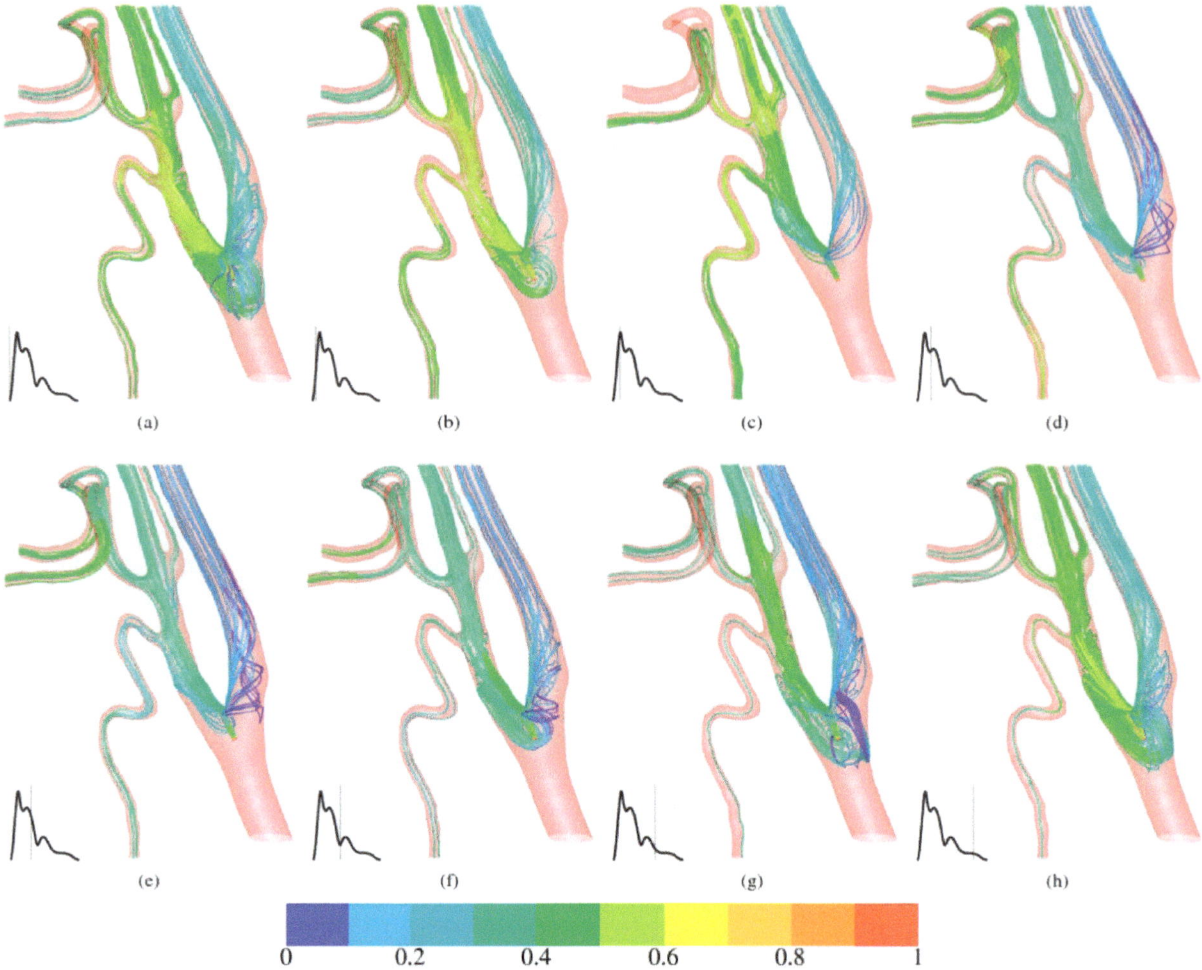

Figure 5: Path lines with origin cisplatin inlet coloured by volume fraction of cisplatin during one cardiac cycle. The catheter is hidden. The blood flow condition at the CCA inlet is shown in each picture on the bottom left.

The mass weighted average cisplatin volume fraction for all 21 saved time steps at all model outlets is presented in Figure 6. The discussed delay of cisplatin concentration between ICA and LA with FA is good to observe. In particular, the functions of ICA, ECA, STA and APA show a qualitatively similar shape contrary to the functions of LA and FA. Apart from the ICA all arteries have an approximately equal cisplatin concentration range between 0.3 and 0.5. It is not expected that a variation in cisplatin dosage will change this uniform distribution, rather a catheter position further downstream could have a positive effect in this matter.

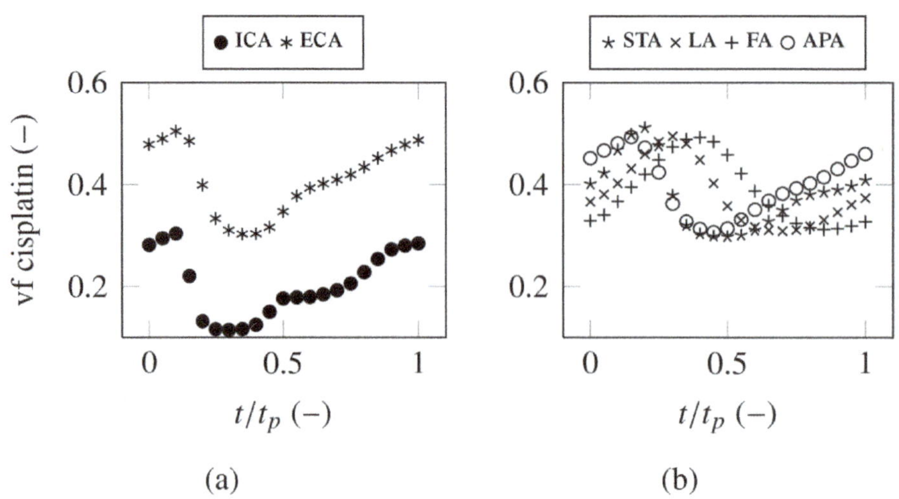

Figure 6: Mass weighted average cisplatin volume fraction at the outlets over one cardiac cycle.

Additionally, the time dependent cisplatin concentration functions, observed for all outlets, indicate that the time continuous boundary condition at the cisplatin inlet is not the most efficient technique for applying high chemical agent dosages. It is possible that an unsteady injection at a particular beneficial time range on higher dosage rate could enhance the cisplatin concentration and reduce side effects, such as local inflammation at artery walls, during the remaining period. An advantageous time range for injection would be between t/t_p = 0-0.3 during peak systole. Furthermore, it is possible that recirculation and the exposure time on high turbulence could be reduced by time dependent injection during systole. Future simulations are required to provide evidence for this assumptions. In absolute values, Figure 7 represents the rate of flow of the chemical. The prescribed flow rate weighting is the reason for the low level of STA, LA, FA and APA compared with ECA and ICA.

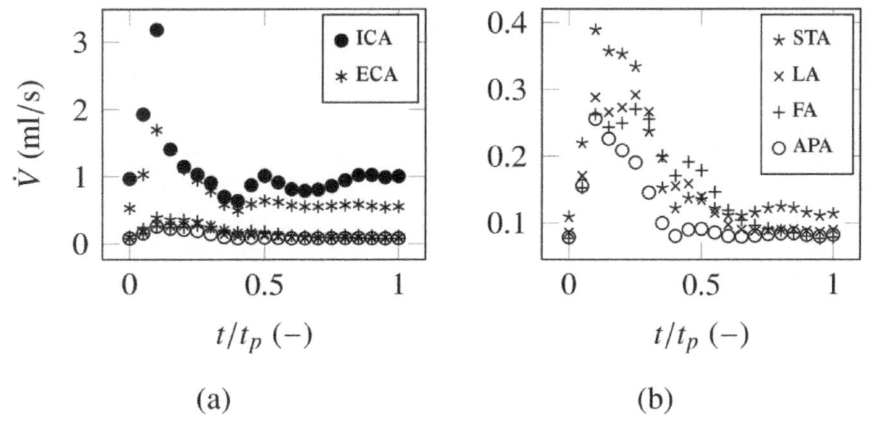

Figure 7: Cisplatin rate of flow at the outlets. Figure (b) shows the outlets STA, LA, FA and APA of Figure (a) separately with a finer ordinate resolution.

CONCLUSIONS

The distribution and concentration of the chemical agent cisplatin have been investigated in a realistic three dimensional patient specific geometry of human head and neck arteries. The used injection position gives an ineffective cisplatin division. This means that the chemical agent is distributed in all arteries and not in one specific tumour feeding artery, a fact of great importance. High wall shear stress and turbulence near the CCA bifurcation inner walls can cause serious harm, for instance acute vascular endothelial changes or haemolysis. It should be expected that an injection position in the core of the artery further downstream would reduce this high exposure. The time continuous cisplatin inlet condition seems to be ineffective during diastole, when upstream recirculation in the CCA mainly occurs. It should be investigated if a time dependent injection during the advantageous systole would lead to an equal cycle averaged cisplatin concentration with reduced side effects caused by the catheter inflow.

In blood flow CFD simulation using multiphase models, neither mass flow nor velocity outlets are practicable, due to required outlet data for each phase. This data is unknown a priori for the secondary phase cisplatin and is the aim of this simulation. An assumed flow rate weighting was used in this model based on measured data for the two main outlets ICA and ECA. It is assumed that the general blood flow division remains unchanged in comparison with measured data of healthy volunteers in the normal state. An injection position further downstream in a side branch of the ECA would probably change the blood flow division. An alternative proper methodology would be to use the pressure inlet and outlet boundary conditions that require only the volume fraction of the secondary phase at the inlets, and adjust the pressure to a correct level. However, time dependent pressure functions for each model boundary were not available and are difficult to assess for small arteries, such as the LA and FA. The implementation of an arterial one dimensional model seems to be the superior way for future CFD simulations dealt with in [26, 27, 28, 29, 30, 31].

List of Abbreviations Used

APA: ascending pharyngeal artery
BCA: brachiocephalic artery
CAD: computer aided design
CCA: common carotid artery
CFD: computational fluid dynamics
CT: computer tomography
ECA: external carotid artery
FA: facial artery
ICA: internal carotid artery
LA: lingual artery
PISO: pressure implicit with splitting of operators
RSA: right subclavian artery
SST: shear stress transport
STA: superior thyroid artery
VA: vertebral artery

ACKNOWLEDGEMENTS

We acknowledge support by Deutsche Forschungsgemeinschaft and Open Access Publishing Fund of Karlsruhe Institute of Technology.

AUTHORS' CONTRIBUTIONS

SR, MCP and EM carried out the CFD studies while DFC provided CT images and clinical inputs. All the authors contributed to the preparation of the manuscript and then approved its final version.

REFERENCES

1. Myers J (Ed): Oral Cancer Metastasis. New York: Springer Verlag; 2010.
2. Alkureishi L, De Bree R, Ross G: RADPLAT: An Alternative to Surgery? The Oncologist 2006, 11(5):469. 10.1634/theoncologist.11-5-469
3. Kovács A: Intraarterial chemotherapy and chemoembolization in head and neck cancer. Establishment as a neoadjuvant routine method. Cancer Therapy 2003, 1: 1–9.
4. Tohnai I: Chemotherapy using intra-arterial infusion for oral cancer. Nagoya Journal of Medical Science 2006, 68(3–4):101–108. [http://ir.nul.nagoya-u.ac.jp/jspui/bitstream/2237/6962/1/p101_108.pdf]
5. Fuwa N, Kodaira T, Furutani K, Tachibana H, Nakamura T, Nakahara R, Tomoda T, Inokuchi H, Daimon T: Intra-arterial chemoradiotherapy for locally advanced oral cavity cancer: analysis of therapeutic results in134 cases. British journal of cancer 2008, 98(6):1039–1045. 10.1038/sj.bjc.6604272
6. Li M, Beech-Brandt J, John L, Hoskins P, Easson W: Numerical analysis of pulsatile blood flow and vessel wall mechanics in different degrees of stenoses. Journal of biomechanics 2007, 40(16):3715–3724. 10.1016/j.jbiomech.2007.06.023
7. Roy A, Back L, Banerjee R: Guidewire flow obstruction effect on pressure drop-flow relationship in moderate coronary artery stenosis. Journal of biomechanics 2006, 39(5):853–864. 10.1016/j.jbiomech.2005.01.020
8. Paul M, Mamun Molla M, Roditi G: Large-Eddy simulation of pulsatile blood flow. Medical Engineering & Physics 2009,31: 153–159. 10.1016/j.medengphy.2008.04.014
9. Paul M, Larman A: Investigation of spiral blood flow in a model of arterial stenosis. Medical Engineering & Physics2009, 31: 1195–1203. 10.1016/j.medengphy.2009.07.008
10. Taylor C, Figueroa C: Patient-Specific Modeling of Cardiovascular Mechanics. Annu Rev Biomed Eng 2009, 11:109–134. 10.1146/annurev.bioeng.10.061807.160521
11. Steinman D, Milner J, Norley C, Lownie S, Holdsworth D: Image-Based Computational Simulation of Flow Dynamics in a Giant Intracranial Aneurysm. American Journal of Neuroradiology 2003, 24(4):559.http://www.ajnr.org/cgi/reprint/24/4/559
12. Cebral J, Löhner R, Soto O, Choyke P, Yim P: Patient-specific simulation of carotid artery stenting using computational fluid dynamics. In Medical Image Computing and Computer-Assisted Intervention-MICCAI 2001. Springer; 2010:153–160.
13. Steinman D: Image-based computational fluid dynamics modeling in realistic arterial geometries. Annals of Biomedical Engineering 2002, 30(4):483–497.
14. Schenkel T, Krittian S, Spiegel K, Höttges S, Perschall M, Oertel H: The Karlsruhe Heart Model KaHMo: A modular framework for numerical simulation of cardiac hemodynamics. In IFMBE Proceedings Volume 25/4, World Congress on Medical Physics and Biomedical Engineering, September 7–12, 2009, Munich, Germany. Springer; 2009:615–618.
15. Simpleware: ScanIP™, + ScanFE™ and + ScanCAD™ Reference Guide . In . Simpleware LTD, Innovation Centre Rennes Drive, Exeter EX4 4RN, UK; 2010.
16. Fluent: Fluent 6.3 User Guide. Fluent Incorporated, Centerra Resource Park 10, Cavendish Court Lebanon, NH 03766; 2007.
17. Rhode S, Paul M, Martens E, Campbell D: Blood flow dynamics in patient specific arterial network in head and neck.2nd International Conference on Mathematical and Computational Biomedical Engineering - CMBE2011 2011, 486–489. . [ISBN: 9780956291417] http://eprints.gla.ac.uk/50432/
18. Holdsworth D, Norley C, Frayne R, Steinman D, Rutt B: Characterization of common carotid artery blood-flow waveforms in normal human subjects. Physiological Measurement 1999, 20: 219. 10.1088/0967-3334/20/3/301
19. Vermeersch S, Rietzschel E, Buyzere M, Bacquer D, Backer G, Bortel L, Gillebert T, Verdonck P, Segers P: Determining carotid artery pressure from scaled diameter waveforms: comparison and validation of calibration techniques in 2026 subjects. Physiological Measurement 2008, 29: 1267. 10.1088/0967-3334/29/11/003
20. Zhao S, Xu X, Hughes A, Thom S, Stanton A, Ariff B, Long Q: Blood flow and vessel mechanics in a physiologically realistic model of a human carotid arterial bifurcation. Journal of Biomechanics 2000, 33(8):975–984. 10.1016/S0021-9290(00)00043-9
21. Hoi Y, Wasserman B, Xie Y, Najjar S, Ferruci L, Lakatta E, Gerstenblith G, Steinman D: Characterization of volumetric flow rate waveforms at the carotid bifurcations of older adults. Physiological Measurement 2010, 31: 291. 10.1088/0967-3334/31/3/002
22. Formaggia L, Quarteroni A, Veneziani A (Eds): Cardiovascular Mathematics: Modeling and simulation of the circulatory system. MS&A; 1SpringerLink: Bücher, Milano: Springer-Verlag Milan; 2009.
23. Laurien E, Oertel H (Eds): Numerische Strömungsmechanik: Grundgleichungen und Modelle - Lösungsmethoden - Qualität und Genauigkeit. Wiesbaden: Vieweg+Teubner Verlag/GWV Fachverlage GmbH, Wiesbaden, 3., vollständig überarbeitete und erweiterte auflage edition; 2009.
24. Sutera S, Mehrjardi M: Deformation and fragmentation of human red blood cells in turbulent shear flow. Biophysical Journal 1975, 15: 1–10. 10.1016/S0006-3495(75)85787-0
25. Fry D: Acute vascular endothelial changes associated with increased blood velocity gradients. Circulation Research1968, 22(2):165. http://circres.ahajournals.org/cgi/reprint/22/2/165.pdf

26. John L: Forward electrical transmission line model of the human arterial system. Medical and Biological Engineering and Computing 2004, 42(3):312–321. 10.1007/BF02344705

27. Snyder M, Rideout V, Hillestad R: Computer modeling of the human systemic arterial tree. Journal of Biomechanics1968, 1(4):341–353. 10.1016/0021-9290(68)90029-8

28. Chellappan K, Zahedi E, Mohd Ali M: Age-related Upper Limb Vascular System Windkessel Model using Photoplethysmography. In 3rd Kuala Lumpur International Conference on Biomedical Engineering 2006. Springer; 2007:563–566.

29. Avolio A: Multi-branched model of the human arterial system. Medical and Biological Engineering and Computing1980, 18(6):709–718. 10.1007/BF02441895

30. Olufsen M, Peskin C, Kim W, Pedersen E, Nadim A, Larsen J: Numerical simulation and experimental validation of blood flow in arteries with structured-tree outflow conditions. Annals of biomedical engineering 2000, 28(11):1281–1299.

31. Olufsen M: Structured tree outflow condition for blood flow in larger systemic arteries. American Journal of Physiology- Heart and Circulatory Physiology 1999, 276: H257.

32. Marshall I, Papathanasopoulou P, Wartolowska K: Carotid flow rates and flow division at the bifurcation in healthy volunteers. Physiological Measurement 2004, 25: 691. 10.1088/0967-3334/25/3/009

33. Deane C, Markus H: Colour velocity flow measurement: in vitro validation and application to human carotid arteries. Ultrasound in medicine & biology 1997, 23(3):447–452. 10.1016/S0301-5629(96)00224-4

CHAPTER 13

NUMERICAL SIMULATION ON THE EFFECTS OF DRUG ELUTING STENTS AT DIFFERENT REYNOLDS NUMBERS ON HEMODYNAMIC AND DRUG CONCENTRATION DISTRIBUTION

Yu Chen[1], Yan Xiong[2], Wentao Jiang[1], Fei Yan[1], Meng Guo[3], Qingyuan Wang[1] and Yubo Fan[3]

[1] Department of Applied Mechanics, Sichuan University, Chengdu, 610065, China

[2] School of Manufacturing Science & Engineering, Sichuan University, Chengdu, 610065

[3] School of Biological Science and Medical Engineering, Beihang University, Beijing, 100191, China.

ABSTRACT

Background

The changes of hemodynamics and drug concentration distribution caused by the implantation of drug eluting stents (DESs) in curved vessels have significant effects on In-Stent Restenosis.

Methods

A 3D virtual stent with 90°curvature was modelled and the distribution of wall shear stress (WSS) and drug concentration in this model were numerically studied at Reynolds numbers of 200, 400, 600, 800.

Results

The results showed that (1) the intensity of secondary flow at the 45° cross-section was stronger than that at the 90° cross-section; (2) As the Reynolds number increases, the WSS decreases. When the Reynolds number reaches 600, the low-WSS region only accounts for 3% of the total area. (3) The effects of Reynolds number on drug concentration in the vascular wall decreases in proportionally and then the blood velocity increased 4 times, the drug concentration in the vascular wall decreased by about 30%. (4) The size of the high drug concentration region is inversely proportional to the Reynolds number. As the blood velocity increases, the drug concentration in the DES decreases, especially at the outer bend.

Citation: Yu Chen, Yan Xiong, Wentao Jiang, Fei Yan, Meng Guo, Qingyuan Wang and Yubo Fan, "Numerical simulation on the effects of drug eluting stents at different Reynolds numbers on hemodynamic and drug concentration distribution," BioMedical Engineering OnLine201514(Suppl 1):S16 DOI: 10.1186/1475-925X-14-S1-S16.

Conclusions

It is beneficial for the patient to decrease vigorous activities and keep calm at the beginning of the stent implantation, because a substantial amount of the drug is released in the first two months of stent implantation, thus a calm status is conducive to drug release and absorption; Subsequently, appropriate exercise which increases the blood velocity is helpful in decreasing regions of low-WSS.

INTRODUCTION

Atherosclerosis is a specific form of arteriosclerosis in which an artery wall thickens as a result of the accumulation of cholesterol and triglyceride [1]. Drug eluting stents (DESs) in cardiovascular surgery significantly reduce the rate of restenosis caused by in-stent intimal hyperplasia [2] and have been considered and accepted as one of the most promising treatment methods for preventing restenosis [3].

The blood flow in humans and animals is always laminar, and only in abnormal conditions is there turbulent blood flow for a long time [4]. Qu et al.[5] found that with age the fluctuation ranges of normal value increased and the mean blood velocity decreased. Yin et al.[6] found that the blood velocity in the basilar and carotid arteries in normal humans decreased with age, with the blood velocity in the range 0.2 m-0.8 m/s. Normally, the peak of diastolic velocities in a human coronary artery is 500 ± 170 mm/s, with a mean blood velocity of 370 ± 120 mm/s. With cardio intervention, the average peak velocity (APV) at the proximal and distal increases and the coronary flow reserve (CFR) improves [7]. In addition, the blood velocity for humans when in motion increases up to 30% even higher [8]. From the viewpoint of biomechanics, factors such as blood flow velocity, impact flow, pressure and wall shear stress (WSS) have important effects on atherosclerosis. Seo et al.[9] discussed that how to reduce erratic flow at different Reynolds numbers for different curvature stents. Sukavaneshvar et al.[10] analyzed the hemodynamics on the two restenosis positions at two different Reynolds numbers. Kolachalama et al.[11] discussed the changes of drug concentration, the Area Under Curve(AUC) and the recirculation length at 2, 4, 6 and 8 times Reynolds number. Chen et al.[12] analyzed the effects of drug-coating positions on the drug concentration in curved stents and found that the contribution of the contacting surface to the drug concentration in the vascular wall was underestimated in 2D numerical simulation. The implantation of stents into coronary arteries can influence the fluid dynamics in the regions adjacent to the arterial wall, and consequently cause damage to the regional arterial wall and a change of WSS. Especially when a drug eluting stent (DES) is implanted in a curved vessel, secondary flow arises due to the curvature. Secondary flows occur in the cross-stream direction caused by the base primary flow in the streamwise direction [13]. Secondary flow structures may affect the wall shear stress in arteries, and is closely related to atherogenesis[14]. So the relationships between stent design and placement, drug release, deposition concentration, distribution of low-WSS, and secondary flow are important issues in research on stents. Autumn et al.[15] compared the change of secondary flow, with and without stents, through PIV experiments. Jung et al.[16] thought that the accumulation of red blood cells on the inner curvature was caused by the secondary flow and higher residence times. Bioron et al.[17] analysed the behaviour of unsteady flows in a bend with experimental and numerical simulation. Sun et al.[18] discussed the flow characteristics in a 90° bent pipe at different Reynolds numbers, and put forward the view that the higher the Reynolds number, the larger the pressure on the pipe wall. Stoesser et al.[19] compared the results of numerical simulation and experimental data for the flow and wall shear stress distributions in a meandering open channel. The emphasis of the aforementioned literature is on secondary flow in a curved channel or pipe, but research on the implantation of a stent in a curved vessel is sparse, therefore it is necessary to study the effects of secondary flow on the hemodynamics and drug distribution in curved vascular walls.

This paper applies the method of computational fluid dynamics (CFD) to investigate the effects on hemodynamics and drug concentration in a 3D 90° DES. At different Reynolds number the relationship between hemodynamics, drug concentration and in-stent restenosis was also studied. The results can provide a theoretical guidance for optimization of the design of DES.

METHODS

Models

The configuration of the 90° stent, with six stent struts is shown in Figure 1). The radius of curvature is 10 mm. The diameter of the stent model is set as 3 mm, the length of stent is 15 mm, the thickness of the vascular wall is 0.5 mm, and the cross-section of the stent struts is square (0.10 mm × 0.10 mm). It was assumed that the stent makes full contact with the vascular wall. The Reynolds number are set as 200, 400, 600 and 800, with blood velocities of 0.22 m/s, 0.44 m/s, 0.66 m/s and 0.88 m/s, respectively. In order to ensure fully developed blood flow, a 30-mm length of straight tube was added at the inlet and outlet of the model.

GAMBIT 2.3 (ANSYS, Inc., USA) software was used to build the model. Unstructured meshes were applied to the model. Grids were refined at the stent and vascular surface to improve computational accuracy. The refinement on the meshes continued until the computational results reached grid-independence. A mesh number of about 3 million cells of the 90°curved stent was finally adopted in the following investigation.

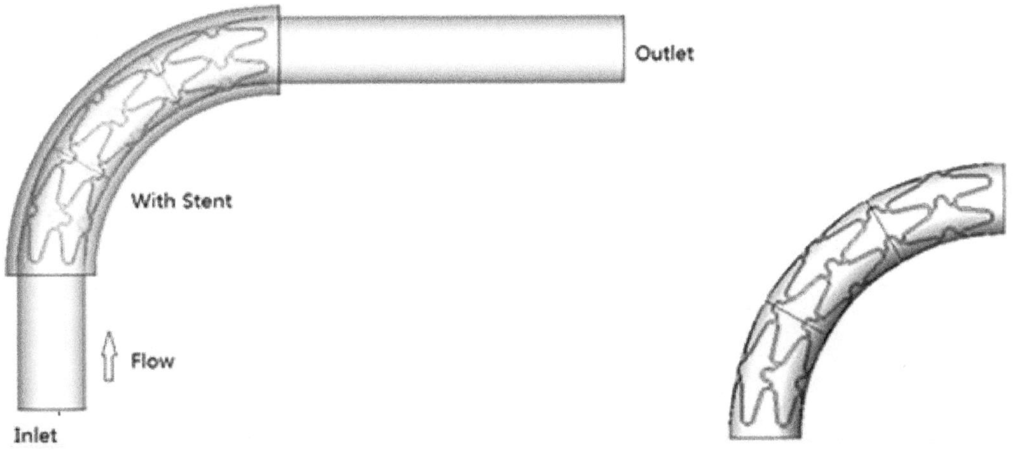

Figure 1: Curved stent model.

Boundary Condition

Blood was assumed to be an isotropic and incompressible Newtonian fluid with constant density and viscosity and in steady flow. The vascular wall was assumed to be impermeable and rigid, and although not thoroughly correct, wall elasticity is of somewhat less importance in regard to the gross features of the flow. Previous studies strongly indicated that the effect of non-Newtonian fluid behavior appears to be negligible[20, 21]. Accordingly, the governing equations are as follows:

$$\begin{cases} \dfrac{\partial u_j}{\partial x_j} = 0 \cdots\cdots\cdots (i,j = 1,2,3) \\ \rho\dfrac{\partial u_j u_i}{\partial x_j} = -\dfrac{\partial P}{\partial x_i} + \mu\dfrac{\partial}{\partial x_j}\left(\dfrac{\partial u_i}{\partial x_j} + \dfrac{\partial u_j}{\partial x_i}\right) \end{cases} \quad (1)$$

where ρ and μ are the density and dynamic viscosity of blood, respectively, $\rho = 1055\ kg/m^3$, $\mu = 3.5 \times 10^{-3} kg/ms$; P is pressure, and u_i is the velocity vector of blood.

Boundary conditions were set as follows. The parabolic profile with a centerline velocity was applied at the inlet; the range of inlet velocity is from 0.22 m/s to 0.88 m/s:

$$\begin{cases} V_z|_{inlet} = V_m \times \left(1 - \dfrac{x^2 + y^2}{r^2}\right) = 0.22\text{m/s},\ 0.44\text{m/s},\ 0.66\text{m/s and } 0.88\text{m/s} \\ V_r|_{inlet} = 0 \end{cases} \quad (2)$$

V_z, V_r are the velocities in the axial and radial directions of the blood flow, respectively. r = 0.0015m.

A zero pressure boundary condition was applied at the outlet.

$$P|_{outlet} = 0 \quad (3)$$

Luminal drug transport was described by the steady-state convection-diffusion Eq. (4) and drug transport through the vascular wall was modelled by the steady state diffusion equation.

$$\begin{cases} \dfrac{\partial}{\partial x_i}\left(u_i\phi_f - \tau_{kf}\dfrac{\partial \phi_f}{\partial x_i}\right) = 0 \\ \dfrac{\partial}{\partial x_i}\left(-\tau_{kt}\dfrac{\partial \phi_t}{\partial x_i}\right) = 0 \end{cases} \quad (i = 1,2,3) \quad (4)$$

Φ_f, Φ_t are the drug concentrations in the blood and in the vascular wall, respectively. The diffusion coefficient of the drug through the blood (τ_{kf}) is $10^{-7}m^2/s$[21], and the diffusion coefficient of the drug through the tissue (τ_{kt}) is $10^{-12}\ m^2/s$[21].

$$\phi_f\Big|_{inlet} = 0,\ \dfrac{\partial \phi_f}{\partial z}\Big|_{outlet} = 0 \quad (5)$$

$$\dfrac{\partial \phi_t}{\partial r}\Big|_{perivascular\ wall} = 0 \quad (6)$$

$$\tau_{kf}\dfrac{\partial \phi_f}{\partial r}\Big|_{blood} = \tau_{kt}\dfrac{\partial \phi_t}{\partial r}\Big|_{tissue} \ \text{(Tissue – blood interface)} \quad (7)$$

$$\dfrac{\partial \phi_t}{\partial z} = 0 \ \text{(Upstream and downstream boundaries of the tissue)} \quad (8)$$

$$\phi = 1 \ \text{(1 indicates the stent surfaces assigned for drug coating)} \quad (9)$$

The drug concentration was set to zero at the luminal inlet (Eq. 5), and an open boundary condition was applied at the distal boundary. The drug transport within the tissue was modelled as a simple diffusion process, with an impermeable boundary condition on the perivascular wall (shown in Eq. 6). The boundary condition of the continuity flux in drug transport is shown in Eq.7, which allows the entire drug in the blood at the tissue-blood interface to be transported into the arterial tissue. The drug release of the stents was simulated as a Dirichlet boundary condition (Eq. 9), with a drug concentration of unity at the strut surfaces.

The computations were conducted by using a commercial CFD package, FLUENT 6.3 (ANSYS Inc.), in which a finite volume method was used to discretize the governing equations. 3D single-precision format and a segregated solver were used. The SIMPLEC algorithm was applied to complete the velocity-pressure correction. The standard format was chosen for the pressure discretization and 2nd-order upwind for the momentum equations. The residual error convergence threshold was set as 0.0001.

To have a better observation on the drug concentration in the vascular wall for the curved stent, three planes were defined as the reference planes at 0°, 45°, 90° cross-sections from the × negative axis, clockwise(as shown in Figure 2(a)). Two lines in the middle segment were chosen for the investigation (as shown in Figure 2(b)). Line A is the reference line in the blood at a 1.25 mm radius, line B is another reference line at a 1.75 mm radius in the vascular wall.

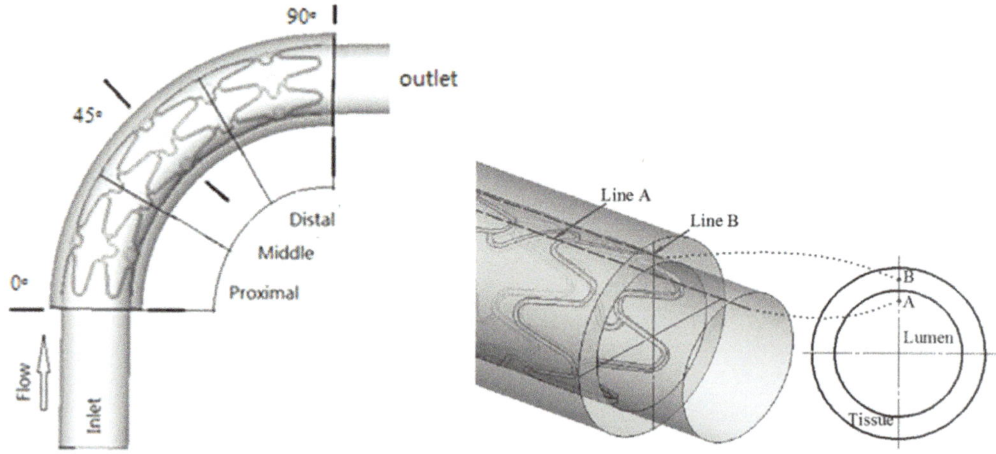

Figure 2: Reference location (a)0°,45°,90° cross-sections (b) Reference Line A and Line B.

RESULTS

Secondary Flow

In the case of Newtonian fluids, Dean Vortices appear as a pair of vortices in curved ducts or bends because of flow inertia. When fluid is directed around a curve under a pressure driven flow, the high velocity streams in the centre of the channel experience a greater centripetal force and so are deflected outward [22]. The Dean number (Dn) is a dimensionless group in fluid mechanics, which occurs in the study of flow in curved pipes and channels [23].

$$Dn = Re \times (r/R)^{1/2}$$

Re --- Reynolds number

r --- The diameter of the pipe

R --- Bend radius

When the Dean number is greater than 36, the fluid appears unstable phenomenon. The Reynolds numbers are 200, 400, 600, 800. In this paper, thus the Dean numbers are calculated as 77.5, 155, 232, and 310, respectively.

The distributions of streamlines on the 45°and 90°cross-sections are shown in Figure 3, where the Dean vortex can be observed clearly. On the 45° cross-section, the high velocity regions are mainly concentrated on the outer bend and the relatively low velocity regions are on the inner bend at small Reynolds numbers. With increasing Reynolds numbers, the high velocity regions reduce gradually in the middle of the pipeline, and close to both sides. Comparing the velocity on the 45°and 90° cross-sections, the trend is consistent, but the velocity on the 90°cross-section is slightly smaller than that on the 45° cross section.

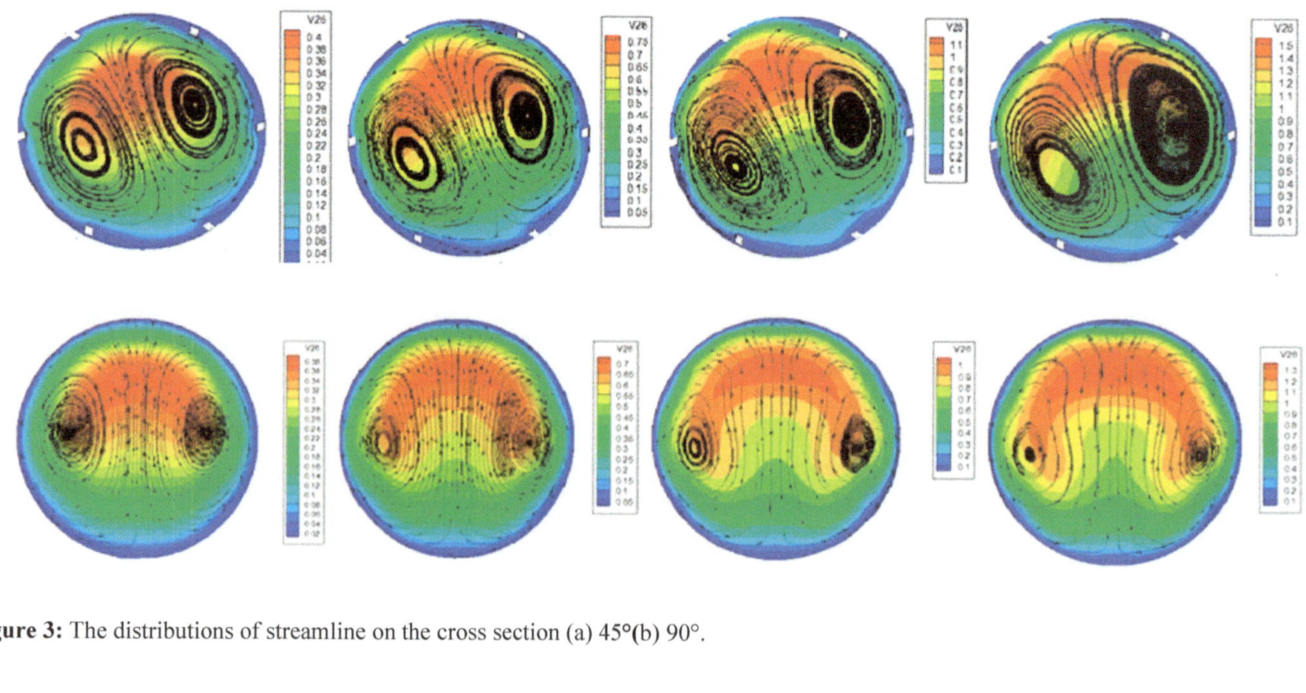

Figure 3: The distributions of streamline on the cross section (a) 45°(b) 90°.

Comparison of the peak values of the velocity on the 45°and 90° cross-sections at different Reynolds numbers is shown as Figure 4. With increasing Reynolds numbers, the peak values have a linear growth. At a low Reynolds number, such as Reynolds number 200, the peak values are similar; the peak values on the 45°cross section are slightly higher than for on 90° at high Reynolds number.

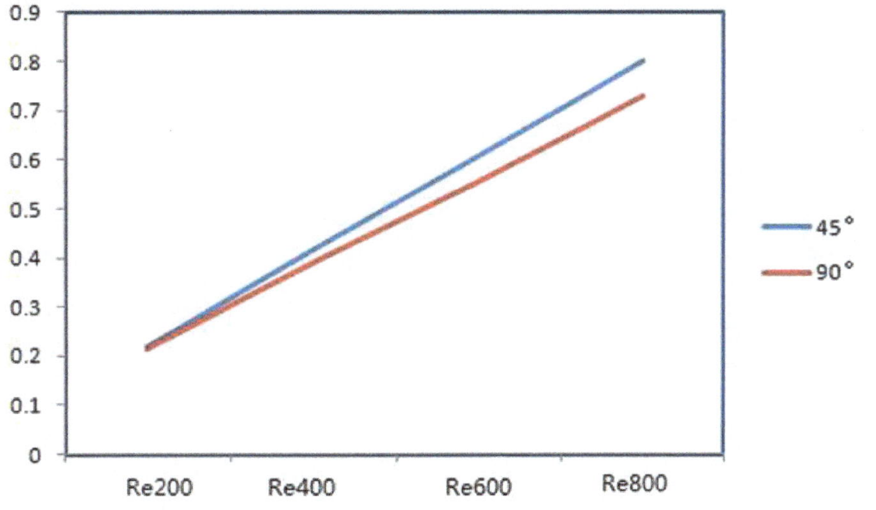

Figure 4: Comparison of the peak values of velocity on the cross-sections.

Velocity in Cross-Section

The distributions of the velocity on the 45°and 90° cross-sections for the model at different Reynolds numbers are shown in Figure 5. On the 45 ° cross-section, the high velocity region occurs in the center of the axis and the relative area is large, and with increasing Reynolds number, the high velocity region moves to the outer bend side. On the 90° cross-section, when the Reynolds number is low, the high velocity region is small, and with the increase of the Reynolds number, the high velocity region slowly appears close to the two sides.

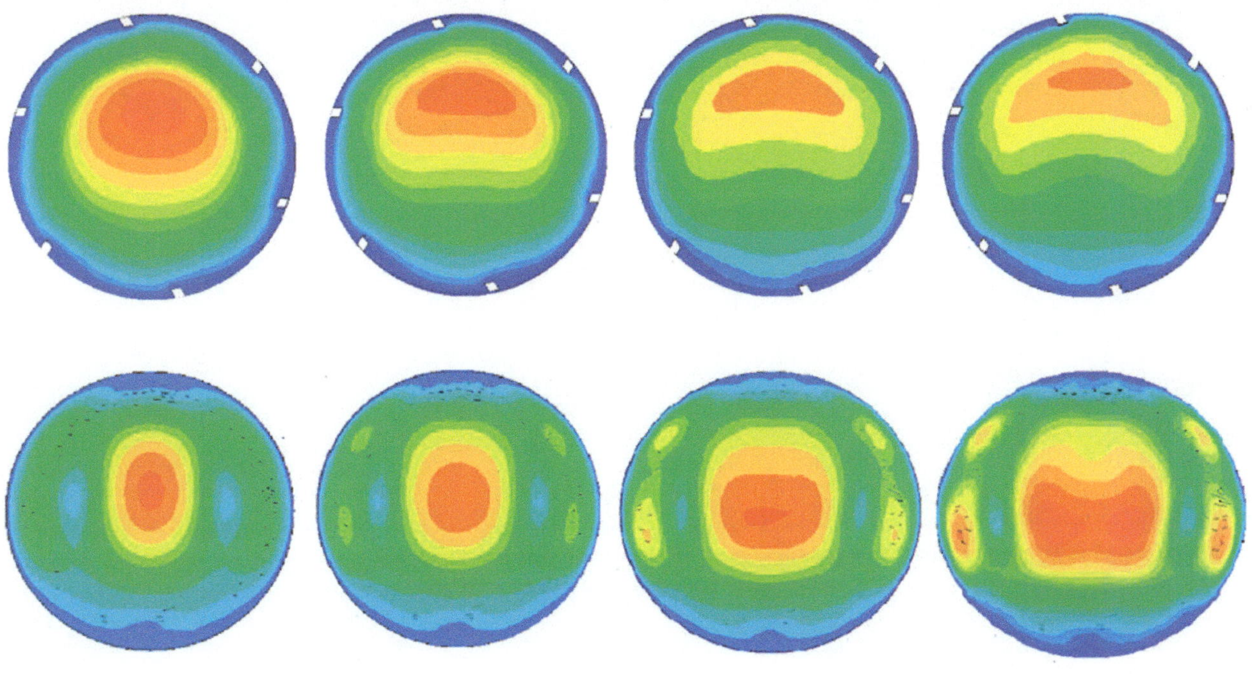

Figure 5: The velocity on the cross sections (a) 45°(b) 90°.

The peak values of velocity for these two-cross sections are shown in Figure 6. The peak values of velocity show the linear increases for both cross-sections and the values are bigger on 45°cross-section than that on 90°with increase of Reynolds number.

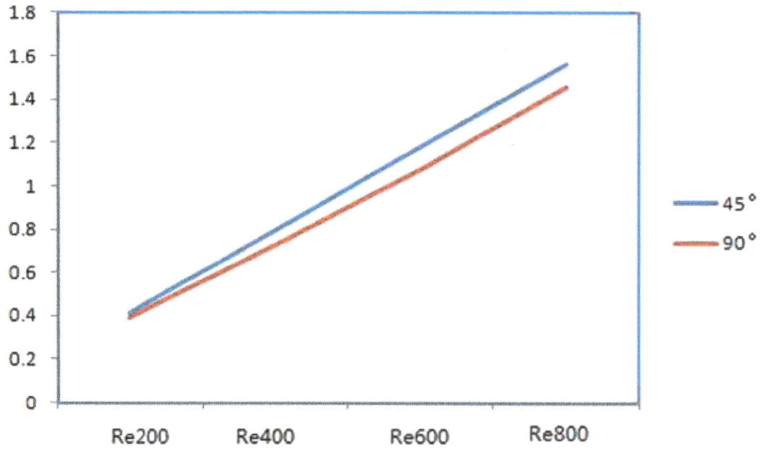

Figure 6: The peak values of velocity on the cross sections.

WSS

The distributions of WSS for the model at different Reynolds numbers are shown in Figure 7. With increasing Reynolds number, low-WSS regions are fewer. A WSS of < 0.5 Pa is defined as a low WSS [20], shown as like dark blue region in Figure 7. The ratios of the low-WSS region to the whole model surface of the model at different Reynolds numbers are shown in Figure 8.

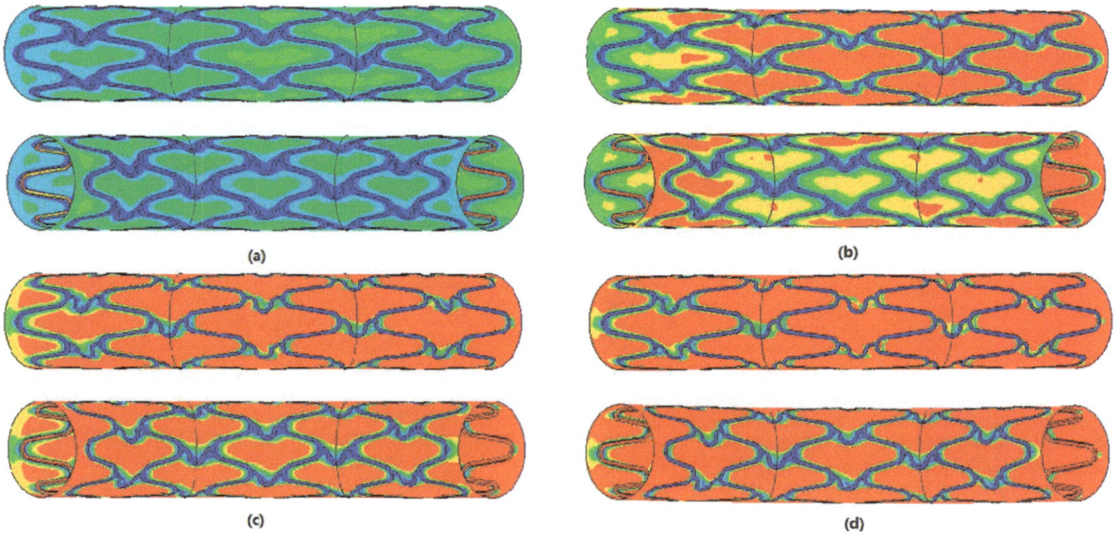

Figure 7: The distributions of WSS for the model at different Reynold number.

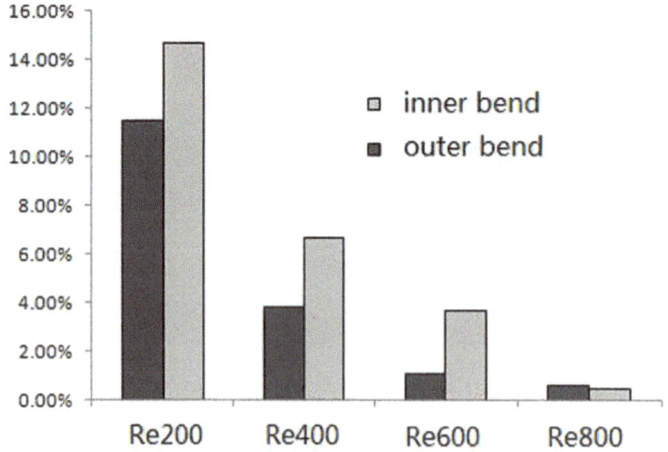

Figure 8: The percentage of the low-WSS regions to the vascular wall surfaces for stent models with different Reynold number.

Obviously, the curved stent model at smaller Reynolds numbers has more low-WSS regions, which implies that it might be the most susceptible to in-stent restenosis. There are more low-WSS regions on the inner bend that on the outer bend. When the Reynolds number is larger than 600, the low-WSS ratio is less than 3%, which helps improve the in-stent restenosis.

Drug Concentration

The drug concentration distributions on the inner/outer bends in Line A in the blood at different Reynolds numbers are shown in Figure 9. The peak value of drug concentration on the inner bend is 5 times higher than that on the outer bend and the average value on the inner bend is 10 times higher than that on the outer bend. The drug concentration on the inner bend reduces in proportion with the increase of blood flow velocity; the drug concentration at a Reynolds number of 200 on the outer bend is significantly higher than that at other Reynolds numbers. The phenomena of "Proximal low and distal high" are very obvious on the inner bend, with the value of drug concentration rising in each stent structs. The phenomena "Proximal low and distal high" is not obvious on the outer bend, where the drug concentration is flat between the stent struts.

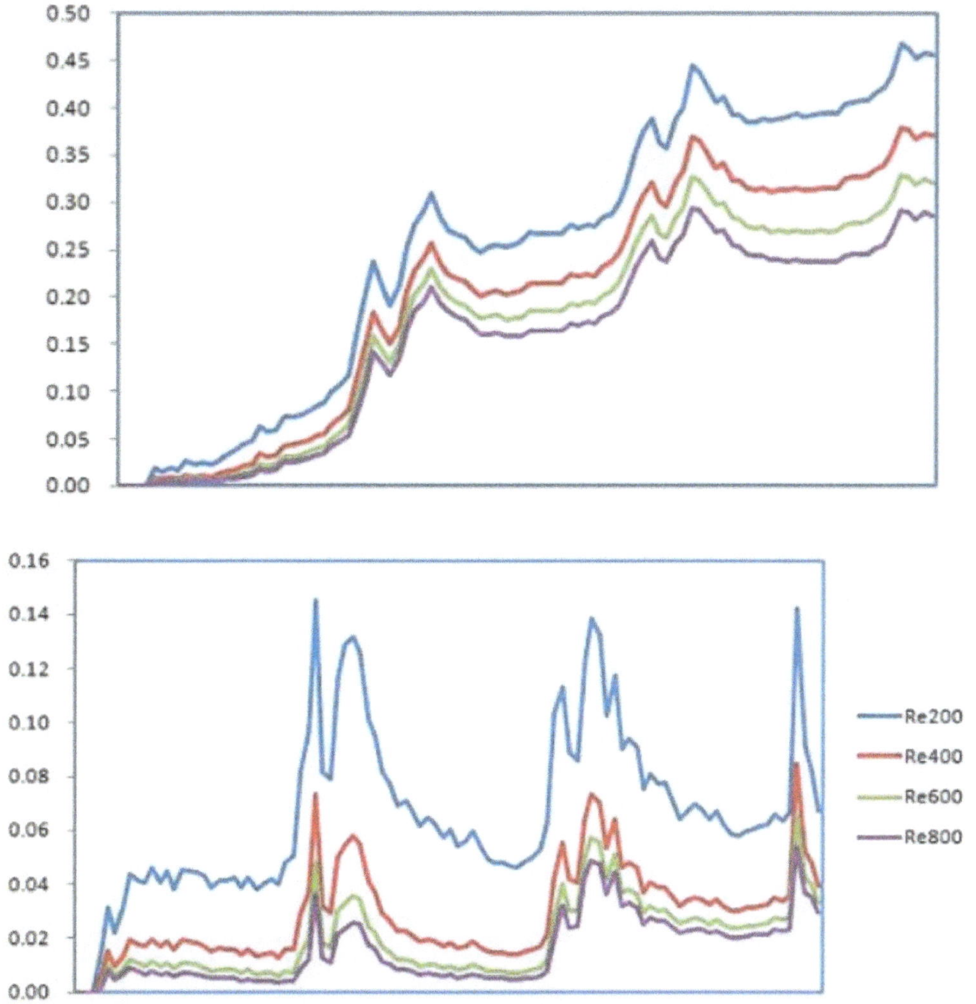

Figure 9: The curves of drug concentration in blood (a) Inner bend (b) Outer bend.

The drug concentration distributions on the inner/outer bends in Line B in the vascular walls at different Reynolds number are shown in Figure 10. The peak value of the drug concentration on the inner bend is 15% higher than that on the outer bend and the average value on the inner bend is 40% higher than that on the outer bend. The drug concentration on the inner bend doesn't obviously decreases compared with that on the outer bend in each stent structs. The drug concentration reduces proportionally with increasing Reynolds number, but the ratio of the drug concentration doesn't change much with Reynolds number.

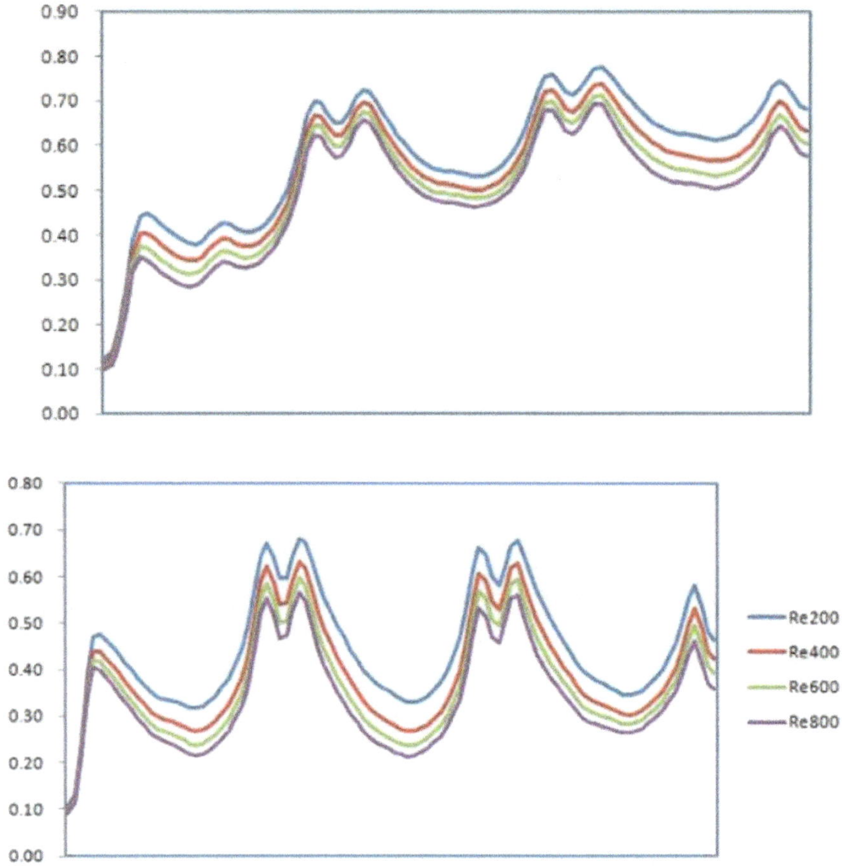

Figure 10: The curves of drug concentration in the vascular walls (a) Inner bend (b) Outer bend.

Comparisons of the mean values of the drug concentration in the proximal, middle and distal segments are shown in Figure 11. The drug concentration increases gradually with blood flow direction, and the drug concentration in the distal is higher than that in the proximal and middle in a straight pipe [21]. For the curved pipe, secondary flow is a problem that can't be avoided. Due to the effects of secondary flow in the curved stents, the velocity in the circumferential direction helps gather and accumulate the drug in the blood in the middle and distal segments of the stent. The drug concentration in middle segment at low Reynolds number is slightly higher than that in distal on the inner bend. On the contrary, the drug concentration in the distal is slightly higher than that in the middle at high Reynolds number; the drug concentrations in the distal and in the middle are almost uniform on the outer bend. It can be seen that the drug concentration is more obvious on the outer bend in the middle and distal of the stent with increase of Reynolds number.

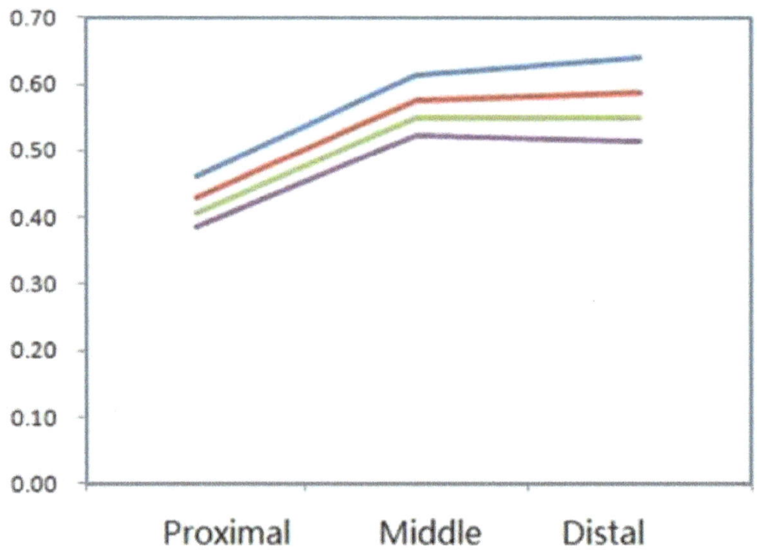

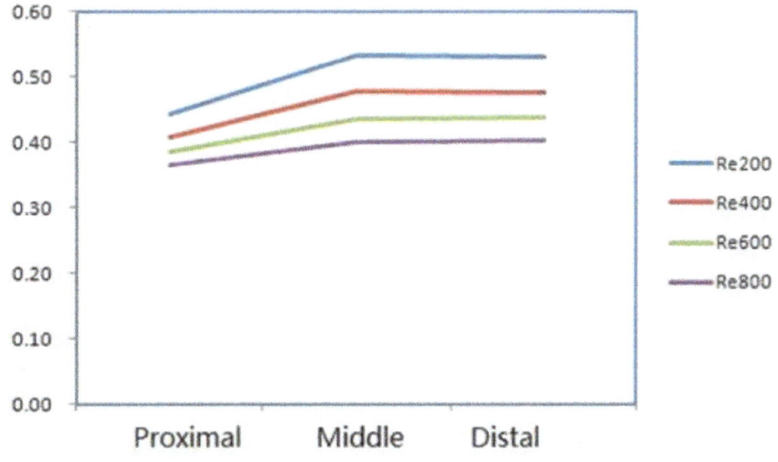

Figure 11: The curves of drug concentration in proximal, middle and distal segments of the vascular walls (a)Inner bend (b)Outer bend.

DISCUSSION

Atherosclerosis in carotid arteries and coronary arteries, abdominal aorta, renal artery and aortic arch, after DES implanted in blood vessels, causes the velocity of the blood flow to be different in different parts of the human body. There is a lack of quantitative researches on the effects of the WSS and drug concentration in the blood and vascular wall at different Reynolds number.

Some diseases may make the blood flow velocity faster or slower, dynamic changes in the blood viscosity would lead to changes in the blood velocity. The elderly are susceptible to hyper viscosity, and there are many factors that affect the blood viscosity: blood cell factors, such as blood cells number, size, shape, RBC deformability, platelet function, etc.; plasma reasons, such as plasma proteins (especially fibrinogen, immune globulin), blood glucose, blood lipid, fibrin lytic activity, etc.; vascular factors, such as length, diameter and lining smoothness, etc.; other factors, such as emotions, life style, smoking, drinking, etc. Common features of these diseases, such as atherosclerosis, moyamoya disease, vasculitis, partial recanalization, inflammation, and tumor thrombus, are caused by vascular stenosis. Radioactive damage caused by artery stenosis, dissecting aneurysm, and vasospasm is one of the reasons causing artery stenosis. For healthy human beings there are similar situations, the velocity of the blood flow changes according to the different statuses of the body: when undertaking sports, the velocity of the blood flow will be faster; when at rest and the velocity of the blood flow will be relatively slower.

We can see clearly the effects of Reynolds number on the WSS and drug concentration. When the Reynolds number increases, the low-WSS region gradually decreases, for a Reynolds number of 600, the percentage of the low-WSS region is less than 3%, and the Reynolds number of 800, the percentage decreases to 1%, with the effect of low-WSS almost negligible. Therefore, as the Reynolds number increases, the low-WSS region decreases.

Increasing blood flow velocity leads to a decrease of the drug concentration in the vascular walls, for example, increasing Reynolds number to 400 from 200, the drug concentration drops 8% approximately, but increase to 800 from 200 that is four times of original value 200, causes the drug concentration to decrease by 30% approximately. The effects of decreasing drug concentrations on the outer bend are greater than that on the inner bend.

CONCLUSIONS

In this study, the effects of different Reynolds number on the hemodynamics, drug concentration and secondary flow were analyzed with 3D DES models. The results showed that:

1. The intensity of secondary flow on the 45° cross-section was stronger than that on the 90° cross-section, meaning that the enhancement of flow in the circumferential direction leads to a decrease of the low-WSS region;
2. With increasing Reynolds number, the region of low-WSS decreased. When the Reynolds number is larger than 600, the region of low-WSS is smaller than 3%.
3. The effects of Reynolds number on drug concentration in the vascular wall decreased proportionally and as blood velocity increased 4 times, the drug concentration in the vascular wall decreased by 30%.
4. As the blood velocity increased, the drug concentration in the DES decreased, especially on the outer bend.

When the DES is implanted in the blood vessel, it is beneficial for the patient to decrease vigorous activities and keep calm at the beginning of stent implantation, because a substantial amount of drug is released in the first two months of stent implantation, and calmness is conducive to drug release and absorption. Later, appropriate exercise can achieve higher blood velocity which is helping in decreasing regions of low-WSS.

ACKNOWLEDGEMENTS

This work is supported by Grants-in-Aid from the National Natural Science Research Foundation of China (No. 61190123, 51205262, 10872138) and Applied Basic Research Programs of science & technology department of Sichuan Province (No. 2014JY0148).

AUTHORS' CONTRIBUTIONS

YC and YX conceived of the study, developed and implemented the modelling, calculation and analysis, and wrote most of the manuscript. WJ conceived the study, contributed to the whole frame, and wrote parts of the manuscript. FY and MG supplied the data and rigorously revised the manuscript. QW and YF thoroughly reviewed the manuscript.

Yu Chen and Yan Xiong contributed equally to this work and should be considered as co-first authors.

REFERENCES

1. Kumar V, Cotran RS, Robbins SL: Basic Pathology. WB Saunders; 1992.
2. Chen J, Ni Z, Gu X: Survey of coronary stents development for restenosis prevention. Chinese journal of medical instrumentation 2009, 33(6):429–434.
3. Acharya Ghanashyam, Park Kinam: Mechanisms of controlled drug release from drug-eluting stents. Advanced Drug Delivery Reviews 2006, 58: 387–401. 10.1016/j.addr.2006.01.016
4. Wang XQ: The application of Reynolds number in biology. Journal of Qinghai University 2003, 21(5):67–69.
5. Qu SB: Transcranial doppler ultrasound instrument detection of carotid artery blood flow velocity normal reference range. Journal of harbin medical university 1994, 28(3):211–213.
6. Yin AY: Transcranial doppler ultrasound measurement in normal human brain artery and carotid artery blood flow velocity. Chinese Journal of Ultrasound in Medicine 1996, 12(12):60–64.
7. Wei M: Doppler blood flow velocity measurement evaluation coronary angioplasty and curative effect of stenting PCI.Chinese Journal of Cardiology 1999, 27(5):337–339.
8. Du W: The effects of exercise on human artery blood flow velocity. Journal of Beijing Sport University 1996, 8(2):14–17.
9. Taewon Seo, Schachter LevantoG: Computational Study of Fluid Mechanical Disturbance Induced by Endovascular Stents. Annals of Biomedical Engineering 2005, 33(4):444–456. 10.1007/s10439-005-2499-y
10. Sukavaneshvar Sivaprasad, Gesse M: Rosa Enhancement of Stent-Induced Thromboembolism by Residual Stenoses: Contribution of Hemodynamics. Annals of Biomedical Engineering 2000, 28: 182–193.
11. Kolachalama VijayaB, Tzafriri AbrahamR: Luminal flow patterns dictate arterial drug deposition in stent-based delivery.Journal of Controlled Release 2009, 13: 24–30.
12. Chen Yu, Yan Fei, Jiang Wen-tao, Wang Qing-yuan, Fan Yu-bo: Numerical Study on the Effects of the Drug-coating Positions of Drug-eluting Stents on Drug Concentration, Journal of Medical and Biological Engineering. 2013.
13. Zhan H, Zhang X, Dai C: Influence of Secondary Flow in Water-only Cyclone on its Separation Mechanism. Mining and Metallurgical Engineering 2002, 2: 017.
14. Glenn AL, Bulusu KV, Shu F, et al.: Secondary flow structures under stent-induced perturbations for cardiovascular flow in a curved artery model. International Journal of Heat and Fluid Flow 2012, 35: 76–83.
15. Glenn AL, Bulusu KV, Shu F: Secondary flow structures under stent-induced perturbations for cardiovascular flow in a curved artery model. International Journal of Heat and Fluid Flow 2012, 35: 76–83.
16. Jung J, Lyczkowski RW, Panchal CB: Multiphase hemodynamic simulation of pulsatile flow in a coronary artery. Journal of biomechanics 2006, 39(11):2064–2073. 10.1016/j.jbiomech.2005.06.023
17. Boiron O, Deplano VER, Pelissier R: Experimental and numerical studies on the starting effect on the secondary flow in a bend. Journal of Fluid Mechanics 2007, 574: 109–130.
18. Sun Y, Hu S, Zhao J: Numerical study on flow characteristics of 90 bend pipe under different reynolds number. Journal of University of Shanghai for Science and Technology 2010, 6: 005.
19. Stoesser T, Ruether N, Olsen NRB: Calculation of primary and secondary flow and boundary shear stresses in a meandering channel. Advances in Water Resources 2010, 33(2):158–170. 10.1016/j.advwatres.2009.11.001
20. Chen Y, Jiang W, Chen X, Zheng T, Wang Q, Fan Y: Numerical simulation on the effects of drug-eluting stents with different links on hemodynamics and drug concentration distribution. Journal of Mechanics in Medicine and Biology2013., 13(04):
21. Balakrishnan B, Dooley JF, Kopia G: Intravascular drug release kinetics dictate arterial drug deposition, retention, and distribution. Journal of Controlled Release 2007, 123(2):100–108. 10.1016/j.jconrel.2007.06.025
22. Howell PB Jr, Mott DR, Golden JP: Design and evaluation of a Dean vortex-based micromixer. Lab on a Chip 2004, 4(6):663–669. 10.1039/b407170k
23. Ali S: Pressure drop correlations for flow through regular helical coil tubes. Fluid Dynamics Research 2001, 28(4):295–310. 10.1016/S0169-5983(00)00034-4

CHAPTER 14 ADVANCES IN MODELLING OF BIOMIMETIC FLUID FLOW AT DIFFERENT SCALES

Sujoy Kumar Saha[1] and Gian Piero Celata[2]

[1] Mechanical Engineering Department, Bengal Engineering and Science University, Shibpur, Howrah, West Bengal 711 103, India

[2] ENEA Casaccia Research Centre, Institute of Thermal Fluid Dynamics, Office Building F-20, Via Anguillarese 301, S. M. Galeria, Rome 00123, Italy

ABSTRACT

The biomimetic flow at different scales has been discussed at length. The need of looking into the biological surfaces and morphologies and both geometrical and physical similarities to imitate the technological products and processes has been emphasized. The complex fluid flow and heat transfer problems, the fluid-interface and the physics involved at multiscale and macro-, meso-, micro- and nano-scales have been discussed. The flow and heat transfer simulation is done by various CFD solvers including Navier-Stokes and energy equations, lattice Boltzmann method and molecular dynamics method. Combined continuum-molecular dynamics method is also reviewed.

INTRODUCTION

Human knowledge is getting enriched from the four billion years' worth of R & D in the natural world of plants and animals and other lower level living creatures and microorganisms, which have evolved through the ages to nicely adapt to the environment. Man has now drawn his attention to soil creatures like earthworms, dung beetle, sea animals like shark and plants and trees like lotus leaf and pastes like termites. In the nature, we see examples of effortless and efficient non-sticking movement in mud or moist soil, high-speed swimming aided by built-in drag-reduction mechanism, water repellant contaminant-free surface cleaning mechanism and natural ventilation and air conditioning, [1, 2, 3, 4, 5, 6, 7, 8]. By nature, feather of the penguin shows staying warm naturally, Figure 1 [4]. The leaf of the lotus is hydrophobic to the extent that water running across the surface of the leaf retains particles of dirt caused by a thick layer of wax on the surface and the sculpture of that surface, Figure 2 [9, 10, 11]. This forces the droplets of water to remain more or less spherical when in contact with the leaf, and reduces the tendency of other contaminants to stick to the leaf. It has been proved that water repellency causes an almost complete surface purification (self-cleaning effect): contaminating particles are picked up by water droplets or they adhere to the surface of the droplets and are then removed with the droplets as they roll off the leaves. This characteristic has been utilized in exterior-quality paint, 'Lotusan', which makes surfaces self-cleaning. Hooks occur in nature as a vast array of designs and in a diversity of animals and plants. The commercial application of this technology of 'Nature' can be found in Velcro [5] having the cheapest and most reliable bur hook-substrate combination. There are now thousands of patents quoting Velcro. This is how the subject of biomimetics has developed. Biomimetics is the application and abstraction of biological methods, systems and good designs found in nature to the study and design of efficient and

Citation: Sujoy Kumar Saha and Gian Piero Celata, "Advances in modelling of biomimetic fluid flow at different scales," Nanoscale Research Letters20116:344. DOI: 10.1186/1556-276X-6-344

sustainable engineering systems and modern technology. The transfer of technology between life forms and manufactures is desirable because evolutionary pressure typically forces living organisms, including fauna and flora, to become highly optimized and efficient. Generally there are three areas in biology after which technological solutions can be modelled.

Figure 1: Feather of a penguin to stay warm naturally in a cold climate. (From [4]).

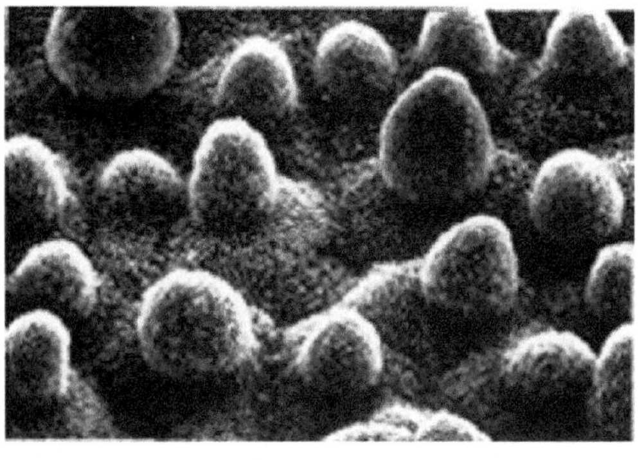

Figure 2: The epidermal structure at the heart of the lotus effect. (From [11]).

- Replicating natural manufacturing methods as in the production of chemical compounds by plants and animals.
- Mimicking mechanisms found in nature such as Velcro and Gecko tape.
- Imitating organizational principles from social behavior of organisms like ants, bees and microorganisms.

Russia has developed a systematic means for integrating the natural knowledge into humankind's technology using 'Teoriya Resheniya Izobretatelskikh Zadatch (TRIZ)', i.e. the theory of inventive problem solving, which provides an objective framework based on functionality for accessing solutions from other technologies and sciences. TRIZ also prevents waste of time trying to find a solution where none exists. The four main tools of TRIZ are a knowledge database arranged by function, analysis of the technical barriers to progress (contradictions), the way technology develops (ideality) and the maximization of resource usage. The biology-based technology 'Biomimetics' suggests new approaches resulting in patents and some into production:

- Strain gauging based on receptors in insects [7],
- Deployable structures based on flowers and leaves [12],
- Tough ceramics based on mother-of-pearl [13],
- Drag reduction based on dermal riblets on shark skin [14],
- Tough composites based on fibre orientations in wood [15],
- Underwater glues based on mussel adhesive [16],
- Flight mechanisms based on insect flight [2],
- Extrusion technology based on the spinneret of the spider [3],
- Self-cleaning surfaces based on the surface of the lotus leaf [17].

The importance of Biomimetics will increase as the incidence of genetic manipulation increases and the genetic manufacturing is developed. In the result, the area between living and non-living materials, where biology interacts with engineering, e.g. bioengineering and biomechatronics, is benefited.

There are innumerable examples of interactions with the environment and balanced and efficient heat, mass, momentum and species transfer through the microstructures in the fluid flow in the manifested living world of plants, animals and other living creatures. Biomimetics involve mimicking these interactions across the functional surfaces with the surrounding environments in the technological design. The physical nature is numerically modelled and simulated using computational fluid dynamics (CFD). Geometrical analogy as well as physical similarity is to be studied to design technological functional surfaces imitating microstructural and biological functional surface morphologies. CFD at micro- or meso-scales and other numerical methodologies are necessary for this [18, 19, 20, 21, 22, 23, 24].

The meso- and micro-scale methods are also being developed in parallel with the continuum theory-based conventional CFD techniques-using finite volume method (FVM) and finite element method (FEM). In the mesoscopic lattice Boltzmann method (LBM), fluid flow is simulated by tracking the development of distribution functions of assemblies of molecules. It is difficult to capture the interfacial dynamics, which is essential for multiphase flow, at the macroscopic level. LBM captures the interaction of fluid particles and is, therefore, helpful for multiphase flow with phase segregation and surface tension. Also, the LBM is computationally more efficient than molecular dynamics (MD) method since it does not track individual molecules; the solution algorithm is explicit, easy to implement and parallel computation can be done. Micro/nano-scale simulations in micro/nano-scale geometries and micro time scales are done in MD method and direct simulation of Monte Carlo (DSMD) method. Coupled macro-scale simulation is being done using high performance computer (HPC). This article makes a review of the advances in multiscale biomimetic fluid flow modelling and simulation of difficult physics problems with complex biological interfaces.

Macroscopic Biomimetic Flow Modelling

The locomotion, power and manoeuvring of aquatic animals like swimming fish having superior and efficient utilization of propulsion through a rhythmic unsteady motion of the body and fin resulting in unsteady flow control has been engineered for the transportation in the underwater vehicles. The fish senses and manipulates large-scale vortices and repositions the vortices through tail motion. The timing of formation and shedding of vortices are important. CFD application by mimicking the swimming of fish and underwater dolphin kicking has been utilized to understand active drag and propulsive net thrust and this has resulted in better sailing performance, Olympic ski jumping, Formula 1 racing, Speedo's new Fastskin FSII swimsuit and an optimal kick profile in swim starts and turns. The undulatory propulsion in aquatic vertebrates is achieved by sending alternating waves down the body towards the tip of the tail and causing sinusoidal oscillation of the body, a jet in the wake and a forward thrust. Two modes of propulsive technique utilized by fish are anguilliform and carangiform, Figure 3 [25]. The carangiform mode is also termed as 'lunate-tail swimming propulsion'.

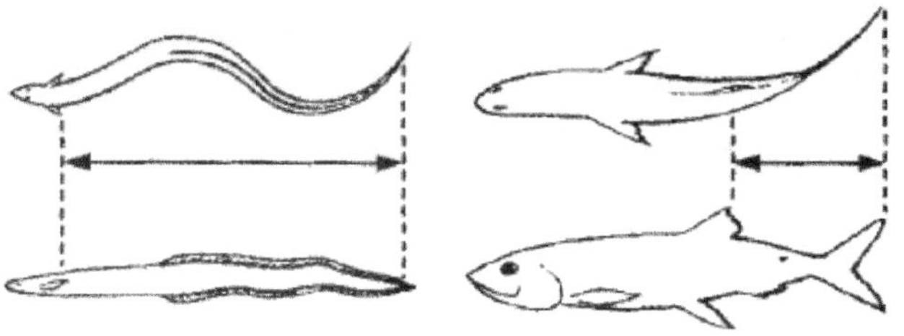

Figure 3: The modes of swimming of fishes. (a) The anguilliform motion of an eel. (b) The carangiform motion of a tuna. (From [25]).

The unsteady incompressible Navier-Stokes equations of turbulent flow are solved in the simulation by applying the Reynolds-averaged Navier-Stokes (RANS) equations with usual boundary conditions to obtain the fluctuating velocity fields. The equations in Cartesian tensor form are:

$$\frac{\partial \rho}{\partial t} + \frac{\partial}{\partial x_i}(\rho u_i) = 0 \tag{1}$$

$$\frac{\partial}{\partial t}(\rho u_i) + \frac{\partial}{\partial x_i}(\rho u_i u_j) = -\frac{\partial p}{\partial x_i} + \frac{\partial}{\partial x_j}\left[\mu\left(\frac{\partial u_i}{\partial x_j} + \frac{\partial u_j}{\partial x_i} - \frac{2}{3}\delta_{ij}\frac{\partial u_l}{\partial x_l}\right)\right] + \frac{\partial}{\partial x_j}\left(-\rho\overline{u'_i u'_j}\right) \tag{2}$$

$$-\rho\overline{u'_i u'_j} = \mu_t\left(\frac{\partial u_i}{\partial x_j} + \frac{\partial u_j}{\partial x_i}\right) - \frac{2}{3}\left(\rho k + \mu_t\frac{\partial u_i}{\partial x_i}\right)\delta_{ij} \tag{3}$$

$$\frac{\partial}{\partial t}(\rho k) + \frac{\partial}{\partial x_i}(\rho k u_i) = \frac{\partial}{\partial x_j}\left[\left(\mu + \frac{\mu_t}{\sigma_k}\right)\frac{\partial k}{\partial x_j}\right] + G_k - \rho\varepsilon \tag{4}$$

$$\frac{\partial}{\partial t}(\rho\varepsilon) + \frac{\partial}{\partial x_i}(\rho\varepsilon u_i) = \frac{\partial}{\partial x_i}\left[\left(\mu + \frac{\mu_t}{\sigma_\varepsilon}\right)\frac{\partial \varepsilon}{\partial x_j}\right] + C_{1\varepsilon}\frac{\varepsilon}{k}G_k - C_{2\varepsilon}\rho\frac{\varepsilon^2}{k} \tag{5}$$

$$G_k = -\rho\overline{u'_i u'_j}\frac{\partial u_j}{\partial x_i}\infty \tag{6}$$

$$\mu_t = \rho C_\mu \frac{k^2}{\varepsilon} \tag{7}$$

where x and u are Cartesian coordinates and velocities, respectively, and t is time. Velocity u, density ρ, viscosity μ and other solution variables represent ensemble-averaged (or time-averaged) values. Reynolds stress, $-\rho\overline{u'_i u'_j}$ is modelled and related to the mean velocity gradients by Boussinesq hypothesis. k is the turbulence kinetic energy, ε the kinetic energy dissipation rate and μ_t the turbulent viscosity. C is constant, σ the Prandtl number. G_k represents the generation of turbulence kinetic energy due to the mean velocity gradients. μ_t is the turbulent viscosity.

The turbulent flow induced by the fish-tail oscillation is characterized by fluctuating velocity fields. The instantaneous governing equations are time averaged to reduce the computational time and complexity which is done in the form of turbulence models like the semi-empirical k-ε work-horse turbulence model for practical engineering flow calculations.

To calculate the flow field using the dynamic mesh, the integral form of the conservation equation for a general scalar φ on an arbitrary control volume V with moving boundary is employed:

$$\frac{d}{dt}\int_V \rho\varphi dV + \int_{\partial V} \rho\varphi\left(\vec{u} - \vec{u}_g\right)\cdot d\vec{A} = \int_{\partial V} \Gamma\nabla\varphi\cdot d\vec{A} + \int_V S_\varphi dV \tag{8}$$

where $\vec{u}$ is the flow velocity vector, $\vec{u}_g$ is the grid velocity of the moving mesh, Γ is the diffusion coefficient, S_φ is the source term of φ and ∂V is the boundary of the control volume V.

The flow is characterized by spatially travelling waves of body bound vorticity. The mix between longitudinal and transverse flow features varies with the phase of oscillation and the unsteady velocity field varies throughout an oscillation cycle. The dynamic pressure distribution contour and the effect of the tail movement on the unsteady flow field of the fish-like body will show that there are high pressure zones at the rear of the body indicating strong vortex and turbulence. The kinematic parameters like Strouhal number, wavelength and oscillating frequency are based on the forward locomotion in a straight line with constant speed in the cruising direction. Figure 4 shows the computational geometric forms of (a) the Robo Tuna, (b) tuna with dorsal/ventral finlets and (c) giant danio [26]. Fish swimming kinematic data shows that the non-dimensional frequencies are close to the value predicted by the instability analysis. Figure 5, from Rohr et al. [27], shows Strouhal number as a function of the Reynolds number for numerous observations of trained dolphins with good agreement between theory and experiment.

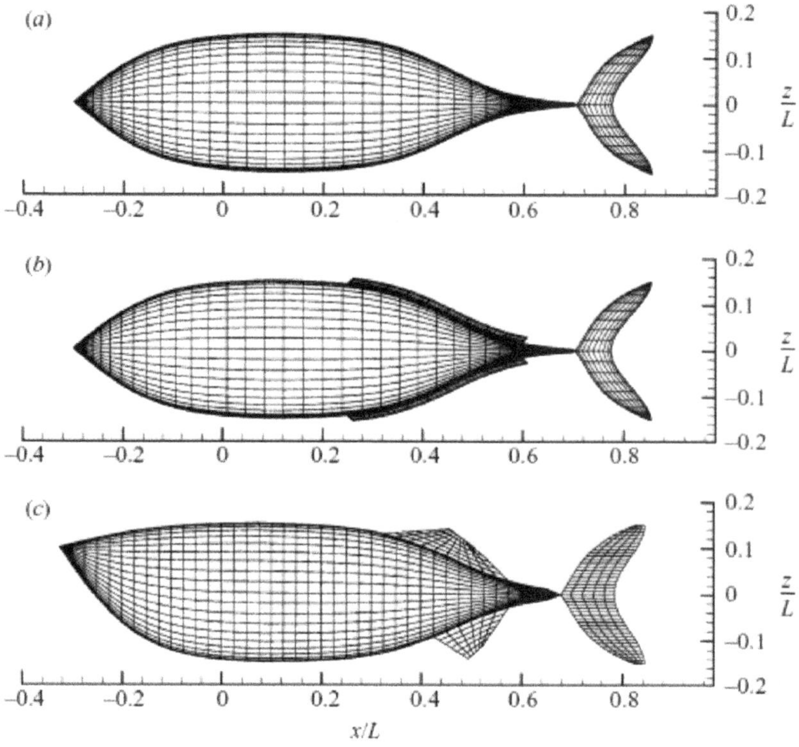

Figure 4: Computational geometric forms of (a) the Robo Tuna, (b) tuna with dorsal/ventral finlets and (c) giant danio. (From [26]).

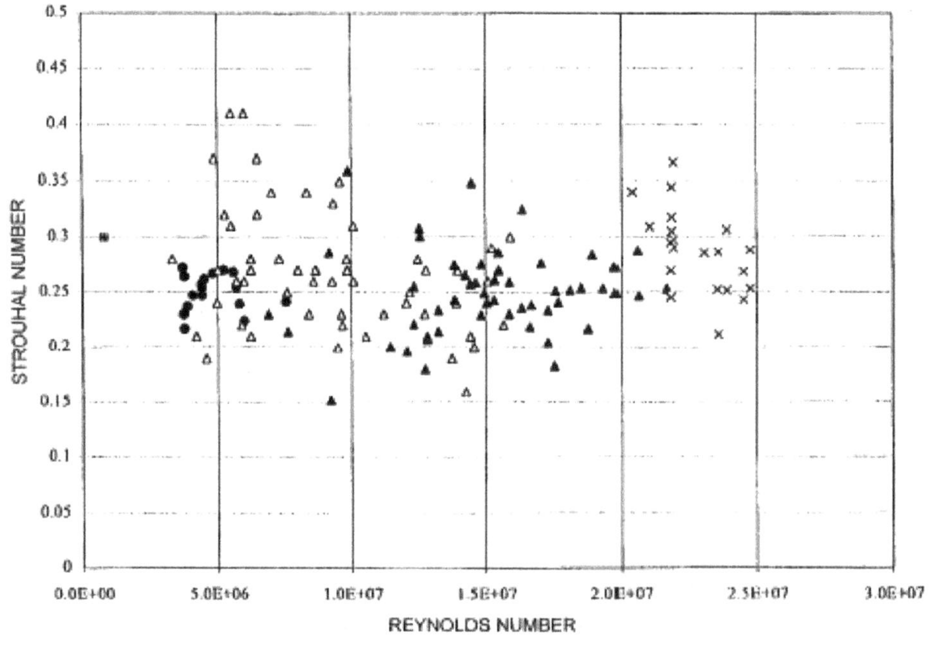

Figure 5: Strouhal number for swimming dolphins as a function of Reynolds number. (From Rohr et al. [27]).

Other example of using CFD to study biomimetic fluid flow problems include simulation of air flow around flapping insect wings, numerical simulation of electro-osmotic flow near earthworm surface and simulation of explosive discharge of the bombardier beetle.

Kroger [28] made a CFD simulation study of air flow around flapping insect wings. The interest in the flapping-wing technique [29, 30] is growing recently due to the fact, that the developments in micro-technology permit people to think about building very small and highly manoeuvrable micro-aircraft that could be used for search and rescue missions or to detect harmful substances or pollutants in areas that are not accessible by or too dangerous for humans. There are three basic principles that contribute to unsteady flapping-wing aerodynamics: delayed stall, rotational circulation and wake capture. However, the exact interactions between them are still subject to ongoing research by CFD simulation. Figure 6 shows surface mesh on fly body.

Figure 6: Surface mesh on fly body. (From [28]).

The dynamic mesh CFD model is used to examine critical flight simulations of normal aircraft, like the undercarriage lowering at low air speed, or the movement of sweep wings of fighter jets at high air speed. Next to flight applications, the dynamic mesh model can also simulate moving heart valves in the biomedical area, or small flapping membrane valves in micro-fluidics or the flow around any arbitrary moving part in other industry or sports applications.

The electro-osmotic flow controlled by the Navier-Stokes equations near an earthworm surface has been simulated by Zu and Yan [31] numerically to understand the anti-soil adhesion mechanism of earthworm. A lattice Poisson method (LPM), which is a derived form of LBM, has been employed to solve externally applied electric potential φ and charge distributions in the electric double layer along the earthworm surface. The external electric field is obtained by solving a Laplace equation. The simulation [32, 33, 34, 35] showed that moving vortices, contributing to the anti-soil adhesion, are formed near earthworm body surface by the non-uniform and variational electric force acting as lubricant. Figure 7 shows the electro-osmotic flow field between the surfaces of soil and earthworm.

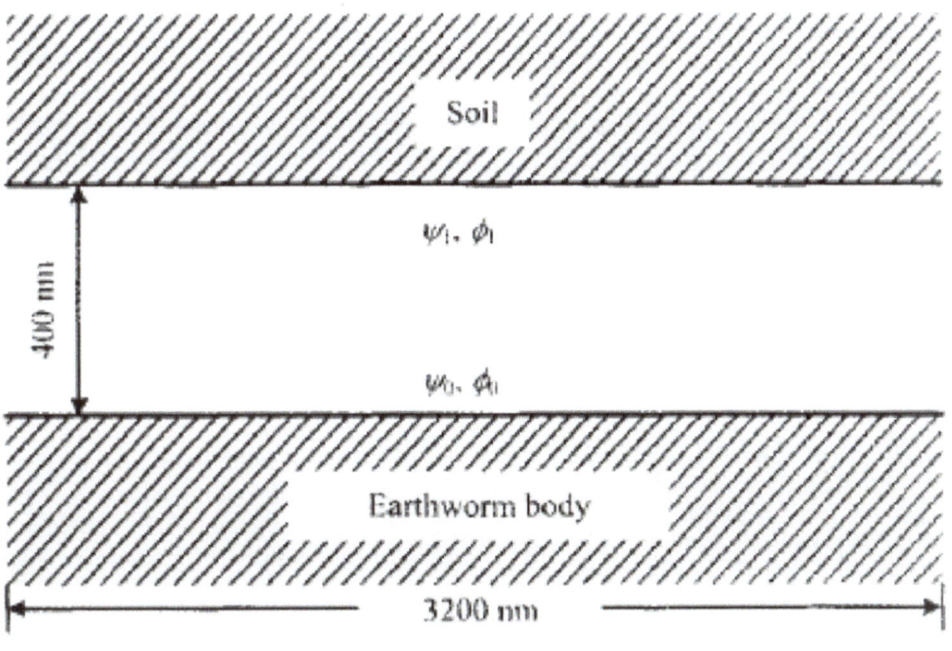

Figure 7: Electroosmotic flow field between the surfaces of soil and earthworm. (From [31]).

A biomimetic CFD study [36, 37, 38, 39] of the bombardier beetle's explosive discharge apparatus and unique natural 'combustion' technique in its jet-based defense mechanism helps designing a short mass ejection system and a long range of spray ejection pertinent to reigniting a gas turbine aircraft engine which has cut out, when the cold outside air temperature is extremely low. The beetle can eject a hot discharge to around 200 to 300 times the length of its combustor. Figure 8 shows a bombardier beetle (brachina) ejecting its water-steam jet at 100°C forward from the tip of its abdomen (from left to right).

Figure 8: A bombardier beetle ejecting its water-steam jet. (From [36]).

Hybrid Molecular-Continuum Fluid Dynamics Simulation

Nanoscale systems such as GaAsMESFETs and SiMOSFETs semiconductor devices, ultra-fast (picoseconds or femtoseconds) pulsed lasers do not conform to the classical Fourier heat diffusion theory in which the mean free path of the energy carriers becomes comparable to or larger than the characteristic length scale of the particle device/system or the time scale of the processes becomes comparable to or smaller than the relaxation time of the energy carriers. Although numerical techniques like Boltzmann transport equation (BTE) or atomic-level simulation (MD) and Monte Carlo simulation (MCS) can capture the physics in this regime, they require large computational resources. The C-V hyperbolic equation, which is not subject to the Fourier law assumption of infinite thermal propagation speed, is also not free from anomalies.

Limitations of Continuum Description of a System

Finite difference and finite element methods serve well for continuum description of a system governed by a set of differential equations and boundary conditions. However, the problem arises when the system has atomic fabric of matter such as in the case of friction problems and phase-change problems of fluid freezing into a solid or dynamic transition such as intermittent stick-slip motion [40].

The Molecular Dynamics (MD) Method

When a system is modelled on the atomic level such as in case of MD, the motion of individual atoms or molecules is approximated. The particle motion is controlled by interaction potentials and equations of motion. MD is used for systems on the nanometre scale.

Coupling MD-Continuum

Coupling two very different descriptions of fluids at MD-continuum interface is a serious issue. The overlapping region of two descriptions must be coupled over space as well as time giving consistent physical quantities like density, momentum and energy and their fluxes must be continuous. Quantities of particles may be averaged locally and temporally to obtain boundary conditions of continuum equations. Getting microscopic quantities from macroscopic non-unique ensembles is, however, difficult.

Coupling Schemes

Several coupling schemes [40, 41, 42, 43, 44] have been developed and the two solutions relax in a finite overlap region before they are coupled. Equations of motion are the language of particles and these are coupled with the continuum language, i.e. the differential equations. The coupling mechanism transmits mass flux, momentum flux and energy flux across the domain boundary. If the remaining boundaries are sealed, i.e. the simulated system is closed; the coupling ensures conservation of mass, momentum and energy.

The two domains are coupled to each other by ensuring that the flux components normal to the domain boundary match. If particles flow towards the boundary, a corresponding amount of mass, momentum and energy must be fed into the continuum. Conversely, any transport in the vicinity of the boundary on the part of the continuum must provide a boundary condition for transport on the part of the particles.

Figure 9 shows the velocity and temperature profiles observed in a simulation using Lennard-Jones particles and a Navier-Stokes continuum.

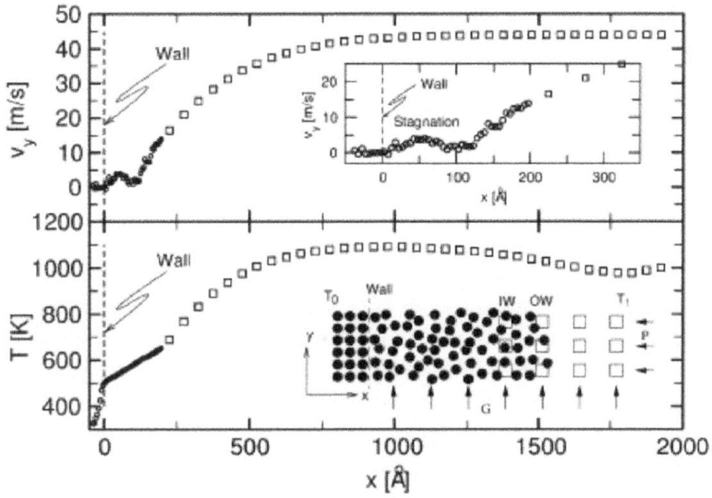

Figure 9: Plot of velocity parallel to a macroscopically flat wall and of temperature as a function of wall distance. Spheres and squares represent the particle and the continuum domain, respectively. (From [40]).

Smoothed Particle Hydrodynamics

Sousa [45] presented a scientific smoothed particle hydrodynamic (SPH) multiphysics simulation tool applicable from macro to nanoscale heat transfer. SPH [45] is a meshless particle based Lagrangian fluid dynamic simulation technique; the fluid flow is represented by a collection of discrete elements or pseudo particles. These particles are initially distributed with a specified density distribution and evolve in time according to the fluid heat, mass, species and momentum conservation equations. Flow properties are determined by an interpolation or smoothing of the nearby particle distribution with the help of a weighting function called the smoothing kernel. SPH is advantageous in (1) tracking problems dealing with multiphysics, (2) handling complex free surface and material interface, (3) parallel computing with relatively simple computer codes, (4) dealing with transient fluid and heat transport.

Following the original approach of Olfe [46] and Modest [47] in case of radiative heat transfer, Sousa [45] made the SPH numerical modelling for the ballistic-diffusive heat conduction equation. In this method, the heat carriers inside the medium are split into two components: ballistic and diffusive. The ballistic component is determined from the prescribed boundary condition and/or nanoscale heat sources and it experiences only outscattering; the transport of the scattered and excited heat carriers inside the medium is treated as diffusive component.

Intrinsic Complex Issues in Hybrid Method

The development and optimization of the performance of micro and nano fluidic devices requires numerical modelling of fluid flow inside micro and nanochannels. The nature of the phenomena involved in these devices invariably and predominantly has the interfacial interactions because of high surface-to-volume ratio and is characterized by an inherent multiscale nature [48, 49, 50, 51, 52, 53, 54, 55, 56, 57, 58, 59, 60, 61, 62]. The traditional continuum models do not capture the flow physics inside the micro and nano scale systems because they neglect the microscopic mechanisms at these scales. The MD is a microscopic model and this can be used where macroscopic constitutive equations and boundary conditions are inadequate. Figure 10 [48] shows the schematic representation of a molecular region in a hybrid simulation. The MD are well suited for the study of slip generation in the solid-fluid interface and other surface properties like nanoroughness and wettability and the boundary conditions. However, high computational cost restricts the molecular simulations to their applications to nanoscale systems and time scales below microseconds. This disparity of spatial and temporal scales is overcome in the hybrid atomistic-continuum multiscale frameworks where the molecular description models only a small part of the computational domain, since the physics of this part of the system cannot be represented by the continuum model. The boundary condition is transferred accurately and efficiently between the atomistic and continuum description in the hybrid methods. Since the microscopic description requires more degrees of freedom than the macroscopic one, the transfer of macroscopic information on a molecular simulation becomes all the more a challenging task.

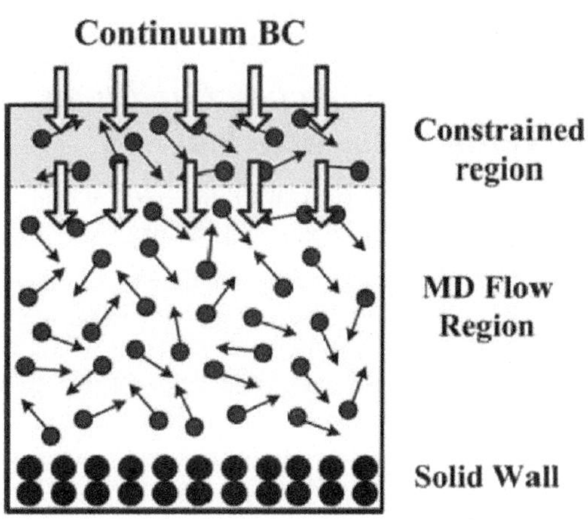

Figure 10: Schematic representation of a molecular region in a hybrid simulation. (From [48]).

MD Model and the Maxwell-Boltzmann Velocity Distribution

The MD atomistic model in the micro-scale framework is a deterministic method. In this model, the evolution of the molecular system is obtained by computing the trajectories of the particles based on the classical molecular model. The continuum conditions can be applied to molecular domain either by the method based on continuous rescaling of atomic velocities or by the periodic resampling method of atomistic velocities that employs velocity distribution functions such as Maxwell-Boltzmann or Chapman-Enskog distribution for non-equilibrium situations of hybrid simulations in dilute gases employing geometrical decomposition and state coupling. The Maxwell-Boltzmann velocity distribution is the natural velocity distribution of an atomic or molecular system in an equilibrium state defining the probability of one-dimensional velocity components of an atom assuming a specific value based on temperature and the atomic mass. The reflective plane placed at the upper boundary of the boundary condition transfer region maintains every particle inside the molecular domain. This scheme is simpler than the velocity reversing scheme, but this can be applied only to incompressible flows because the normal pressure is a result of the reflected atoms.

Rescaling Techniques

In the rescaling techniques, in addition to the velocity restrictions, the continuum pressure applies to the atomistic region. The normal pressure is applied through external forces generating a potential energy field. Energy is decreased because of the reduction of potential energy of the atoms moving towards the continuum boundary. The resulting energy oscillations in the molecular system are reduced by velocity reversing of the outermost atoms. This scheme is simple and robust because of uncontrolled transfer of energy. The continuum temperature to the molecular system is accomplished by an energy transfer scheme. The energy is added or removed from the microscopic system to parallel the macroscopic temperature without modifying the mean velocity of the particles. The energy transfer takes place independent of each dimension and is accomplished by the velocity vectors of the atoms [42, 61, 62, 63, 64, 65, 66, 67, 68].

Issues Related to Boundary Conditions in Hybrid Multiscaling Modelling

Drikakis and Asproulis [69] applied macroscopic boundary conditions in hybrid multiscale modelling. MD microscopic simulation was employed. They employed the methods for various liquid and gas flows with heat transfer and identified specific parameters for accuracy and efficiency. Their work has shown that knowledge about boundary conditions development and application is needed in multiscale computational frameworks. Continuum temperature and velocity as well as macroscopic pressure constrain molecular domain. Inconsistent pressure can shrink the simulation domain and the particles may drift away generating errors and instabilities in the hybrid procedure. Also, the size of the regions for the application of velocity constrains is important to avoid unrealistic heat transfer across the computational domain and inconsistencies between the molecular and continuum state. Resampling frequency and the termination of the atomistic region have significant impact in the resampling techniques and these can influence trapping of particles in the constrained region and may cause deviations between the macroscopic and microscopic velocities. The domain termination needs correct continuum pressure application.

Challenge in Biomimetic Flow Simulation

The task of imitating biological functional surfaces with variety of complex three-dimensional micro- and nano-structures is very challenging in biomimetic flow simulation. The transfer of biological morphologies of plants and animals by imitating both geometrical and physical similarity to technological applications is to be identified [70, 71,72, 73, 74, 75, 76, 77, 78, 79, 80, 81, 82, 83, 84, 85, 86, 87, 88, 89, 90, 91, 92, 93, 94, 95, 96, 97, 98, 99, 100, 101,102, 103, 104, 105, 106, 107, 108, 109, 110, 111, 112, 113, 114, 115, 116, 117, 118, 119, 120, 121, 122, 123, 124,125, 126, 127]. Studies on micro surface structures of different species are to be made by scanning electron microscope (SEM) and atomic force microscope (AFM) to imitate engineering functional surfaces. The mesoscopic LBM has been applied in studying electro-osmotic driving flow within the micro thin liquid layer near an earthworm body surface [128]. The moving vortices give the effect of anti-soil adhesion. Few multiphase LBM models are the pseudo-potential model, the free energy model and the index-function model [129, 130, 131, 132]. In LBM, effective interaction potential describes the fluid-fluid interaction. Interface is introduced by modelling the Boltzmann collision operator imposing phase separation. Also, the fluid-fluid interactions are represented by a body force term in Boltzmann equation. In this case, second-order terms in the pressure tensor are removed and more realistic interfacial interactions are produced.

Hard spheres fluids, square well fluids and Lennard-Jones fluids are model fluids in MD. The fluid flow and heat transfer in micro-scale and nano-scale systems get microscopic and nanoscopic insight from MD [133].

CONCLUSIONS

A comprehensive and state-of-the-art review of CFD techniques for numerical modelling of some biomimetic flows at different scales has been done. Fluid-fluid interfaces contacting with functional solid surfaces have been discussed. The multiphysics modelling at different scales by Navier-Stokes and energy equations, mesoscopic LBM, MD method and combined continuum-MD method with appropriate coupling schemes have been dealt with in detail.

ABBREVIATIONS

AFM: atomic force microscope

BTE: Boltzmann transport equation

CFD: computational fluid dynamics

DSMD: direct simulation of Monte Carlo

FEM: finite element method

FVM: finite volume method

HPC: high performance computer

LBM: lattice Boltzmann method

LPM: lattice Poisson method

MCS: Monte Carlo simulation

MD: molecular dynamics

RANS: Reynolds-averaged Navier-Stokes

SEM: scanning electron microscope

SPH: smoothed particle hydrodynamic

TRIZ: Teoriya Resheniya Izobretatelskikh Zadatch.

REFERENCES

1. Ball P: Life's lessons in design. Nature 2001, 409: 413–416. 10.1038/35053198
2. Ellington CP: The novel aerodynamics of insect flight: applications to micro-air vehicles. J Exp Biol 1999, 202: 3439–3448.
3. Knight DP, Vollrath F: Liquid crystals and flow elongation in a spider's silk production line. Proc R Soc Lond B 1999, 266: 519–523. 10.1098/rspb.1999.0667
4. Dawson C, Vincent JFV, Jeronimidis G, Rice G, Forshaw P: Heat transfer through penguin feathers. J Theoret Biol 1999, 199: 291–295. 10.1006/jtbi.1999.0959
5. Velcro SA: Improvements in or relating to a method and a device for producing a velvet type fabric. Swiss patent no. 721338 1955.
6. Vincent JFV, Mann DL: Systematic technology transfer from biology to engineering. Philos Trans R Soc Lond A 2002, 360: 159–173. 10.1098/rsta.2001.0923
7. Skordos A, Chan PH, Vincent JFV, Jeronimidis G: A novel strain sensor based on the campaniform sensillum of insects.Philos Trans R Soc Lond A 2002, 360: 239–253. 10.1098/rsta.2001.0929
8. Vincent JFV, King MJ: The mechanism of drilling by wood wasp ovipositors. Biomimetics 1996, 3: 187–201.
9. Barthlott W, Neinhuis C: The Lotus-effect: non-adhesive biological and biomimetic technical surfaces. In Proceedings of 1st International Industrial Conference BIONIK. Hannover, Germany; 2004:211–214.
10. Barthlott W, Neinhuis C: The lotus-effect: Nature's model for self-cleaning surfaces. Int Text Bull 2001, 8: 10–12.
11. Barthlott W, Neinhuis C: Purity of the sacred lotus, or escape from contamination in biological surfaces. Planta 1997, 202: 1–8. 10.1007/s004250050096
12. Guest SD, Pellegrino S: Inextensional wrapping of flat membranes. In Proceedings of International Seminar Structure Morphology. LMGC Universite Montpellier II, Montpellier; 1992:203–215.
13. Jackson AP, Vincent JFV, Turner RM: A physical model of nacre. Compos Sci Technol 1989, 36: 255–266. 10.1016/0266-3538(89)90024-9
14. Bechert DW, Bruse M, Hage W, Meyer R: Biological surfaces and their technological application--laboratory and flight experiments on drag reduction and separation control. In Proceedings of 28th AIAA Fluid Dynamics Conference. Snowmass Village, CO, USA; 1997.
15. Chaplin RC, Gordon JE, Jeronimidis G: Development of a novel fibrous composite material. US patent no. 4409274 1983.
16. Holl SM, Hansen D, Waite JH, Schaefer J: Solid-state NMR analysis of cross-linking in a mussel protein glue. Arch Biochem Biophys 1993, 302: 255–258. 10.1006/abbi.1993.1207
17. Barthlott W, Neinhuis C: Lotusblumen und Autolacke: Ulstrastruktur, Pflanzucher, Grenzflachen und biomimetische unverschmutzbare Werkstoffe. In BIONA Report 12. Edited by: Nachtigall W, Wisser A. Gesellschaft fur Technische Biologie und Bionik, Universitat des Saarlandes; 1999:281–293.
18. McIntosh A, Forman M: The efficiency of the explosive discharge of the bombardier beetle with possible biomimetic application. In Design and Nature II--Comparing Design in Nature with Science and Engineering. Edited by: Collins MW, Brebbia CA. Southampton & Boston: WIT Press; 2004:227–236.
19. Amon CH: Advances in computational modeling of nano-scale heat transfer. In Proceedings of 12th International Heat Transfer Conference. Grenoble, France; 2002:41–53.
20. Chen S, Doolen GD: Lattice Boltzmann method for fluid flows. Annu Rev Fluid Mech 1998, 30: 329–364. 10.1146/annurev.fluid.30.1.329
21. Maruyama S: Molecular dynamics methods in microscale heat transfer. In Heat Transfer and Fluid Flow in Microchannels. Edited by: Celata GP. New York: Begell House Inc; 2002.
22. Bird GA: Molecular gas Dynamics and Direct Simulation of Gas Flows. New York: Oxford Univ. Press; 1994.
23. Wagner G, Flekkoy E, Fedder J, Jossang T: Coupling molecular dynamics and continuum dynamics. Comput Phys Commun 2002, 147: 670–673. 10.1016/S0010-4655(02)00371-5
24. Prizjev NV, Darhuber AA, Troian SM: Slip behavior in liquid films on surfaces of patterned wettability: Comparison between continuum and molecular dynamics simulations. Phys Rev E 2005, 71: 041608. 10.1103/PhysRevE.71.041608
25. Adkins D, Yan YY: CFD simulation of fish-like body moving in viscous liquid. J Bionic Eng 2006, 3: 147–153. 10.1016/S1672-6529(06)60018-8
26. Zhu Q, Wolfgang MJ, Yue DKP, Triantafyllou MS: Three-dimensional flow structures and vorticity control in fish-like swimming. J Fluid Mech 2002, 468: l-28. 10.1017/S002211200200143X
27. Rohr JJ, Hendricks EW, Quigley L, Fish FE, Gilpatrick JW, Scardina-Ludwig J: Observations of dolphin swimming speed and Strouhal number. In Space and Naval Warfare Systems Center Technical Report 1769. San Diego; 1998.
28. Kroger R: Simulation of airflow around flapping insect wings. In Proceedings of 1st International Industrial Conference BIONIK. Hannover, Germany; 2004:185–190.
29. Dickinson MH, Lehmann FO, Sane SP: Science. 1999,

284: 1954–1960. 10.1126/science.284.5422.1954

30. Nachtigall W: Biona-Report 11, Biology and Related Natural Sciences. Stuttgart: Fischer Verlag; 1997:115–156.
31. Zu YQ, Yan YY: Numerical simulation of electroosmotic flow near earthworm surface. J Bionic Eng 2006, 3: 179–186. 10.1016/S1672-6529(07)60001-8
32. Ren L, Tong J, Li J, Cheng B: Soil adhesion and biomimetics of soil-engaging components: a review. J Agricult Eng 2001, 79: 239–263. 10.1006/jaer.2001.0722
33. Ma J: Creatures and Bionics. Tianjin: Tianjin Science and Technology Press; 1984.
34. Yan YY, Hull JB: The concept of electroosmotically driven flow and its application to biomimetics. J Bionic Eng 2004, 1: 46–52.
35. Yan YY, Ren L, Li J: The electroosmotic driven flow near an earthworm surface and the inspired bionic design in engineering. Int J Des Nat 2007, 1: 135–145. 10.2495/D&N-V1-N2-135-145
36. Beheshti N, Mcintosh AC: A biomimetic study of the explosive discharge of the bombardier beetle. Int J Des Nat 2007, 1: 61–69. 10.2495/D&N-V1-N1-61-69
37. Aneshansley DJ, Eisner T: Spray aiming in the bombardier beetle: photographic evidence. Proc Natl Acad Sci USA 1999, 96: 9705–9709. 10.1073/pnas.96.17.9705
38. Aneshansley DJ, Eisner T, Widom M, Widom B: Biochemistry at 100°C: explosive secretory discharge of bombardier beetles (brachinus). Science 1969, 165: 61–63. 10.1126/science.165.3888.61
39. Dean J, Aneshansley DJ, Edgerton H, Eisner T: Defensive spray of the bombardier beetle: a biological pulse jet. Science1990, 248: 1219–1221. 10.1126/science.2349480
40. Wagner G, Flekkoy E, Feder J, Jossang T: Coupling molecular dynamics and continuum dynamics. Comput Phys Commun2002, 147: 670–673. 10.1016/S0010-4655(02)00371-5
41. O'Connell ST, Thompson PA: Molecular dynamics-continuum hybrid computations: a tool for studying complex fluid flows.Phys Rev E 1995, 52: 792–795.
42. Priezjev NV, Darhuber AA, Troian SM: Slip behavior in liquid films on surfaces of patterned wettability: comparison between continuum and molecular dynamics simulations. Phys Rev E 2005, 71: 041608/1–041608/11. 10.1103/PhysRevE.71.041608
43. Qian T, Wang X: Hydrodynamic slip boundary condition at chemically patterned surfaces: a continuum deduction from molecular dynamics. Phys Rev E 2005, 72: 022501. 10.1103/PhysRevE.72.022501
44. Hadjiconstantinou NG: Combining atomistic and continuum simulations of contact-line motion. Phys Rev E 1999, 59: 2475. 10.1103/PhysRevE.59.2475
45. Sousa ACM: Multiphysics modeling with SPH: from macro to nanoscale heat transfer. Proceedings of IHTC-13, KN 29, Sydney, Australia 2006.
46. Olfe DB: A modification of the differential approximation for radiative transfer. AIAA J 1967, 5: 638–643. 10.2514/3.4041
47. Modest MF: The modified differential approximation for radiative transfer in general three-dimensional media. J Thermophys Heat Transf 1989, 3: 283–288. 10.2514/3.28773
48. Asproulis N, Drikakis D: Nanoscale materials modelling using neural networks. J Comput Theoret Nanosci 2009, 6: 514–518. 10.1166/jctn.2009.1062
49. Wijesinghe HS, Hadjiconstantinou NG: A hybrid atomistic-continuum formulation for unsteady, viscous, incompressible flows. CMES 2004, 5: 515–526.
50. Wijesinghe HS, Hornung RD, Garcia AL, Hadjiconstantinou NG: Three dimensional hybrid continuum-atomistic simulations for multiscale hydrodynamics. J Fluids Eng 2004, 126: 768–777. 10.1115/1.1792275
51. Nie XB, Robbins MO, Chen SY: Resolving singular forces in cavity flow: multiscale modeling from atomic to millimeter scales. Phys Rev Lett 2006, 96: 1–4.
52. De Fabritiis G, Delgado-Buscalioni R, Coveney PV: Modeling the mesoscale with molecular specificity. Phys Rev Lett 2006, 97: 134501. 10.1103/PhysRevLett.97.134501
53. Schwartzentruber TE, Scalabrin LC, Boyd ID: A modular particle-continuum numerical method for hypersonic non-equilibrium gas flows. J Comput Phys 2007, 225: 1159–1174. 10.1016/j.jcp.2007.01.022
54. Bhattacharya DK, Lie GC: Nonequilibrium gas flow in the transition regime: a molecular-dynamics study. Phys Rev 1991, 43: 761–767. 10.1103/PhysRevA.43.761
55. Delgado-Buscalioni R, Coveney P: Hybrid molecular-continuum fluid dynamics. Philos Trans R Soc Lond A 2004, 362: 1639–1654. 10.1098/rsta.2004.1401
56. Hadjiconstantinou NG: Discussion of recent developments in hybrid atomistic continuum methods for multiscale hydrodynamics. Bull Pol Acad Sci Tech Sci 2005, 53: 335–342.
57. Kalweit M, Drikakis D: Coupling strategies for hybrid molecular continuum simulation methods. Proc Inst Mech Eng C J Mech Eng Sci 2008, 222: 797–806. 10.1243/09544062JMES716
58. Hadjiconstantinou NG, Patera AT: Heterogeneous atomistic-continuum representations for dense fluid systems. Int J Mod Phys 1997, 8: 967–976. 10.1142/S0129183197000837
59. Cao BY: Non-maxwell slippage induced by surface roughness for microscale gas flow: a molecular dynamics simulation.Mol Phys 2007, 105: 1403–1410. 10.1080/00268970701361322
60. Delgado-Buscalioni R, Coveney PV: Continuum-particle hybrid coupling for mass, momentum and energy transfers. Phys Rev E 2003, 67: 046704. 10.1103/PhysRevE.67.046704
61. Liu J, Chen SV, Nie XB, Robbins MO: A continuum-atomistic simulation of heat transfer in micro- and nano-flows. J Comput Phys 2007, 227: 279–291. 10.1016/j.jcp.2007.07.014
62. Nie XB, Chen SY, Robbins MO: A continuum and

molecular dynamics hybrid method for micro- and nanofluid flow. J Fluid Mech 2004, 500: 55–64. 10.1017/S0022112003007225

63. Schwartzentruber TE, Scalabrin LC, Boyd ID: Hybrid particle-continuum simulations of non-equilibrium hypersonic blunt-body flow fields. J Thermophys Heat Transf 2008, 22: 29–37. 10.2514/1.30216
64. Kalweit M, Drikakis D: Multiscale methods for micro/nano flows and materials. J Comput Theoret Nano Sci 2008, 5: 1923–1938. 10.1166/jctn.2008.906
65. Ren W, Weinan E: Heterogeneous multiscale method for the modeling of complex fluids and micro-fluidics. J Comput Phys2005, 204: 1–26. 10.1016/j.jcp.2004.10.001
66. Schwartzentruber TE, Scalabrin LC, Boyd ID: Hybrid particle-continuum simulations of hypersonic flow over a hollow-cylinder-flare geometry. AIAA J 2008, 46: 2086–2095. 10.2514/1.36681
67. Werder T, Walther JH, Koumoutsakos P: Hybrid atomistic-continuum method for the simulation of dense fluid flows. J Comput Phys 2005, 205: 373–390. 10.1016/j.jcp.2004.11.019
68. Yan YY, Lai H, Gentle CR, Smith JM: Numerical analysis of fluid flow inside and around a liquid drop using an incorporation of multi-block iteration and moving mesh. Trans IChE Chem Eng Res Des 2002, 80: 325–331. 10.1205/026387602753582132
69. Drikakis D, Asproulis N: Multi-scale computational modeling of flow and heat transfer. Int J Num Methods Heat Fluid Flow2010, 20: 517–528. 10.1108/09615531011048222
70. Ahlborn B, Chapman S, Stafford R, Blake RW, Harper D: Experimental simulation of the thrust phases of fast-start swimming of fish. J Theor Biol 1997, 200: 2301–2312.
71. Drucker EG, Lauder GV: Locomotor forces on a swimming fish: three-dimensional vortex wake dynamics quantified using digital particle image velocimetry. J Exp Biol 1999, 202: 2393–2412.
72. Ahlborn B, Harper D, Blake R, Ahlborn D, Cam M: Fish without footprints. J Theor Biol 1991, 148: 521–533. 10.1016/S0022-5193(05)80234-6
73. Bandyopadhyay P, Castano J, Rice J, Philips R, Nedderman W, Macy W: Low speed maneuvering hydrodynamics of fish and small underwater vehicles. J Fluids Eng 1997, 119: 136–144. 10.1115/1.2819099
74. Cheng HK, Murillo LE: Lunate-tail swimming propulsion as a problem of curved lifting line in unsteady flow. Part 1. Asymptotic theory. J Fluid Mech 1984, 143: 327–350. 10.1017/S0022112084001373
75. Ellington CP: The aerodynamics of hovering insect flight. V. A vortex theory. Philos Trans R Soc Lond B 1984, 305: 115–144. 10.1098/rstb.1984.0053
76. Dewar H, Graham J: Studies of tropical tuna swimming performance in a large water tunnel--I, energetics. J Exp Biol 1994, 192: 13–31.
77. Dewar H, Graham J: Studies of tropical tuna swimming performance in a large water tunnel--III, kinematics. J Exp Biol 1994, 192: 45–59.
78. Domenici P, Blake RW: The kinematics and performance of fish fast-start swimming. J Exp Biol 1997, 200: 1165–1178.
79. Ellington CP, van den Berg C, Thomas A: Leading edge vortices in insect flight. Nature 1996, 384: 6–26. 10.1038/384626a0
80. Lighthill MJ: A quatic animal propulsion of high hydromechanical efficiency. J Fluid Mech 1970, 44: 265–301. 10.1017/S0022112070001830
81. Lighthill MJ: Large-amplitude elongated-body theory of fish locomotion. Proc R Soc Lond B 1971, 179: 125–138. 10.1098/rspb.1971.0085
82. Fish F: Power output and propulsive efficiency of swimming bottlenose dolphins. J Exp Biol 1993, 185: 179–193.
83. Barrett DS, Triantafyllou MS, Yue DKP, Grosenbaugh MA, Wolfgang M: Drag reduction in fish-like locomotion. J Fluid Mech1999, 392: 183–212. 10.1017/S0022112099005455
84. Fish F, Hui CA: Dolphin swimming--a review. Mammal Rev 1991, 21: 181–195. 10.1111/j.1365-2907.1991.tb00292.x
85. Gray J: The locomotion of nematodes. J Exp Biol 1964, 13: 135–154.
86. Kayan YP, Pyatetskiy VY: Kinematics of bottlenose dolphins swimming as related to acceleration mode. Bionika 1977, 11: 36–41.
87. Kato N: Locomotion by mechanical pectoral fins. J Mar Sci Technol 1998, 3: 113–121. 10.1007/BF02492918
88. Gray J: Studies in animal locomotion. VI. The propulsive powers of the dolphin. J Exp Biol 1936, 13: 192–199.
89. Lighthill MJ: Note on the swimming of slender fish. J Fluid Mech 1960, 9: 305–317. 10.1017/S0022112060001110
90. Lighthill MJ: Hydromechanics of aquatic animal propulsion. Annu Rev Fluid Mech 1969, 1: 413–445. 10.1146/annurev.fl.01.010169.002213
91. Liu H, Ellington CP, Kawachi K, van den Berg C, Wilmlmott AP: A computational fluid dynamic study of hawkmoth hovering.J Exp Biol 1998, 201: 461–477.
92. Liu H, Wassenberg R, Kawachi K: The three-dimensional hydrodynamics of tadpole swimming. J Exp Biol 1997, 200: 2807–2819.
93. Maxworthy T: Experiments on theWeis-Fogh mechanism of lift generation by insects in hovering flight. Part I. Dynamics of the fling. J Fluid Mech 1979, 93: 47–63. 10.1017/S0022112079001774
94. Muller U, van den Heuvel B, Stamhuis E, Videler J: Fish foot prints: morphology and energetics of the wake behind a continuously swimming mullet (Chelon labrosus risso **).** J Exp Biol 1997, 200: 2893–2896.
95. Weihs D: The mechanism of rapid starting of slender fish. Biorheology 1973, 10: 343–350.
96. Wolfgang M, Anderson JM, Grosenbaugh MA, Yue DKP, Triantafyllou MS: Near body flow dynamics in swimming fish. J Exp Biol 1999, 202: 23032327.
97. Newman JN, Wu T: A generalized slender-body theory for fish-like forms. J Fluid Mech 1973, 57: 673–693.

10.1017/S0022112073001953

98. Wu T: Hydromechanics of swimming fishes and cetaceans. Adv Appl Math 1971, 11: 1–63.
99. Rome L, Swank D, Corda D: How fish power swimming. Science 1993, 261: 340–343. 10.1126/science.8332898
100. Triantafyllou GS, Triantafyllou MS, Grosenbaugh MA: Optimal thrust development in oscillating foils with application to fish propulsion. J Fluids Struct 1993, 7: 205–224. 10.1006/jfls.1993.1012
101. Triantafyllou MS, Triantafyllou GS: An efficient swimming machine. Sci Am 1995, 272: 64–70. 10.1038/scientificamerican0395-64
102. Weihs D: A hydrodynamical analysis of fish turning maneuvers. Proc R Soc Lond Ser B 1972, 182: 59–72. 10.1098/rspb.1972.0066
103. Triantafyllou MS, Barrett DS, Yue DKP, Anderson JM, Grosenbaugh MA: A new paradigm of propulsion and maneuvering for marine vehicles. Trans Soc Naval Architects Mar Eng 1996, 104: 81–100.
104. Fierstine H, Walters V: Studies in locomotion and anatomy of scombroid fishes. Mem Soc South Calif Acad Sci 1968, 6: 1–31.
105. Videler JJ, Muller UK, Stamhuis EJ: Aquatic vertebrate locomotion: wakes from body waves. J Exp Biol 1999, 202: 3423–3430.
106. Kagemoto H, Wolfgang M, Yue D, Triantafyllou M: Force and power estimation in fish-like locomotion using a vortex-lattice method. Trans ASME J Fluids Eng 2000, 122: 239–253. 10.1115/1.483251
107. Newman J: The force on a slender fish-like body. J Fluid Mech 1973, 58: 689–702. 10.1017/S0022112073002429
108. Lan C: The unsteady quasi-vortex-lattice method with applications to animal propulsion. J Fluid Mech 1979, 93: 747–765. 10.1017/S0022112079002019
109. Triantafyllou MS, Barrett DS, Yue DKP, Anderson JM, Grosenbaugh MA, Streitlien K, Triantafyllou GS: A new paradigm of propulsion and maneuvering for marine vehicles. Trans Soc Naval Architect Mar Eng 1996, 104: 81–100.
110. Pedley TJ, Hill SJ: Large-amplitude undulatory fish swimming: fluid mechanics coupled to internal mechanics. J Expl Biol1999, 202: 3431–3438.
111. Triantafyllou G, Triantafyllou M, Yue DKP: Hydrodynamics of fish swimming. Annu Rev Fluid Mech 2000, 32: 33–53. 10.1146/annurev.fluid.32.1.33
112. Liu H, Wassersug RJ, Kawachi KA: Computational fluid dynamic study of tadpole swimming. J Exp Biol 1996, 199: 1245–1260.
113. Cheng JY, Zhuang LX, Tong BG: Analysis of swimming three-dimensional waving plates. J Fluid Mech 1991, 232: 341–355. 10.1017/S0022112091003713
114. Anon: Biomimetics: Strategies for Product Design Inspired by Nature. Bristol: Department of Trade and Industry; 2007.
115. Barthelat F, Tang H, Zavattieri PD, Li CM, Espinosa HD: On the mechanics of mother-of-pearl: a key feature in the material hierarchical structure. J Mech Phys Solids 2007, 55: 306–337. 10.1016/j.jmps.2006.07.007
116. Bhushan B, Jung YC: Wetting, adhesion and friction of superhydrophobic and hydrophilic leaves and fabricated micro/nanopatterned surfaces. J Phys Condens Matter 2008, 20: 225010. 10.1088/0953-8984/20/22/225010
117. Bhushan B, Sayer RA: Surface characterization and friction of a bio-inspired reversible adhesive tape. Microsyst Technol2007, 13: 71–78. 10.1007/s00542-006-0256-2
118. Elbaum R, Gorb S, Fratzl P: Structures in cell wall that enable hygroscopic movement of wheat awns. J Struct Biol 2008, 164: 101–107. 10.1016/j.jsb.2008.06.008
119. Gorb S, Varenberg M, Peressadko A, Tuma J: Biomimetic mushroom-shaped fibrillar adhesive microstructure. J R Soc Interface 2007, 4: 271–275. 10.1098/rsif.2006.0164
120. Goswami L, Dunlop JWC, Jungniki K, Eder M, Gierlinger N, Coutand C, Jeronimidis G, Fratzl P, Burgert I: Stress generation in the tension wood of poplar is based on the lateral swelling power of the G-layer. Plant J 2008, 56: 531–538. 10.1111/j.1365-313X.2008.03617.x
121. Koch K, Bhushan B, Barthlott W: Multifunctional surface structures of plants: an inspiration for biomimetics. Prog Mater Sci2009, 54: 137–178. 10.1016/j.pmatsci.2008.07.003
122. Luz GM, Mano JF: Biomimetic design of materials and biomaterials inspired by the structure of nacre. Philos Trans R Soc A2009, 367: 1587–1605. 10.1098/rsta.2009.0007
123. Mueller T: Biomimetics design by natures. Natl Geogr 2008, 2008: 68–90.
124. Reed EJ, Klumb L, Koobatian M, Viney C: Biomimicry as a route to new materials: what kinds of lessons are useful? Philos Trans R Soc A 2009, 367: 1571–1585. 10.1098/rsta.2009.0010
125. Stegmaier T, Linke M, Planck H: Bionics in textiles: flexible and translucent thermal insulations for solar thermal applications. Philos Trans R Soc A 2009., 367: 10.1098/rsta.2009.0019
126. van der Zwaag S, van Dijk N, Jonkers H, Mookhoek S, Sloof W: Self healing behaviour in man-made engineering materials: bio-inspired but tasking into account their intrinsic character. Philos Trans R Soc A 2009., 367: 10.1098/rsta.2009.0020
127. Youngblood JP, Sottos NR: Bioinspired materials for self-cleaning and self-healing. MRS Bull 2008, 33: 732–738. 10.1557/mrs2008.158
128. Yan YY: Recent advances in computational simulation of macro-, meso-, and micro-scale biomimetics related fluid flow problems. J Bionic Eng 2007, 4: 97–107. 10.1016/S1672-6529(07)60021-3
129. Shan X, Chen H: Lattice Boltzmann model for simulating flows with multiple phases and components. Phys Rev E 1993, 47: 1815–1819. 10.1103/PhysRevE.47.1815
130. Shan X, Chen H: Simulation of non-ideal gases and liquid-gas phase transitions by a lattice Boltzmann equation. Phys Rev E 1994, 49: 2941–2948. 10.1103/PhysRevE.49.2941
131. Swift MR, Osborn WR, Yeomans JM: Lattice Boltzmann

simulation of non-ideal fluids. Phys Rev Lett 1995, 75: 830–833. 10.1103/PhysRevLett.75.830

132. He XY, Chen SY, Zhang RY: A lattice Boltzmann scheme for incompressible multiphase flow and its application in simulation of Rayleigh-Taylor instability. J Comput Phys 1999, 152: 642–663. 10.1006/jcph.1999.6257

133. Ji CY, Yan YY: A molecular dynamics simulation of liquid-vapor-vapor-solid system near triple-phase contact line of flow boiling in a microchannel. Appl Therm Eng 2008, 28: 195–202. 10.1016/j.applthermaleng.2007.03.029

SECTION III: CFD APPLICATIONS TO ARCHITECTURE DESIGN AND OPTIMIZATION IN TERMS OF ENERGY CONSUMPTION, VENTILATION AND COOLING SYSTEMS

CHAPTER 15

CFD MODELING AND SIMULATION OF A HYDROPOWER SYSTEM IN GENERATING CLEAN ELECTRICITY FROM WATER FLOW

Oladapo S. Akinyemi[1] and Yucheng Liu[2]

[1] Department of Mechanical Engineering, University of Louisiana at Lafayette, Lafayette, LA 70504, USA

[2] Department of Mechanical Engineering, Mississippi State University, Mississippi State, MS 39762, USA

ABSTRACT

In this study, electric generation capacity of a hydropower system was evaluated through computational fluid dynamics (CFD) modeling and simulation. Performance of paddle wheels in producing hydropower out of running water under different speeds was evaluated, and effects of side and bottom fins and paddle wheel shape on power generation were discussed based on the CFD results. It was also found that, compared to the regular straight paddles, the curved paddles can increase the electricity generation capacity of the entire system by 10–20 % at different water speeds. A preliminary economic analysis revealed the great prospects of applying the presented hydropower technology in regions with rich water resource to produce and supply low-cost, clean, and renewable electricity for a large amount of households and industries.

INTRODUCTION

In our previous study [1, 2], hydroelectric generation capacity of a paddle wheel had been evaluated through both analytical calculations and computational fluid dynamic (CFD) simulations. Based on the outcomes, a hydropower system was fully developed, which consists of three pairs of paddle wheels and may include side and bottom fins to produce more hydropower out of running water. CFD modeling and simulation approach presented in [1, 2] was applied in this study to evaluate the performance of the developed system, and, based on the results, the potential impact of such system on energy industry and local economy was also estimated.

Citation: Oladapo S. Akinyemi and Yucheng Liu, "CFD modeling and simulation of a hydropower system in generating clean electricity from water flow," Int J Energy Environ Eng (2015) 6:357–366 DOI 10.1007/s40095-015-0180-2.

Because of its important role in generating electric power, the paddle wheel has been extensively employed in a rich variety of power plants to provide renewable and sustainable power generation. For example, paddle wheels will be used in a series of large concrete hydroelectric stations being constructed in the Belle Isle Strait, Canada [3]. Paddle wheel can also be found from the Rance River tidal power plant to generate electricity out of ocean energy [4]. In a microalgal biomass production station, paddle wheel is even mixing of ponds to achieve a required water velocity (20–25 cm/s) [5]. Several inventors also used paddle wheel as an important component in their designs of generating hydroelectricity out of water [6, 7].

In the present hydropower system, all the hydroelectricity is generated from paddle wheels. In real application, the full-sized hydropower system will be anchored in a river, and when the water passes through the equipment, it will rotate the paddle wheels to create very high torque at their axes. The rotating paddle wheels will then drive conventional electromagnetic generators to produce electric power. Such system includes three pairs of paddle wheel, which are arranged in series. The dimensions of original design are huge: 55 × 36 × 9 m^3, and two CFD models, the original model and a down scale (1:30) model, were created and used for simulations. In next phase, a scaled prototype will be manufactured and used for experiments; the experimental results will then be compared to the simulation results to validate the employed computational approach.

CFD techniques have been extensively applied in design and assessment of hydro equipments, such as fluidized-bed reactors [8], structured packings [9], double turbine stirred tank [10], and reservoir [11]. However, in spite of the many CFD models created for other hydro equipments, only a few models have been created to simulate and assess paddle wheels. Peterson et al. [12] presented a methodology to simulate combinations of paddle wheels and propeller-aspirators in a single pond and showed from their simulations that a paddle wheel imparts more circulation into a pond than a propeller-aspirator of the same motor horsepower. Kang et al. [13] surveyed and simulated paddlewheel-driven circulation in rectangular shrimp culture ponds using a 2D depth-integrated numerical model. Compared to the previous paddle wheel models, the modeling and simulation work presented in this study for the first time predicts the role of paddle wheels in a clean energy system using ANSYS CFD software packages. In addition to that, the computational results generated from this study provide a solid benchmark for the future experimental validation of the presented hydropower system.

Modeling and Simulation of a Single Paddle Wheel

In this section, the power generation capacity of an entire paddle wheel is calculated through the computer simulation. Software package ANSYS FLUENT and ANSYS Design-Modeler were used for modeling, analysis, and simulation. The paddle wheel consists of 16 paddles, and only 7 paddles are submerged into water at the same time.

Computer Model

2D Computer-aided design (CAD) models for the paddle wheel and the moving water were first created, as shown in Fig. 1 (without bottom fin) and Fig. 2 (with bottom fin). In the generated computational environment, the physical domain was considered to be only the fluid part and the heat and mass transfer between the paddle wheel and water was neglected. The model consists of stationary and rotational reference frame, which were separated by circular boundary around the paddles.

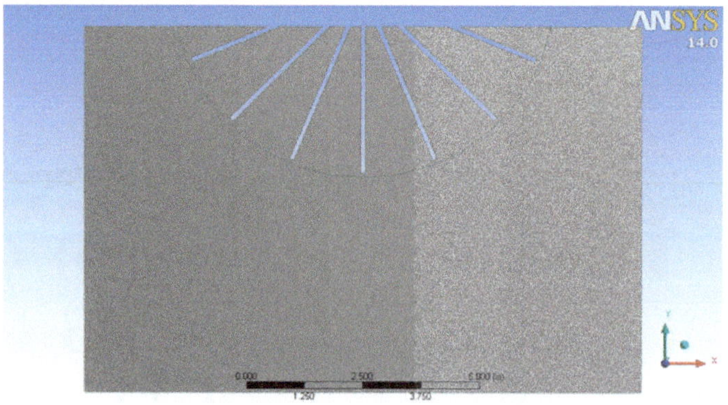

Figure 1: Geometry of a paddle wheel under water.

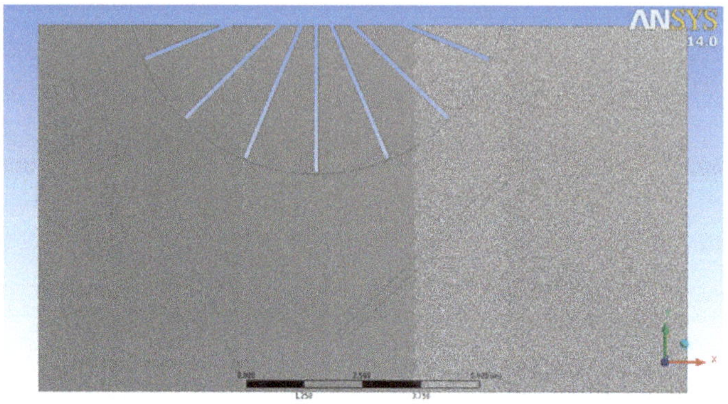

Figure 2: Geometry of a paddle wheel under water with bottom fin.

The CAD model was then meshed, and cells were generated in the entire domain to create the CFD model. Quadrilateral mesh method was employed for meshing the model, and refinement was applied on edges of the paddles because it was expected that the results in those regions would vary very quickly and fine meshes were therefore necessary to accurately capture the variations [14]. Figures 3 and 4 display meshed paddle wheel model without and with bottom fins, which have 4668 cells and 7243 cells, respectively. In the meshed CFD model, the essential fluid flows were described through these cells, which would be solved numerically, so that the discrete values of the flow properties such as the velocity, pressure, temperature, and other transport parameters of interest can be determined [15].

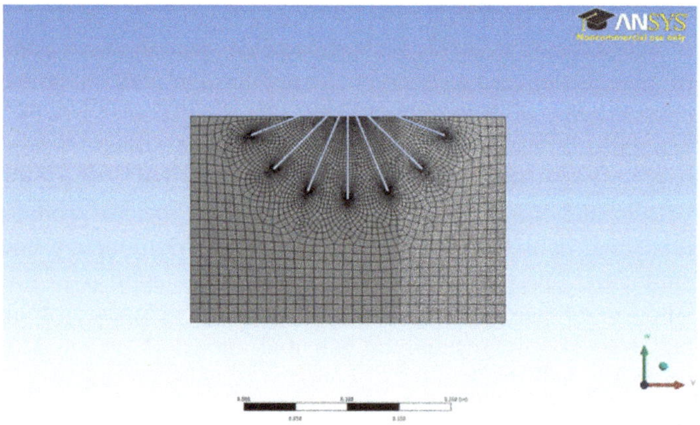

Figure 3: Meshes of a paddle wheel without bottom fin.

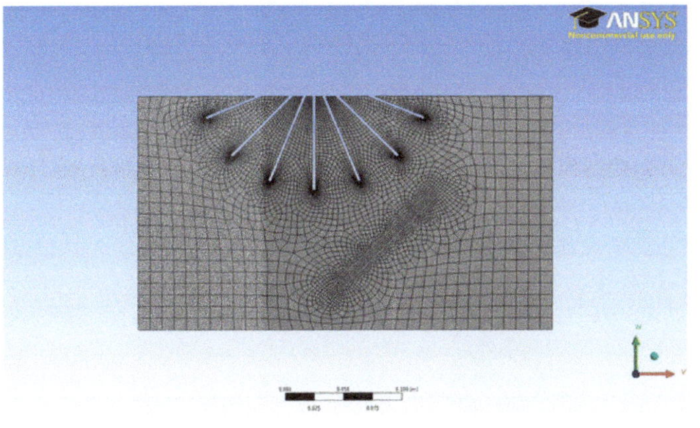

Figure 4: Meshes of a paddle wheel with bottom fin.

CFD Analysis

The maximum angular velocity is calculated with respect to that the net generated torque on the wheel shaft becomes zero. In another word, when the wheel is rotating by maximum angular velocity, the torque exerting on the paddles is zero. Following this manner, overall 30 simulations are performed. Table 2 illustrates important algorithms, approximations, and other settings.

Water speeds of 1.8, 2.2, 2.7, 3.6, and 4.5 m/s were defined in this model and used for CFD analyses to find the hydropower electricity generation capacity of the paddle wheel. Those speeds were selected as typical water speeds attainable at Wax Lake in State of Louisiana, United States (a candidate site for deployment of the presented system), and those values were provided by a local agency. At each water speed, six equally increased angular velocities from 0 to a maximum value were applied on the paddle wheel and used for simulation. The maximum angular velocity was calculated as when the net generated torque on the wheel shaft became zero. Overall 30 simulations were carried out. Table 1 illustrates important algorithms, approximations, and other settings used for the analyses.

Table 1: Important CFD analysis settings

Solver type	Pressure based, double precision, steady state, 2D
Viscose model	k–ε realizable with standard wall function
Fluid	Water with density of 998 kg/m^3
Rotation	Multiple reference frame model
Pressure–velocity coupling	Coupled
Gradient discretization	Least square cell based
Pressure discretization	Standard
Momentum discretization	Second-order upwind
Turbulent kinetic energy discretization	Second-order upwind
Turbulent dissipation energy discretization	Second-order upwind
Boundary condition	Velocity inlet and ambient pressure outlet

Simulation Results

Figures 5, 6, 7, and 8 display the distribution of the pressure and water velocity around a stationary paddle wheel with and without the bottom fin at a water speed of 1.8 m/s,. From these figures, it can be seen that the majority of the force is applied on the left four paddles through the flow, while other paddles were indispensable in making a continuous rotation. It is also obvious that the left four paddles created most power because of the high pressure and high velocity around them.

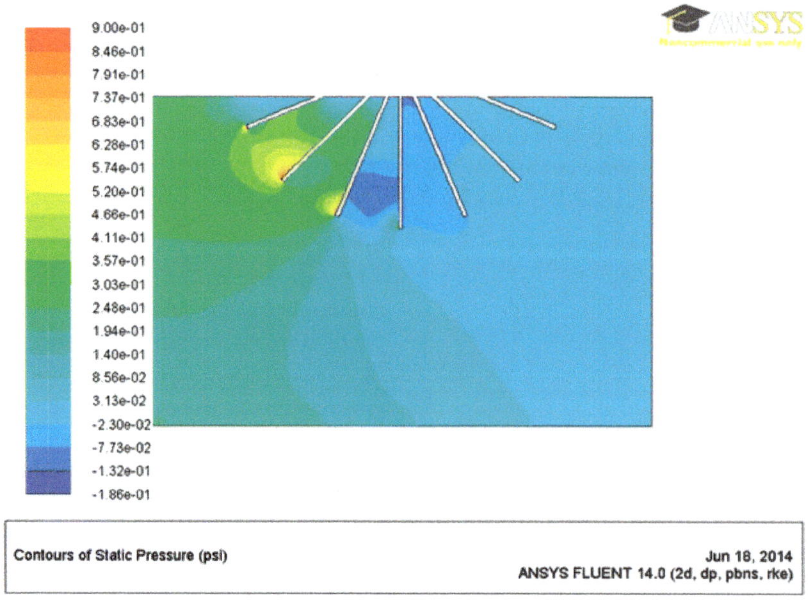

Figure 5: Pressure distribution around the paddle blades without bottom fin, water speed = 1.8 m/s.

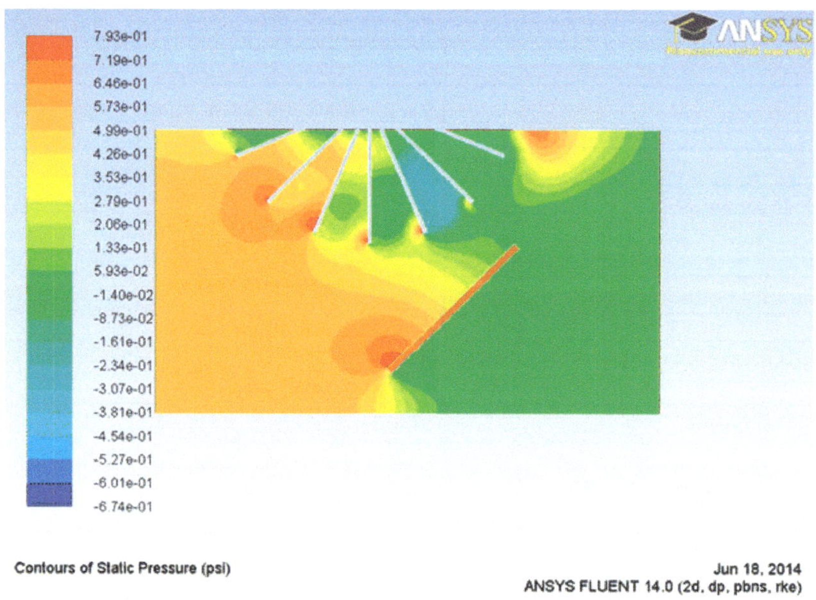

Figure 6: Pressure distribution around the paddle blades and bottom fin, water speed = 1.8 m/s

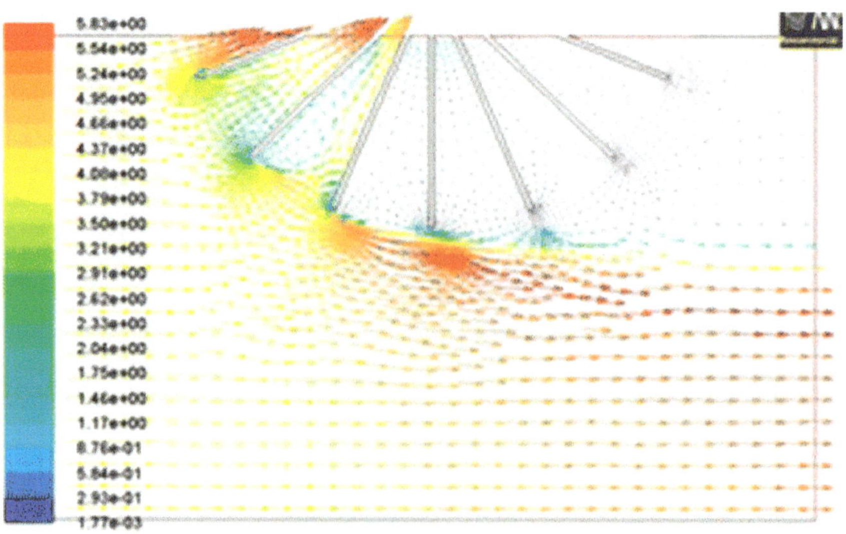

Figure 7: Velocity profile of water around the paddle blades, water speed = 1.8 m/s.

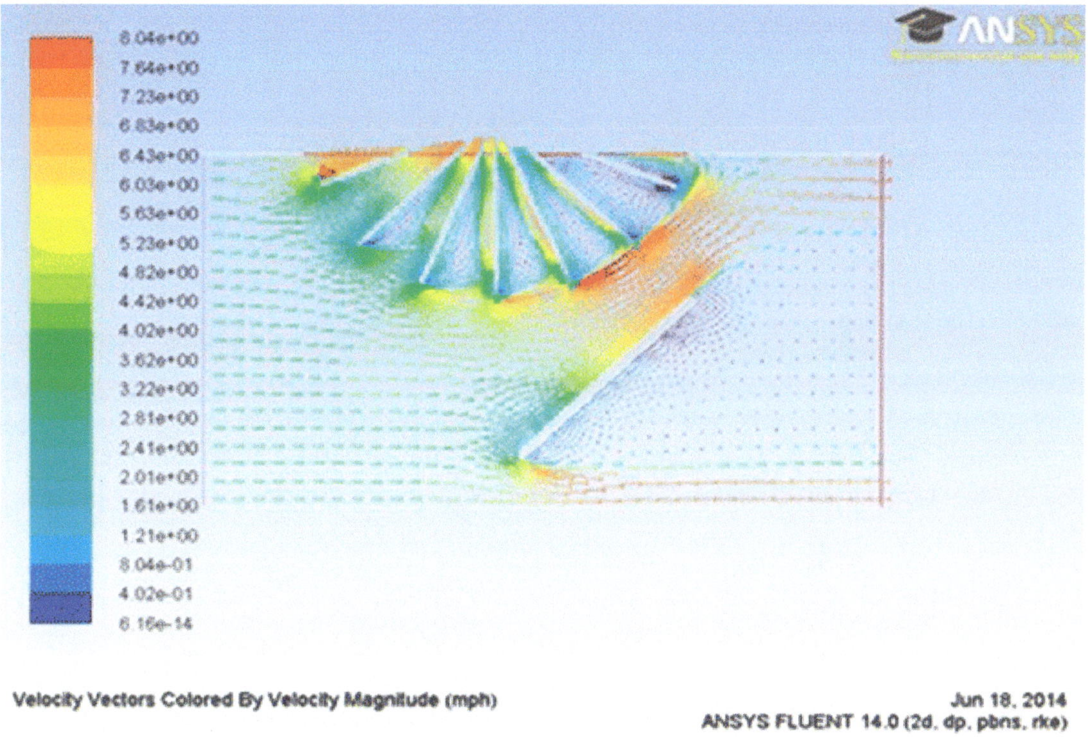

Figure 8: Velocity profile of water around the paddle blades and fin, water speed = 1.8 m/s.

CFD Simulation of the Hydropower System

After successfully running CFD simulation for a single paddle wheel, the entire hydropower system CFD model was created, on which simulations were performed (Fig. 9). During the simulation, water would pass through three pairs of paddle wheels in sequence; therefore, the water speed profile would change after passing a pair of paddle wheels and would then affect the power generation of the next pair of paddle wheels. As shown in Fig. 10, the water speed profile at outlet of the first paddle wheel is not uniform. Such speed profile will then be applied at inlet of the second pair of paddle wheels. A flowchart (Fig. 11) is presented here to show the procedure of assessing the entire hydropower system.

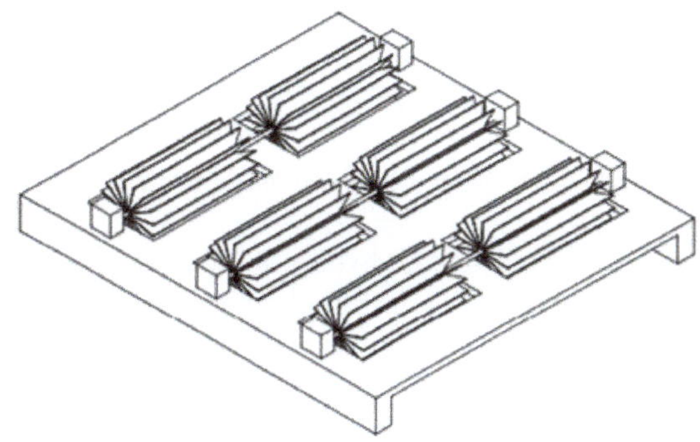

Figure 9: 3D sketch of the entire hydropower system.

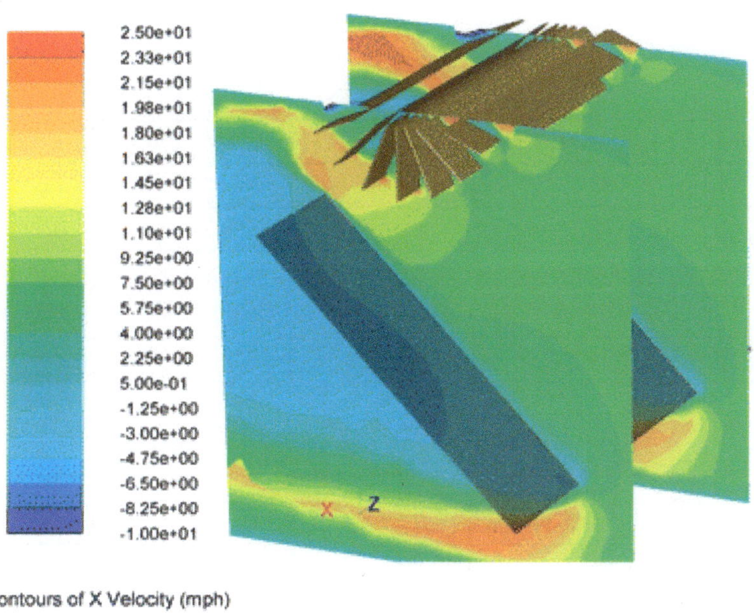

Figure 10: Water speed contour at outlet of the first paddle wheel, water speed = 3.6 m/s.

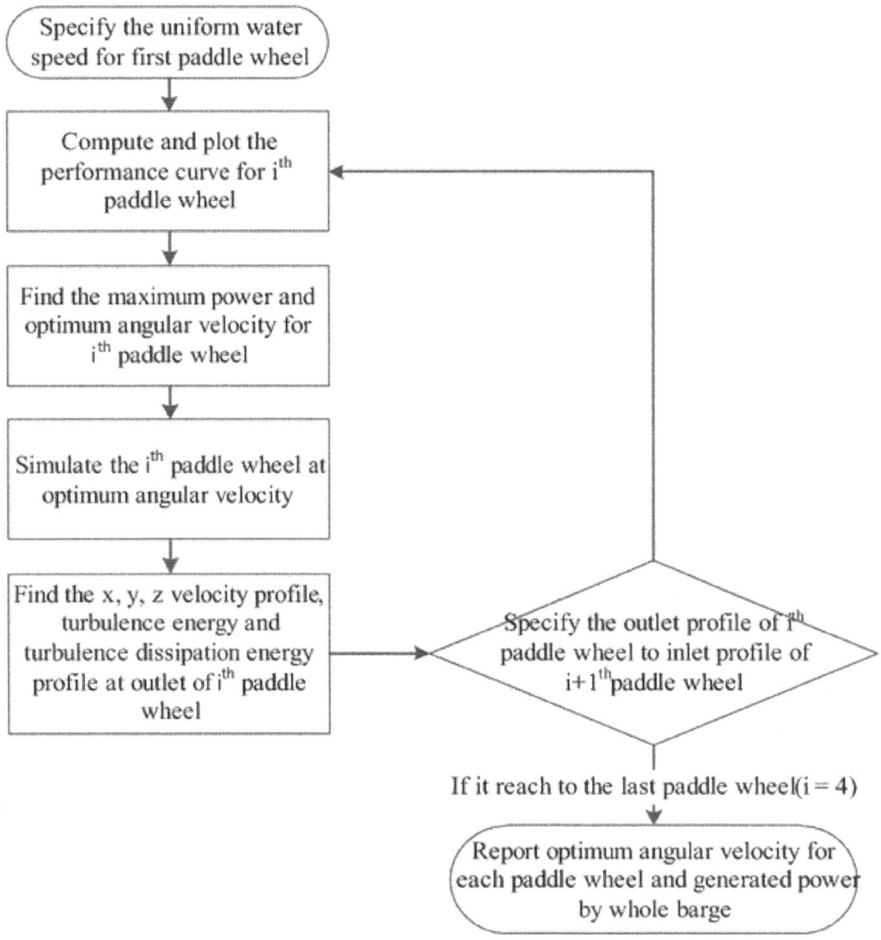

Figure 11: Hydropower system simulation algorithm.

Power generated by the hydropower system at different water speeds is listed in Table 2 (without bottom fin) and Table 3 (with bottom fin), along with the contribution from each pair of paddle wheels. From both tables, it can be found that the first pair of paddle wheels produces majority of hydropower and the power generated by the following two pairs of paddle wheels decreases subsequently. The results also confirm significant effect of bottom fins on improving the power generation capacity of the paddle wheels. During the simulation, the bottom fins captured energy in deep water and carried that energy directly to the paddle wheels. As shown in Tables 2and 3, without bottom fins, only first two pairs of paddle wheels generated hydropower, but after applying bottom fins, even the third pair of paddle wheels produced 8–10 % of hydropower at different water speeds. Moreover, the application of bottom fins dramatically improved the overall produced hydropower by 2.7 to about 4 times at different water speeds (as seen in Table 4).

Table 2: Barge-produced power and contribution of each paddle wheel (without bottom fin)

Water speed (m/s)	Power (kW)	First pair (%)	Second pair (%)	Third pair (%)
1.8	185.4	69.4	29.3	1.3
2.2	318.3	71.3	27.8	0.9
2.7	502.5	71.7	27.4	0.9
3.6	1025.0	70.9	28.3	0.8
4.5	1740.0	71.7	27.5	0.8

Table 3: Barge-produced power and contribution of each paddle wheel (with bottom fin)

Water speed (m/s)	Power (kW)	First pair (%)	Second pair (%)	Third pair (%)
1.8	506.9	63.2	26.7	10.1
2.2	943.0	64.9	25.3	9.8
2.7	1589.8	65.2	24.9	9.9

3.6	3620.2	65.2	26.0	8.8
4.5	6838.1	66.0	25.8	8.2

Table 4: Effects of bottom fins in hydropower generation

Water speed (m/s)	No bottom fin (kW)	With bottom fins	Optimized model
1.8	185.4	506.9 kW (173.4 %)	605.5 kW (226.6 %)
2.2	318.3	943.0 kW (196.3 %)	1098.8 kW (245.2 %)
2.7	502.5	1589.8 kW (216.4 %)	1823.1 kW (262.8 %)
3.6	1025.0	3620.2 kW (253.2 %)	4041.6 kW (294.3 %)
4.5	1740.0	6838.1 kW (293.0 %)	7482.0 kW (330 %)

It is evident that the bottom fins enhance the overall performance of the entire hydropower system.

Evaluation of Curved Paddles

From the simulation results of the present design, it was found that paddle wheels are the elements that convert the flow energy to mechanical energy and bottom fins have the major effect in hydropower generation. Originally, the paddles are flat plates and it was hypothesized that the shape of paddle wheels can be curved to bring more water flow running through the hydropower system to generate more power and also to reduce the energy loss as a result of sudden change in flow direction [16]. In optimizing the shape of the fins, Bejan's constructal law as well as Lorenzini's work was also referred to [17–20].

As suggested by Bansal [16], the optimized blade was developed such that the curvature begins at the 3″ length radius of the blade from the center of the paddle wheel with a radius of curvature of 1.07″. The starting position of 3″ length was chosen in order to guarantee that there are at least seven paddles submerged into water at any time. Specifically, the curvature was made along the direction of the water flow because it would increase the overall torque and therefore improve the generated hydropower. Figure 12 illustrates the optimized CFD model. The optimized model was then validated through the same CFD simulation, and as compared in Table 5, comparing to the original model with flat paddles and fins, the optimized model evidently enhanced the hydropower generation from about 10–20 % at different water speeds.

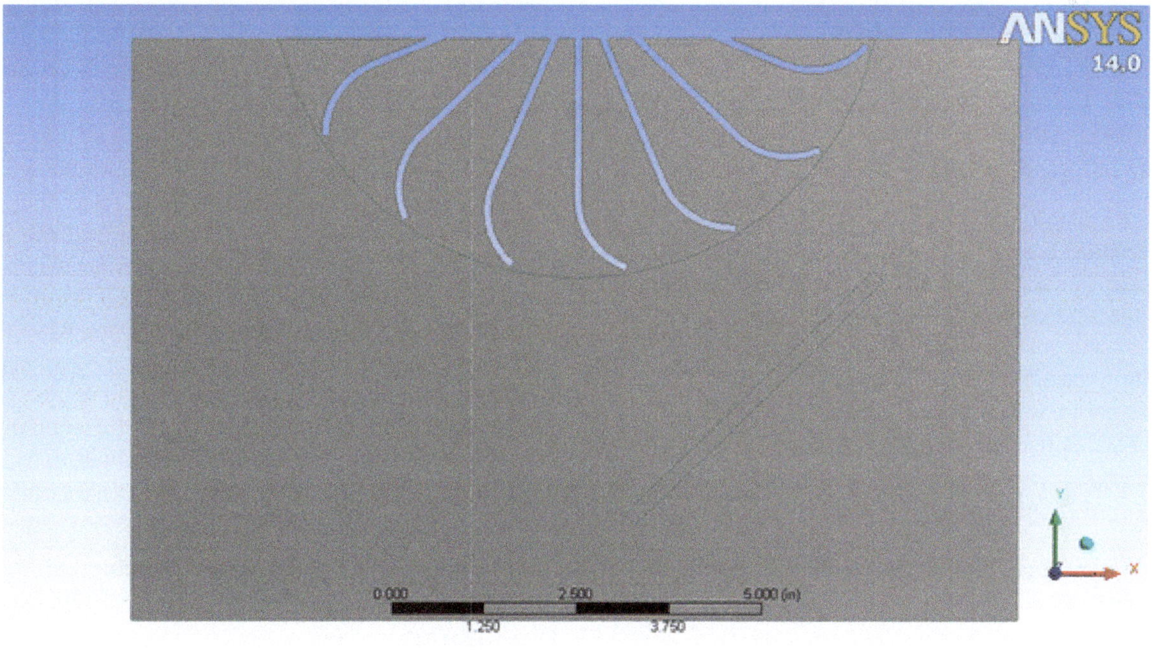

Figure 12: Optimized CFD hydropower system model with curved paddles.

Table 5: Comparison of original hydropower system and optimized model (with bottom fins)

Water speed (m/s)	With bottom fins (kW)	Optimized model
1.8	506.9	605.5 kW (19.5 %)
2.2	943.0	1098.8 kW (16.5 %)
2.7	1589.8	1823.1 kW (14.7 %)
3.6	3620.2	4041.6 kW (11.6 %)
4.5	6838.1	7482.0 kW (9.42 %)

This work only represents an initial effort in shape optimization of the blade paddles. To achieve an optimal shape of the blade paddle that maximizes the hydropower generation capacity, a full-scale optimization study has to be conducted, where the blade number, blade length, and curvature of the blades should be defined as design variables, whose influence on the overall performance of the present hydropower system needs to be examined.

CFD Simulation of the Down Scale Model

The full-size design of the hydropower system is too huge to be used for experimental validation. Thus, 1:30 down scale models were created and used for the same simulations, and the results will be compared to the experimental data measured from tests to validate the employed computational approach as well as to verify the potential of implementation and commercialization of the proposed hydropower technology. Three CFD models were created for simulations: the original scaled hydropower system without bottom fins, with bottom fins, and the optimized system with bottom fins. The overall produced hydropower from the three models at five different water speeds was recorded and compared, as shown in Table 6. As can be seen from that table, the relationship among the power generated from the three down scale models is very close to that from the three original-size models.

Table 6: Comparison of hydropower generation capacity of three down scale models

Water speed (m/s)	No bottom fin (W)	With bottom fins	Optimized model
1.8	1.42	3.89 W (174.0 %)	4.66 W (228.2 %)
2.2	2.74	8.12 W (196.4 %)	9.53 W (247.8 %)
2.7	4.54	14.37 W (216.5 %)	16.51 W (263.7 %)
3.6	8.84	31.24 W (253.4 %)	34.99 W (295.8 %)
4.5	15.55	61.09 W (292.9 %)	67.19 W (332.1 %)

Economic Analysis

A primary economic analysis was performed based on the computer results. Based on the results listed in Table 3, the total electricity generated annually from one such system under different water speeds was estimated and displayed in Table 7. The initial investment for manufacturing the system, purchasing and installing generators, controllers, and inverters was estimated at about $8.2 million. Assuming an annual average water speed of 3.2 m/s, the total generated hydroelectricity may reach 25 GWh, and considering the renewable energy credits of $11 for every 1000 kWh renewable electricity generated, which is offered by the federal government, the total annual income is close to $1.8 million at the average water speed if the generated electricity is sold at $0.06/kWh. Thus, the assumed payback period is about four and half years. After the payback period, the cost for the electricity will be reduced to $0.02/kWh to cover basic maintenance and labor fee. Considering the household electricity consumption, the amount of 25 GWh electricity can benefit about 1645 households.

Table 7: Electric generation capacity of the facility and its economic and social impact

Water speed (m/s)	Electricity (kWh)	Annual income ($0.06/kWh)	Renewable energy credit $11/1000 kWh	Total annual income	# influenced households
1.8	6,797,710	$407,862	$74,767	$482,629	443
2.7	11,746,443	$704,786	$129,206	$833,992	767
3.6	27,483,420	$1,670,605	$306,273	$1,976,878	1794
4.5	54,381,679	$3,262,900	$598,191	$3,861,091	3551
3.2 (avg)	25,192,313	$1,511,538	$277,112	$1,788,650	1645

Economic analysis results suggest the great prospects of applying the presented hydropower technology in regions with rich water resource to produce and supply low-cost, clean, and renewable electricity for a large amount of households and industries. Full implementation of such technology will lead to the establishment of many run-of-the-river hydroelectric power stations, in turn will create thousands of manufacturing, construction, supply, and maintenance jobs.

CONCLUSIONS

In this study, a hydropower system which employs paddle wheels and bottom fins to capture hydropower out of water flow was developed and optimized. CFD simulations were performed to validate the power generation capacity of the system, and its economic impact was evaluated based on the CFD results. In CFD simulations, effects of water speed, bottom fins, and shape of paddles and bottom fins on the system power generation capacity were illustrated. Contribution of each pair of paddle wheels in the hydropower generation was analyzed. The CFD simulation results and economic analysis data showed that, averagely, the presented hydropower equipment can generate and provide electricity for 1645 households at \$0.06/kWh during the first four and half years (payback period) and \$0.02/kWh after that period. Part of the present CFD simulation results had been verified through analytical approach [1, 2], and in order to further verify the applied computational approach and confirm the economic potential and commercialization prospect of the developed hydropower technique, an experimental analysis is required. To prepare for the future experimental validation, a 1:30 down scale CFD model was created and used for simulations. The results correlated very well with the CFD results obtained from the full-size model and will be compared with the experimental data in next phase.

REFERENCES

1. Peymani, Y.F., Liu, Y.-C.: Development and computational verification of an analytical model to evaluate performance of paddle wheel in generating electricity from moving fluid. Distrib. Gener. Altern. Energy J. **30**, 58–80 (2015)
2. Peymani, Y.F., Liu, Y.-C.: Development of an analytical model to predict the performance of paddle wheel in generating electricity and its validation using computational fluid dynamics (CFD). In: IEEE Green Technologies Conference, Tulsa, OK, 19–20 April 2012
3. Derradji-Aouat, A., Lau, M.: Ice loads on electric power generating stations to harness ocean energy in the Belle Isle Strait. In: Proceedings of 18th International Conference on Port and Ocean Engineering under Arctic Conditions, Potsdam, New York (2004)
4. Charlier, R.H.: Forty candles for the Rance River TPP tides provide renewable and sustainable power generation. Renew. Sustain. Energy Rev. **11**(9), 2032–3057 (2007)
5. Brune, D.E., Lundquist, T.J., Benemann, J.R.: Microalgal biomass for greenhouse gas reductions: potential for replacement of fossil fuels and animal feeds. J. Environ. Eng.**135**(11), 1136–1144 (2009)
6. Rogers, E.R.: Water wheel electric generation device. US4268757 A, 19 May 1981
7. Winius, H.C.: Paddle wheel electric generator. US7969034 B2, 28 June 2011
8. Zimmermann, S., Taghipour, F.: CFD modeling of the hydrodynamics and reaction kinetics of FCC fluidized-bed reactors. Ind. Eng. Chem. Res. **44**(26), 9818–9827 (2005)
9. Petre, C.F., Larachi, F., Iliuta, I., Grandjean, B.P.A.: Pressure drop through structured packings: breakdown into the contributing mechanisms by CFD modeling. Chem. Eng. Sci.**58**(1), 163–177 (2003)
10. Kerdouss, F., Bannari, A., Proulx, P.: CFD modeling of gas dispersion and bubble size in a double turbine stirred tank. Chem. Eng. Sci. **61**(10), 3313–3322 (2006)
11. Andersson, A.G., Andreasson, P., Lundstrom, T.S.: CFD-modeling and validation of free surface flow during spilling of reservoir in down-scale model. Eng Appl Comput Fluid Mech**7**(1), 159–167 (2013)
12. Peterson, E.L., Harris, J.A., Wadhwa, L.C.: CFD modeling pond dynamic processes. Agric Eng **23**(1–3), 61–93 (2000)
13. Kang, Y.H., Lee, M.O., Choi, S.D., Sin, Y.-S.: 2-D hydrodynamic model simulating paddlewheel-driven circulation in rectangular shrimp culture ponds. Aquaculture **231**(1–4), 163–179 (2004)
14. Liu, Y.-C., Peymani, Y.F.: Evaluation of paddle wheels in generating hydroelectric power. In: ASME 2012 International Mechanical Engineering Congress and Exposition, IMECE2012-85121, pp. 675–684, Houston, TX, 9–15 Nov 2012
15. Tu, J.-Y., Yeoh, G.-H., Liu, C.-Q.: Computational Fluid Dynamics: A Practical Approach. Butterworth-Heinemann, Oxford (2008)
16. Bansal, R.K.: A Textbook of Fluid Mechanics and Hydraulic Machines. LAXMI Publications, Ltd., New Delhi (2005)
17. Lorenzini, G., Moretti, S.: A CFD application to optimize T-shaped fins: comparisons to the constructal theory's results. ASME J. Electron. Packag. **129**(3), 324–327 (2007)
18. Lorenzini, G., Moretti, S.: Numerical performance analysis of constructal I and Y finned heat exchanging modules. ASME J. Electron. Packag. **131**(3), 031012-1–031012-10 (2009)
19. Lorenzini, G., Moretti, S.: A Bejan's constructal theory approach to the overall optimization of heat exchanging finned modules with air in forced convection and laminar flow condition. ASME J. Heat Transf. **131**(8), 081801-1–081801-18 (2009)
20. Lorenzini, G., Moretti, S.: Bejan's constructal theory analysis of gas-liquid cooled finned modules. ASME J. Heat Transf. **133**(7), 071801-1–071801-10 (2011)

CHAPTER **16**

IMPACT OF URBAN MORPHOLOGY ON INFILTRATION-INDUCED BUILDING ENERGY CONSUMPTION

Andrius Jurelionis[1] and Demetri G. Bouris[2]

[1]Faculty of Civil Engineering and Architecture, Kaunas University of Technology, Studentu str. 48, 51367 Kaunas, Lithuania

[2]School of Mechanical Engineering, National Technical University of Athens, 15780 Zografou, Greece

ABSTRACT

External air movement within built neighborhoods is highly dependent on the morphological parameters of buildings and surroundings, including building height and street cavity ratios. In this paper, computational fluid dynamics (CFD) methods were applied to calculate surface pressure distributions on building surfaces for three city models and two wind directions. Pressure differences and air change rates were derived in order to predict the heating load required to cover heat losses caused by air infiltration. The models were based on typical urban layouts for three cities, and were designed of approximately equal built volumes and equal air permeability parameters. Simulations of the three analyzed building layouts resulted in up to 41% differences in air change rates and heat losses caused by air infiltration. In the present study, wind direction did not have a significant impact on the relative difference between the models, however sideward wind direction caused higher air change rates and heat losses for all simulated layouts.

INTRODUCTION

Urban morphology parameters, such as urban plan area density, frontal area density, geometry of the buildings, and topographical features influence airflows in and around buildings and, ultimately, energy consumption on a regional scale [1,2]. Airflow patterns in urban areas, referred further as neighborhoods, are especially important for buildings with natural or hybrid ventilation. However, air infiltration, urban heat island formation and airborne pollutant accumulation can affect air quality (IAQ) on the neighborhood or building scale, the coefficient of performance (COP) of heating, ventilation and air conditioning (HVAC) systems, and the heating and cooling demand of mechanically ventilated buildings as well [1]. Spatial arrangement of the neighborhoods influences energy transfer via convection, infiltration and conduction.

Several studies have provided integrated approaches for combining urban airflow simulations with energy performance tools. Indoor-outdoor building energy simulator TUF3D was one of the first three-dimensional fully-coupled indoor-outdoor building energy simulators which allows analysis of urban energy use based on urban geometry, material modifications and the interaction between buildings and their surroundings [3]. Yang et al. established an integrated simulation method capable of quantifying the effects of various microclimatic factors on building energy performance under given urban contexts [4].

Among the many types of energy related interactions between a building and its surroundings, air infiltration

Citation: Jurelionis, A.; Bouris, D.G. Impact of Urban Morphology on Infiltration-Induced Building Energy Consumption. Energies 2016, 9, 177. doi:10.3390/en9030177.

can be responsible for a significant portion of a building's energy consumption, depending on construction and design parameters. It is proved [5] that the time-averaged wind pressure coefficient C_p is one of the best indicators of indoor-outdoor environment interaction due to air infiltration. It is defined as follows:

$$C_p = \frac{p_x - p_0}{p_d};\ p_d = \frac{\rho \cdot U_{ref}^2}{2} \quad (1)$$

where p_x is the static pressure at a given point on the building façade (Pa), p_0 is the static reference pressure (Pa), p_d is the dynamic pressure, ρ is the air density (kg/m³) and U_{ref} is the reference wind speed at building height h in the windward undisturbed flow (m/s) [5].

C_p values are determined according to orientation and height of the component, building and zone characteristics, shielding and building location [6]. It is common practice to use surface-averaged C_p values for air infiltration and ventilation studies. However, using such values may lead to significant errors in the airflow calculations compared to using local C_p values at the exact coordinates of the building where ventilation equipment is located [5]. This is especially true for natural ventilation cases. Uncertainties of the air change rate calculations can also be increased by neglecting the surroundings of the analyzed buildings or neighborhoods [7].

Van Moeseke et al. demonstrated changes in air flow inside buildings when horizontal as well as vertical pressure coefficient gradients on buildings' sides are considered [8]. CFD has proven to be an effective tool to predict air movement and air temperature distribution for solving complex problems within urban neighborhoods [9,10]. Experiments and CFD simulations performed by Hang et al. showed different wake flows and even airflow patterns for round and square idealized city models. The overall city form, the configuration of streets and street orientation relative to the approaching wind direction was found to have a great influence on the airflow within the street cavity. Weaker wind was observed in the street network of the square city model than that in the round city model [11]. The interaction of the external wind flow and the internal thermally-driven flow depends upon the ratio of the building height to the urban canyon width [12].

This study is based on the hypothesis that the urban morphology parameters can either increase or decrease wind impacts on buildings, depending on the building type and the aim of urban planners. Actually, heat loss due to air infiltration and leakage depends heavily on a buildings' plan layout and construction techniques. Modern, airtight and mechanically ventilated buildings are expected to be rather insensitive to wind effects on infiltration so the main focus in this study is on existing buildings without mechanical ventilation, which represent a large portion of the European building stock and even more so in the case of residential buildings. However, apart from the effect of each building's parameters, the purpose of this study is to implement urban airflow simulations in order to identify the possible effect of urban scale morphology on the potential for infiltration. This is considered as the driving force that interacts with the building scale construction and plan parameters. The pressure distribution on building surfaces was used as an indicator in order to estimate the potential impact of urban morphology on air infiltration and energy use. Three city models, each at two wind directions, were analyzed by means of computational fluid dynamics (CFD) simulations. Air speed and turbulence within the street cavity is examined and used as an indicator of general neighborhood aeration. The results of the simulations showed up to 41% increase in both air change rates and heat losses caused by air infiltration for the analyzed city models. Sideward wind direction resulted in higher overall air change rates for all neighborhoods compared to perpendicular wind direction.

RESULTS

Urban Morphologies Selected for the Study

Two wind directions and three urban morphologies were analyzed in this study, which are further defined as UM-1, UM-2 and UM-3. The morphologies were designed to accommodate approximately the same built volume (600,000 m³) within the same area of the neighborhood (300 m × 300 m) and the surroundings were imitated by placing buildings of a smaller size around the main domain of interest. The geometry of the surrounding buildings was the same for all cases. The morphology of the analyzed regions was designed according to layouts of real cities, namely Zürich (Switzerland), Kaunas (Lithuania) and Athens (Greece). The geometries and views of the cities they represent are provided in Figure 1.

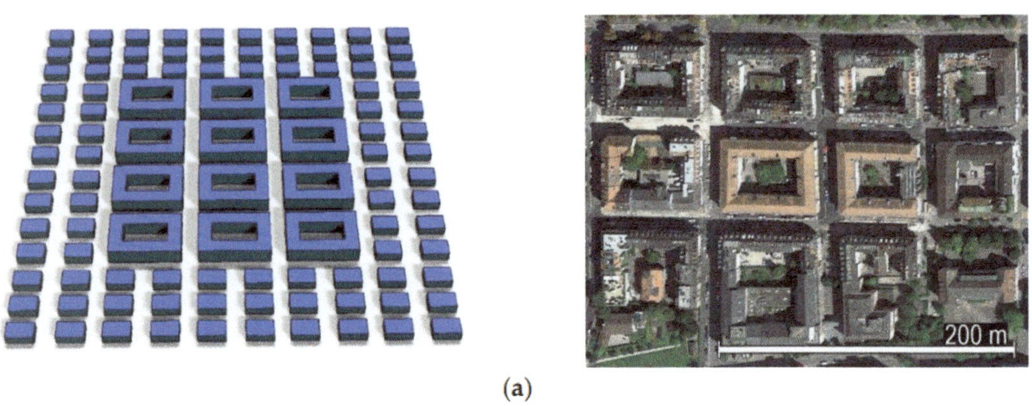

(a)

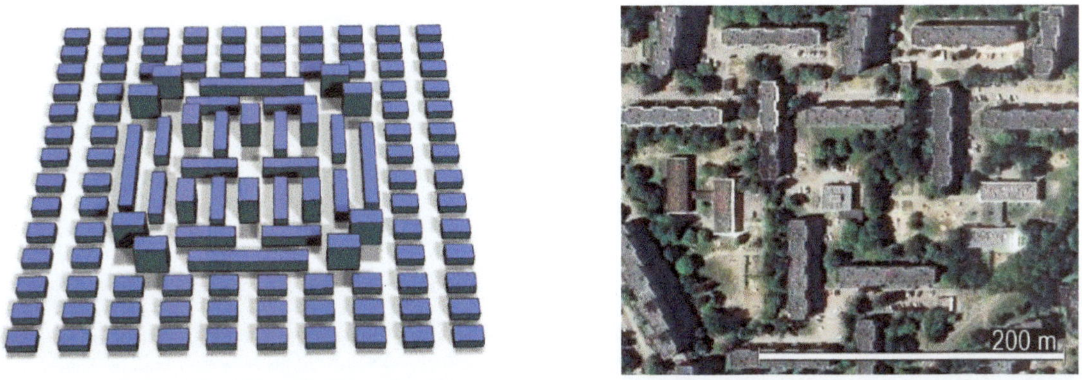

(b)

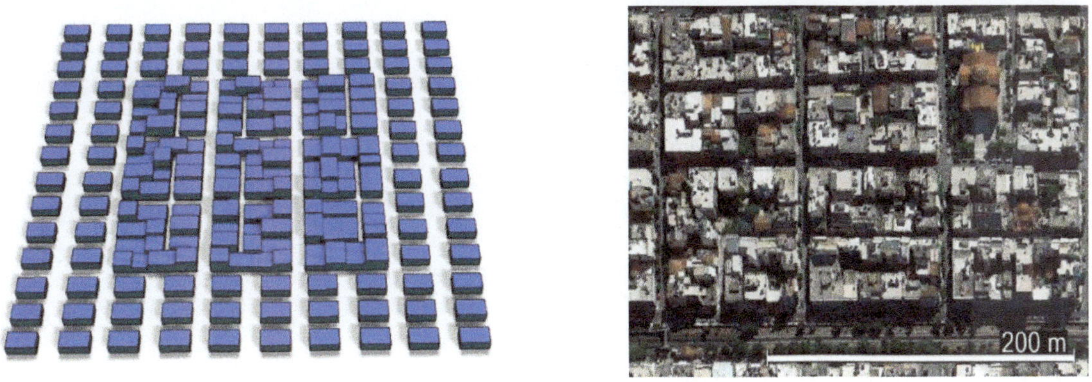

(c)

Figure 1. Urban morphologies with uniform surrounding buildings used for CFD analysis in comparison to the real city images: (**a**) UM-1—Zürich, Switzerland; (**b**) UM-2—Kaunas, Lithuania; (**c**) UM-3—Athens, Greece. Map data ©2015 Google.

The real city layouts and building heights were modified to satisfy the requirements of equal built volumes and neighborhood area. The building heights, average building height and street cavity width ratios used for the simulations are presented in Table 1.

Table 1. Height of buildings and its ratio to street width for selected urban morphologies

Urban Morphology	Height of Buildings, m	H_b/W [1]
UM-1	18 m	1.00
UM-2	16 and 36 m	1.73
UM-3	8 to 16 m	0.76

[1] Ratio between average building height (m) and average distance between buildings or blocks (m).

Pressure differences on building surfaces were used to predict air infiltration and the potential impact on energy consumption due to air infiltration in the built neighborhood. Detailed descriptions of the simulation methods, validation of the CFD model and calculation procedures are provided in the Methods section.

CFD Prediction Results for Air Speed and Building Surface Pressure Distribution

CFD predictions proved the hypothesis that urban morphology is a critical factor, significantly affecting pressure distribution on building surfaces (Figure 2), which in turn determines the pressure differences that drive infiltration. Results of the simulations with perpendicular wind (0°) revealed that the highest pressure differences appear on the windward side building surfaces for morphologies UM-1 and UM-3. However, higher buildings (up to 36 m) were present in UM-2 case, and therefore, relatively high pressure differences on the leeward building surfaces were observed as well. Effects of irregularities of building heights can also be observed in UM-3 cases. CFD predictions proved the hypothesis that urban morphology is a critical factor, significantly affecting pressure distribution on building surfaces (Figure 2), which in turn determines the pressure differences that drive infiltration. Results of the simulations with perpendicular wind (0°) revealed that the highest pressure differences appear on the windward side building surfaces for morphologies UM-1 and UM-3. However, higher buildings (up to 36 m) were present in UM-2 case, and therefore, relatively high pressure differences on the leeward building surfaces were observed as well. Effects of irregularities of building heights can also be observed in UM-3 cases.

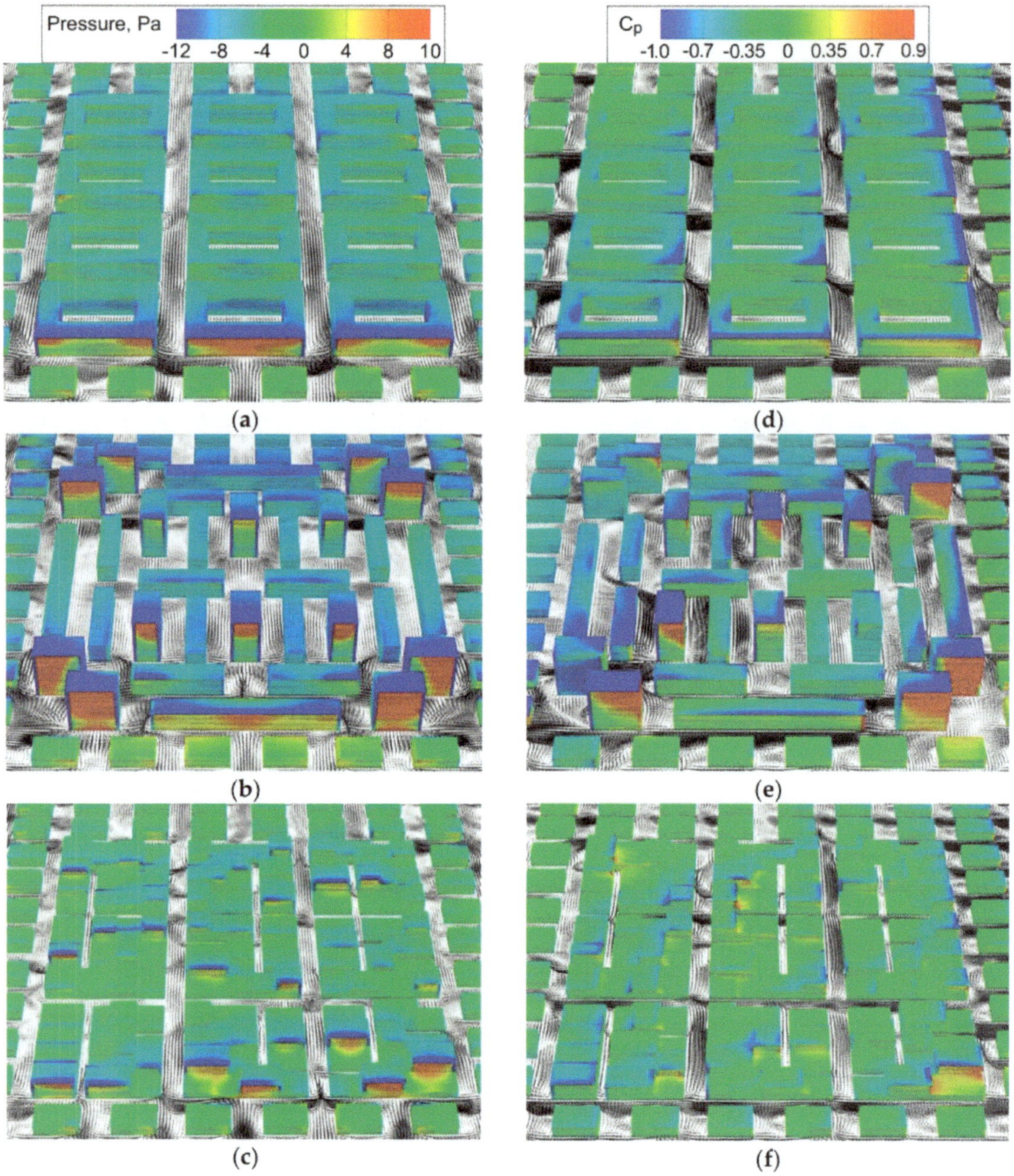

Figure 2. CFD prediction results of air pressure and wind pressure coefficient distribution on building surfaces and air speed vectors at the height of 2 m: (**a**,**d**) Urban morphology UM-1; (**b**,**e**) Urban morphology UM-2; (**c**,**f**) Urban morphology UM-3; (**a**–**c**) Perpendicular wind direction; (**d**–**f**) Sideward wind direction, 45°.

Air speed contours at three heights (2 m, 10 m and 15 m) and turbulent kinetic energy contours at 10 m height for both wind directions are provided in the Supplementary Materials of this paper as Figures S1 and S2, respectively.

Also shown in Figure 2 is a scale of the commonly used wind pressure coefficient (1). As the buildings of different heights are present in different models, the reference velocity has been taken at 10 m height i.e., U_{ref} = 4.5 m/s. It can be observed from Figure 2a,d that homogeneous urban morphology resulted in better wind shading effects. Highest pressure differences were present on surfaces of the buildings located on the windward side (UM-1). On the other hand, irregularities in building heights caused higher pressure differences on the leeward building surfaces for UM-2 and UM-3 cases. In the street canyons, higher values for turbulent kinetic energy were observed for UM-2 and UM-3 models, although air speed within the neighborhood was the highest for the UM-1 model.

Results of Air Infiltration and Expected Impact on Energy Consumption Calculations

Pressure differences on building surfaces were obtained by post-processing the CFD simulation results. The obtained values are presented in Figure 3 (median values, inter-quartile range as well as minimum and maximum values). The highest standard deviation and range of values were obtained for urban morphology UM-2 at both wind directions. As can be observed from Table 2, this layout also resulted in the highest air change rates and energy consumption required to cover air infiltration heat losses, subject to assumptions for leakage area and discharge coefficient.

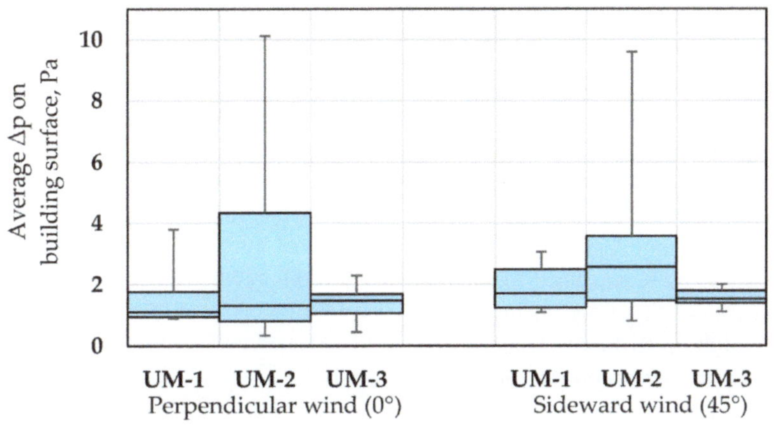

Figure 3. Pressure difference (Δp) distribution on building surfaces for urban morphologies UM-1, UM-2, UM-3 and two wind directions (0° and 45°). Horizontal lines within bars indicate median values, bars denote inter-quartile ranges and whiskers define the full range of values.

Table 2. Results of air infiltration and expected impact on energy consumption calculations

Wind Direction	Urban Morphology	Average Speed in Street Cavities at 2 m, m/s	Weighted Average Δp, Pa	Weighted Air Change Rates [1]	Heat Losses at Δt=10K, kW	Percentile Increase Compared to Lowest Result
0°	UM-1	2.30	1.62	0.273	546	0%
	UM-2	1.60	3.30	0.384	768	40.7%
	UM-3	1.40	1.47	0.303	607	11.1%
45°	UM-1	2.52	1.87	0.300	601	0%
	UM-2	2.11	3.55	0.424	848	41.0%
	UM-3	2.05	1.55	0.314	628	4.5%

[1] Specific air leakage area (A_L) and discharge coefficient (C_D) were considered 4 cm²/m² and 0.6, respectively.

In Table 2, both weighted Δp values and weighted air change rates were calculated considering the volume of buildings. In all cases, potential heat losses were estimated by Equation (6), presented in the Section 4.4 by using the same typical 10 K air temperature difference between indoors and outdoors. In Table 2 heat losses are presented for the whole analyzed neighborhood. The results of this study prove that urban morphology has a significant impact on air infiltration and air movement within street cavities.

With regard to energy demand, this impact may or may not be significant, depending on other factors such as airtightness, building plan depth etc. The results presented here refer to the change in relative energy consumption due to infiltration alone. The following conclusions can be drawn:

- The highest values of average speed and the lowest for turbulent kinetic energy within street cavities were observed in the case of urban morphology UM-1. However, it seems that this layout exhibits better aerodynamic properties that allow a reduction of wind induced energy consumption by approx. 41% compared to urban morphology UM-2. Therefore, designing built neighborhoods according to this overall spatial shape should be considered to achieve lower energy consumption of smart cities.
- The highest wind induced air change rates and pressure differences on building surfaces were observed for urban morphology UM-2. Mixed building height and highest average building height and street width ratio resulted in a significant increase of the estimated heat load required to cover the heat losses despite the wind direction.
- A relatively small difference between the results of the layouts UM-1 and UM-3 was found. These models however, were similar in that buildings were joined into blocks in both cases. This leads to overall lower values of air leakage areas as, in this study, air leakage areas were calculated from specific leakage area (A_L, cm²/m²) according to the exposed surface areas of the buildings.

DISCUSSION

This study focused on finding relative differences of air change rates and energy consumption caused by wind effects for different urban layouts. Therefore, isothermal conditions were simulated by means of CFD and the stack effect was neglected as a driving force. This assumption is not expected to influence the results and conclusions since pressure effects on infiltration rates are additive [13]. Air change rates were estimated by using differential pressure on building surfaces as an input and considering similar air leakage area per total surface area of the buildings for all cases. Inclusion of the stack effect would be more realistic, but under the relatively high wind speed being examined, the wind induced pressure difference is expected to be the dominant mechanism. Furthermore, inclusion of the stack effect would increase the number of degrees of freedom for the design of the numerical experiment, such as indoor-outdoor air temperature difference and air leakage within the buildings itself.

It is also worth noting that urban morphologies analyzed in this study were selected as representative cases for Northern, Central and Southern Europe and were only used to build the initial geometries. The simulations were carried out with the same wind profiles used to define boundary conditions and the same temperature difference of 10 K was used for heating load calculations. The results should therefore be interpreted in a way that optimal solution might be different depending on the geographical location of the neighborhood. Higher values of air speed within the street cavities can be chosen as a goal while planning neighborhoods in warm climates and the aim to reduce pressure differences on building surfaces can be adopted for cold climate cities.

METHODS

Urban Morphologies and Boundary Conditions

Three urban morphologies were analyzed in this study. Details of the models UM-1, UM-2 and UM-3 are provided inSection 2.1. The analyzed built neighborhood was surrounded by buildings of a smaller height. The width, length and height of the overall solution domain for UM-1 and UM-3 cases was 750 m × 750 m × 125 m. In UM-2 case, buildings of 36 m were present in the model, therefore the domain was extended to 750 m × 800 m × 180 m. The wind profile was assumed to correspond to neutral atmospheric stability and followed a log-law:

$$u = \frac{u_*}{\kappa} ln\left(\frac{z}{z_0}\right) \quad (2)$$

with a roughness length of z_0 = 0.5 m and the Von Karman constant κ = 0.41. The friction velocity u_* was calculated by considering wind speed equal to 4.5 m/s at the height of 10 m and the full profile is presented in Figure 4. A Cartesian grid was adopted with a vertical discretization of 1 m up to a height equal to two building heights above ground and a horizontal discretization of 2 m (width and length). The size of the grid cells was increased closer to the boundaries of the domain. Total number of grid cells varied between 5.2 million (UM-1 and UM-3) and 9.2 million (UM-2). An example of the generated grid for UM-2 case is presented in Figure 5.

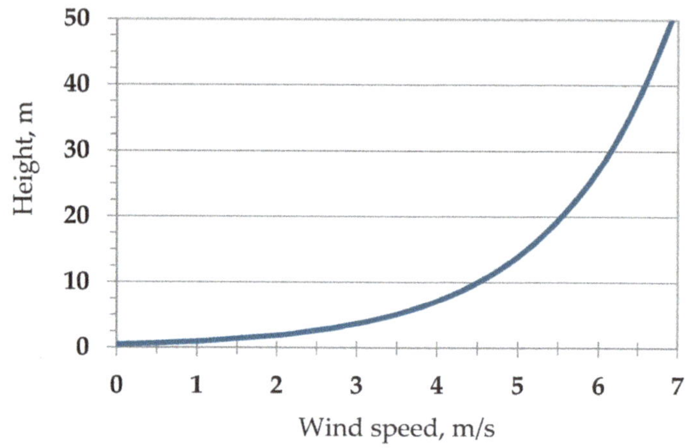

Figure 4. Wind profile used for CFD simulations.

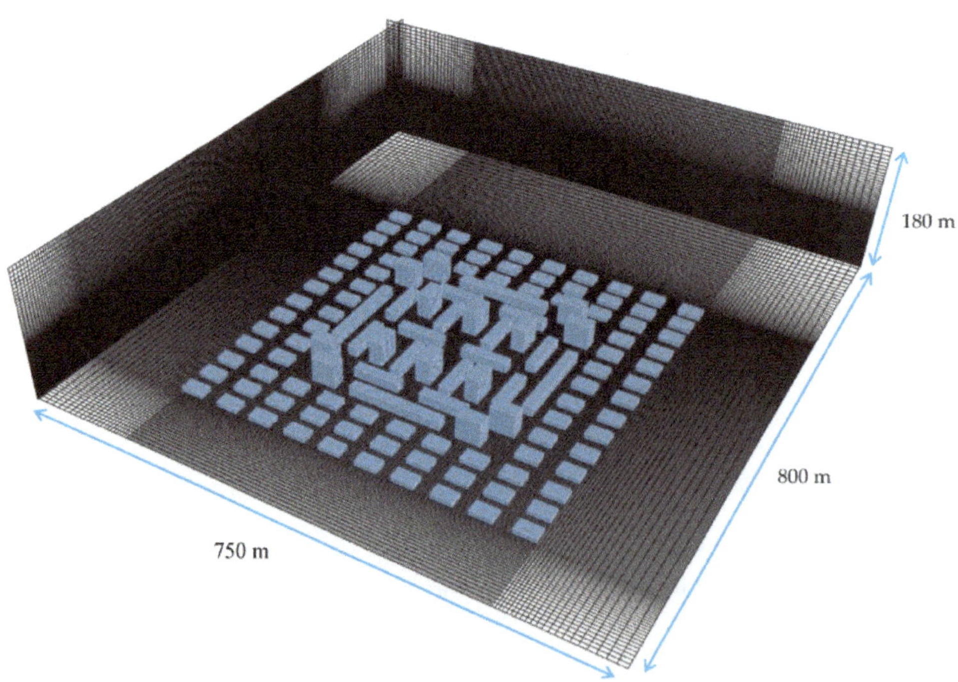

Figure 5. Example of the calculation grid used for CFD simulations (UM-2 case).

Tools and Procedures Used for CFD Simulations

Geometries for the CFD models were created through 3D design software (Sketchup, 2015, Trimble Navigation, Ltd., Sunnyvale, CA, USA) and then processed using an in-house algorithm for the definition of solid regions on a Cartesian grid. Sufficient distance from the computational domain boundaries ensured minimal blockage and boundary effects. The numerical modeling involved solution of the 3D volume averaged Reynolds equations with SIMPLE algorithm for pressure velocity coupling [14] and the Rhie and Chow corrections [15] for the collocated grid arrangement. A bounded second order upwind discretization scheme was used for the convective terms and central differences for the rest.

The k-ε turbulence model with wall functions was used for turbulence modeling [16]. The profiles for turbulence kinetic energy and its dissipation rate were calculated assuming local equilibrium [17]:

$$k = \left(\frac{u_*}{C_\mu^{0,25}}\right)^2, \varepsilon = \frac{u_*^3}{\kappa y} \quad (3)$$

where u* is the friction velocity, κ is the von Karman constant (=0.40–0.42) and C_μ is a model constant of the standard k-ε model (=0.09). The velocity and turbulence quantities were considered constant along the top of the computational domain. These boundary conditions correspond to neutral atmospheric conditions and a moderately rough upstream fetch and are in accord with the COST 732 guidelines for CFD simulation of flows in urban environments [18]. Tecplot 360 EX software was used for visualization of the CFD results (2015, Tecplot, Inc., Bellevue, WA, USA).

Validation of the CFD model

Validation of the CFD methodology in the prediction of surface pressure distribution has been performed in the previously published studies dealing with the flow past a single building [19,20]. The solution domain was extended in this study to a neighborhood scale. Therefore, validation of the computational methodology was performed through a simulation of the experimental study by Davidson et al. [21]. This study was selected for the validation of the model due to well documented boundary conditions, including upstream wind profiles. Staggered array configuration of building imitating cubes were analyzed with the distance between the blocks double the building height. During the wind tunnel experiments, both vertical and horizontal air velocity profiles were measured at X and Y axis by means of pulsed-wire and hot-wire anemometers in between the street cavities. The study demonstrated the reduction in magnitude of the velocity within the array including the far-wakes of individual obstacles spreading and merging with those of neighboring obstacles, reducing the mean flow through the array [21].

Similar boundary conditions were used as an input for CFD simulations. The geometries of the model and the comparison of the velocity profiles are presented in Figure 6. Air velocity profiles are presented at the same coordinates as documented by Davidson et al. and horizontal profiles along the X axis are compared within the same region of interest, excluding the areas shaded in Figure 6.

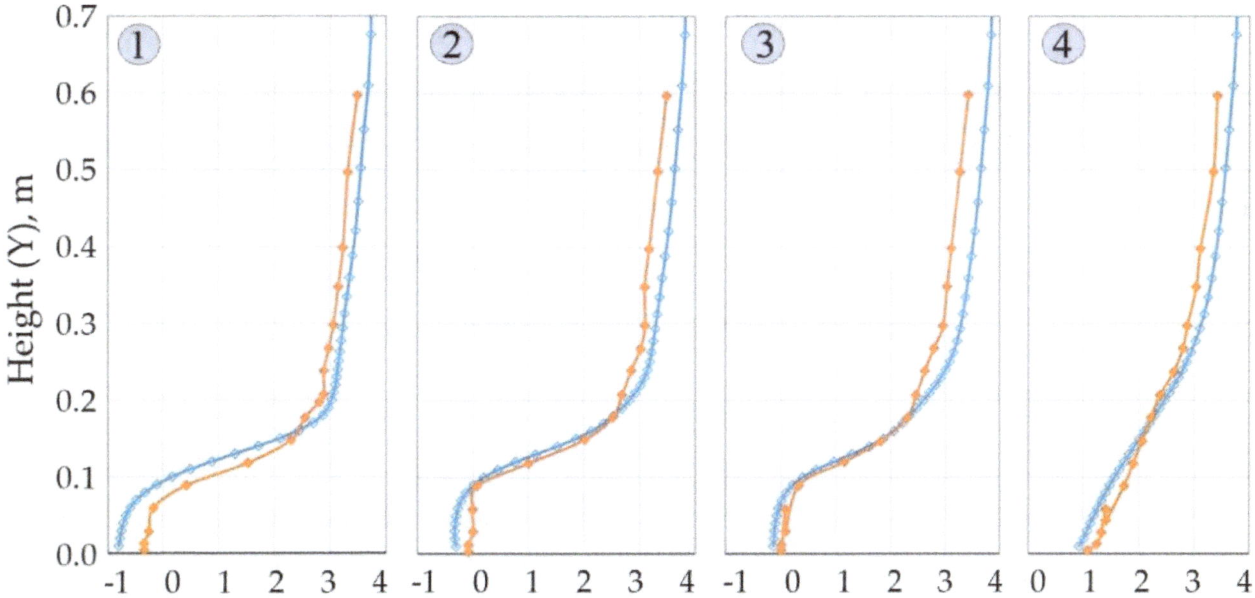

(a)

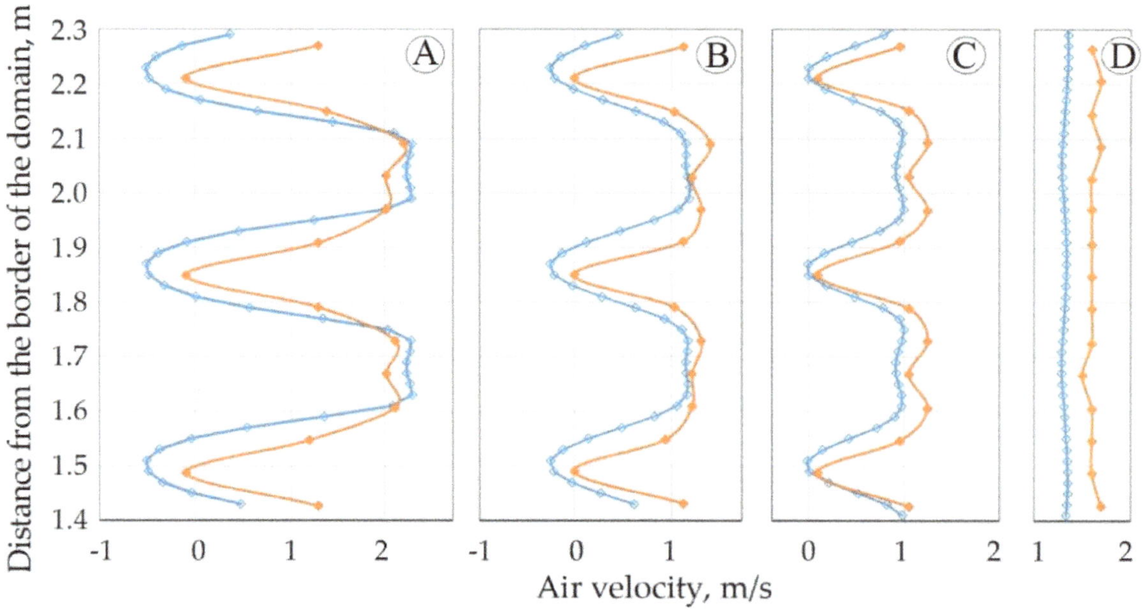

(b)

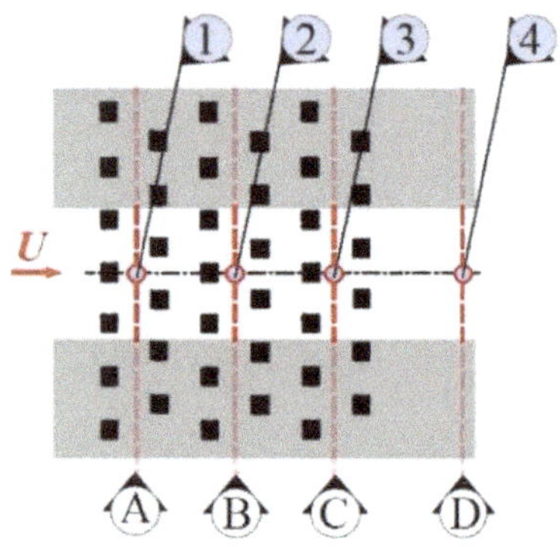

(c)

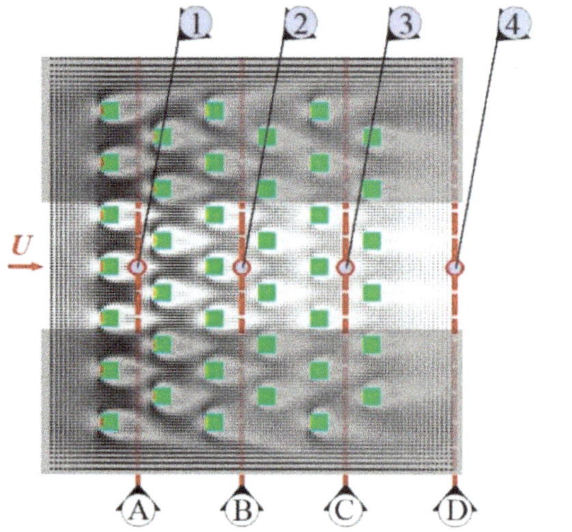

(d)

Figure 6. Vertical (**a**) and horizontal (**b**) air velocity profiles obtained by the CFD validation model (**d**) based on the wind tunnel study (**c**) by Davidson et al. [21] (wind tunnel experiment results, CFD prediction results).

As it can be observed from Figure 6a,b, vertical profiles obtained from CFD simulations were in good agreement with the experimental results. Horizontal profiles indicated that CFD tends to overestimate the building impact on the flow and shows slightly faster reduction in magnitude of the velocity deeper within the building array. However, considering that the main aim of this study was to observe differences between the models, the level of agreement between the models was considered satisfactory. Surface pressure distribution was not measured by Davidson et al., but given the strong relation between pressure distribution on building surfaces and air velocity patterns within street cavities, as well as the previous validation studies for pressure distributions on single buildings [19,20], the computational model's performance was considered adequate in order to proceed with the present study.

Air Infiltration Estimation

The technique of equivalent leakage area (ELA) was adopted to obtain the volume flow rates resulting from pressure differences on building surfaces. It is important to note, that stack effect in buildings was not considered in this study and the air change rates were assumed to be generated solely by wind as a driving force. However, the (ELA) technique is based on overall pressure differences between the interior and the exterior of the building and calculation based on detailed local surface pressures is not straightforward. Here, air change rate caused by wind was calculated based on the assumption that the building's inner pressure is determined by the mean pressure on its exposed surfaces. Therefore, infiltration and exfiltration are determined by independently calculating the positive and negative differences of the local external surface pressure to the building's inner pressure. The following steps were performed during post processing:

- Reading the pressure values for each grid cell of the building's exposed surfaces obtained by CFD, and calculating the surface weighted mean external pressure. This defines the building's inner pressure.
- Determining overall pressure difference by using the difference between the sum of the values which are higher than the inner pressure on the building surface, indicating air infiltration and the sum of the values which are lower than the inner pressure, indicating exfiltration.
- The pressure difference thus obtained was used for calculating the air flow rate by applying the ELA equation [22]:

$$Q = C_d \cdot A \cdot \left(\frac{2 \cdot p}{\rho}\right)^{0.5} \quad (4)$$

where: Q—air flow rate, m³/h; C_d—the discharge coefficient (0.6 value was used in this study), A—equivalent leakage area, m²; Δp—pressure difference across building surface, Pa (obtained within the step 2); ρ—air density. In this study the values were used as follows:

- Discharge coefficient—0.6 (i.e., the discharge coefficient for a sharp-edged orifice) [13];
- Equivalent leakage area was calculated by using the specific leakage area i.e., the ratio of (A_L) and exposed surface area of the building. This ratio was considered 4 cm² per 1 m² of the building surface area [23];
- Air density—1.16 kg/m³.
- Building air change rate was calculated as follows:

$$ach = \frac{Q}{V} \quad (5)$$

where: Q—air flow rate, m³/h (obtained within the step 3), volume of the building, m³.

- The total energy consumption of the built neighborhood was estimated by calculating total air flow rates using the average built volume between the simulated cases and air change rates of each particular case. Heating load required to cover air infiltration heat losses was calculated as follows [13]:

$$q = \Sigma Q \cdot \rho \cdot c_p \cdot \Delta t \quad (6)$$

where: q—sensible heat load, W; Q—air flow rate, m³/s; ρ—air density, kg/m³; c_p—specific heat of air, J/kgK; Δt—temperature difference between indoors and outdoors, K.

10 K air temperature difference between indoors and outdoors was used for this study as an estimate of average air temperature difference during the heating season. Example of the infiltration induced air change rate and heating load calculation results are presented for one example building in Appendix A.

ACKNOWLEDGMENTS

This paper was supported by COST (European Cooperation in Science and Technology) by a STSM Grant from COST Action TU1104 "Smart Energy Regions".

AUTHOR CONTRIBUTIONS

The authors contributed equally to this work. CFD code and post processor used for this study was developed by Demetri G. Bouris; Andrius Jurelionis and Demetri G. Bouris conceived and designed the simulation cases and post processing tools; Demetri G. Bouris performed CFD calculations; Andrius Jurelionis and Demetri G. Bouris analyzed the data; Andrius Jurelionis wrote the paper.

ABBREVIATIONS

The following abbreviations are used in this manuscript:

CFD	Computational fluid dynamics
ELA	Equivalent Leakage Area
IAQ	Indoor air quality
UM	Urban morphology

APPENDIX A

In this appendix, the results of air change rate and infiltration induced energy consumption calculation for a single building are presented. The building on the windward side of 36 m height was extracted from urban morphology UM-2 (perpendicular wind direction case, 0°).

The location of the building within the neighborhood and CFD simulation results of pressure distribution on building surfaces are presented in Figure A1. The results of the calculations are presented in Table A1 and follows the structure described in Section 4.4.

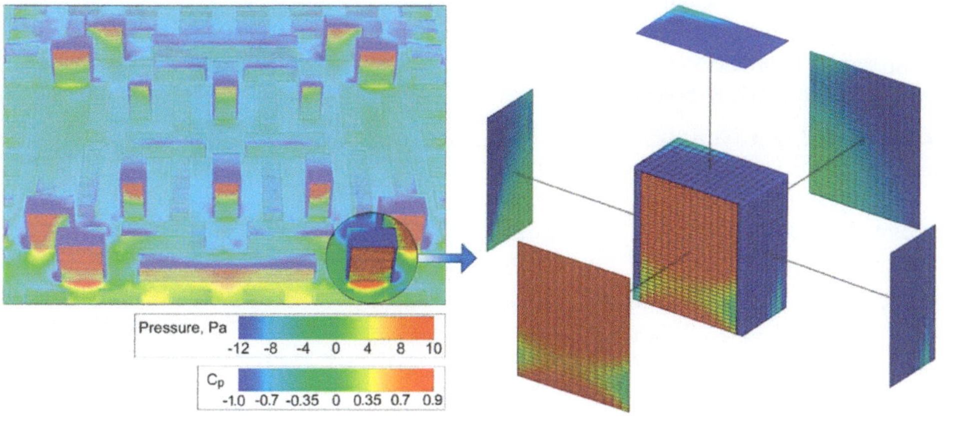

Figure A1. Air pressure distribution and wind pressure coefficient distribution on the surfaces of one building (UM-2 case, perpendicular wind), used as an example.

Table A1. Steps and results of air change rate and heat losses calculations for the example building

Step	Description and Known Values		Values Obtained
1-2	Calculating the mean pressure on building surfaces and determining the pressure difference	$(p-p_{avg})_{>0} = 6.88$ Pa $(p-p_{avg})_{<0} = -3.24$ Pa	$\Delta p = 10.12$ Pa
3	Air flow rate calculation	Building volume—20979 m^3 Air leakage area—1.69 m^2	Q = 15222 m^3/h
4	Building air change rate calculation		ACH = 0.726
5	Heating load required to cover air infiltration heat losses at $\Delta t = 10$ K		q = 40.3 kW

REFERENCES

1. Srebric, J.; Heidarinejad, M.; Liu, J. Building neighborhood emerging properties and their impacts on multi-scale modeling of building energy and airflows. Build Env. **2015**, 91, 246–262.
2. Rasheed, A.; Robinson, D.; Clappier, A.; Narayanan, C.; Lakehal, D. Representing complex urban geometries in mesoscale modeling. Int.J. Climatol. **2011**, 31, 289–301.
3. Yaghoobian, N.; Kleissl, J. An indoor–outdoor building energy simulator to study urban modification effects on building energy use—Model description and validation. Energy Build **2012**, 54, 407–417.
4. Yang, X.; Zhao, L.; Bruse, M.; Meng, Q. An integrated simulation method for building energy performance assessment in urban environments. Energy Build **2012**, 54, 243–251.
5. Cóstola, D.; Blocken, B.; Ohba, M.; Hensen, J.L.M. Uncertainty in airflow rate calculations due to the use of surface-averaged pressure coefficients. Energy Build **2010**, 42, 881–888.
6. EUROPEAN STANDARD STORE. Ventilation for Buildings—Calculation Methods for the Determination of Air Flow Rates in Buildings Including Infiltration; EUROPEAN STANDARD STORE: Pilsen, Czech Republic, 2007.
7. Van Hooff, T.; Blocken, B. On the effect of wind direction and urban surroundings on natural ventilation of a large semi-enclosed stadium. Comput. Fluid. **2010**, 39, 1146–1155.
8. Van Moeseke, G.; Gratia, E.; Reiter, S.; De Herde, A. Wind pressure distribution influence on natural ventilation for different incidences and environment densities. Energy Build **2005**, 37, 878–889.
9. Ramponi, R.; Blocken, B.; de Coo, L.B.; Janssen, W.D. CFD simulation of outdoor ventilation of generic urban configurations with different urban densities and equal and unequal street widths. Build Env. **2015**, 92, 152–166.
10. Toparlar, Y.; Blocken, B.; Vos, P.; van Heijst, G.J.F.; Janssen, W.D.; van Hooff, T.; Montazeri, H.; Timmermans, H.J.P. CFD simulation and validation of urban microclimate: A case study for Bergpolder Zuid, Rotterdam. Build Env. **2015**,83, 79–90.
11. Hang, J.; Sandberg, M.; Li, Y. Effect of urban morphology on wind condition in idealized city models. Atmos Env. **2009**,43, 869–878.
12. Syrios, K.; Hunt, G.R. Passive air exchanges between building and urban canyon via openings in a single façade. Int. J. Heat Fluid Flow **2008**, 29, 364–373.
13. ASHRAE TC 4.3. Ventilation and infiltration. In ASHRAE Handbook-Fundamentals; American Society of Heating, Refrigerating and Air-Conditioning Engineers, Inc. (ASHRAE): Atlanta, GA, USA, 2013.
14. Patankar, S.; Spalding, D. A calculation procedure for heat, mass and momentum transfer in three dimensional parabolic flows. Int. J. Heat Mass Transf. **1972**, 15, 1787–1806.
15. Rhie, C.; Chow, W. Numerical study of the turbulent flow past an airfoil with trailing edge separation. AIAA J. **1983**,21, 1525–1532.
16. Launder, B.E.; Spalding, D.B. The numerical computation of turbulent flows. Comput. Method. Appl. Mech. Eng. **1974**, 3, 269–289.
17. Blocken, B.; Stathopoulos, T.; Carmeliet, J. CFD simulation of the atmospheric boundary layer: Wall function problems. Atmos Env. **2007**, 41, 238–252.
18. Franke, J.; Hellsten, A.; Schlünzen, H.; Carissimo, B. Best Practice Guideline for the CFD Simulation of Flows in the Urban Environment, COST Action 732 Quality Assurance and Improvement of Microscale Meteorological Models; Meteorological Institute: Hamburg, Germany, 2007.
19. Barmpas, F.; Bouris, D.; Moussiopoulos, N. 3D Numerical simulation of the transient thermal behaviour of a simplified building envelope under external flow. J. Sol. Energy Eng. **2009**, 131.
20. Petridou, M.; Bouris, D. Experimental and numerical study of the effect of openings on the surface pressure distribution of a hollow cube. WSEAS Trans. Fluid Mech. **2006**, 1, 655–662.
21. Davidson, M.J.; Snyder, W.H.; Lawson Jr, R.E.; Hunt, J.C.R. Wind tunnel simulations of plume dispersion through groups of obstacles. Atmos. Env. **1996**, 30, 3715–3731.
22. ATTMA. Technical standard 1. Measuring Air Permeability of Building Envelopes, Air Tightness

Testing and Measurement Association. Available online: http://www.attma.org (accessed on 1 October 2015).

23. Urquhart, R.; Richman, R.; Finch, G. The effect of an enclosure retrofit on air leakage rates for a multi-unit residential case-study building. Energy Build **2015**, 86, 35–44.

CHAPTER 17 ADVANCES IN COMPUTATIONAL FLUID DYNAMICS APPLIED TO THE GREENHOUSE ENVIRONMENT

Jorge Flores-Velázquez[1], Guillermo De la Torre-Gea[2], Enrique Rico-García[2], Irineo L. López-Cruz[1] and Abraham Rojano-Aguilar[1]

[1]Department of Agricultural Engineering, University of Chapingo, Chapingo, México

[2]Department of Biosystems, School of Engineering Queretaro State University Querétaro, México

INTRODUCTION

In order to perform an accurate study of the climate inside a greenhouse, it is necessary to use models based on heat and mass transfer (Rico-García et al., 2008). Mathematical models based on mass and energy balances assume a homogeneous greenhouse environment. These models generate a set of nonlinear ordinary differential equations without an analytical solution. However, a more detailed monitoring of the environment inside the greenhouses reflects a two-dimensional and three-dimensional variability of climatic variables. Recently, this problem has been tackled using the fundamental equations of fluid dynamics. The set of numerical methods applied in order to solve those equations are called Computational Fluid Dynamics (CFD). Computational Fluid Dynamics (CFD) provides a numerical solution from an energy balance of a controlled volume, which in comparison with other methods and expensive technologies allows an efficient study of the climate inside the greenhouse. CFD techniques consider the values of the independent variables as primary unknowns in a finite number of places inside the domain, and then a set of algebraic equations are derived from the fundamental equations applied to the domain and can be solved by pre-establish algorithms.

In spite of, the greenhouse is a very complex bio-system, in which there are several physical, chemical and biological interacting process and phenomena, during the last decade, due to the development of computer simulation tools and the increase in computational processing power, it is possible to develop numerical models for the greenhouse environment such as more accurate models for transport phenomena and energy exchange inside the greenhouse. As a consequence, these studies have led improvements in the design of greenhouses (Norton et al., 2007).

According to Boulard et al. (2002), CFD is a branch of fluid mechanics that uses numerical methods and algorithms to solve and analyze problems involving fluids flow. Therefore, it is possible with the use of computers to perform millions of calculations to simulate the interaction of liquids and gases with surfaces defined by the boundary conditions. In recent studies the modeling of air flow, CFD has deepened to test their effectiveness in relationships of climatic factors (Bournet and Boulard, 2010). Computational parametric studies on greenhouse structures can help to the identification of design factors that affect greenhouse ventilation under specific climatic conditions (Romero-Gómez et al, 2008; Romero-Gómez et al., 2010).

Citation: Jorge Flores-Velázquez, Guillermo De la Torre-Gea, Enrique Rico-García, Irineo L. López-Cruz and Abraham Rojano-Aguilar (2012). Advances in Computational Fluid Dynamics Applied to the Greenhouse Environment, Applied Computational Fluid Dynamics, Prof. Hyoung Woo Oh (Ed.), InTech, DOI: 10.5772/26477.

In the last years, many studies have used CFD to investigate the climatic conditions inside greenhouses. CFD has been able to increase the degree of realism by taking into account insect-proof screens and simulation of the crop effect, considering it as a porous medium, among others in 3D models. The results have been able to improve our understanding of the phenomenon of greenhouse ventilation. Therefore, this chapter discusses significant recent studies to understand how the use of CFD has evolved.

FUNDAMENTAL CFD EQUATIONS AND METHODOLOGY

Finite Element Models using CFD

Computational fluid dynamics is based on the governing fluid dynamics equations (continuity, momentum and energy). The set of equations obtained directly from the volume or fixed element in space is known as "conservative form" Euler type. The equations obtained directly from the volume or movement with the fluid element are called "nonconservative form" Lagrange type (Anderson, 1995).

Substantial Derivative

The substantial derivative physically is the exchange rate of any substance that moves with a fluid element. It consists of two parts, where the first part is called the local derivative, which means the rate of change over time in a fixed point. The second part is called the convective derivative, which physically is the exchange rate due to movement of the fluid from one point to another in the field of fluid, where the fluid properties are spatially different. The resulting material can be applied to any field variable fluid, for example: velocity (u), pressure (p) or temperature (T) (Anderson, 1995).

$$\nabla \equiv i\frac{\partial}{\partial x} + j\frac{\partial}{\partial y} + k\frac{\partial}{\partial z} \tag{1}$$

$$V \equiv (u, v, w)$$

$$\frac{\partial}{\partial t}\iiint_V \rho dV + \iint_S \rho V \cdot dS = 0 \tag{2}$$

$$\rho\frac{Du}{Dt} = -\frac{\partial p}{\partial x} + \frac{\partial t_{xx}}{\partial x} + \frac{\partial t_{yx}}{\partial x} + \frac{\partial t_{zx}}{\partial x} + \rho f_x$$

$$\rho\frac{Dv}{Dt} = -\frac{\partial p}{\partial x} + \frac{\partial t_{xy}}{\partial x} + \frac{\partial t_{yy}}{\partial x} + \frac{\partial t_{zy}}{\partial x} + \rho f_y$$

$$\rho\frac{Dw}{Dt} = -\frac{\partial p}{\partial x} + \frac{\partial t_{xz}}{\partial x} + \frac{\partial t_{yz}}{\partial x} + \frac{\partial t_{zz}}{\partial x} + \rho f_z \tag{3}$$

$$\rho\frac{Dw}{Dt}\left(e + \frac{V^2}{2}\right) = \rho q + \frac{\partial}{\partial x}\left(k + \frac{\partial T}{\partial x}\right) + \frac{\partial}{\partial y}\left(k + \frac{\partial T}{\partial y}\right) + \frac{\partial}{\partial z}\left(k + \frac{\partial T}{\partial z}\right) - \frac{\partial(up)}{\partial x} - \frac{\partial(vp)}{\partial y} - \frac{\partial(wp)}{\partial z}$$
$$+ \frac{\partial(ut_{xx})}{\partial x} + \frac{\partial(ut_{yx})}{\partial y} + \frac{\partial(ut_{zx})}{\partial z} + \frac{\partial(vt_{xy})}{\partial x} + \frac{\partial(vt_{yy})}{\partial y} + \frac{\partial(vt_{zy})}{\partial z}$$
$$+ \frac{\partial(wt_{xz})}{\partial x} + \frac{\partial(wt_{yz})}{\partial y} + \frac{\partial(wt_{zz})}{\partial z} + pf \cdot V \tag{4}$$

- Continuity equation
- Momentum equation (a non-conservative)
- Components in x, y and z
- Energy equation (a non-conservative)

The fundamental CFD equations form a coupled system of nonlinear partial differential equations. So far no analytical solution has been found. It is commonly assumed that the fluid is an ideal gas where the intermolecular forces can be neglected. For an ideal gas, the state equation is:

$$p = \rho RT \tag{5}$$

Where R is the specific gas constant. For a calorically ideal gas we have:

$$e = CvT, \tag{6}$$

Where C_v is the specific heat at constant volume (Rodríguez, 2006, Norton et al, 2007).

CFD Procedure applied to the Greenhouse Environment

Domain setting and grid generation are the first steps in the CFD modeling process. This implies the choice of a computational domain that is large enough to correctly assess the main mechanisms that occur in the system and to avoid interference with artificial Boundaries (Bournet and Boulard, 2010). The CFD modeling process encompasses three stages: preprocessing, solution and post-processing.

The preprocessing is the most time-consuming activity because of mesh generation that is the basis for an accurate simulation, does require heavy calculations. Therefore, keeping an adequate strategy should allow reliability in the calculations and physically consistent results. The meshing process is based on a serie of activities that can be summarized as follows:

- Geometry definition (Figure 1)
- Geometry decomposition
- Computational mesh generation (Figure 2), refining, giggling, quality
- Functional and quality meshing
- Definition of the boundary conditions and
- Export mesh

One of the most important problems currently in CFD modeling of the greenhouse environment is the time-consuming for a simulation to converge, due to the high number of cells that come from the process of meshing. Thus, for many years was chosen to model only in two dimensions (2D) (Flores-Velázquez, 2010). However, air distribution inner greenhouse has a third-dimensional (3D) pattern, for that, nowadays 3D CFD models are more commonly developed. According to Bournet and Boulard (2010) the calculation process is an iterative procedure that requires the definition of convergence criteria, and the user of the CFD codes must to decide on an appropriate level of convergence, of the solution. Generally, 10–4 of the value of a variable at all of the nodes is used. Faster convergence may also be reached by optimizing the grid shape or by assuming the Boussinesq model as density dependent rather than by setting up the problem on the basis of the ideal gas theory. However, this choice may not be applicable in the case of large thermal gradients and may also overlook the influence of temperature on air viscosity.

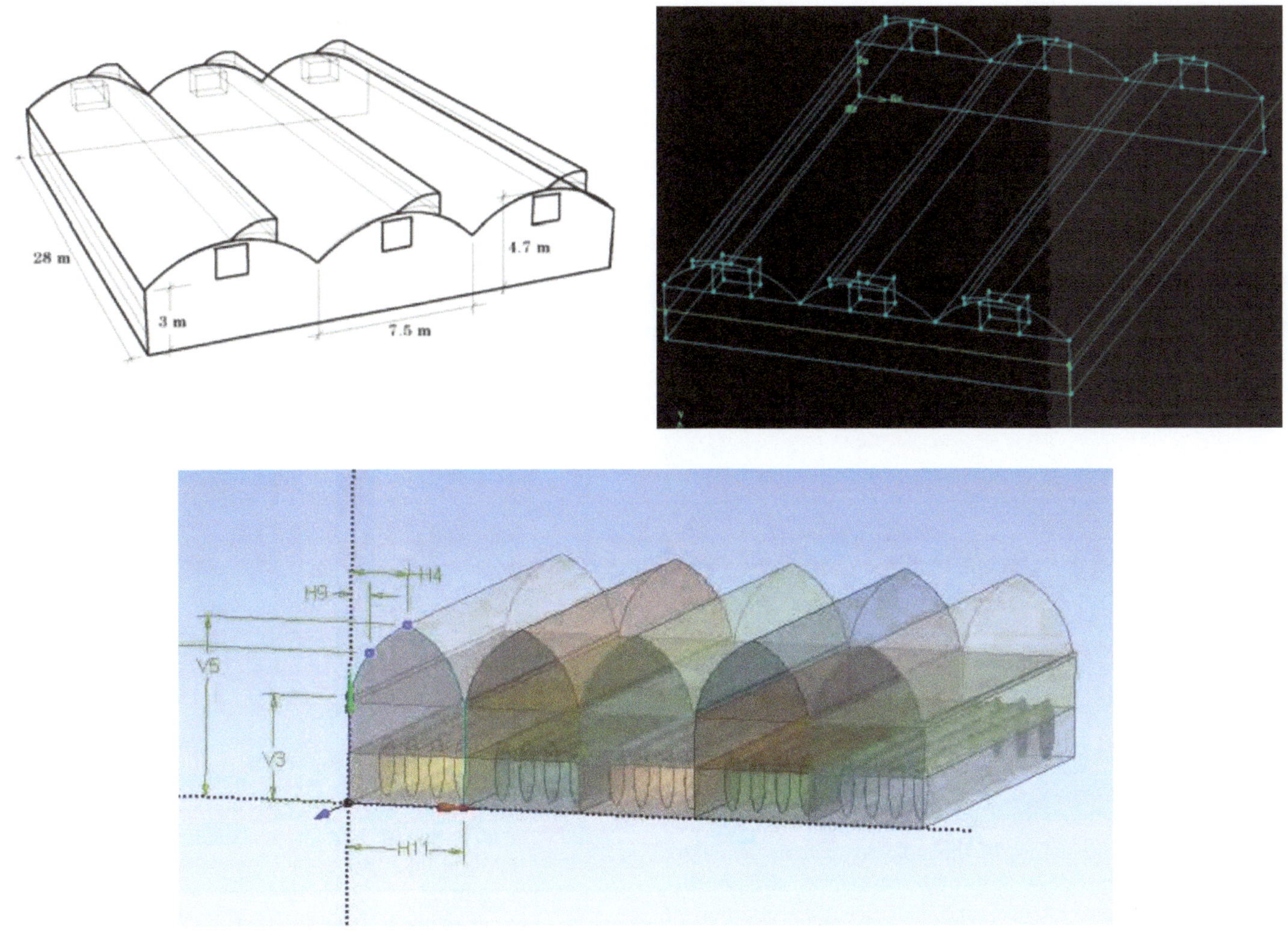

Figure 1: Geometry generation for a greenhouse.

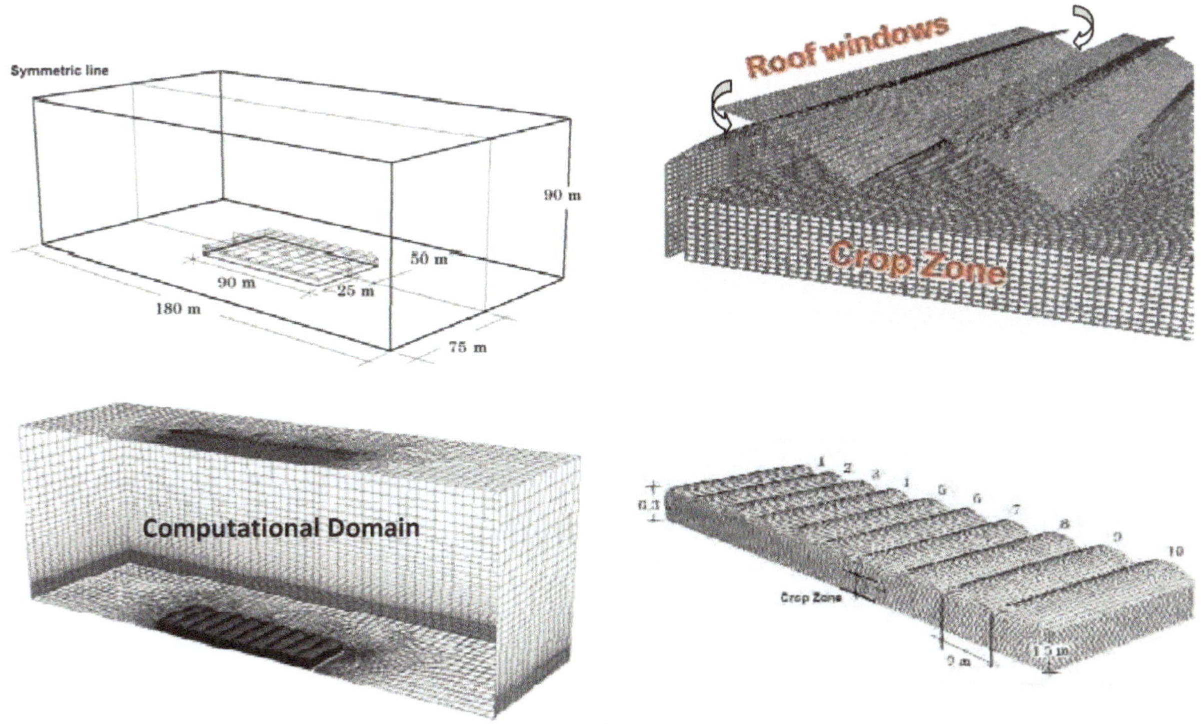

Figure 2: Computational mesh generation for a greenhouse (Flores-Velázquez, 2010).

The presence of turbulence in a fluid is indicated by the fluctuating velocity components and the quantities carried out by the flow, even when the boundary conditions for the problem under study are kept constant. These fluctuations determine the difference between laminar flow and turbulent flow (Figure 3). For most situations, ventilation (effect of temperature, wind or both) measurements and visualization experiments have demonstrated the turbulent air flow inside and outside the greenhouse. Therefore, the phenomenon of turbulence must be taken into account (Norton et al. 2007).

Figure 3: Velocity vectors on a multi-hood greenhouse of four spans. CFD model considers anti-insect mesh vents (Rico-García, 2008).

If inertial effects are large enough with respect to viscous effects, then the flow can be turbulent. Turbulence means that the instantaneous velocity varies at each point of the flow field. The turbulent nature of the velocity can be explained considering that the rate consists of the sum of two components, a main component (stable) and a fluctuating component. Depending on Reynolds number, laminar or turbulent flow can be modeled. For instance, most turbulence models, such as standard k-ε Model, and Re-Normalized Group Turbulence Model (RNG), to name a few (Rico-García, 2008).

The post-processing stage allows the user to visualize and search for the solution. Figures contours, vectors and graphs can be obtained from analyzing the solution. It is remarkable that figures allow us to observe the full distribution of temperature, speed, pressure and so on the whole flow field (Figure 4).

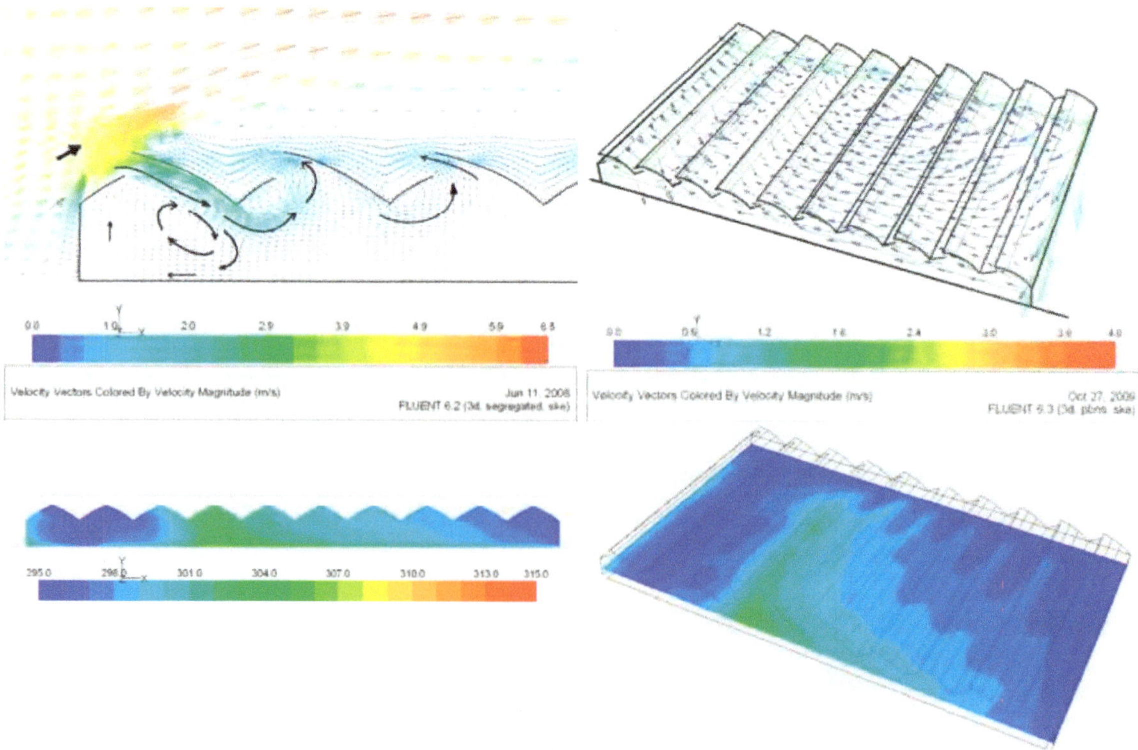

Figure 4: Wind velocity (m s-1) and temperature (K) representative post processing characteristics at different greenhouses sceneries (Flores-Velázquez, 2010).

As soon as a CDF model has been tested, the computational greenhouse environment can become a powerful climate analysis tool. Nowadays, it is possible to visualize, for instance, the wind distribution along the greenhouse when the income windows are up or down, and also the consequent temperature profiles, among many other possibilities (Figure 5). Even though, in the last decade research on wind behavior inside the greenhouse has been enormous, still, as a fundamental part of the greenhouse environment modeling process, it is necessary to take into consideration the physical verification in order to provide accuracy on the results obtained by numerical simulation (Flores-Velázquez, 2010). Scale models, water and wind tunnels and direct measurements of the climatic variables are some of the main options for verification of the CFD models of the greenhouse climate (FloresVelázquez, et al., 2011).

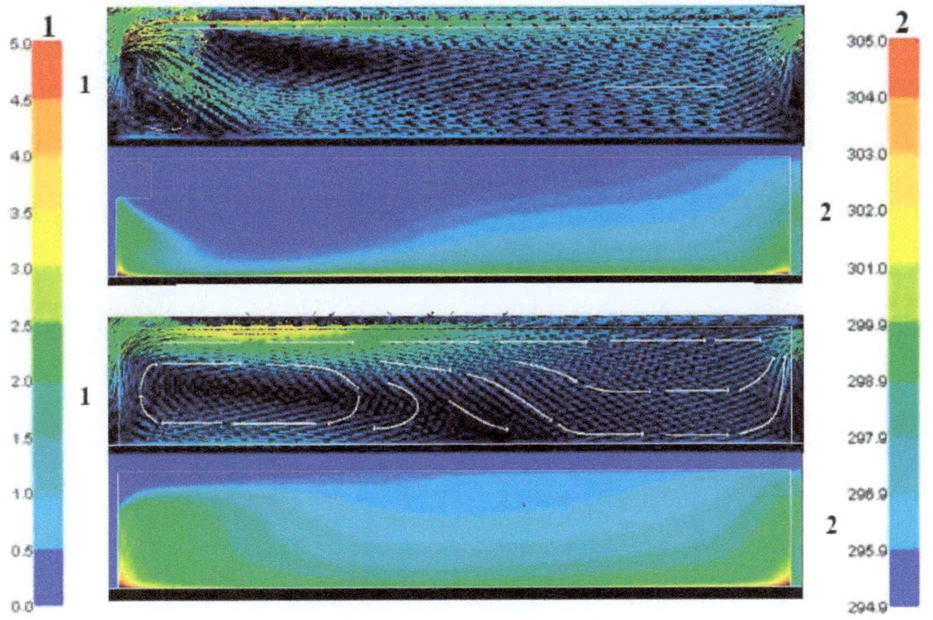

Figure 5: Comparison of wind speed (ms^{-1}) (1) and profiles of air temperature (K) (2) in the greenhouse with and without the projection of the rectangular window. Wind speed exterior 4 ms^{-1}, soil heat flux 315 Wm^{-2} (Flores-Velázquez, 2010).

APPROACHES USED IN THE APPLICATION OF CFD TO THE GREENHOUSE ENVIRONMENT

CFD modeling is used to design facilities that provide suitable climatic conditions for the crops. According to Sase (2006), within a mild climate, appropriate design and control of ventilation are required to ensure effective cooling and uniformity of the environment. It is possible to design an optimal greenhouse by calculating its area, volume and vents area as well as the material properties of the roof (Impron et al., 2007). Rico-García et al. (2006), comparing two different greenhouses, showed the importance of its geometry and found that the ventilation rate for a greenhouse with larger vertical roof and windows was better than a multi-span greenhouse. Omer (2009) describes several designs of low energy greenhouses. In agreement with Baeza et al. (2008), design changes in the greenhouse, such as size and shape of vents, can improve air movement in the area of crops. Bakker et al. (2008) investigated energy balance, and determined that the amount of energy used per unit of output is defined by improvements in energy conversion, environmental control to reduce energy consumption and efficiency of agricultural production. In a study of outdoor areas using the turbulence model Reynolds-averaged Navier-Stokes equations (RANS), van Hoff (2010) found that small geometric modifications can increase the ventilation rate up to 43%. The performance of ventilation in enclosed spaces is affected by the flow of outside air, type of cover, height of the installation and the ventilation opening (Kim et al., 2010). Computational parametric studies on greenhouse structures can help to identify design factors that affect greenhouse ventilation under specific climatic conditions (Romero-Gómez et al., 2008; Romero-Gómez et al., 2010; Flores-Velázquez et al., 2008).

Windward and Leeward Wind Directions

The wind direction outside the greenhouse is an important factor in defining the flow of air and climate inside the greenhouse system. The boundary conditions of wind speed distribution are deduced from experimental data and wind direction with respect to the longitudinal axis of the greenhouse, which can range from 0° to 90°. Roy and Boulard (2005) simulated the impact of wind at 45 ° and 90 °, showing the influence of wind direction in the air velocity, temperature and humidity distributions inside the greenhouse; a similar result was found by Campen (2003). Rico-García et al. (2006) also showed that a greenhouse with larger vertical roof windows works better with a windward condition, whereas the multi-span greenhouse works better with a leeward condition. Therefore, wind direction affects the degree of ventilation. In a experiment carried out by Khaoua et al. (2006), four different openings of roof vents obtained ventilation rates from 9 to 26.5 air exchanges per hour for the windward and 3.7 to 12.5 on the leeward wind condition, respectively, which can maintain acceptable and uniform climate conditions for particular cases where the wind is perpendicular to the main axis of the greenhouse. Overhead ventilation to the windward and leeward directions represents a reduction in the ventilation rate by 25% to 45%, compared with only opening to the windward direction (Bournet et al., 2007). Openings to the windward direction generate the highest rate of ventilation; however, the greatest homogeneity of the temperature and wind speed arises from combining windward and leeward roof vents (Bournet and Khaoua, 2007).

Kacira et al. (2008) showed that the air temperature inside the greenhouse was higher on the windward side than on the leeward side when roof vents were used. Wind speed had a linear influence on air exchange rates, while the wind direction did not affect them. Majdoubi et al. (2009) observed a strong wind air current above a tomato canopy that was fed by a windward side vent and a slow air stream flowing within the tomato canopy space. The first third of the greenhouse, until the end of the leeward side, was characterized by a combination of wind and buoyancy forces, with warmer and more humid inside air that was removed through upper roof vents. There may be a conflict between increasing ventilation and improving uniformity because there is little information on air movement affecting the cooling efficiency and the uniformity of the environment (Sase, 2006). According to Rico-García (2008) the relationship between the thermal gradient and ventilation of gases shows a linear behavior, while the relationship between the combined effect of temperature and wind greenhouse ventilation presents a piecewise linear behavior. The wind pattern in a greenhouse is strongly affected not only by the outside wind velocity but also by the number of greenhouse spans (Flores-Velázquez, 2010). It was found recently, that as the greenhouse has three or four spans roof windows orientation is independent, however, when the greenhouse has five or more spans, side ventilation is dominant over the roof ventilation (Figure 6).

By increasing the surface of the front windows, opening windows and increasing roof capacity exhaust fans can ventilate properly larger greenhouses (Figure 7). It is important to calibrate the fan power, but also to determine a representative inlet area (Flores-Velázquez et al., 2009; Flores-Velázquez, 2010).

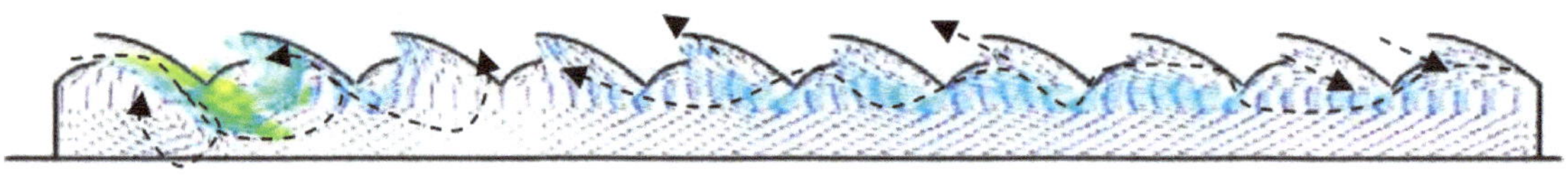

A) Roof windows windward open

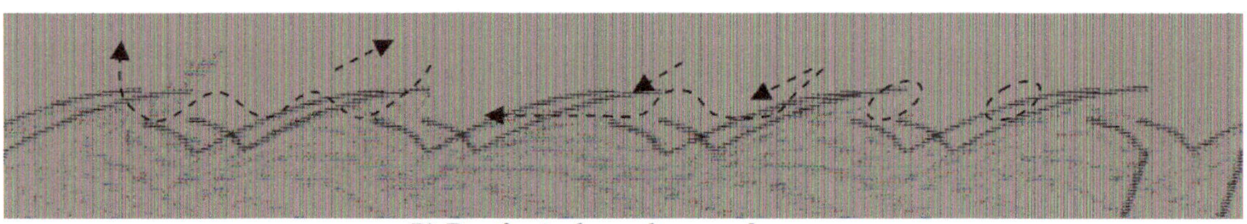

B) Roof windows leeward open

C) Roof and side windows windward open

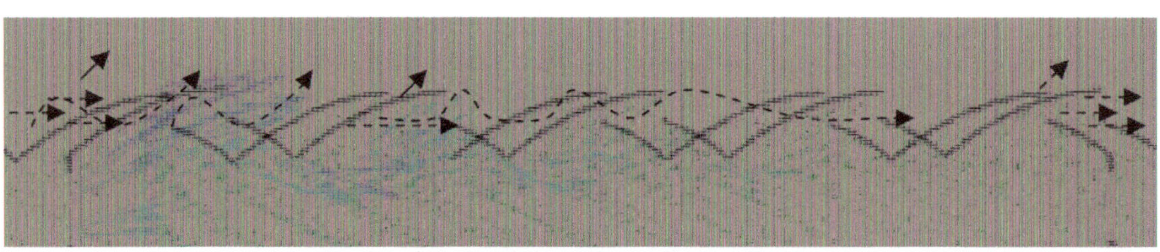

D) Roof and side windows leeward open

Figure 6: Wind velocity interior vectors with 5 m s-1 wind velocity outside the greenhouse, on four open windows sceneries tested.

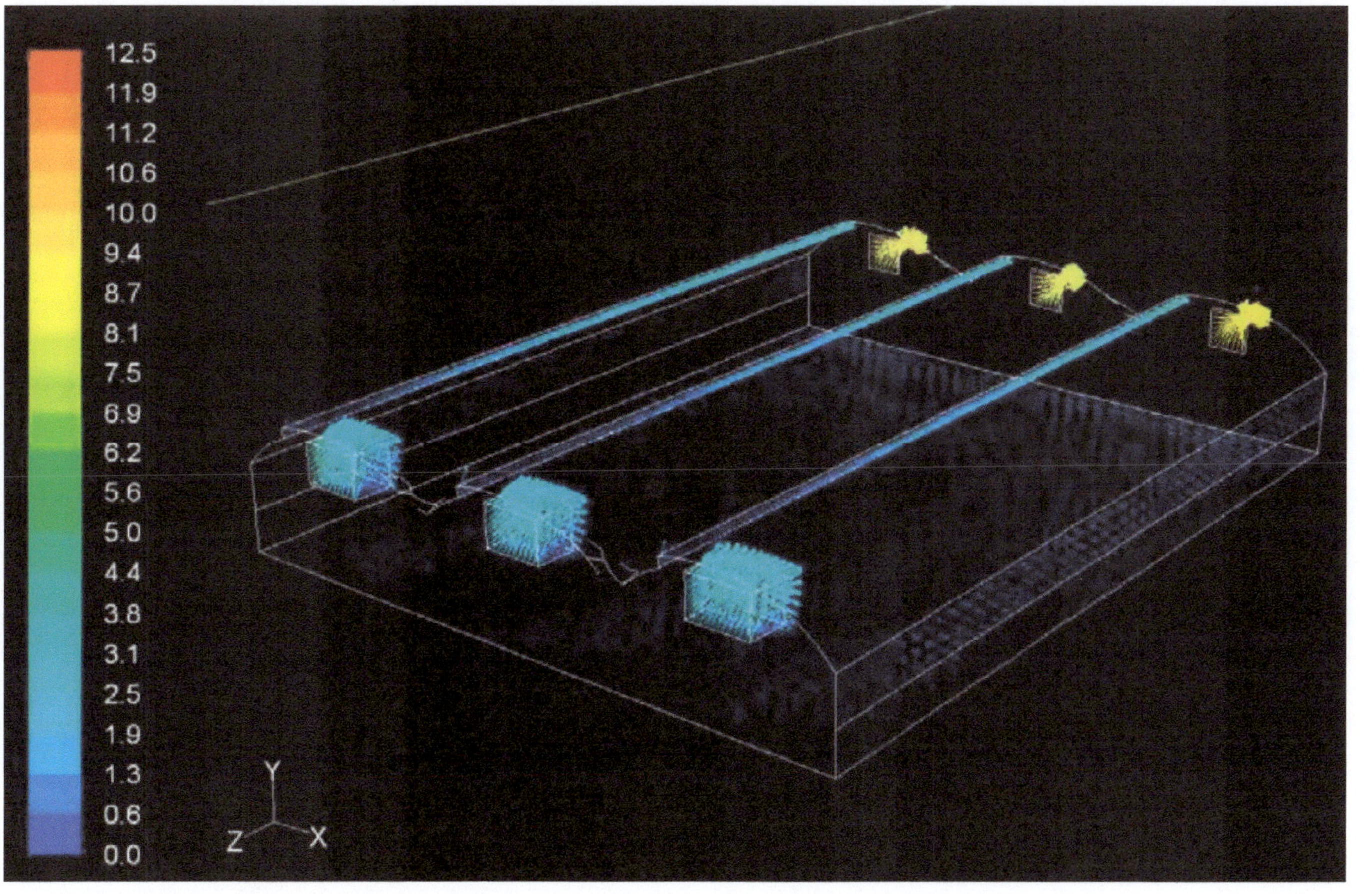

Figure 7: Wind velocity vectors in a inlet area and outside fans in a three span greenhouse (Flores-Velazquez, 2010).

Heat Exchange Tube and Natural Ventilation

Rouboa and Monteiro (2007) simulated, by using a CFD model, the effects on temperature and wind speed of the introduction of hot water pipes along a greenhouse on nighttime conditions, under three scenarios: natural convection heating (case A), artificial heat pipes (case B) and artificial heat pipes and natural ventilation (case C) by using the turbulence model. Re-Normalization Group (RNG) observed an average increase in air temperature to 2.2 ° C, 6.7 ° C and 3.5 ° C; the turbulence was lower for case A, slightly increasing with the heating system for case B and higher for case C, due to the effect of natural ventilation.

Forced Ventilation

The study of fluid dynamics in ventilation systems application provides elements of natural ventilation. A numerical investigation by Rousseau (2008) on a prototype air-forced unit for crop growth chambers obtained simulations that show a nonlinear relationship between airflow rate and opening vents, showing the mixing zone. Dayan et al. (2004) developed another simplified model to demonstrate the calculation of plant temperature when applying forced ventilation for climate control in greenhouses. The use of a device air flow deflector below the roof vents proved to increase air exchange in the area of cultivation effectively. According to the CFD simulations, the combination of the side vent dual configuration has little effect on overall air exchange; nonetheless it increases air movement in the crops and homogenizes temperatures (Baeza et al., 2008). Another investigation by Hughes and Abdul (2010) took into account the effect of the external angle of the ventilation device (wind vent) louvers against the internal pressure and velocity to optimize the device performance. The optimum angle was 35° to 40° with a wind velocity of 4.5 ms-1. Forced ventilation is an excellent option to abate the high temperature, but mainly in small greenhouses as fans are designed properly (Figure 8). As the length of the greenhouse increases, the overhead natural ventilation becomes a positive and more important complement to mechanical ventilation (Flores-Velazquez, 2010; Flores-Velázquez et al., 2011).

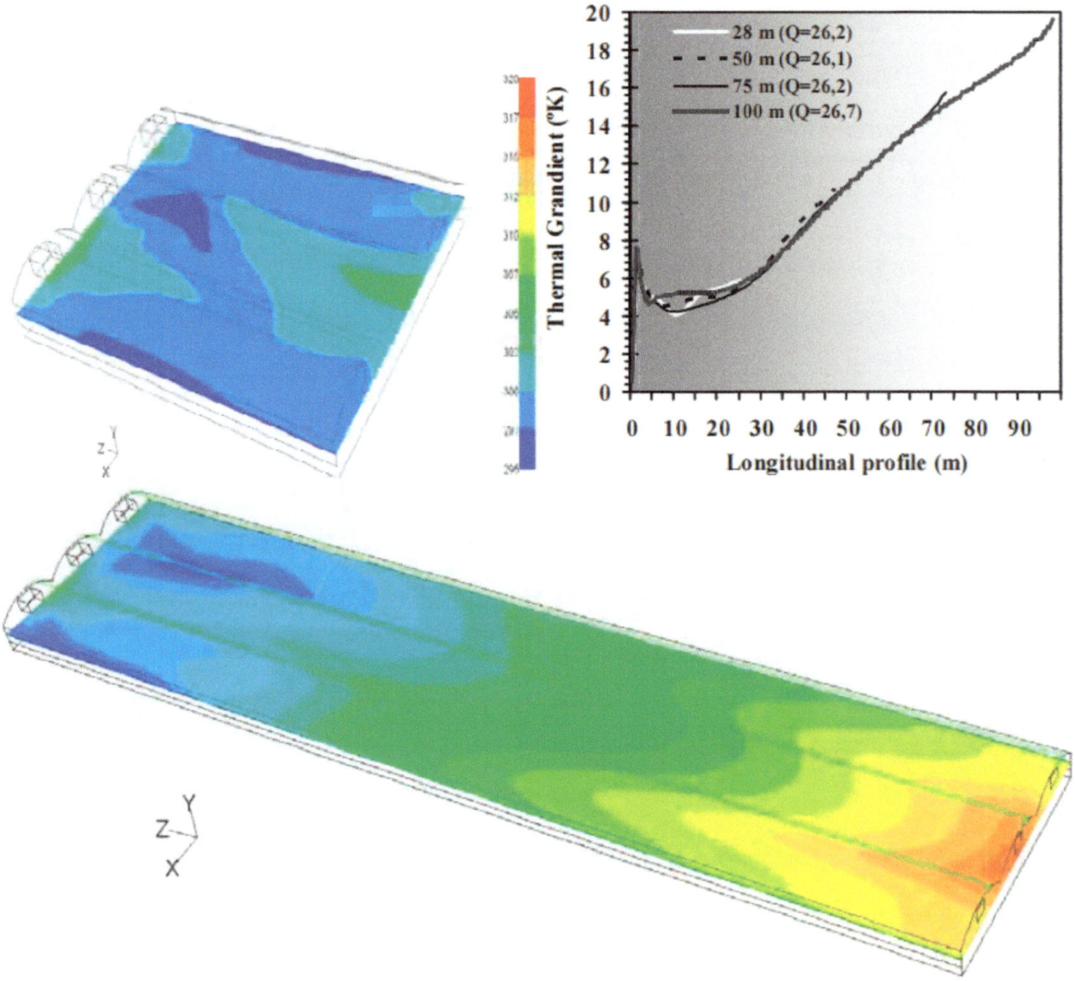

Figure 8: Three spans mechanically ventilated greenhouse, power fan 25 Pa, Heat flux convection 315 W, top view 2 m high.

Fog-Cooling System

According to Sase (2006), in a fog-cooled greenhouse in combination with natural ventilation, air cooled by fogging above the plants is likely to go down. Kim et al. (2007) developed a CFD model to simulate air temperature and relative humidity distribution in a greenhouse with fog-cooling systems, regardless of the presence of plants. Air temperatures and simulated measures ranged from 0.1° to 1.4° C and relative humidity differences were 0.3-6.0%. The results showed that the best cooling system performance occurs when fog nozzles are within 2.3 meters of the floor and 1.9 m of the side walls with a uniform spacing of 3.7 m and the best location for the injectors is at the entrance of the side openings of the greenhouse.

Eave Cladding Type

In order to understand the air mixing properties of the installation, Norton et al. (2010b) studied airflow and buoyancy, determining the relationship between eave openings and resistance to airflow and indoor air mixing for ventilation. They quantified the domain based on the effect of opening conditions, eave cladding type, porosity and height and found that the porosity of the eave opening cladding system significantly increases the efficiency of ventilation. They developed another CFD model incorporating shipping space and the cladding of the eave under conditions of opening and modifying its height, to determine the effects and characteristics of ventilation inside. It was found that the cladding of the eave influences the efficiency of ventilation and thermal comfort. They also found that the strength and the height of the eave determine whether it opens to leeward, acting as an air inlet (Norton et al., 2010c).

Screens and Vents

Recent research using CFD models includes further refinement in adaptive meshing areas, in order to maintain a high level of accuracy during modeling making the simulations more reliable (Norton and Sun, 2006). Screens reduce ventilation rate by 33%, according to a study carried out by Kittas et al. (2005). In agreement with Harmanto et al. (2006), using different screen's sizes over the vent opening has a significant effect, reducing 50% to 35% mesh 40, 78 and 52 and giving rise to a temperature gradient of 1 to 3 ° C with a mesh of 52 as optimal for a tropical greenhouse. Majdoubi et al. (2007) found that insect-proof screens significantly reduced airflow, increasing thermal gradients inside the greenhouse by 46%. Using a wind tunnel with screens of different porosity (0.62, 0.52 and 0.4), Teitel et al. (2008a) showed that a screen inclined by airflow reduces drag compared to a flow perpendicular to the screen, allowing an increase of 15-30% and 25% in the upper compared with a flat screen. Also, Teitel et al. (2009)

found that higher speed screens are inclined at 45 ° and decreased to 135° tilt.

Ali et al. (2009) investigated the effect of roof vents on the temperature and coefficient of heat transfer in naturally ventilated facilities. Better flow patterns and heat transfer from the heated ceiling are observed when the front opening is located closest to the ceiling and the rear opening is located closest to the center. The increase of temperature and humidity as a result of insect-proof screens is particularly evident in the vicinity of the crop canopy (Majdobi et al., 2009). Also it was found that a larger roof vent area can greatly enhance ventilation, while the extension of an insect-proof screen on side walls hardly changes the air exchange rate (Romero-Gómez et al, 2008; Romero-Gómez et al, 2010). In hot regions a common practice is the substitution of the plastic cover of the greenhouse by a screen cover. This structure is called a screenhouse or a shadehouse. Although the climate control is rather difficult in such structure, it is an interesting and important system for growing crops. In some respect a screenhouse is better than a greenhouse as it is required to avoid hot air temperatures during specific hours of a day or a season, which is feasible because of the roof ventilation lets a higher air exchange (Figure 9).

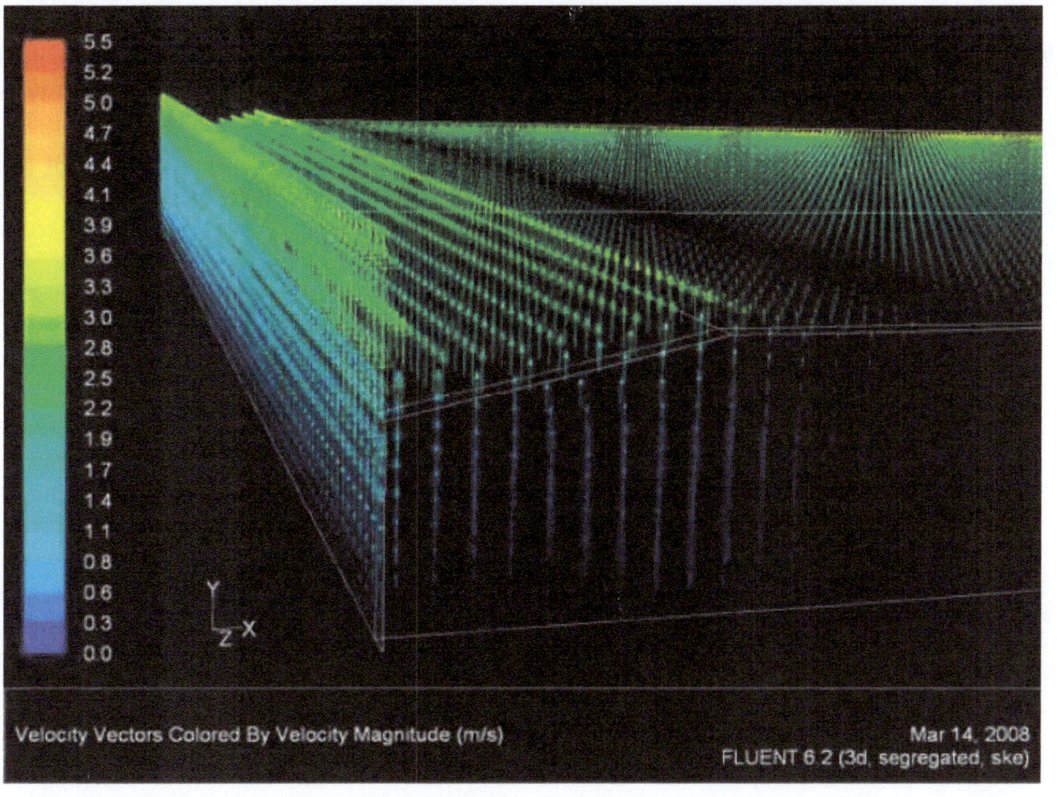

Figure 9: Y-component wind velocities showing the wind exchange by the roof at entrance of a screenhouse (Flores-Velazquez, et al., 2008).

When several screen porosities are located on side/roof screenhouses a similar behavior is observed, regardless the kind of screen, and a strong reduction of the wind velocity is predicted as the crop effect is taken into account to simulate the CFD model of the screenhouse (Figure 10).

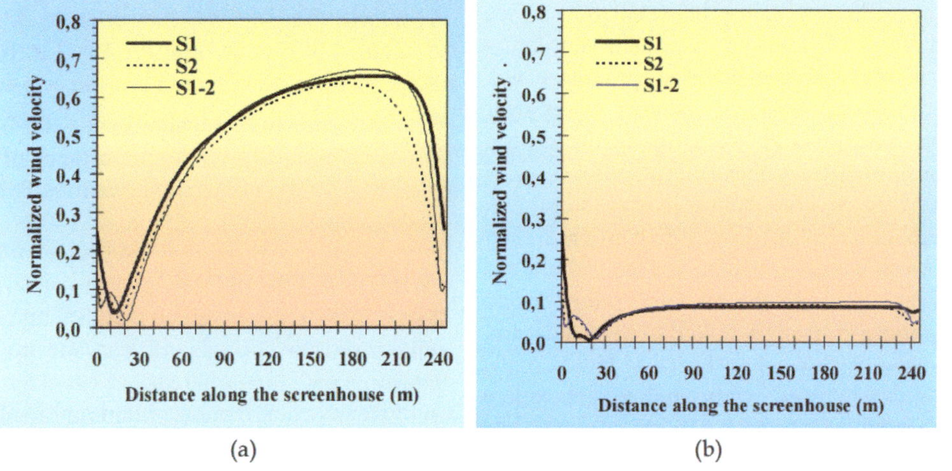

Figure 10: Comparison of wind velocities between screenhouses with the same screen (S1 and S2) and screen on the side (S1) and screen on the roof (S2). (a) With crop effect simulated. (b) Without crop effect simulated.

Solar Radiation and Temperature

Some studies have used solar radiation and transpiration models based on the heat and water balances of the crop, to investigate the distributions of air temperature and humidity and also the interactions between the crop and the air, in addition to the airflow distribution (Sase, 2006). According to Tablada (2005), the factor of solar protection plays a crucial role in maintaining stable thermal conditions indoors, even if the outside air temperature is higher. The slightly higher air speed on the top floor is insignificant in view of reducing the negative effect of the solar radiation over the roof and facade. The temperature of the greenhouse cover is an essential parameter needed for any analysis of energy transferred in the greenhouse. A sub-model developed by Impron et al. (2007) calculated the transmission of radiation through the greenhouse, including the reduction of NIR transmission through the roof. Tong et al. (2009) developed a numerical model to determine time-dependent temperature distributions based on hourly measured data for solar radiation, indoor air, soil and outside temperature, taking into account variable solar radiation and natural convection inside the greenhouse during the winter in northern China.

Temperature and Air Exchange

The effect of solar and thermal radiation is often taken into account by setting specific wall or heat fluxes at the physical boundaries of the greenhouse. Radiation transfer within the crop itself is still the major concern since it determines the two main physiological crop processes: transpiration and photosynthesis. This challenge is now launched and will probably receive more attention within the next few years (Bournet and Boulard, 2010). Pontikakos et al. (2006) analyzed data obtained from a CFD model, and showed that the external boundary temperature is a critical parameter in the pattern of internal greenhouse temperatures and that for specific external temperatures and wind directions, airspeed becomes the crucial parameter. According to Molina et al. (2006), opening vents affect the air flow, the ventilation rate and the air temperature distribution in a greenhouse; where the mean air temperature at the middle varied from 28.2 to 32.9°C with an outside air temperature of 26°C, there were regions inside the greenhouse that were 13°C warmer than the outside air. Nebbali et al. (2006) used a semi-analytical method to determine the ground temperature profile from weather parameters and other characteristics, to help in evaluating heat flux exchange between the surface and the air. Rico-García et al. (2008) showed that ventilation in greenhouses due to the temperature effect produces high air exchange rates; however, those air patterns occur near the openings, causing almost no air exchange in the central zone of the greenhouse due to a stagnant effect that reduces the wind effect throughout the greenhouse. In agreement with the results of Majdoubi et al. (2009), convection and radiation are the dominant forms of heat transfer. The measurements show that the difference between the air temperature inside and outside the greenhouse is strongly linked to solar radiation and secondly to wind speed. However, Chow and Hold (2010) obtained the following conclusions from studying buoyancy forces from thermal gradients:

1. Thermal radiation without air involvement changes air temperature distribution by radiating upper zone thermal energy in the wall towards the lower zone wall, which then affects air temperature through conduction and convection;
2. The inclusion of air absorption increases the effect of radioactive thermal redistribution by allowing air to absorb and radiate heat, reducing temperature gradients further;
3. Thermal boundary conditions and heat loads affect the predicted absolute temperature bounds, but do not affect the temperature distribution.

Radiation conditions play an important role in redistributing heat. Atmospheric conditions, especially relative humidity, are important for the calculation of radiation and heat transfer. Flores-Velázquez (2010) recently found that without overhead natural ventilation, it is possible to find linear relationships between temperature increase and the length of the greenhouse respectively, and the fan power that determines the slope of the regression line.

Turbulence and Buoyancy

As computing power has increased the complexity and sophistication of CFD models also have increased. According to Norton and Sun (2006), the standard k-ε turbulence model commonly used in CFD models for greenhouses, in some cases provides inadequate results, and the choice of turbulence models must be based on the phenomena involved in the simulation. Different turbulence models give rise to differences in speed, temperature and humidity patterns, confirming the importance of choosing the model that most closely matches the actual conditions of turbulence (Roy and Boulard, 2005). Teitel and Tanny (2005) showed that the output of the turbulent heat flux is mainly due to cold air entering the greenhouse, which produces hot and cold eddies coming in and out the greenhouse. Roy and Boulard (2005) showed that the effects of wind direction on climate parameters inside the greenhouse are usually simulated by using different turbulence models available, to determine the energy balance between the flow of perspiration and the flow of radiation. Under ventilation parameters based on Bernoulli's theorem, Majdoubi et al. (2007), showed that bad ventilation performance is not a result of the low value of the greenhouse wind–related ventilation efficiency coefficient, but rather, that the low rate of discharge due to pressure drop in air flow is generated both by the use of anti-insect screens with small openings as an obstruction due to the orientation of the rows of crops. Moreover, Rouboa and Monteiro (2007) note that the RNG turbulence model is best suited to simulate microclimates in arc-shaped greenhouses.

According to Baxevanou et al. (2007), the circulation of air buoyancy effect shows the importance of internal temperature gradients, forced convection resulting from natural ventilation predominates. Rico-García et al. (2008) found that applying temperatures as the main driven forces for the buoyancy effect provides a simple way to study ventilation and inner air patterns. Vera et al. (2010a) observed that differences in temperature and ventilation rates strongly influence the movement of air, pushing it through openings where space is colder, while creating rising air currents when it is hot. Majdoubi et al. (2009) showed that the buoyancy forces induced by air temperature and increased humidity result in loops of air between the crop and the roof windows, which in turn tend to accelerate the pace of removal of heat and water vapor, enhancing indoor climate. Fidaros et al. (2010) studied turbulence in Greek greenhouses and found that external temperature variation is very important because internal temperature is determined by convection induced by the input current. The housing area had a higher circulation in the center of the greenhouse near the deck and in the corners of the ground, where the effect of the input current is weak. Defraeye et al. (2010) used a RANS turbulence model in CFD simulations to evaluate heat transfer by forced convection at the surface of a cube immersed in a turbulent boundary layer for applications in the atmospheric boundary layer (ABL), where wind speed is not disturbed at a height of 10 m. In a study of airfoil wakes, three turbulence models were simulated by Roberts and Cui (2010); the Reynolds Stress Model (RSM) is superior over the k-ε model, and when a time-dependent solution is necessary, Large Eddy Simulation (LES) is the desired option. However, LES does require the airfoil geometry to be included in the domain because it performs poorly when given only inlet velocities, turbulence kinetic energy and eddy dissipation at the trailing edge of the airfoil. According to Bournet and Boulard (2010), although they have been used for a long time in both the agriculture and environment studies, less empirical approaches to turbulence based on the use of LES have never been applied to greenhouse climate modeling and might perhaps be used to look for a solution to this complex situation.

Inside buildings, it is difficult to maintain a thermally stratified space with low ceilings, such as in offices and houses. Vera et al. (2010b) studied buoyancy in enclosed spaces, drawing the following conclusions:

- Rising air currents and the exchange of humidity are closely related to the temperature difference between the lower and upper space. Low temperature in the upper space promotes the exchange of humidity and air flow through the opening; the hotter you are, the greater the restriction of air and humidity transport.
- The existence of upward air currents when the space is warmer than the bottom is caused by local conditions such as non-uniform temperature distributions in the upper space and convective warm currents of the base and the humidity source.
- Compared with conditions without mechanical ventilation, ventilation severely restricts the flow of air through the opening.

The main difficulty in the choice of the model is that greenhouse systems cover a range of length and velocity scales that generally require different modeling approaches (Bournet and Boulard (2010).

Incorporation Crop Effects and Crop Modeling

The effect of plants on greenhouse ventilation has also been studied in the past. For instance; Bournet et al. (2007), based on studies by Nederhoff (1985) and Lee and Short (1998), assumed that a crop of 90 cm high and low density decreases between 12 and 15% greenhouse ventilation. Dayan et al. (2004) built a representative model of a greenhouse for three vertical segments, horizontally oriented to the directions of energy and vapor transfer between the segments containing plants, considering the external weather. They concluded that Representative Plant Temperatures (RPTs) can be calculated instead of measured. Roy and Boulard (2005) developed a 3D CFD model for the characterization of climatic conditions in a greenhouse, incorporating five rows of ripe tomatoes as a porous medium where the buoyancy, heat and moisture transfer between the crop and air flow inside were considered. The heat and moisture transfer coefficients are deduced from the characteristics of the laminar boundary layer of the leaf, which are calculated with the velocity of flow in the crop. Khaoua et al. (2006) found that under external conditions of 1 ms-1 air velocity and 30° of temperature, wind speed at crops' height varies according to the modalities of ventilation from the windward 0.1 and 0.5 ms-1 for the leeward side, while temperature differences ranged from 2.0 to 6.1 ° C. In a study with tomatoes, Majdoubi et al. (2007) found that crop rows oriented perpendicular to air movement reduce the rate of airflow through the cultivation in a greenhouse by 50%. According to Baeza et al. (2008), a greenhouse with natural ventilation efficiency must combine an enough number of air changes to remove excess of heat, with good circulation of air through the crop. The effect of the crop was evaluated by Impron et al. (2007) using a sub-model to determine its effects on ventilation, the properties of the cover, and crop transpiration. In agreement with Kruger and Pretorius (2007), the temperature and velocity at the plant level are influenced by the arrangement and number of windows. A study carried out by Sapounas et al. (2007) simulated a tomato crop as a porous medium, taking into account the addition of buoyancy to develop a model of the pressure drop of air flow due to the crop. The model depended on the area leaf stage of growth and cultivation, under the RANS turbulence model together with the RNG k-ε turbulence model. The results, validated with experimental measurements obtained at 1.2 m inside the canopy; show that the evaporative cooling system is effective with numerical parameters, providing a useful tool to improve system efficiency. A study performed by Roy et al. (2008) on leaf level through an experimental setup based on Münger cells measured the temperature, relative humidity and different heat flows to the leaves of soybeans, obtaining minimum stomatal resistance values ranging from 66 to 200 sm^{-1}.

Teitel et al. (2008b) built a small-scale model and found that wind direction significantly affects the ventilation rate and temperature distribution in crops. A study by von Elsner et al. (2008) on the effect of near-infrared (NIR) reflecting pigments in microclimate and plant growth found that a temperature drop up to 4° C in a young crop is the result of a 18% reduction in the transmission of global radiation in spring. At the same time, during the rainy season, minimizing transpiration differences in temperature and shading reduces water requirements in the plants, and they observed parthenocarpic fruit rot and yieldreducing crop. In a tunnel-type greenhouse, a tomato crop was modeled by Bartzanas et al. (2008) by designing a porous medium, where they emphasize the influence of the heating system on greenhouse microclimate. The climatic behavior of the rows of the tomato crop is taken into account using external user defined functions (Baxevanou et al., 2007). According to Majdoubi et al. (2009), reorienting crop rows in simple ways improved climatic conditions. Endalew et al. (2009) performed CFD modeling of a plant with leaves and branches of the canopy, using turbulent energy equations in porous sub-domains created around the branches. Fidaros et al. (2010)

simulated a greenhouse tomato crop as a porous medium so as to model radiation transport by discrete ordinates (DO). According to Teitel et al. (2010a), when applying the porous medium approach, the Forchheimer equation is often used, which gives rise to erroneous results with respect to the pressure drop through screens. An alternative way to calculate it through several panels of porous media used to simulate screens with realistic geometries. Moreover, the crop exerts a mechanical strain (drag force) on the flow just above but also interacts through the transpiration process with the temperature and humidity distributions (Bournet and Boulard, 2010). A simple model of transpiration of a crop was developed by Sun et al. (2010), who related it to the characteristics of ventilation in a greenhouse in eastern China, obtaining a good approximation. In general, there have been enormous efforts devoted to the analysis of ventilation in greenhouses (Norton, 2007); each new study provides new elements not only in the movement of air in the greenhouse but also in the forms it takes due to interactions occurring in the environment, such as position, shape and size of windows, and (one of the most important), the presence of a crop (Flores-Velázquez, 2010).

Humidity

Roy and Boulard (2005) simulated wind directions of 0°, 45° and 90° with respect to the orientation of the greenhouse ridge to determine wind speed, temperature and humidity distributions inside the greenhouse, getting a good approximation for the humidity. In agreement with Demrati et al. (2007), models allow estimation, with better accuracy, of water requirements for a banana crop under cover and improved water saving in regions where water is the main limiting factor for agriculture. Roy et al. (2008) studied moisture on the surface of leaves at low light levels; crop transpiration and air flow were integrated into a single parameter model of leaf stomatal response to air flow and radiation. Campen (2008) showed that climate through a ventilation system is more homogeneous and the control is more efficient than with the conventional method of steam extraction. Dehumidifiers and cooling reduce the overall difference in humidity between the middle and lower areas of a greenhouse, as demonstrated by Kim (2008) using a 3D model could identify the heterogeneous distribution of relative humidity in a greenhouse. According to Majdobi et al. (2009), an increase in air temperature precedes a more moderate increase in specific humidity.

Main Results of CFD Models in Greenhouses

Many CFD studies are focused on defining the conditions for a suitable environment. There has been less work on automation and control variables. Investigations that seek for a greater understanding of the interactions among climatic variables are increasing. Studies such as those of Hooff, 2010; Teittel, 2010 and Fidaros, 2010, evaluating geometries, have increased in the last year. Figure 11 shows the frequency of climatic variables studied during the period from the year 2005 to 2009 in the studies of CFD models in greenhouses.

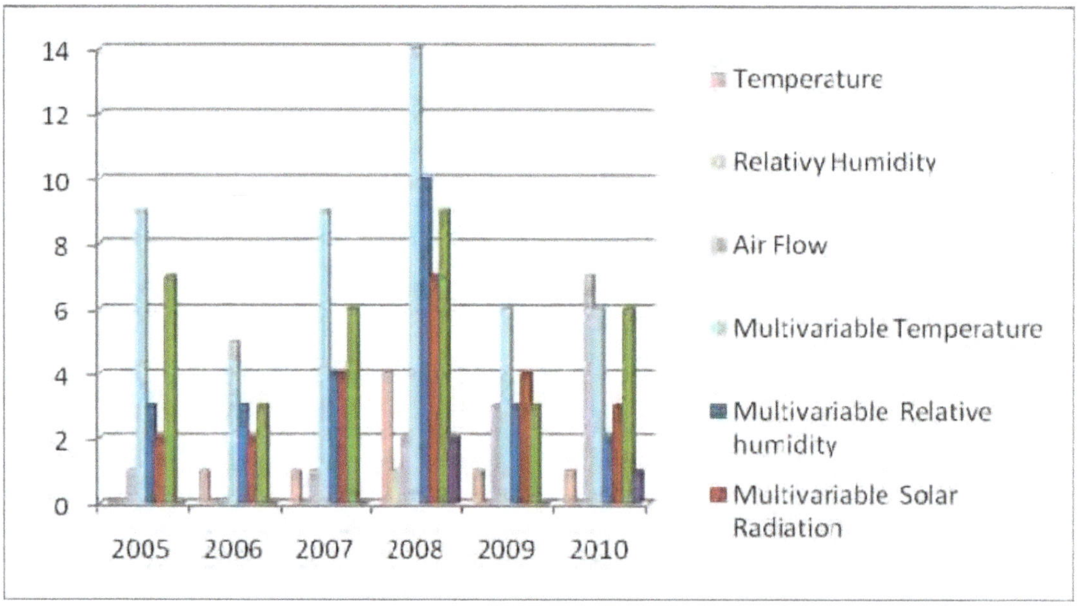

Figure 11: Frequency of climatic variables analyzed by CFD models applied to greenhoses.

Most studies show multi-variable relationships, of which temperature and air flow are predominant. Humidity has been linked to temperature, while there are still few CO2 distribution models. Solar radiation is the subject of investigations that evaluate housing, and is also related to the temperature in simulations with a greater degree of realism. Studies to determine the influence of windward and leeward wind direction indicate that roof vents are of great importance for air renewal, where aperture settings that maximize the number of air renewals are obtained only with windows open to the windward roof (Bournet et al., 2007; Kacira, 2008 and Majdoui, 2009). However, the combination of openings to the windward and leeward sides homogenizes the temperature inside the greenhouse better (Bournet and Khaoua, 2007). There have been many studies to determine a kind of optimal geometric design of greenhouses that encourage improvements in weather conditions and the use of new technologies such as monitoring systems in real time, allowing improvements in automation using Web technology (Pontikakos, 2005). Other studies have focused on the evaluation of misting systems (Kim et al., 2007; Gázquez et al. 2008), forced ventilation (Dayan et al., 2004, Baeza et al., 2008 and Hughes and Abdul 2010) looking for energy savings. As simulation technology and computing power have improved, accuracy and realism in research based on CFD models, has increased as well, by defining more detailed models and by the use of textures that define the materials of the facilities.

The use of insect-proof screens in commercial greenhouses is very important as a means of crop protection; even though they reduce natural ventilation, by this, there have been many research efforts to reduce its negative influence (Kittas et al., 2005; Harmanto et al., 2006; Majdoubi et al., 2007; Teitel et al., 2008a). These studies tested different designs in size of the box and tilt and determined the most affected areas within the greenhouse, where the use of porous media allowed its CFD simulation.

Several studies have investigated the influence of solar radiation on temperature and relative humidity (Tablada et al., 2005; Impron et al., 2007, Tong et al., 2009), and the result in crop response (Baxevanou et al., 2007). Other studies evaluated the use of pigments (Elsner et al., 2008) taking into account the convection, and thermal gradients.

Most of the recent studies developed 3D CFD models, some of which reported the use of models of turbulence and buoyancy, which appear more often during the past two years (Fidaros, 2010; Defraeye, 2010; Norton, 2010; Majdoubi, 2009). By taking into account turbulence, CDF models can make simulations more accurate, in turn increasing the processing and memory requirements for computing resources. Norton and Sun (2006) and Roy and Boulard (2005) discuss the importance of choosing the turbulence model that best meets the conditions of the study. Moreover, the concept of buoyancy appears frequently in order to incorporate the effects of growing space on the air flow and temperature gradients into the models (Figure 12).

Many studies consider the growing space, some of which are designed to measure phenomena based on their influence on the development and crop yield. Other studies are focused on the influence of crops on the other elements, such as temperature, relative humidity, CO_2 concentration and air flow, where it is necessary to model the space occupied by the crop by using porous media approach (Fidaros et al ., 2010). Other investigations measure biological phenomena such as evapotranspiration and Photosynthetically Active solar Radiation (PAR) by using indirect measures of climatic variables (Baxevanou, 2007; Sun, 2008). However, some studies do not mention an experimental phase aimed at validating the numerical model. In studies of air flow, the experimental methods mostly used are scaled models and unidirectional anemometry; the tracer gas technique is used less often, as well as three-dimensional anemometry, which is considerably more expensive. Studies that have used new methods to assess ventilation systems, such as those by Lu (2009), Molina (2010), Endalew (2009), van Henten (2008), Mikulka (2010) and Defraeye (2010), have been increasing in the past three years. The main question is the validation of these studies because they mainly concern to real scale greenhouses, whereas the measurements and characterizations have merely been done on scale models.

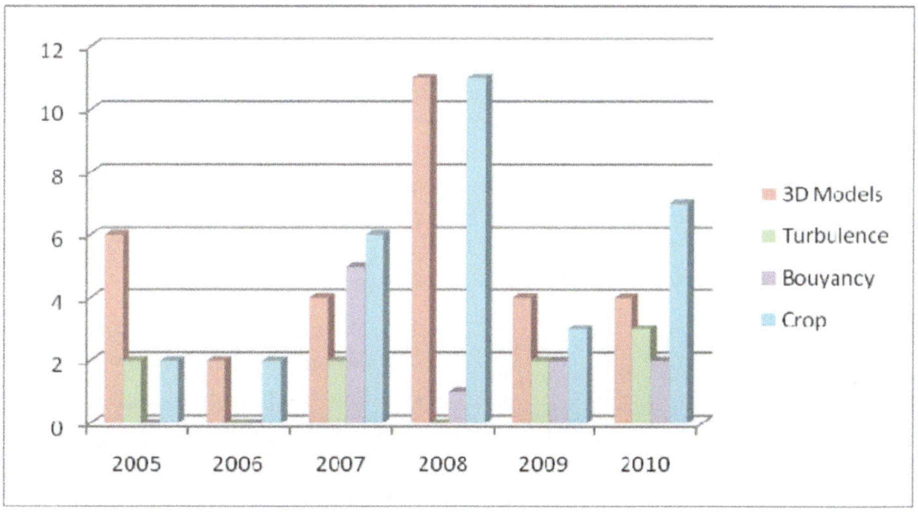

Figure 12: Degree of realism and accuracy in the CFD models for greenhouse climate.

CFD modeling is an area of knowledge that in recent years has developed enormously through the development of software and hardware, which has contributed to research on natural ventilation a greater understanding of the interactions between the variables that make up the climate inside greenhouses. In the past five years, CFD simulation has become increasingly realistic and detailed, obtaining more accurate solutions. However, their use requires depth and extensive knowledge of climatic variables, fluid dynamics and turbulence. Simulating more accurately requires more processing power, so research tends to use CFD models together with other tools. Further studies are required to incorporate more realistic crops beyond a porous medium, taking into account the role of gas exchange, which is necessary for an understanding of the physiology and phenology of crops. There is still a need to develop high-precision systems in greenhouses, and CFD is a powerful tool for defining parameters with high precision, in order to control better the greenhouse environment.

VALIDATION PROCEDURES FOR CFD MODELS OF GREENHOUSE ENVIRONMENT

Models and Experimental Validation

According to Sase (2006), recent progresses in CFD techniques have accelerated a more detailed analysis of air movement in combination with verification tests. However, studies in this area are required in order to address the detailed design of each element involved in the greenhouse climate, highlighting the difficulty involved in the analysis of air movement inside a greenhouse (Flores-Velázquez, 2010). The quality of the CFD models predictions is often evaluated from the agreement with experimental data. Nevertheless, no standard procedure exists yet in order to properly assess the accuracy of the simulations, and the type of comparison often differs from one study to the next (Bournet and Boulard, 2010). Figure 13 summarizes the main approaches used in the recent past to validate the CFD models of the greenhouse environment.

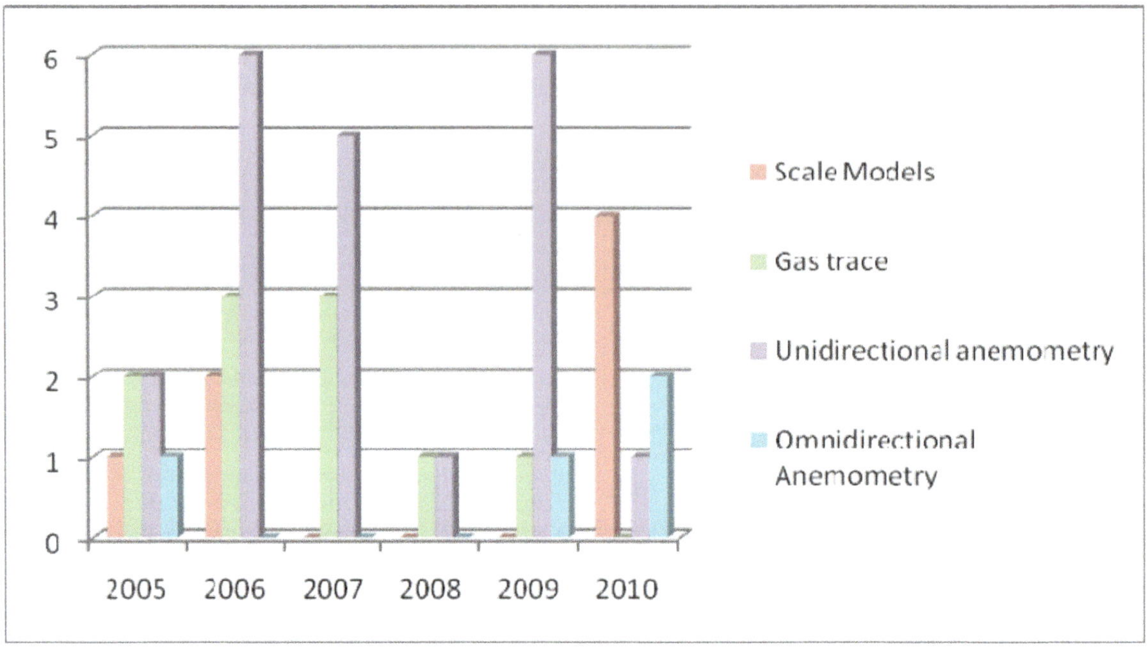

Figure 13: Frequency of validation procedures used in CFD models applied to greenhouses.

The use of porous media models to simulate the pressure drop across flow boundaries, such as insect screens and fences, is very popular and must be validated experimentally. New technologies such as particle image velocimetry have worked properly to complement predictions; field solutions may include biological responses increasing the realism of the simulations (Norton et al., 2007). According to Rouboa and Monteiro (2007), improvements could be achieved by incorporating nighttime transpiration and optimizing the size of the mesh elements to lower computation time. Recent progress offers the opportunity to build a grid that fits the physical boundaries of the structures studied much more realistically than a Cartesian structured grid, which closely follow the contour of the solid boundaries. However, they require verification of the meshing quality to obtain accurate data and an appropriate computational convergence (Bournet and Boulard, 2010).

Statistical Models

The statistical models developed by Pontikakos et al. (2006) are less computationally expensive than the original CFD model, and therefore, they could be used for real-time estimates of temperature and flow rate in a greenhouse.

Model Types

In a small-scale model developed by Teitel et al. (2008b), wind direction significantly affected the ventilation rate and temperature distribution in crops. Chen et al. (2010) evaluated seven types of models (analytical, empirical, experimental small-scale, pilot-scale, multi-area network and CFD to predict the ventilation rate in crops, obtaining the following conclusions:

- The analysis of the model can give an overall assessment of a ventilation system if flow can be approximated to obtain a solution.
- The empirical model has similar abilities to the model of analysis, but develops a database.
- The small-scale model can be useful to examine the complex problems of ventilation if a similarity of flow can be maintained between the model and reality.
- The large-scale model is the most reliable in predicting the efficiency of ventilation, but is expensive and time consuming.
- The multi-zone model is a useful tool for the design of the ventilation of an entire building, but cannot provide detailed flow information in a room.
- The zone model can be useful when a user has prior knowledge of the flow in a room.
- The CFD model provides more detailed information on the performance of ventilation and is the most sophisticated.

However, the model must be validated by corresponding experimental data and the user must have depth knowledge of fluid mechanics and numerical technique. Therefore, the choice of an appropriate model depends on the problem to be solved.

Finite Element vs. Finite Volume

In a study by Molina et al. (2010), on the effectiveness of the Finite Element Method (FEM) and Finite Volume Method (FVM) for two-dimensional incompressible turbulent flow in ventilation rates, it was found that the FEM requires twice the computation time and 10 times more memory storage than FVM. FVM software (ANSYS/FLUENT v 6.3.) is the most frequently used CFD package in ventilation research, and only few papers using FEM software (ANSYS/FLOTRAN v. 11.0) have been published. CFD simulations have been compared to experimental data for 12 cases corresponding to three greenhouse types. The experimental greenhouses were chosen to represent a large range of ventilation situations: buoyancy effect in a mono-span greenhouse with adiabatic walls, as well as buoyancy and wind effect in a multi-span greenhouse and ventilation.

LOOKING INTO THE FUTURE

Advances in telecommunications such as wireless networking and Internet technology (TCP / IP) facilitate the monitoring of environmental conditions in greenhouses. Pontikakos et al. (2005) designed a Web-based application for real-time predictive modeling of temperature and air velocity patterns, which consists of a user interface, interpolation process data generated by CFD and an output interface.

A lighting systems model with different optical properties was developed by Mikulka et al. (2010) who shows various settings for the R-FEM method in the CFX environment.

CFD ventilation space still tends to be a slow process today, while the computation time for the ventilation system and control simulation strategy is negligible. Sun and Wang (2010) found that the test method is more effective than the simplified numerical models, which require more powerful computers. Stavrakakis et al. (2010) concluded that Artificial Neural Networks coupled with CFD models are a powerful computational tool to evaluate the energy savings of various architectural designs.

Currently, CFD studies that mainly considered natural ventilation, increasing the realism of simulations by adding features such as the modeling of the crop inside greenhouse. However, it is still necessary to model crop physiological processes such as respiration and transpiration that define the terms of CO2 and relative humidity mainly in crop space (De la Torre-Gea and Rico-García, 2010).

REFERENCES

1. Ali, A.A.A.; Nabil, M.G. & Mahmoud, A.H (2009). Effect of natural ventilation and wind direction on the thermal performance of a building ceiling, Proceedings Of ISES World Congress 2007,Vol. I-V, Springer, No. 2, pp. 436-440.
2. Anderson, J.D. (1995). Computational Fluid Dynamics. ISBN: 978-3-540-85055-7, McGrawHill, US.
3. Baeza, E.J.; Perez-Parra, J.J.; Lopez, J.C. & Montero, J.I. (2008). CFD simulation of natural ventilation of a parral greenhouse with a baffle device below the greenhouse vents. Acta Horticulturae, No. 801, pp. 885-892.
4. Bakker, J.C.; Adams, S.R.; Boulard, T. & Montero, J.I. (2008). Innovative technologies for an efficient use of energy. Acta Horticulturae, No. 801, pp. 49-62.
5. Bartzanas, T.; Tadj, N.; Draoui, B. & Kittas, C. (2008). Numerical and experimental analysis of convective heat transfer in a heated greenhouse. Acta Horticulturae, No. 801, pp. 847-854.
6. Baxevanou, C.; Bartzanas, T.; Fidaros, D. & Kittas, C. (2008). Solar radiation distribution in a tunnel greenhouse. Acta Horticulturae, No. 801, pp. 855-862.
7. Bournet, P.E.; Ould Khaoua, S.A.; Boulard, T.; Migeon, C. & Chassériaux, G. (2007). Effect of roof and side opening combinations on the ventilation of a greenhouse using computer simulation. American Society of Agricultural and Biological Engineers, No. 50, pp. 201–212.
8. Bournet, P.E. & Ould Khaoua, S.A. (2007). Predicted effects of roof vent combinations on the climate distribution in a glasshouse considering radioactive and convective heat transfers. Acta Horticulturae, No. 801, pp. 925-932.
9. Bournet, P.E. & Boulard, T. (2010). Effect of ventilator configuration on the distributed climate of greenhouses: A review of experimental and CFD studies. Computers and Electronics in Agriculture, No. 74, pp. 195–217.
10. Campen, J.B. (2008). Vapor removal from the greenhouse using forced ventilation when applying a thermal screen. Acta Horticulturae, No. 801, pp. 863-868.
11. Chen, Q. (2009). Ventilation performance prediction for buildings: a method overview and recent applications. Building and Environment, No. 44, pp. 848-858.
12. Chen, Q; Lee, K.; Mazumdar, S.; Poussou, S.; Wang, L.; Wang, M. & Zhang, Z. (2010). Ventilation performance prediction for buildings: model assessment. Building and Environment, No. 45, pp. 295–303.
13. Chow, K.; & Hold, A.E. (2010). On the influence of boundary conditions and thermal radiation on predictive accuracy in numerical simulations of indoor ventilation. Building and Environment, No 45, pp. 437–444.
14. Dayan, J.; Dayan, E.; Strassberg, Y. & Presnov, E. (2004). Simulation and control of ventilation rates in greenhouses. Mathematics and Computers in Simulation, No. 65, pp. 3-17.
15. Defraeye, T.; Blocken, B. & Carmeliet, J. (2010). CFD analysis of convective heat transfer at the surfaces of a cube immersed in a turbulent boundary layer. Int. J. Heat Mass Tran., pp.297-308.
16. De la Torre-Gea, G. & Río-García, E. (2010). Crop Under Greenhouse Simulations as Porous Media in Computational Fluid Dynamics Models. Proceedings of The 5th symposium of engineering, Queretaro State University, Queretaro, Mexico, November 2010.
17. Endalew, A.M.; Hertog, M.; Gebrehiwot, M.G.; Baelmans, M.; Ramon, H.; Nicolaï, B.M. & Verboven, P. (2009). Modelling airflow within model plant canopies using an integrated approach. Computers and Electronics

in Agriculture, No. 66, pp. 9-24.

18. Fidaros, D.K.; Baxevanou, C.A.; Bartzanas, T. & Kittas, C. (2010). Numerical simulation of thermal behavior of a ventilated arc greenhouse during a solar day. Renewable Energy, No. 35, pp. 1380-1386.

19. Flores-Velázquez, J. (2010). Climate analysis in the main models of greenhouses in Mexico (mesh shade and Baticenital multitunnel) using CFD. PhD Tesis, Almeria University, Spain. Flores-Velázquez, J. & Montero J.I. (2008). Computational fluid dynamics CFD study of large-scale screenhouse. Acta Horticulturae, No. 797, pp. 117-122.

20. Flores-Velázquez, J.; Montero, J.I; Baeza, E.J.; Lopez, J.C.; Bonachela, S. & Pérez-Parra, J.J. (2009). Analysis of mechanical ventilation in a three span greenhouse using computational fluid dynamics (CFD). Acta Horticulturae No. 893, pp. 653-660.

21. Flores-Velázquez, J.; Mejía, E.; Montero, J.I.; & Rojano, A. (2011). Numerical analysis of the inner climate in a mechanically-ventilated greenhouse with three spans. Acepted in Agrociencia. Gan, G. (2009). CFD modelling of transparent bubble cavity envelopes for energy efficient greenhouses. Building and Environment, No. 44, pp. 2486–2500.

22. Gan, G. (2010). Impact of computational domain on the prediction of buoyancy-driven ventilation cooling. Building and Environment, No. 45, pp. 1173–1183.

23. Hughes, B.R. & Ghani, S.A.A. (2010). A numerical investigation into the effect of windvent louvre external angle on passive stack ventilation performance. Building and Environment, No. 45, pp. 1025–1036.

24. Iglesias, N.; Montero, J.I.; Muñoz, P. & Antón, A. (2009). Study for the night weather and the use of double cover roof as a passive alternative to increase temperature in greenhouses through a model based in computational fluid dynamics (CFD). Horticultura Argentina, No. 28, pp. 18-23.

25. Impron, I.; Hemming, S. & Botb, G.P.A. (2007). Simple greenhouse climate model as a design tool for greenhouses in tropical lowland. Biosystems Engineering, No. 98, pp. 79-89.

26. Kacira, M.; Sase, S.; Ikeguchi, A.; Ishii, M.; Giacomelli, G. & Sabeh, N. (2008). Effect of vent configuration and wind speed on three-dimensional temperature distributions in a naturally ventilated multi-span greenhouse by wind tunnel experiments. Acta Horticulturae, No. 801, pp. 393-401.

27. Khaoua, S.A.O.; Bournet, P.E.; Migeon, C.; Boulard, T. & Chassériaux, G. (2006). Analysis of greenhouse ventilation efficiency based on computational fluid dynamics. Biosystems Engineering, No. 95, pp. 83–98.

28. Katsoulas, N.; Bartzanas, T.; Mermier, M. & Kittas, C. (2006). Effect of vents opening and insect screen on greenhouse ventilation. Biosystems Engineering., No. 94, pp. 427- 436.

29. Kim, K.; Giacomelii, G.A.; Yoon, J.Y.; Sase, S.; Son, J.E.; Nam, S.W. & Lee, I.B. (2007). CFD modeling to improve the design of a fog system for cooling greenhouses. Japan Agricultural Research Quarterly, No. 41, pp. 283-290.

30. Kim, K.; Yoona, J.Y.; Kwonb, H.J.; Hana, J.H.; Sonc, J.E.; Namd, S.W.; Giacomelli, G.A. & Lee, I.B. (2008). 3-d cfd analysis of relative humidity distribution in greenhouse with a fog cooling system and refrigerative dehumidifiers. Biosystems Engineering, No. 100, pp. 245-255.

31. Kim, T.; Kim, K. & Kim, B.S. (2010). A wind tunnel experiment and cfd analysis on airflow performance of enclosed-arcade markets in korea. Building And Environment, No. 45, pp. 1329-1338.

32. Kruger, S. & Pretorius, L. (2007). The effect of ventilator configurations in naturally ventilated greenhouse applications. Proceedings of 10th Uk Heat Transfer National Conference, Edinburgh, Scotland, September, pp. 10-11.

33. Majdoubi, H.; Boulard, T.; Hanafi, A.; Bekkaoui, A.; Fatnassi, H.; Demrati, H.; Nya, M. & Bouirden, L. (2007). Natural ventilation performance of a large greenhouse equipped with insect screens. American Society of Agricultural and Biological Engineers, No. 50, pp. 641-650.

34. Majdoubi, H.; Boulard, T.; Fatnassi, H. & Bouirden, L. (2009). Airflow and microclimate patterns in a one-hectare canary type greenhouse: an experimental and CFD assisted study. Agricultural and Forest Meteorology, No. 149, pp. 1050–1062.

35. Mercado-Luna, A.; Rico-García, E.; Lara-Herrera, A.; Soto-Zarazúa, G.; Ocampo-Velázquez, R.; Guevara-González, R.; Herrera-Ruiz, G. & Torres-Pacheco, I. (2010). Nitrogen determination on tomato (Lycopersicon esculentum Mill.) seedlings by color image analysis (RGB). African Journal of Biotechnology, No. 9,Vol. 33, pp. 5326-5332.

36. Mikulka, J; Kriz, T.; Kroutilova, E. & Fiala, P. (2010). Numerical modeling of light sources with R-Fem method in CFX environment. Piers Proceedings, March 22-26, pp. 424-427.

37. Molina-Aiz, F.D.; Valera, D.L.; Pena, A.A.; Alvarez, A.J. & Gil, J.A. (2006). Analysis of the effect of rollup vent arrangement and wind speed on Almería-type greenhouse ventilation performance using Computational Fluid Dynamics. Acta Horticulturae, No. 719, pp. 173-180.

38. Molina, F.D.; Fatnassi, H.; Boulard, T.; Roy, J.C. & Valera, D.L. (2010). Comparison of finite element and finite volume methods for simulation of natural ventilation in greenhouses. Computers and Electronics in Agriculture, No. 72, pp. 69-86.

39. Nebbali, R.; Makhlouf, S.; Boulard, T. & Roy, J.C. (2006). A dynamic model for the determination of thermal boundary conditions in the ground of a greenhouse. Acta Horticulturae, No. 719, pp. 295-302.

40. Norton, T. & Sun, D.W. (2006). Computational Fluid Dynamics (CFD) an effective and efficient design and analysis tool for the food industry: a review. Trends in Food Science & Technology, No. 17, pp. 600-620.

41. Norton, T.; Sun, D.W.; Grant, J.; Fallon, R. & Dodd, V. (2007). Applications of Computational Fluid Dynamics (CFD) in the modeling and design of ventilation systems

in the agricultural industry: a review. Bioresource Technology, No. 98, pp. 2386-2414.

42. Norton, T.; Grant, J.; Fallon, R. & Sun, D.W. (2009). Assessing the ventilation effectiveness of naturally ventilated livestock buildings under wind dominated conditions using Computational Fluid Dynamics. Biosystems Engineering, No. 103, pp. 78–99.

43. Norton, T.; Grant, J.; Fallon, R. &, Sun, D.W. (2010a). Optimizing the ventilation configuration of naturally ventilated livestock buildings for improved indoor environmental homogeneity. Building And Environment, No. 45, pp. 983–995.

44. Norton, T.; Grant, J.; Fallon, R. & Sun, D.W. (2010b). Assessing the ventilation performance of a naturally ventilated livestock building with different eave opening conditions. Computers and Electronics in Agriculture, No. 71, pp. 7-21.

45. Norton, T.; Grant, J.; Fallon, R. & Sun, D.W. (2010c). A Computational Fluid Dynamics study of air mixing in a naturally ventilated livestock building with different porous eave opening conditions. Biosystems Engineering, No. 106, pp. 125-137.

46. Norton, T.; Grant, J.; Fallon, R. & Sun, D.W. (2010d). Improving the representation of thermal boundary conditions of livestock during CFD modeling of the indoor environment. Computers and Electronics in Agriculture, article in press. Omer, A.M. (2009) Constructions, applications and the environment of greenhouses. African Journal of Biotechnology, No. 8, Vol. 25, pp. 7205-7227.

47. Pontikakos, C.; Ferentinos, K.P. & Tsiligiridis, T.A. (2005). Web-based estimation model of natural ventilation efficiency in greenhouses using 3D Computational Fluid Dynamics. In Proceedings of the International Congress on Information Technologies in Agriculture, Food and Environment, October: 12-14.

48. Pontikakos, C.; Ferentinos, K.P.; Tsiligiridis, T.A. & Sideridis, A.B. (2006). Natural ventilation efficiency in a twin-span greenhouse using 3D Computational Fluid Dynamics. Of The 3rd International Conference on Information and Communication Technologies in Agriculture, September: 20-23.

49. Rico-García, E.; Reyes-Araiza, J.L. & Herrera-Ruiz, G. (2006). Simulations of the climate in two different greenhouses. Acta Horticulturae, No. 719, pp. 325-332.

50. Rico-García, E.; López-Cruz, I.L.; Herrera-Ruiz, G.; Soto-Zarazua, G.M. & CastanedaMiranda, R. (2008). Effect of temperature on greenhouse natural ventilation under hot conditions: Computational Fluid Dynamics simulations. Journal of Applied Sciences, No. 8, pp. 4543-4551.

51. Rico-García, E. (2008). Climate Modeling Greenhouse: Natural Ventilation. PhD Tesis, Queretaro State University, México. Roberts, R.A. & Cui, J. (2010). Selection of the simulation domain of turbulent flow around and airfoil. Engineering Aplications of Computational Fluid Mechanics, No. 4, pp. 441-449.

52. Rodríguez, N.Y. (2006). Fundamental Equations of Fluid Dynamics. Universidad de la Habana, Facultad de Matemática y Computación, Ciudad de la Habana, Cuba.

53. Romero-Gómez, P.; Choi, C.Y. & Lopez-Cruz, I.L. (2010). Enhancement of the greenhouse air ventilation rate under climate conditions of central Mexico. Agrociencia No. 44, pp. 1-15.

54. Romero-Gómez, P.; López-Cruz, I.L. & Choi, C.Y. (2008). Analysis of greenhouse natural ventilation under the environmental conditions of central México. Transactions of the ASABE, Vol. 51(5), pp. 1753-1761.

55. Rousseau, T. (2008). Structure design and indirect adaptive general predictive temperature control of a class of passive HVAC. WSEAS Transactions On Systems and Control, No. 3, pp. 493-502.

56. Rouboa, A. & Monteiro, E. (2007). Computational Fluid Dynamics analysis of greenhouse microclimates by heated underground tubes. Journal of Mechanical Science and Technology, No. 21, Vol 12, pp. 2196-2204.

57. Roy, J.C. & Boulard, T. (2005). CFD prediction of the natural ventilation in a tunnel-type greenhouse: influence of wind direction and sensibility to turbulence models. Acta Horticulturae, No. 691, pp. 457-464.

58. Roy, J.C.; Vidal, C.; Fargues, J. & Boulard, T. (2008). CFD based determination of temperature and humidity at leaf surface. Computers and Electronics in Agriculture, No. 61, pp. 201-212.

59. Sapounas, A.A.; Nikita-Martzopoulou, C.; Bartzanas, T. & Kittas, C. (2007). Aspects of CFD modeling of fan and pad evaporative cooling system in a greenhouses. International Journal of Ventilation. No. 6, pp. 379-388.

60. Sase, S. (2006). Air movement and climate uniformity in ventilated greenhouses. Acta Horticulturae, No. 719, pp. 313-324.

61. Seo, I.H.; Lee, I.B.; Moon, O.K.; Kim, H.T.; Hwang, H.S.; Hong, S.W.; Bitog, J.P.; Yoo, J.I.; Kwon, K.S.; Kim, Y.H. & Han, J.W. (2009). Improvement of the ventilation system of a naturally ventilated broiler house in the cold season using computational simulations. Biosystems Engineering, No. 104, pp. 106-117.

62. Stavrakakis, G.M.; Zervas, P.L.; Sarimveis, H. & Markatos, N.C. (2010). Development of a computational tool to quantify architectural-design effects on thermal comfort in naturally ventilated rural houses. Building and Environment, No. 45, pp. 65–80.

63. Sun, H.; Yang, J. & Wang, X. (2010). CFD based determination of crop transpiration in local greenhouses in eastern china. On 4th International Conference, Bioinformatics and Biomedical Engineering, pp. 1-4.

64. Sun, Z. & Wang, S. (2010). A CFD-based test method for control of indoor environment and space ventilation. Building and Environment, No. 45, pp. 1441–1447.

65. Tablada, A.; Blocken, B.; Carmeliet, J.; De Troyer, F. & Verschure, H. (2005).The influence of courtyard geometry on air flow and thermal comfort: CFD and thermal comfort simulations. Proceedings of 22nd Conference on Passive and Low Energy Architecture, No.1, pp. 75-80.

66. Teittel, M.; & Tanny, J. (2005). Heat fluxes and airflow patterns through roof windows in a naturally ventilated

enclosure. Flow, Turbulence and Combustion, No. 74, Vol. 1, pp.: 21-47.

67. Teitel, M.; Liron, O; Haim, Y. & Seginer, I. (2008a). Flow through inclined and concertinashape screens. Acta Horticulturae, No. 801, pp. 99-106.

68. Teittel, M.; Ziskind, G.; Liran, O.; Dubovsky, V. & Letan, R. (2008b). Effect of wind direction on greenhouse ventilation rate, airflow patterns and temperature distributions. Biosystems Engineering, No. 101, pp. 351-369.

69. Teitel, M.; Dvorkin, D.; Haim, Y.; Tanny, J. & Seginer, I. (2009). Comparison of measured and simulated flow through screens: effects of screen inclination and porosity. Biosystems Engineering, No. 104, pp. 404-416.

70. Teittel, M. (2010). Using Computational Fluid Dynamics Simulations to determine pressure drops on woven screens. Biosystems Engineering, No. 105, pp. 172-179.

71. Tong, G.; Christopher, D.M. & Li, B. (2009). Numerical modelling of temperature variations in a chinese solar greenhouse. Computers and Electronics in Agriculture, No. 68, pp. 129-139.

72. Van Hoff, T. & Blocken, B. (2010). Coupled urban wind flow and indoor natural ventilation modelling on a high-resolution grid: a case study for the amsterdam arena stadium. Environmental Modelling & Software, No. 25, pp. 51 – 65.

73. Vera, S.; Fazio, P. & Rao, J. (2010a). Interzonal air and moisture transport through large horizontal openings in a full-scale two-story test-hut: part 2 – CFD study. Building and Environment No. 45, pp. 622–631.

74. Vera, S.; Fazio, P. & Rao, J. (2010b). Interzonal air and moisture transport through large horizontal openings in a full-scale two-story test-hut: part 1 – experimental study. Building and Environment, No. 45, pp. 1192–1201.

CHAPTER 18

CFD SIMULATION AND OPTIMISATION OF A LOW ENERGY VENTILATION AND COOLING SYSTEM

John Kaiser Calautit, Dominic O'Connor, Polytimi Sofotasiou and Ben Richard Hughes

Department of Mechanical Engineering, University of Sheffield, Sheffield S10 2TN, UK

ABSTRACT

Mechanical Heating Ventilation and Air-Conditioning (HVAC) systems account for 60% of the total energy consumption of buildings. As a sector, buildings contributes about 40% of the total global energy demand. By using passive technology coupled with natural ventilation from wind towers, significant amounts of energy can be saved, reducing the emissions of greenhouse gases. In this study, the development of Computational Fluid Dynamics (CFD) analysis in aiding the development of wind towers was explored. Initial concepts of simple wind tower mechanics to detailed design of wind towers which integrate modifications specifically to improve the efficiency of wind towers were detailed. From this, using CFD analysis, heat transfer devices were integrated into a wind tower to provide cooling for incoming air, thus negating the reliance on mechanical HVAC systems. A commercial CFD code Fluent was used in this study to simulate the airflow inside the wind tower model with the heat transfer devices. Scaled wind tunnel testing was used to validate the computational model. The airflow supply velocity was measured and compared with the numerical results and good correlation was observed. Additionally, the spacing between the heat transfer devices was varied to optimise the performance. The technology presented here is subject to a patent application (PCT/GB2014/052263).

INTRODUCTION

Commercial wind towers are passive ventilation systems adapted from vernacular architecture of Middle Eastern cultures that date back hundreds of years [1]. Wind towers are capable of delivering ventilation to buildings by the manipulation of pressure differences caused by wind velocity and buoyancy effects of thermally stratified air. Mechanical Heating Ventilation and Air-Conditioning (HVAC) systems commonly used to ventilate and condition air for buildings account for 60% of the total energy consumption in the operation and maintenance of buildings [2]. As a sector, the construction, operation and maintenance of buildings contributes 40% of the total global energy demand [3]. By using new technology coupled with passive ventilation from wind towers, significant amounts of energy can be saved, reducing the emissions of greenhouse gases.

Citation: John Kaiser Calautit, Dominic O'Connor, Polytimi Sofotasiou and Ben Richard Hughes, CFD Simulation and Optimisation of a Low Energy Ventilation and Cooling System, doi:10.3390/computation3020128.

The development of commercial wind towers can be seen to run parallel to the development of Computational Fluid Dynamics (CFD). As improvements in the commercial codes capable of being processed and advancements in mesh detail have pushed CFD analysis forward, the computing power capable in CFD analysis had resulted in the design of commercial wind towers in more efficient and compacts approaches capable of meeting recommended guideline levels of ventilation without causing significant architectural provocations. Early work in fluid analysis was completed by hand calculation, resulting in slow progress and long calculation time with only very simple and small grids being analysed. Early analysis of natural ventilation did not espouse the use of CFD, instead relying on numerical methods to provide guidance for design [4].

More recently, CFD analysis had become vital to design engineers and researchers interested in obtaining fast and accurate data, used to inform design decisions for prototype technology. This was equally true for those interested in commercial wind towers [5]. The speed at which modifications can be introduced and tested in design through the use of CFD far outstrips the practicality of experimental testing though it should be noted that both methods of analysis complement each other for complete validation of a design [6].

In this study, the development of CFD analysis in aiding the development of wind towers was explored. Initial concepts of simple wind tower mechanics to detailed design of wind towers which integrate modifications specifically to improve the efficiency of wind towers were detailed. From this, using CFD analysis, heat transfer devices were integrated into a wind tower to provide cooling for incoming air, thus negating the reliance on mechanical HVAC systems. Furthermore, the spacing between the heat transfer devices was varied to optimise the cooling and natural ventilation performance.

PREVIOUS RELATED WORK

CFD Simulation of Wind Towers

Wind towers provide passive ventilation of buildings by supplying external air and extracting indoor air by manipulating pressure differences around buildings. Areas of positive pressure are created on the windward faces of objects as wind velocity increases, areas of negative pressure are created on the leeward faces. This difference in pressure causes the movement of air from the areas of positive pressure to the areas of negative pressure to attempt to equalise the pressure. This is one of the principles which wind towers are able to manipulate to provide ventilation to buildings. Negative pressure on the leeward faces of the wind tower create suction forces, drawing air out of the building, this is replaced by air from the windward faces of the wind tower forced through by the positive pressure. This is shown schematically in Figure 1.

Wind towers also operate by a secondary action of the stack effect; the density of air decreases as the temperature increases, causing warmer air to rise. The warm air is passes through the wind tower and escapes to the environment. Cooler, fresh air replaces the outgoing air, either through the wind tower or open doors and windows lower in the structure. Linden [7] provided comprehensive models of the fluid flow involved in natural and passive ventilation, much of the work completed helped with the development of CFD analysis for wind towers.

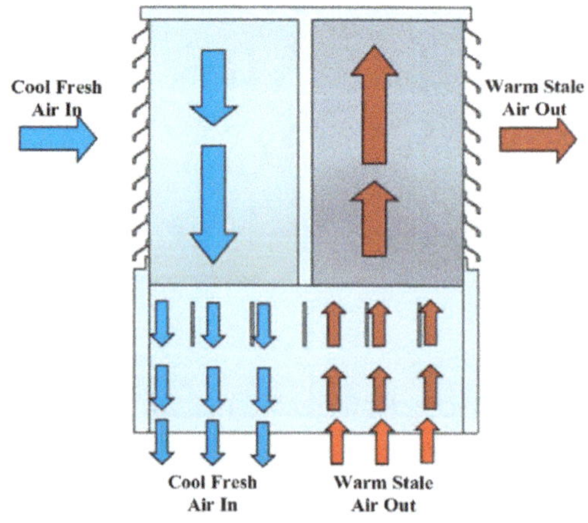

Figure 1: Schematic of typical wind tower operation.

Prior to the development of modern commercial wind towers, Bahadori [8] investigated traditional wind towers. The aim of this was to better understand the operation and ventilation techniques of traditional wind towers and determine if modifications could be made. The aims of the modifications were to reduce the dust infiltration into the building through the wind tower and provide cooling of the incoming air to improve thermal comfort to occupants. Following the design modifications proposed by Bahadori, it was found that the flow rate through the wind tower increased whilst the dust infiltration was reduced. Furthermore, the ability of the wind tower to cool the incoming air was significant when evaporative cooling sprinklers were integrated into the design. Yaghoubi et al. [9] provided similar analysis which aligned with the measurements and conclusion drawn by Bahadori.

Computational Fluid Dynamics (CFD) analysis provided designers and engineers with a quicker method of processing modifications to wind towers than numerical methods allow. In addition to this, dependence on long and expensive experimental testing can be minimised to a stage when final modifications had been agreed upon. The advancements in processing power of work stations had resulted in greater detail in results due to increased mesh quality and better reliability due to the number of models commercial CFD codes were capable of running depending on the conditions required.

CFD had been used by a wide range of researchers interested in the analysis and design of wind towers. Hughes and Mak [10] compared the efficiency of the two primary methods of ventilation through wind towers; wind driven flow and buoyancy driven flow. A model of a wind tower integrated into a building to provide ventilation air was designed, air flow could be provided by a velocity inlet boundary condition or by heating of the internal building air to create density stratification. The results from the CFD analysis showed that wind driven air flow was able to provide 76% more supply air than the buoyancy effect. A flow visualisation of the wind driven flow can be seen in Figure 2a.

(a)

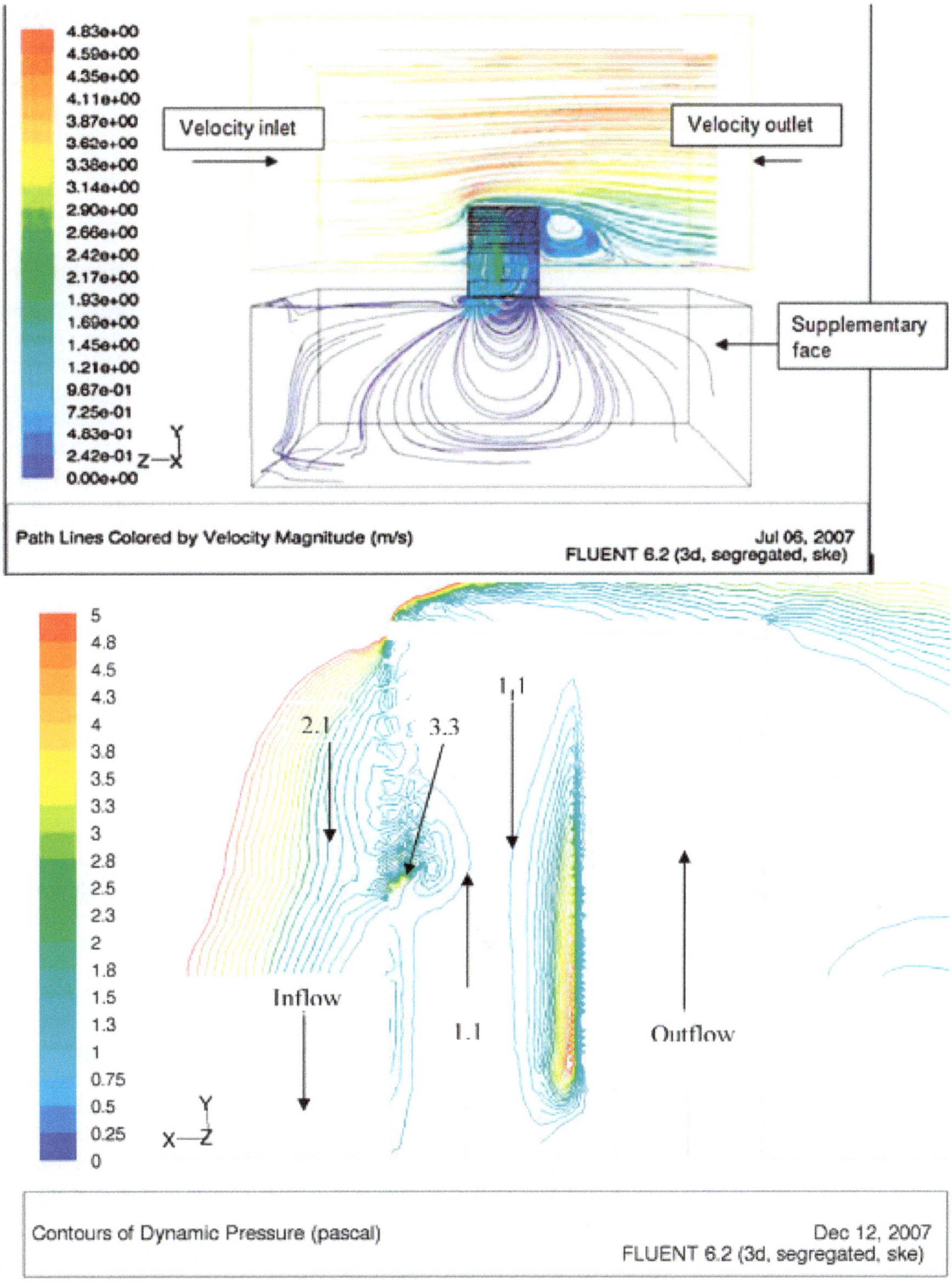

(b)

Figure 2: (**a**) Velocity vectors simulating the con-current flow through the wind tower [10]; (**b**) Computational Fluid Dynamics (CFD) calculated pressure contours at louvre angle of 35° [11].

Due to the ability to change dimensions and design quickly through the use of computer-aided design (CAD) programs which integrated into CFD simulations, on the fly modifications was made to concepts. This approach had been readily employed in the development of wind towers. A large section of the research that had been conducted in relation to the development of wind towers was to improve the air velocity through the wind tower and provide the maximum amount of supply air. Liu et al. [12] focused on determining the optimum length and number of louvers in a wind tower. Louvers were used on wind towers to accelerate air flow through the wind tower whilst also preventing the ingress of unwanted particles such as pollution, rain and snow. By increasing the number of louvers on the wind tower, it was found that the flow through the wind tower increased. A 12.7% increase in air flow rate was observed by the addition of 2–3 louvers to a maximum total number of 6–8 louvers. Beyond this number of louvers the air flow rate only increased by 1.5%. Furthermore, short-circuiting, air entering and exiting the wind tower without circulating around the building, was observed in the upper section of the wind tower due to the high number of louvers.

In addition to increased number of louvers used in the study, the length of the louvers protruding from the wind tower was altered. It was also noted that a single louver length for wind towers of varying sizes would not be sufficient due to the likeliness of varying performance. Therefore a length ratio was applied $l = L/L_{ref}$. L is the nominal length of the louver and L_{ref} is the louvers length when the dimension of louver projection on the corresponding quadrant equates with the gap between two adjacent louvers, L_{gap}. Results from simulations showed that the optimum value for louver length is when the length ratio is close to 1. The performance of the wind tower for ventilation is influenced by the louver length.

As the design of louvers was a critical factor in the efficiency of wind towers, Hughes and Ghani [11] conducted a similar study. The angle of the louvers was varied by increments of 5° within a range of 10°–45°. This was done to analyse the effect on the air flow rate and pressure distribution within the building. It was established that the maximum internal air velocities were measured at a louver angle of 35°. Above 35°, flow separation occurs at the louvers reducing the ventilation rate of the wind tower. A pressure contour plot can be seen in Figure 2b for the louver angle designed at 35°.

The geometry of a wind towers was a significant factor in the overall performance for ventilation. The majority of analysis of wind towers was conducted using a square or rectangular shape with flat surfaces and sharp edges. Multi-direction wind towers were tested by Montazeri [13] to determine the effect of increasing the number of internal section within the wind tower on the behaviour of the air flow. Five different shapes of wind towers were tested experimentally in a wind tunnel and validated by numerical modelling. Two sided, three sided, four sided, six sided and twelve sided variations were all tested by measuring the air flow rate into a test room at different air incident angles [13].

This research was driven by the understanding that for a typical square four sided wind tower, maximum efficiency was possible at a wind incident angle of 45°. In addition to this, a large portion of a four sided wind tower was used to extract internal air; Montazeri proposed that a more appropriate use of sections of the wind tower would be for supply; this was tested using a circular cross section for the wind tower with increasing number of divisions. Experimental testing and numerical modelling proved this idea to be incorrect; a rectangular wind tower was 13% more efficient than a circular design. Furthermore, at an incident angle of 0° increasing the number of partitions within the wind tower decreased the induced air flow rate within the test room. However, as the number of partitions was increased the influence of a changing incident angle of the air flow was reduced.

CFD Simulation of Heat Transfer Devices

Despite the innovations that had been accomplished through the use of CFD for the development of wind towers, regulating the temperature of supply air through a low energy process still remains incomplete. Heat transfer devices technology could provide the link between passive ventilation and low energy temperature regulation [14]. It had been noted previously that heat transfer devices integrated into other ventilation systems do not create a significant pressure drop compared to other technology, this was critical in the integration into passive ventilation systems as pressure losses should be kept to a minimum. Gan and Riffat [15] conducted CFD analysis to understand the causes of pressure loss within a heat recovery system with heat transfer devices. It was found that pressure loss was a result of the blockage caused by the physical presence of the heat transfer devices; which was further increased by the addition of fins and smaller spacing. An indirect cause of pressure loss came from the reduced temperature difference between the supply and exhaust air streams after the heat recovery.

Hughes et al. [16] completed CFD modelling to determine the most effective arrangement of heat transfer devices to allow air streams to flow through. Both the pitch and distance between heat transfer devices were varied along with the horizontal or vertical arrangement in order to determine which arrangement resulted in the greatest heat recovery (Figure 3). The optimal arrangement of heat transfer devices was determined as a pitch of 0.035 m between centres, pre-cooling recovery was calculated at 15.6 °C in hot, dry climates. Pre-heating was calculated at 3.3 °C across the heat transfer devices for a milder climate. Calautit et al. [17] used a similar CFD methodology when modelling heat transfer devices in a vertical and horizontal arrangement. This analysis yielded similar results in terms of temperature drop.

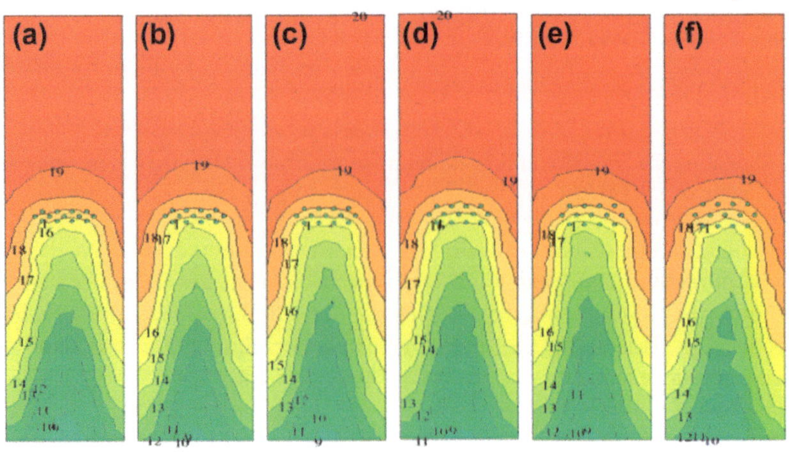

Figure 3: Effect of different pitch spacing on the temperature profile **(a)** 0.020 m **(b)** 0.025 m **(c)** 0.030 m **(d)** 0.035 m **(e)** 0.040 m **(f)** 0.045 m [16].

CFD Optimisation of Natural Ventilation Systems

Due to demanding need for holistic ventilation design approaches, researchers employed optimization techniques, leading to optimal or improved design solutions. Stavrakakis et al. [18] employed CFD techniques to investigate the most reliable turbulence model for the study of naturally cross-ventilated building model. In sequence, for the same model, Stavrakakis et al. [19] performed coupled CFD-Artificial Neutral Network (ANN) techniques to evaluate the effect of different opening configurations and building directions on the ventilation rate, minimizing simultaneously the percentage of dissatisfaction. Similar simulation-based optimization techniques were employed by Zhou and Haghighat [20], who investigated the result of various influential parameters, such as temperature, airflow, position and number of ventilation systems, etc., in an office environment, in order to determine the most energy efficient ventilation systems configuration for optimal comfort conditions.

Shen et al. [21] employed Response Surface Methodology (RSM) techniques to investigate the impact effect of the wind speed and direction on the ventilation rate of a naturally ventilated building. They examined several Design of Experiments (DoE) methods to determine the quadratic response surface that shows better performance on the prediction of ventilation rate. In a more recent work, Shen et al. [22] employed similar RSM techniques to study the ventilation efficiency on a livestock building, under variable window opening sizes and outdoor wind conditions. The results obtained from CFD simulations revealed that the Box-Bohnken Design (BBD) and Central Composite Rotation Design (CCRD) methods can accurately predict the response behavior. However, the performance of the selected DoE methods may vary depending on the case study.

CFD techniques were also used for conducting parametric studies, intending to identify design solutions that would achieve increased ventilation rates, improving comfort conditions. Lal [23] studied different geometrical characteristics for a solar chimney integrated into a building structure. CFD simulations enabled the identification of an enhanced chimney design, reducing the internal temperature up to 4 °C. Calautit et al. [24] determined the spacing and arrangement for wind towers for optimum natural ventilation performance.

Yin et al. [25] evaluated 16 CFD models of different ventilation scenarios and exhaust location, in a patient ward. The simulation study revealed that the location of the exhaust, depending on the ventilation system (displacement or mixing), was of great importance in the air quality and contaminant dispersion. Aliabadi et al. [26] used CFD to design and optimise the performance of a hybrid ventilation scheme for a dining hall. The influence of occupancy, heating and cooling modes, ceiling fans operation low level bypass exhaust on the ventilation energy consumption and thermal comfort were investigated.

CFD METHODOLOGY

A commercial general purpose CFD code ANSYS FLUENT was used in this study to simulate the airflow inside the wind tower model and into to the small room connected to it. The simulation was conducted at steady state and with a three-dimensional computational domain. The CFD code uses the Finite Volume Method (FVM) with the Semi Implicit Method for Pressure Linked Equations (SIMPLE) velocity-pressure coupling algorithm. The Finite Volume Method (FVM) is a discretisation technique for solving partial differential equations that computes the values of the conserved variables averaged across a domain. The first step was to subdivide the domain into a number of elements. At the centroid of each of the elements, the variable values were computed. The next step was to transform the governing partial differential equations in a conservative form and solved over discrete elements. The resulting equation is called the discretised equation. As the effect of heat transfer across the heat pipes was of high importance in this study, the SIMPLE algorithm was chosen. Furthermore, the as wind driven flow was the dominant driving force for airflow, SIMPLE was deemed the most appropriate solution [27]. The turbulent component of the airflow was modeled using the k-epsilon turbulence model [28,29]. The governing equations for the mass conservation (Equation (1)), momentum conservation (Equation (2)), energy conservation (Equation (3)), turbulent kinetic energy (Equation (4)), and energy dissipation rate (Equation (5)) are detailed below:

$$\frac{\partial \rho}{\partial t} + \nabla \times (\rho u) = 0 \tag{1}$$

where ρ is density, t is time and u refers to fluid velocity vector.

$$\frac{\partial (\rho u)}{\partial t} + \nabla \times (\rho u u) = -\nabla p + \rho g + \nabla \times (\mu \nabla u) - \nabla \times \tau_t \tag{2}$$

where p is the pressure, g is vector of gravitational acceleration, μ is molecular dynamic viscosity and τ_t is the divergence of the turbulence stresses which accounts for auxiliary stresses due to velocity fluctuations.

$$\frac{\partial (\rho e)}{\partial t} + \nabla \times (\rho e u) = \nabla \times (k_{eff} \nabla T) - \nabla \times \left(\sum_i h_i j_i \right) \tag{3}$$

where e is the specific internal energy, k_{eff} is the effective heat conductivity, T is the air temperature, h_i is the specific enthalpy of fluid and j_i is the mass flux.

$$\frac{\partial (\rho k)}{\partial t} + \nabla \times (\rho k u) = \nabla \times [a_k \mu_{eff} \nabla k] + G_k + G_b - \rho_\varepsilon \tag{4}$$

$$\frac{\partial (\rho \varepsilon)}{\partial t} + \nabla \times (\rho \varepsilon u) = \nabla \times [a_\varepsilon \mu_{eff} \nabla \varepsilon] + C_{1\varepsilon} \frac{\varepsilon}{k} (G_k + C_{3\varepsilon} G_b) - C_{2\varepsilon} \rho \frac{\varepsilon^2}{k} \tag{5}$$

where G_k stands for source of turbulent kinetic energy due to average velocity gradient, G_b is source of turbulent kinetic energy due to buoyancy force, α_k and α_ε are turbulent Prandtls numbers, $C_{1\varepsilon}$, $C_{2\varepsilon}$ and $C_{3\varepsilon}$, are empirical model constants.

Design Geometry

The wind tower geometry (Figure 4a) was created using Solid Edge CAD software and then imported into ANSYS Geometry (FLUENT pre-processor) to create a computational model. The solid parts of the wind tower and test room geometry were created using the CAD software. However, replicating the physical geometry of the wind tower does not represent a computational domain. To achieve this, the fluid volume was defined before generating a computational mesh. The fluid volume was extracted from the solid model as shown in Figure 5a. The fluid volume or domain was separated into three parts: the macro-climate (external), wind tower and micro-climate (test room). The macro-climate was created to simulate the airflow around the wind tower model. The macro-climate consisted of an inlet at the left hand side of the domain, and an outlet on the opposing boundary wall of the macro climate. The distance between the wind tower and the wall of the macro-climate was determined by calculating the blockage caused by the wind tower model. The macro-climate dimensions was 5 m × 5 m × 10 m. According to the dimensions of the wind tower (1 m × 1 m × 1 m), the model produced a maximum blockage of 4.8%, therefore no corrections were made to the measurements obtained with these configurations [13].

The wind tower device was incorporated to a micro-climate with the height, width, and length of 3 m × 5 m and 5 m, representing a small room [30,31]. The wind tower was modelled with seven louvres angled at 45° [11]. The wind tower was assumed to be supplying at 100% (fully open), therefore the volume control dampers was not added to the model [30]. The cylindrical heat transfer devices, each with an outer diameter of 0.02 m, were integrated in to the lower part of the wind tower channel. Figure 4b illustrates the arrangement of the heat transfer devices inside the wind tower. In order to determine the optimum cooling and ventilation performance of the wind tower, the horizontal spacing of the heat transfer devices were varied. The horizontal spacing (x) between the heat transfer devices was varied between 50 mm and 100 mm.

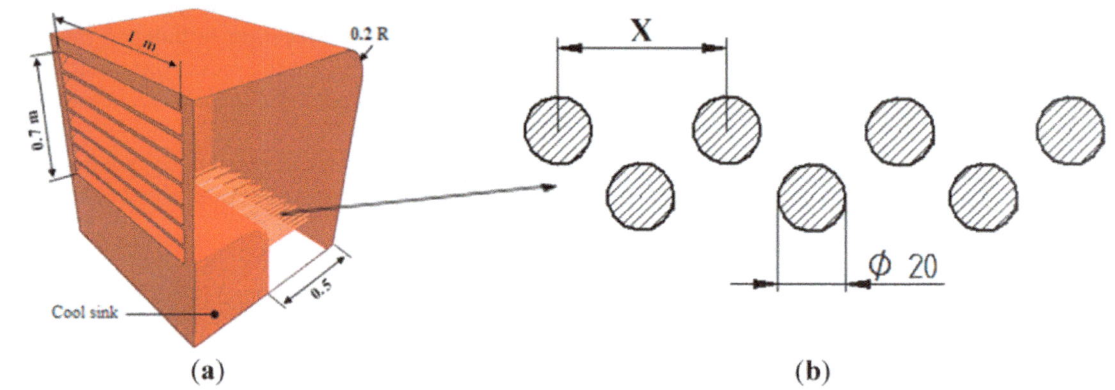

Figure 4: (**a**) Illustration of the computer-aided design (CAD) geometry of the wind tower with heat transfer devices; (**b**) Cross-sectional diagram of the heat transfer device in staggered arrangement.

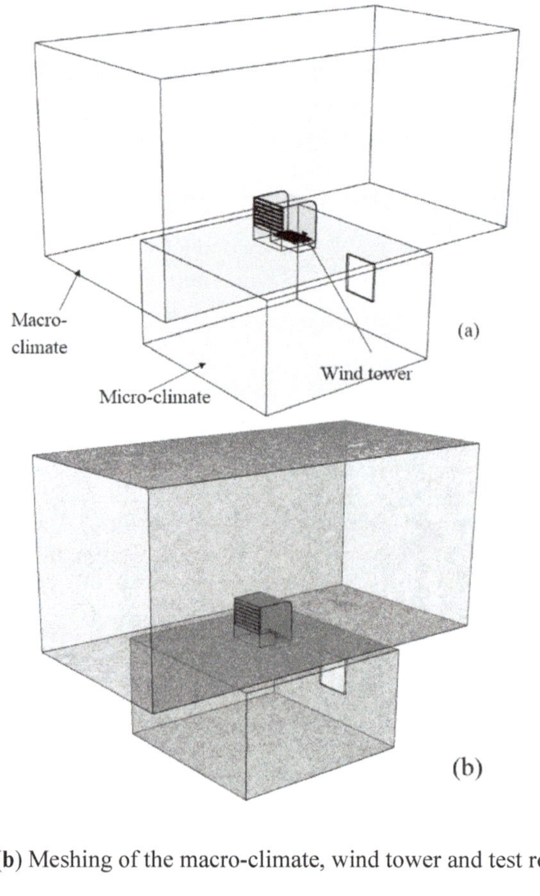

Figure 5: (**a**) Illustration of the geometry; (**b**) Meshing of the macro-climate, wind tower and test room.

Mesh Generation and Verification

Due to the geometrical complexity, an unstructured mesh was used to discretise the surface and volumes of the computational domains [32]. To be able to capture properly the flow fields near the critical areas of interest in the simulation such as the heat transfer devices and louvers, size functions were applied in those surfaces. The size of the mesh element was extended smoothly to resolve the sections with high gradient mesh and to improve the accuracy of the results of the velocity and temperature fields [33]. The mesh element size was varied from 0.005 m for the mesh near the heat transfer devices and louvers to 0.07 m for the middle of the space. Figure 5b shows the meshing of the modeled wind tower and room using ANSYS Mesh. The total number of the mesh elements used for the CFD analysis was equal to 7,249,235.

Mesh verification method was used to validate the programming and computational operation of the model. The numerical grid was refined and locally enriched using the h-method grid verification. The h-method (mesh refinement) refers to the targeting of areas of the mesh which have high levels of skewness and replacing them with an increased number of elements, or a different element type. The mesh was evaluated and refined until the posterior estimate error becomes insignificant between the number of elements and the posterior error indicator [33,34].

The mesh verification method starts with a coarse mesh and gradually refines it until the variation observed between the results were smaller than the predefined acceptable error. The accuracy of the results was improved by using successively smaller cell sizes for the computation [33,35]. The set boundary conditions remained fixed throughout the simulation process to ascertain precise comparison of the results. The initial coarse grid consisted of 2,674,484 elements which was refined over five stages until an acceptable error of 0.01% was achieved with a total of 7,249,235 elements.

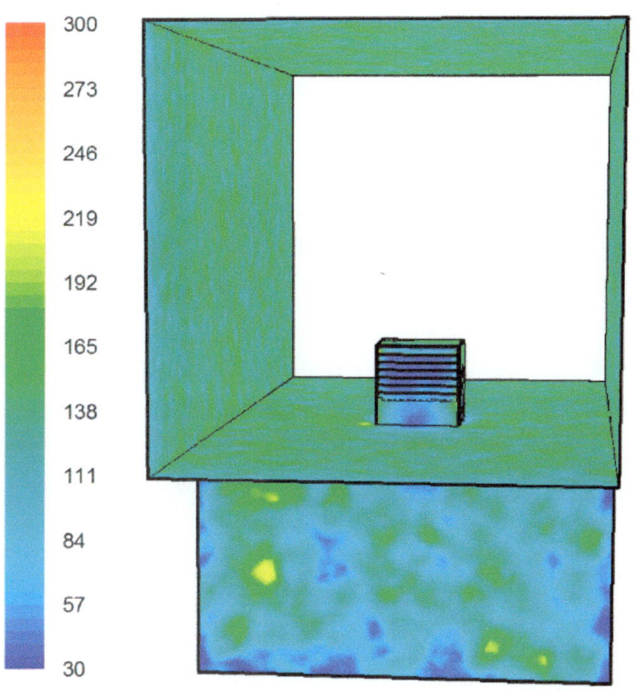

Figure 6: Contour of wall y-plus value.

Furthermore, the grid resolution was determined taking into account an acceptable value for the wall y+. The log-law, which is valid for equilibrium boundary layers and fully developed flows, provides upper and lower limits of the acceptable distance between the near-wall cell centroid and the wall. The distance is usually measured in the dimensionless wall units, y+. For standard wall functions, each wall-adjacent cell's centroid should be located within the log-law layer, 30 < y+ < 300 [28]. Figure 6 shows the wall y+ analysis for the model.

Boundary Conditions

Figure 7 shows the physical domain containing the macro- and micro-climate fluid volumes. A wall boundary condition was used to create a boundary between each region. The macro-climate fluid volume, used to simulate the external velocity of the flow field, generated velocity into the wind tower. A horizontal plane was used as a velocity inlet, while the opposite boundary wall was set as the pressure outlet (atmospheric). The velocity inlet was varied between 0 and 5 m/s. The external air temperature was set to 318 K to simulate a hot external environment. The external air temperature was also varied between 318 K and 302 K. In order to cool the air that was supplied to the room, the surface temperature of the heat transfer devices was set to 293 K. An opening located at the leeward side of the room was set as pressure outlet (atmospheric) to extract the air out of the room. The influence of the internal gains (occupants, equipment, etc.) and heat gain through walls were not investigated in this work.

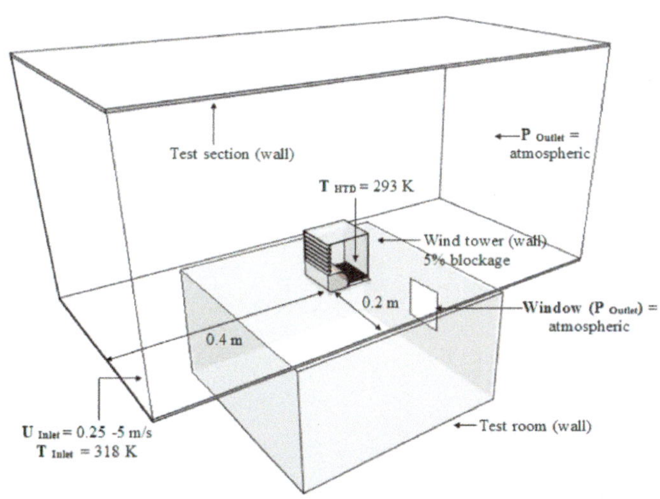

Figure 7: Flow domain of the computational model showing the boundary conditions.

Solution Convergence

Solution convergence is the term for a numerical method using iterations to produce a solution of the grid, whereby the error approaches zero. Solutions are based on iterations against pre-defined convergence criterion [28]. The default convergence criterion in FLUENT is 10-6 for the energy equation and 10-3 for all other equations. When the set convergence criterion is met the iteration process is complete. In this study, the convergence checking was disabled and the convergence was monitored by examining the residual levels (Figure 8). In addition, relevant integrated quantities such as the indoor and supply air velocity (Figure 9) and temperature (Figure 10) were also monitored.

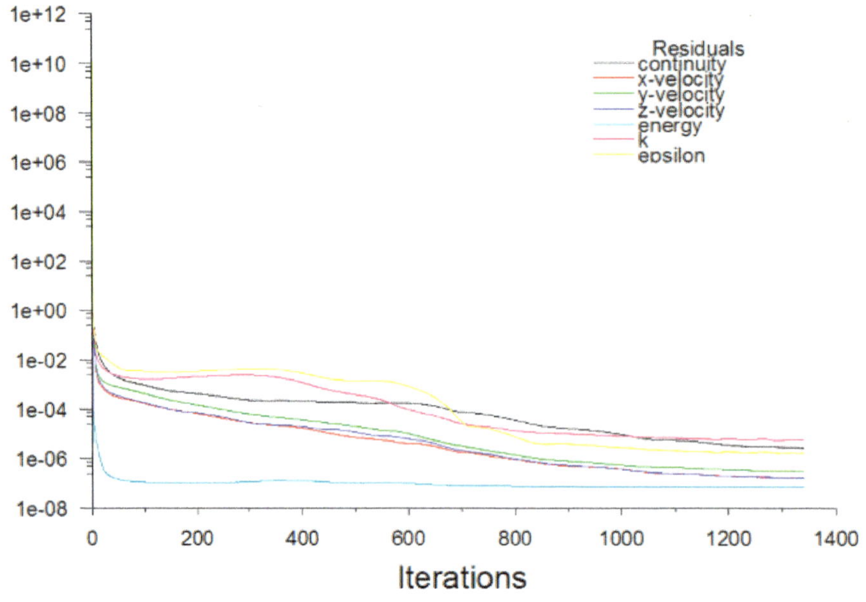

Figure 8: CFD solution residuals.

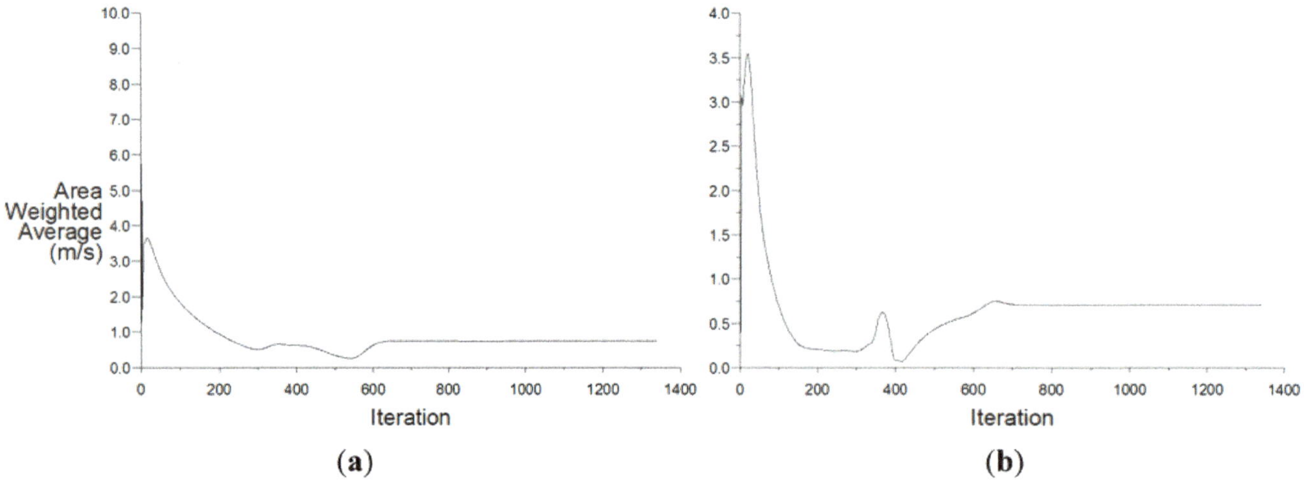

Figure 9: (**a**) Indoor velocity; and (**b**) air supply velocity convergence.

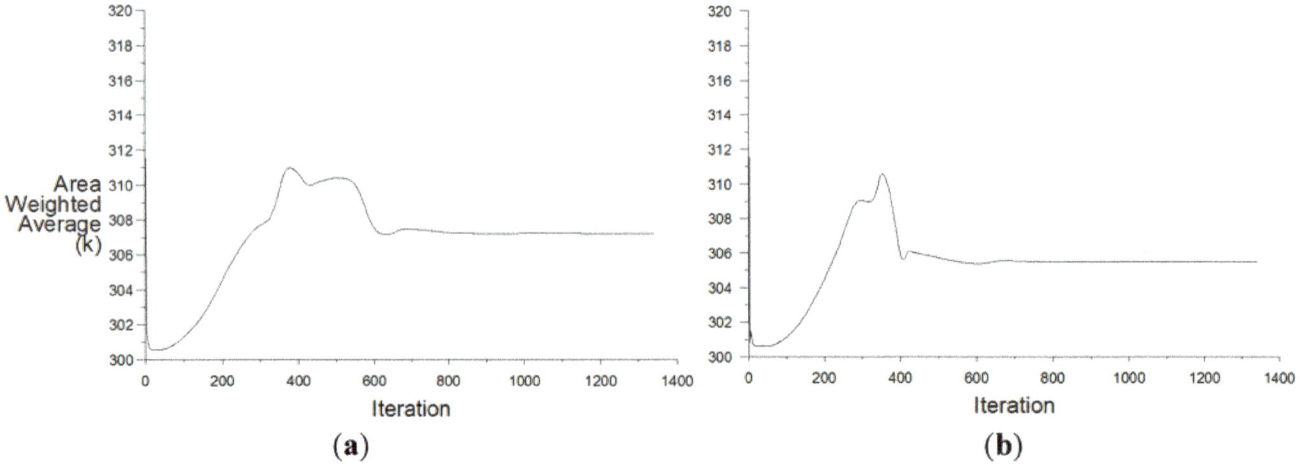

Figure 10: (**a**) Indoor temperature; and (**b**) air supply temperature convergence.

EXPERIMENTAL METHODOLOGY

The experimental test was carried out in a low speed wind tunnel detailed in [36]. The wind tunnel had a test section of the height, width, and length of 0.5 m, 0.5 m, and 1 m as shown in Figure 11. A 1:10 scale model of the wind tower and test room was used in the experimental study. According to the dimensions of the 1:10 model and the wind tunnel cross-section, the wind tower scale model produced a maximum wind tunnel blockage of 4.8%, and no corrections were made to the measurements obtained with these configurations [30,35]. Maintaining the geometric similarity between the model and prototype dimensions is necessary for accurate comparison between the two analysis methods. In order for the data measured from the scale model to be reliable, similarity is required for geometric and kinematic parameters such that the Reynolds numbers are maintained between both scales and analysis methods [37]. The geometric scale of the model of the wind tower was selected to maintain, as close as possible, equality of model and prototype ratios of overall dimensions to the important meteorological lengths of the simulated wind. All the relevant dimensions of the prototype were equally scaled down by the appropriate factor. Maintaining the ratio between the inertial and viscous forces, commonly known as the Reynolds number, for the flow has been shown to be not possible when air is used at the working fluid [38]. However, as long as the Reynolds number exceeds the critical value (10,000) for fully developed turbulent flow then the analysis is reliable [38]. The critical Reynolds number can be calculated for each scale, provided that the scale used is greater than this, fully developed turbulent flow is realised.

Figure 11: Closed-loop subsonic wind tunnel system for wind tower testing [36].

Due to the size of the required experimental model and complexity of the geometry, the wind tunnel prototype was created using a 3D printer. The wind tower model (Figure 12a) was connected to a 0.5 m × 0.5 m × 0.3 m room (Figure 12b), which was mounted underneath the wind tunnel test section (Figure 7). A 0.1 m × 0.1 m opening at the leeward side of the room represents the outlet. The room model was made of clear perspex sheets to allow the accurate positioning of the measurement sensors inside the room. The top plate of the room was constructed that it could be rotated in the test section in order to test different approaching wind directions.

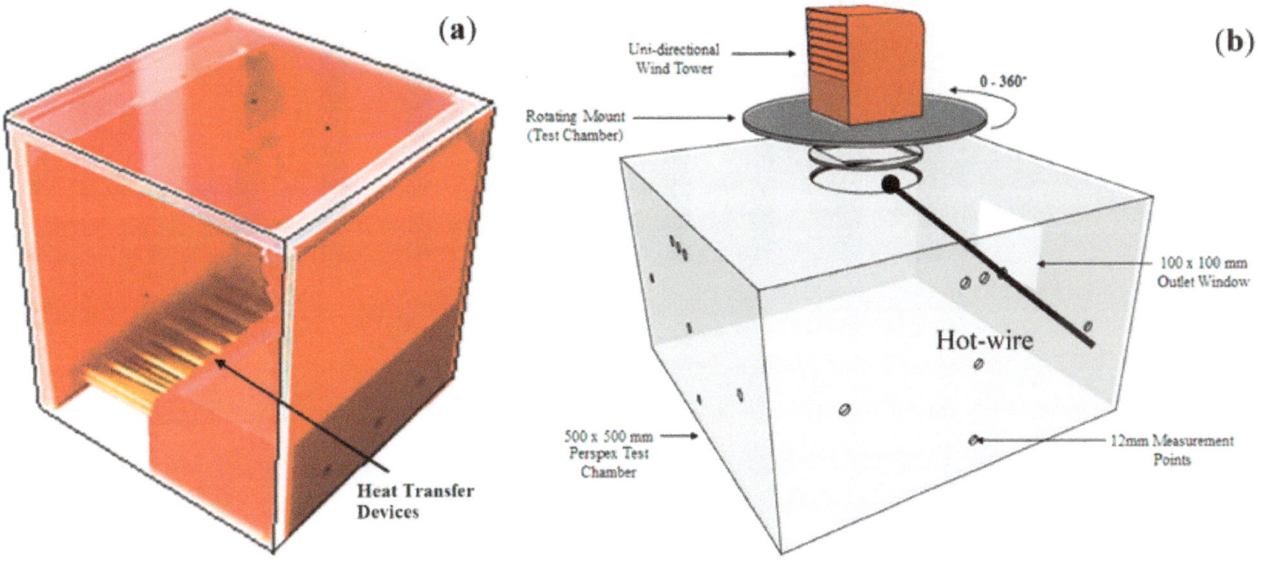

Figure 12: (**a**) 1:10 wind tower prototype; (**b**) Test room prototype.

The airflow into the room was measured using a hot-wire anemometer, which was positioned below the channels of the wind tower. The cross-sectional area of the wind tower channel was divided into several portions and the supply rate through the channel was calculated. Figure 13a shows the position of the measurement points inside the channel at a height of 0.27 m from the base of the room. The hot-wire sensor gave airflow velocity measurements with uncertainty of ±1.0% rdg. at speeds lower than 8 m/s and uncertainty of ±0.5% rdg. at higher speeds (8–20 m/s).

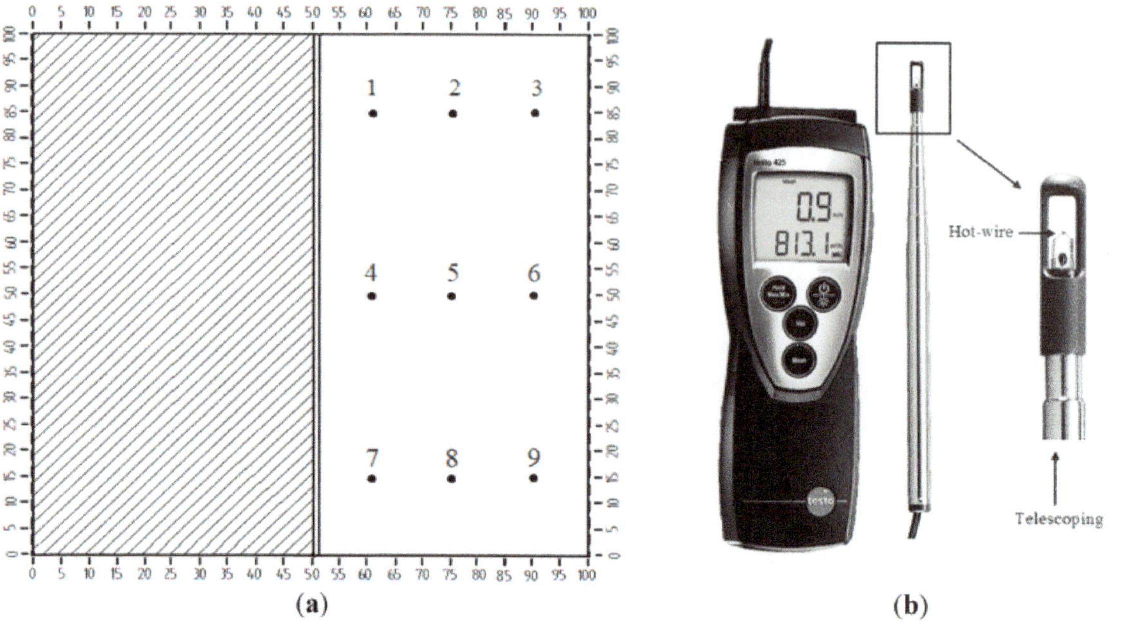

Figure 13: (**a**) Measurement points; (**b**) Hot-wire anemometer.

RESULTS AND DISCUSSION

Validation of Computational Model

Figure 14a shows the comparison between the predicted and experimental results for the airflow velocity measurements below the wind tower channel. An uneven airflow distribution was observed below the supply channel. The airflow speed on the right corner of the channel (Points 1, 4 and 7) was 30%–60% higher than the left corner (Points 3, 6 and 9). This was due to the flow separation created by the sudden change in direction (90° bend). Good agreement was observed between the CFD results and measurement, with the error below 10% for all the points except for Point 6. The average error across the points was 7.2%. Figure 14b compares the CFD and experimental results of the volumetric airflow through the wind tower channel. In this figure, the direction of the airflow in the supply channel is recognised by positive and negative values of airflow. A volumetric airflow rate of 0.6 m^3/s was achieved through the supply channel at 0° for an external wind speed of 3 m/s.

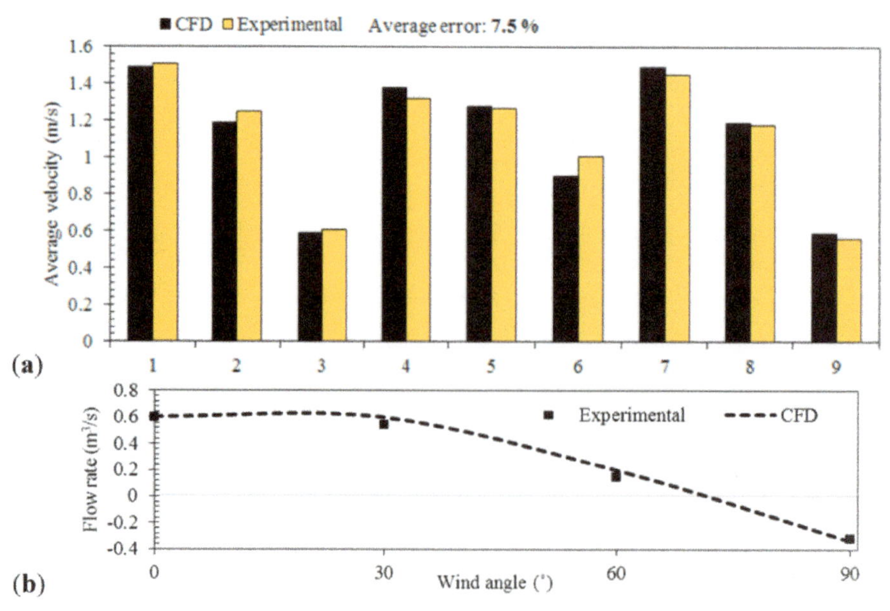

Figure 14: (**a**) Comparison between the prediction of the CFD and the measurement of the airflow speed at different points; (**b**) Effect of variation of wind direction on airflow rate.

Due to the nature of the single-sided wind tower, the dominant direction of the prevailing wind is important in obtaining maximum flow rate. At a 0° wind angle, the flow rate is at the maximum value. As the wind angle increases, the flow rate reduces to a minimum at 75°. Above this angle, negative pressure cross the opening of the wind tower causes a suction force, drawing air out shown by the negative flow rate values. Using a wind rose calculated for a specific location, the wind tower can be positioned to utilise the wind flow and direction most effectively. Should the wind the direction be 180° from the opening of the wind tower, negative pressure at the opening of the wind tower would cause air to be drawn out of the building. Though no air entering the building would pass over the heat transfer devices (HTD), ventilation would still occur as outdoor air is induced into the building through other openings.

Airflow Distribution

Figure 15a illustrates a cross-sectional plot of the velocity contours in the room and wind tower with heat transfer device. The right hand side of the plot shows the scale of airflow velocity (m/s). The contour plot in the fluid domain is colour coded and related to the CFD colour map, ranging from 0 to 3.7 m/s. As observed, the airflow passed around the wind tower, parts of it entered the wind tower and parts of it exited through the pressure outlet on the right side of the domain. After entering the wind tower, the airflow was accelerated as it hit the rear wall of the channel and re-directed downward towards the room. Separation zone was observed near the lower edge of the opening, which caused a sharp variation in velocity in this region and reduced the maximum efficiency of the wind tower. High draft speeds were observed at the centre of the room reaching up to 0.83 m/s. The air stream was circulated inside the structure and exited the opening located on the leeward side of the room. The average air velocity inside the room model was 0.28 m/s. Figure 15b shows the relation between the varying external wind speeds and performance of the wind tower in terms of the internal air distribution.

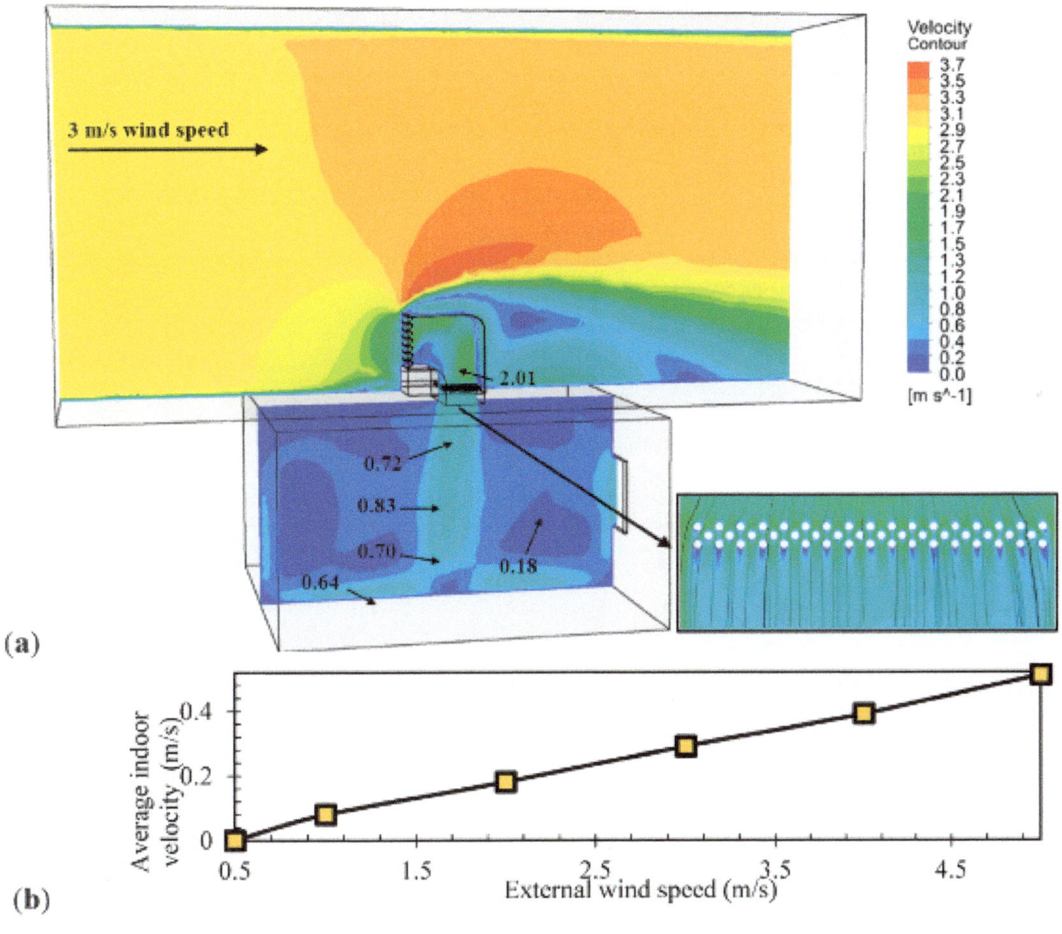

Figure 15: (**a**) Velocity contour plot of a cross-sectional plane; (**b**) Effect of variation of external wind speed on average indoor airflow speed.

Airflow Supply Rate

Since people nearly spend 90% of their time indoors, concern over human exposure to the indoor pollutants and its potential adverse effects on the occupant's health and productivity is therefore growing. Achieving good indoor air quality in buildings, hence, depends on reducing the influence of indoor sources and also decreasing pollutant entry by effective design and operation of the ventilation system. The Building Regulation's Approved Document F1A suggests that an air supply rate of 8–10 L/s per occupant is recommended for office spaces and classrooms [39]. A number of studies have identified that ventilation rates below 10 L/s/person affect occupants' health and perceived air quality. In contrast, ventilation rates above 10 L/s/person are related to lower risk of Sick Building Syndrome (SBS) [40]. As shown in Table 1, the device did not meet this recommendation for an external wind speed of 1 m/s and lower. However, the system surpassed the recommended supply rates as the external velocity increased. For comparison with a standard wind tower device, the model was simulated without the heat transfer devices and the results are presented in Table 2.

Table 1: Summary of the supply rates of the wind tower with heat transfer device

Inlet Speed [m/s]	**CFD Supply Rate** [L/s]	**CFD Supply Rate** [L/s/occupant]	**Building Regulation** [L/s/occupant]	**CFD** [L/s/m^2] *Area = 25 m^2*
0.50	65.00	4.33	10.00	2.60
1.00	145.00	9.67	10.00	5.80
2.00	325.00	21.67	10.00	13.00
3.00	485.00	32.33	10.00	19.40
4.00	695.00	46.33	10.00	27.80
5.00	840.00	56.00	10.00	33.60

Table 2: Summary of the supply rates of the wind tower without heat transfer device

Inlet Speed [m/s]	**CFD Supply Rate** [L/s]	**CFD Supply Rate** [L/s/occupant]	**Building Regulation** [L/s/occupant]	**CFD** [L/s/m^2] *Area = 25 m^2*
0.50	100.00	6.67	10.00	4.00
1.00	200.00	13.33	10.00	8.00
2.00	430.00	28.67	10.00	17.20
3.00	610.00	40.67	10.00	24.40
4.00	870.00	58.00	10.00	34.80
5.00	1090.00	72.67	10.00	43.60

Temperature Distribution

Figure 16 displays a cross-sectional plot of the temperature distribution inside the room with a wind tower. The right hand side of the contour plot shows the scale of static temperature (K). The contour plot in the fluid domain is colour coded and related to the CFD colour map, ranging from 293 K (heat transfer device temperature) to 318 K (external temperature). The average temperature inside the room was 310.4 K when the temperature of the external wind was set at 318 K. The temperature was reduced further at the immediate downstream of the heat transfer device with a supply temperature value of 309 K.

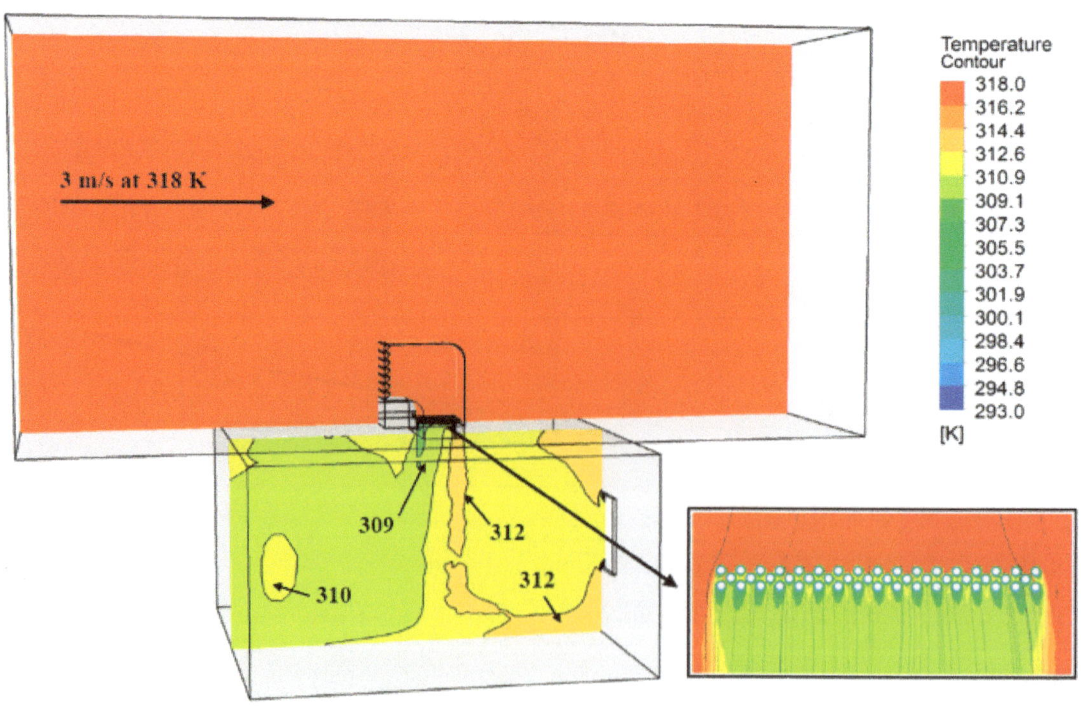

Figure 16: Cross-sectional plot of the temperature contour.

Figure 17 shows the effect of increasing the external wind speeds (1–5 m/s) on the thermal performance of the wind tower with heat transfer devices. As observed, the cooling performance of the heat transfer device decreased as the speed of the airflow increased. For instance, at 5 m/s external wind speed, the air temperature was only reduced by 5 K. Significant reduction in temperature was observed at lower wind speeds (1–2 m/s), up to 9.5–12 K reduction. Figure 18 shows the effect of varying the outdoor air temperature on the supply air temperature. The highest temperature reduction was achieved when the outdoor temperature was at 318 K and the wind speed was low (1 m/s). When the outdoor temperature was set at 302 K, the temperature reduction was only 4 K and 5 K (3 m/s and 1 m/s wind speed).

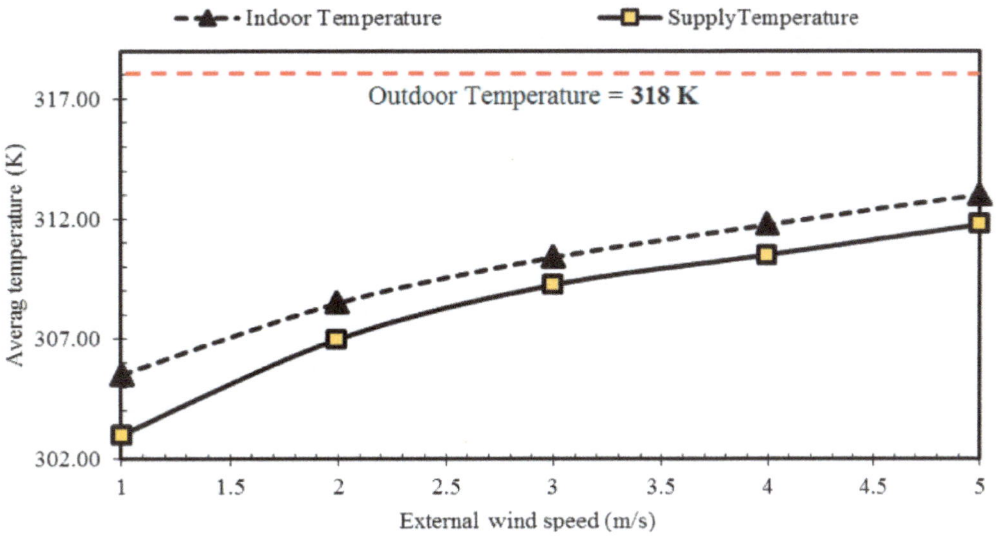

Figure 17: Effect of the wind speed on the indoor and supply air temperature.

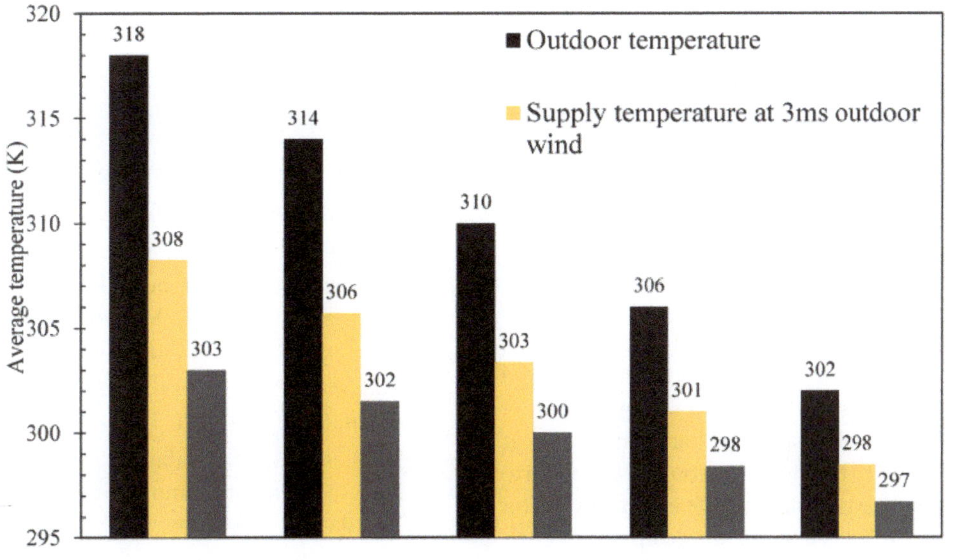

Figure 18: Effect of the outdoor temperature on the supply air temperature.

Determining the Optimum Heat Transfer Device Spacing

Figure 19 illustrates the combined response of the velocity and temperature due to the variation of the horizontal spacing between the heat transfer devices. It was observed that the effect of the horizontal spacing on the airflow temperature was more significant compared to the effect on the airflow velocity. Reducing the spacing from 100 mm to 50 mm improved the temperature drop by 3.48 K while the supply airflow velocity was reduced by only 0.12 m/s. The shaded region represents the optimum (in terms of a balance between the temperature and air velocity) horizontal spacing for the heat transfer device, and appears at 62.5–75.0 mm. U_{avg} and T_{avg} are the average predicted U and T values.

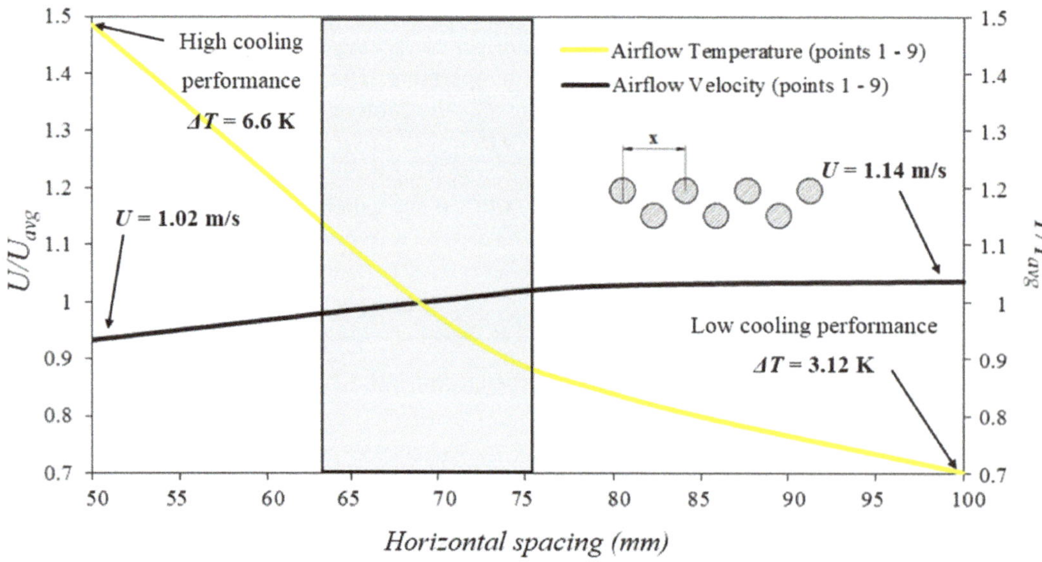

Figure 19: Effect of varying the horizontal spacing on the ventilation and thermal performance.

CONCLUSIONS

The integration of natural ventilation wind towers as a low energy alternative to HVAC systems has the potential to improve the thermal comfort of occupants, the indoor air quality and reduce energy consumption and greenhouse gas emissions. A comprehensive review of the advancement of Computational Fluid Dynamics (CFD) in aiding the development of wind towers was presented. From the review, it was concluded that regulating the temperature of supply air through a low energy process still remains incomplete. Therefore, a wind tower incorporating heat transfer devices was proposed to provide the link between passive ventilation and low energy temperature regulation. CFD modelling and scaled wind tunnel testing were used to evaluate the performance of the wind tower device. The CFD simulations were generally in good agreement with the experimental measurements. Simulation of different external wind speeds (1–5 m/s) showed that the cooling performance of the heat transfer device was indirectly proportional with the air supply rate. At 5 m/s external air speed, the air temperature was only reduced by 6 K. Significant reduction in the temperature was observed at lower wind speeds (1–2 m/s), up to 9.5–12 K reduction. Furthermore, the spacing between the heat transfer devices was varied to optimise the cooling and natural ventilation performance. It was found that the effect of varying the horizontal spacing on the supplied air temperature was more significant compared to the effect on the airflow velocity. Reducing the spacing from 100 mm to 50 mm improved the temperature drop by 3.48 K while the supply airflow velocity was reduced by only 0.12 m/s.

ACKNOWLEDGMENTS

The support by the University of Sheffield (Department of Mechanical Engineering) is gratefully acknowledged. The statements made herein are solely the responsibility of the authors. The technology presented here is subject to a patent application (PCT/GB2014/052263).

AUTHOR CONTRIBUTIONS

The main author (J.K.C.) conducted the CFD modelling and wind tunnel testing of the wind tower incorporating the heat transfer devices. The co-authors reviewed the development of CFD analysis in aiding the development of wind towers (D.O.C.) and the optimisation of natural ventilation systems (P.S.). B.R.H. participated in the analysis of the CFD and experimental data.

REFERENCES

1. Bouchahm, Y.; Bourbia, F.; Belhamri, A. Performance analysis and improvement of the use of wind tower in hot dry climate. Renew. Energy 2011, 36, 898–906.
2. WBCSD. Energy Efficiency in Buildings: Facts & Trends; Atar Roto Presse: Satigny, Switzerland, 2008.
3. Sofotasiou, P.; Hughes, B.R.; Calautit, J.K. Qatar 2022: Facing the FIFA World Cup climatic and legacy challenges.Sustain. Cities Soc. 2015, 14, 16–30.
4. Bahadori, M.N. Passive cooling systems in Iranian architecture. Sci. Am. 1978, 238, 144–154.
5. Hughes, B.R.; Calautit, J.K.; Ghani, S.A. The Development of Commercial Wind Towers for Natural Ventilation: A Review. Appl. Energy 2012, 92, 606–627.
6. O'Connor, D.; Calautit, J.K.; Hughes, B.R. A Study of Passive Ventilation Integrated with Heat Recovery. Energy Build.2014, 82, 799–811.
7. Linden, P.F. The fluid mechanics of natural ventilation. Ann. Rev. Fluid Mech. 1999, 31, 201–238.
8. Bahadori, M.N. An improved design of wind towers for natural ventilation and passive cooling. Sol. Energy 1985, 35, 119–129.
9. Yaghoubi, M.A.; Sabzevari, A.; Golneshan, A.A. Wind Towers—Measurement and Performance. Sol. Energy 1991, 47, 97–106.
10. Hughes, B.R.; Mak, C.M. A study of wind and buoyancy driven flows through commercial wind towers. Energy Build.2011, 43, 1784–1791.

11. Hughes, B.R.; Ghani, S.A. A numerical investigation into the effect of Windvent louvre external angle on passive stack ventilation performance. Build. Environ. 2010, 45, 1025–1036.
12. Liu, S.C.; Mak, C.M.; Niu, J.L. Numerical Evaluation of Louver Configuration and Ventilation Strategies for the Windcatcher System. Build. Environ. 2011, 46, 1600–1616.
13. Montazeri, H. Experimental and numerical study on natural ventilation performance of various multi-opening wind catchers. Build. Environ. 2011, 46, 370–378.
14. Calautit, J.K.; Chaudhry, H.N.; Hughes, B.R.; Ghani, S.A. Comparison between evaporative cooling and a heat pipe assisted thermal loop for a commercial wind tower in hot and dry climatic conditions. Appl. Energy 2013, 101, 740–755.
15. Gan, G.; Riffat, S.B. Naturally ventilated buildings with heat recovery: CFD simulation of thermal environment. Build. Serv. Eng. Res. Technol. 1997, 18, 67–75.
16. Hughes, B.R.; Chaudhry, H.N.; Calautit, J.K. Passive energy recovery from natural ventilation air streams. Appl. Energy 2014, 113, 127–140.
17. Calautit, J.K.; Hughes, B.R.; Chaudhry, H.N.; Ghani, S.A. CFD analysis of a heat transfer device integrated wind tower system for hot and dry climate. Appl. Energy 2013, 112, 576–591.
18. Stavrakakis, G.; Koukou, M.; Vrachopoulos, M.; Markatos, N. Natural cross-ventilation in buildings: Building-scale experiments, numerical simulation and thermal comfort evaluation. Energy Build. 2008, 40, 1666–1681.
19. Stavrakakis, G.; Zervas, P.; Sarimveis, H.; Markatos, N. Development of a computational tool to quantify architectural-design effects on thermal comfort in naturally ventilated rural houses. Build. Environ. 2010, 45, 65–80.
20. Zhou, L.; Haghighat, F. Optimization of ventilation system design and operation in office environment, Part I: Methodology. Building and Environment. Build. Environ. 2009, 44, 651–656.
21. Shen, X.; Zhang, G.; Bjerg, B. Investigation of Response Surface Methodology for Modelling Ventilation Rate of a Naturally Ventilated Building. Build. Environ. 2012, 54, 174–185.
22. Shen, X.; Zhang, G.; Bjerg, B. Assessments of Experimental Designs in Response Surface Modelling Process: Estimating Ventilation Rate in Naturally Ventilated Livestock Buildings. Energy Build. 2013, 62, 570–580.
23. Lal, S. Experimental, CFD simulation and parametric studies on modified solar chimney for building ventilation. Appl. Sol. Energy 2014, 50, 37–43.
24. Calautit, J.K.; O'Connor, D.; Hughes, B.R. Determining the optimum spacing and arrangement for commercial wind towers for ventilation performance. Build. Environ. 2014, 82, 274–287.
25. Yin, Y.; Xu, W.; Gupta, J.; Guity, A.; Marmion, P.; Manning, A.; Gulick, B.; Zhang, X.; Chen, Q. Experimental Study on Displacement and Mixing Ventilation Systems for a Patient Ward. HVAC&R Res. 2010, 15, 1175–1191.
26. Aliabadi, A.; Faghani, E.; Tjong, H.; Green, S. Hybrid Ventilation Design for a Dining Hall Using Computational Fluid Dynamics (CFD). In Proceedings of the CSME International Congress 2014, Toronto, ON, Canada, 2014.
27. Chen, Q. Using computational tools to factor wind into architectural environment design. Energy Build. 2004, 36, 1197–1209.
28. ANSYS® Academic Research. ANSYS FLUENT User's Guide Release 14.0. Available online: https://www.ansys.com(accessed on 1 January 2014).
29. Calautit, J.K.; Hughes, B.R.; Ghani, S.A. A Numerical Investigation into the Feasibility of Integrating Green Building Technologies into Row Houses in the Middle East. Archit. Sci. Rev. 2013, 56, 279–296.
30. Calautit, J.K.; Hughes, B.R. Wind tunnel and CFD study of the natural ventilation performance of a commercial multi-directional wind tower. Build. Environ. 2014, 80, 71–83.
31. Calautit, J.K. Integration and Application of Passive Cooling within a Wind Tower. Ph.D. Thesis, University of Leeds, Leeds, UK, 2013.
32. Calautit, J.K.; Hughes, B.R.; Ghani, S.A. Numerical investigation of the integration of heat transfer devices into wind towers. Chem. Eng. Trans. 2013, 34, 43–48.
33. Chung, T. Computational Fluid Dynamics; Cambridge University Press: Cambridge, UK, 2002.
34. Calautit, J.K.; Hughes, B.R. Measurement and prediction of the indoor airflow in a room ventilated with a commercial wind tower. Energy Build. 2014, 84, 367–377.
35. Calautit, J.K.; Hughes, B.R. Integration and application of passive cooling within a wind tower for hot climates. HVAC&R Res. 2014, 20, 722–730.
36. Calautit, J.K.; Chaudhry, H.N.; Hughes, B.R.; Sim, L.F. A validated design methodology for a closed-loop subsonic wind tunnel. J. Wind Eng. Ind. Aerodyn. 2014, 125, 180–194.
37. Cermak, J.E.; Isyumov, N. Wind Tunnel Studies of Buildings and Structure; ASCE Publications: Virginia, USA, 1999.
38. Linden, P.F.; Cooper, P. Multiple sources of buoyancy in a naturally ventilated enclosure. J. Fluid Mech. 1996, 311, 177–192.
39. Building Regulations 2000: Approved Document F1A: Means of Ventilation. Available online: https://www.planningportal.gov.uk (accessed on 1 January 2014).
40. Seppänen, O.A.; Fisk, W.J.; Mendell, M.J. Association of ventilation rates and CO_2 concentrations with health and other responses in commercial and institutional buildings. Indoor Air 1999, 9, 226–252.

SRCTION IV: CFD APPLICATION TO THE DESIGNING AND OPTIMIZATION OF TURBINES AND CENTRIFUGES

CHAPTER 19

3D CFD ANALYSIS OF A VERTICAL AXIS WIND TURBINE

Andrea Alaimo, Antonio Esposito, Antonio Messineo, Calogero Orlando and Davide Tumino

Faculty of Engineering and Architecture, Kore University of Enna, Cittadella Universitaria, 94100 Enna, Italy

ABSTRACT

To analyze the complex and unsteady aerodynamic flow associated with wind turbine functioning, computational fluid dynamics (CFD) is an attractive and powerful method. In this work, the influence of different numerical aspects on the accuracy of simulating a rotating wind turbine is studied. In particular, the effects of mesh size and structure, time step and rotational velocity have been taken into account for simulation of different wind turbine geometries. The applicative goal of this study is the comparison of the performance between a straight blade vertical axis wind turbine and a helical blade one. Analyses are carried out through the use of computational fluid dynamic ANSYS® Fluent® software, solving the Reynolds averaged Navier–Stokes (RANS) equations. At first, two-dimensional simulations are used in a preliminary setup of the numerical procedure and to compute approximated performance parameters, namely the torque, power, lift and drag coefficients. Then, three-dimensional simulations are carried out with the aim of an accurate determination of the differences in the complex aerodynamic flow associated with the straight and the helical blade turbines. Static and dynamic results are then reported for different values of rotational speed.

INTRODUCTION

In recent decades, the prospect of the depletion of fossil fuels has directed attention toward renewable energy sources. Among the renewable energy sources, onshore wind power is one of the most attractive because of its low cost of maintenance of installed systems. Although the concept of vertical axis wind turbines (VAWTs) was proposed by Darrieus [1] as early as 1931, the research and development in this area are still of interest and in progress nowadays. Among the several categories of wind generators, small wind turbines have gained more and more interest for their excellent adaptability to the urban environment in terms of visual impact and noise pollution. Being axisymmetric, they are omnidirectional turbines, which respond well to changes in wind direction. In order to determine the performance of the VAWT, analytical and numerical approaches, as well as experimental tests are generally used [2–11]. Among others, Sutherland et al. [2] report on the streamtube model and vortex models that allow one to analyze the aerodynamic response of VAWTs. Islam et al. [3] analyzed three aerodynamic models used for the performance prediction and design of a straight blade Darrieus turbine. These three models are the double-multiple streamtube model, free vortex model and cascade model, and we highlight

Citation: Andrea Alaimo, Antonio Esposito, Antonio Messineo, Calogero Orlando and Davide Tumino, 3D CFD Analysis of a Vertical Axis Wind Turbine, doi:10.3390/en8043013.

their strengths and weaknesses. The first simplest approach developed by Templin [4] is the single streamtube numerical model. Templin calculates the performance for a single blade whose chord equals the sum of the chords of the actual rotor's blade; airfoil characteristics are introduced through the application of blade element theory. This approach gives predictions of rotor performance in terms of average torque per revolution only. Strickland [12] extends Templin's approach into a multiple single streamtube model by considering a number of adjacent smaller streamtube, applying the conservation of momentum and blade element theory to each streamtube. This allows one to determine the performance considering the local Reynolds number rather than an average. Further improvement in the multiple streamtube model has been reported by Paraschivoiu [5], namely the double multiple streamtube model (DSMT); the difference with respect to Strickland's model is that upwind and downwind passages through each streamtube are considered separately. Low solidity with a low chord-radius ratio ($c/r \leq 0.1$) can be modeled well using a DMST analysis, such as the one developed by Paraschivoiu [5]. Brusca et al. [6], using a calculation code based on the multiple stream tube model, analyze the link between the aspect ratio of a VAWT with straight blade; it was highlighted that the power coefficient is influenced by both rotor solidity and the Reynolds number, concluding that a low aspect ratio has several advantages compared to higher ratio values. A second class of models is those based on vortex representation of the blades and their wakes. Strickland et al.'s model [7] is based on replacement of the rotor blade by a bound vortex filament, called a lifting line, whose strength is a function of the azimuthal position. Wang et al.[8] present a potential flow 2D vortex panel model (VPM2D) applied to the vertical axis straight blade variable pitch turbine; however, this model does not include the consideration of the phenomenon of dynamic stall. Experimental tests can be performed by means of wind tunnel or open-field measurement in order to evaluate the real performance of wind turbines. Sheldahl [9] tested a 2-m diameter Darrieus VAWT with NACA-0012 airfoil blades, in the field at the Sandia National Laboratories, and operated it to determine if the data obtained from these tests are comparable with wind tunnel data obtained under ideal conditions. Bedon et al. [10] describe the results of an experimental campaign of measurements conducted on a VAWT with helical blade having a radius of 1 m with a NACA-0018 airfoil section, tested in an open field. With the advent of computers, many authors have investigated the feasibility and accuracy of different computational fluid dynamics (CFD) approaches. Analytical and numerical investigations were conducted by Biadgo et al. [11] on straight blade VAWT using NACA0012 as a blade profile. The analytical investigation is performed using the double-multiple streamtube model, while the numerical investigation of two-dimensional unsteady flow is carried out using CFD. Wolfe et al. [13] use CFD to investigate the two-dimensional aerodynamic performance of typical wind turbine airfoils with the standard κ − □ [14] turbulence model, concluding that this model is not appropriate at angles of attack with flow separation. Carrigan et al. [15], fixing the tip speed ratio (TSR) of the wind turbine, developed an iterative design system to maximize the torque for different airfoil cross-sections and solidities in two-dimensional CFD simulations. The TSR, which is also referred to as λ in the manuscript, is defined as U/V_∞, where U is the tip peripheral velocity of the rotor, namely ωR, ω is the angular velocity of the blade and R the rotor radius, while V_∞ is the free stream velocity. A 2.5D model [16] is also presented in the literature, which differs from a 3D simulation, because only a segment of the airfoil blades is modeled with periodic boundaries at the extremities of the domain. The limitation of this model is that it can only be used with Darrieus-type straight-bladed VAWT. Furthermore, although this modeling approach produces a more realistic three-dimensional vortex diffusion after flow separation, it cannot capture the effect of tip vortex and vertical flow divergence in VAWT. These two processes are the major cause of over-prediction of VAWT power in the two-dimensional model [17]. A CFD performance model based on blade element momentum theory was presented by Raciti Castelli et al. [18] for evaluating the aerodynamic forces of straight blade VAWT. Most of the authors have the adopted Reynolds averaged Navier–Stokes equations (RANS) approach with different turbulence models [19,20]. However, Iida et al. [21] used a model with large eddy simulations (LES), proving that the as-obtained results are in good agreement at high TSR with that of momentum theory, but the model was revealed to be not accurate for low TSR cases. Ferreira et al. [22] use detached eddy simulation (DES) focusing on 2D dynamic stall and validate the results with particle image velocimetry data. It is to be noted that in the Darrieus-type VAWT, a cyclic phenomenon of dynamic stall and vortex separation is present [8,22,23]. Amet et al. [23] provided a detailed numerical analysis of the physical phenomena that occur during dynamic stall where the compressible RANS $\kappa - \omega$ model and multi-block mesh structure were used. Furthermore, studies of dynamic stall show that the Shear Stress Transport (SST) $\kappa - \omega$ turbulence model performs better than the $\kappa - \omega$model, as reported by Ekaterinaris and Platzer [24] and Qian et al. [25]. Only recently have some researchers carried out three-dimensional CFD simulations [26,27]. Howell et al. [26] provided a three-dimensional CFD model based on the Re-Normalisation Group (RNG) κ − □ turbulence model, and the effects of turbine solidity, roughness and tip vortices were considered. Zhang et al. [27], analyze a three-bladed straight wind turbine, using a NACA 0015 airfoil with a chord length of 0.4 m and a blade length of 2 m with 4 m in diameter. The aim of the present paper is to numerically compare different Darrieus turbine configurations that differ for the twist angle of the blades. In Section 2, a two-dimensional dynamic numerical model is explained, and the results are: (i) the influence of the rotational domain mesh; (ii) the airfoil behavior; and (iii) the ideal dynamic torque coefficient. In Section 3, a three-dimensional Darrieus turbine configuration and computational domain are described. Then, static analyses are carried out to compute the static performance of the studied configurations, namely the torque produced by the still turbine. Dynamic simulations are also performed to compare the straight blade turbines with the select twisted blade ones in terms of torque and power coefficients during operative cycles. It is to note that, in order to perform the simulations, the κ − □ model is employed; this model does not befit analyzing strong adverse pressure gradients. Nevertheless, its strength is given by the reduced calculation times required to perform a simulation with good levels of convergence [28]. Last, conclusions are given in Section 4.

TWO-DIMENSIONAL MODEL AND RESULTS

In this section, the geometrical features of the analyzed Darrieus turbine are described first. Then, the adopted control volumes and the corresponding boundary conditions are described. The mesh generation parameters are also given. The solid geometry was created with a CAD modeler and imported into a "Mesh Component Systems" in ANSYS®Workbench™, to create the mesh model. The discretized model generated was then read into FLUENT® for numerical solution. The geometric characteristics for the turbine are a NACA 0021 blade airfoil, with a chord length c = 0.3 m and a rotor radius R = 0.99 m. The length of the control volume along the free-stream direction is equal to Lx = 100 R, while its width in the y direction is Ly = 60 R. The circular rotating domain has an overall diameter of D_r = 3 m; see Figure 1. At the velocity inlet, the velocity distribution is uniform, directed along the x-axis and equal to 7 m/s. The pressure outlet is set equal to 101,325 Pa. The no slip boundary condition is applied on the turbine wall blades. An interface wall is introduce between the fixed and rotating domain. The origin of the reference frame is the center of the rotor.

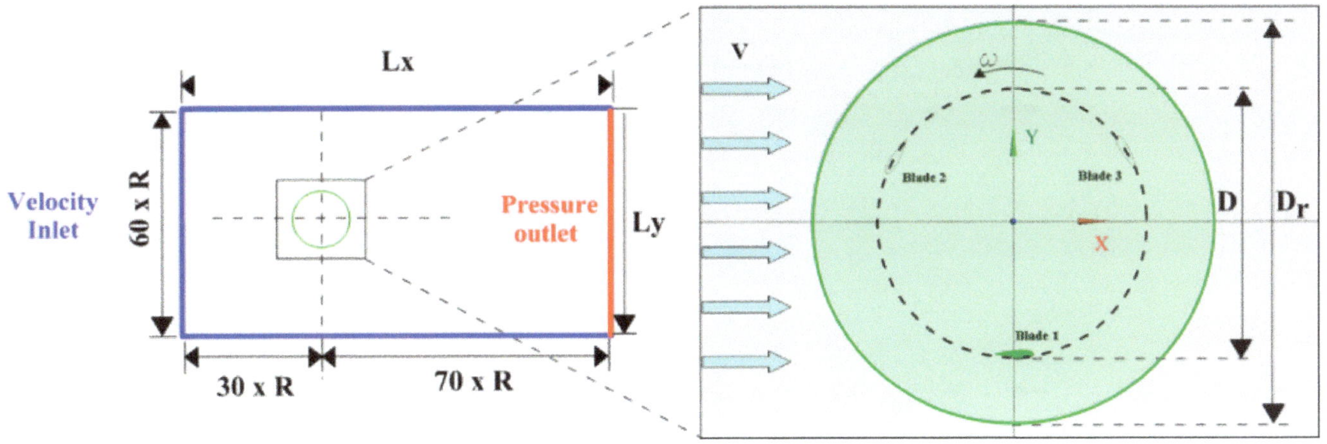

Figure 1: Two-dimensional computational domain.

In the present section, the results obtained from the two-dimensional analyses are shown. Transient analyses are carried out in this section to characterize the performances of the investigated airfoil. Performances are described in terms of torque and power coefficients, which are computed using procedures described in [29]. The total torque vector about the rotor axis is computed by summing the elemental cross products of the pressure and viscous force vectors times the vector r⃗ r→, that is the distance between the element centroid and the rotor axis; see [30], page 692. Equation (1) represents the pressure and viscous torque vectors:

$$\vec{T}(\theta) = \vec{r} \times \vec{F}_p(\theta) + \vec{r} \times \vec{F}_v(\theta) \tag{1}$$

where $\vec{F}_p$ is the pressure force vector and $\vec{F}_v$ is the viscous force vector. The instantaneous torque coefficient C_t is defined as:

$$C_t(\theta) = \frac{T(\theta)}{0.5 \rho V_\infty^2 A_S R} \tag{2}$$

where ρ is the density, A_S is the size of turbine equal to the product of diameter D and rotor height H and V_∞ is the free stream velocity. For two-dimensional simulation, H is unitary. The average torque coefficient $C_{t,ave}$ is computed by averaging C_t (θ) over a whole revolution. A minimum of fifteen revolutions of the rotor have been simulated for each λ. The average power coefficient is computed as $C_{p,ave} = C_{t,ave}\ \lambda$, with λ ranging from 0.44 to 2.22, taking into account that for this study, the considered design tip speed ratio is about 1.4.

Mesh Configurations

The mesh is characterized by triangular elements; in particular, the maximum element size for the fixed domain is set to 0.5 m, whereas the length of the element at the interface between the fixed and the rotating domain is equal to 0.05 m. The maximum length of the element size close to the airfoils varies between 0.0005 m and 0.005 m for convergence analysis purposes.

The solution model is set using the least-squares cell-based gradient option with second order interpolation method for face pressure. The sliding mesh model [29] is used to model the turbine blade rotation. A multiple reference frame solution [30] is used to compute a flow field as an initial condition for the transient sliding mesh calculation. This eliminates the need for a start-up calculation. The κ − ϵ approach with enhanced wall treatment [31] is adopted to model the flow turbulence.

When using a sliding mesh, it is known [32] that to have a good numeric convergence characteristic, the value of the time step should not be larger than the time it takes for a moving cell to advance of a characteristic length L_c. An estimate for the time step in the case of translational motion can be calculated as Equation (3):

$$\Delta t = \frac{1}{3}\frac{L_c}{V} \tag{3}$$

where the step size Δt is proportional to the cell characteristic size L_c and inversely to velocity V. On the other hand, for turbomachinery, the following relation is suggested, Equation (4) [32]:

$$\Delta t = \frac{1}{10}\frac{Number\ of\ blades}{rotational\ velocity} \tag{4}$$

Taking into account that the radius of the turbine is equal to that of the rotational sliding computational domain and assuming an angular velocity of 120 RPM, from Equation (3), a value of Δt about equal to 0.001 s is obtained. In applyingEquation (3), the characteristic length is $L_c = 0.05$ m and V = ω R = 18, 85 m/s; whereas from Equation (4), Δt = 0.016 s is obtained. For all two-dimensional simulations, the step size is set equal to 0.0005 s to increase the accuracy of the solution.

At first, the behavior of the two-dimensional turbine as a function of the mesh discretization along the profile is analyzed. In Table 1 are listed the four selected maximum elements lengths, the number of nodes and the number of elements of the control volume and on the airfoil edge, whereas in Figure 2, the mesh grids close to the airfoils are shown.

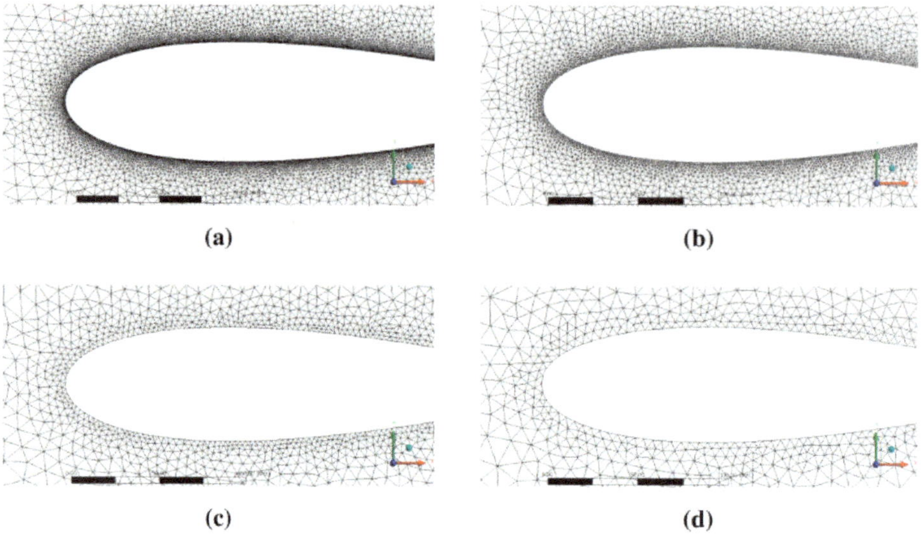

Figure 2: Mesh discretization along the two-dimensional turbine blade airfoils. (**a**) Mesh 0; (**b**) Mesh 1; (**c**) Mesh 2; (**d**) Mesh 3.

Table 1: Mesh discretization values

Name	Length (m)	Airfoil Elements	Domain Node	Domain Elements
Mesh 0	0.0005	1,304	32,901	61,405
Mesh 1	0.001	650	19,081	35,699
Mesh 2	0.003	216	9,732	18,277
Mesh 3	0.005	130	7,614	14,301

The influence of discretization is assessed by investigating the effect of the characteristics of the maximum element length close to the airfoil on the torque coefficients C_t. More particularly, it appears from Figure 3 that the torque coefficient for the three blades is only slightly affected by the chosen mesh size values, as well as by the y+ (see [30], page 109), which is lower then 10 for Mesh 0, Mesh 1 and Mesh 2.

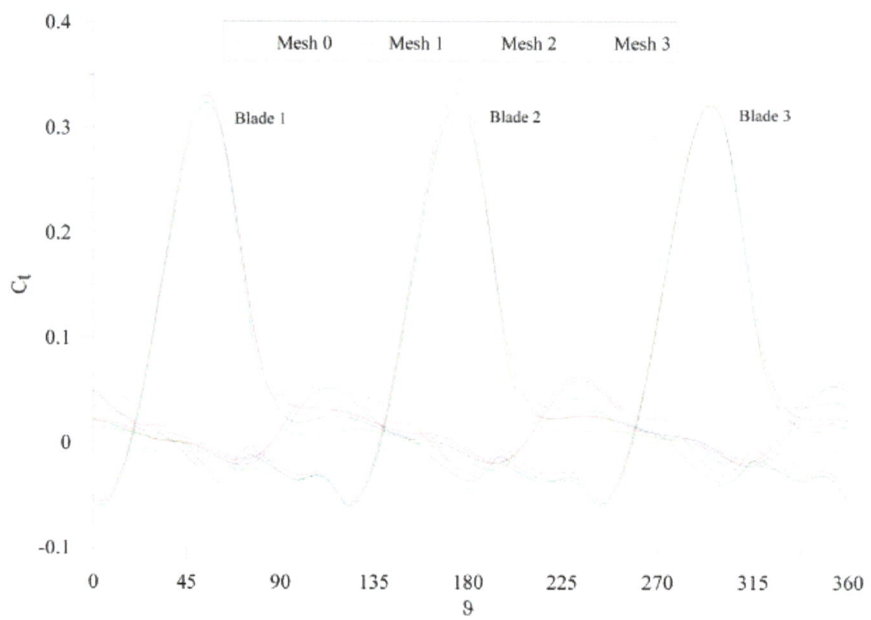

Figure 3: Influence of mesh discretization on blade torque coefficients.

The trend of curve for the individual blade, in a time interval corresponding to one revolution, is shown in Figure 3, for a rotational speed of 60 RPM. With reference to Figure 3, the Mesh 3 and Mesh 2 curves show an overprediction with respect to the Mesh 1 and Mesh 0 data. However, the average discrepancy between Mesh 3 and Mesh 1 is equal to 8.09%, over one revolution, while the value of Mesh 2 with respect to Mesh 1 is equal to 3.91%. Mesh 1 has an average discrepancy of 1.27% with respect to Mesh 0. For this reason, to reduce the computational times, Mesh 2 is selected to carry out the simulations.

Dynamic Stall Angle

Once the discretization of the mesh along the airfoil is set, two-dimensional simulations with increasing rotational speeds of the blades are carried out. Six different rotational speeds corresponding to λ from 0.44 to 2.22 are chosen; the selected range of velocity allows one to analyze the characteristic bell shape curve that describes the behavior of a Darrieus turbine in terms of the power coefficient [33]. The aim is to investigate the aerodynamic performance behavior.

In order to compute the lift and drag coefficients of the blade airfoil, the angle of attack has to be considered, and it is directly computed by the CFD analysis. Taking into account Figure 4, it appears that the angle of attack α is the angle between the airfoil cord and the relative velocity V_r. The relative velocity V_r of the air mass deflected by the airfoils is given by the vectorial composition of the wind velocity V_a with the tangential relative velocity $U = -\omega R$.

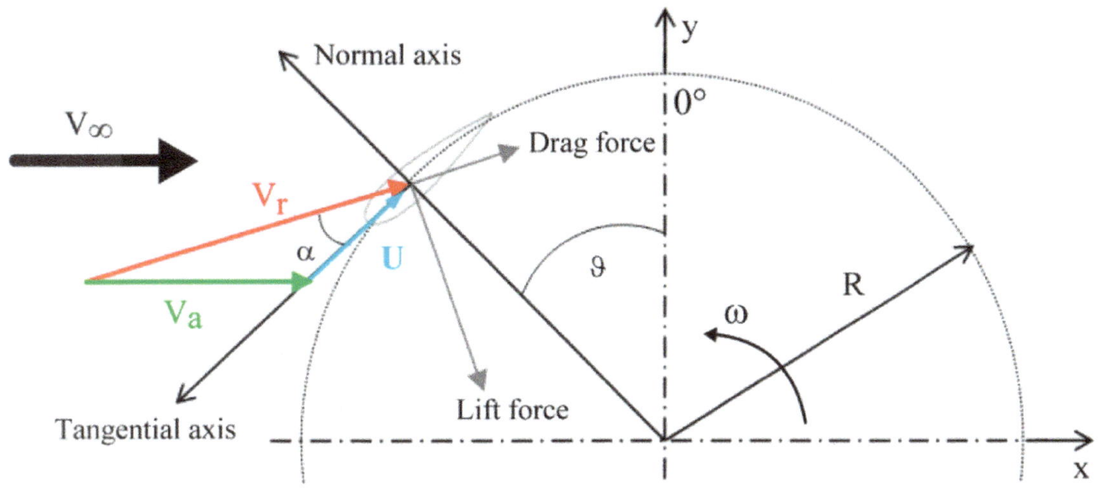

Figure 4: Azimuthal coordinate of the blade midsection's center of pressure and blade velocity vectors.

By using numerical results for C_x and C_y, the lift and drag coefficients are computed according to:

$$C_l = C_y\left[\cos\left(\vartheta + \alpha\right)\right] - C_x\left[\sin\left(\vartheta + \alpha\right)\right] \tag{5}$$

$$C_d = C_y\left[\sin\left(\vartheta + \alpha\right)\right] + C_x\left[\cos\left(\vartheta + \alpha\right)\right] \tag{6}$$

The lift and drag coefficients' variation with respect to azimuthal position are reported for different values of the tip speed ratio in Figure 5. It is seen that as the λ increases, both the maximum absolute value of the lift and drag coefficients produced by airfoil increase.

Summary results are given in Table 2 in terms of airfoil stall condition and maximum torque coefficients. As λ increases, the azimuthal position where the airfoil stall increases, while the stall angle of attack decreases. Similar results are obtained for the maximum torque coefficient, both in terms of azimuthal position and angle of attack. According to Sheldahl et al. [9], it is found that for a Reynolds number of 355,000, which corresponds to the case λ = 2.22, the stall angle of attack is about 12 degrees, which compares well with our results.

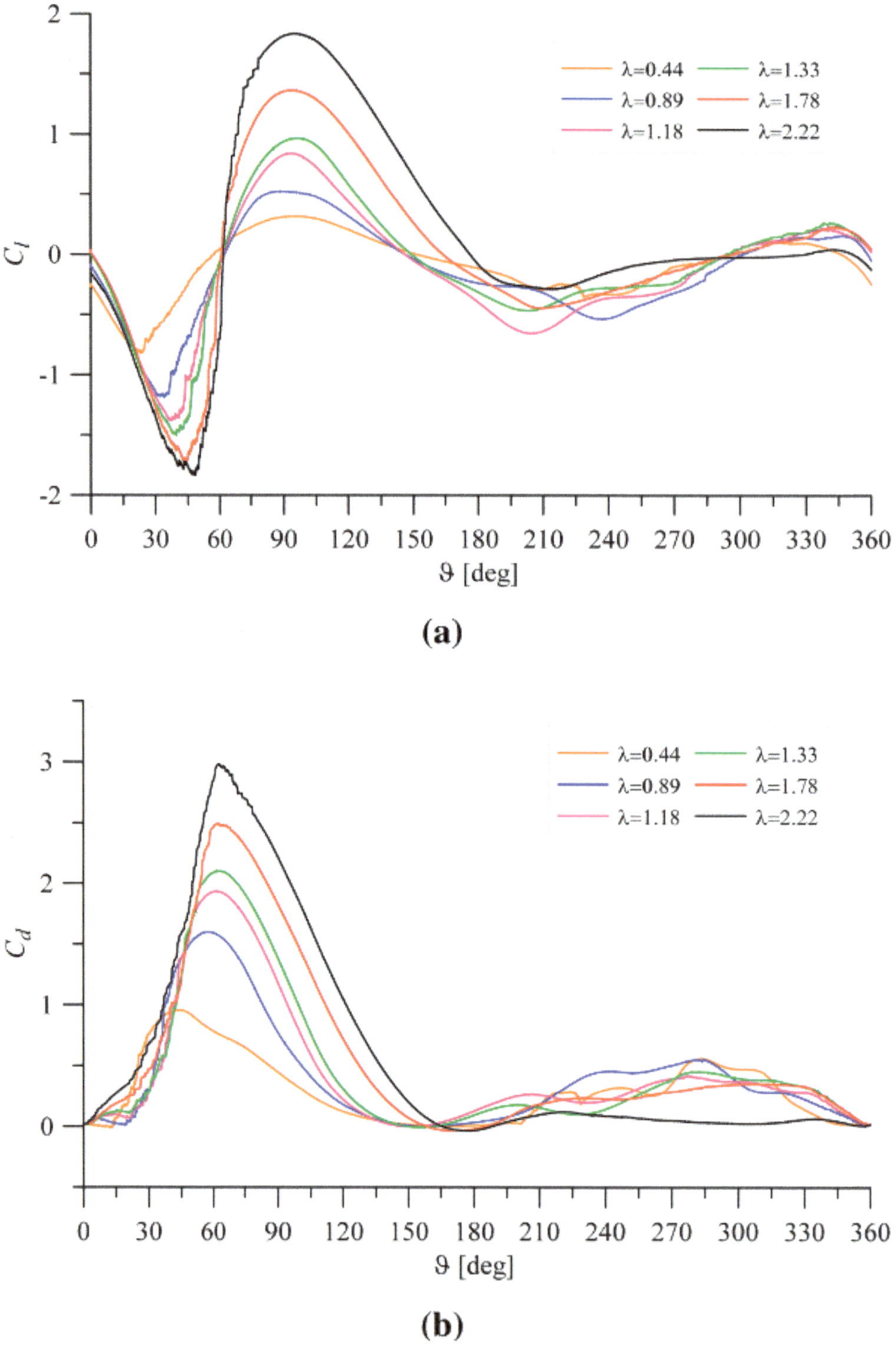

Figure 5: Lift and drag coefficients of airfoil over one turbine revolution for different λ. (**a**) Lift coefficient (C_l) with respect to the azimuthal position; (**b**) drag coefficient (C_d) with respect to the azimuthal position.

However, by looking at Figure 6a, it appears that the maximum value of torque coefficients is obtained for λ equal to 1.78. This trend is also confirmed by the average distribution of both torque and power coefficients, as appears from Figure 6b.

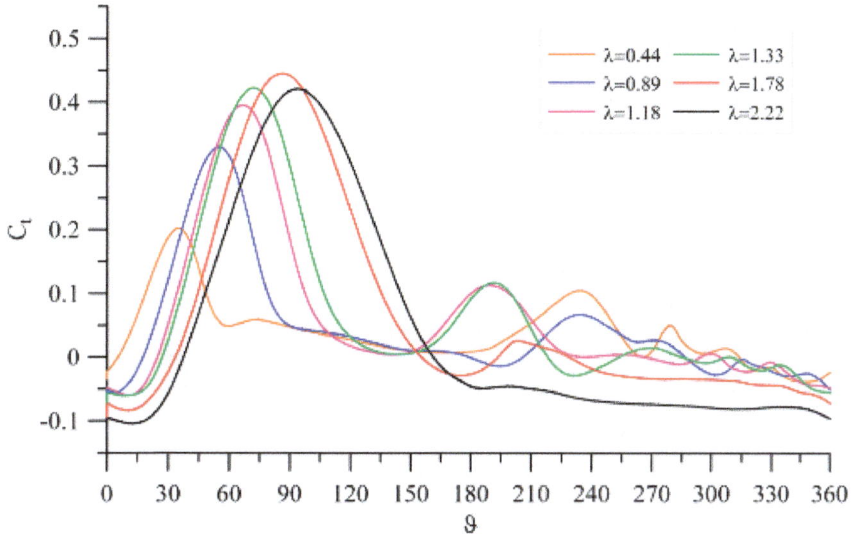

(a)

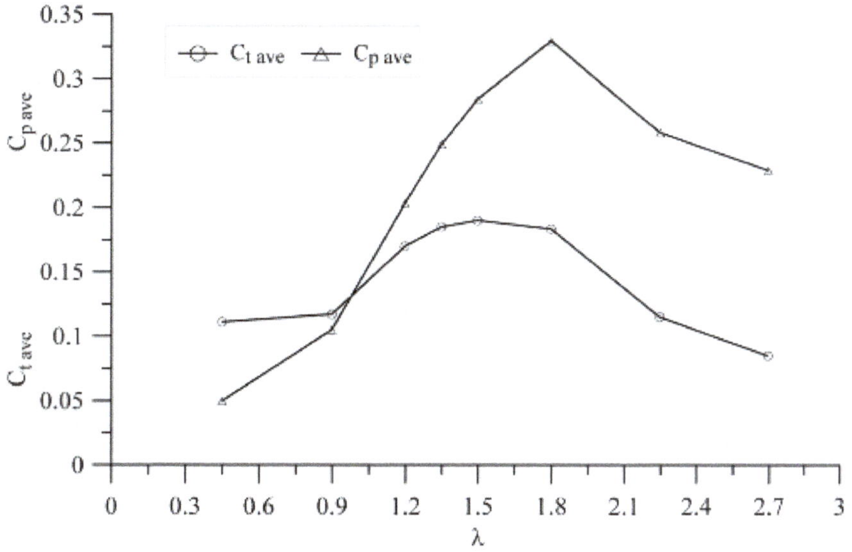

(b)

Figure 6: Torque and power coefficients as a function of azimuthal position and λ. (**a**) Instantaneous torque coefficient (C_t) for a single blade with respect to the azimuthal position (θ) at various tip speed ratios (λ); (**b**) rotor average torque coefficient (C_t) and average power coefficient (C_p) with respect to tip speed ratio (λ).

Table 2: Stall and maximum torque data at different λ values

RPM	λ	α_{Stall}	$C_{l,Stall}$	θ_{Stall}	$\alpha_{C_{t,max}}$	$C_{t,max}$	$\theta_{C_{t,max}}$
30	0.44	1.124	−0.819	23.940	28.905	0.201	35.028
60	0.89	4.088	−1.188	33.948	35.759	0.329	55.044
80	1.18	9.016	−1.381	37.056	40.304	0.395	66.960
90	1.33	11.529	−1.499	39.528	42.622	0.422	71.874
120	1.78	11.760	−1.719	42.912	46.419	0.445	86.904
150	2.22	12.081	−1.834	48.420	44.422	0.420	93.600

THREE-DIMENSIONAL MODELS AND RESULTS

Three-dimensional models are now considered to take into account the finite aspect ratio effects. Four distinct wind turbine configuration are taken into account, namely the Straight blades VAWT and three Helical blades VAWT with different twist angles. The considered twist angles for the Helical blade VAWT are ψ = 30°, 60°, 90°; see Figure 7. The airfoils' characteristics and rotor diameters are the same for the two-dimensional analyses (NACA 0021, R = 0.99 m), and the rotor height is H = 1.15 m.

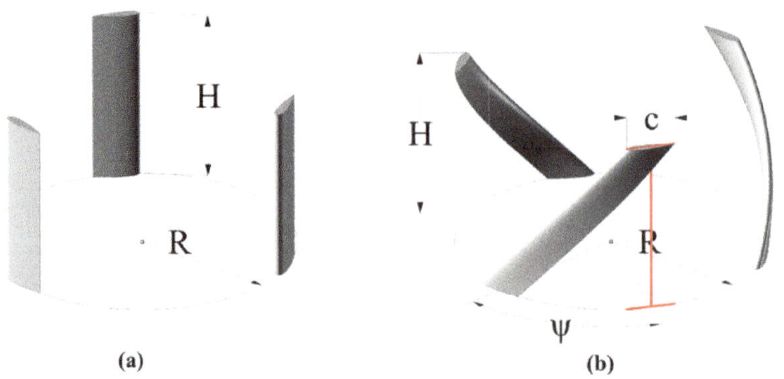

Figure 7: Turbine configurations. (**a**) Straight blade; (**b**) Helical blade with ψ = 30°, 60°, 90°.

The length of the control volume along the free-stream direction is equal to Lx = 32 R, while its width in the y and zdirection is Ly = Lz = 12 R; see Figure 8. The origin of the reference frame is set on the center of the rotor, 8 m from the inlet. The boundary conditions are defined as a velocity inlet on surface S1 in Figure 8; the inflow velocity is directed along the xaxis only and equals 7 m/s. Surface S2 is set as a pressure outlet, keeping the pressure constant equal to 101,325 Pa. In Figure 8, surface S3 is the interface between the rotating and the fixed volumes. The no slip boundary condition is applied on the turbine wall blades. The rotational speed of the turbines, for the sake of dynamic analyses, ranges from 30 RPM to 120 RPM. The origin of the reference frame is the center of the rotor; on surface S4 in Figure 8.

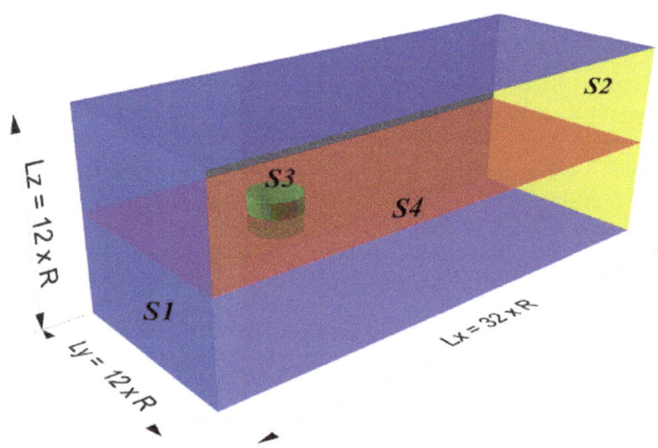

Figure 8: Three-dimensional computational domain.

The mesh is characterized by tetrahedral elements. In particular, the maximum element size for the fixed domain is set to 0.5 m; the length of the element at the interface is equal to 0.05 m, whereas the minimum element size is 0.003 m and is used to discretize the fluid domain on the turbine wall; see Figure 9. The number of elements and nodes for the three-dimensional mesh are given in Table 3.

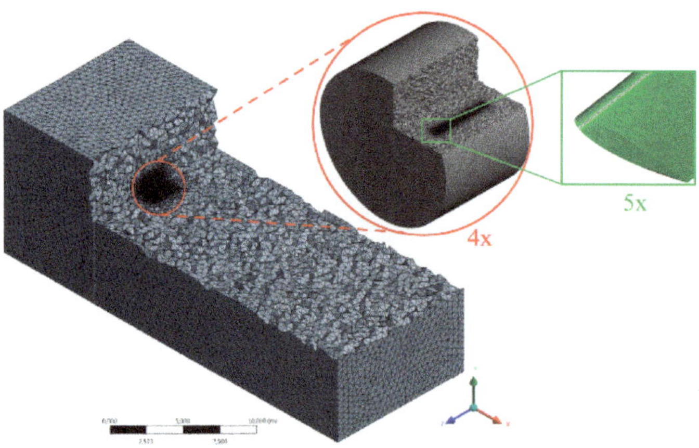

Figure 9: Three-dimensional mesh discretization.

Table 3: Number of three-dimensional grid nodes and elements

	Straight ($\psi = 0°$)	**Helical** ($\psi = 30°$)	**Helical** ($\psi = 60°$)	**Helical** ($\psi = 90°$)
Mesh Node	1,736,475	1,809,898	1,885,205	1,990,686
Mesh Element	9,503,471	9,933,335	10,367,297	10,967,904

In the following, results obtained from the three-dimensional analyses are shown. Static and transient analyses are carried out in this section to characterize the performances of the investigated wind rotors. The performance of Darrieus wind turbine is described in terms of moment coefficients, which are computed using procedures described in [29]. On the other hand, dynamical wind turbine performances are obtained by computing the average torque coefficients, over an operational cycle of the rotor.

Static 3D Comparison

Initially, static simulations are carried out to characterize the behavior of the turbine blade considering rigid rotations with steps of 10°. Because of the rotational symmetry of the geometry, only one blade is considered. Static results for Blade 2 and Blade 3 differ from Blade 1 data for a phase shift of +120° and −120°, respectively. The curves corresponding to the single blade are shown in Figure 10b, respectively for the Straight and Helical configurations. Figure 10a instead shows the three-blade curve trend. It can be seen that the curve of the Helical blade is smoother than that of the Straight blade. This is due to the fact that the Straight blade has an optimal angular position with respect to the relative wind corresponding to 30°, while the Helical blade always presents a section in the optimal position with respect to the relative wind. This becomes more noticeable with increasing twist angle. Moreover, the mean torque for the Helical blade results are greater than the one for the Straight blade for the Helical(ψ = 60°) and Helical(ψ = 90°) configurations. Instead, there is a decrease in mean performance for Helical(ψ = 30°), as shown in Table 4.

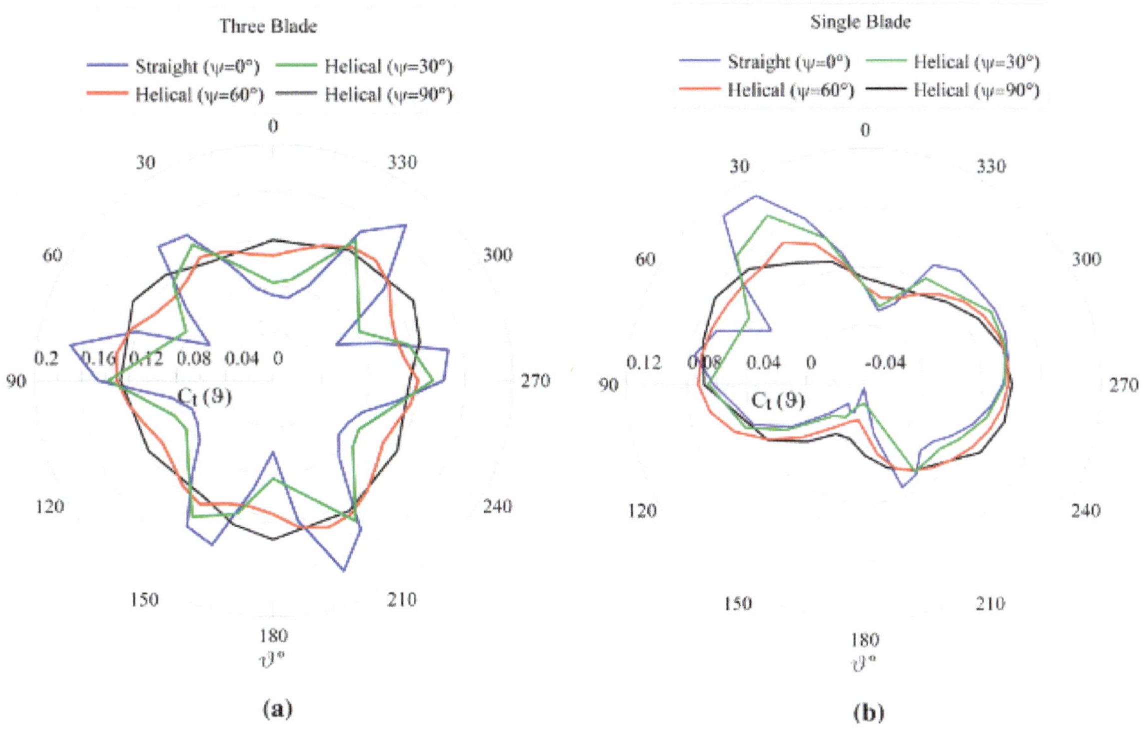

Figure 10: Static performance behavior for different azimuthal position θ values. (**a**) Turbine static three-dimensional comparison; (**b**) single-blade static three-dimensional comparison.

Table 4: Comparison of the average torque coefficient between ψ = 0°, 30°, 60°, 90°

Configuration	**Turbine C_t Average**	**Variance**	**Blade C_t Average**	**Variance**	**100 $\left(\frac{C_{tH}}{C_{tS}} - 1\right)$** [1]
Straight ($\psi = 0°$)	0.107	0.001282	0.0359	0.001347	-
Helical ($\psi = 30°$)	0.105	0.000500	0.0352	0.001109	−2.09
Helical ($\psi = 60°$)	0.117	0.000077	0.0391	0.000701	8.75
Helical ($\psi = 90°$)	0.124	0.000049	0.0415	0.000588	15.53

[1] C_{tH} refers to *Helical* C_t average; C_{tS} refers to *Straight* C_t average.

To decide which one of the helical configurations would be the best, not just consideration of the numerical performances should be done. In fact, although a comparison between the ψ = 60° and ψ = 90° Helical VAWT could lead to choosing the ψ = 90° as the best one, the technological point of view must be kept in mind also when manufacturing and installing such turbines. Complications arise in manufacturing highly twisted blades as the ψ = 90°, especially if a fiber composite is used as the bulk material. Furthermore, as the twist angle increases, also the weight increases: this means difficult installing procedures and higher loads on rotating supports. The improvements in performance given in Table 4 from ψ = 60° to ψ = 90° are not so significant to justify the complications previously mentioned. The goal of the paper is to perform a preliminary comparative numerical analysis to be confirmed with a future experimental campaign on turbine prototypes. For these reasons, it has been decided to choose the ψ = 60° turbine as the one to be studied in the following dynamic three-dimensional analysis. Hereinafter, the ψ = 60° VAWT will be the only Helical turbine compared to the Straight one.

Dynamic Behavior

In order to define the time step size for three-dimensional simulations, Equations (3) and (4) are invoked for a rotational speed of 60 RPM resulting in Δt = 0.0017 s and Δt = 0.03 s, respectively. Three different values of step sizes are selected, namely Δt = 0.002, 0.005, 0.01 s. The performance characteristics for the selected time step size are thus investigated. From the results shown in Figure 11, it can be seen that time step equal to 0.01 s allows one to obtain matching results with the finer step sizes. Moreover, the choice of Δt = 0.01 s allows one to reduce the calculation time considerably.

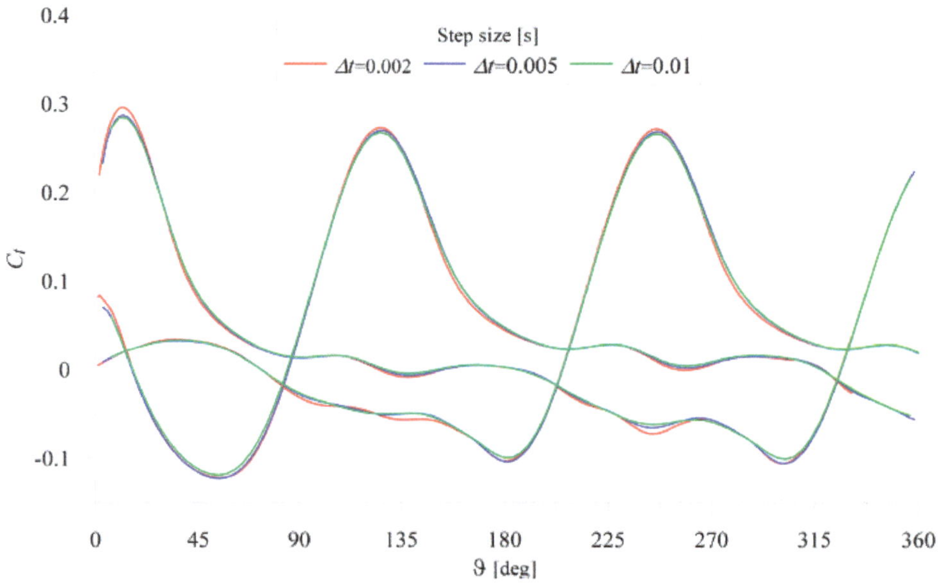

Figure 11: Torque coefficient for different time step sizes.

Dynamical wind turbine performances are obtained by computing the instantaneous torque coefficient C_t. Moreover, average torque and power coefficients over an operational cycle of the rotor are shown at the end of this subsection.

Figures 12 and 13 show the trend of the moment coefficient over a full operational cycle. The polar distributions refer to the sum of the three blades and to three different values of RPM, namely 60, 90 and 120.

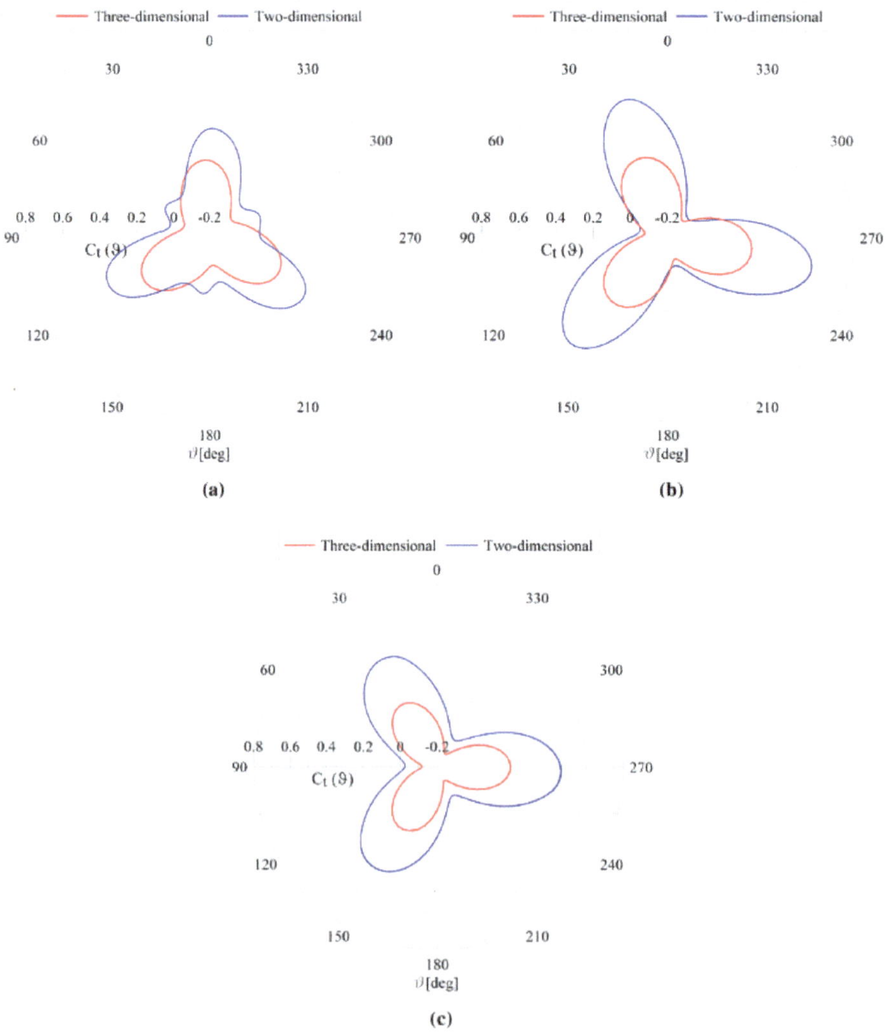

Figure 12: Two-dimensional and three-dimensional dynamic behavior for a Straightblade. (**a**) C_t 60 RPM; (**b**) C_t 90 RPM; (**c**) C_t 120 RPM.

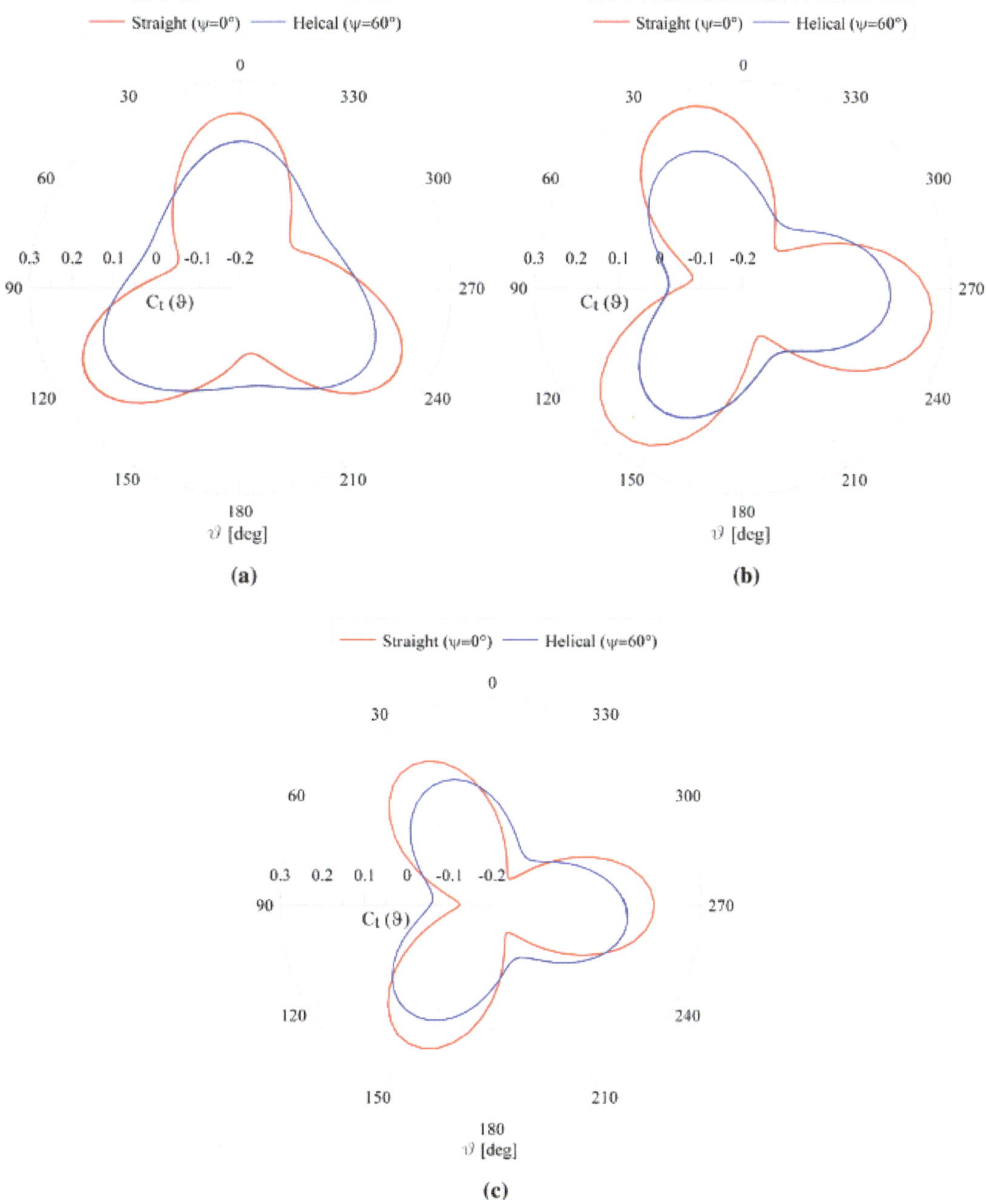

Figure 13: Instantaneous torque coefficient Straight blades with respect to Helicalblades. (**a**) Rotor polar behavior in terms of $C_t(\theta)$, corresponding to a 60-RPM value; (**b**) rotor polar behavior in terms of $C_t(\theta)$, corresponding to a 90-RPM value; (**c**) rotor polar behavior in terms of $C_t(\theta)$, corresponding to a 120-RPM value.

In Figure 12, the comparison between two-dimensional simulations with respect the three-dimensional Straight blade data is presented, whereas a comparison between the three-dimensional Straight blade and Helical($\psi = 60°$) blade is given inFigure 13. From the results reported in Figure 12, it appears that a finite turbine three-dimensional blade has a performance reduction linked to vortex diffusion after flow separation; this is in accordance with the literature [17].

For the sake of completeness and to better highlight this effect, the vortex core regions are shown in Figure 14. It appears that as the rotational speed increases, vortex core regions extend more and more, influencing the upcoming blade. This phenomenon is more evident in the Straight blade geometry (Figure 14), where vortices accumulate mainly at blade tips. On the contrary, in the Helical blade geometry, vortices are evenly distributed along the blade trailing edges. Considering a three-dimensional blade, the Helical blade has a smoother behavior with respect to the Straight blade; as shown in Figure 13, the maximum value of $C_t(\theta)$ is obtained at about the same azimuthal position. This effect reduces with the increase of the angular velocity.

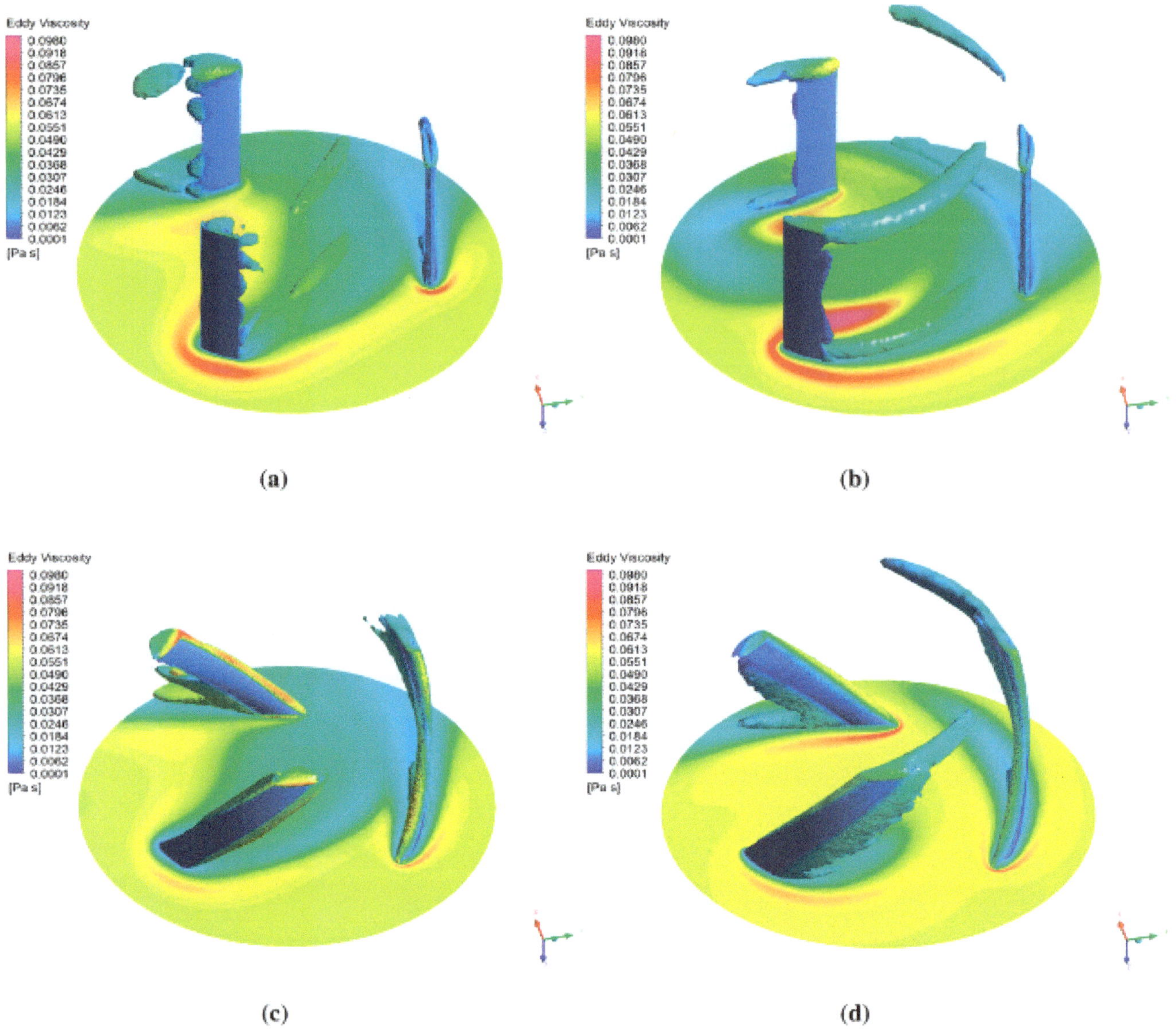

Figure 14: Tip vortex diffusion. (**a**) Straight ($\psi = 0°$) 60 RPM; (**b**) Straight ($\psi = 0°$) 120 RPM; (**c**) Helical ($\psi = 60°$) 60 RPM; (**d**) Helical ($\psi = 60°$) 120 RPM.

For a given free stream velocity, the moment coefficient of Helical blades is greater than the Straight blade configuration, as shown in Figure 15. For higher values of the RPM, the tip vortex diffusion leads to a deterioration of blade performances on the advancing blade with respect to the straight configuration.

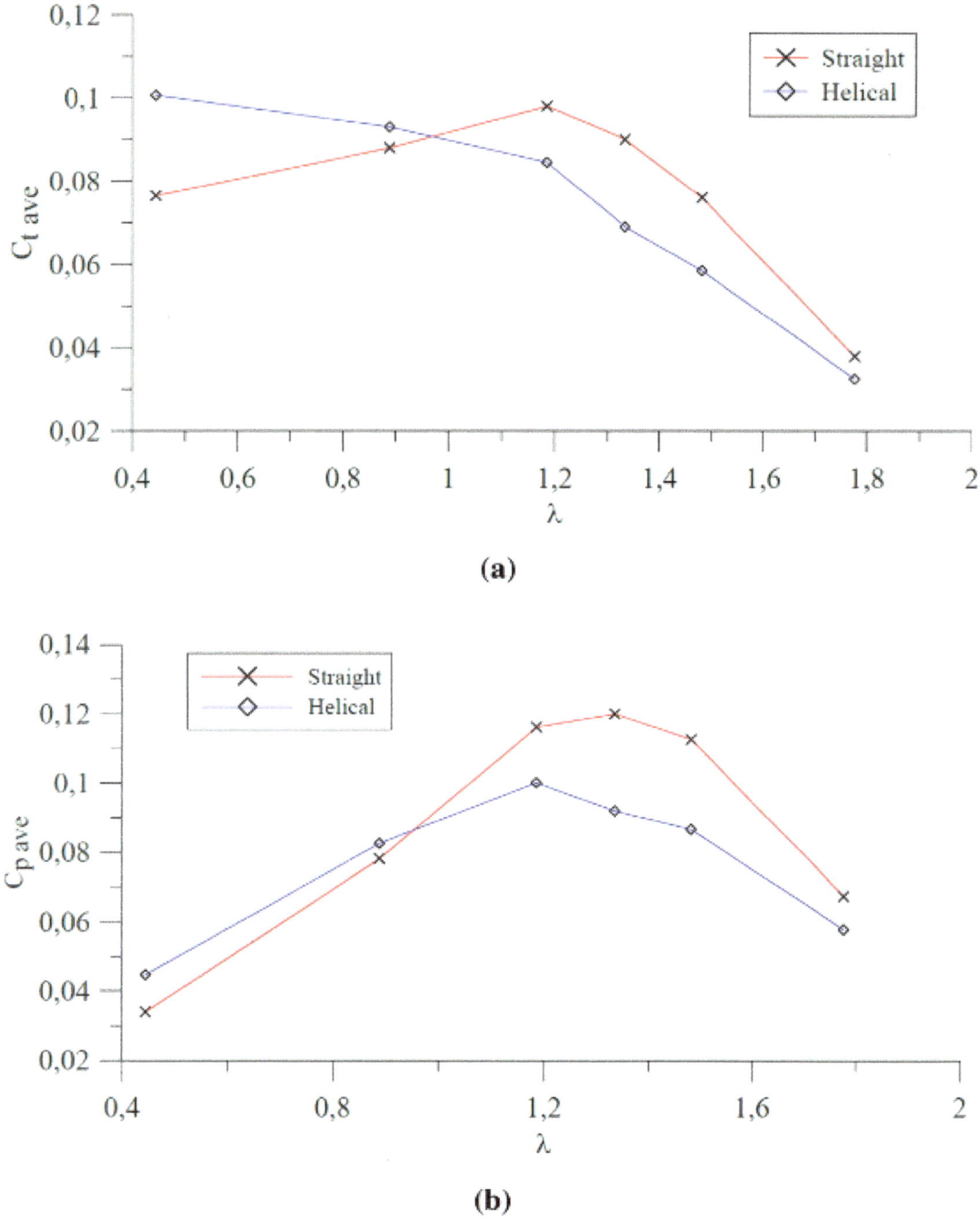

Figure 15: Comparison of the average torque and power coefficient for the Straight andHelical($\psi = 60°$) three-dimensional turbines. (**a**) Torque coefficient for Straight andHelical configurations; (**b**) power coefficient for Straight and Helical configurations.

At low rotational speed, the average C_{tave} for the helical blade turbine is higher with respect to the straight blade up to $\lambda = 1$, where the curves intersect. The straight blade configuration produces higher C_{tave} as the rotational speed increases beyond this value. In accordance with static results for the given conditions, the helical turbine shows a lower cut-in speed.

CONCLUSIONS

In the present paper, a comparative evaluation of a small Helical-VAWT and of a Straight-VAWT has been performed. Preliminary two-dimensional and subsequent three-dimensional CFD analyses have been carried out. Dynamic two-dimensional simulation data have been collected at different tip speed ratio values, and the results show the torque, lift and drag coefficient behavior of the rotor blade over a revolution with respect to the azimuthal position. Subsequently, three-dimensional static analyses of four different configurations, $\psi = 0°$, 30°, 60°, 90°, have been performed. Among these, only two configurations, Straight ($\psi = 0°$) and Helical ($\psi = 60°$), were used in the dynamic analyses. From steady simulations, over a complete cycle, it has been observed that the Helical blades have an average torque coefficient increment equal to 8.75%. This leads the Helical VAWT to a low speed of start-up (this is confirmed by dynamic results at low rotational speed) and then to a higher number of operational hours in similar environmental conditions. Furthermore, it has been shown, by comparing three-dimensional with two-dimensional results, that the generation of the tip vortices greatly reduces the aerodynamic performance. However, for a three-dimensional comparison, it is shown that this effect is less relevant in theHelical VAWT than in the Straight one.

AUTHOR CONTRIBUTIONS

All authors contributed equally to this work. All authors were involved in the design of the model and discussion of the results. The manuscript is written by A. Esposito and revised and commented by Andrea Alaimo and Calogero Orlando. The simulation were done by Antonio Esposito with the supervision of Antonio Messineo and Davide Tumino.

REFERENCES

1. Darrieus, G. Turbine Having Its Rotating Shaft Transverse to the Flow of the Current. U.S. Patent 1,835,018, 8 December 1931.
2. Sutherland, H.J.; Berg, D.E.; Ashwill, T.D. A Retrospective of VAWT Technology; Sandia Report, SAND2012-0304; Sandia National Laboratories: Albuquerque, NM, USA; January; 2012.
3. Islam, M.; Ting, D.S.K.; Fartaj, A. Aerodynamic models for Darrieus-type straight-bladed vertical axis wind turbines.Renew. Sustain. Energy Rev. 2008, 12, 1087–1109.
4. Templin, R.J. Aerodynamic Performance Theory for the NRC Vertical-Axis Wind Turbine; Technical Report N-76-16618; LTR-LA-160; National Aeronautical Establishment: Ottawa, ON, Canada; 1; June; 1974.
5. Paraschivoiu, I. Double-multiple streamtube model for studying vertical-axis wind turbines. J. Propuls. Power. 1988, 4, 370–377.
6. Brusca, S.; Lanzafame, R.; Messina, M. Design of a vertical-axis wind turbine: How the aspect ratio affects the turbine's performance. Int. J. Energy Environ. Eng. 2014, 5, 333–340.
7. Strickland, J.; Webster, B.; Nguyen, T. A vortex model of the Darrieus turbine: An analytical and experimental study.J. Fluids Eng 1979, 101, 500–505.
8. Wang, L.; Zhang, L.; Zeng, N. A potential flow 2-D vortex panel model: Applications to vertical axis straight blade tidal turbine. Energy Convers. Manag. 2007, 48, 454–461.
9. Sheldahl, R.E. Comparison of field and wind-tunnel Darrieus wind-turbine data. J. Energy. 1981, 5, 254–256.
10. Bedon, G.; Castelli, M.R.; Benini, E. Experimental tests of a vertical-axis wind turbine with twisted Blades, Proceedings of the 15th International Conference on Mechanical, Industrial, and Manufacturing Engineering, Istanbul, Turkey, 5–6 December 2013.
11. Biadgo, M.A.; Simonović, A.; Komarov, D.; Stupar, S. Numerical and analytical investigation of vertical axis wind turbine. FME Trans 2013, 41, 49–58.
12. Strickland, J.H. Darrieus Turbine: A Performance Prediction Model Using Multiple Streamtubes; Technical Report; Sandia Labs.: Albuquerque, NM, USA, 1975.
13. Wolfe, W.P.; Ochs, S.S. Predicting Aerodynamic Characteristic of Typical Wind Turbine Airfoils Using CFD; Technical Report; Sandia National Labs.: Albuquerque, NM, USA, 1997.
14. Launder, B.; Spalding, D. Lectures in Mathematical Models of Turbulence; Academic Press: London, UK, 1972.
15. Carrigan, T.J.; Dennis, B.H.; Han, Z.X.; Wang, B.P. Aerodynamic shape optimization of a vertical-axis wind turbine using differential evolution. ISRN Renew. Energy. 2012.
16. Li, C.; Zhu, S.; Xu, Y.L.; Xiao, Y. 2.5 D large eddy simulation of vertical axis wind turbine in consideration of high angle of attack flow. Renew. Energy. 2013, 51, 317–330.
17. Hofemann, C.; Ferreira, C.S.; Dixon, K.; van Bussel, G.; van Kuik, G.; Scarano, F. 3D Stereo PIV study of tip vortex evolution on a vawt, Proceedings of the European Wind Energy Conference and Exhibition (EWEC), Brussels, Belgium, 31 March–3 April 2008.
18. Raciti Castelli, M.; Englaro, A.; Benini, E. The Darrieus wind turbine: Proposal for a new performance prediction model based on CFD. Energy 2011, 36, 4919–4934.
19. Castelli, M.R.; Benini, E. Effect ofblade inclination angle on a Darrieus wind turbine. J. Turbomach. 2012, 134, 031016.
20. Lanzafame, R.; Mauro, S.; Messina, M. 2D CFD modeling of H-darrieus wind turbines using a transition Turbulence Model. Energy Procedia 2014, 45, 131–140.
21. Iida, A.; Kato, K.; Mizuno, A. Numerical simulation of unsteady flow and aerodynamic performance of vertical axis wind turbines with LES, Proceedings of the 16th Australasian Fluid Mechanics Conference (AFMC), Gold Coast, Australia, 2–7 December 2007; pp. 1295–1298.
22. Ferreira, C.S.; Bijl, H.; Van Bussel, G.; van Kuik, G. Simulating dynamic stall in a 2D VAWT: Modeling strategy, verification and validation with particle image velocimetry data. J. Phys. Conf. Ser. 2007, 75.
23. Amet, E.; Maître, T.; Pellone, C.; Achard, J.L. 2D numerical simulations of blade-vortex interaction in a darrieus turbine. J. Fluids Eng 2009, 131.
24. Ekaterinaris, J.A.; Platzer, M.F. Computational prediction of airfoil dynamic stall. Progress Aerosp. Sci. 1998, 33, 759–846.
25. Qian, W.; Fu, S.; Cai, J. Numerical study of airfoil dynamic stall. Acta Aerodyn. Sin. 2001, 19, 427–433.
26. Howell, R.; Qin, N.; Edwards, J.; Durrani, N. Wind tunnel and numerical study of a small vertical axis wind turbine.Renew. Energy. 2010, 35, 412–422.
27. Zhang, L.; Liang, Y.; Liu, X.; Jiao, Q.; Guo, J. Aerodynamic performance prediction of straight-bladed vertical axis wind turbine based on CFD. Adv. Mech. Eng. 2013, 2013.
28. Alaimo, A.; Esposito, A.; Milazzo, A.; Orlando, C.; Trentacosti, F. Slotted blades savonius wind turbine analysis by CFD. Energies 2013, 6, 6335–6351.
29. FLUENT Manual—ANSYS Release Version 14.5, User's Guide; ANSYS Inc.: Canonsburg, PA, USA, 2012.

30. FLUENT Manual—ANSYS Release Version 14.5, Theory Guide; ANSYS Inc.: Canonsburg, PA, USA, 2012.

31. Chen, H.; Patel, V. Near-wall turbulence models for complex flows including separation. AIAA J. 1988, 26, 641–648.

32. FLUENT Training Material—ANSYS Release Version 14.5, Lecture Transient Flow Modeling, Introduction to ANSYS Fluent; ANSYS Inc.: Canonsburg, PA, USA, 2012.

33. Paraschivoiu, I. Wind Turbine Design: With Emphasis on Darrieus Concept; Polytechnic International Press: Montreal, QC, Canada, 2002.

CHAPTER
20

NUMERICAL AND EXPERIMENTAL ANALYSIS OF MICRO HAWTS DESIGNED FOR WIND TUNNEL APPLICATIONS

R. Lanzafame[1], S. Mauro[1], M. Messina[1]

[1]Department of Industrial Engineering, University of Catania, Viale A. Doria, 6, 95125 Catania, Italy

ABSTRACT

In this paper the authors describe a design and optimization process of micro HAWTs using a numerical and experimental approach. An in-house 1D BEM model was used to obtain a first geometrical draft. It allowed to quickly optimize blade geometry to maximize energy production as well. As these models are quite sensitive to airfoil coefficients, above all at low Reynolds numbers, an accurate 3D CFD model was developed to support and validate the 1D BEM design, analyzing and fixing the discrepancies between model output. The 3D CFD model was developed and optimized using ANSYS Fluent solver and a RANS transition turbulence model. This allowed to correctly reproduce the transition and stall phenomena that characterize the aerodynamic behavior of micro wind turbines, solving the issues related to low Reynolds flows. The procedure was completed, thus building two micro HAWTs with different scales, testing them in the subsonic wind tunnel of the University of Catania. Wind tunnel features, experimental set-up and testing procedures are presented in the paper. Through the comparison of numerical CFD and experimental test results, a good compatibility was found. This allowed the authors to analyze and compare numerical calculation results and verify blockage effects on the prototypes as well.

INTRODUCTION

The growing need for sustainable energy generation has led to the development of huge wind turbine rotors. However, this is not the only way to generate more electrical power from wind. The design and optimization of mini and micro wind turbines is also becoming an essential element to this field of research due to the increasing interest in smart grid generation and widespread generation of electrical power. It means that it is possible to generate great electrical power through large rotors or through the use of numerous mini and micro rotors, which are designed and optimized for local electrical power generation. Basically, the widespread generation of electrical power is a very smart way to produce energy from wind because it can reduce the visual and environmental impact, maximizing annual energy production in the same way as the large rotors do.

However, design and optimization of mini and micro wind turbines require an in-depth study of fluid dynamics due to low Re effects like transition, laminar separation and laminar bubbles.

These phenomena are not of the utmost importance for the large rotors due to operation in high turbulent flow conditions; however, they are quite important in low Re flows, because

Citation: R. Lanzafame, S. Mauro, M. Messina, Numerical and experimental analysis of micro HAWTs designed for wind tunnel applications, DOI: 10.1007/s40095-016-0202-8.

they greatly influence performance and thus have to be taken into account in the aerodynamic modeling of micro rotors.

In this paper the authors describe the development strategy for high accurate design of micro wind turbines, operating in low Re flows, using 1D BEM modeling, 3D CFD computations and wind tunnel experiments. Following previous works [1–6], the 1D BEM model was used to generate a first geometrical draft, defined as the twisting and tapering action of the blade to maximize energy production, based on wind distributions. The widely known problems with 1D BEM model are the necessity for accurate experimental data of airfoil aerodynamic coefficients and the limitations of the 1D modeling in reproducing 3D effects.

To overcome the limitations of 1D modeling, the authors developed a 3D CFD model, using the ANSYS Fluent solver and a four equations RANS transition turbulence model, therefore optimizing it for wind turbine applications [7]. The CFD model was validated, showing excellent compatibility with experimental data. Subsequently, this model may be used to analyze the effects of the aerodynamic coefficients on the 1D model. Moreover, the wind tunnel experiment results were compared to the numerical results in terms of torque, power and power coefficient. Therefore, in designing the rotors with the use of both models, it will then be possible to greatly reduce the prototyping phase and wind tunnel experimentation, thus saving time and costs.

The entire procedure was validated through the design of two micro rotors to scale, using the 1D BEM model and analyzing them with the 3D CFD model. The larger prototype was made with wood whereas the smaller one was made with ABS in a 3D printer. Both prototypes were tested in the (0.5 × 0.5 × 1.2 m) wind tunnel using an AC brushless servo system to measure torque and rotational speed with a pc—Labview PCB interface. A hot-wire anemometer and a five hole Pitot probe were used to measure the flow speed in the open test section wind tunnel. In this way it was possible to verify the blockage effects on rotor performance as well.

MATHEMATICAL MODELS

Review of the in-house 1D BEM model

The 1D BEM based models are powerful tools for obtaining design drafts in terms of diameter, twist, taper and rotational speed thus quickly optimizing rotor performance of HAWTs.

The in-house 1D code developed by the authors (see [1–6]), is based on BEM theory and Glauert propeller theory with suitable modifications for application to wind turbines.

In particular, 3D effects like centrifugal pumping were taken into account using a suitable mathematical representation of the lift and drag coefficients [6] whereas, for the tip losses, a modified Prandtl function was implemented [1, 2, 4, 5].

The correct evaluation of axial and tangential induction factors was obtained comparing experimental measurements and numerical simulations as well [1].

The code was validated using experimental data from literature [9], comparing generated and simulated power and torque as a function of wind speed [1]. In Fig. 1 a power comparison is presented for the NREL Phase VI wind turbine. Comparing numerical and experimental data [9] an excellent compatibility can be seen.

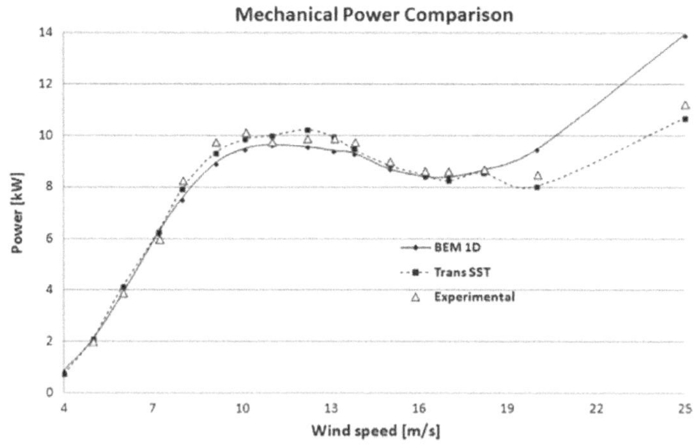

Figure 1: NREL phase VI. Comparison between experimental and simulated power.

As the 1D model was validated, it was used in this paper in a reverse way, defining a power target, based on a hypothetical PDF wind distribution. First of all the design was optimized maximizing the C_l/C_d ratio, it was then refined to optimize the power curve and the performance for maximizing annual energy production [5]. The final design was obtained in terms of twist and taper, choosing the optimal airfoil as well.

The most significant problem was the lack of airfoil experimental data in scientific literature, above all at very low Reynolds numbers. As BEM models are strictly dependent on the experimental airfoil coefficients, it was not a simple task to adequately scale-down the prototype fluid dynamic design for the wind tunnel size requirements. This issue was successfully resolved thorough the development of both a CFD and experimental approach. The 3D CFD model was optimized for transitional flows at very low Re and the results were tested in the wind tunnel.

Review of the CFD 3D model

CFD modeling is widely used in industrial and scientific research, especially with the superior performance of calculators in modern technology. This has led to the possibility of simulating numerous important aerodynamic phenomena related to wind turbines, thus aiding in the improvement of researchers' designs and optimizations.

However, it is well known that the best results are obtained when large rotors are simulated. This is because RANS "full turbulent" modeling works very well at higher Reynolds numbers, reproducing fully developed turbulence [10–16].

As the mini and micro rotors operate at low Reynolds numbers, the fully turbulent modeling lacks the capability to adequately simulate this kind of flow-field and rotor performance. Laminar bubble, laminar separation and boundary layer transition are quite important phenomena that must be captured by the CFD model to reproduce the actual physical aerodynamic behavior. In general, the traditional full turbulent modeling leads to an overestimation of fluid-dynamic torque and power. This is due to the earlier separation of the laminar boundary layer in incipient stall region compared to a fully developed turbulent flow [17–22]. Moreover, the need to scale-down the prototypes for wind tunnel testing, led to the necessity to further examine the study of laminar and transitional flows.

For these reasons the authors developed a strategy to generate a HAWT high accurate 3D CFD model [7], using the correlation based transition model of Menter [20, 23, 24], calibrated for airfoils which operate at low Re number flows.

To compare the results and to validate the procedure, a model of the NREL PHASE VI wind turbine was first implemented [7], calibrating the model using experimental data found in scientific literature [9]. In this way it was also possible to qualitatively compare the results with the 1D BEM calculations (Fig. 1).

A 3D CAD was used to accurately reproduce the rotor geometry, the file was then imported into the ANSYS Workbench multiphysics platform. The computational domain was generated using the ANSYS CAD interface 'Design Modeler'.

Spatial discretization was optimized doing a grid independence study, focusing on turbulence model limitations [23, 24] and geometric complexity. Specifically a $y^+ < 1$ and a maximum skewness value of 0.85 were reached. The mesh was later converted by Fluent solver from a tetrahedral to a polyhedral geometry. This led to a considerable reduction in the number of cells. As this type of cell considerably reduces mesh skewness, better alignment of the flow with the cell faces were obtained. Interpolation errors and false numerical diffusion decreased as well.

The innovative concept was the use of the transition SST turbulence model. To calibrate the transition model for wind turbine applications, a long process of optimizing the local correlation variables was carried out. Several simulations on typical wind turbine airfoils (S809, NASA ls421, NACA 4415) were performed using Fluent 2D. Inlet and outlet turbulence boundary conditions were also optimized (turbulent intensity and turbulent viscosity ratio).

In detail, the local correlation variables F_{length}, $Re_{\theta c}$, $Re_{\theta t}$ were modified through the use of a UDF, written in C language and interpreted by the Fluent solver [7].

A moving reference frame model was used for rotation in a steady state, pressure based, coupled solver. This was found to be the best balance between reasonable calculation time and accuracy of results. This was an important achievement in the obtaining of a quick response in design process and scientific research.

As the CFD 3D calculation results were in close proximity to experimental data, the methodology of generating and optimizing the CFD 3D model was considered valid. In Fig. 1 a comparison between Transition CFD calculations, 1D BEM results and experimental data [7] of generated power for the NREL PHASE VI turbine is also reported.

The NREL PHASE VI operates at Reynolds numbers of approximately 1 million. To demonstrate the validity of the methodology even at lower Reynolds numbers and validate the process of designing micro rotor for wind tunnel testing, two micro rotors with different scales were designed. They were subsequently analyzed using the proposed CFD modeling. The rotors were finally constructed and tested in the wind tunnel, owned by the University of Catania. In this way, it was possible to identify both the advantages and the limitations of the different numerical approaches, as related to the absence of experimental data for low Re flows, as well.

DESIGN AND NUMERICAL ANALYSIS OF MICRO HAWTS

Geometrical design of prototypes

The objective of this paper was to design and analyze the performance of two scaled micro rotors to fully examine both the capabilities and the limitations of 1D BEM and CFD 3D modelling for low Re applications. The rotors were then tested in the wind tunnel to evaluate the blockage effects as well.

Through previous works by the authors [3, 4], it was possible to choose the geometric characteristics of the turbines (rotor diameter, number of blades, airfoils, chord, pitch, twist and rotational speed). It was also possible to evaluate the forces acting on the blades, torque and power at the rotor shaft.

Choosing a NACA 4415 airfoil, the rotors were first optimized maximizing the C_l/C_d ratio, and hence the power coefficient. To make the machines suitable for a wide range of wind conditions, the twist was modified to flatten the power curve, thus increasing annual energy production [3,4].

One of the objectives was to examine the effects of the use of aerodynamic coefficients related to higher Reynolds numbers. As the only complete available experimental data were those reported in [25], these aerodynamic coefficients were used in the design process. These data refer to a Reynolds number of approximately 1 million while the prototypes operate at Re < 80,000.

Final features of the rotors are presented in Table 1. The rotors were named as rotor A, for the largest one, and rotor B, for the smallest one.

Table 1: Experimental rotors features

Rotor type	**Rotor A**	**Rotor B**
Number of blades	3	3
Blade airfoil	NACA 4415	NACA 4415
Rotor diameter (*D*) (m)	0.45	0.225
Hub diameter (*d*) (m)	0.108	0.054
Rotational speed (*n*) (r/min)	1000	2450
Wind speed (V_w) (m/s)	5–30	5–30

Tip speed ratio (λ)	1–7	1–7
On design power (W)	25	10
Chord (c) (m)	Variable	Variable
Twist (°)	Variable	Variable

Rotor A was made with wood, while rotor B was built in a 3D printer with ABS material. The construction clearly followed the twist and taper laws, defined by 1D BEM calculations, as reported in Fig. 2. Although rotor B was half the size of rotor A, the differences in twist and taper laws were related to the different operative conditions, defined in 1D BEM calculations. In this way it was possible to further examine different fluid dynamic behaviors.

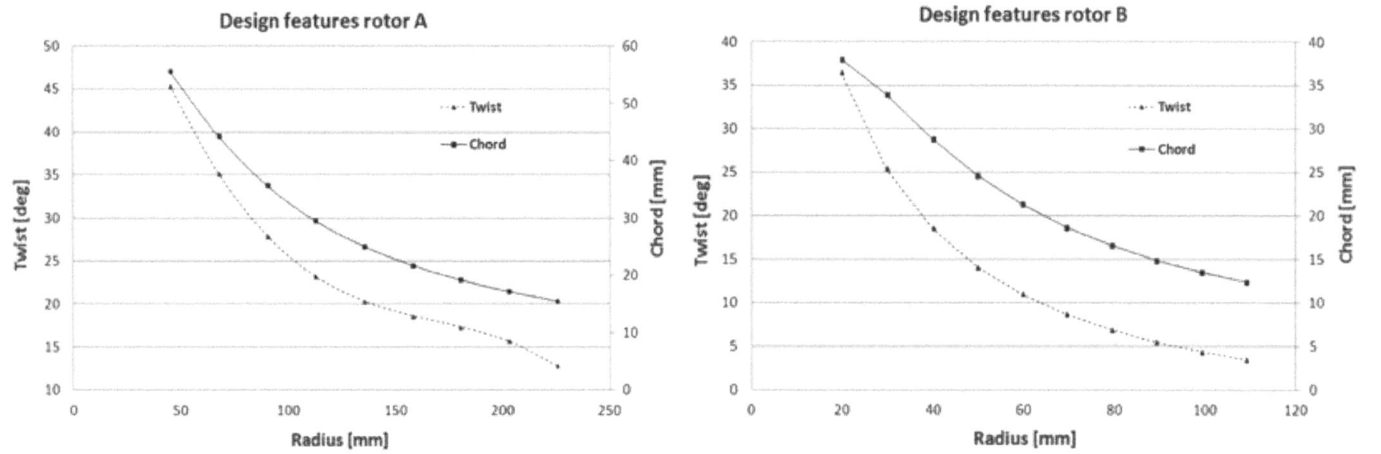

Figure 2: Twist and taper design: rotor A (left) and rotor B (right).

The prototypes are presented in Fig. 3. Both rotors were built with three separated blades, assembled with a central pentahedron as a shaft connection system and two plates for longitudinal structural connections (exploded view Fig. 3, right). An aerodynamic nose cone completed the assembly.

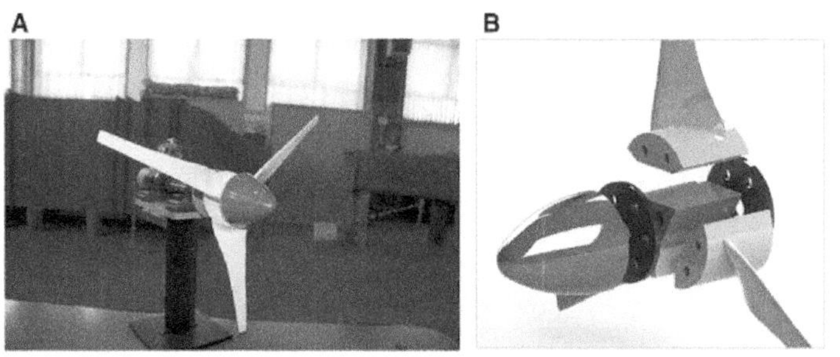

Figure 3: Fully assembled rotor A (left) and exploded view of rotor B (right).

Numerical analysis of rotor performance

Once the features of the rotors were defined, they were analyzed both with 1D BEM and CFD 3D models, comparing the results in terms of generated fluid dynamic power.

1D BEM performance simulation analysis was quick and simple to implement. Once the geometric characteristics and the rotational speed were known, wind speed was simulated from 5 to 25 m/s, thus obtaining power curves and power coefficient curves in very few seconds. The results are presented in Fig. 4 for both types of rotors along with the CFD 3D calculation results for immediate comparison.

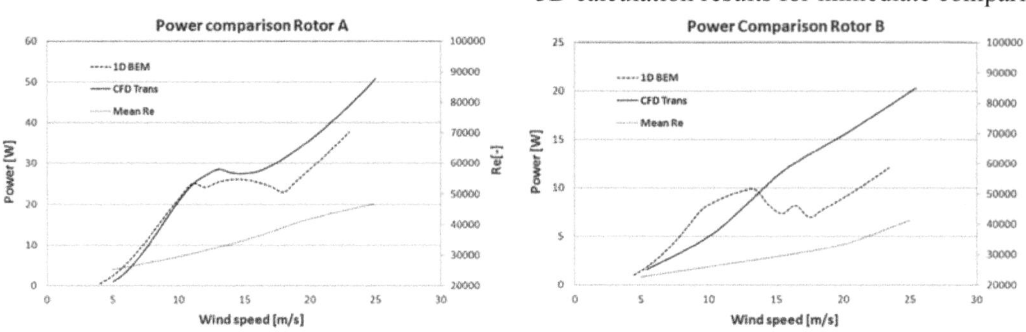

Figure 4: Calculated fluid dynamic power comparison and Reynolds number trends for rotor A (left) and rotor B (right).

The 3D CFD models instead, require all the steps presented in paragraph 2.2 as reported in a previous work by the authors [7]. Specifically, an accurate grid independence study was carried out in both cases, reaching a $y^+ < 1$ in all simulations, as required by the transition turbulence model.

The optimal local correlation variables for very low Re flows were found after a lengthy calibration process. Several simulations on NACA 4415 airfoil were performed using Fluent 2D and the local variables were modified until the numerical aerodynamic coefficients matched the experimental one. Inlet and outlet turbulence boundary conditions were optimized as well (turbulent intensity and turbulent viscosity ratio) [7]. As complete experimental data forRe $< 10^5$ were not available in scientific literature, the calibration process results were compared to the data reported in [26].

All of the simulations were performed on a workstation HP z820, with 2 Intel Xeon E5-2695 and 6 cores for each, 128 GB of RAM memory and a NVIDIA TESLA K20c for GPU computing. To speed up calculation time, the grid was partitioned using a METIS technique on the 24 threads available. The use of the coupled solver allowed Fluent to enable the GPU computing, speeding up the calculations as well. The main features of the CFD models are summarized in Table 2.

Table 2: CFD 3D setup features

	Rotor A	**Rotor B**
Grid elements	~2 million	~900,000
Grid features	Polyhedral cells $y^+ < 1$ Skewness <0.85	
Turbulence boundary conditions		
Inlet	Tu = 0.1	
Outlet	TVR = 10	
Solution methods		
Solver	Steady state Pressure based Coupled	
Discretization method	Least squares cell based method for gradients Second order upwind discretization for equations	
Solution controls		
Relaxation factors	0.2	0.25
Courant number	50	100
Rotational model	Moving reference frame	
Time to convergence (h)	~1.5	~40 min

The simulation results are presented in Fig. 4, as a generated fluid dynamic power comparison between both models. The substantial differences are quite obviously due to the use of inadequate aerodynamic coefficients and limitations of 1D modelling. Specifically, a general overestimation of the BEM calculated power at lower wind speed and an underestimation at higher wind speed are evident.

It was possible to evaluate both of these effects on performance overcoming the 1D modelling limitations using a more powerful 3D approach. This will be fully assessed in the following paragraphs where experimental and numerical results will be compared to validate procedure and hypotheses.

EXPERIMENTAL SETUP: WIND TUNNEL FEATURES AND MEASURING CHAIN

Experimental tests on both types of rotors were carried out in the subsonic wind tunnel of the Department of Industrial Engineering (University of Catania, Italy). The wind tunnel was designed and calibrated by the authors [27], the main geometrical features are presented in Fig. 5.

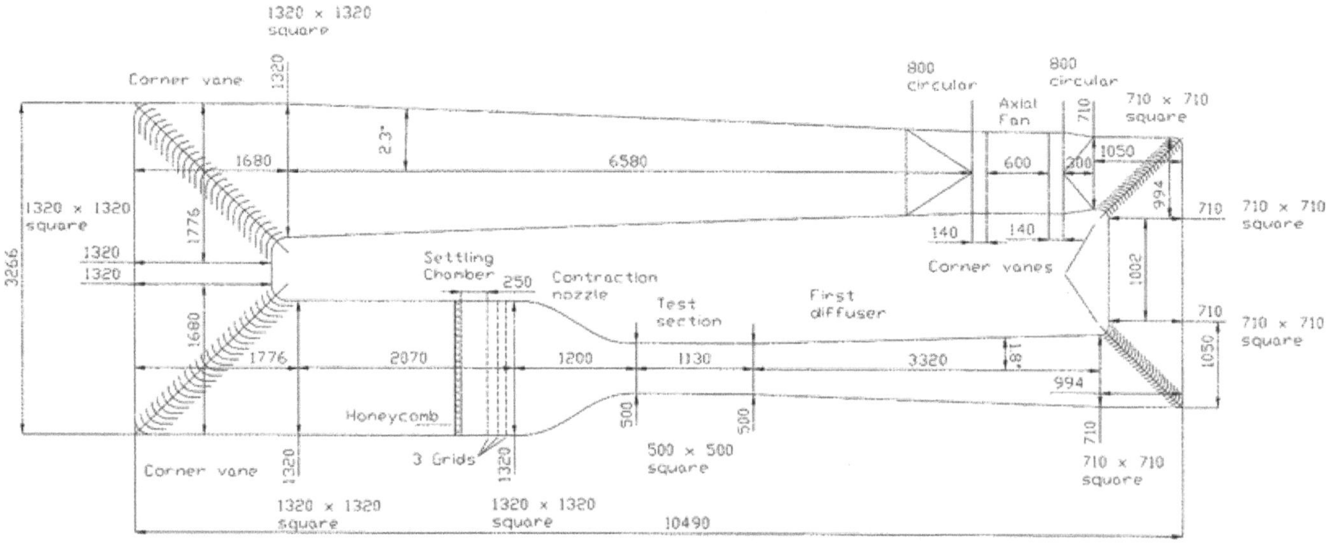

Figure 5: Wind tunnel geometrical features (units in mm).

It is a closed circuit wind tunnel with a test section of 0.5 × 0.5 × 1.2 m and it is possible to perform tests with an open or closed test section. There are four corner vanes to correctly guide the flow. An axial fan of 5.5 kW power peak at a maximum rotational speed of 1400 r/min is used to generate the flow. The fan speed is regulated using an inverter from 0 to 50 Hz. A settling chamber and a honeycomb with three screen levels are used to reduce turbulence in the test section. The tunnel was tested and calibrated by the authors following what is reported in [28–33]. The maximum flow speed possible is 30 m/s with closed test section and 27 m/s in the open test section configuration. Turbulence Intensity, measured at the contraction nozzle section was lower than 0.5 % at maximum speed [27]. The misalignment of the flow-field was also studied to adequately align the flow-field with the rotational axis of the rotors.

The measuring chain included: the use of a five hole Pitot probe to determine the real flow speed and direction; a hot wire anemometer to confirm the measured wind speed; a thermometer and a barometer to define the environmental conditions before each test. A Sanyo Denki AC brushless servo system was used for torque, power and rotational speed measurement with a National Instruments PCB–LabVIEW interface for acquisition and signal conditioning [8].

Specifically, the Sanyo Denki servo motor R2AA04010FXH1CM was controlled by the servo amplifier RS1A01AA through the National Instruments DAQ NI—6008 USB PCB to connect the servo amplifier with a pc where the LabVIEW interface allowed generation control signal and data acquisitions. A regenerative resistor was connected to the servo amplifier to dissipate braking energy. In Fig. 6 an outline of the control system is reported.

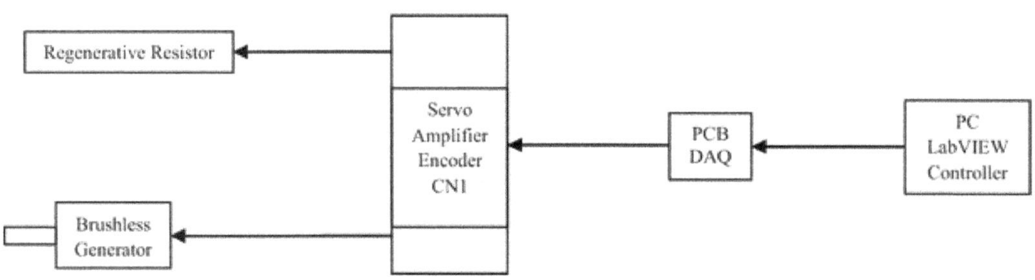

Figure 6: Torque, power and rotational speed control system.

Essentially, in Fig. 6, the control strategy was generated in LabVIEW. The signal was passed through the PCB DAQ to CN1 connection of the Servo Amplifier, that directly controlled the brushless generator in terms of rotational speed and braking torque. In this way the motor worked like a dynamic brake (generator) and the encoder dissipated the braking energy in the regenerative resistor. As the rotors were directly locked into the shaft, knowing the conversion efficiency of the brushless generator, the fluid dynamic torque was obtained in the PC LabVIEW Controller [8].

The control strategy was to fix the rotational speed and to vary the wind speed. In this way the system reacted by braking continuously to maintain the dynamic balance, dissipating surplus energy in the resistor, indicating the relative torque and hence generated power.

An example of the connection of rotor B to the generator shaft is presented in Fig. 7.

Figure 7: Connection between micro rotor B and generator shaft.

ANALYSIS OF THE RESULTS: EXPERIMENTAL VALIDATION

Experiments were first carried out for rotor A in open test section configuration. Five tests were carried out to verify the dispersion of the measurements. For each test, the rotational speed was fixed to the on-design value (n = 1000 r/min), through the servo system control while the flow speed was gradually increased from 0 to 26 m/s and then decreased. Resistive torque was obtained directly on the LabVIEW interface and therefore generated power was calculated.

The experimental and numerical power data were compared and reported in Fig. 8.

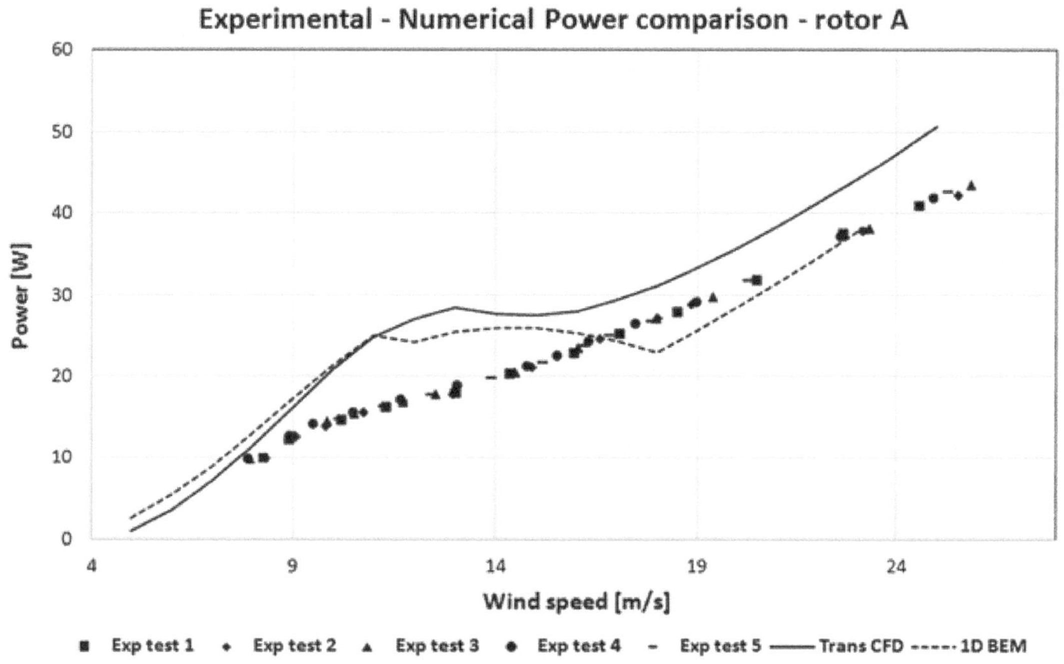

Figure 8: Experimental, CFD 3D and 1D BEM power comparison for rotor A.

It was possible for the rotor A measurements to be affected by blockage effects. This was due to an overly large blockage ratio, $B_r = \frac{A_{wt}}{A_{ts}} = 0.318$. To further examine this issue and validate numerical models, the experiments were replicated for rotor B $(B_r = \frac{A_{wt}}{A_{ts}} = 0.159)$, using the same methodology. The results are presented in Fig. 9 where the same generated and calculated power comparison is proposed.

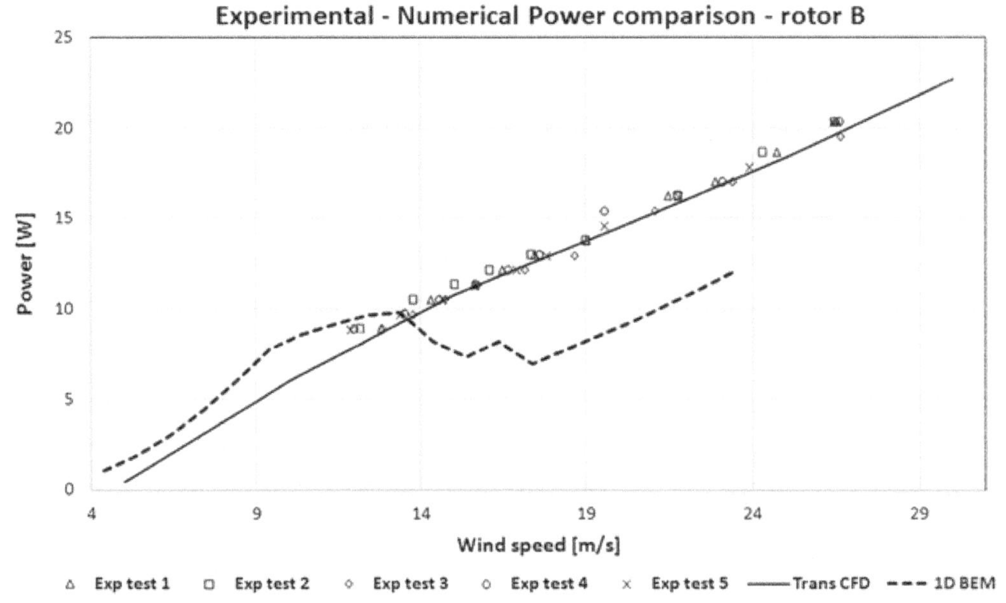

Figure 9: Experimental, CFD 3D and 1D BEM power comparison for rotor B.

Due to blockage effects, the experimental results for rotor A underestimated the actual power, as evidenced in Fig. 8. In an open test section, in fact, the divergence of the flow field, occurring at a short distance from the contraction nozzle and the wake expansion effects caused by the size of the larger rotor (A), lead to a slowing of the fluid flow and to a misalignment of the flow field [28]. Hence, the experiment for rotor A was useful in the verification of the limitations of rotor dimensions to avoid blockage effects.

The above is supported by the results presented in Fig. 9. In this case, in fact, CFD results were in close proximity to experimental data and blockage effects were definitely of minor importance as rotor B had a diameter which was half the diameter of rotor A. The optimized transition CFD 3D model was therefore considered valid even at very low Reynolds numbers and it can be used to analyze, and finally optimize, rotor geometry in a wide range of transitional flow conditions. It was useful in assisting wind tunnel testing in the proposed design process as well.

To further confirm the validation, a power coefficient comparison between CFD calculations and experimental data is presented in Fig. 10. The chart refers to rotor B at n = 2450 r/min.

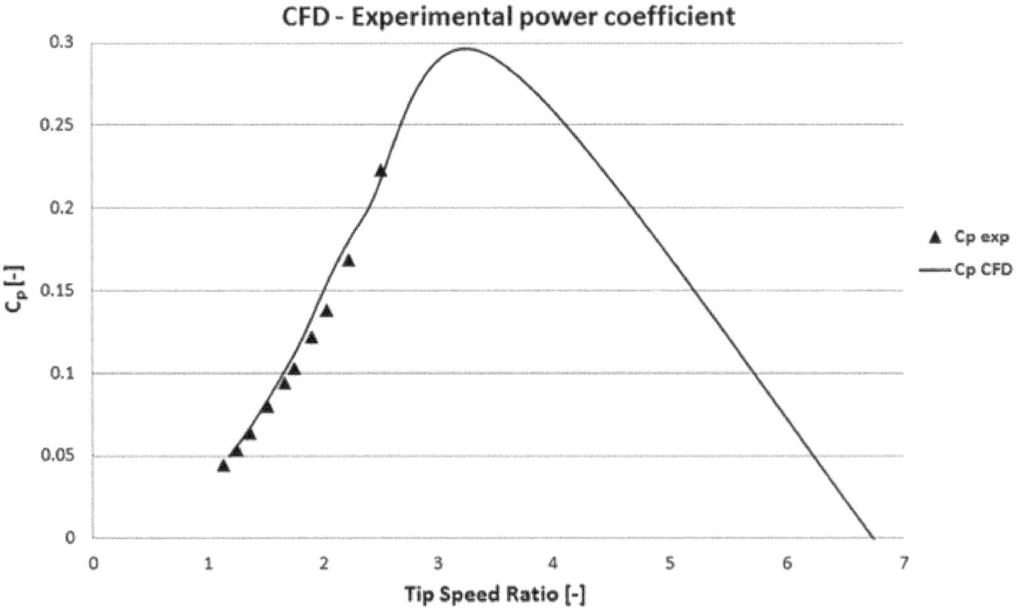

Figure 10: Power coefficient vs λ comparison between CFD and experimental results (rotor B).

The range of validation was limited by the maximum possible rotational speed of the servo system (n = 4500 r/min). For this reason, in Fig. 10, the experimental data are related to low λ. However, all the following considerations were made in the validation range of the CFD model.

Specifically, through the use of CFD post-processing, referring to the validated rotor B results, it was possible to analyze the discrepancies between both numerical models.

A comparison of calculated flow-field at different radial stations and different wind speeds was obtained using ANSYS CFD post-processor and is presented in Fig. 11. The sections were obtained using cylindrical surfaces at a distance of 0.026 m (near the hub), 0.06 m (medium radius) and 0.1 m (near the tip). Three different wind speeds were examined as well: low (10 m/s), medium (20 m/s) and high (30 m/s). By doing so, it was possible to analyze a wide range of fluid dynamic rotor behaviors.

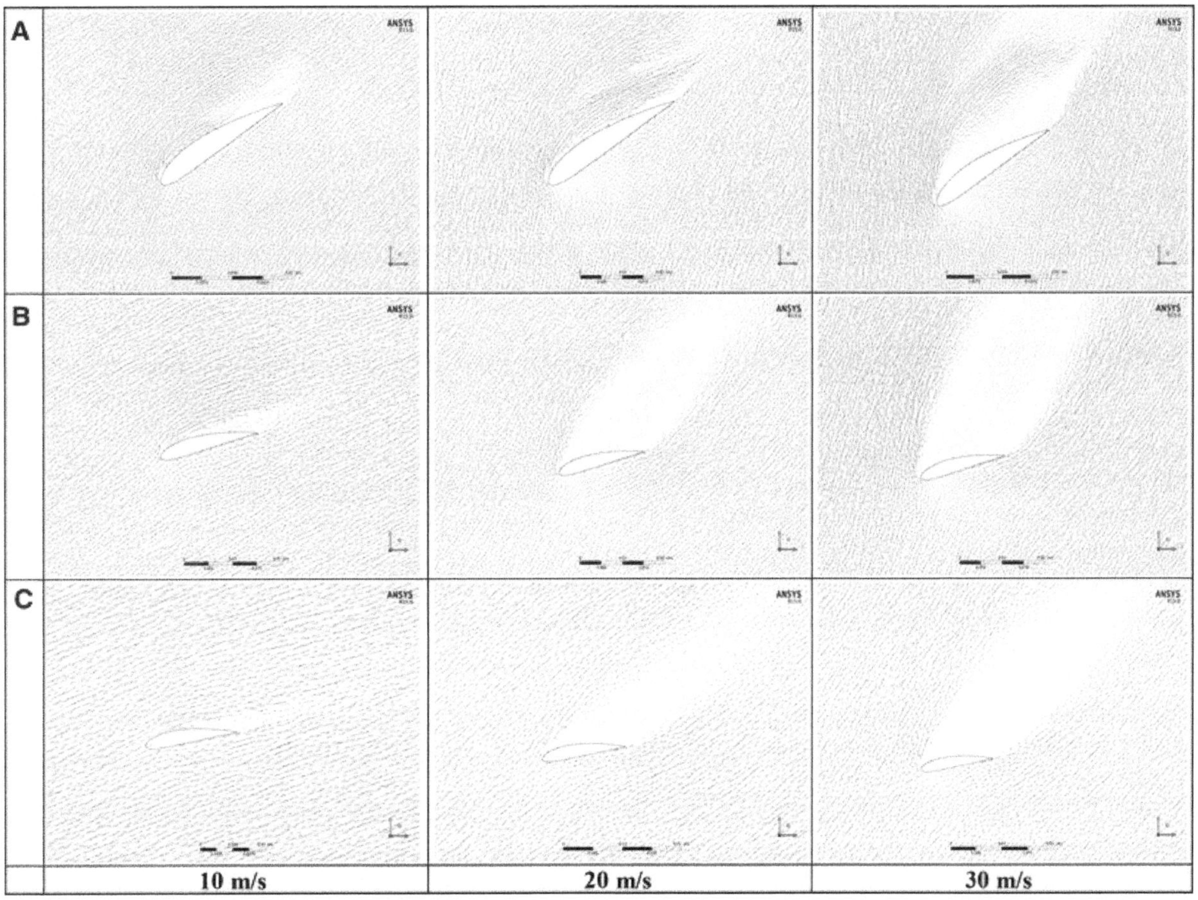

Figure 11: Calculated flow-field at different radial stations r = 0.026 m (**a**), r = 0.06 m (**b**), r = 0.1 m (**c**) for different wind speeds.

At a lower wind speed (Fig. 11, 10 m/s) the flow was mostly attached along the blades. Only laminar separation bubbles were detectable and this transitional behavior was the reason why the 1D BEM calculations overestimated power at lower wind speeds, as seen in Fig. 9. The excellent capabilities in both transitional modeling and behavior of the proposed CFD 3D model were clearly evident.

More interesting was the fluid dynamic behavior at medium and higher wind speeds. The flow in fact is fully separated along the blade, already at 15 m/s. The phenomenon is shown in Fig. 11(20; 30 m/s). The transition between attached and separated flow is abrupt as the leading edge stall characterizes low Reynolds flows. This is well captured by the CFD model. Specifically, the influence of the hub at r = 0.026 m, where the flow is accelerated and the stall vortex is flattened, is also evident. Moreover, extrapolating the sectional AoAs, post-processing CFD calculations [34, 35], it was possible to highlight a particular fluid dynamic behavior that influences rotor performance, as evidenced in Fig. 9.

In Fig. 12, a comparison between numerical predictions of AoA trends along the blade, for three different wind speeds, is presented. The way in which the 1D BEM model underestimates angles of attack along the entire blade has been highlighted. It is therefore possible to conclude that lift and drag coefficients will also be different than the actual ones.

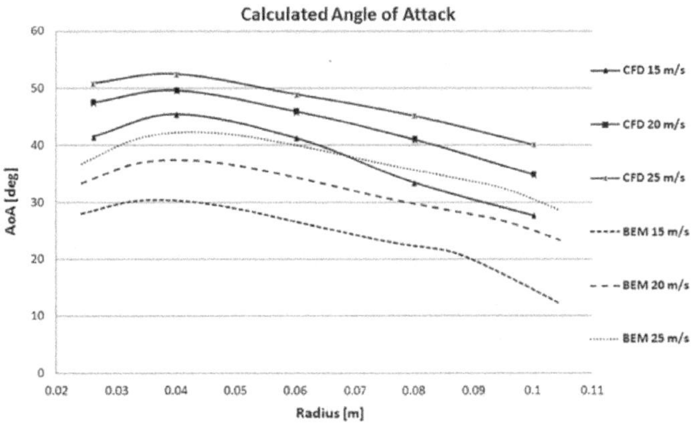

Figure 12: Comparison of CFD 3D and BEM calculated AoAs at different wind speeds (rotor B).

In fact, through the calculations of the 3D aerodynamic coefficients, from CFD simulations, and comparing them with 2D data, the effects of the centrifugal pumping can be evaluated. As is known from literature [35], the 3D coefficients are much higher than the 2D values even in deep stall conditions.

For example, at 20 m/s, r = 0.06 m (medium radius) and AoA = 46°, the calculated 3D aerodynamic coefficients were C_l = 1.49 and C_d = 1.42 (medium Reynolds number along the blade was 33,000) while corresponding experimental

2D values at Re = 10^6 are C_l = 1.39 and C_d = 1.32. This indicates that centrifugal pumping effects are clearly evident in deep stall conditions as well, increasing forces on the blades compared to 2D conditions.

In this way, two issues were highlighted. This is of the utmost importance because 1D BEM code is a fundamental tool in wind turbine design and optimization. Therefore, the joint use of a validated CFD 3D model and wind tunnel experiments will allow the BEM code to be refined.

Certainly, the errors introduced through the use of inadequate aerodynamic coefficients lead to errors in the optimal design and modeling. The authors will try to resolve this issue in future works, generating a database of 2D aerodynamic coefficients at very low Re numbers through the use of experimental and CFD calculations. However, the increase of aerodynamic coefficients, due to centrifugal pumping effects, even in deep stall conditions, suggests that it is necessary to further study this phenomenon. It is also important to take this phenomenon into account in the BEM code, modifying the proposed empirical equations [6] for centrifugal pumping in low Re conditions. This will also be examined in future works.

An example of radial flow along a blade was obtained using the ANSYS CFD-Post processing and is presented in Fig. 13 for a wind speed of 20 m/s, in deep stall condition. A centrifugal vortex is highlighted along the entire blade and this supports the hypothesis that centrifugal pumping heavily influences fluid dynamic behavior, even in fully stalled conditions.

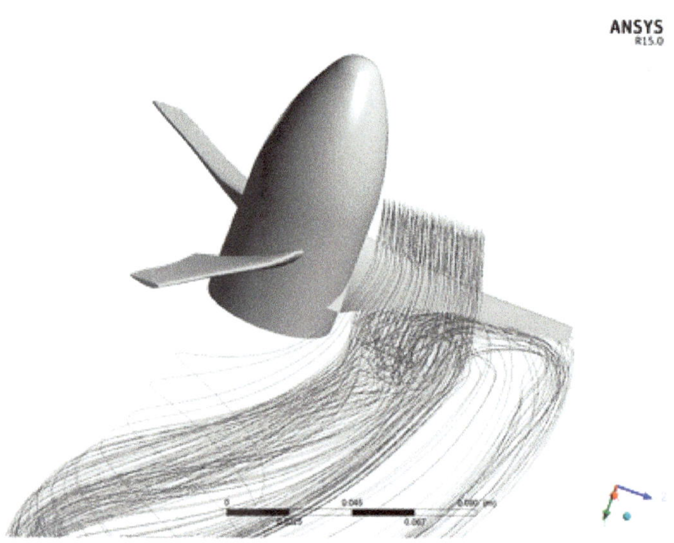

Figure 13: Streamlines of velocity highlighting radial flows along a blade (rotor B; V_w = 20 m/s).

CONCLUSIONS

In this paper a design strategy for low Re HAWTs is proposed and verified. Through the joint use of 1D BEM code, 3D CFD modeling and wind tunnel experiments, the authors examined the advantages and limitations of numerical modeling, thus resolving issues related to very low Reynolds number flows, leading the way for future research.

Using results from previous works, an in-house 1D BEM model was applied to obtain a first geometrical draft of two micro rotors for wind tunnel applications, thus optimizing the twist and taper of the blades and fixing the on-design rotational speed.

To fully examine the effects of the Reynolds number on performance, the authors developed a 3D CFD model, using and optimizing a SST transition turbulence model. The CFD model was validated in a previous work and was applied to verify its capabilities in reproducing very low Reynolds number flows as well.

The rotors were simulated with both of the numerical codes and the results were compared to wind tunnel experiments, carried out in the subsonic wind tunnel, belonging to the University of Catania.

In addition, rotor dimensions were chosen to verify the widely known blockage effects in the open test section of the wind tunnel. The largest rotor (rotor A), had a diameter which was twice the diameter of the other smaller rotor (rotor B). From the experimental results, an underestimation of power for rotor A, due to high blockage ratio, was verified. For rotor B, however, the power comparison showed a good correlation with 3D CFD computations, demonstrating good predictive capabilities of the modified transition turbulence model, even at very low Re conditions.

As highlighted by the CFD computation results, errors were induced in the 1D model through the use of unsuitable aerodynamic coefficients, related to higher Reynolds numbers. This led to errors in the optimal design and modeling. This is an important issue with the 1D BEM modeling that must be resolved to improve the good reliability of this code at low Reynolds number ranges as well. With the support of the proposed CFD model and wind tunnel experiments, a database of aerodynamic coefficients at low Re will be generated for implementation into the BEM model.

CFD results also showed that centrifugal pumping effects heavily influence fluid dynamic behavior along blades, increasing the 3D aerodynamic coefficients, even in deep stall conditions. This proved to be true not only in incipient stall regions (stall delay). In fact, comparing calculated angles of attack and the 3D aerodynamic coefficient values, it was possible to deduce that the combination of the widely known sensitivity to flow separation at low Re with the centrifugal effects increase the aerodynamic forces more than at higher Reynolds numbers.

Both of the aforementioned effects will be further examined in future works to define new equations. This definition is necessary to accurately take into account the 3D radial flow inside the 1D numerical code, based on BEM theory.

REFERENCES

1. Lanzafame, R., Messina, M.: Fluid dynamics wind turbine design: critical analysis, optimization and application of BEM theory. Renew. Energy-Elsevier Sci. 32(4), 2291–2305 (2007)
2. Lanzafame, R., Messina, M.: Design and performance of a double-pitch wind turbine with non-twisted blades. Renew. Energy-Elsevier Sci. 34(5), 1413–1420 (2009)
3. Lanzafame, R., Messina, M.: Optimal wind turbine design to maximize energy production. Proc. IMechE Part A: J. Power Energy 223(A2), 93–101 (2009)

4. Lanzafame, R., Messina, M.: Power curve control in micro wind turbine design. Energy Elsevier Sci. 35(2), 556–561 (2010). ISSN 03605442. doi:10.1016/j.energy.2009.10.025
5. Lanzafame, R., Messina, M.: Horizontal axis wind turbine working at maximum power coefficient continuously. Renew. Energy 35, 301–306 (2010)
6. Lanzafame, R., Messina, M.: BEM theory: how to take into account the radial flow inside of a 1D numerical code. Renew. Energy 39, 440–446 (2012)
7. Lanzafame, R., Mauro, S., Messina, M.: Wind turbine CFD modeling using a correlation-based transitional model. Renew. Energy 52, 31–39 (2013). (Elsevier)
8. Torresi, M., De Benedittis, A.F., Fortunato, B., Camporeale, S.M.: Performance and flow field evaluation of a Savonius rotor tested in a wind tunnel. Energy Proc. 45, 207–216 (2014). (Elsevier)
9. Lindenburg, C.: Analysis of the stationary measurements on the UAE phase-VI rotor in the NASA-Ames wind tunnel. ECN-C-03-025
10. Hansen, M.O.L., Sørensen, J.N., Voutsinas, S., Sørensen, N., Aa, H.: Madsen State of the art in wind turbine aerodynamics and aeroelasticity Elsevier -. Prog. Aerosp. Sci. 42, 285–330 (2006)
11. Sumner, J., Watters, C.S., Masson, C.: CFD in wind energy: the virtual, multiscale wind tunnel. Energies 3, 989–1013 (2010). doi:10.3390/en3050989
12. Hartwanger, D., Horvat, A.: 3D modelling of a wind turbine using CFD. In: Proceedings of NAFEMS Conference; NAFEMS: Cheltenham, UK (2008)
13. Gòmez-Iradi, S., Steijl, R., Barakos, G: Development and validation of a CFD technique for the aerodynamic analysis of HAWT. ASME J. Sol. Energy Eng 131(3), 031009 (2009). doi:10.1115/1.3139144
14. Bechmann, A., Sørensen, N.: CFD simulation of the MEXICO rotor wake. In: Proceedings of EWEC 2009; EWEA: Marseille, France (2009)
15. Benjanirat, S., Sankar, L.N.: Evaluation of turbulence models for the prediction of wind turbine aerodynamics. AIAA-2003-0517
16. Miller, A., Chang, B., Issa, R., Chen, G.: Review of computer-aided numerical simulation in wind energy. Renew. Sustain. Energy Rev. Elsevier 25, 122–134 (2013)
17. Wilcox, D.C.: Turbulence Modeling for CFD. DWC Industries, La Canada, California, pp. 143–163
18. Davidson, P.A.: Turbulence, 1st edn. pp. 107–145. Oxford University Press, Oxford (2004)
19. Villalpando, F., Reggio, M., Ilinca, A.: Assessment of turbulence models for flow simulation around a wind turbine airfoil. Model. Simul. Eng. 2011, 714146 (2011). doi:10.1155/2011/714146
20. Menter, F.R., Langtry, R., Völker, S.: Transition modelling for general purpose CFD codes. Flow Turbulence Combust. 77, 277–303 (2006). doi:10.1007/s10494-006-9047-1MATH
21. Langtry, R.B., Gola, J., Menter, F.R.: Predicting 2D airfoil and 3D wind turbine rotor performance using a transition model for general CFD codes. In: 44th AIAA Aerospace Sciences Meeting and Exhibit, Reno, Nevada, 9–12 Jan 2006 (2006)
22. Sørensen, N.N.: 3D CFD computations of transitional flows using DES and a correlation based transition model. Risø-R-1692(EN). ISBN 978-87-550-3749-6
23. Langtry, R.B., Menter, F.R., Likki, S.R., Suzen, Y.B., Huang, P.G., Völker, S.: A correlation-based transition model using local variables—part I: model formulation. In: Vienna, ASME Paper No. ASME-GT2004-53452 (2006)
24. Langtry, R.B., Menter, F.R., Likki, S.R., Suzen, Y.B., Huang, P.G., Völker, S.: A correlation-based transition model using local variables—part II: test cases and industrial applications. ASME J. Turbomach. 128(3), 423–434 (2006)
25. Reuss, R.L., Hoffmann, M.J., Gregorek, G.M.: Effects of Surface Roughness and Vortex Generators on the NACA 4415 Airfoil. The Ohio State University, Columbus, NREL/TP-442-6472 (1995)
26. Saliveros, E.: The Aerodynamic Performance of the NACA 4415 Aerofoil Section at Low Reynolds Numbers. Department of Aerospace Engineering, University of Glasgow (1988)
27. Brusca, S., Lanzafame, R., Messina, M.: Low speed wind tunnel: design and build. In: Pereira, J.D. (ed.) Wind Tunnels: Aerodynamics, Models and Experiments, chap 7, pp. 189–220. Nova Science Publishers, Inc., Hauppauge, NY. ISBN: 978-1-61209-204-1 (2011)
28. Barlow, J.B., Rae, W.H., Pope, A.: Low Speed Wind Tunnel Testing, 3rd edn. Wiley-Interscience, New York (1999)
29. Sørensen, J.N., Shen, W.Z., Mikkelsen, R.: Wall Correction Model for Wind Tunnels with Open Test Section. AIAA J. 44(8), 1890–1894 (2006)
30. Bell, J.H., Metha, R.D.: Contraction design for small low-speed wind tunnels. In: NASA—CR—182747, April (1988)
31. Zell, P.: Performance and test section flow characteristics of the national full-scale aerodynamics complex 80-by 120-foot wind tunnel. In: NASA TM 103920 (1993)
32. Eckert, W., Mort, K., Pope, J.: Aerodynamic design guidelines and computer program for estimation of subsonic wind tunnel performance. In: NASA TN D—8243, Washington, D.C., October (1976)
33. Metha, R.D., Bradshaw P.: Design rules for small low speed wind tunnels. Aeronaut. J. R. Aeronaut. Soc. 442–449 (1979)
34. Shen, W.Z., Hansen, M.O.L., Sørensen, J.N.: Determination of the angle of attack on rotor blades. Wind Energy 12, 91–98 (2009). doi:10.1002/we.277
35. Johansen, J., Sørensen, N.N.: Aerofoil characteristics from 3D CFD rotor computations. Wind Energy 7, 283–294 (2004). doi:10.1002/we.127

CHAPTER 21 DEVELOPMENT OF A MODEL COUNTER-ROTATING TYPE HORIZONTAL-AXIS TIDAL TURBINE

B Huang, K Yoshida, T Kanemoto

Faculty of Engineering, Kyushu Institute of Technology

ABSTRACT

In the past decade, the tidal energies have caused worldwide concern as it can provide regular and predictable renewable energy resource for power generation. The majority of technologies for exploiting the tidal stream energy are based on the concept of the horizontal axis tidal turbine (HATT). A unique counter-rotating type HATT was proposed in the present work. The original blade profiles were designed according to the developed blade element momentum theory (BEMT). CFD simulations and experimental tests were adopted to the performance of the model counter-rotating type HATT. The experimental data provides an evidence of validation of the CFD model. Further optimization of the blade profiles was also carried out based on the CFD results.

INTRODUCTION

Ocean energy resources have attracted great attention to get sustainable society with accompanying hydro, wind and solar resources, and then many kinds of the power unit have been provided for the tidal stream [1]-[4]. The unique tidal stream power unit called "Counter-Rotating Type Tidal Stream Power Unit" has been proposed in the previous studies [5, 6]. As shown in Fig.1, tandem propellers counter-drive double rotational armatures respectively, in keeping the rotational torque counter-balance between both propellers/armatures. The counter-rotating type tidal stream power unit has fruitful advantages as follows: (1)The tandem propellers may be easy to suppressthe cavitation rather than the single propeller, because the individual rotational speed can be reduced by about half while the armature diameter and the induced voltage are the same as those of the traditional power unit with single propeller. (2)The individual rotational speed can be designed freely in response to the tidal circumstances. (3) The rotational moment hardly acts on the casing and/or the pile because the rotational torque between both propellers/ armatures is counter-balanced in the unit. That is, the unit can be mounted easily on the mono-pile. (4)The flow has scarcely the swirling component and the disturbed wave downstream of the unit, because the angular momentum change, namely the rotational torque, through the front propeller is almost the same as that through the rear propeller. That is, the flow discharges in the axial direction while the tidal stream has no swirling component.

In the present work, the front and rear propellers of counter-rotating type tidal stream power unit were designed according to BEMT. And then CFD simulations and experimental tests

Citation: B Huang, K Yoshida and T Kanemoto, Development of a model counter-rotating type horizontal-axis tidal turbine, doi:10.1088/1757-899X/129/1/012057.

were adopted to the performance of the model counter-rotating type HATT.

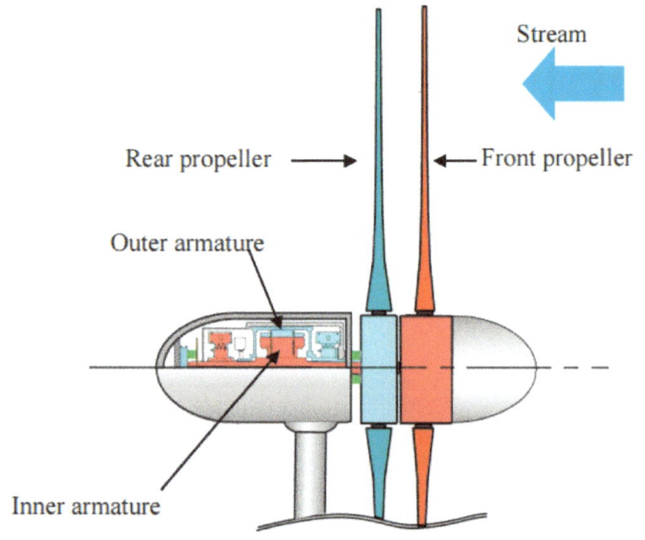

Figure 1: Concept of counter-rotating type tidal stream power unit (upstream type).

TURBINE DESIGN

A model counter-rotating type HATT with the diameter of 500 mm was designed as a trade-off between maximizing the Reynolds number and not incurring excessive tunnel blockage correction [7, 8]. The blade profiles were composed of the blade element KIT001 with three different relative thicknesses as shown in Fig. 2. The blade profiles of the model counter-rotating type HATT designed by BEMT are illustrated in Fig.3.

The diameter of the front propeller is d_F = 500mm and the rear propeller is d_R = 420 mm, namely the diameter ratio [$=d_R/d_F$] is 0.84. The number of blades for upstream propeller and downstream propeller is Z_F = 3 and Z_R = 5, respectively. The axial distance between the front and the rear propellers is set to 80mm, and the diameter of the hub is set to 90 mm.

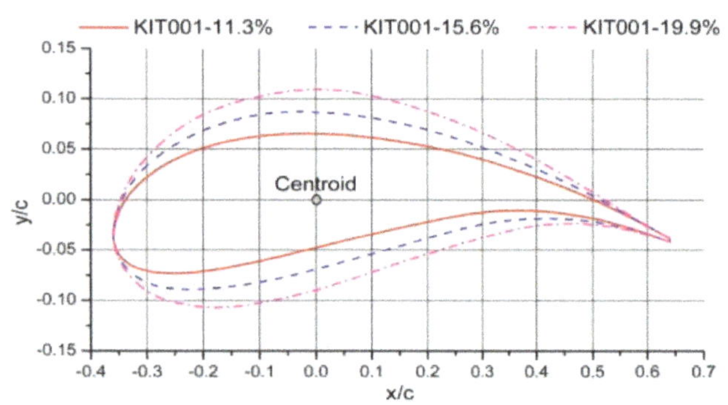

Figure 2: Blade element KIT001 with three different relative thicknesses.

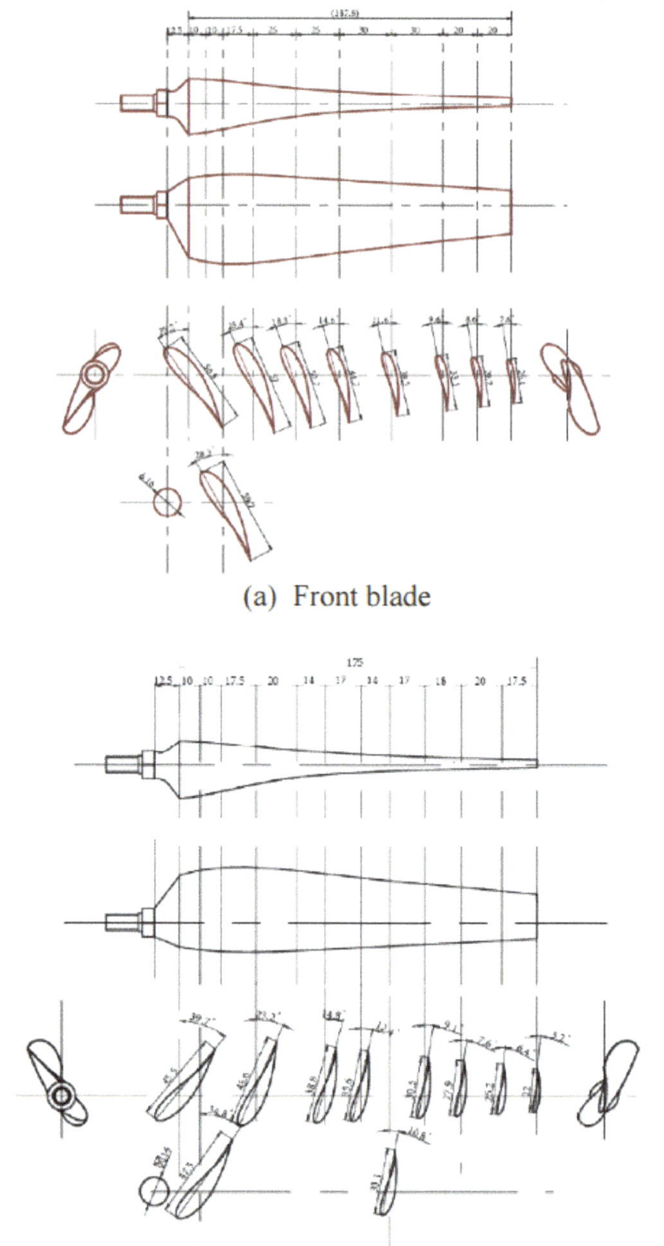

(a) Front blade

(b) Rear blade

Figure 3: Blade profiles of model turbine.

PERFORMANCE OBTAINED BY CFD AND EXPERIMENTAL TESTS

In order to obtain the hydrodynamic performance of the model counter-rotating type HATT designed in the previous section, CFD simulations and experimental tests were carried out in this section. The detailed introduction of CFD simulations can refer to the published works [5, 6]. And at this stage the experimental tests were conducted in the wind tunnel in Kyushu Institute of Technology as shown in Fig.4.

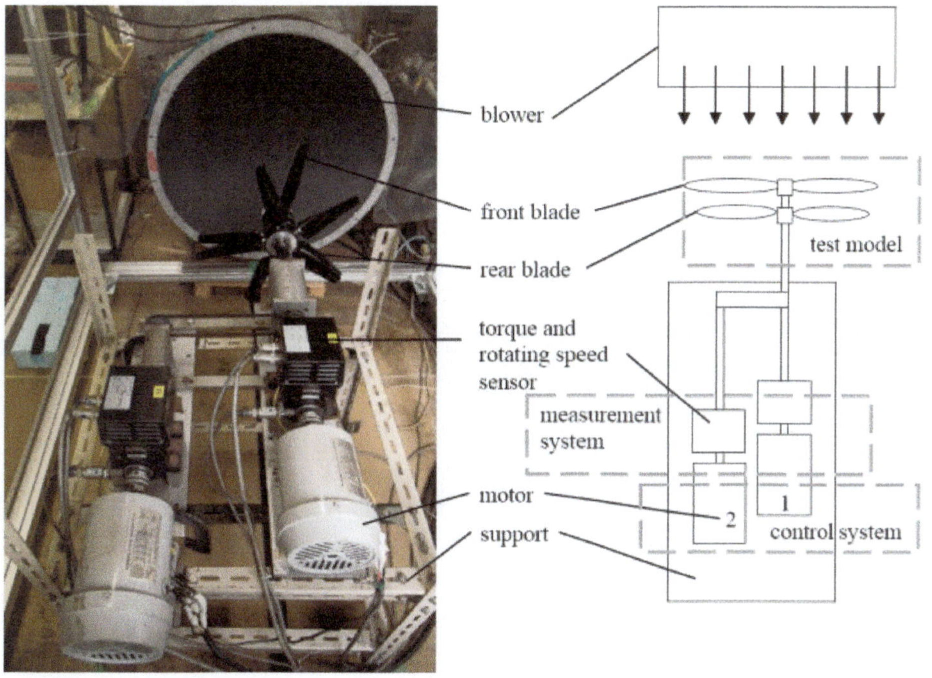

Figure 4: Schematic diagram of test rig in a wind tunnel.

Fig.5 shows the performance of the designed model counter-rotating type HATT in terms of power coefficient. It can be seen that the trend of the power coefficient shows good agreement over a range of tip speed ratios. However, the CFD over predicts the power coefficient in the high tip speed ratio region. It should be noted that in the wind tunnel tests, there is no blockage effect and lower Reynolds number than that in the water tunnel. These differences lead to the deviation between CFD predictions and experimental results.

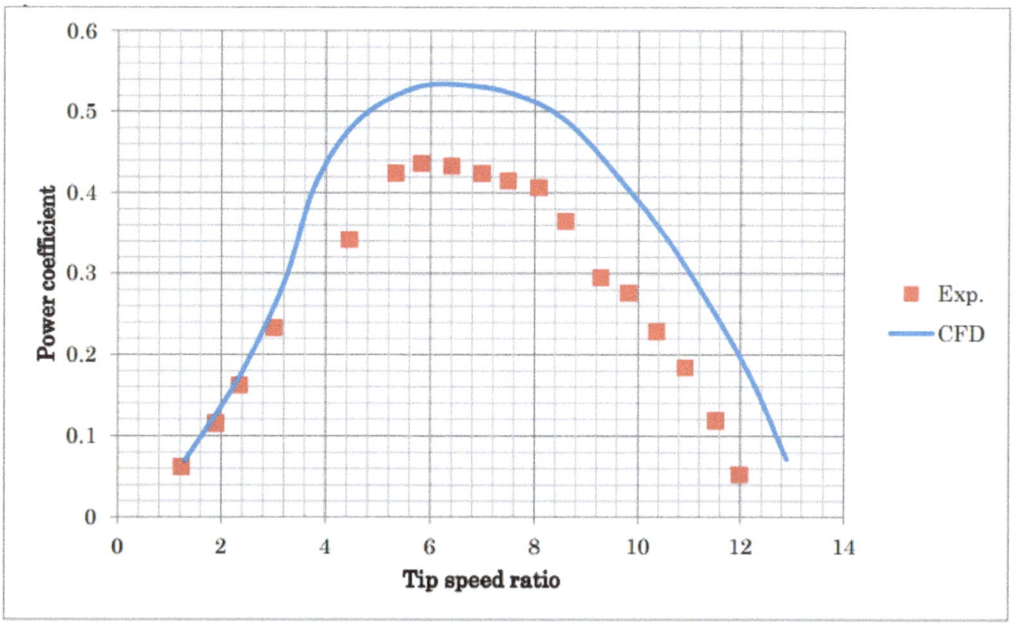

Figure 5: Comparison between CFD simulations and experimental tests.

EFFECTS OF REYNOLDS NUMBER

Consideration was tried about the way to presume the output coefficient of the actual machine size from model test outcome. When converting from model test to actual performance, it is necessary to know the difference from the similarity law (size effect) due to the fact that the Reynolds number is different. The efficient conversion formula to predict efficient difference between a model and an actual machine has been announced from the old days. Among these the next Moody 1/5th power formula obtained empirically to a water turbine has been also used much to a pump as well as a water turbine about the total efficiency.

$$\eta = 1 - (1 - \eta')(D/D')^{-1/5} \quad (4\text{-}1)$$

Here η: The total efficiency and the D: representative size and the symbol ' indicates a model. Formula (4-1) can be transformed into following equation.

$$(1-\eta)/(1-\eta') = (Re/Re')^{-1/5} \quad (4\text{-}2)$$

Here -1/5 of the index corresponds to a relation between friction coefficient λ on a flat smooth board and Reynolds number Re (λ∝ Re -1/5).

Based on the above, we examined the effect of Reynolds number based on the flow analysis around the two-dimensional blade profile. For a two-dimensional blade profile, we designed the camber blade profile (Basic thickness form: NACA65-010, Mean line: NACA Mean Line 65, Chord length: 50mm, Trailing edge radius: 0.5mm) to show in figure 4-1. With analysis code, ANSYS 14.5 was used .

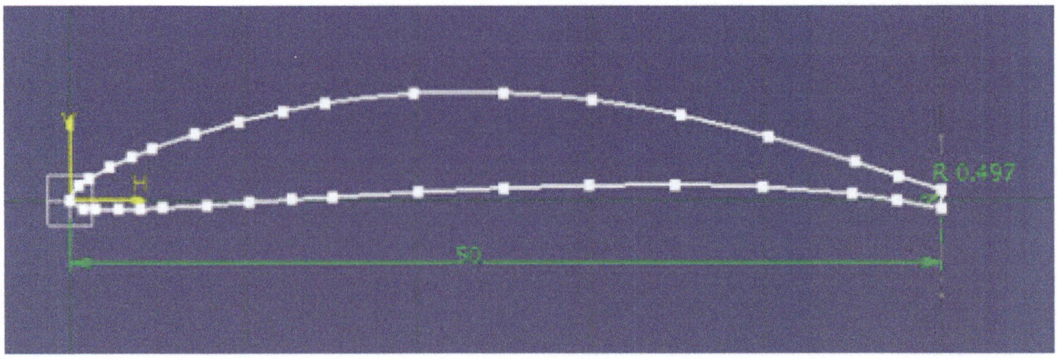

Figure 6: Blade profile used for analysis.

The analysis condition did the attack angle (α=0∼12 deg) number (Re=10 and Reynolds 4∼ Re=10^4 ∼ 10^7 10 ^{5} was). equivalent to the former model and Re=10^6 ∼ 10^7 test range [9], assumed a future actual power unit (Propeller diameter is on the order of 10m).

An analysis result of lift coefficient Cl, drag coefficient Cd, and pressure coefficient Cp=ΔP/ (0.5 ρ v ^{2}) is indicated in Fig.7.

As shown in Fig.7(a), Cl of Re=10^4 is the small value extremely compared with other Re . According to the analysis result of the streamline, in the case of Re = 10^4 , separation is seen near trailing edge in an analysis range of α.But this separated region greatly decreases if it becomes more than Re=10^5 . Further, when it'll be the actual power unit size level of Re=10^6∼10^7 , the curves are overlap each other, and influence of Reynolds number doesn't show.

As shown in Fig.7(b) and (c), Cd and Cp of Re=10^4 is the big value extremely compared with other Re. Cp also shows the same trend as Cd, the decrease amount becomes smaller as the Reynolds number increases. It has a tendency to approach saturation.

Here, we consider the relation between above-mentioned CFD result and Moody 1/5th power formula. The left-hand side of (1-η)/(1-η') in equation (4-2) represents a loss ratio of actual and model. So this parameter is considered to be physically associated with pressure coefficient ratio Cp/Cp' of actual and model. We show a rearranging result in Fig.7(d). In addition, Cp/Cp' varies a little according to an attack angle, but the tendency is almost the same. By the two dimensional blade profile which simplified that effects of the centrifugal force are not considered at all, the CFD result is similar of Moody 1/5th power formula as shown in Fig.7 (d). From this ,efficiency improvement by the scale effect of the future actual power unit (Re = 10^6∼10^7) is expected.

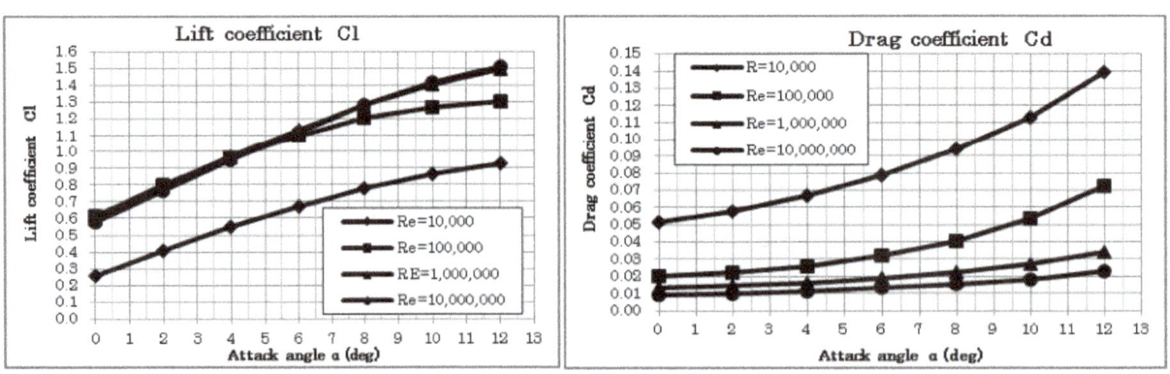

(a) (b)

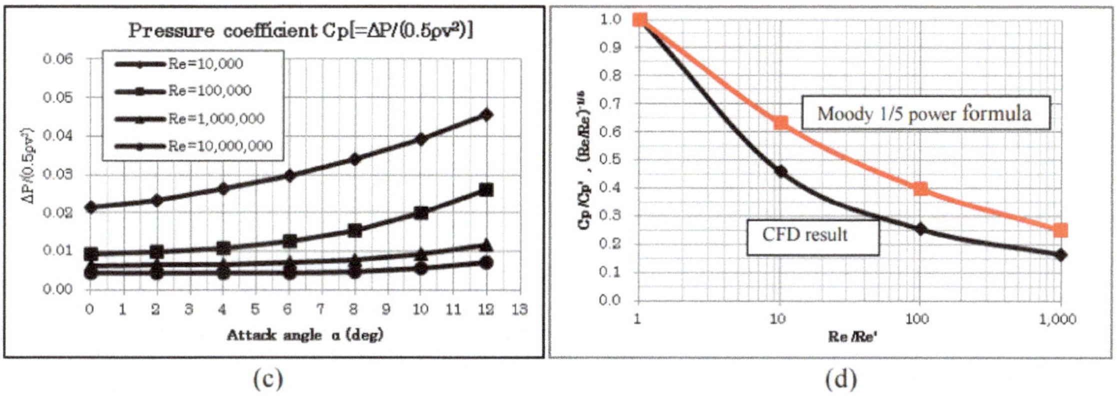

Figure 7: Results of CFD analysis.

CONCLUSIONS

In the present study, a unique counter-rotating type HATT was proposed and the blade profiles were designed based on the BEMT. Both of CFD simulations and experimental tests in the wind tunnel have been carried out to verify our design. Effects of Reynolds number have also been conducted on a 2D hydrofoil, which provides a reference for the efficiency conversion from model to prototype turbine.

REFERENCES

1. Fraenkel P.L. 2007 Marine current turbines: pioneering the development of marine kinetic energy converters Journal of Power and Energy 221 159-169
2. Bahaj A.S., Molland A.F., Chaplin J.R. and Batten W.M.J. 2007 Power and thrust measurements of marine current turbines under various hydrodynamic flow conditions in a cavitation tunnel and a towing tank Renewable Energy 32 407-426
3. Clarke J., Connor G., Grant A., Johnstone C. and Ordonez-Sanchez S. 2008 Contra-rotating marine current turbines: performance in field trials and power train developments Proceedings of 10th World Renewable Energy Congress (Glasgow, Scotland, UK,)
4. Hiraki K., Wakita R. and Kanemoto T. 2012 Demonstrative power generation by twin-runner darrieus turbine in Kanmon strait Proceedings of the 22nd International Offshore and Polar Engineering Conference (Rhodes, Greece,) 725-729
5. Huang B. and Kanemoto T. 2015 Multi-objective numerical optimization of the front blade pitch angle distribution in a counter-rotating type horizontal-axis tidal turbine Renewable Energy 81 837-844
6. Huang B. and Kanemoto T. 2015 Performance and internal flow of a counter-rotating type tidal stream turbine Journal of Thermal Science 24 1-7
7. Bahaj A.S., Molland A.F., Chaplin J.R. and Batten W.M.J. 2007 Power and thrust measurements of marine current turbines under various hydrodynamic flow conditions in a cavitation tunnel and a towing tank Renewable Energy 32 407-426
8. Bahaj A.S., Batten W.M.J. and McCannb G. 2007 Experimental verifications of numerical predictions for the hydrodynamic performance of horizontal axis marine current turbines Renewable Energy 32 2479-2490
9. Usui Y., Kanemoto T., Takaki k. and Hiraki K. 2014 Counter-rotating type tidal stream power unit mounted on a mono-pile Journal of Energy and Power Engineering 1748-1755

CHAPTER 22

THE NUMERICAL SIMULATION BASED ON CFD OF HYDRAULIC TURBINE PUMP

X H Duan[1], F Y Kong[1], Y Y Liu[1], R J Zhao[1], Q L Hu[1]

[1]National Research Center of Pumps, Jiangsu University, Zhenjiang, Jiangsu 212013, China

ABSTRACT

As the functions of hydraulic turbine pump including self-adjusting and compensation with each other,it is far-reaching to analyze its internal flow by the numerical simulation based on CFD, mainly including the pressure field and the velocity field in hydraulic turbine and pump.The three-dimensional models of hydraulic turbine pump are made by Pro/Engineer software;the internal flow fields in hydraulic turbine and pump are simulated numerically by CFX ANSYS software. According to the results of the numerical simulation in design condition , the pressure field and the velocity field in hydraulic turbine and pump are analyzed respectively .The findings show that the static pressure decreases systematically and the pressure gradient is obvious in flow area of hydraulic turbine ; the static pressure increases gradually in pump . The flow trace is regular in suction chamber and flume without spiral trace . However , there are irregular traces in the turbine runner channels which contrary to that in flow area of impeller . Most of traces in the flow area of draft tube are spiral .

INTRODUCTION

Computational fluid dynamics (CFD), which is theoretical and practical, describes the numerical solution of the flow field quantitatively on the time and space by computer numerical calculations and a method of image displayed[1]. The relationship of performance parameters between pumps and PAT was predicted by theoretical analysis and the empirical formulas based on the experimental data[2]. However, turbine Pump has some advantages, such as low cost and the ability of simulating complicated or desirable fluid motion. As the functions of hydraulic turbine pump including selfadjusting and compensation with each other , it is far-reaching to analyze its internal flow.

Citation: X H Duan, F Y Kong, Y Y Liu, R J Zhao and Q L Hu, The numerical simulation based on CFD of hydraulic turbine pump, doi:10.1088/1757-899X/129/1/012042.

COMPUTATION OF PARAMETERS AND CONDITIONS

Table 1: The main parameters of the pump and the turbine

Pump flow $/m^3 \cdot h^{-1}$	Pump head /m	N $/r \cdot min^{-1}$	T /℃	Turbine flow $/m^3 \cdot h^{-1}$	Turbine head /m	P /KW	ρ $/kg \cdot m^{-3}$
60	11	1475	20	90	41.5	1.8	998.2

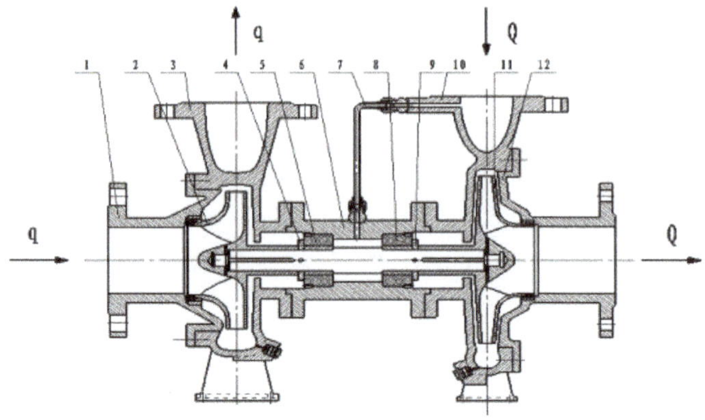

1.pump lid 2.impeller 3.Pump body 4.Balance disc 5.Sliding bearing 6.Connector 7.Rushing pipette 8.Sliding bearing 9.Balance disc 10.Turbine body 11.Runner 12.Turbine lid

Figure 1: Structure diagram of hydraulic turbine.

THE GEOMETRIC MODEL AND NUMERICAL SIMULATION

Solid model of fluid region

Using Pro/Engineer software, the three-dimensional solid models are made. The models are presented in figure2 to figure3[3].

Figure 2: Hydraulic Turbine.

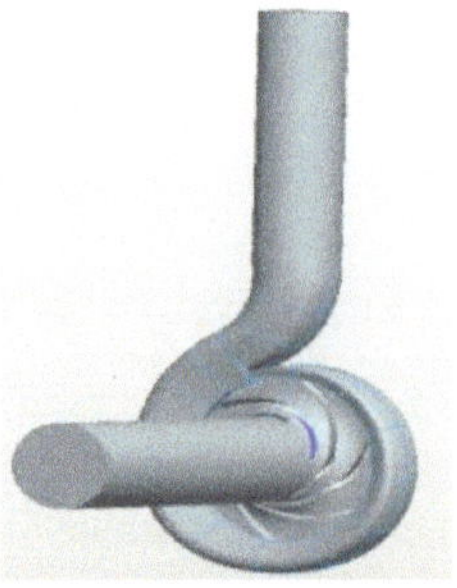

Figure 3: Pump.

Mesh generation of fluid region

The number of grid of hydraulic turbine is 1158877 and pump is 1458943. The grids are presented in figure4 to figure5[4].

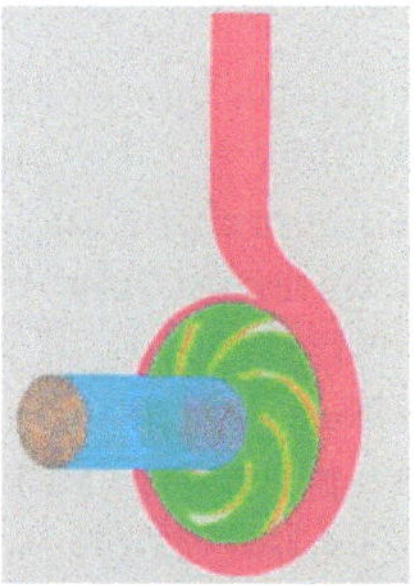

Figure 4: Meshes in Hydraulic Turbine.

Figure 5: Meshes in Pump.

Turbulence Model

At present , the standard k-ε model is the most basic one of the double equations model being the most widely used the eddy viscosity model. The equations[5] are as follows :

$$\frac{\partial}{\partial t}(\rho k)+\frac{\partial}{\partial x_i}\left(\rho k u_i\right)=\frac{\partial}{\partial x_j}\left[\left(\mu+\frac{\mu_t}{\sigma_k}\right)\frac{\partial k}{\partial x_j}\right]+G_k-\rho\varepsilon-S_k \tag{1}$$

$$\frac{\partial}{\partial t}(\rho\varepsilon)+\frac{\partial}{\partial x_i}(\rho\varepsilon u_i)=\frac{\partial}{\partial x_j}\left[\left(\mu+\frac{\mu_t}{\sigma_\varepsilon}\right)\frac{\partial\varepsilon}{\partial x_j}\right]+C_{1\varepsilon}\frac{\varepsilon}{k}G_k-C_{2\varepsilon}\rho\frac{\varepsilon^2}{k}+S_\varepsilon \tag{2}$$

Where: μ_t —turbulent viscosity, $\mu_t=\rho C_\mu\frac{k^2}{\varepsilon}$; G_k —turbulent kinetic energy, $G_k=\mu_t\left(\frac{\partial u_i}{\partial x_j}+\frac{\partial \mu_j}{\partial x_i}\right)\frac{\partial u_i}{\partial x_j}$; $\sigma_k,\sigma_\varepsilon$ — turbulence Prandt1 constant of k -ε equation ; S_k —vicious stress S_k =1.0 ; S_ε —turbulence stress S_ε =1.3; $C_m=0.09; C_1=1.44; C_2=1.92$

The simulation and analysis of internal flow field in pump and hydraulic turbine

The inner flow field in pump and hydraulic turbine are simulated by ANSYS CFX ,and then the velocity profile and pressure profile are analyzed to get some conclusions through the pictures from the simulation.

The analysis of pressure profile in hydraulic turbine

The pressure value compared to 101325 Pa is a relative value in this model[6].

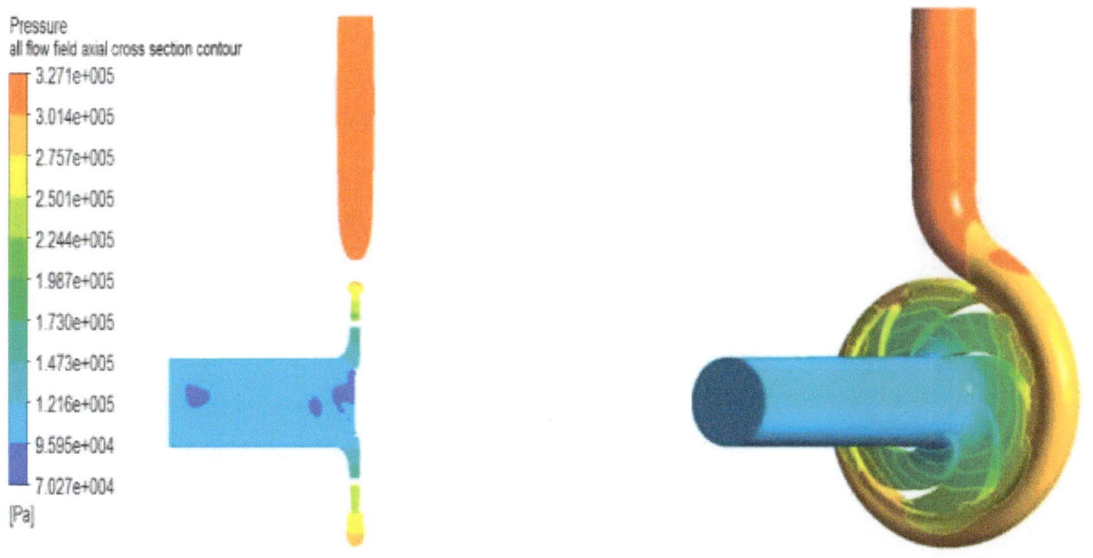

Figure 6: Static pressure distribution diagram of hydraulic turbine.

From Figure6 ,the static pressure is lower and lower as the order of intake pipe , turbine flume ,impeller and draft tube. The dividing line is obvious among the different pressure fluid ,and the static pressure value is rather low in the draft tube. The static pressure contours of the whole flow passage of hydraulic turbine are shown in Figure6. As can be seen from the picture, the static pressure of the water diversion chamber, the runner and the draft tube is decreasing in order .The indoor static pressure of the water diversion chamber begins to decrease from the IX section, which is decreasing as the diversion chamber flow section area reducing. Some regions appears high-pressure. The static pressure of the runner is decreasing from the inlet to the outlet and the pressure gradient is significant. The static pressure of draft tube is basically constant.

The analysis of pressure field of the internal flow region of the pump.

The pressure values in the picture is relative to the reference value of the 101325Pa of the model. And the values in the following static pressure contours are relative. The static pressure contours of the axial section of the whole passage of pump is shown in Figure7.As can be seen from the picture, the static pressure of the water inlet pipe, the water diversion chamber, the impeller, the water pressing chamber and the water outlet pipe is gradually decreasing in order. And the pressure gradient is significant.

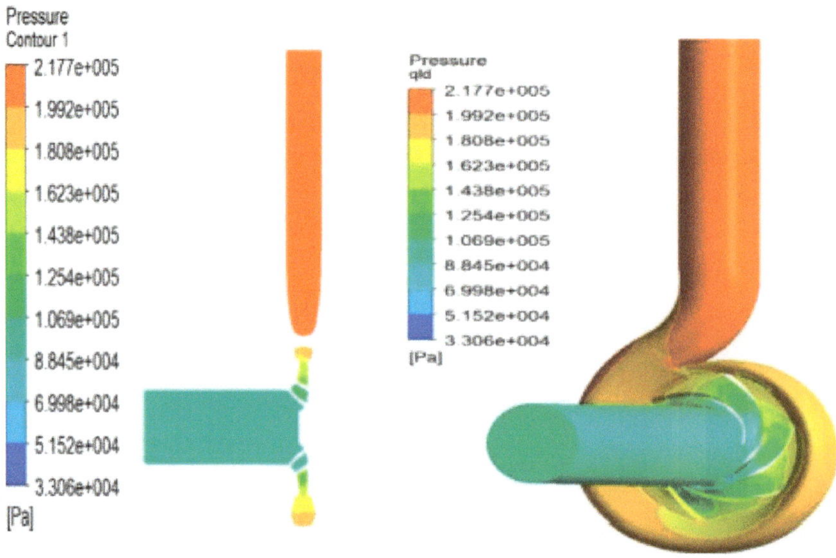

Figure 7: Static pressure contours of the whole passage of pump.

The static pressure contours of the whole passage of pump is shown in Figure7. As can be seen from the picture, the pressure of the suction chamber, the impeller, the pump chamber and the outlet pipe is gradually increasing .The the static pressure gradient of liquid impeller is significant, which is increasing gradually. The indoor static pressure of the pumping chamber is decreasing from I section to XIV section. Some regions appeared high-pressure. The static pressure of impeller passage increased gradually from the inlet to the outlet direction, the adverse pressure gradient. And the pressure gradient is significant. The static pressure of outlet pipe is basically constant.

The analysis of velocity field of the internal flow region of the hydraulic turbine.

The velocity vector distribution of the whole flow passage of pump is shown in Figure8.The velocity vector distribution of the axial section of the whole flow passage is shown in Figure8. As can be seen from the picture, the velocity of the inlet pipe, the suction chamber, the impeller and the pumping chamber increase first ,then decrease. The liquid velocity value reached the maximum in the impeller. And it reached the minimum in the draft tube. The velocity field of inlet liquid in the tube is regular, which is spiral in the tail pipe and the outlet pipe.

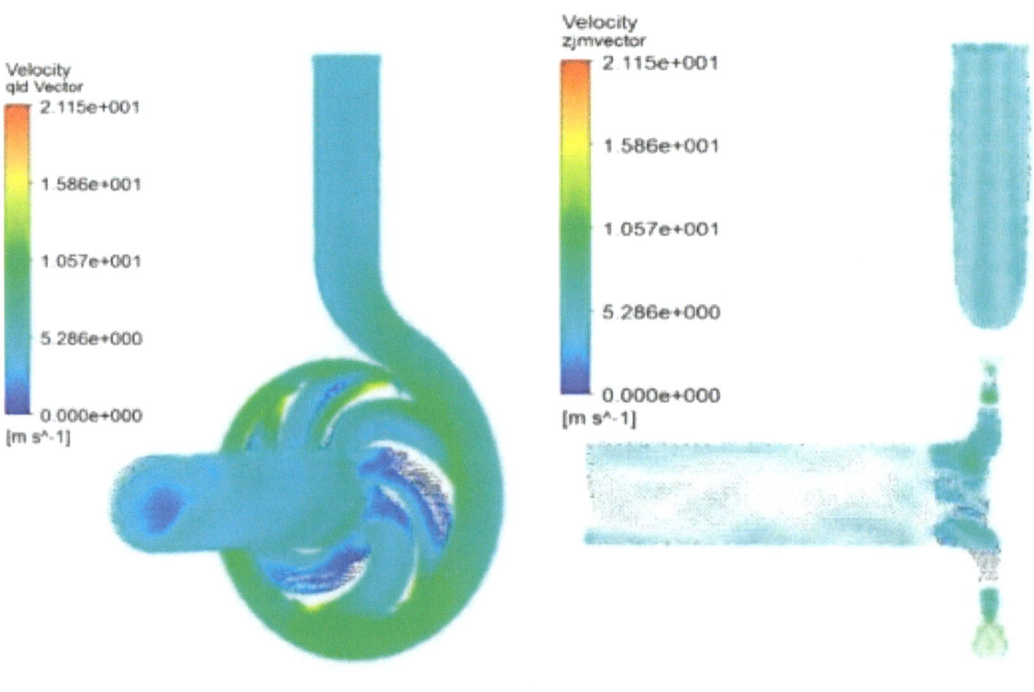

Figure 8: Velocity vectors contours of hydraulic turbine.

The trace diagram of the whole flow passage of pump is shown in Figure9.As can be seen from the picture, the trace in the inlet pipe and the water diversion chamber is regular. There is no spiral trace appearing. There is irregular trace appearing in the flow passage of impeller. Most of trace in the tail pipe is spiral.

The analysis of velocity field of the internal flow region of pump.

The velocity vector distribution of the whole flow passage of pump is shown in Figure10.As can be seen from the picture, the trace in the suction chamber and the impeller is regular. And the trace of the outlet pipe is bending up.

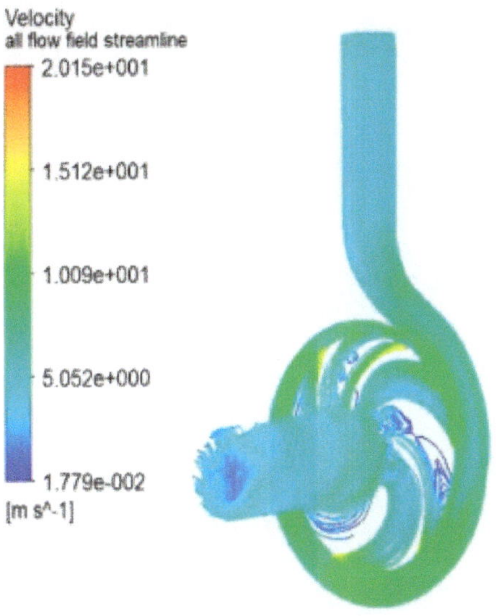

Figure 9: Trace diagram of the whole passages in hydraulic turbine.

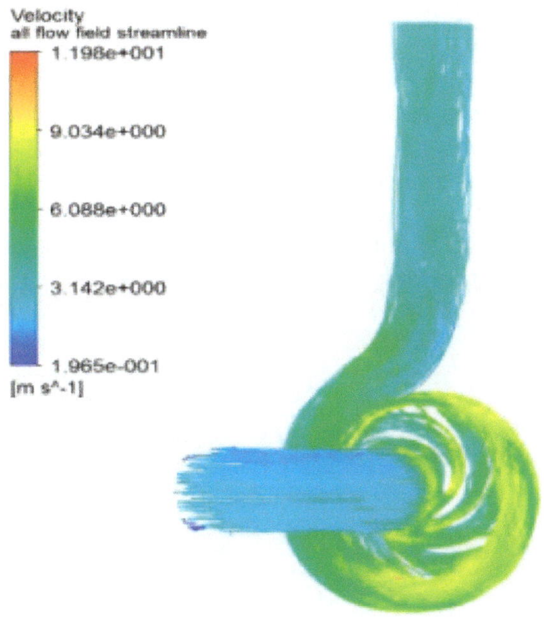

Figure 10: Trace diagram of whole passage in pump.

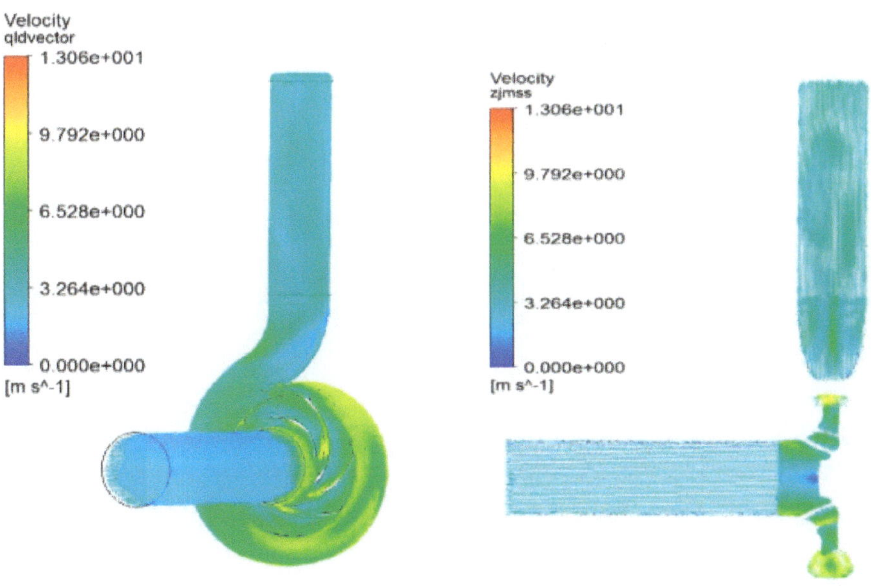

Figure 11: Velocity vectors contours of pump.

The velocity vector distribution of pump is shown in Figure11.As can be seen from the picture, the relative velocity of the suction chamber, the impeller, the volute and the outlet pipe increases first ,then decreases. The liquid relative velocity value reached the maximum in the flow passage of the impeller. And it reached the minimum in the center of back cover plate. The velocity field in the suction chamber and the impeller is regular. And there is vortex of the velocity field in the volute. The velocity field in the outlet pipe is regular.

CONCLUSION

In the hydraulic turbine, the static pressure of the suction chamber, the impeller, the tail pipe is lower and lower, the pressure gradient is significant. In the pump, the pressure of the suction chamber, the impeller, the pumping chamber and the outlet pipe is gradually increasing . The trace in the inlet pipe and the diversion chamber is regular, There is no spiral trace appearing. There is irregular trace appearing in the flow passage of the turbine impeller which contrary to in the follow passage of pump impeller. Most of trace in the tail pipe is spiral.

REFERENCES

1. Wang F J 2004 The Analysis of Computer Fluid Dynamics (Beijing: Tsinghua University Press)
2. Wang T, Kong F Y, He Y Y and Yang S S 2013 The current research status of centrifugal pump for turbine Journal of Drainage and Irrigation Machinery Engineering 08 674-680
3. Yang S S, Kong F Y, Su X H et al 2012 Numerical simulation and external characteristic of pumps and pump as turbine Academic Journal of Xi'an Jiaotong University 3 36-41 f
4. Yang S S, Kong F Y and Shao F 2010 Numerical simulation and comparison of pump and pump as turbine ASME Fluids Engineering Summer meeting (Montreal, Canada,) 1-10
5. Xie L H and Xinyu Zhao 2013 ANSYS CFX analysis and simulation of fluid (Beijing: Electronic Industry Press)
6. Yang S S, Derakhshan Shahram and Kong F Y 2012 Theoretical numerical and experimental prediction of pump as turbine performance Renewable Energy 507-513

CHAPTER 23

COMPUTATIONAL FLUID DYNAMICS (CFD) AND DISCRETE ELEMENT METHOD (DEM) APPLIED TO CENTRIFUGES

Xiana Romani Fernandez, Lars Egmont Spelter and Hermann Nirschl

Karlsruhe Institute of Technology, Campus South, Institute for Mechanical Process Engineering and Mechanics, Karlsruhe Germany

INTRODUCTION

The simulation of fluid flow in e.g. air channels, pipes, porous media and turbines has been the emphasis of many research projects and optimization processes in the academic and industrial community during the last years. A key advantage of the calculation of fluid flow through a given geometry when compared to the experimental determination is the efficient identification of disadvantages of the current state and the fast cycle of change in geometry and analysis of its effects. This saves time and costs and furthermore opens up new fields of research and development. There are, however, few drawbacks of the new optimisation and research methods. The numerically obtained results need validation before it can be stated whether they are correct or misrepresent the true state. In some cases, such as the optimisation of existing systems, the experience of the user may be sufficient to evaluate the calculated flow pattern. In the case of the prediction of flow in new geometries or process conditions, the numerical result can be validated only by measuring the flow pattern at representative process conditions (flow velocity, pressure, and temperature). For nonmoving geometries, many measurements have been conducted in order to determine the flow behaviour in various geometries and for different process conditions. These measurements are readily available for the evaluation of new simulations. However, only little research has been conducted to evaluate the flow conditions in centrifuges, especially at high rotational speeds. The reason for the lack of experimental data lies in the difficult conditions for the measurement in the centrifugal field with high circumferential velocities and pressure gradients. For low rotational speeds, the flow patterns have been determined by an electrolytic technique (Glinka, 1983) and visualised by adding ink to the flowing liquid (Bass, 1959; 1959; 1962).

The emphasis of the work presented is the evaluation of the efficiency of the acceleration geometry and the prediction of the axial flow profile in the centrifuge. In centrifugation technology, the prediction of the separation efficiency is often restricted to the cut size, the smallest particle size that can settle on the bowl wall at certain operating conditions. In a centrifuge, the highest velocities occur in tangential direction but, due to the throughput, a secondary flow in the axial direction appears. The tangential velocity creates the centrifugal force acting on the particles and the flow in the axial direction determines the residence time of particles in the bowl. Thus it is necessary to know the flow patterns in a centrifuge in order to calculate its separation capability. Often the predicted cut size of a centrifuge diverges from the real separation efficiency due to the fact that the assumptions of the analytical models are not strictly valid for industrial process conditions. The deviation

Citation: Xiana Romani Fernandez, Lars Egmont Spelter and Hermann Nirschl, Computational Fluid Dynamics (CFD) and Discrete Element Method (DEM) Applied to Centrifuges, DOI: 10.5772/27176.

is caused by the complex shape and internal assemblies of industrial solid bowl centrifuges, which create complex flow patterns. Hence estimating these flow patterns, as well as measuring them, is complicated.

In rotating machinery, the fluid flow is mostly turbulent, but in centrifuges the degree of turbulence is difficult to determine. Furthermore it may vary throughout the different zones of the geometry. For example, at the inlet of a continuous centrifuge, high shear rates and intensive mixing occur due to the high tangential velocity gradient. Hence a solver that incorporates equations to solve turbulent flow has to be used to obtain accurate predictions. Behind the inlet zone, the liquid flows undisturbed until the continuous phase leaves the centrifuge via an overflow weir or another discharge system. In this zone, the degree of turbulence depends on the geometry of the outlet system and the velocity gradients at the outlet. This example shows how complicated it is to gain the basis for the selection of an appropriate simulation method. Thus it is advisable to choose a model that is as precise as possible, but with an affordable computational demand. Various models have been developed to predict turbulent flow, but the codes were initially not designed to solve the Navier-Stokes equation in a rotating reference frame. Modifications were made over the years to include rotating systems and phenomena, e.g. the modified k-ε model "k-ε-RNG". Since then, only little work has been conducted to evaluate the accuracy of the various turbulence models when applied to centrifuges. The work presented compares several models such as the k-ε RNG, the k-ω, the Reynolds Stress Model (RSM) and the large-eddy simulation (LES) for predicting flow pattern in centrifuges. Furthermore the computational demand is evaluated and compared for different mesh sizes and types. The nomographs presented help to estimate the computational time for a given problem and computer system depending on the model and mesh.

Since with increasing fill level the sediment interacts with the flow pattern and reduces the capacity of the centrifuge, the sediment build-up must be included in the calculation of the separation efficiency and the determination of the flow pattern for tubular centrifuges. For calculation of the settling particles, diverse approaches are available. In this study, a Lagrange formulation has been chosen for calculation of the particle trajectories. In order to investigate the sediment build-up, particle–particle interactions must be considered. For this purpose, it is appropriate to apply the Discrete Element Method (DEM) (Cundall & Strack, 1979), originally developed for calculating the flow of bulk granular materials and then extended to the entire particle processing technology. This method accounts for particleparticle and particle-wall interactions by means of various contact models. However this method is limited because it does not account for hydrodynamic forces (drag, lift, torque), which play a vital role in e.g. centrifuges. Hence the simulation of the multiphase flow in centrifuges requires a coupling of the DEM with computational fluid dynamics (CFD). The flow pattern and particle trajectories must be calculated in an alternating scheme. First, the flow pattern is determined using CFD and afterwards, the motion of the particles is calculated using DEM adding the hydrodynamic forces to the contact forces.

The advantage of the numerical simulation of the complex multiphase flow in centrifuges is that the flow, the particle trajectories and their deposition are represented accurately. It allows a detailed description of the flow and particle behaviour for various operating conditions, which is of high importance for the design and optimization of centrifuges and other apparatus in solid-liquid separation. This work focuses on the prediction and validation of the results of flow patterns in solid bowl centrifuges of different kind. The centrifuges operate within a range of 1000 up to 40000 rpm and with throughputs between 0.1 and 6 l/min. The article highlights the issues that will occur if flow behaviour has to be calculated in a rotating confinement using commercial software. Detailed results show the variation of the flow patterns achieved using different models and the possibilities to run a sanity check for the calculated flow pattern.

SOLID BOWL CENTRIFUGES

The centrifugal separation of particles in a suspension is one of the most common problems appearing in industrial processes such as waste water, mining and mineral processing, biotechnology, solid-fuel, pharmaceutical, chemical and food industry. Solid bowl centrifuges are widely used in the pharmaceutical and fine chemicals industry. The tubular bowl centrifuge has been providing the basis for all centrifuge designs. A scheme of a typical state of the art tubular bowl centrifuge is shown in Figure 1.

Starting with a rotating cylinder, modifications were made, leading to the disc stack centrifuge, screen scroll centrifuge and various other types. The basic design still has its applications. Due to the lean cylinder, high rotational speeds are possible. One of the first high speed centrifuges was the Sharples superfuge (Saunders, 1948; Taylor, 1946) that has been modified and adapted by various companies. Applications range from laboratory purposes to the production of pharmaceuticals in an industrial scale. New centrifuge designs were developed allowing the extension of the techniques to the industrial manufacture of new products such as vaccines (Anderson et al., 1969; Brantley et al., 1970; Perardi & Anderson, 1970; Perardi et al., 1969). The application of most high-speed centrifuges is limited by the solids capacity of the rotor. Due to the lean design, the solids capacity usually does not exceed a few kilograms. For most products that are processed in such centrifuges, the small capacity does not increase the production costs significantly because the value of the final product exceeds the cost of labour and machinery by an order of magnitude. This is the reason why most research has been conducted in the field of process optimization, e.g. in sterilisation, product recovery, and operator safety. Optimisation of the flow pattern in the tubular centrifuge has not yet been the focus of research.

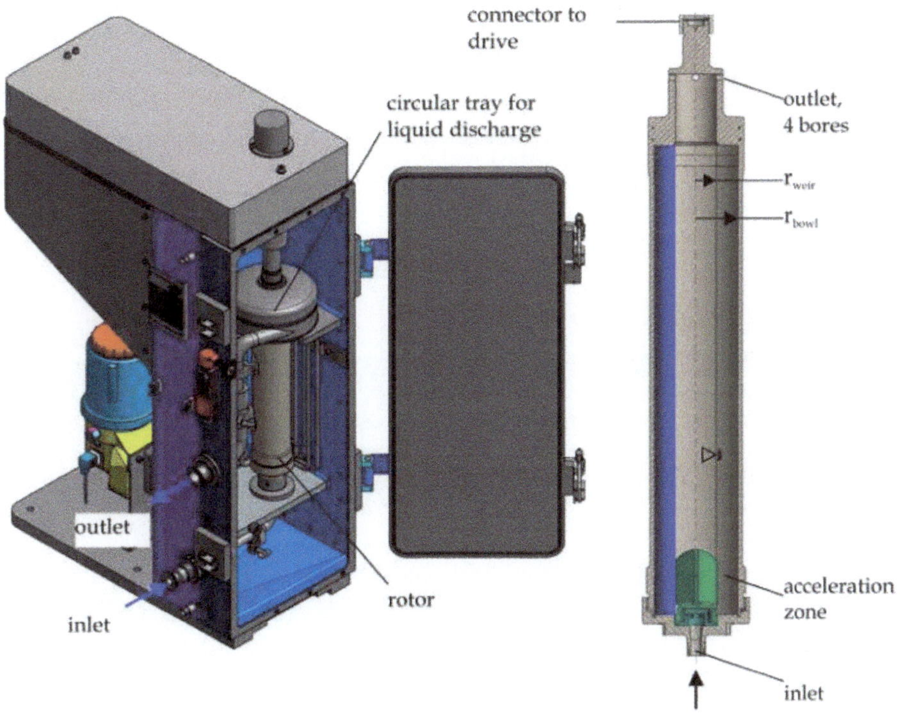

Figure 1: Tubular bowl centrifuge (CEPA Z41G).

The advantages of this type of centrifuge are the high centrifugal numbers and the simple geometry as compared to e.g. disc stack centrifuges. Tubular bowl centrifuges reach centrifugal numbers up to 40000 g. Their stability is due to their relatively high length to diameter aspect ratio in the range of approximately 4 to 7. There are other types of solid bowl basket centrifuges with conical or tubular bowls but smaller length to diameter aspect ratios, usually below 0.75 (Leung, 1998) such as overflow centrifuges. Sometimes radial or vertical compartments as well as blades are included to improve the separation efficiency. Up to now, the flow pattern in existing centrifuges has been analysed and compared to some theories describing the fluid flow in rotating geometries (Glinka, 1983; Golovko, 1969; Gösele, 1968; 1974; Schaeflinger & Stibi, 1991; Schaeflinger, 1991; Trawinski, 1959). Using the determined axial flow pattern, the authors developed various analytical models for calculation of the cut size. Since it was not possible to measure or predict the tangential velocity, corrections were made to take into account insufficient tangential acceleration. These empirical corrections were often only valid for a specific centrifuge. Moreover, the cut size depends not only on the tangential and axial flow but also on the density difference $\Delta\rho$ between the solids and the liquid, the viscosity of the suspension μ and the geometry of the rotor as stated in Equation (1). The geometric parameters are the length of the rotor L, the radius of the overflow weir r_{weir} and the radius of the bowl r_{bowl}:

$$x = \sqrt{\frac{18 \cdot \mu \cdot v_{ax} \cdot \ln\left(\frac{r_{bowl}}{r_{weir}}\right)}{\Delta\rho \cdot L} \cdot \frac{2 \cdot \pi \cdot n}{60}} \qquad (1)$$

Modifications were made because the observed separation efficiency of the test rig often was not in good agreement with simple theoretical approaches. These modifications included certain levels of turbulence, boundary layer flow and inhomogeneous flow in the acceleration and discharge zones (Bass, 1959; 1959; 1962; Sokolow, 1971; Zubkov & Golovko, 1968; 1969). The different flow patterns in centrifuges are explained in detail in Chapter 2.1. The assumption that the entire length L is available for the deposition of particles is only valid for an ideal tangential acceleration of the feed so that no turbulent inlet area is present and a rotor without any settled particles that may interact with the flow pattern. In fact, the length L is reduced by up to 30 % due to the effects of the inlet. With increasing fill level, the sediment interacts with the flow pattern and reduces the free length L so that the cut size of the centrifuge increases (Spelter et al., 2010). Therefore the sediment build-up has to be included in the calculation of the separation efficiency and the determination of the flow pattern for tubular centrifuges.

Flow pattern

There are two limiting cases for the flow pattern in solid bowl centrifuges. Either the entire volume is passed by the feed or the main axial fluid flow takes place in a thin layer on top of a stagnant but rotating pool. The two different flow patterns are schematically shown in Figure 2. When a boundary layer flow is present, the geometric constant r_{bowl} in Equation 1 has to be replaced by the radius where the stagnant pool begins ($r_{boundary}$ layer).

When starting building centrifuges and evaluating the separation performance, it was assumed that the proportion of the volume in which significant axial fluid velocities occur, further referred to as active volume V_{ac} or area A_{ac}, is nearly as high as the entire volume of the centrifuge, V_{ac}=100 % (Horanyi & Nemeth, 1971). Other researchers realised that the active volume of the centrifuge is significantly smaller than the capacity of the centrifuge (V_{ac}<<100%) and developed the boundary layer theory (Bass, 1959). Today, most researchers who work with solid bowl centrifuges support the boundary layer theory.

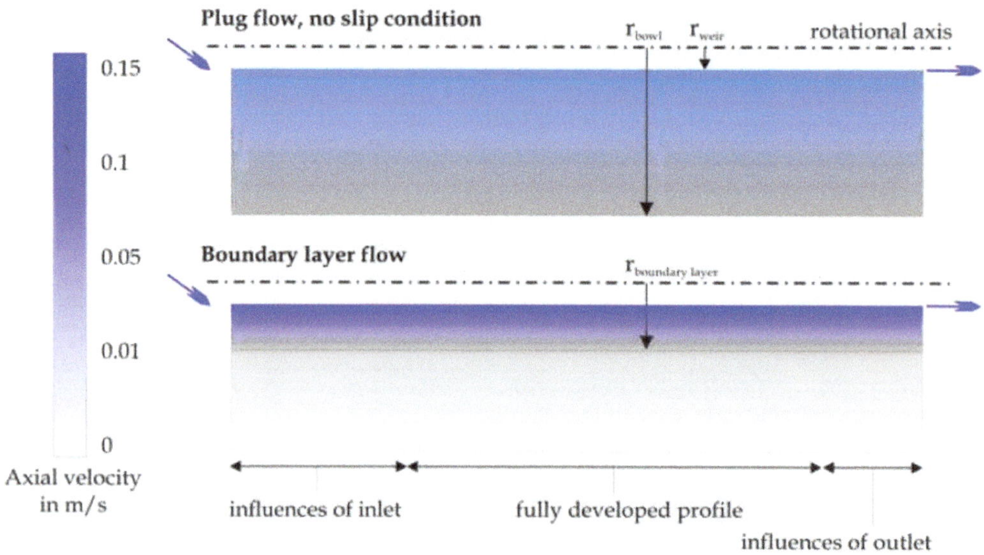

Figure 2: Flow pattern in solid bowl centrifuges.

Recent studies by the authors have confirmed the validity of this theory for industrial solid bowl and tubular bowl centrifuges (Romani Fernandez & Nirschl, 2009; Spelter & Nirschl, 2010; Spelter et al., 2010). Nevertheless it is important to pay attention to the acceleration geometry of the centrifuge and the influence of the settled particles on the flow pattern. In many centrifuges, the feed enters the rotor via a cone or a jet hitting a distributor plate. The liquid is not accelerated to the tangential velocity of the pool surface and thus hits the surface with a certain velocity gradient. The rotating pool accelerates the incoming liquid to the angular velocity. The pressure gradient in the rotating liquid stabilises the flow pattern and reduces the mixing of the feed and the rotating pool in radial direction. Thus a stratification of the liquid in radial layers occurs from the inlet and the liquid flows axially at the liquid-gas interface in a boundary layer. The flow is eased by the reduced friction at the interface between liquid and gas. The flow pattern will be different if the liquid is not fed to the pool surface or if intensive mixing occurs at the inlet. This is the case for some tubular centrifuges, as long as no sediment is present. Figure 3 a) shows the inlet geometry of a tubular bowl centrifuge as it was used in previous studies. The metal structure accelerates the incoming liquid to a certain rotational speed and significant mixing of incoming and rotating liquid occurs. The rigid body rotation is reached after a few centimetres behind the inlet. Figure 3 b) shows the acceleration geometry after 5 minutes of centrifugation. The settled yeast cells block a significant section of the inlet geometry and thus impede intensive mixing and acceleration. This has been observed for various minerals and biological products (Stahl et al., 2008) and is detectable by measuring the residence time and separation efficiency in the centrifuge at different fill levels. The growth of the sediment will change the flow pattern significantly if it reaches the active area. This may not be the case for industrial scale solid bowl centrifuges but has to be included in the calculations of the separation efficiency in tubular bowl centrifuges.

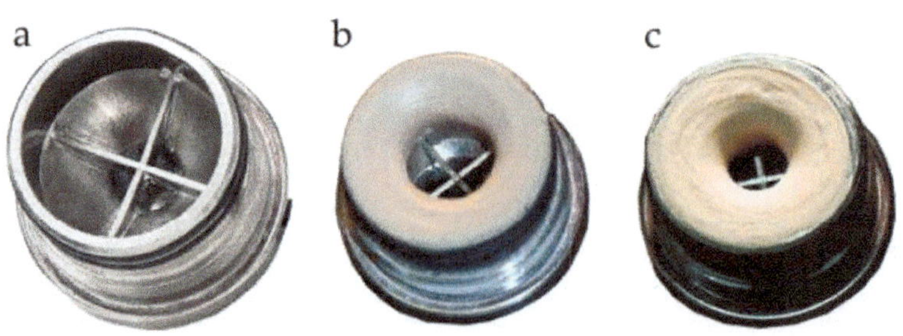

Figure 3: Inlet geometry of tubular bowl centrifuge – a) empty rotor; b) with sediment after 5 min of centrifugation at 15000 rpm; c) as "b" but at 40000 rpm.

Rotor geometry

Other inlet systems have been analysed in a recent research project (Spelter, et al. 2011). The liquid is accelerated by rotating inlet pipes and the outlet of these pipes is below the pool surface. This minimises the lag of the tangential velocity of the incoming liquid significantly and the trapping of air in the fluid. This feeding system may be used in existing centrifuges in order to enhance the separation efficiency by reducing the axial section where the tangential velocity gradient reduces the effective centrifugal force on the settling particles. Figure 4 shows a simplified scheme of the rotor of the test rig. The test rig is operated with two different rotor setups. Either the centrifuge runs using both cylinders, further referred to as tandem setup, or without the inner glass cylinder, further referred to as overflow setup. The tandem setup is similar to the one applied in high-performance centrifuges used for the production of vaccines. In these centrifuges, the outer rotor is often made of titanium and the inner one of plastic, e.g. Noryl®. The liquid enters the cylinder halfway between inner and outer cylinder at a radius of 42 mm at the top. The inlet is marked with an arrow and the height of the inlet is further referred to as H=0 mm. Both cylinders and the feed pipes rotate with the same revolutions per minute. There is no air in the system as the void is entirely filled with water. The liquid leaves the centrifuge at two outlet bores at a radius of 36 mm. In the overflow setup, only the outer cylinder is used and thus an air core is present. The liquid leaves the centrifuge via the two outlet bores at 36 mm and via two additional outlet bores at a radius of 30 mm, which adjust the minimum pool level. The two bores at 36 mm can be sealed. The height of the outlet is at H=170 mm, the overall length of the cylinder is 200 mm. Depending on the pressure drop at the outlet bores, the water level rises above the radius of the weir. The superelevation is the equilibrium of pressure drop, rotational speed and throughput. With higher rotational speed and lower throughput the elevation is close to zero. With increasing throughput and lower rotational speeds, the centrifuge can be entirely filled. In the test rig, the maximum elevation is limited by a bore (radius 12.5 mm) that is necessary for the measurement of the flow pattern with Laser Doppler Anemometry (Spelter, et al. 2011). Once this radius is reached by the liquid, it overflows the circular weir without hindrance.

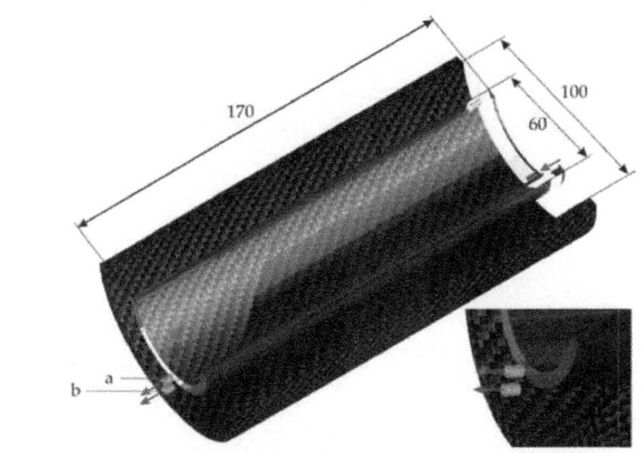

Figure 4: Simplified scheme of the setups of the centrifuge test rig - outer rotor made of carbon fibre reinforced plastic, inner one made of tempered glass - a) additional outlet for overflow setup; b) outlet for tandem setup.

MATHEMATICAL FORMULATION OF THE FLUID

To describe the flow pattern, it is necessary to determine the velocity components in all of the spatial directions considered as well as the pressure and, in the case of compressible flow, the density at any time and position in the domain. This chapter summarizes the mathematical methodology used to calculate these parameters. The conservation equations of the fluid mechanics, conservation of mass, momentum and energy, can be obtained by means of a balance around an infinitesimally small fluid element. The continuity equation, which is derived from the mass conservation, describes the temporal change of mass in the volume element as the net rate of flow mass into the element across its boundaries. In the case of the incompressible flow considered here, the continuity equation simplifies to

$$\nabla \cdot \vec{v} = 0 \,. \tag{2}$$

In an analogous way, the momentum equation can be determined based on the conservation of momentum in all of the spatial directions considered. Newton's second law states that the rate of change of momentum of a fluid element equals the sum of the forces acting on the volume element. These forces can be surface forces, such as viscous forces and shear and normal stresses, or mass forces, such as gravity and centrifugal forces. For incompressible and Newtonian fluids in turbulent regime, the resulting momentum conservation equation follows to be

$$\rho\left(\frac{\partial \vec{v}}{\partial t}+(\vec{v}\cdot\nabla)\vec{v}\right)=-\nabla p+\mu\,\nabla^2\vec{v}+\nabla\cdot\tau_t+\vec{F}\,, \tag{3}$$

where τ_t represents the tensor of turbulences and F other additional forces that can be included expressed as volumetric forces. All terms from Equation (3), the Reynoldsaveraged Navier-Stokes equation (RANS), are discretized and calculated for each volume cell with the exception of τ_t, which is modelled. The RANS equation together with the continuity equation form an equation system of four differential equations with four unknown variables, the flow velocities u, v, w in x, y and z direction, respectively and the pressure p, which can be numerically calculated. This equation will be used for simulation of the centrifugal flow when only one phase is considered. For the cases where the liquid pool rotates around an air core, a multiphase approach must be used in order to simulate both, the air and the liquid. The presence of solid particles is ignored for flow simulation purposes, which is an acceptable assumption for low solid concentrations of the suspensions, as is the case in the present study.

Two-phase approach

The Volume of Fluid method (VOF) (Hirt & Nichols, 1981) is an interface tracking method that simulates the gas-liquid multiphase flow. This method is a simple and efficient formulation designed to track the interface of two phases that do not interpenetrate with a relatively small number of interfaces. Hence it is often used to simulate the gas-liquid multiphase flow in industrial devices (Brennan, 2003; Brennan et al., 2007; Li et al., 1999; Mousavian & Najafi, 2009). VOF introduces a variable a which takes values from zero to one and represents the volume fraction of one of the phases in each cell. A continuity conservation equation is solved for each phase

$$\frac{\partial}{\partial t}\left(\alpha_q \rho_q\right)+\nabla\cdot\left(\alpha_q \rho_q \vec{v}_q\right)=0 \tag{4}$$

Furthermore the volume fraction of each phase α_q in any cell has to obey

$$\sum_{q=1}^{n}\alpha_q = 1\,. \tag{5}$$

Using the VOF method, a single momentum equation, the Reynolds-averaged Navier-Stokes equation, is solved throughout the domain, and the resulting velocity field is shared among the phases. The dependency on the volume fraction is implemented by using volumeaveraged values for the density $\bar{\rho}$ and viscosity $\bar{\mu}$, as explained in Equations 6 - 8.

$$\bar{\rho}\left(\frac{\partial \vec{v}}{\partial t} + (\vec{v}\cdot\nabla)\vec{v}\right) = -\nabla p + \bar{\mu}\,\nabla^2\vec{v} + \nabla\cdot\tau_t + \vec{F} \tag{6}$$

$$\bar{\rho} = \sum_{q=1}^{n} \alpha_q\, \rho_q \tag{7}$$

$$\bar{\mu} = \sum_{q=1}^{n} \alpha_q\, \mu_q \tag{8}$$

Turbulence models

Due to high velocities and complex flow areas, most of the flow patterns of interest in process engineering, aviation, and automotive engineering become unstable above certain Reynolds numbers and, thus turbulences emerge. Turbulent eddies appear at very different time and length scales and are always three-dimensional. The kinetic energy stored in the rotational movement of the large eddies is passed to smaller eddies, until they disappear by energy dissipation. A turbulent flow is characterized by a chaotic and random fluctuation of the flow properties velocity and pressure. The conservation equations are able to capture the physics of the fluctuation motion, but in order to solve all the vortex structures of a turbulent flow, an extremely fine grid and very small time steps would be necessary. The memory and the computing power required for the socalled Direct Numerical Simulation (DNS) can only be provided by supercomputers. For that reason, turbulence is usually modelled with different mathematical approaches. A large-eddy simulation (LES) performs a filtering of the fluctuating flow quantities. The small eddies have a common behaviour and are nearly isotropic, while large eddies are more anisotropic and their behaviour is strongly influenced by the geometry of the domain, the boundary conditions and the body forces. Thus larger eddies are directly calculated, while fine structures are modelled. As a result, the computational demand decreases regarding to the DNS but it is nonetheless high in comparison with the approaches modelling all the turbulence scales. The flow variables can be represented as the sum of a time-averaged size and an additional fluctuation:

$$u = \bar{u} + u'; \quad v = \bar{v} + v'; \quad w = \bar{w} + w'; \quad p = \bar{p} + p'. \tag{9}$$

The substitution of these expressions in the momentum conservation equation for incompressible flows and the subsequent temporal averaging yield to the Reynoldsaveraged conservation equations. In the modified continuity equation, the velocity component is just replaced by the time-averaged variable. In the Reynolds-averaged NavierStokes equation (RANS), the fluctuation velocities appear in an additional tensor, the Reynolds stresses tensor τ_t :

$$\tau_t = -\rho \begin{pmatrix} \overline{u'u'} & \overline{u'v'} & \overline{u'w'} \\ \overline{v'u'} & \overline{v'v'} & \overline{v'w'} \\ \overline{w'u'} & \overline{w'v'} & \overline{w'w'} \end{pmatrix}. \tag{10}$$

The description of a turbulent flow with the RANS equation requires modelling of these tensor terms. The most complex classical turbulence model is the Reynolds Stress Model, RSM, (Launder et al., 1975), which uses transport equations to model each of the elements of the stress tensor and in addition an equation for the dissipation rate of the turbulences. This means that five additional transport equations are required in two-dimensional flows and seven in three-dimensional flow. This way, the directional effects of the Reynolds stress field can be taken into account. This model is recommended for flows with complex strain fields, such as highly swirling flows, or significant body forces. However, the computational cost of RSM is often unaffordable.

An effective method to reduce the modelling effort is to apply the Boussinesq approach, Equation (11). This approach introduces the eddy viscosity or turbulent viscosity μ_t, which represents the momentum and energy transport by diffusion by the eddies or turbulent fluctuations. Thus the Reynolds stresses can be calculated as the viscous stresses and only this new variable μ_t must be modelled.

$$-\rho\,\overline{u'v'} = -\rho\,\overline{v'u'} = \mu_t\left(\frac{\partial \bar{u}}{\partial y} + \frac{\partial \bar{v}}{\partial x}\right) \tag{11}$$

One of the most common turbulence models for the determination of μ_t is the k- ε model (Launder & Spalding, 1974). This model introduces two variables to calculate μ_t, the turbulent kinetic energy k, and its dissipation rate ε :

$$\mu_t = \rho C_\mu \frac{k^2}{\varepsilon}, \tag{12}$$

where C_μ represents a semi-empirical dimensionless constant. The temporal and spatial changes of both variables are described with a transport equation which includes not just the convection and diffusion but also the creation of turbulence and its dissipation by using several source and sink terms. This advantage justifies the additional expense of the transport equation turbulence models compared to algebraic turbulence models, which presume that the turbulence only depends on local conditions. The disadvantage of the k-ε model is that it assumes the turbulences to be isotropic, which is not strictly true in most of the flows. This model, well established and most widely validated, has already been used to simulate the flow in cyclones and centrifuges; although in its development this model assumes a fully turbulent flow, which is only partially applicable to centrifuges.

The k-ε renormalization group model, k-ε RNG, (Yakhot & Orszag, 1986) is an extension of the standard k-ε model that takes into account the effect of swirl on the turbulence by means of an extra source term in the transport equations. Thus the k-ε RNG model exhibits a higher accuracy for swirling flows. The RNG procedure systematically removes the small scales of motion from the governing equations by expressing their effects in terms of larger scale motions and a modified viscosity (Versteeg & Malalasekera, 2007). While the standard k- ε model is appropriate for high-Reynolds-number flows, the RNG theory accounts for lowReynolds-number effects. Thus this model is more reliable for a wider class of flows than the standard k- ε model. The k- ε RNG model has been succeeding for previous simulations of the flow in centrifuges by other groups and in our own research group (Romani Fernandez & Nirschl, 2009; Spelter & Nirschl, 2010).

The k-ω model (Wilcox, 1988), like the k- ε model, uses two equations to describe turbulence. Instead of the dissipation rate ε, the specific dissipation or turbulent frequency $\omega = k/\varepsilon$ is introduced in this model. The turbulent viscosity is then given by Equation (13). The accuracy of the model depends on the undisturbed velocity of the fluid outside the boundary layer which is subject to large fluctuations. This dependence can lead to significant errors in the calculation (Bardina et al., 1997). Due to the strong dissipation of the turbulence on the wall, no special treatment of the boundary layer at the wall is necessary. This feature is desired for the exact calculation of the flow near the wall.

$$\mu_t = \rho \frac{k}{\varpi} \tag{13}$$

Comparing the turbulence models, the k-ω model, as a two-equation model, has a similar range of strengths and weaknesses as the k- ε model. RSM is complex, but it is generally accepted as the simplest kind of model with the potential to describe all mean flow properties and Reynolds stresses without case-by-case adjustment. RSM is by no means as comprehensively validated as the k- ε model and due to the high computational cost it is not as widely used in industrial flow simulations as the k- ε model. LES, due to the inherent unsteady nature, is much more computationally expensive than the k- ε and k-ω models. However, compared to RSM it was proved that it requires only twice the computational demand of RSM (Ferziger, 1977) as cited in (Versteeg & Malalasekera, 2007). This is not a big difference taking into account the solution accuracy and the ability of LES to resolve certain time-dependent features. In order to investigate the applicability of the different turbulence models to the simulation of the flow pattern in centrifugal field, different approaches, the k- ε RNG model, the k-ω model, the RSM, and the LES, were used in this work.

Boundary conditions, discretization schemata and solver

In order to simplify the simulation, the air is considered as an incompressible gas. This is a reasonable assumption for the operation conditions of atmospheric pressure and for a nontemperature dependent problem. The fluid is defined as an incompressible Newtonian one with the density and viscosity of water. The inlet velocity is set to match the desired volume flux. Using a turbulence model, also the turbulence quantities of the inlet flow must be specified. The input of fixed values for k and ε or k and ω are difficult to estimate. Thus these parameters are usually obtained as a function of the turbulence intensity and a characteristic length which must be set at the boundaries (Paschedag, 2004). For the multiphase simulations, a value of one was given to the volume fraction of water at the inlet. A static pressure of 101325 Pa (1 atmosphere) is set as ambient condition for the outlet, defined as a pressure outlet. Sometimes, there is a backflow through the pressure outlet in order to obey the mass conservation. Therefore it is recommended to set realistic conditions for a possible backflow to avoid convergence problems. For the cases presented, only backflow of air is allowed. It is possible to reduce the computational demand by dividing the geometry into periodic sections. The flow pattern is calculated in one of these segments with periodic conditions at the intersecting planes. The periodic surfaces were defined with a periodic boundary condition which uses the flow conditions at one of the periodic surfaces as its cells were the adjacent cells of the other periodic plane. The rotational periodicity must be taken into account and the rotation axis must be defined. Due to the rotational nature of the system, it is helpful to use a rotating reference frame to perform the calculations. The side and top walls of the bowl, as well as the inlet accelerator, all defined with no-slip condition, rotate with the same angular velocity as the reference frame.

Another problem emerging from the use of turbulence models is the treatment of turbulent quantities at the walls. Despite the turbulent flow, there always exists a laminar sublayer at the wall followed by a transient region until the flow can be considered as turbulent. A common approach is to create a first layer of cells next to the wall with a width including the entire viscous sublayer and the transition layer. In this layer, standard wall functions (Paschedag, 2004), most widely used for industrial flows, can be applied. The wall functions define values for the velocity field and the turbulent quantities suitable as boundary conditions for the solution in the second cells layer, where the turbulence is fully developed. This approach is particularly suitable for flows with larger spatial extent in which the thickness of the boundary layer at the wall is small compared to the extension of the whole domain. To choose the thickness of the first layer, the criterion of the y+ value (Equation 14) is often used. The value of y^+ is a function of the first layer thickness y, the shear stress at the wall τ_w, and the physical properties of the fluid density r and viscosity μ. Values of y^+ between 30 and 100 allow the use of standard wall functions. For details concerning the exact formulation of the y^+ and standard wall functions, the reader is referred to (ANSYS, 2009) which is based on (Launder & Spalding, 1974).

$$y^+ = \frac{\rho y}{\mu} \sqrt{\frac{|\tau_w|}{\rho}} \tag{14}$$

The k- ε model and the RSM need these wall functions in order to calculate the turbulent properties near the wall. The k-ω model does not require wall-damping functions, because the value of turbulence kinetic energy at the wall is set to zero and for the turbulent frequency ω a hyperbolic variation at the near-wall grid point is applied (Wilcox, 1988). The discretization schema used for the velocity and the turbulent variables is set to firstorder upwind, due to the convective nature of the flow in centrifugal field and to avoid convergence problems. For the volume fraction, the Geo-Reconstruct method (ANSYS, 2006) is applied in order to obtain a sharp interface between both phases. For the interface between two fluids, this method assumes a linear slope, which is calculated with a piecewise linear approach. The variable pressure was discretized with the PRESTO! (PREssure STaggering Option) scheme recommended for high speed rotating flows (ANSYS, 2009). If velocities and pressure values are both defined at the centres of the cells, a highly non-uniform pressure field can act like a uniform field in the discretized momentum equations (Versteeg & Malalasekera, 2007). A solution to this problem is to use a staggered grid to calculate the pressure via a discrete continuity balance. In the staggered grid, the values of pressure in the centre are known and these are the values at the interfaces in the normal grid.

The temporal discretization of the transport properties was performed using a first-order implicit method. However, using the VOF model, the time step for the variable volume fraction is refined with respect to the general time step based on the maximal Courant number allowed for this variable. The Courant number, also called Courant-Froudel-Lewy number (CFL), Equation 15, is a dimensionless number that compares the time step of the calculation to the characteristic time needed by a fluid element to cross a cell. In some explicit time discretization methods, the criterion CFL<1 (Oertel & Laurien, 2003) is imposed to achieve stability in the numerical calculation. An explicit schema limited by a Courant number of 0.25 was applied for the volume fraction.

$$\mathrm{CFL} = \frac{\Delta t}{\Delta x_{\mathrm{cell}} / v_i} \qquad (15)$$

The simulations were performed with the commercial software ANSYS FLUENT with the versions 6.3.26, 12.0.16, 12.1.2 and 13.0, which are comparable. An overview of the type of solver used by ANSYS can be found in (Kelecy, 2008). The solver used is an unsteady, segregated pressure-based solver. The segregated pressure-based algorithm solves the conservation equations sequentially. Because these equations are coupled, the solution loop must be carried out iteratively in order to obtain a converged numerical solution. Pressure and velocity were coupled either with the SIMPLE algorithm, Semi-Implicit Method for Pressure Linked Equations, or with the PISO schema, which stands for Pressure Implicit with Splitting of Operators and is available for unsteady calculations. The PISO pressurevelocity coupling algorithm has one predictor step and two corrector steps, extending the pressure correction procedure of the commonly used SIMPLE algorithm with a further corrector step (Versteeg & Malalasekera, 2007). The SIMPLE algorithm is a guess-andcorrect procedure which uses a Poisson's equation, developed from the continuity condition, for the pressure correction.

SIMULATED FLOW IN SOLID BOWL CENTRIFUGES

The previously published simulations of a tubular bowl centrifuge were carried out twodimensionally (Spelter & Nirschl, 2010); hence it was not possible to include the real acceleration geometry in the model. For the work presented, all simulations were run threedimensionally, including the acceleration and discharge geometry, and were compared with experimental data. The flow pattern was predicted for a tubular bowl centrifuge with a feeding system similar to the one used in a recent published work (Spelter, et al. 2011). The experimental data has been obtained by Laser Doppler Anemometry, a non-invasive technique for the determination of flow velocities.

Prediction of the flow pattern (k-ε RNG model, transient solution) - tandem setup

Figure 5 shows the tangential velocity profiles for four different rotational speeds at a constant throughput of 1 l/min. There is no significant deviation between the predicted tangential velocity and the rigid body rotation (RBR). The measurement of the tangential velocity for various rotational speeds and throughputs at different heights was in good agreement with the rigid body rotation as well (Spelter, et al. 2011). Thus it can be stated that the simulation is in good agreement with the experimentally determined tangential velocity profiles. Therefore it is possible to optimise existing acceleration geometries by using a feeding system similar to the one proposed. By calculating the present state and modifying the geometry, the changes in the acceleration efficiency can be evaluated by means of CFD. Furthermore the calculation will help in the case of a scale-up of the centrifuge. The acceleration efficiency and the shear in the feed zone can be calculated before a pilot machine or a full scale centrifuge is built. With the knowledge of the flow behaviour, the geometry can be optimised, leading to an improved apparatus with less cost in development and testing.

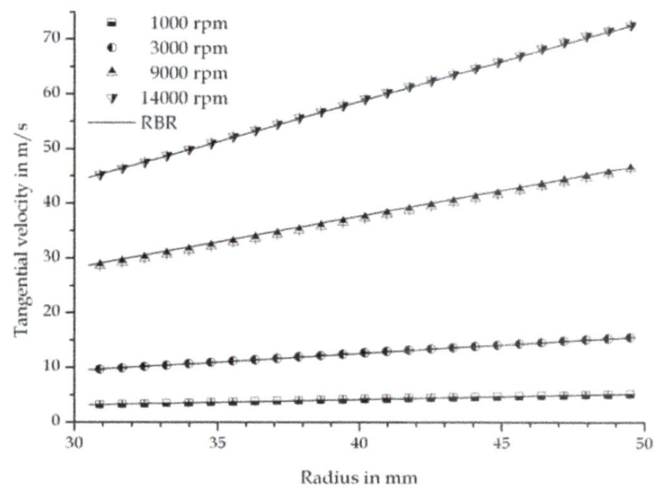

Figure 5: Comparison of calculated tangential velocity and rigid body rotation, 1 l/min.

The tangential velocity of the fluid creates a pressure in the cylinder in addition to the stress in the rotor created by the rotating cylinder itself. For the case that the rotor is designed as an ideal cylinder, the pressure distribution and thus the tensile stress can be calculated analytically (Sokolow, 1971; Stahl, 2004). However in many industrial centrifuges, the rotor has a non-cylindrical shape. For these designs, the stress in the rotor is determined via finite element methods (FEM), but the arbitrary load by the rotating liquid has to be calculated separately. The geometry of the rotor can be imported from various CAD programs such as ProEngineer, Autodesk Inventor, Catia, etc.. The pressure on the different planes in the bowl is then accessible by calculating the pressure distribution for the highest rotational speed that occurs in the centrifuge. The combination of FEM and CFD leads to a complete virtual development and saves time and costs. The simulations will never substitute completely experimental evaluation of the separation efficiency and mechanical integrity of the centrifuge, but the scale-up factor may be significantly increased. The predicted pressure distributions in the cylinder are shown in Figure 6 for 9000 rpm (a) and for a range of rotational speeds in (b). The outlet is located at the top end of the cylinder. The predicted pressures in Figure 6 b) are compared with the analytically calculated pressure distribution for the different rotational speeds. All values are given in bar.

The gauge pressure must be taken into account for dimensioning the rotor. Furthermore it is possible to increase the density of the liquid to model the arbitrary load of a sediment on the bowl wall.

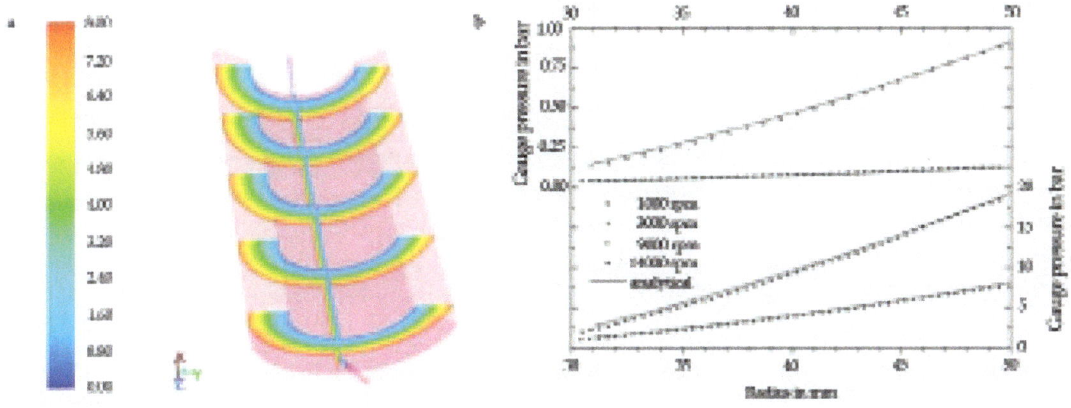

Figure 6: Pressure distribution, all values in bar: a) 9000 rpm, pressure b) 1000-14000 rpm and comparison to analytical values.

The axial velocity profiles are displayed as contour plots in Figure 7 for 1000 rpm (a) and 3000 rpm (b) at a constant throughput of 1 l/min. The inlet is located at the bottom while the liquid leaves the centrifuge at the top. The incoming liquid enters the centrifuge as a jet that does not widen up significantly towards the outlet. At 1000 rpm, the main axial flow takes place at a small circular cross-section that is not displaced in angular direction on the path towards the outlet bores. With increasing rotational speed, a small angular displacement of the axial flow path is predicted, as shown in Figure 7 b). This effect has been observed by Laser Doppler Anemometry as well.

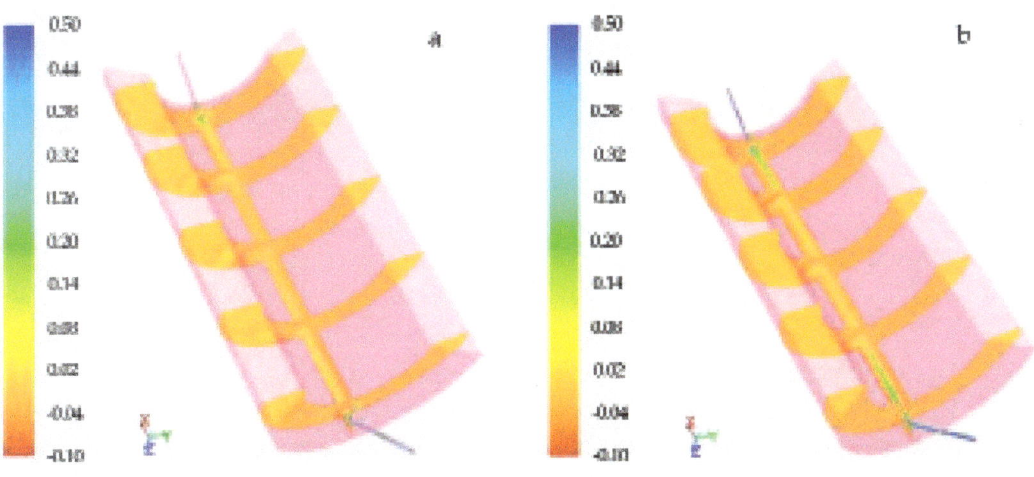

Figure 7: Axial velocity in m/s, throughput 1 l/min: a) 1000 rpm b) 3000 rpm.

However, the predicted axial flow pattern does not depict the experimentally determined flow behaviour accurately. It has been observed that the inlet jet widens within the first 30 % of the rotor (Spelter, et al. 2011) until a parabolic profile is developed that is distributed over the whole circumference. The predicted axial flow profile close to the inlet is in good agreement with the experimental data but the change of flow pattern in direction of the angle of rotation is not correct. The axial flow profiles averaged over the length of the centrifuge between 1000 and 14 000 rpm for two angles of rotation and a constant throughput of 1 l/min are shown in Figure 8. The standard deviation is plotted in both diagrams. The scatter of the data is very high for the angle where the inlet is located, which is 0°. The diagram on the right side of Figure 8 shows the axial flow profiles for an angle of rotation of ±45° displacement to the inlet. There is no significant flow towards the exit of the rotor, only recirculating currents are predicted. The standard deviation of the data is small when compared to the angle of 0°. The origin of the high standard deviation of the data displayed in the diagram on the left in Figure 8 is caused by the change in flow pattern from the feed towards the liquid discharge. The radius of the inlet is higher than the radius of the outlet. Hence the main flow is shifted towards the rotational axis. These effects are displayed in Figure 9 for 1000 rpm (left) and 3000 rpm (right).

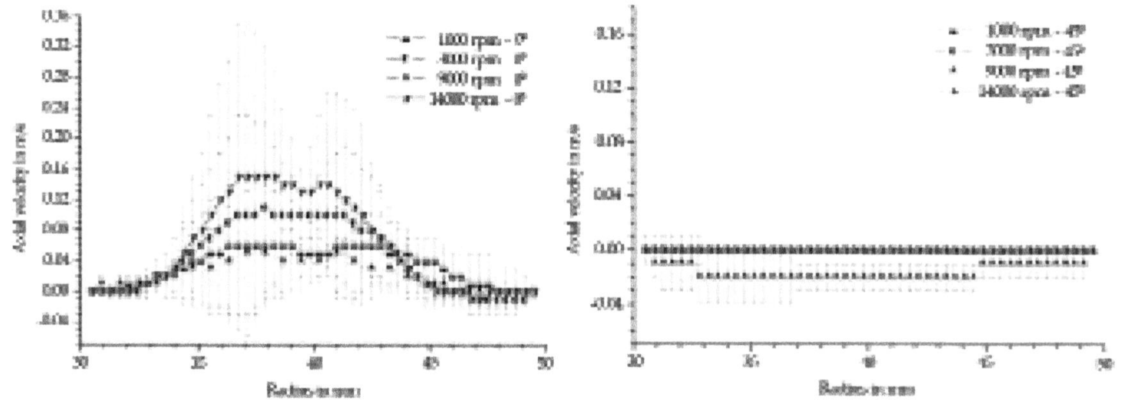

Figure 8: Axial velocity, mean values from throughout the rotor: left: radius of inlet; right: angular position ± 45°.

The maximum of the axial velocity is close behind the inlet (H=0mm) at a radius of 42 mm, which corresponds to the inlet radius. Towards the liquid discharge, the maximum shifts to a radius of 37 mm, where the outlet pipes are located. The change in flow pattern is indicated by the arrows in Figure 9. This behaviour was detected for all rotational speeds and different throughputs. The changing flow profile explains the high standard deviation of the averaged profiles shown in the diagram on the left in Figure 8. The deviation between simulation and experimental data with respect to the flow profile over the full angle of rotation may be explained by the turbulent model used. The modelling does not calculate the turbulences in detail, but estimates the degree of turbulence and thus extenuates the fluctuations at the inlet and outlet. Furthermore the flow between inlet and outlet may not be entirely turbulent, as mentioned in the introduction.

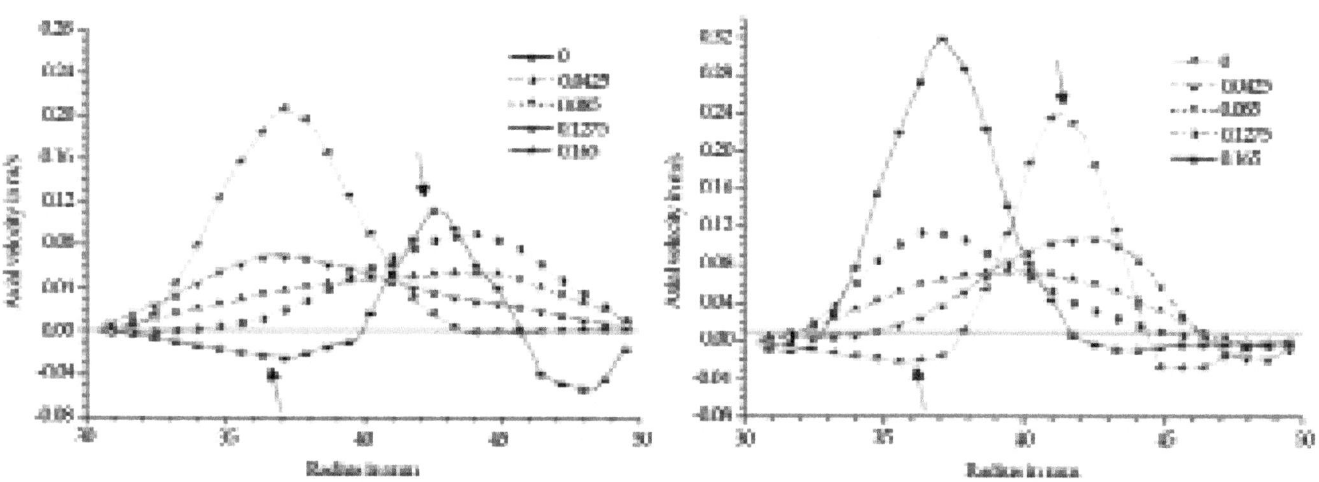

Figure 9: Axial velocity, radius of inlet for different distances from inlet: left: 1000 rpm; right: 3000 rpm.

Comparison of different turbulence models

It is presumed that the deviations between observed flow pattern and predicted flow behaviour are caused by the turbulence model used. Some authors already compared different models that simplify turbulent effects by correlation factors and additional equations like the k-ε RNG model. They stated that the k-ε RNG model is a reasonable compromise between computational demand and validity of the results (Wardle et al., 2006). Other authors relied on the more extensive Reynolds Stress Model (RSM) or the large-eddy simulation (LES) (Mainza et al., 2006; Wang & Yu, 2006). Nowakowski et al. reviewed the milestones in the application of computational fluid dynamics in hydrocyclones (Nowakowski et al., 2004). The most common models used are the k-ε and recently the RSM and LES model. For the evaluation of the differences in the prediction of the flow behaviour in centrifuges between several turbulence models, the flow pattern was calculated for various rotational speeds and throughputs, using the k-ε RNG, k-ω, Reynolds Stress Model and large-eddy simulation. A representative comparison of the obtained results by using the different models is shown in Figure 10. The plots illustrate the differences in the calculated tangential and axial velocity halfway between inlet and outlet at the angular position of the inlet.

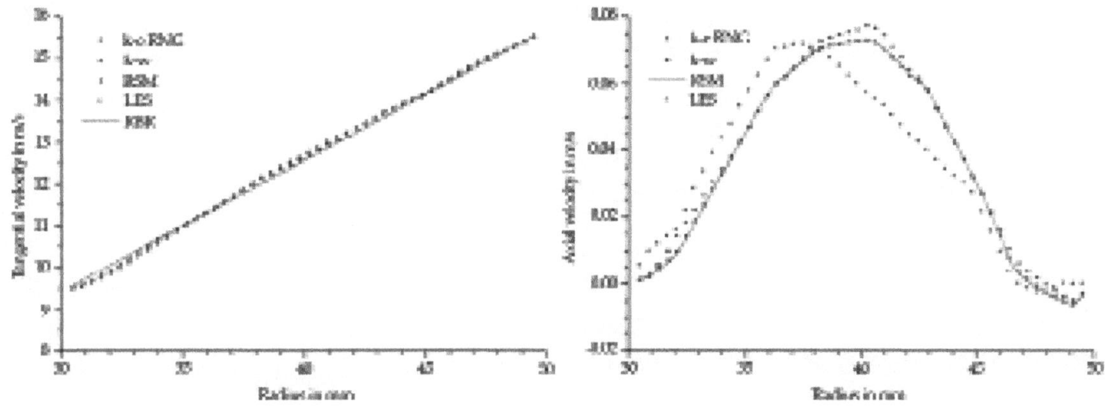

Figure 10: Tangential (left) and axial velocity profiles (right), throughput 1 l/min, k-ε RNG, k-ω, RSM and LES turbulence model.

The comparison reveals no significant deviation of the predicted flow pattern. The differences in the tangential velocity are barely detectable. All calculations are in good agreement with the rigid body rotation and thus with the experimentally determined profiles (Spelter, et al. 2011). Using the LES model, the obtained axial flow profile differs from the other models by predicting an earlier shift of the main axial flow towards the radius of the outlet at 36 mm. Because of the high computational demand and instability of the RSM, the simulation with this turbulence model was started with a solution obtained with the k-ε RNG model. Nevertheless the differences are small and all calculated axial flow pattern are in disagreement with the measured flow profile. The mixing and expanding of the inlet jet is significantly understated. By not questioning and validating the simulations, the user of the CFD software is led to false conclusions which may result in low performance of the centrifuge due to disadvantageous design. The classical evaluation of the simulation does not indicate a computational error. The mass balance between the inlet and the outlet is fulfilled as stated in Table 1, the residuals are sufficiently small (10^{-3} to 10^{-5}) and the flow patterns do not change with increasing flow time. Furthermore the predictions are reproducible with different turbulence models. All these facts suggest an accurate calculation. Only the comparison of the main velocities with experimentally determined data shows that the prediction misrepresents the true case in respect to the axial flow profile.

Table 1: Deviation of mass balance dV for different turbulence models in % at 3000 rpm and 1 l/min

Model	k-ε RNG	k-ω	RSM	LES
% dV	0.0395	0.0015	0.0022	0.0172

Comparison of two and three dimensional modelling

The analysis of the acceleration geometry is only possible in three dimensional modelling, because, as mentioned in the introduction, the inlet geometry cannot be correctly reproduced in the two dimensional case. The results presented in chapter 4.1 and 4.2 show a short-circuit flow between inlet and outlet. The comparison between the results obtained when calculating three and two dimensional are shown in Figure 11. The predicted tangential velocity profiles are in good agreement with the rigid body rotation for both cases at various rotational speeds. The water is fed through an annulus in the two dimensional model and the tangential acceleration takes place via the friction at the wall at both sides of the annulus. According to the prediction, this feeding geometry is sufficient for an acceleration of the incoming liquid up to 9000 rpm. The differences between the cases are distinguishable in the axial flow profiles which are shown in the diagram on the right side of Figure 11.

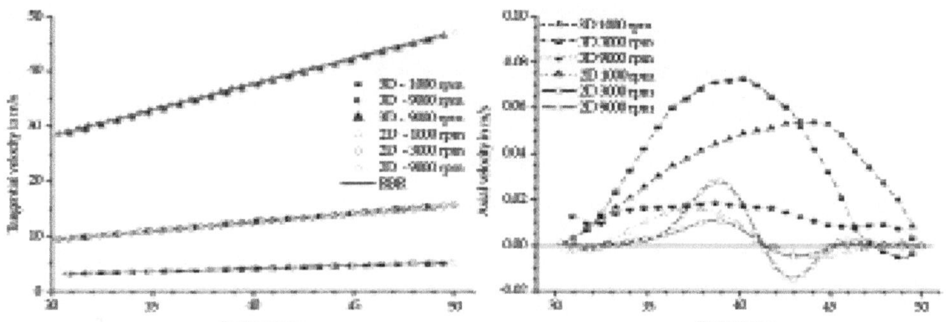

Figure 11: Tangential and axial velocity profiles, throughput 1 l/min, k-ε RNG, k-ω, RSM and LES turbulence model.

The profiles presented are the values halfway between inlet and outlet. Due to the symmetry condition, the flow is homogeneously distributed along the whole circumference for a specific radius. Therefore the maximum axial velocity is lower than in the three dimensional case due to the conservation of mass. There are recirculating currents next to the inlet in the two dimensional cases which increase in magnitude with rotational speed. In the three dimensional case, recirculating flows are located next to the inlet as well, but displaced few degrees in the angle of rotation.

Nevertheless the predicted axial velocity profiles in the two dimensional cases are comparable to the ones obtained in the three dimensional ones in respect to the radial distribution of the flow. The inlet jet remains until the water leaves the centrifuge via the circular weir, so that no plug-shaped profile - resulting in a high active volume -, as observed in the experiments, is formed.

Calculation of the flow pattern (k-ε RNG model, VOF method, transient solution) - overflow setup

The overflow setup was simulated by using the previously described Volume of Fluid method. The centrifuge is filled with water at the beginning of the simulation. As an initial condition, the water is already spinning with the rigid body rotation. This assumption is justifiable because in the experiments, the centrifuge was slowly filled with water and ran for at least 30 minutes in steady state prior to the measurements. The initial condition reduces the necessary time until the steady state is reached significantly. Figure 12 shows the distribution of the two phases for 1000 (a) und 9000 rpm (b) for a throughput of 1 l/min. Due to the gravitational and centrifugal force, a conical shape of the water surface is developed at 1000 rpm. The centrifugal number outbalances the gravitational force with increasing rotational speed, thus the water phase approaches the shape of a cylinder, as shown in Figure 12 b).

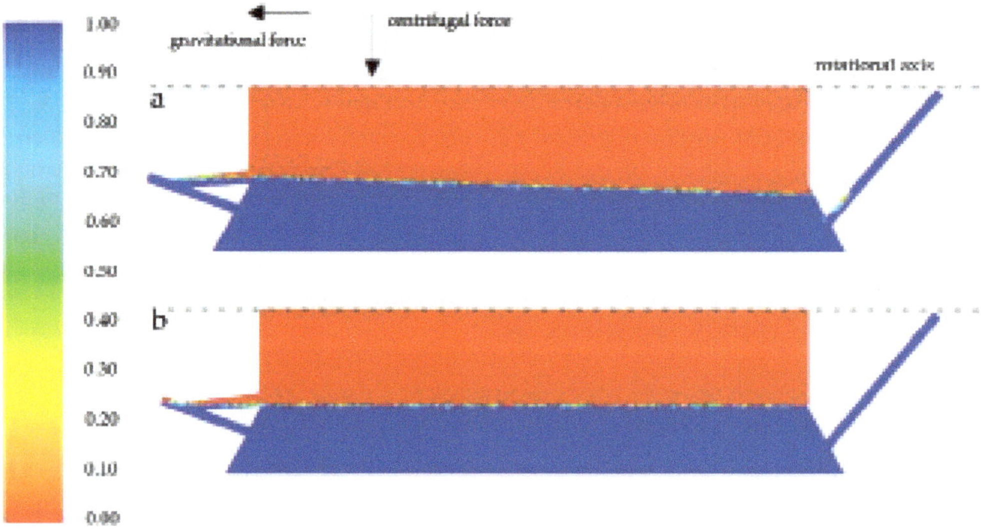

Figure 12: Profiles of water-air interface for a throughput of 1 l/min: a) 1000 rpm; b) 9000 rpm.

The tangential velocity profiles for 1000 and 9000 rpm are shown in Figure 13. The diagram on the right side shows the tangential velocity values averaged over the length of the centrifuge for the water phase only. The computed values are compared to the rigid body rotation. The contours of the tangential velocity at 9000 rpm are shown on the left side of Figure 13. The calculated velocities are in good agreement with the analytical values and with the measurements. At 9000 rpm, the tangential velocity of the water exhibits a small lag when compared to the rigid body rotation. The magnitude of the deviation is within the accuracy of the measurements, so that no statement whether this difference is the true case or not, is possible.

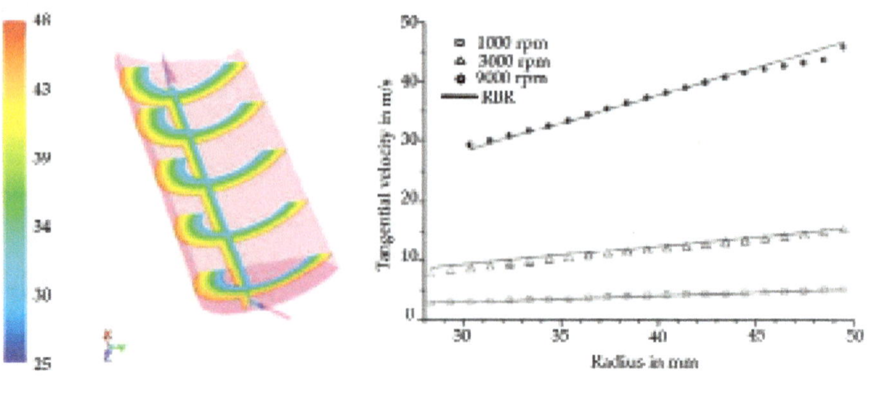

Figure 13: Tangential velocity profiles, water phase, throughput of 1 l/min; left: contours at 9000 rpm; right: for different rotational speeds and different axial positions.

Figure 14 shows the axial velocity profiles for 9000 rpm at a constant throughput of 1 l/min. The axial flow pattern is similar to the one predicted in the single-phase model. The inlet jet does not widen so that a short-circuit flow towards the outlet is formed, as shown in the contour plot on the left side of Figure 14. There is no significant expansion of the flow profile neither in angular nor in radial direction. The peak of the axial velocity shifts from a radius of 42 mm close to the inlet towards a radius of 36 mm. This behaviour was also predicted in the single-phase model. Nevertheless the simulation misrepresents the true case, in which a significant widening of the inlet jet was observed.

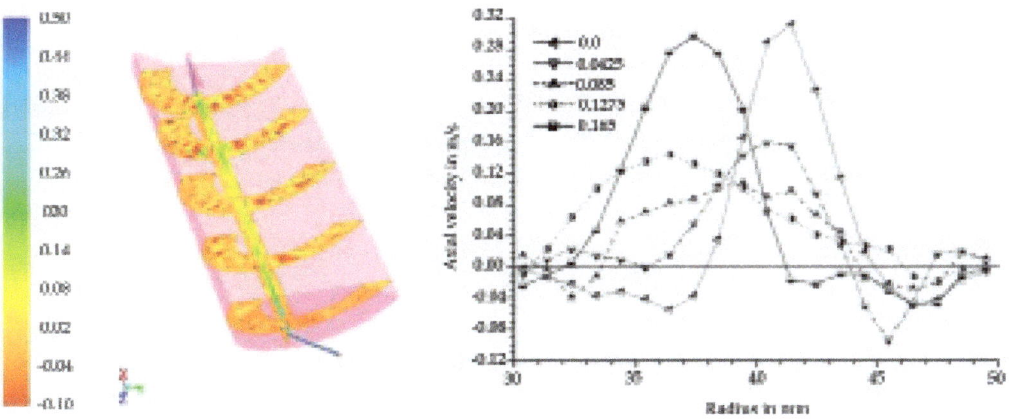

Figure 14: Axial velocity profiles at 9000 rpm and 1 l/min throughput; left: contours; right: different distances from the inlet.

Influence of the inlet geometry on the flow pattern

In order to investigate how the acceleration of the feed affects the flow pattern, a centrifuge with a simpler geometry and a low performance feed accelerator was simulated. The main differences when compared to the tubular bowl centrifuge are the lower length-to-diameter ratio of 0.8 and the feed accelerator. The feed enters a rotating disc and leaves it radially with a certain angular velocity. The water exits the centrifuge through eight boreholes distributed along the circumference at the top of the bowl. Figure 15 shows a scheme of the centrifuge geometry on the left and the results of the volume fraction of air on the right. Water is fed axially through the inlet to the accelerator; there, the water changes its direction and gains in tangential velocity. The transfer of rotational movement occurs just at the plate surfaces, defined as no-slip boundaries, via momentum diffusion. Then, the water exits the accelerator as streams that travel through the air core and enter the rotating liquid pool. The radius of the interface changes 35 % along the height of the centrifuge from the inlet to the upper part of the centrifuge because of the low rotational speed (454 g).

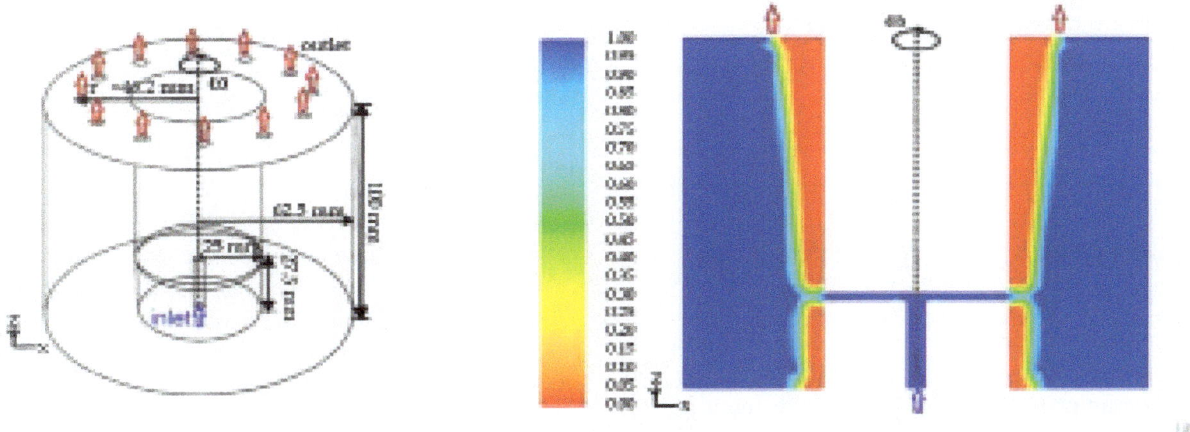

Figure 15: Left: geometry of the centrifuge; right: contours of volume fraction of air at 2550 rpm and 5.8 l/min.

The main flow occurs in the direction of the rotation of the bowl. Both, the liquid pool and the air core rotate. The tangential velocity of the water jet from the inlet accelerator (height 0.0275 m) is lower than the tangential velocity of the bowl. This causes a drag of the whole rotating liquid pool regarding the angular velocity of the bowl. The values of the tangential velocity of the liquid averaged over the circumference are plotted for different heights in Figure 16. The averaged velocity of the liquid in this geometry is approximately 50 % below the rigid body rotation. The values at the wall have to match the rigid body rotation of the bowl to fulfil the no-slip boundary condition. The steep decrease of the tangential velocity near the wall was found to be smoother for a 6 times finer grid. However, the computational effort for the multiphase simulation including the particles with the finer grid was unaffordable. The deviation with respect to the rigid body rotation decreases for lower flow rates because a smaller amount of liquid has to be accelerated. The simulation of this geometry shows how important is to accelerate the feed properly to the angular velocity of the liquid pool before entering it in order to reach the angular velocity of the bowl in the rotating liquid and thus the highest centrifugal acceleration possible. In contrast to the other centrifuge, a layer with high axial velocity at the gas-liquid interface, as expected by the boundary layer model, has been observed in the simulations as depicted in the diagram on the right in Figure 16. Near the impact height of the inlet water jet two whirls with opposed directions are formed. This can be seen at the negative axial velocity values at the interface for the height 0.02 m, just below the inlet, and the positive axial velocities at he interface for the height 0.0275 m, just above the inlet. Upsides the height of the inlet, a homogeneous axial boundary layer is formed at the interface with a recirculation layer at the bowl wall. This layer exhibits a similar averaged thickness δ as the one proposed by the theoretical model of Gosele (Gösele, 1968) but much thicker as the one calculated following the theory of Reuter (Reuter, 1967) as seen in Table 2. Indeed, it occupies the whole liquid pool until a recirculation layer is formed close to the bowl wall. This is due to the lower tangential velocities which cause a small pressure gradient in the radial direction reducing the stratification of the flow and allowing the fluid to move in the axial direction. The velocity values oscillate with the angle, especially near the inlet at 0.02 m and 0.0275 m and the outlet at 0.0975 m, respectively. Thus the standard deviation is higher at these positions.

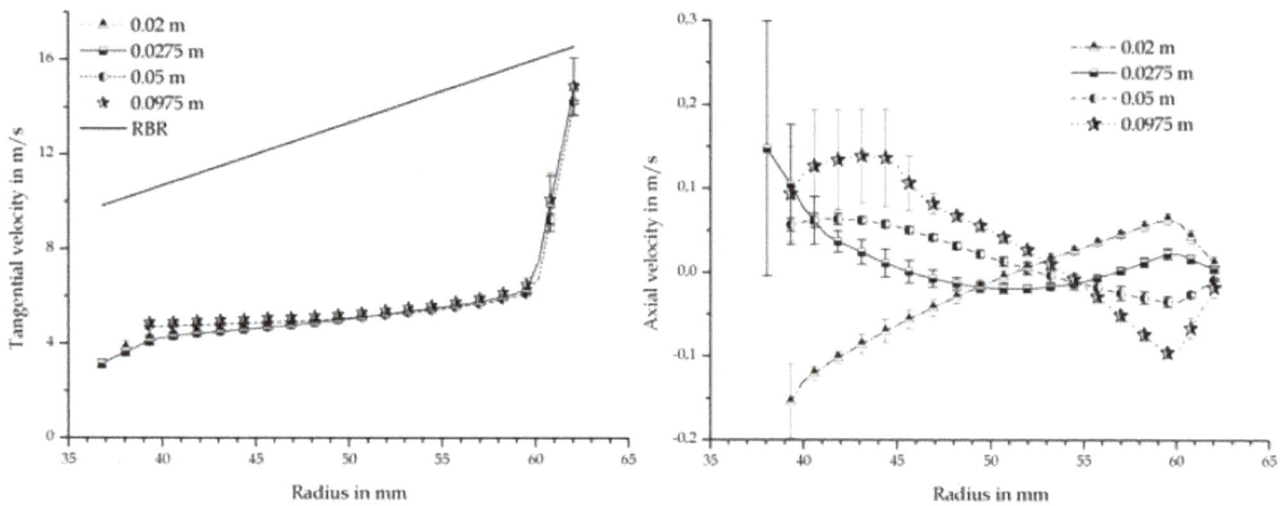

Figure 16: Left: tangential velocity versus radial position; right: axial velocity versus radial position at 2550 rpm and 5.8 l/min.

Table 2: Thickness of the axial boundary layer δ obtained in the simulations and calculated analytically at 2550 rpm and a throughput of 5.8 l/min

$\delta_{simulated}$ (mm)	δ_{Reuter} (mm)	$\delta_{Gösele}$ (mm)
14.1	1.71	15.8

MATHEMATICAL FORMULATION OF THE PARTICLES

There are several multiphase models available for simulation of a particulate phase. A summary of the different simulation methods for multiphase flows in CFD can be found in (van Wachem & Almstedt, 2003). The multiphase models can be divided into two groups; the Euler-Euler and the Euler-Lagrange methods.

The Euler-Euler method considers all phases as continuous ones. For each phase, a momentum conservation equation is solved. In the momentum equation of the dispersed phases, the kinetic theory for granular flows (Lun & Savage, 1987) is used. The intensity of the particle velocity fluctuations determines the stresses, viscosity, and pressure of the solid phase. The interaction between phases is considered by means of momentum exchange coefficients. This numerical method has the advantage that it is suitable for high volume concentrations of the disperse phase. The disadvantage is that the properties of each dispersed particle cannot be simulated and a particle size distribution cannot be considered. Moreover, the stability of the calculation depends on the choice of model parameters and often convergence problems appear. For the simulation of sedimentation and thickening processes there are also some Euler-Euler models based on the Kynch's kinematics sedimentation theory (Kynch, 1952), which were reviewed by Bürger in (Bürger & Wendland, 2001). These models are all based on the measurement of fundamental material properties of the suspension to be separated as explained by (Buscall, 1990; Buscall et al., 1987; Landman & White, 1994). These multiphase models are based on mass balances for the fluid and the disperse phase. The fluid flow is not calculated but they consider the relative velocity between fluid and particles. On this basis, these models are unsuitable for studying the particle behaviour in solid bowl centrifuges where the liquid flow plays a crucial role in particle settling. These models usually describe the sedimentation and consolidation in batch processes in simple-geometry centrifuges. An extension of these models is necessary to describe the sediment build-up in two and three dimensions correctly.

The Euler-Lagrange method solves the continuum conservation equations just for the characterization of the continuous phase. It describes the dispersed phase as mass points and determines the particle trajectories by integrating a force balance for each individual particle. This way, instead of a partial differential equation, only an ordinary differential equation must be solved for the dispersed phase. The weakness of the Euler-Lagrange method is that it is only valid for diluted suspensions if a one-way or a two-way coupling method between continuous and disperse phase is applied. With the one-way coupling, the motion of the discrete phase is calculated based on the velocity field of the continuous phase but the particles have no effect on the pressure and velocity field of the continuous phase. By using a two-way coupling, the equations of continuous and discrete phase are calculated alternatively. The momentum conservation equation of the continuous phase has a source term to consider the momentum exchange with the particulate phase. The maximum value for the volume concentration of the disperse phase is between 5% (Schütz et al., 2007) and 10-12% (ANSYS, 2009), above this value the interaction between particles should be considered. Thus there is a third method, four-way coupling, which involves the interaction within the disperse phase: impacts between particles, particle-wall interactions, van der Waals forces, electrostatic forces, etc. Thus the restriction for the volume fraction is not valid anymore. However, if the number of particles increases, the computational effort rises significantly. Usually the characteristics of the problem and the information that should be obtained from the simulation decide which multiphase model and coupling methodology should be used.

The Discrete Element Method (DEM) (Cundall & Strack, 1979) is a Lagrangian method originally developed for the simulation of bulk solids and then extended to the entire field of particle technology. This method solves the Newton's equations of motion for translation and rotation of the particles. In these equations, impact forces of particle-particle and particle-wall interactions are considered. Other forces such as electrostatic or Brownian forces can also be implemented if necessary. In order to describe the collisions of particles, the soft-sphere approach was chosen. When two particles collide, they actually deform. This deformation is described by an overlap displacement of the particles. A schematic representation of the forces on the particles in the soft-sphere model is depicted in Figure 17. This approach uses a spring-dashpot-slider system to calculate the behaviour of the particles during the contact time. The soft-sphere model allows contacts of a particle with more neighbouring particles during a time step. The net contact force acting on a particle is added in the case of multiple overlapping. This model for contact forces was first developed by Cundall (Cundall & Strack, 1979), who proposed a parallel linear spring-dashpot model for the normal direction and a parallel linear spring-dashpot in a series with a slider for the tangential direction. A comparison of the performance of several soft-sphere models can be found in (Di Renzo & Di Maio, 2005; Stevens & Hrenya, 2005). There, also the Hertz-Mindlin (Mindlin, 1949) non-linear soft-sphere model used in the present work is analysed. This model proposes a normal force as a function of the normal overlap δ_n, the Young's modulus of the particle material and the particle radii. The damping force in normal direction depends on the restitution coefficient, the normal stiffness, and the normal component of the relative velocity between the particles. The tangential force depends on the tangential overlap δ_t, on the shear modulus of the particle material, and on the particle radii; while in the damping force in tangential direction the tangential stiffness and the tangential component of the relative velocity between the particles are considered.

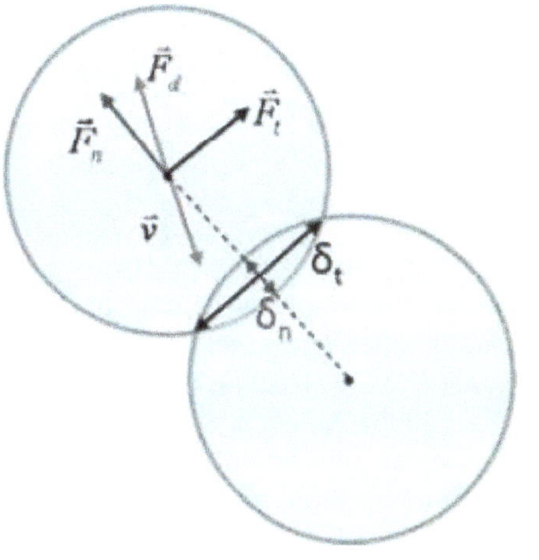

Figure 17: Contact forces between two particles following the soft-sphere model.

While the DEM model was originally used in environments without fluid, increasing efforts have been made to extend this model to the mixture of fluid and particles. For that, a coupling of the DEM calculations with the fluid flow simulation is necessary to be able to include the hydrodynamic forces in the particle simulation. An important aspect in the calculation of particles in a fluid is the energy dissipation into the surrounding fluid. In viscous fluids, there are, in contrast to air, two additional effects that lead to energy losses. The drag force leads to lower particle velocities and when a particle is about to hit a surface or another particle, a deceleration occurs. This is due to the drainage of the fluid, which is located in the gap between the two particles or the particle and the surface. In his work (Gondret et al., 2002), Gondret presents the coefficient of restitution as a function of the Stokes number St (Equation 16), which describes the ratio of the dynamic response time of a particle to flow changes and the characteristic flow time. He determined empirically a critical Stokes number of about 10. Below this value, a particle does not bounce when it hits a wall and instead remains in contact with it.

$$St = \frac{1}{9}\frac{\rho_s v x}{\mu} \tag{16}$$

In a centrifugal field, the hydrodynamic forces (drag, lift, torque) play a crucial role and must be included in the calculation of the particle trajectories. Therefore the fluid flow and particles movement have to be computed alternatively. A scheme of the coupling between the CFD software FLUENT and the DEM software EDEM can be seen in Figure 18. At first, the flow will be determined with CFD, then for each position in the computational domain, the velocity, pressure, density and viscosity of the flow are transferred to DEM in order to calculate the hydrodynamic forces and include them in the balance. Afterwards, the fluid flow will be computed again considering the influence of the particles as an additional sink term in the momentum equations. This sink term include the force transmitted to the particles by the fluid.

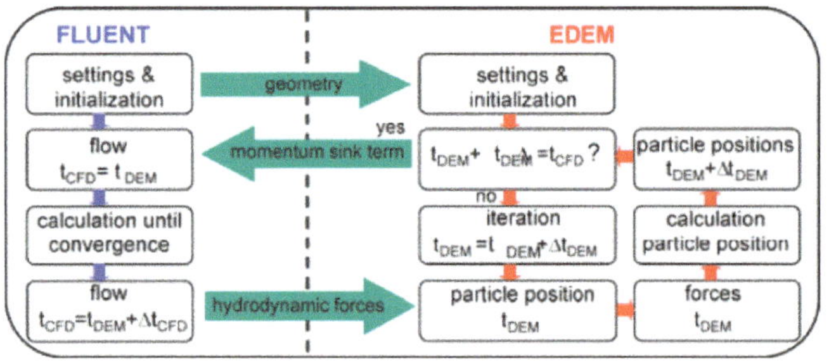

Figure 18: Coupling between CFD and DEM.

Some authors have already simulated such complex multiphase flows in other separation equipment such as filters and hydrocyclones (Chu et al., 2009; Chu & Yu, 2008; Dong et al., 2003; Ni et al., 2006). Nevertheless due to the high velocity gradients and complex interactions between the phases, the simulation of multiphase flow and in particular the sediment behaviour in centrifugal field is still an unsolved challenge. The advantage of this approach is that an accurate description of the flow, the particle paths and the sediment behaviour can be obtained. However, the amount of simulated particles and its size is limited due to the computational demand.

Particles simulation parameters

The calculation in CFD must be run as an unsteady case when coupled with DEM. The DEM time steps are typically smaller than the CFD time steps in order to correctly capture the contact behaviour. The choice of time steps in DEM simulation is of great importance in order to achieve stability in the calculation but does not perform very long calculations, although still representing the real particle behaviour. Therefore it is necessary to analyse the forces acting on the particles to determine the different time scales. In the case presented, the Hertzian contact time - the total time of contact in Hertz theory of elastic collision -, the particle sedimentation time - the time required for a particle to settle one diameter -, and the particle relaxation time - the time for a particle to adapt to the surrounding fluid flow - are considered. On the basis of these, the smallest time step must be chosen so that all physical phenomena are represented. In the simulations presented, the Hertzian contact time is the smallest time scale and, thus a time step $\Delta t_{DEM} = 10^{-6}$ s was chosen in the DEM software, while in the CFD a time step $\Delta t_{CFD} = 10^{-4}$ s was sufficient.

The size of the particles was defined as a logarithmic normal distribution with a mean diameter of 200 µm and a standard deviation of $_{\sigma}$ = 10 µm. Simulations with smaller particles lead to divergence of the calculation despite of the small time steps set. Thus simulations with lower densities of the particles and rotational speeds were performed to obtain similar settling velocities as with small quartz particles (mean diameter of 2.07 µm, $_{\sigma}$=2.15 µm) which were used in the experiments. In order to reproduce the experimental concentration of approximately 0.5 vol. %, 100,000 particles per second were injected at the inlet of the centrifuge for the simulation. In future work, the comparison of predicted and experimentally determined sediment build-up will show whether the simulation methodology leads to correct predictions.

For the contact models, a parameter setting is needed to represent the real particle behaviour. The information needed is mainly about the physical properties of the particle system because the spring and damper constants are calculated as a function of the shear modulus and the Poisson's ratio of the particle and wall materials. The coefficient of restitution depends on the Stokes particle number. The friction coefficient is an empirically determinable property, which can be found in the literature for various material combinations. Some authors have already determined this friction coefficient for sand and glass particles for DEM (Asaf et al., 2007; Li et al., 2005). Here, shear experiments of a sediment made of quartz particles were performed in a Jenike shear cell under normal pressure in order to calculate the friction coefficient as was done by (Hartl & Ooi, 2008).

Table 3: Parameters used in the DEM simulation

Parameter	Value
Particle radius (µm)	200
Particle density (kg/m^3)	1016
Shear modulus particle (Pa)	$2.2 \cdot 10^8$
Shear modulus wall (Pa)	$7.0 \cdot 10^{10}$
Poisson's ratio particle (-)	0.25
Poisson's ratio wall (-)	0.30
Restitution coefficient (-)	0.010
Static friction coefficient (-)	0.787
Rolling friction coefficient (-)	0.100

PREDICTION OF THE SEDIMENT BUILD-UP BY MEANS OF DEM

The geometry of the centrifuge in which the particle trajectories and the sediment build-up were simulated is shown in Figure 15. For the simulation purpose of predicting the sediment build-up it is not possible to use periodic boundaries as was done for the simulation of the fluid. Moreover, we chose this geometry with the VOF model for air and liquid in order to investigate the interaction between the multiphase model and the use of DEM to calculate the particles. The calculations were made with the software ANSYS FLUENT 12.0.16 and EDEM 2.3, which include a coupling feature for both programs. The coupled simulation was performed with a rotational speed of 2550 rpm. For this rotational speed, the Stokes settling velocity of the particles is by two orders of magnitude higher than the Stokes settling velocity of the quartz particles used in the experiments. The results of this simulation are depicted in Figure 19.

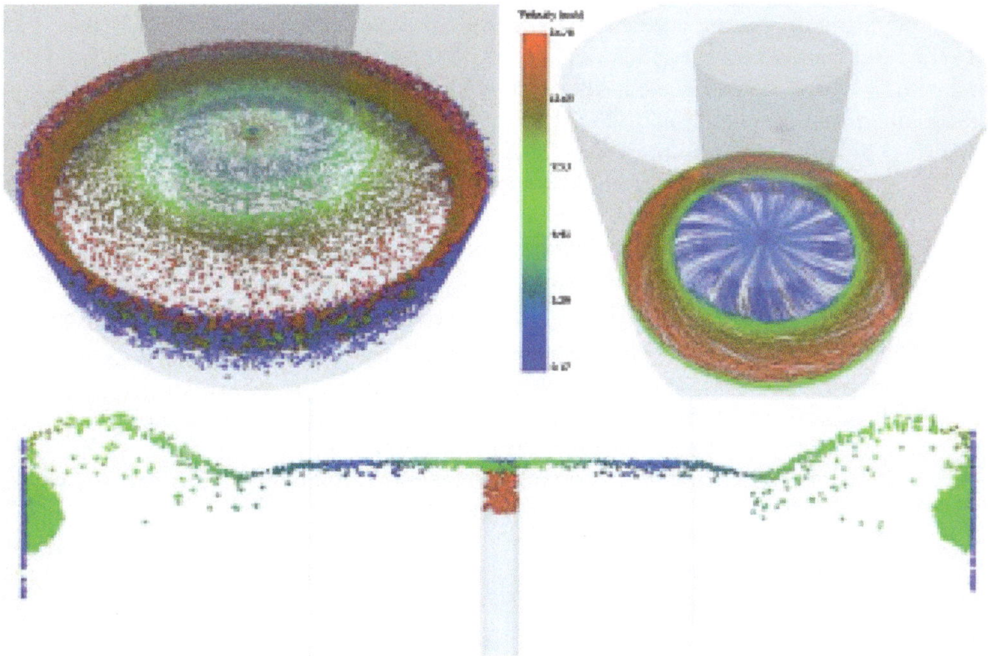

Figure 19: Particles and path lines simulated with EDEM coupled with a two-phase simulation of a centrifuge in FLUENT and coloured by velocity magnitude.

The particles are injected at the inlet pipe with the same axial velocity as the fluid. They hit the top of the feed accelerator and gain in radial velocity. Afterwards, they move through the air core until they hit the liquid pool. There, under the effect of the centrifugal force, the particles perform spiral path lines, as shown in Figure 19 on the right, until reaching the wall or already settled particles. The lower plot in Figure 19 shows a vertical plane of the centrifuge with the particles in the region near the feed accelerator where the sediment is built. Due to the drag force of the axial flow in the rotating pool, the particles move first to higher axial positions. Then, near the wall, they move backwards because of the recirculation layer. The evolution of the sediment can be observed in Figure 20 as a function of the calculated particles for two different rotational speeds of the liquid pool. For the case with lower rotational speed, 686 rpm, the first settled particles deposit at the wall and are pushed to lower axial positions because of the recirculation layer near the wall and due to the lower normal pressure, 3 kPa, acting on the sediment, which allows the influence of the gravity. Then, particles deposit on the sediment and it growths from the wall to the inner part of the centrifuge in radial direction. The sliding of the sediment in this case is due to the relationship of low normal pressure and high shear stress acting on it. In the case of higher rotational speed in the liquid pool, 1937 rpm, the particles travel radially direct to the bowl wall and the sediment formed is flat. The particles settling in the next time steps slide to the side of the sediment which grows along the wall rather than in the radial direction.

In this study the value of the friction coefficient between particles and between particles and wall was constant. Changes of this parameter would also lead to changes in the sediment behaviour under certain conditions of normal pressure and shear stress. A further investigation in this field, including the experimental determination of the friction coefficient for different materials and particle size distributions, is needed. Here only the basis of the coupling methodology was studied.

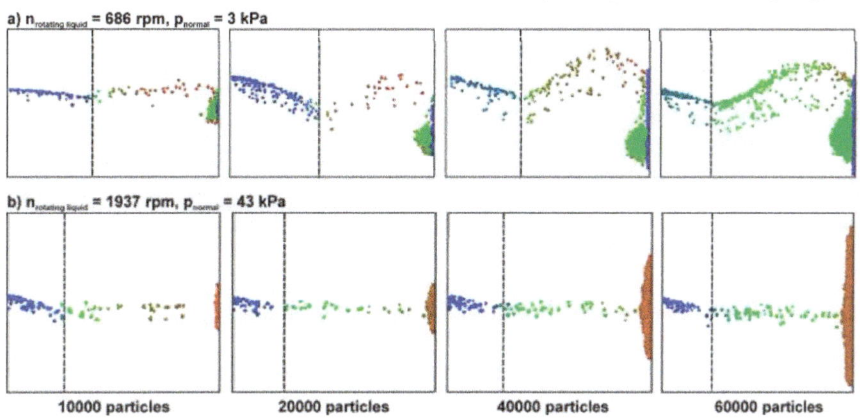

Figure 20: Sediment built at the bowl wall as a function of the particles simulated with EDEM coupled with a two-phase simulation of a centrifuge in FLUENT and coloured by velocity magnitude for two different rotational speeds.

In order to ensure that the coupling between both programs works and that the hydrodynamic forces are correctly passed from FLUENT to EDEM, the particle trajectories calculated with the coupling and with FLUENT were evaluated. Although the time steps and the mathematical method used in FLUENT to calculate the particles is different from the time steps and algorithm used in EDEM, the results are comparable as shown in Figure 21. Another different is that in FLUENT, the particles were calculated using a static flow field at a certain time step whereas in EDEM, the flow field was calculated alternating to the position calculation of the particles. Thus in Figure 21 on the left, the particles calculated in FLUENT acquire higher axial positions when compared to the particles calculated with EDEM. The reason for this disagreement is the difference in the axial flow pattern, where the velocities oscillate until a quasi-steady state is achieved. However, qualitatively both particle trajectories are in good agreement. The tangential velocity of the particles is represented in Figure 21 on the right. The tangential velocity of the particles is almost zero in the regions where there is no liquid. Then, they are accelerated in tangential direction until reaching the bowl wall. There, a deviation between FLUENT and EDEM appears. The particles in EDEM hit the sediment and the tangential velocity diminishes because the wall rotation is not taken into account here, but in FLUENT, no particle accumulation is considered and, thus all particles are calculated until they reach the wall. Once they have reached the wall, they are taken from the simulation domain.

The location and quantity of the particles are stored as a variable in FLUENT representing the volume fraction of the particles, as can be seen in Figure 22 on the left. However, for the flow calculations they are considered as points where a momentum exchange occurs. The drag force of the flow acting on the particles is subtracted from the momentum equation diminishing the flow in regions where the particles accumulate. Figure 22 on the right represents the volume fraction of air in the centrifuge. The effect of the particles on the flow can be seen as the water jet coming from the inlet accelerator falls down in the axial position until it reaches the liquid pool due to the particles weight and the momentum sink term.

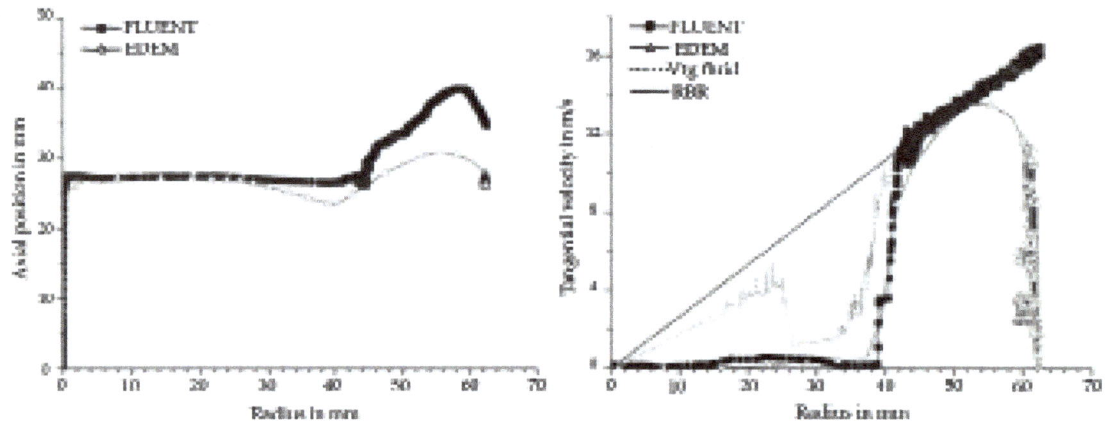

Figure 21: Comparison of the particle trajectories obtained in FLUENT and EDEM: left: axial position versus radius; right: tangential velocity of the particles and the fluid versus radius.

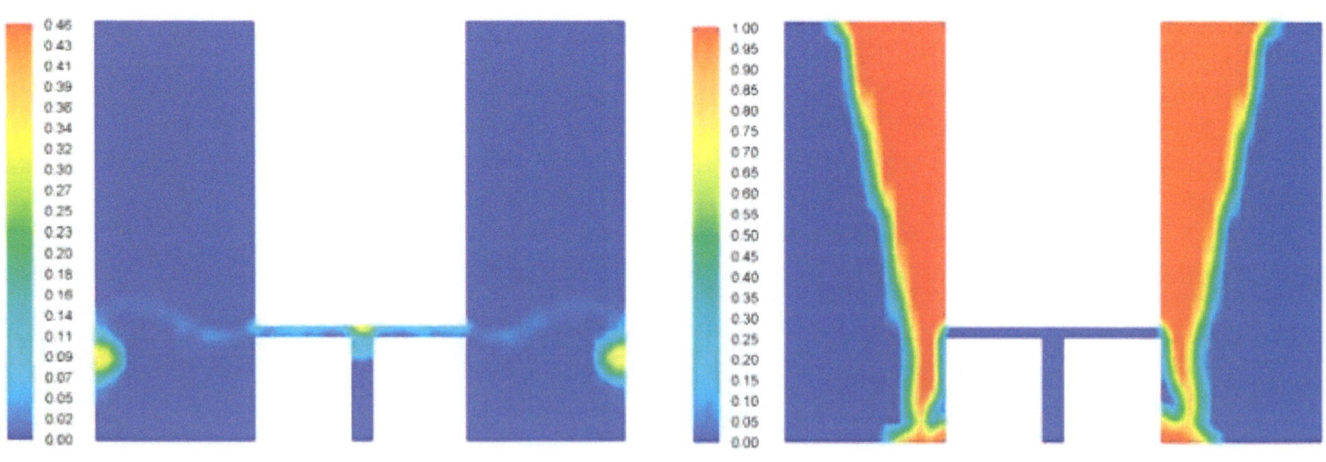

Figure 22: Contours of the volume fraction in the FLUENT simulation domain; left: particles; right: air.

The effect of the sediment on the flow pattern is shown in Figure 23, where the tangential and axial velocity values are represented. The tangential velocity values are lower than in the simulation without particles because of the momentum exchange between the fluid and the particles. Furthermore the tangential velocity is lower in the lower part of the centrifuge, where most of the particles can be found. The interface radius is not constant over the height of the centrifuge, as can be observed in Figure 22 on the right. Therefore the values of the tangential and axial velocity of the water in Figure 23 begin for lower radii in the lower part of the centrifuge. The form of the axial flow pattern does not change qualitatively except for the height where the particles accumulate, around 0.02 m. There, the axial velocity takes approximately the value zero due to the exchange of momentum between fluid and particles. At the wall, where the sediment is located, there is no axial flow in contrast to the simulation without particles. This fact shows how important is to consider the sediment build-up in the multiphase flow simulation of a centrifuge. However this also carries an increase of the computational effort.

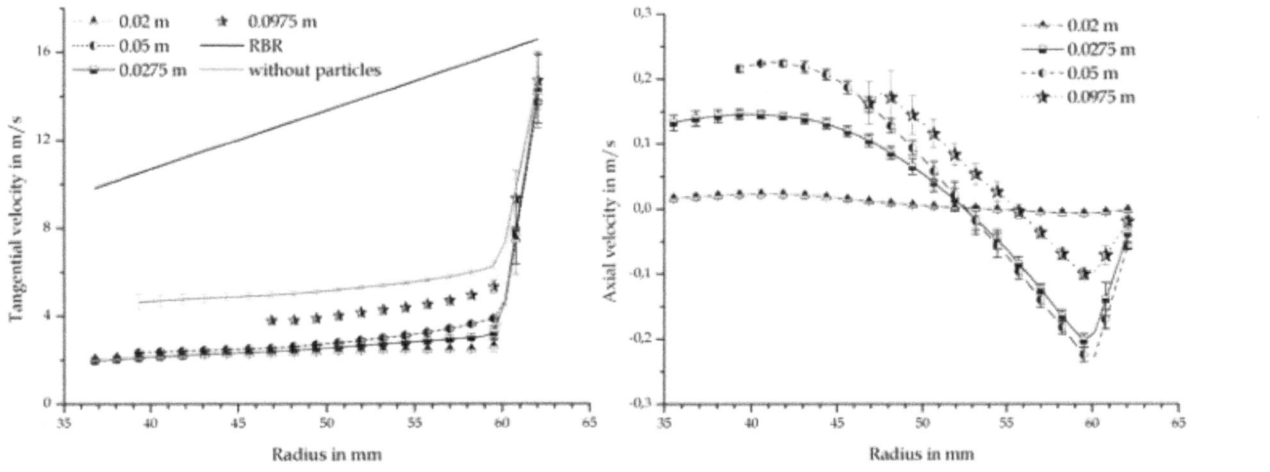

Figure 23: Averaged values over the circumferential position for different axial positions of the centrifuge obtained in FLUENT coupled with EDEM, left: tangential velocity values versus radius; right: axial velocity versus radius.

EVALUATION OF THE COMPUTATIONAL DEMAND FOR PREDICTING FLOW IN CENTRIFUGES

The prediction of flow pattern in centrifuges requires a high computational demand that increases with the maximum velocity in the system. Accordingly to Equation 17, the highest velocity is reached at the bowl wall and increases linearly with the rotational speed.

$$v = \frac{2 \cdot \pi \cdot n}{60} \cdot r \qquad (17)$$

Figure 24 shows the computational demand in $h_{sim}/(h_{flow}$ time 3.6 10^{-3}) (dimensionless) against circumferential velocity in m/s for different simulations which include two and three dimensional problems, one and two-phase simulations, and different models to predict turbulent flow. The simulations were performed with Win xp 32 bit desktop computers using an Intel Core 2 Duo E4500 2.2 GHz processor, 2 GB DDR 2 memory running at 266 MHz (2 x 1GB Kingston) and a ASrock motherboard (dual channel memory). All data are averaged values. For reasons of readability, the standard deviation is not stated, because it is too small to be included in the plot. Evaluating the diagram on the left side of Figure 24, it is remarkable how costly the VOF method for simulating two-phase flow for the two and three dimensional cases is. Moving from one to two phases causes the same increase in computational demand as moving from a two to a three dimensional grid and simultaneously increasing the cell number by one order of magnitude. However, the size of the grid does not necessarily increase the computational time. The convergence in a fine grid is faster and the influence of the rotating geometry overcomes the one of the grid when simulating centrifuges. This is concluded by comparing the three dimensional, single-phase calculations with a grid size of 215000 cells with the simulations with a grid of 715000 cells. The same effect is observed when comparing the three dimensional, two-phase simulations with different cell sizes and different size of geometry. All data is well fitted with a single function for each of the two compared sets of calculations. Looking at the differences between the turbulent models, all simulations exhibit a comparable computational demand. The RSM is the model with the highest computational demand whereas the large-eddy simulation needs less computational resources than all other models evaluated. This is unexpected since the LES is usually rated as more demanding than the RSM. This is explainable due to the single equation modelling of turbulence in the subgrid, the low degree of turbulence in the case presented and the fine mesh that was constant in all cases presented in the diagram on the right side of Figure 24. The k-ε RNG and the k-ω simulations require comparable computational resources. This is in agreement with the theory, because both models are similar to each other except of the modelling of the near wall region.

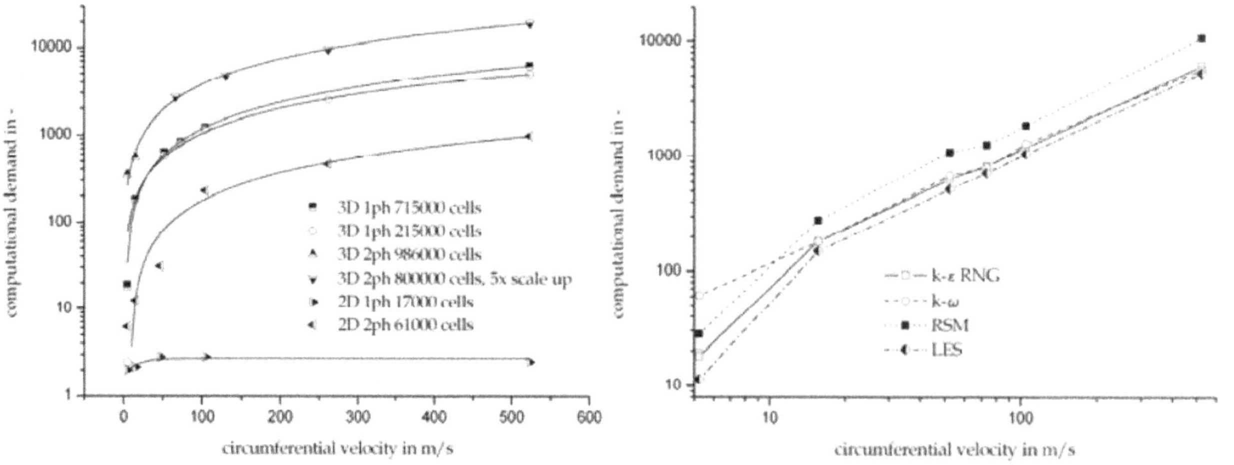

Figure 24: Correlation of computational demand in $h_{sim}/(h_{flow}$ time 3.6 $10^{-3})$ and circumferential velocity: for one and two-phase modelling, two and three-dimensional geometries, and different sizes of the geometry (left); dependence on turbulent model used (right), 715000 cells, three-dimensional and single-phase model.

CONCLUSIONS

The results presented show that the prediction of the flow pattern in centrifuges by means of computational fluid dynamics give good results for evaluation of the tangential acceleration efficiency. The predictions are in good agreement with the measurements conducted with Laser Doppler Anemometry. Nevertheless the axial fluid profiles do not depict the true case accurately. The calculated axial velocity profiles give reasonable results close behind the inlet, but the widening of the inlet jet is not distinguishable in the simulations. The axial velocity profile is essential for the calculation of the cut size and the sediment distribution in the centrifuge, thus any significant deviation from the true case is not acceptable. It was demonstrated that the Reynolds Stress Model (RSM) and the large-eddy simulation (LES) do not increase the accuracy of the predictions when compared to the results obtained using the k-ε RNG and k-ω model, but increase the computational demand in case of the RSM. All models predict a short-circuit flow between inlet and outlet. The flow pattern does not change with increasing rotational speed or throughputs, only the maximum fluid velocity increases with higher throughputs as expected. This behaviour was observed in single and two-phase simulations.

The geometry of the feed accelerator is decisive for the axial flow pattern developed in the centrifuge. A radial feed impacting on the surface of the liquid pool leads to an axial boundary layer flow similar to the one predicted and experimentally verified by other authors such as Glinka (Glinka, 1983). The design of the feed accelerator and its efficiency is essential for a good centrifugal performance. If the liquid is not accelerated to the angular velocity of the bowl, the rotating pool exhibits a significant lag in tangential velocity when compared to the rigid body rotation. The proposed feeding system via rotating inlet bores leads to an efficient tangential acceleration of the fed suspension.

Particles with a diameter of 200 μm were calculated for the first time in a centrifuge using a coupling between the simulation of the flow in FLUENT and the simulation of the particles in EDEM. The results show a good agreement between particles simulated in FLUENT only considering the hydrodynamic forces and the results from EDEM, which implies that the hydrodynamic forces are correctly transferred from one program to the other. Furthermore it proves that the coupling method can be applied for rotating geometries such as centrifuges. The particle trajectories and the sediment build-up agree qualitatively with the expected results. Particles travel through the air core and enter the liquid pool, where they describe spiral paths until they reach the wall or already settled particles to form a sediment. The volume occupied by the particles has not been considered in the calculation of the fluid but the effect of the particles is taken into account by a sink term in the momentum equation. Thus in the flow pattern obtained in FLUENT, the fluid velocity is approximately zero where the sediment is located. The consideration of the particle volume in the calculation of the fluid is a gap in the methodology. A simulation with three Euler phases - air, liquid and particles - calculated in EDEM could be a possible but computationally very expensive solution. The method is complex and there is still some effort needed to reduce the particle size without significantly increasing computing time. A possibility to reduce the computational time is to divide the computational domain into regions with different time steps according to the controlling forces and phenomena.

There is a substantial increase in computational demand with the increase in rotational speed. This limits the predictions to low rotational speeds when standard workstation computers are used for the calculations. If supercomputers are accessible as it will be the case in research at universities, predictions for high rotational speed are possible. But before advancing to higher rotational speeds it is important to improve the accuracy of the simulations of flow in centrifugal force fields and validate the models with experimental data, especially in terms of the axial velocity profile. In order to run a sanity check for the simulations in FLUENT or EDEM, it is not sufficient to evaluate the residuals and mass balance. For new methodologies or boundary conditions, for example when simulating centrifuges, it is necessary to compare different turbulent models, transient and stationary calculation methods and most important to validate the simulation methodology proposed with experimental data. Once validated, geometry and parameters can be changed and the advantage of CFD, which is a fast change to evaluation cycle, can be used.

ACKNOWLEDGMENTS

We acknowledge support by Deutsche Forschungsgemeinschaft and Open Access Publishing Fund of Karlsruhe Institute of Technology.

REFERENCES

1. Anderson, N. G., Waters, D. A., Nunley, C. E., Gibson, R. F., Schilling, R. M., Denny, E. C., Cline, G. B., Babelay, E. F. & Perardi, T. E. (1969). K-Series Centrifuges I. Development of the K-II continuous-sample-flow-with-banding centrifuge system for vaccine purification. Analytical Biochemistry, 32, 3, 460-494
2. ANSYS. (2006). Chapter 23.3.2 Volume Fraction Equation, In: User's Guide Ansys Fluent 6.3
3. ANSYS. (2009). Chapter 4.12.2 Standard Wall Functions, In: Theory Guide Ansys Fluent 12.0
4. ANSYS. (2009). Chapter 18.4.1 Discretization of the Momentum Equation, In: Theory Guide Ansys Fluent 12.,
5. ANSYS. (2009). Chapter 23. Modeling Discrete Phase, In: User's Guide Ansys Fluent 12.0
6. Asaf, Z., Rubinstein, D. & Shmulevich, I. (2007). Determination of discrete element model parameters required for soil tillage. Soil & Tillage Research, 92, 1-2, 227-242
7. Bass, E. (1959). Strömungen im Fliehkraftfeld I. Periodica Polytechics chem. Ingenieurwes, 3, 321-340
8. Bass, E. (1959). Strömungen im Fliehkraftfeld II - Absetzsicherheit von Röhrenzentrifugen. Periodica Polytechics chem. Ingenieurwes, 4, 41-61
9. Bass, E. (1962). Strömungs- und Absetzvorgänge in Röhrenzentrifugen. 249-264
10. Brantley, J. N., Willis, D. D., Breillatt, J. P., Gibson, R. F., Patrick, L. C. & Anderson, N. G. (1970). K-series centrifuges : IV. Measurement and control of temperature. Analytical Biochemistry, 36, 2, 434-442
11. Brennan, M. S., M., N. & Holtham, P. N. (2007). Multiphase modelling of hydrocyclones – prediction of cut-size. Minerals Engineering, 20, 395-406
12. Bürger, R. & Wendland, W. L. (2001). Sedimentation and suspension flows: Historical perspective and some recent developments. Journal of Engineering Mathematics, 41, 2-3, 101-116
13. Buscall, R. (1990). The sedimentation of concentrated colloidal suspensions. Colloids and Surfaces Papers Presented at the Society of Chemical Industry Meeting, 43, 1, 33-53
14. Buscall, R., Mills, P. D. A., Stewart, R. F., Sutton, D., White, L. R. & Yates, G. E. (1987). The rheology of strongly-flocculated suspensions. Journal of Non-Newtonian Fluid Mechanics, 24, 2, 183-202
15. Chu, K. W., Wang, B., Yu, A. B. & Vince, A. (2009). CFD-DEM modelling of multiphase flow in dense medium cyclones. Powder Technology, 193, 3, 235-247
16. Chu, K. W. & Yu, A. B. (2008). Numerical simulation of complex particle-fluid flows. Powder Technology, 179, 3, 104-114
17. Cundall, P. A. & Strack, O. D. L. (1979). Discrete Numerical-Model for Granular Assemblies. Geotechnique, 29, 1, 47-65
18. Di Renzo, A. & Di Maio, F. P. (2005). An improved integral non-linear model for the contact of particles in distinct element simulations. Chemical Engineering Science, 60, 5, 1303- 1312
19. Glinka, U. (1983). Die Strömung in Überlaufzentrifugen - Neue Ergebnisse mit einem elektrolytischen Markierungsverfahren. Verfahrenstechnik, 17, 5, 315-318
20. Golovko, Y. D. (1969). Liquid Lag in Tubular Centrifuge Rotors. Khimicheskoe i Neftyanoe Mashinostroenie, 6, 13-14
21. Gondret, P., Lance, M. & Petit, L. (2002). Bouncing motion of spherical particles in fluids. Physics of Fluids, 14, 2, 643-652
22. Gösele, W. (1968). Schichtströmung in Röhrenzentrifugen. Chemie IngenieurTechnik, 40, 13, 657-659
23. Gösele, W. (1974). Strömungen in Überlauf-Zentrifugen und ihr Einfluß auf die Sedimentation. Chemie Ingenieur Technik, 46, 2, 67
24. Hartl, J. & Ooi, J. Y. (2008). Experiments and simulations of direct shear tests: porosity, contact friction and bulk friction. Granular Matter, 10, 4, 263-271
25. Hirt, C. W. & Nichols, B. D. (1981). Volume of Fluid (VOF) method for the dynamics of free boundaries. Journal of Computational Physics, 39, 1, 201-225
26. Horanyi, R. & Nemeth, J. (1971). Theoretical Investigation of Clarification Process in a Tube Centrifuge. Acta Chimica Academiae Scientarium Hungaricae, 69, 1, 59-75
27. Kelecy, F. J. (2008). Coupling momentum and continuity increases robustness. ANSYS Advantage, 2, 2, 49-51
28. Kynch, G. J. (1952). A theory of sedimentation. Transactions of the Faraday Society, 48, 2, 166- 176
29. Landman, K. A. & White, L. R. (1994). Solid/liquid separation of flocculated suspensions. Advances in Colloid and Interface Science, 51, 175-246
30. Launder, B. E., Reece, G. J. & Rodi, W. (1975). Progress in development of a Reynolds-Stress turbulence closure. Journal of Fluid Mechanics, 68, APR15, 537-566
31. Launder, B. E. & Spalding, D. B. (1974). The numerical computation of turbulent flows. Computer Methods in Applied Mechanics and Engineering, 3, 2, 269-289
32. Leung, W. (1998). Industrial Centrifugation Technology, McGraw.Hill, New York
33. Li, Y., Zhang, J. P. & Fan, L. S. (1999). Numerical simulation of gas-liquid-solid fluidization systems using a combined CFD-VOF-DPM method: bubble wake behavior. Chemical Engineering Science, 54, 21, 5101-5107
34. Li, Y. J., Xu, Y. & Thornton, C. (2005). A comparison

of discrete element simulations and experiments for ‹sandpiles› composed of spherical particles. Powder Technology, 160, 3, 219-228

35. Lun, C. K. K. & Savage, S. B. (1987). A simple kinetic theory for granular flow of rough, inelastic, spherical particles. Journal of Applied Mechanics-Transactions of the Asme, 54, 1, 47-53

36. Mainza, A., Narasimha, M., Powell, M. S., Holtham, P. N. & Brennan, M. (2006). Study of flow behaviour in a three-product cyclone using computational fluid dynamics. Minerals Engineering, 19, 10, 1048-1058

37. Mindlin, R. D. (1949). Compliance of elastic bodies in contact. Journal of Applied MechanicsTransactions of the Asme, 16, 3, 259-268

38. Mousavian, S. M. & Najafi, A. F. (2009). Numerical simulations of gas-liquid-solid flows in a hydrocyclone separator. Archive of Applied Mechanics, 79, 5, 395-409

39. Ni, L. A., Yu, A. B., Lu, G. Q. & Howes, T. (2006). Simulation of the cake formation and growth in cake filtration. Minerals Engineering, 19, 10, 1084-1097

40. Nowakowski, A. F., Cullivan, J. C., Williams, R. A. & Dyakowski, T. (2004). Application of CFD to modelling of the flow in hydrocyclones. Is this a realizable option or still a research challange? Minerals Engineering, 17, 661-669

41. Perardi, T. E. & Anderson, N. G. (1970). K-series centrifuges : III. Effect of core taper on particle capture efficiency. Analytical Biochemistry, 34, 1, 112-122

42. Perardi, T. E., Leffler, R. A. A. & Anderson, N. G. (1969). K-series centrifuges II. Performance of the K-II rotor. Analytical Biochemistry, 32, 3, 495-511

43. Reuter, H. (1967). Sedimentation in der Überlaufzentrifuge. Chemie-Ing. Tech., 39, 9, 548-553

44. Romani Fernandez, X. & Nirschl, H. (2009). Multiphase CFD Simulation of a Solid Bowl Centrifuge. Chemical Engineering & Technology, 32, 5, 719-725

45. Saunders, E. (1948). Nomograph for Particle Size Determination with the Sharples Supercentrifuge. Analytical Chemistry, 20, 4, 379-381

46. Schaeflinger, U. & Stibi, H. (1991). Centrifugal separation of a mixture in a rotating bucket II. Chemical Engineering Science, 46, 8, 2143-2152

47. Schaeflinger, U., Stibi, H. (1991). Centrifugal separation of a mixture in a rotating bucket. The American Society of Mechanical Engineers, 118, 173-182

48. Schütz, S., Piesche, M., Gorbach, G., Schilling, M., Seyfert, C., Kopf, P., Deuschle, T., Sautter, N., Popp, E. & Warth, T. (2007). CFD in der mechanischen Trenntechnik. Chemie Ingenieur Technik, 79, II, 1777-1796

49. Sokolow, W. J. (1971). Moderne Industriezentrifugen, VEB Verlag Berlin, Leipzig

50. Spelter, L. E., Schirner, J. and Nirschl, H., (2011). A novel approach for determining the flow patterns in centrifuges by means of Laser-Doppler-Anemometry. Chemical Engineering Science, In Press, Corrected Proof

51. Spelter, L. E. & Nirschl, H. (2010). Classification of Fine Particles in High-Speed Centrifuges. Chemical Engineering & Technology, 33, 8, 1276-1282

52. Spelter, L. E., Steiwand, A. & Nirschl, H. (2010). Processing of dispersions containing fine particles or biological products in tubular bowl centrifuges. Chemical Engineering Science, 65, 14, 4173-4181

53. Stahl, S., Spelter, L. E. & Nirschl, H. (2008). Investigations on the Separation Efficiency of Tubular Bowl Centrifuges. Chemical Engineering & Technology, 31, 11, 1577-1583

54. Stahl, W. H. (2004). Industrie-Zentrifugen, DrM Press, Männedorf

55. Stevens, A. B. & Hrenya, C. M. (2005). Comparison of soft-sphere models to measurements of collision properties during normal impacts. Powder Technology, 154, 2-3, 99-109

56. Taylor, A. R. (1946). Concentration of the Rabbit Papilloma Virus with the Sharples Supercentrifuge. Journal of Biological Chemistry, 163, 1, 283-287

57. Trawinski, H. (1959). Die äquivalente Klärfläche von Zentrifugen. Chemiker Zeitung Chemische Apperatur, 83, 18, 606-612

58. van Wachem, B. G. M. & Almstedt, A. E. (2003). Methods for multiphase computational fluid dynamics. Chemical Engineering Journal, 96, 1-3, 81-98

59. Wang, B. & Yu, A. B. (2006). Numerical study of particle–fluid flow in hydrocyclones with different body dimensions. Minerals Engineering, 19, 1022-1033

60. Wardle, K. E., Allen, T. R. & Swaney, R. (2006). Computational Fluid Dynamics (CFD) Study of the Flow in an Annular Centrifugal Contactor. Separation Science and Technology, 41, 2225-2244

61. Wilcox, D. C. (1988). Reassessment of the scale-determining equation for advanced turbulence models. Aiaa Journal, 26, 11, 1299-1310

62. Yakhot, V. & Orszag, S. A. (1986). Renormalization-group analysis of turbulence. Physical Review Letters, 57, 14, 1722-1724

63. Zubkov, V. A. & Golovko, Y. D. (1968). Liquid Flow in the Extraction Head of a TubularCentrifuge Rotor. Khimicheskoe i Neftyanoe Mashinostroenie, 12, 18-19

64. Zubkov, V. A. & Golovko, Y. D. (1969). Flow of Liquid in Rotor Outlet of a Tubular Centrifuge. International Chemical Engineering, 9, 3, 403-406

SECTION V: CFD APPLICATIONS TO STUDY OF COMBUSTION, SEDIMENTATION, GRANULATION, OXIDATION PROCESSES IN MANUFACTURING ENGINEERING

CHAPTER 24

POLLUTANT EMISSION VALIDATION OF A HEAVY-DUTY GAS TURBINE BURNER BY CFD MODELING

Roberto Meloni

Department of Astronautics, Electrical and Energy Engineering, University of Rome, "Sapienza", via Eudossiana 18, Rome 00184, Italy

ABSTRACT

3D numerical combustion simulation in a can burner fed with methane was carried out in order to evaluate pollutant emissions and the temperature field. As a case study, the General Electric Frame 6001B system was considered. The numerical investigation has been performed using the CFD code named ACE+ Multiphysics (by Esi-Group). The model was validated against the experimental data provided by Cofely GDF SUEZ and related to a real power plant. To completely investigate the stability of the model, several operating conditions were taken into account, at both nominal and partial load. In particular, the influence on emissions of some important parameters, such as air temperature at compressor intake and steam to fuel mass ratio, have been evaluated. The flamelet model and Zeldovich's mechanism were employed for combustion modeling and NO_x emissions, respectively. With regard to CO estimation, an innovative approach was used to compute the Rizk and Mongia relationship through a user-defined function. Numerical results showed good agreement with experimental data in most of the cases: the best results were obtained in the NO_x prediction, while unburned fuel was slightly overestimated.

INTRODUCTION

In recent decades, stationary gas turbines have become firmly established as prime movers in the gas and oil industry, acquiring new ranges of application in many areas of utility power generation and, in particular, in combined cycle plants [1]. These developments have been accompanied by a continuous increase in pressure ratio, and furthermore, many practices were developed to reduce pollutant emissions and their impact on health and the environment, keeping a high combustion efficiency in a wide range of operating conditions. In general, the combustion efficiency and the performance of a burner depend on the quality of the fuel-air mixture [2]. In systems where the fuel is directly injected into the combustion chamber, it is necessary that the fuel is mixed into the air over a time interval comparable with the burning rate of the mixture. In this system, the Reynolds number characterizes the intensity of the turbulent mixing, while the Damköhler number (Da) characterizes the rate of the mixing against the rate of the chemical reactions. Specifically, this number represents the quotient of a characteristic time scale of turbulence/mixing and a characteristic time scale of chemical reactions: larger Da corresponds to faster reactions [3]. This phenomenon depends on many parameters that can be classified into two

Citation: Roberto Meloni, Pollutant Emission Validation of a Heavy-Duty Gas Turbine Burner by CFD Modeling, doi:10.3390/machines1030081.

distinct groups: those related to the injection system and those related to the environment in which the injected gas evolves. From a numerical point of view, a CFD investigation of the mixture formation processes can be carried out only by making use of meshes whose refinement is able to accurately describe fuel mixing and its dispersion within the region near the injector and the primary zone [4].

In the present article, the numerical study was set with reference to operating conditions of one of the ten burners of the FRAME 6B General Electric gas turbine machine, having the nominal electric power of 40 MW_e. In particular, the simulations referring to both nominal and partial load conditions were investigated, focusing attention on the thermal field and pollutant emissions in order to firstly validate the numerical model through experimental data. Secondly, the model was used to obtain regression laws to predict pollutant emissions out of the simulated ranges or to improve the employed correlations. This information is important, since it represents a starting point of a more complex study for the evaluation of the power plant conversion to bio-liquid. The first step is thus the construction of a numerical model to reproduce the actual configuration and its operating conditions. The experimental data were provided by Cofely GDF SUEZ and related to the combined cogenerative power plant located in Nera Montoro (Narni, Italy), whose characteristics will be described in the following paragraphs.

THE POWER PLANT

The gas turbine we studied is installed within a combined cycle power plant whose steam turbine has a nominal electric power of 12 MW_e (Figure 1). To reduce NO_x pollutant emissions, bled steam from the high pressure turbine is injected within the combustion chambers of the gas turbine. The main operating parameters of the gas turbine and steam injection system are reported in Table 1, and they refer to the nominal operating conditions. The electrical energy is introduced into the grid and remunerated through the CIP6 feed-in tariff law; the steam is sold to the thermal users connected with the power plant.

The power plant can be fed with both methane and off-gas, the latter produced by one of the thermal users. Since the off-gas contribution is minimal, our studies have focused on methane. The most important parameters regarding the fuel injection system are listed in Table 2.

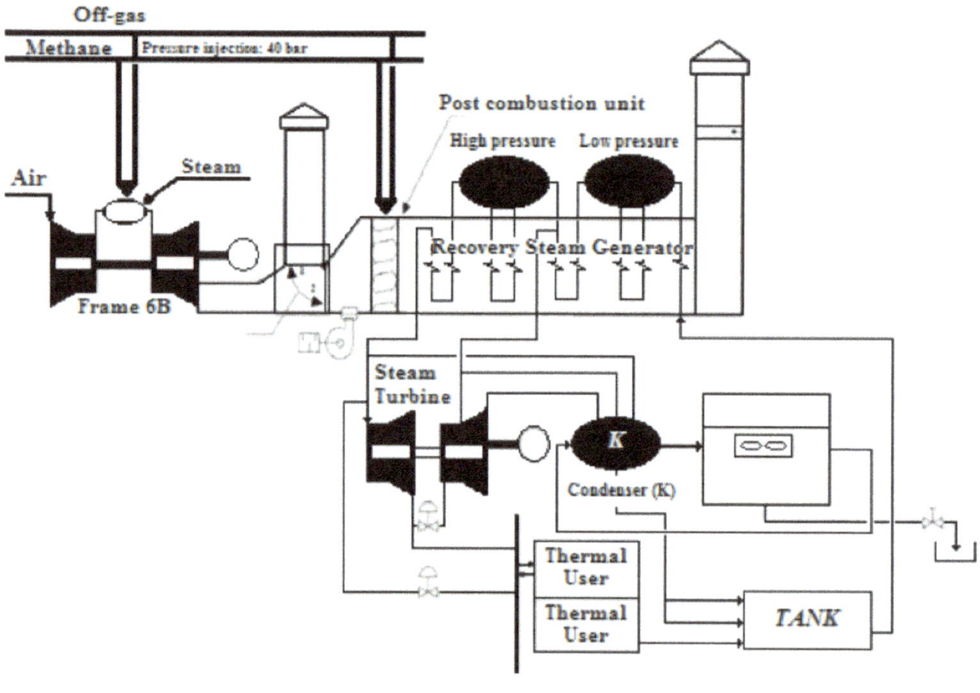

Figure 1: Power plant schematization.

Table 1: Gas turbine specifications at the design condition

Engine Type	Frame 6B (General Electric)
	Air
Compression ratio (β)	11.8:1
Temperature at the end of compression	670 K
Mass flow rate (kg/s)	139 kg/s
Engine speed (at nominal conditions)	5,115 rpm
	Steam
Injection pressure	Variable (up to 40 bar)
Temperature	620 K
Mass flow rate	Minimum: 0.208/Maximum: 0.445 kg/(s·can)
Nozzle hole diameter	2 mm

Table 2: Fuel injector: system specifications at nominal conditions

Injector Type	Pressure swirl Atomizer
Injection pressure	40:1
Temperature	350 K
Mass flow rate	0.28 kg/s
Number of nozzles	4
Fuel to air ratio (FAR)	0.019
Fuel	Methane/off-gas

The steam can be produced by both the heat recovery generator and the post combustion unit: when the steam demand is at the maximum level, these two sections operate in parallel, with the gas turbine working at partial load. Figure 2 reports a typical load diagram for the gas turbine: the blue line represents the produced energy with the gas turbine working at nominal conditions and the yellow one, the energy produced at partial load. For example, during days 1 and 4, the gas turbine produced the same energy (about 500,000 kWh) working at full load and partial load.

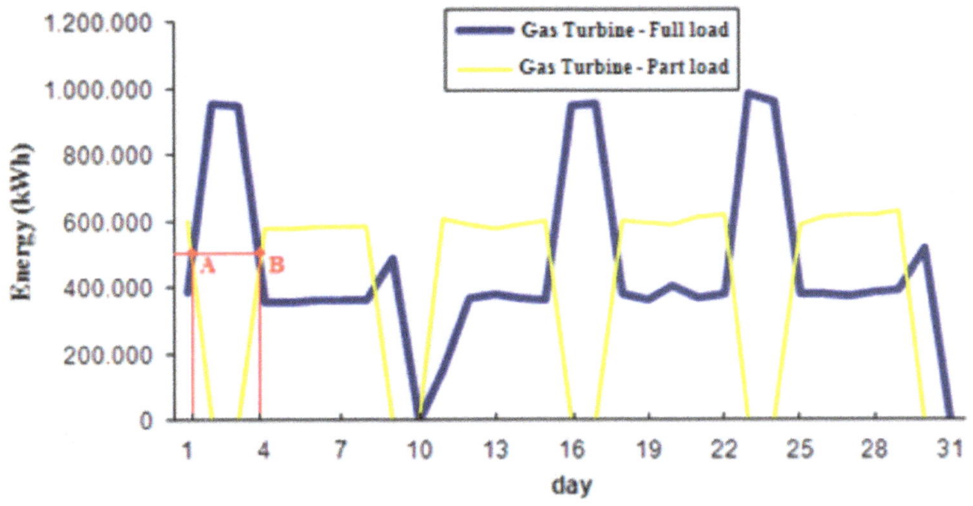

Figure 2: Typical gas turbine load diagram.

Considering the most frequent operating conditions of the plant, simulations for 75, 80, 90 and 100% of the nominal thermal power have been taken into account. Obviously, changes of the air conditions due to compressor regulation have been considered, too. The information about the pollutant emissions will be reported in the following paragraphs, where they will be compared with the numerical results.

NUMERICAL PROCEDURE

Burner Geometry and Grid Discretization

The FRAME 6B can is a reverse-flow combustor, i.e., the air coming from the compressor is passed inside a pre-heating space heading from the base towards the dome. This path allows the compressed air to be heated and to enter the combustion chamber with a higher temperature than that at the end of compression, thus increasing the combustion efficiency.

When the air arrives close to the dome, it is swirled and pushed into the primary combustion zone. The injectors of the fuel gas and of the steam are placed in the terminal part of the dome. The central part of the liner, where the combustion chemical reactions develop, is followed by a dilution zone: the holes on the walls of the liner have the dual function to dilute the exhaust gas flow and cool the walls of the combustion chamber. The total length of the burner and its maximum diameter are about 1.50 m and 0.56 m, respectively (Figure 3); the diameter of the combustion chamber is 0.38 m.

The computational mesh was created using the ACE+ Multiphysics tool. Because of the symmetry of the system and its components, the CFD calculations are performed on 90° sector meshes. Two different domains have been defined: the first is related to fluid volume, the second to the metal walls of the liner. This choice allows the pre-heating of the compressed air that occurs in the interspace to be described, solving concurrently fluid-dynamic and heat transfer equations. In both cases, unstructured meshes have been chosen. With regard to the fluid mesh, more dense discretizations have been located near the outlet sections of the fuel and steam injectors, where the speed of the jets takes on values on the order of hundreds of meters per second and within the primary and secondary zones. Table 3 summarizes the characteristics of the fluid grid in each region of the burner. The adopted resolution was found to give adequately grid independent results.

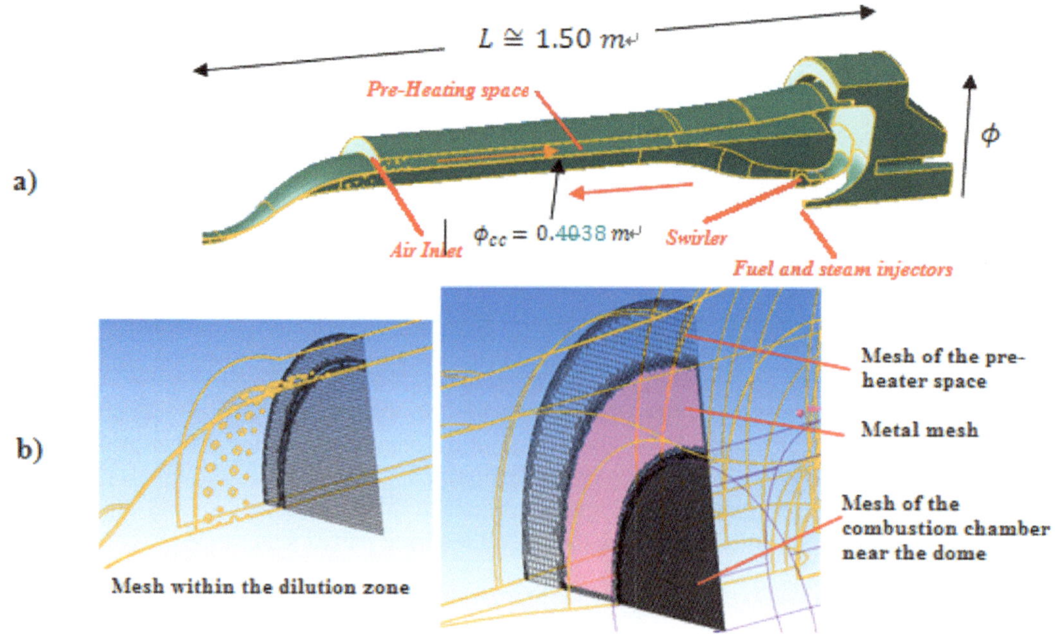

Figure 3: (**a**) Main components of the burner (Computer Aided Drafting model); (**b**) and details of the mesh.

Table 3: Fluid mesh properties

Region	Characteristics
Pre-heater region	Cell type: hexahedral
	Averaged cell resolution: $\phi_{cc}/210$
Injectors region	Axial length: 5 cm
	Cell type: hexahedral
	Averaged cell resolution: $\phi_{cc}/475$
Combustion chamber: primary and secondary zones	Cell type: hexahedral
	Averaged cell resolution: $\phi_{cc}/315$
Combustion chamber: dilution zone	Cell type: hexahedral
	Averaged cell resolution: $\phi_{cc}/210$
Wall region	Two layer of tetrahedral cells
Total number of cells	1.5 million cells

Table 4 provides more details about mesh boundaries. In particular, the air mass flow rate at the inlet of the single burner (cf. Table 1) has been defined through the specification of the velocity and the temperature, so that the experimental data have been directly introduced into the numerical model. This choice allows these parameters to be easily changed to follow the different operating conditions of the burner; for example, at partial load conditions, the effect of the Inlet Guide Vane (IGV) on the flow has been modeled by changing the numerical value of the flow velocity. On the contrary, the turbulence intensity has not been modified during simulations (cf. Table 4).

Table 4: Boundary conditions

Boundary	Type	Value (at Nominal Condition)	Turbulence
Air inlet	Velocity inlet	28 m/s (normal to boundary)	Turbulence intensity: 0.08 Dissipation rate: 0.055 m (hydraulic diameter)
Gas outlet	Pressure outlet	11.8 bar	Back flow kinetic energy: 0.01 m^2/s^2 Dissipation rate: 0.135 m (hydraulic diameter)
Fuel inlet	Mass flow rate inlet	0.07 kg/s (for the single nozzle)	Turbulence intensity: 0.12 Dissipation rate: 0.001 m (hydraulic diameter)
Steam inlet	Mass flow rate inlet	0.032 kg/s (for the single nozzle)	Turbulence intensity: 0.12 Dissipation rate: 0.002 m (hydraulic diameter)

The Turbulent Combustion Model

As to the turbulent closure, the k-ε approach with standard wall function has been used to model the turbulent effects. With regard to the combustion model, a chemistry with 35 species and 177 reactions was employed. This is a homemade chemistry derived from the detailed mechanism, GRIMech 3.0 [5] using the Open Source software, Kintecus, and, specifically, its tool, Atropos [6]. The Atropos software addition to Kintecus allows one to accurately reduce species and reactions out of larger systems that have no bearing on the results. This is realized through a sophisticated principle component analysis (PCA) on the basis of the normalized sensitivity coefficients (NSC) files outputted by Kintecus. The employed mechanisms result in being not so large as to be integrated within a multi-dimensional CFD code for the computing of laminar flamelets.

The flamelet model, used in the non-premixed combustion approach to account for chemical non-equilibrium, regards the turbulent flame as an ensemble of thin, locally one-dimensional flamelet structures. In this work, three flamelets with strain rate χ equal to 0.01, 0.1, one and 57 flamelets with initial strain rate $\chi_0 = 2$ and $\Delta\chi = 1$ have been generated to characterize combustion. Using flamelets, species mass fraction, temperature, as well as the chemical reactions can be computed as a function of the mixture fraction (Z) and the strain rate (χ) only. After translation of the flamelet equations from a physical space to a mixture fraction space, a set of simplified equations can be re-written in the mixture fraction space, including equations for the species mass fraction (Equation (1)) and "solved for temperature" energy (Equation (2)) [7,8]:

$$\frac{\partial}{\partial t}(\rho Y_i) = \frac{1}{2}\rho\chi\frac{\partial^2 Y_i}{\partial Z^2} + S_i \tag{1}$$

$$\rho\frac{\partial T}{\partial t} - \rho\frac{\chi}{2}\frac{\partial^2 T}{\partial Z^2} - \rho\frac{\chi}{2c_p}\frac{\partial T}{\partial Z}\frac{\partial c_p}{\partial Z} - \sum_i^N \rho\frac{\chi}{2Le_i}\frac{c_{p,i}}{c_p}\frac{\partial Y_i}{\partial Z}\frac{\partial T}{\partial Z} + \frac{1}{c_p}\sum_i^N S_i h_i = 0 \tag{2}$$

where Y_i, T, ρ, Z, S_i and h_i represent the mass fraction, temperature, density, mixture fraction, reaction rate and enthalpy of the i-th species, respectively; $c_{p,i}$ and c_p are the specific heat of the i-th species and the specific heat of the mixture. At the stoichiometric condition, the strain rate χ and the mixture fraction Z are related according to the following Equation (3):

$$\chi_{st} = \frac{a_s \exp\left(-2[erfc^{-1}(2Z_{st})]^2\right)}{\pi} \tag{3}$$

where a_s is a characteristic strain rate, Z_{st} and χ_{st} are the mixture fraction and the strain rate at the stoichiometric condition, and erfc is the complementary error function. In the turbulent flame, both Z and χ_{st} are random variables with a joint probability density function (PDF). Therefore, the characteristic scalar (such as density, temperature and mass fraction) in the turbulent diffusion flame may be gained by laminar flamelet statistically:

$$\bar{\phi} = \iint \phi(Z,\chi_{st})p(Z,\chi_{st})dZd\chi_{st} \tag{4}$$

Assuming that Z and χ_{st} have independent distributions, the previous equation can be written as:

$$\bar{\phi} = \iint \phi(Z,\chi_{st})p_Z(Z)p_\chi(\chi_{st})dZd\chi_{st} \tag{5}$$

Equations (1) and (2) for species mass fraction and energy are solved simultaneously with the flow: the flamelet equations are advanced for a fractional time-step using properties computed from the fluid flow, and then, the latter is advanced for the same fractional time-step using properties from the flamelets (Figure 4). The flamelet time-step is computed through the volume-averaged scalar dissipation, pressure, fuel and oxidizer temperatures, which are passed from the flow solver to the flamelet solver.

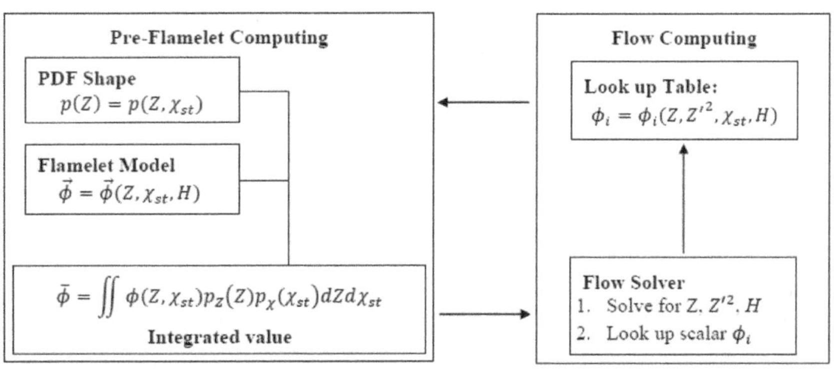

Figure 4: Flamelet model computing. PDF, probability density function.

The Pollutant Emission Models

For the estimation of CO emission, an innovative approach has been used. The following semi-analytical correlation proposed by Rizk and Mongia has been implemented into a user-defined function (UDF) [9]:

$$CO = \frac{a_1 exp(\frac{a_2}{T_{pz}})}{p^{a_3}\tau_r(\frac{\Delta p}{p})^{a_4}} \tag{6}$$

where the values of the constants are reported in the following Table 5 and the burner pressure loss, Δp, has been fixed to 4%. This equation allows carbon monoxide concentration to be related to important operating physical quantities, since there are two fundamental parameters in the previous equation: τ_r represents the residence time of the methane parcels within the combustion zone and T_{pz}, the averaged temperature in the primary zone. Using the software particle tracking, the methane parcels are followed until they have been completely consumed by combustion; an averaged value of the axial distance covered by parcels and their residence time are transferred to the UDF. According to [10], the temperature of the primary zone has been calculated as follows:

$$T_{pz} = A \cdot FAR^{\alpha} \exp(\gamma(FAR + \lambda)^2) \left(\frac{p}{p_{ISO}}\right)^x \left(\frac{T_{out}}{T_{ISO}}\right)^y \psi^z$$

(7)

where:

- A, α, γ and λ are constants whose values are given in [10];
- x, y and z are constants chosen in the function of fuel to air ratio (FAR) [10];
- T_{out} is the averaged numerical temperature at the burner discharge;
- □ the hydrogen to carbon atomic ratio, equal to four for methane.

Table 5: Values of the constants for CO prediction [9]

a_1	a_2	a_3	a_4
66.48×10^6	28.16	2.69	0.5

The extended Zeldovich's mechanism has been taken into account to model reactions for thermal NO_x formation:

$$N + O_2 \Leftrightarrow NO + O;\ N_2 + O \Leftrightarrow NO + N;\ N + OH \Leftrightarrow NO + H$$

The reaction rates are assumed as follows [11]:

$$w^{+}_{NO,Th} = 1.35 \cdot 10^{16} \rho Y_{N_2} Y_{O_2}^{0.5} T^{-1} \exp\left(-\frac{69160}{T}\right)$$

(8)

$$w^{-}_{NO,Th} = 22.5 \rho Y_{N_2}^2 Y_{O_2}^{-0.5} T^{-1} \exp\left(-\frac{47355}{T}\right)$$

(9)

Prompt NO is caused by the presence of radicals in the flame zone; the following equation was adopted to compute its contribution [11]:

$$w_{pro} = 1.9 \cdot 10^3 (Y_{N_2}[Y_{O_2}]_b Y_{O_2}[Y_{CH_4}]_b)^{0.45} \exp\left(-\frac{16000}{T}\right)$$

(10)

where the subscript, "b", denotes the concentration of the species before the start of the combustion.

Heat Transfer Modeling for the Pre-Heating of the Air

Heat transfer modeling involves the wall of the liner. The gas-side (T_{w1}) and the air-side (T_{w2}) wall temperatures of the liner can be computed using the following equilibrium equations, involving the gas-side and air-side radiation (R) and convection (C) contributes:

$$R_1 + C_1 = R_2 + C_2 = \frac{k_w}{t_w}(T_{w1} - T_{w2})$$

(11)

where k_w and t_w are the thermal conductivity and the thickness of the liner, respectively. As a simplifying assumption, the external radiation, R_2, has been neglected, being quite small compared with the external convective heat transfer, C_2. Thus, the remaining contributes are computed by the software as follows [11]:

$$R_1 = 0.5\sigma(1 + \varepsilon_w)(\varepsilon_g T_g^4 - \alpha_g T_{w1}^4)$$

(12)

$$C_1 = h_1(T_g - T_{w1})$$

(13)

$$C_2 = h_2(T_{w2} - T_{air})$$

(14)

where σ is the Stefan-Boltzmann constant, ε_g and α_g are the gas emissivity and absorptivity that are a function of the gas composition, $0.5 \times (1 + \varepsilon_w)$ is the effective wall absorptivity, T_g and T_{air} are the local gas and air temperature and h_1 and h_2 are the gas-side and air-side heat transfer coefficients. The latter have been computed based on the local flow-field conditions, assuming a fully developed turbulent heat transfer within the annulus.

RESULTS AND DISCUSSIONS

Comparison between Experimental and Numerical Results at 100% Load

A comparison of the results with experimental data can be carried out focusing attention only on the discharge temperature of the combustion chamber. The values reported in Table 6 refer to nominal conditions for the gas turbine and Standard conditions (ISO) for the environment and summarize the results that have been obtained in terms of pollutant emissions, too. Their concentration represent volume-weighted average values at the outlet section of the burner. NO_x and CO formation can be estimated, but there is only one measured value; for further assessment of the prediction capability of the model, several simulations with different air temperatures at the inlet of the compressor were performed in order to carry out a sensitivity analysis.

Table 6: Comparison between experimental and numerical result at 100% load and ISO conditions

Firing Temperature		NO_x Pollutant Emission		CO Pollutant Emission	
Experimental measurement	Numerical result and percentage error	Experimental measurement (ppm)	Numerical result and percentage error (ppm) (at 15% O2)	Experimental measurement (ppm)	Numerical result and percentage error (ppm)
1,373 K	1,420 K 3.42%	83.7	84.7 +1.25%	11.8	11.9 +0.82%

These simulations are also able to examine the response of the gas turbine model to variations in operating condition. In particular, six further simulations have been performed, using inlet temperatures ranging from 268 K to 298 K. In Table 7, for both experimental and numerical data, the percentages refer to the experimental value reported in Table 6. The experimental results derive from an averaged process of several experimental conditions close to any operating point taken in consideration. In Figure 5, other intermediate experimental values have been plotted.

Table 7: The effect of the air temperature at the compressor inlet on pollutant emissions

Air Temperature Inlet (K)	NOx (at 15% O_2)		CO	
	Experimental Value in Comparison with ISO Condition (%)	Numerical Value in Comparison with ISO Condition (%)	Experimental Value in Comparison with ISO Condition (%)	Numerical Value in Comparison with ISO Condition (%)
268	90.05	87.15	102.33	102.91
273	94.16	92.22	101.87	102.66
278	96.54	95.41	101.29	102.27
283	98.19	98.63	100.50	101.15
288	100	101.25	100	100.82
293	103.21	104.50	99.76	100.33
298	104.93	106.77	99.55	99.77

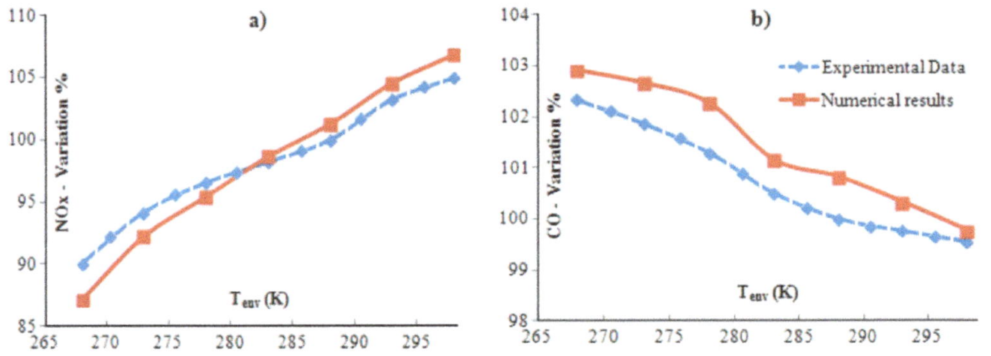

Figure 5: (**a**) NO_x and (**b**) CO percentage variation for different environmental temperatures.

It is evident from the previous figures that the predicted trends are qualitatively similar to the experimental values. In particular, they capture the trend of reduced CO and increased NO_x with air temperature rising. As for CO emissions, although the model tends to overestimate the real data, it is evident how the simulations can completely follow the small variations measured on the real machine. In fact, as a typical characteristic of all heavy-duty machine series, the carbon monoxide emissions increase quickly only if the firing temperature is drastically reduced. Therefore, the stability of the numerical model is based on the ability to appreciate these very small variations. However, the methodical overestimation is due to a systematic error in the evaluation of the main parameters of Equation 6. Two different sources of error are the residence time estimation, since UDF elaborates an averaged value, and the numerical temperature at the exit of the burner. Similarly, the supposed pressure loss may deviate from the real value, because there is no experimental data available for its estimation.

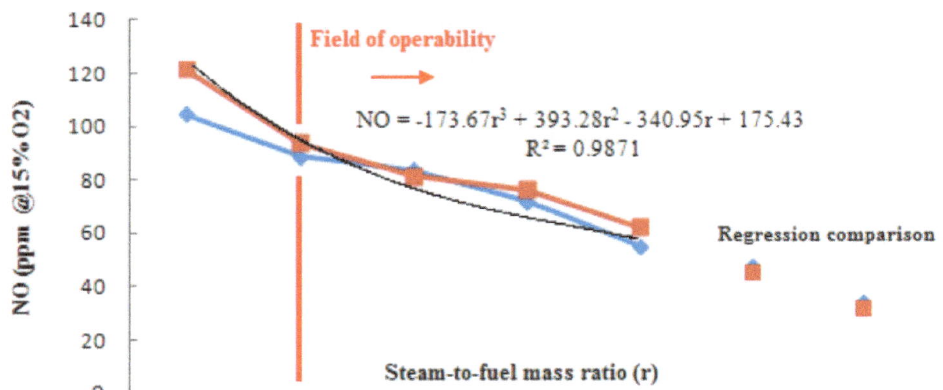

	0.2	0.4	0.46	0.6	0.8	1.1	1.2
Experimental data	104.63	88.72	83.70	71.98	55.24	47.39	33.28
Numerical data	121.37	93.74	81.19	76.17	61.94	45.09	32.51

Figure 6: NO_x emission vs. steam-to-fuel mass ratio.

To completely evaluate the pollutant emissions, the effect of different mass flow rates of steam injected within the combustion chamber has been studied at full load and ISO condition, also. In a STIG (STeam Injected Gas Turbine) cycle, the amount of injected steam represents a sensible parameter to take into account, since there are practical limits to the amount of steam, as well as water that can be injected into the combustor before a serious flame stability problems occur [12]. This has been experimentally determined and must be taken into account in all applications if the combustor is to ensure low NO_x emissions and long hardware life for the gas turbine user. In ordinary operating conditions, the steam-to-fuel mass ratio (r) is between 0.2 and 0.8. The results shown in Figure 6 for NO_x have been plotted taking this range as the reference (r= 0.46 represents the design value).

It is evident how the computational model has a good agreement with experimental results, primarily for intermediate values of r (0.4 □ r □ 0.6). For values close to the minimum, the code tends to produce a higher percentage of errors at about 15%. These discrepancies could be related to experimental uncertainties in input parameters to the computations that can be amplified in the case of extremely low values of this parameter, producing an overestimation of temperature within the flame zone. If this error cannot be considered tolerable, the field of operability is limited to a steam-to-fuel-mass ratio higher than 0.4; in fact, lower error percentages have been obtained for the other operating conditions within the whole simulated range.

The model analysis has been extended for higher values of r, which are usually adopted just when an increase of electrical power is required. A third order regression curve, built through the numerical data, has been used in the cases of r = 1.1 andr = 1.2. The comparison with experimental data provides an excellent fitting, since very limited differences can be noticed. These results confirm that the software prediction capability can be extended to extremely high values for steam mass flow rate.

The effect of steam injection on CO emission is almost negligible. Steam temperature is very similar to the air temperature, so that experimental data at nominal conditions show no appreciable variations with increasing r. From a numerical point of view, the employed UDF cannot consider the influence of r, because there is no parameter that depends on the steam mass flow rate.

Thermal and Fluid Dynamic Internal Fields

Since the previous study demonstrated the validity of the present CFD model, it is interesting to analyze some physical characteristics of the internal fields of the combustion chamber. Figure 7a shows the temperature contour plot on the symmetry plane of the burner and on a cutting plane in the dilution zone. Different peculiarities of a non-premixed combustion can be highlighted. First of all, the flame occupies the central and terminal part of the burner, since the chemical species need time for mixing and complete reactions: the primary zone is characterized by the low temperature of the mixture, since the reaction rate is limited by a high mass fraction of the fuel (Figure 7b).

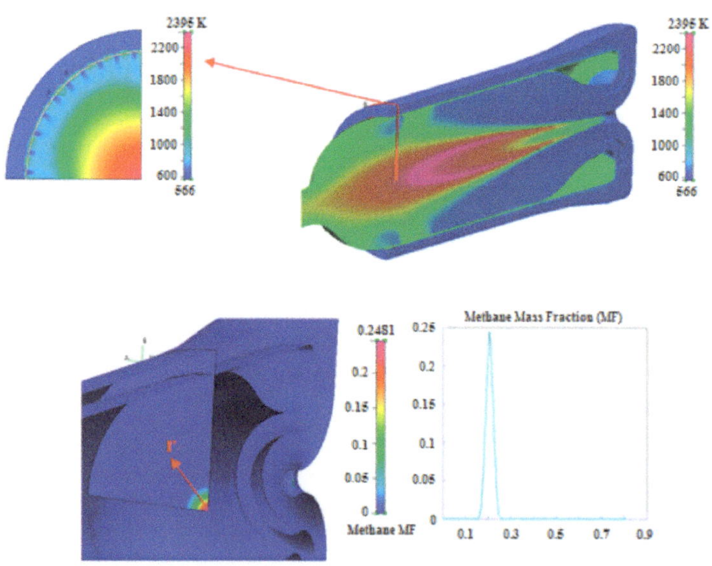

Figure 7: (**a**) Temperature contour plot for a secondary zone cross-section (**Left**) and the symmetry plane (**Right**); (**b**) Methane concentration within the primary zone.

Secondly, the flame develops into a thin interface with the air flux and produces a compact body flame: this structure is typical of the diffusive reaction mechanism. The cutting plane on the dilution zone allows one to verify the influence of the cold air flux in preserving walls and decreasing the temperature of exhaust gases before the first stage of the gas turbine.

The strong connection between the temperature field and the NO_x production rate is evident in Figure 8, where three different cross-sections have been plotted. Numerical simulations reveal that the distribution of maximum values follow the flame shape within the secondary zone of the combustion chamber where temperature range exceeds 2,000 K.

Keeping attention on fluid dynamics, one of the most important parameters that can be verified is the Mach number of the methane jet (Figure 9). These kind of combustion chambers are not designed to operate with supersonic gaseous jets, and simulations confirm that the injection is characterized by a supersonic flow just near the injectors' discharge. The delaying action of the air involves a decreasing of the Mach number in the range between 0.5 and 0.8, already within the primary zone.

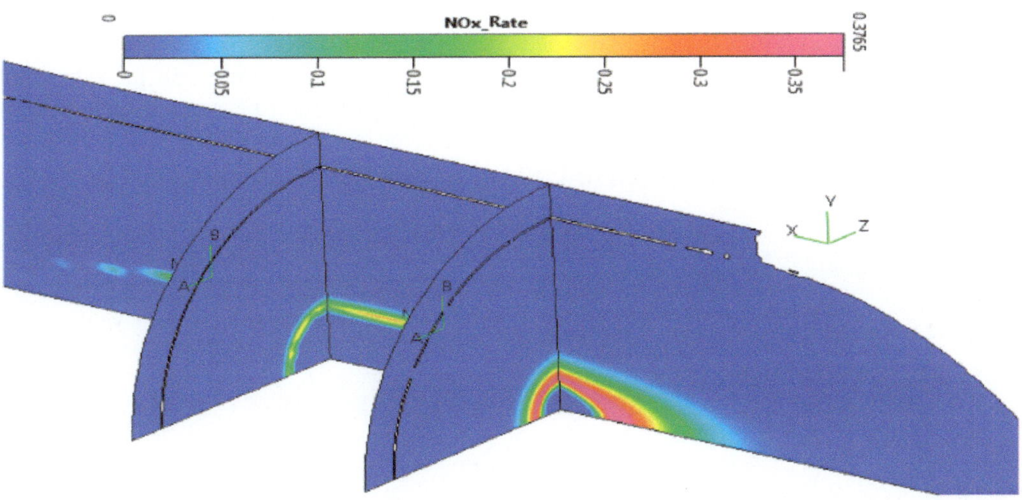

Figure 8: NO_x production rate on the symmetry plane and on two different normal x cuts.

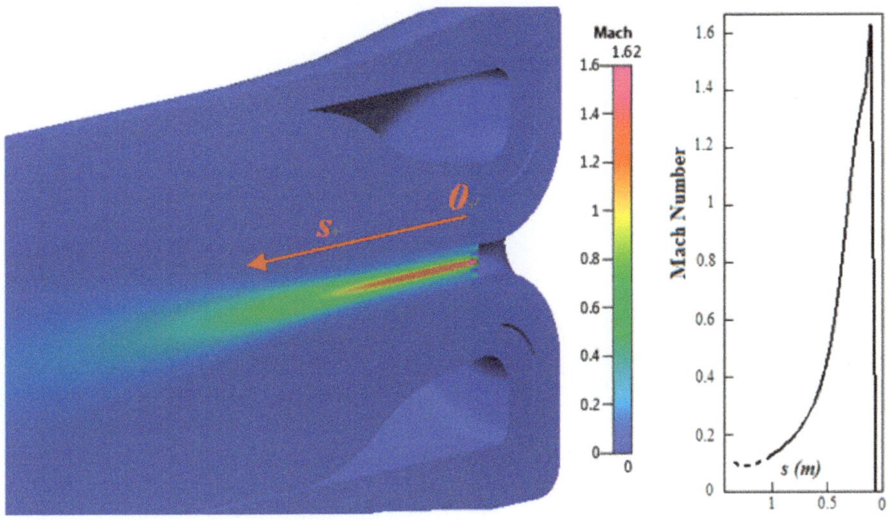

Figure 9: Fuel Mach number: contour plot and chart.

Figure 10 reports the stream lines and the y-direction velocity contour plot in a cross-section at the beginning of the primary zone. It is evident how the CFD model can describe the recirculation flow at the inlet of the combustion chamber. This is an important phenomenon to take into account, because it influences the pressure losses and the mixing process between reactants.

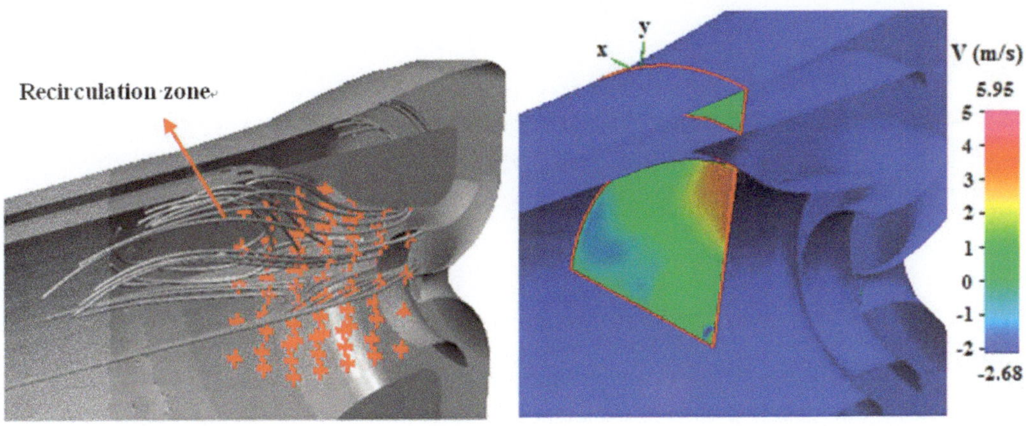

Figure 10: Recirculation zone.

From this point of view, the attention has been focused on the Mach number and recirculation flow to characterize the primary zone, since the employed correlation for CO contains the assessment for T_{pz}, which is directly influenced by fuel distribution and fluid dynamics. Good results in CO estimation are primarily related to the ability of the model to describe these phenomena.

Experimental and Numerical Results at Partial Load

As mentioned earlier, the analysis has been extended to 75, 80 and 90% of the nominal thermal power. Figure 11 shows the comparison between the computed temperatures and the measured values at the discharge of the combustion chamber. The decreasing trend in temperature is the result of the combination of both the air and fuel lower mass flow rate, due to the (IGV) regulation at the compressor intake. The numerical simulations can provide, in any case, a good agreement with the experimental data, with a maximum value of the error lower than 4%.

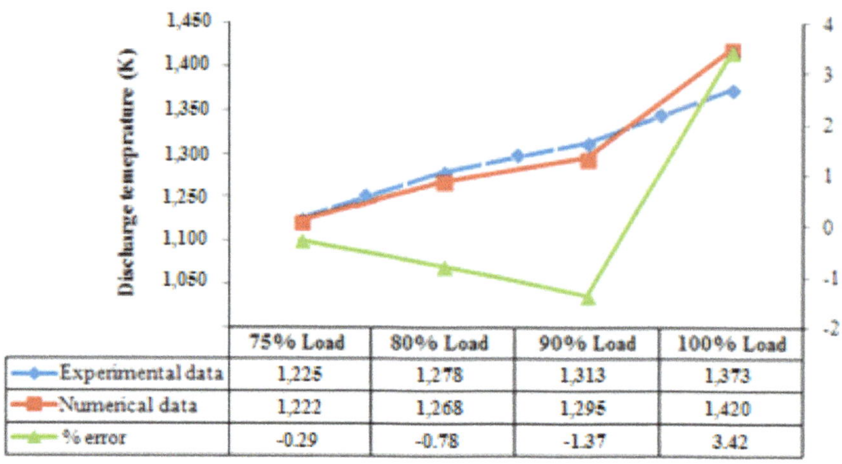

	75% Load	80% Load	90% Load	100% Load
Experimental data	1,225	1,278	1,313	1,373
Numerical data	1,222	1,268	1,295	1,420
% error	-0.29	-0.78	-1.37	3.42

Figure 11: Discharge temperature at different loads.

Figure 12 reports the averaged temperature for each cross-section within the intermediary and secondary zones of the combustion chamber: the rough rising in temperature due to the completion of the fuel oxidation is followed by a fast cooling due to the action of the fresh air entering from the dilution zone and the decreasing of the body flame diameter, as can be seen in Figure 7a. In general, the presence of IGV as a power control system has a slight effect on the NO_x emissions; in fact, the IGV maintains a high exhaust temperature also at partial load conditions, so that the NO_x concentration can vary by a few ppm.

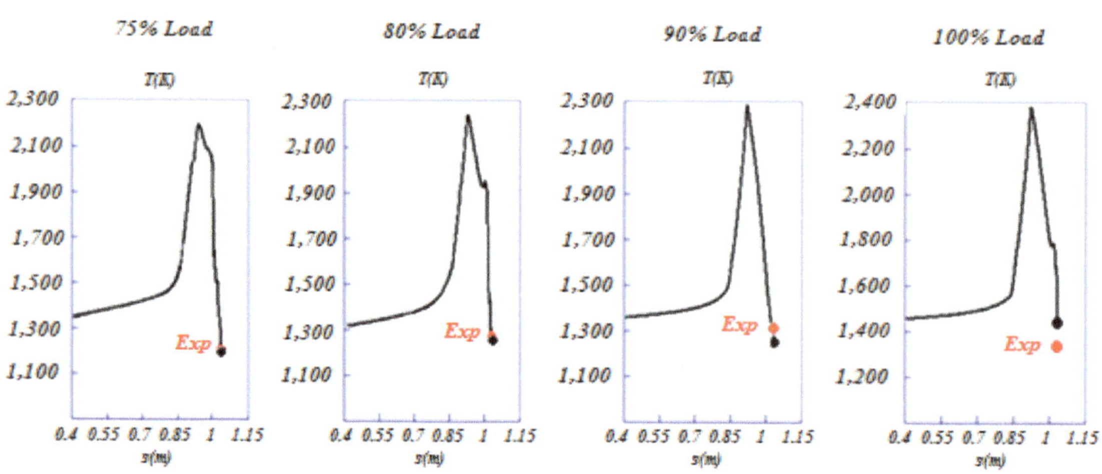

Figure 12: Comparison between numerical and experimental temperatures at different load values.

However, for the studied cases, the decreasing of the temperature peak reached within the combustion chamber determines a reduction in terms of NO_x pollutant emissions. This trend is well described also by the numerical model (Figure 13): as NO_x strongly depends on temperature, this is an indication that the model can provide an averaged internal thermal field close to the real one.

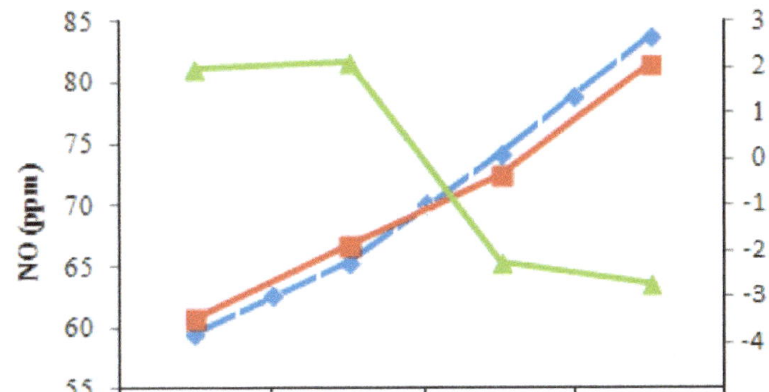

	75% Load	80% Load	90% Load	100% Load
Experimental Data	59.62	65.37	74.21	83.76
Numerical results	60.80	66.75	72.54	81.49
% Error	1.98	2.11	2.25	2.71

Figure 13: NO_x prediction at different load values.

The evaluation of the numerical model performance in terms of CO assessment at partial load has been summarized inFigure 14. First of all, it is important to underline that the model can predict the increasing in CO emissions, due to the reduction of the peak temperatures within the combustion chamber. However, as well as is seen in the analysis dedicated to the full load condition, also in this case, the software tendency is to overestimate the carbon monoxide concentration at the discharge section of the burner.

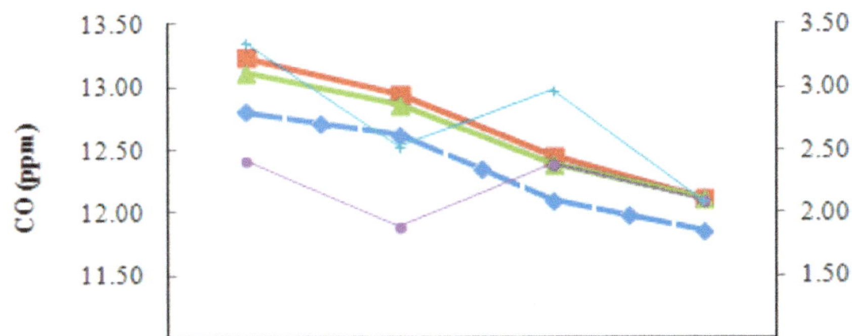

	75% Load	80% Load	90% Load	100% Load
Experimental data	12.81	12.63	12.10	11.87
Numerical data	13.24	12.95	12.46	12.12
Optimized UDF	13.12	12.87	12.39	12.12
Optimized UDF % error	2.42	1.90	2.40	2.11
% Error	3.36	2.53	2.98	2.11

Figure 14: CO prediction at different load values. UDF, user-defined function.

To improve the UDF and minimize the error, a hybrid strategy can be adopted (Figure 15): while the numerical temperature, T_{out}, is directly transferred to the UDF, the experimental value of temperature and experimental CO concentration can be given as the external input. In this way, the code can compute Equation (6) choosing the temperature that minimizes the difference with the experimental value. This approach reduces the error when the estimated temperature is lower than the real one, and it has the advantage of directly joining together the code with real data.

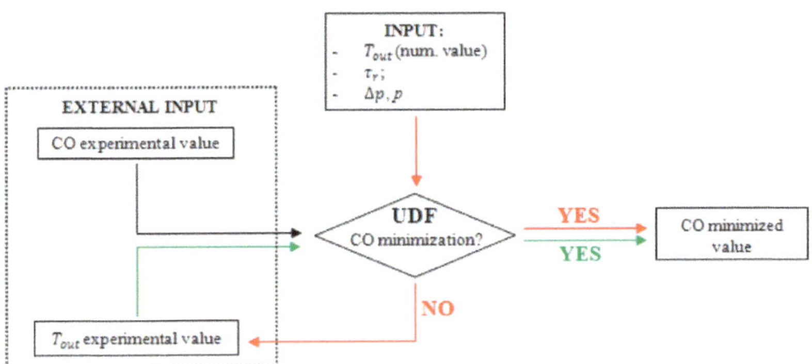

Figure 15: UDF flow chart for the CO minimization.

CONCLUSIONS

In this paper, a numerical model for the prediction of pollutant emissions within a Frame 6B combustion chamber was presented. The analysis involved both nominal and partial load conditions in order to completely evaluate the setting of the models that have been implemented. Furthermore, a sensitive analysis was developed varying the intake temperature of the air and the steam to the fuel mass ratio at full load.

The real discharge temperature was used for a direct comparison with the numerical result: in any case, the differences provided negligible percentage errors. As to the internal temperature field, there is no experimental data available; its numerical analysis confirmed some peculiarities of non-premixed diffusion combustion, such as the body flame structure and the mixing phenomena between reactants. However, an indirect control is possible analyzing the NO_x emissions at the outlet section of the combustion chamber. In fact, excluding the results obtained for extremely low values of the steam-to-fuel mass ratio, NO_x concentrations have a good agreement with experimental results. Therefore, it can be concluded that also the thermal-fluid-dynamic field within the burner reflects the real phenomenon. A regression numerical curve was also obtained to compute NO_x for a high steam mass flow rate, producing an excellent fitting with real data.

Ritz and Mongia correlation was implemented through an innovative UDF for CO computing. Even if the discrepancies with the experimental data are always acceptable, a systematic overestimation was verified at both nominal and partial load. In this latter case, the influence of discharge temperature on calculation was studied: instead of the numerical temperature, the experimental value was used to improve the code and limit the percentage error.

ACKNOWLEDGMENTS

The author wishes to acknowledge Cofely GDF SUEZ S.p.A for experimental data related to the power plant located in Nera Montoro (Narni, Italy) and Esi-Group for technical support.

REFERENCES

1. Ol'Khovskii, G.G. Prospective gas turbine and combined-cycle units for power engineering (A review). Therm. Eng.2013, 60, 79–88.
2. Lefbvre, A.H.; Ballal, D.R. Gas Turbine Combustion—Alternative Fuels and Emissions, 3rd ed.; CRC Press: Boca Raton, FL, USA, 2010; pp. 153–154.
3. Stöhr, M.; Arndt, C.M.; Meier, W. Effects of Damköhler number on vortex-flame interaction in a gas turbine model combustor. Proc. Combust. Inst. 2013, 34, 3107–3115.
4. Strakey, P.A.; Eggnspieler, G. Development and validation of a thickened flame modeling approach for large eddy simulation of premixed combustion. J. Eng. Gas Turbine Power 2010, 132, 1–9.
5. Smith, G.P.; Golden, D.M.; Frenklach, M.; Moriarty, N.W.; Eiteneer, B.; Goldenberg, M.; Bowman, T.C.; Hanson, R.K.; Song, S.; Gardiner, W.C.; et al. Gri-Mech 3.0. Available online: http://www.me.berkeley.edu/gri_mech/ (accessed on 10 October 2012).
6. Ianni, J. A comparison of the Bader-Deuflhard and the Cash-Karp Runge-Kutta. Integrators for the GRI-MECH 3.0 Model Based on the Chemical Kinetics Code Kintecus. In Computational Fluid and Solid Mechanics 2003; Bathe, K.J., Ed.; Elsevier Science Ltd.: Oxford, UK, 2003; pp. 1368–1372.
7. Kong, S.; Han, Z.; Reitz, R.D. The Development and Application of a Diesel Ignition and CombustionModel for Multidimensional Engine Simulation; SAE Technical Paper 950278; SAE International: Detroit, MI, USA, 1995.
8. Baritaud, T. Multi-Dimensional Simulation of Engine Internal Flow, 1st ed.; Technip: Rueil-Malmaison, France, 1998; pp. 234–237.
9. Aïssani, S.; Bouam, A.; Kadi, R. Combustion chamber steam injection for gas turbine performance improvement during high ambient temperature operations. J. Eng. Gas Turbine Power 2008, 130, article number 041701.
10. Gülder, O.L. Flame temperature estimation of conventional and future jet fuels. J. Eng. Gas Turbine Power 1986, 108, 376–380.
11. ACE+ Suite 2011, Theory Guide. 2011.
12. Srinivas, T.; Gupta, A.V.S.S.K.; Reddy, B.V. Parametric simulation of steam injected gas turbine combined cycle. Proc. Inst. Mech. Eng. Part A J. Power Energy 2007, 221, 873–883.

CHAPTER 25

A COMPUTATIONAL FLUID DYNAMICS MODEL OF FLOW AND SETTLING IN SEDIMENTATION TANKS

Ali Hadi Ghawi[1] and Jozef Kriš[2]

[1]Department of Civil Engineering, Faculty of Civil Engineering, AL-Qadisyia University Iraq

[2]Department of Sanitary and Environmental Engineering, Faculty of Civil Engineering, Slovak University of Technology, Bratislava Slovakia

INTRODUCTION

Sedimentation is perhaps the oldest and most common water treatment process. The principle of allowing turbid water to settle before it is drunk can be traced back to ancient times. In modern times a proper understanding of sedimentation tank behavior is essential for proper tank design and operation. Generally, sedimentation tanks are characterized by interesting hydrodynamic phenomena, such as density waterfalls, bottom currents and surface return currents, and are also sensitive to temperature fluctuations and wind effects. On the surface, a sedimentation tank appears to be a simple phase separating device, but down under an intricate balance of forces is present. Many factors clearly affect the capacity and performance of a sedimentation tank: surface and solids loading rates, tank type, solids removal mechanism, inlet design, weir placement and loading rate etc. To account for them, present-day designs are typically oversizing the settling tanks. In that way, designers hope to cope with the poor design that is responsible for undesired and unpredictable system disturbances, which may be of hydraulic, biological or physicochemical origin. To improve the design of process equipment while avoiding tedious and time consuming experiments Computational Fluid Dynamics (CFD) calculations have been employed during the last decades. Fluid flow patterns inside process equipment may be predicted by solving the partial differential equations that describe the conservation of mass and momentum. The geometry of sedimentation tanks makes analytical solutions of these equations impossible, so usually numerical solutions are implemented using Computational Fluid Dynamics packages. The advent of fast computers has improved the accessibility of CFD, which appears as an effective tool with great potential. Regarding sedimentation tanks, CFD may be used first for optimizing the design and retrofitting to improve effluent quality and underflow solids concentration. Second, it may increase the basic understanding of internal processes and their interactions. This knowledge can again be used for process optimization. The latter concerns the cost-effectiveness of a validated CFD model where simulation results can be seen as numerical experiments and partly replace expensive field experiments (Huggins et al. 2005).

Generally, many researchers have used CFD simulations to describe water flow and solids removal in settling tanks for sewage water treatment. However, works in CFD modelling of sedimentation tanks for potable water treatment, rectangular sedimentation tanks, and iron removal by sedimentation tank in

Citation: Ali Hadi Ghawi and Jozef Kriš (2012). A Computational Fluid Dynamics Model of Flow and Settling in Sedimentation Tanks, Applied Computational Fluid Dynamics, Prof. Hyoung Woo Oh (Ed.), InTech, DOI: 10.5772/27160.

surface and groundwater treatment plants have not been found in the literature. Moreover, the physical characteristics of the flocs may not be such significant parameters in the flow field of sedimentation tanks for potable water, due to the much lower solids concentrations and greater particle size distributions than those encountered in wastewater treatment.

Design of sedimentation tanks for water and wastewater treatment processes are often based on the surface overflow rate of the tank. This design variable is predicated on the assumption of uniform unidirectional flow through the tank. Dick (1982), though, showed that many full-scale sedimentation tanks do not follow ideal flow behavior because suspended solids removal in a sedimentation tank was often not a function of the overflow rate. Because of uncertainties in the hydrodynamics of sedimentation tanks, designers typically use safety factors to account for this nonideal flow behavior (Abdel-Gawad and McCorquodale, 1984).

It can be concluded from the discussion that the current ways in which STs are designed and modified could and should be improved. Providing a tool that might lead to sedimentation tank optimization, as well as understanding, quantifying and visualizing the major processes dominating the tank performance, are the main goals of this research.

SCOPE AND OBJECTIVES

This research focuses on the development of a CFD Model that can be used as an aid in the design, operation and modification of sedimentation tanks (Ghawi, 2008). This model represents in a 2D scheme the major physical processes occurring in STs. However, effect of scrapers and inlet are also included, hence the CFD Model definition. Obviously, such a model can be a powerful tool; it might lead to rectangular sedimentation tanks optimization, developing cost-effective solutions for new sedimentation projects and helping existent sedimentation tanks to reach new-more demanding standards with less expensive modifications. An important benefit is that the model may increase the understanding of the internal processes in sedimentation tanks and their interactions. A major goal is to present a model that can be available to the professionals involved in operation, modification and design of sedimentation tanks.. The ultimate goal of the project is to develop a new CFD methodology for the analysis of the sediment transport for multiple particle sizes in fullscale sedimentation tanks of surface and groundwater potable water treatment plants with high iron concentration. The CFD package FLUENT 6.3.26 was used for the case study of the effect of adding several tank modifications including flocculation baffle, energy dissipation baffles, perforated baffles and relocated effluent launders, were recommend based on their field investigation on the efficiency of solids removal. An overview of the outline of the project is given in Figure 1.

The specific objectives of this research include:

- Improve the operation and performance of horizontal sedimentation tank in Iraq which have been identified as operating poorly, by predicting the existing flow, coagulant dose to remove iron and flocculent concentration distribution of the sedimentation tank by means of CFD techniques.
- Develop a mathematical model for sedimentation tanks in 2D;
- Introduce a flocculation submodel in the general ST model,
- Introduce a temperature submodel in the general ST model.
- Design CFD model for simulation of sedimentation tanks, i.e. grids and numerical descriptions.
- Develop a model calibration procedure, including the calibration of the settling properties, and validate the models with experimental data.
- Evaluate the suitability of CFD as a technique for design and research of rectangular sedimentation tanks for drinking water treatment plants and iron removal.
- Use CFD to investigate the effects of design parameters and operational parameters.

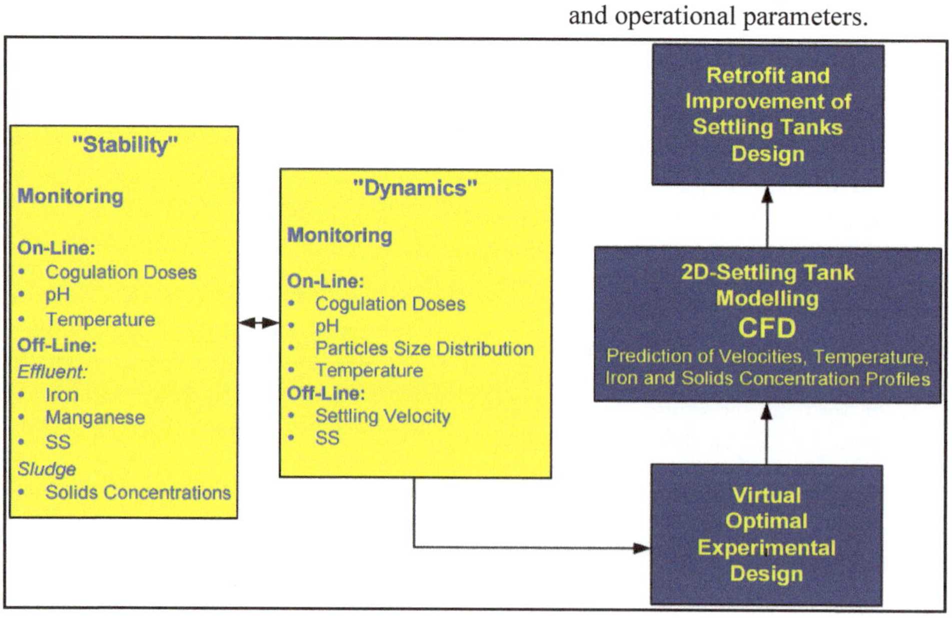

Figure 1: Overview of the settling tank project.

Finally, a CFD model was developed to simulate the full scale rectangular sedimentation tanks at the AL-DEWANYIA purification works in Iraq. The CFD simulations of the ALDEWANYIA tanks were done by setting up standard cases for each, i.e. a configuration and operating conditions that represented the physical tanks as they were built, and then varying different aspects of the configuration or operating conditions one or two at a time to determine the effect. discrete particles in dilute suspension was simulated, as it is the applicable type for the operating conditions in rectangular sedimentation tanks for potable water treatment.

MODELLING THE SETTLING TANK

Figure 2 shows the set-up of the settling tank CFD model which developed in this work. The code predicts fluid flow by numerically solving the partial differential equations, which describe the conservation of mass and momentum. A grid is placed over the flow region of interest and by applying the conservation of mass and momentum over each cell of the grid sequentially discrete equations are derived. In the case of turbulent flows, the conservation equations are solved to obtain time-averaged information. Since the time-averaged equations contain additional terms, which represent the transport of mass and momentum by turbulence, turbulence models that are based on a combination of empiricism and theoretical considerations are introduced to calculate these quantities from details of the mean flow.

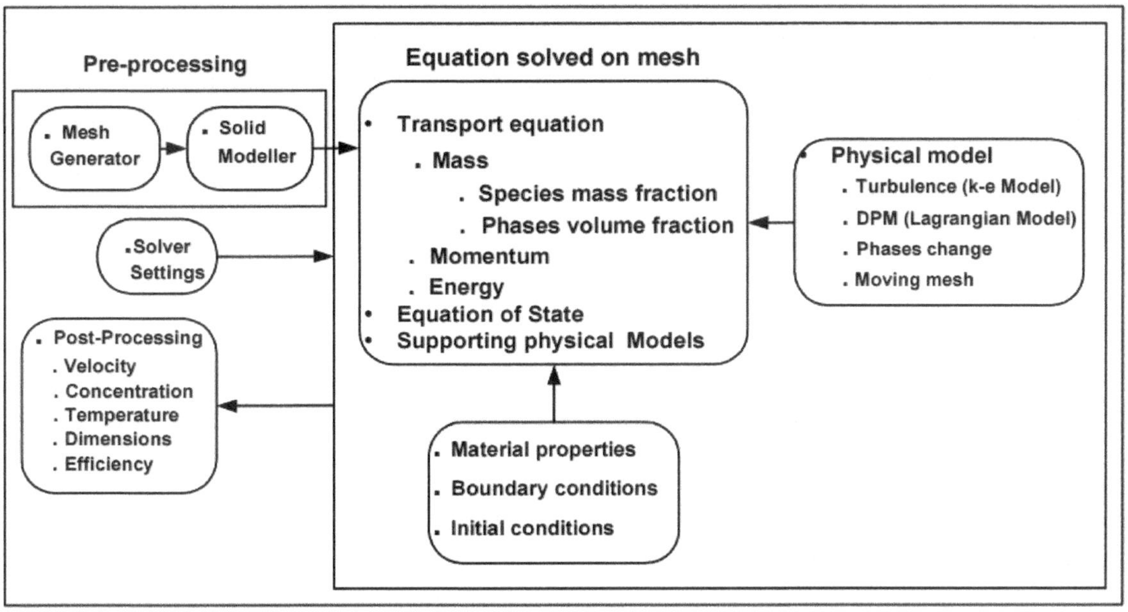

Figure 2: CFD model.

NUMERICAL TECHNIQUES USED IN FLUENT

This section will shortly deal with the methods applied in (Ghawi, 2008). The Fluent software utilises the finite volume method to solve the governing integral equations for the conservation of mass and momentum, and (when appropriate) for scalars such as turbulence and solids concentration. In the work (Ghawi, 2008), the so-called segregated solver was applied; its solution procedure is schematically given in Figure 3. Using this approach, the governing equations are solved sequentially, i.e. segregated from one another. Because the governing equations are non-linear (and coupled), several iterations of the solution loop must be performed before a converged solution is obtained. Concerning the spatial discretisation, the segregated solution algorithm was selected.

The k-ε turbulence model was used to account for turbulence, since this model is meant to describe better low Reynolds numbers flows such as the one inside our sedimentation tank. The used discretisation schemes were the simple for the pressure, the PISO for the pressure-velocity coupling and the second order upwind for the momentum, the turbulence energy and the specific dissipation. Adams and Rodi 1990 pointed out that for real settling tanks the walls can be considered as being smooth due the prevailing low velocities and the correspondingly large viscous layer. Consequently, the standard wall functions as proposed by Launder and Spalding 1974 were used. The water free surface was modeled as a fixed surface; this plane of symmetry was characterized by zero normal gradients for all variables.

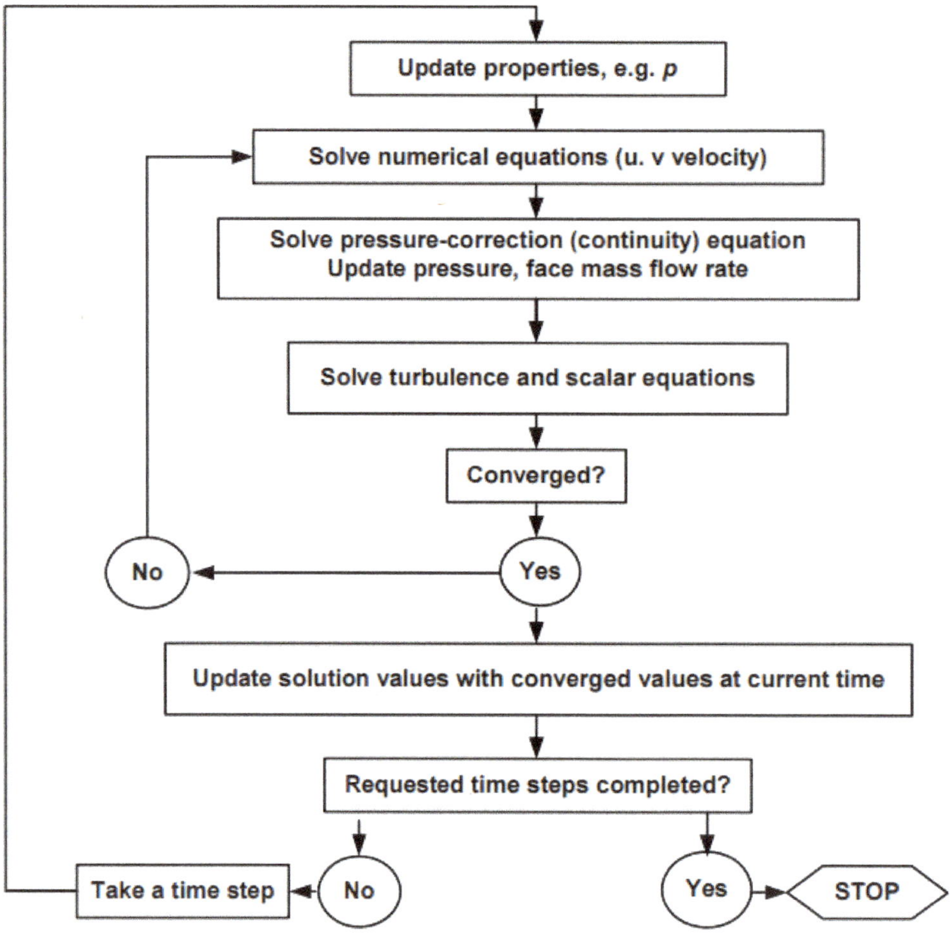

Figure 3: Solution procedure.

EXPERIMENTAL TECHNIQUES FOR MODEL CALIBRATION AND VALIDATION

The process of developing (incl. calibration), verifying, and validating a CFD code requires the use of experimental, theoretical and computational sciences. This process is a closed loop as presented in Figure 4.

The above clearly indicates that good experimental data are indispensable for settling tank model validation; their quality largely depends on the applied experimental technique.

For the purposes of testing the numerical model presented in this thesis on a full scale tank, the data set gathered laboratories, was selected. Here, a comprehensive experimental study of a working settling tank at AL-DEWANYIA in Iraq were carried out. Velocity and concentration profiles were gathered at 7 stations along the length and 3 stations across the width of the tank for a variety of inlet conditions and inlet and outlet geometries. Volumetric flow rates through the inlets and outlets were measured for each test condition studied. Details of the tank geometry and the experimental conditions for which 3D numerical simulations have been made are given in next sections. The following topics are dealt with which measured in the sites:

Flow rate, (2) Settling velocity, (3) Solids concentration (Turbidity), Iron, and Manganese, (4) Particle size distribution, (5) Velocity of liquid, and (6) Temperature.

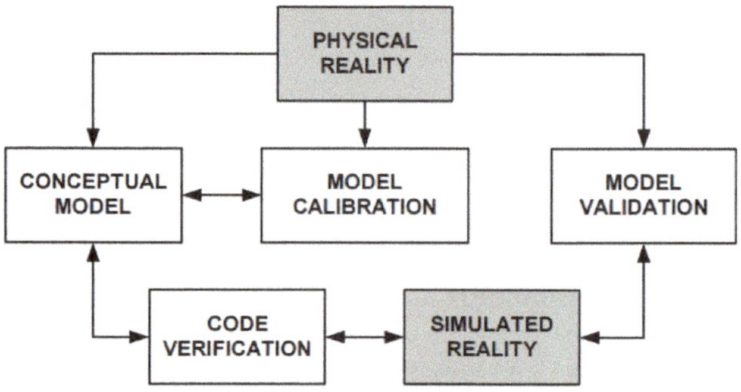

Figure 4: Process of developing CFD code.

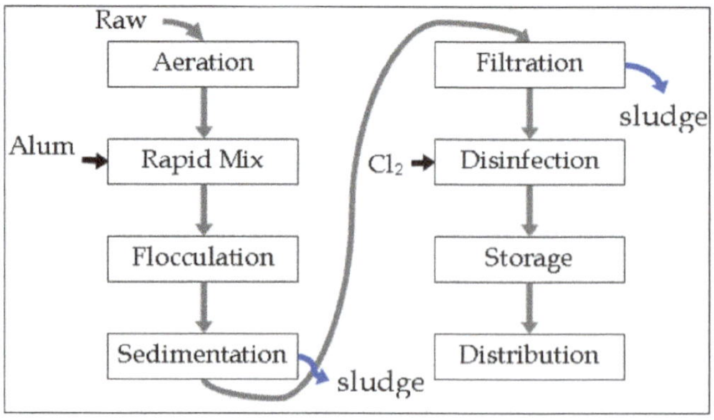

Figure 5: Layout of AL-DEWANYIA WTP.

MODEL DEVELOPMENT, APPLICATIONS AND RESULTS

Introduction

The full-scale horizontal settling tanks at the drinking treatment plant of AL-DEWANYIA were opted for. Most settling tanks of Iraq Water exhibit a horizontal settling tank.

This research was focused on this type of settling tanks. Figure 5 represents treatment of water obtained from a deep well in AL-DEWANYIA WTP a The AL-DEWANYIA WTP were built to remove turbidity and organic material.

Simulation

To limit computational power requirements, the rectangular settling tank was modeled in 2D. The major assumption in the development of the model is that the flow field is the same for all positions; therefore, a 2D geometry can be used to properly simulate the general features of the hydrodynamic processes in the tank. As a first step, a mesh was generated across the sedimentation tank. As a result, the solutions from the grid of 137,814 quadrilateral elements were considered to be grid independent.

For simulation purposes, the range of the suspended solids was divided into thirteen distinct classes of particles based on the discretization of the measured size distribution. The number of classes was selected in order to combine the solution accuracy with short computing time. Two other numbers, 6 and 15, were tested. While the predictions obtained using 6 classes of particles were found to be different from those resulting from the 13 classes, the difference between the predictions made by the 13 and the 15 classes were insignificant. Therefore, a number of 13 classes were selected as a suitable one. Within each class the particle diameter is assumed to be constant (Table 1). As it can be seen in Table 1, the range of particle size is narrower for classes that are expected to have lower settling rates.

Table 1: Classes of particles used to account for the total suspended solids in the STs in ALDEWANYIA STs

Class	Range of particle size (μm)	Mean particle size (μm)	Mass fraction
1	10-30	20	0.025
2	30-70	50	0.027
3	70-90	80	0.039
4	90-150	120	0.066
5	150-190	170	0.095
6	190-210	200	0.115
7	210-290	250	0.126
8	290-410	350	0.124
9	410-490	450	0.113
10	490-610	550	0.101
11	610-690	650	0.077
12	690-810	750	0.057
13	810-890	850	0.040

The Influence of Particle Structure

The settling velocity of an impermeable spherical particle can be predicted from Stokes' law. However, the aggregates in the water not only are porous but it is well known that they have quite irregular shapes with spatial varying porosity. The flow chart of this computations sequence is presented in Figure 6.

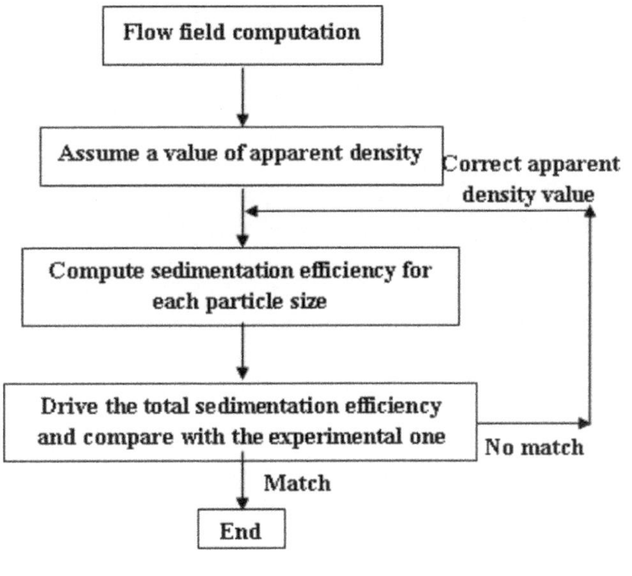

Figure 6: Flow chart of computation sequence.

Simulation of Existing Sedimentation Tanks

The AL-DEWANYIA water treatment plant uses lime, and $(NH_4)_2SO_4$ and $Fe_2(SO_4)_3$ to flocculate the and solid concentrations, respectively before entering the sedimentation tanks. There are 4 rectangular tanks at the AL-DEWANYIA WTP . The Physical and hydraulic data during study periods, and settling tank data for two WTPs are shown in Table 2.

Table 2: Physical and hydraulic data during study periods, and settling tank data.

Geometry	Value
Tank length	30.0 m
Tank width	4.50 m
Hopper depth	2.50 m
Bottom slop	0.00
Weir length	4.50 m
Weir width	0.70 m
Weir depth	0.50 m
loading	Value
SOR	2.7 m/h
Inlet concentration	30-80 mg/l
Density of water	1000 kg/m³
Density of particulate	1066 kg/m³
Tank parameter	Value
Average flow rate	60-80 l/s
Sludge pumping rate	5-15 l/s
Inflow temperature average	4°C -11°C , and 20°C -27°C
Inflow suspended solids	25-80 mg/l
Detention time	2.5-3.6 hr
C_{min}	0.17 mg/l
μ	0.002 N.s/m²

AL-DEWANYIA WTP

Figure 7 shows the velocity profiles of the existing tanks for a flow rate of 80 l/s and an inlet concentration of 50 mg/l (~75 NTU). High velocities are present at the inlet (0.065 m/s). The flow is further accelerated towards the bottom of the hopper due to the density differences as well as the wedge shape of the hopper. The strong bottom current is balanced by a surface return current inside the hopper. The velocities near the effluent weir are very low. The solids concentration profile is shown in Figure 8. Note the high concentration downstream of the ss. The sludge that is supposed to settle in the hopper is washed out of the hopper into the flat section of the tank. Over time a significant amount of sludge accumulates. According to both the field observations and the modeling of the existing process, each of the following reasons (or combination of them) may cause the ST problems, i.e. the flocculant solids blowing out:

- The location of the existing weir (distributed in a range of 1 meter at the very downstream end of the ST) cause very strong upward currents, which could be one of the major reasons that the flocculant solids were blowing out around the effluent area.
- The strong upward flow is not only related to the small area the effluent flow passes through but also to the rebound effect between the ST bottom density current and the downstream wall. The "rebound" phenomenon has been observed and reported by many operators as well as field investigators, especially in ST with small amounts of sludge inventory. A reasonable amount of sludge inventory can help dissipate the kinetic energy of the bottom density current.
- In the existing operation, the bottom density current must be fairly strong due to the lack of proper baffling and the shortage of sludge inventory in the tank.

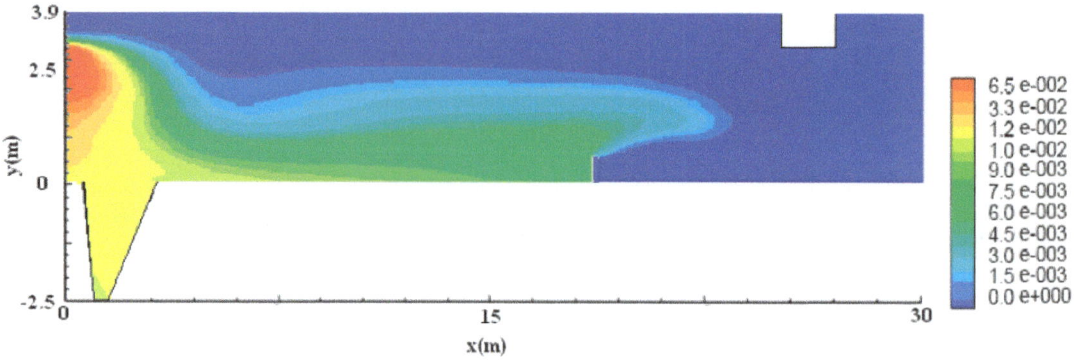

Figure 7: Velocity contours of existing tank (m/s).

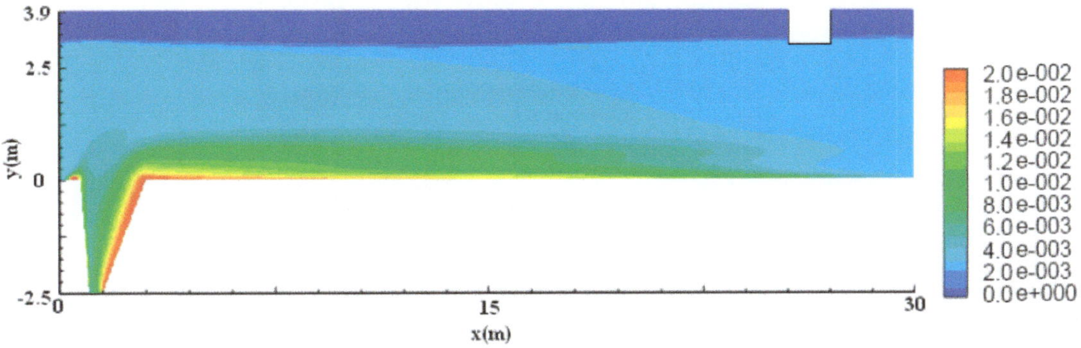

Figure 8: Solids concentration profile for existing tank.

SIMPLE IMPROVEMENTS TO THE EXISTING SEDIMENTATION TANK IN WTP

Field data collected from the laboratories during the last 3 years was used to analyze the tank behavior and to enhance the performance of the settling tanks at the AL-DEWANYIA WT Plant. Several tank modifications including flocculation baffle, energy dissipation baffles, perforated baffles and inboard effluent launders, were recommend based on their field investigation The relationship between the effluent SS and the hydraulic loading is summarised in Table 3 for the existing STs and with different modification combinations. The predicted Effluent SS (ESS) in Table 3 and Figure 9 indicates that the average ESS can be significantly reduced by improving the tank hydraulic efficiency. The comparison of model predictions with the subsequent field data indicates that the significantly improvement of STs performance was obtained by using the minor modifications based on the 2-D computer modeling.

Table 3: Performance data for modelled settling tank

	Q= 50 l/s Influent conc.= 40 mg/l	Q= 70 l/s Influent conc.= 40 mg/l	Q= 80 l/s Influent conc.= 50 mg/l	Q= 80 l/s Influent conc.= 75 mg/l
	Predicted average effluent concentration			
Existing tank	20	30	40	50
Modification 1	12	11	30	22
Modification 2	6	8	12	13
Modification 1 and 2	4	6	7	9
(1) Perforated baffle distance from inlet = 16m; gap above bed = 0.5 m; height above bed = 1.8 m; porosity = 55%				
(2) Length of launder = 12 m.				

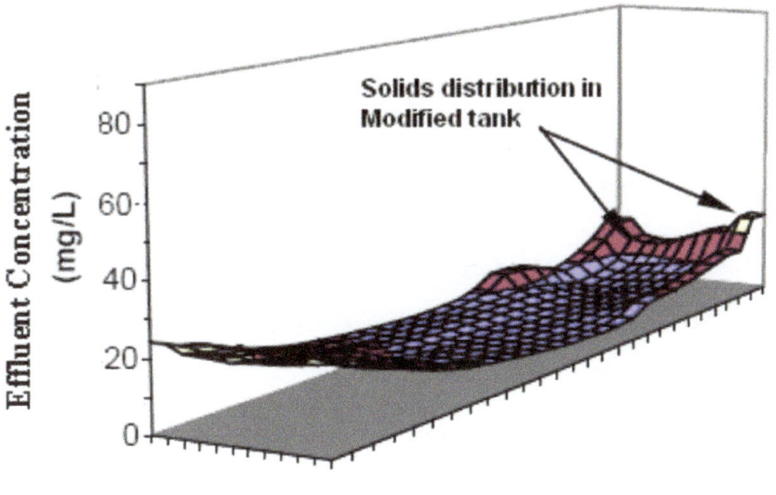

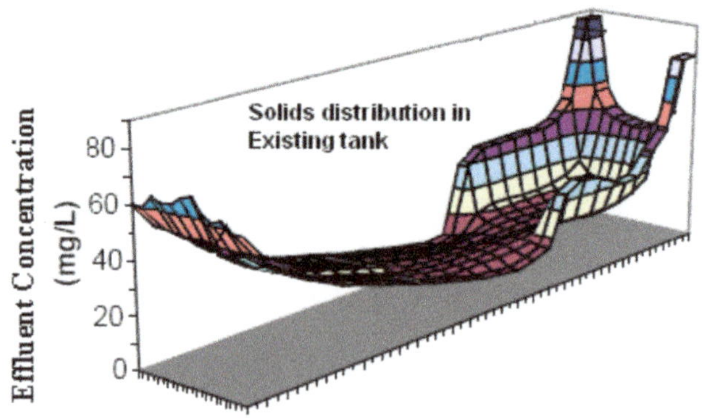

Figure 9: Comparison of solids distributions on surface layer between existing and modified tanks.

MODELLING THE SCRAPER MECHANISM

The gravitational (and laminar) flow along the bottom, which may go up to 8-15 mm/s near the sump, is blocked for 40 minutes of scraper passage. This is clearly seen in Figure 10. The scraper blade thus constrains the bottom flow discharge by counteracting the gravitational force. Near the floor the velocity increases with height in the shear flow region, but is obviously limited by the scraper's velocity.

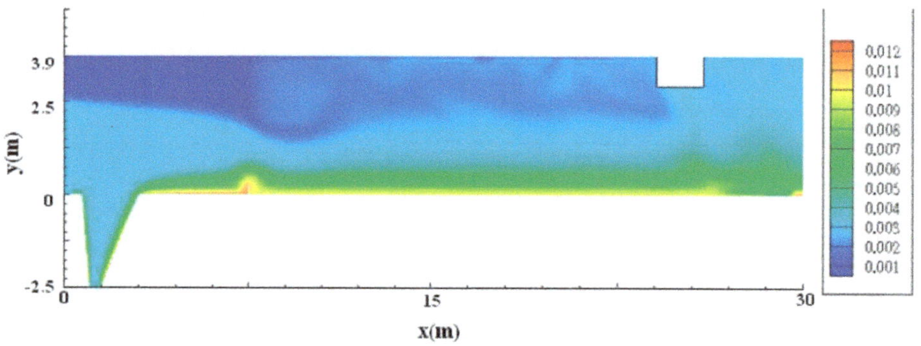

Figure 10: Effect of scraper on solids concentration profiles.

DESIGN PROCEDURES AND GUIDELINES

The design procedures are necessarily based on many assumptions, not normally stated as shortcomings and limitations during the design process. To demonstrate the implications of these assumptions and the way in which these assumption deviate from real tanks. (Ghawi, 2008) tried to improve design procedure as show in Table 4.

Table 4: Proposed CFD enhanced design procedure

Improved design procedure	
Step	description
Step 1	Measurement of settling velocity and sludge density
Step 2	Set up of computational grid
Step 3	Simulate tank
Step 4	Evaluate results and check for evidence of the following: - short circuiting - high velocities zones - high overflow concentration - poor sludge removal
Step 5	If none of the above is present, tank size can be reduced to reduce capital cost. If problems are evident, adjust the design by adjusting the: - inlet - position of sludge withdrawal - position of overflow launders Also consider using perforated, porous and deflecting baffles
Step 6	Repeat until a satisfactory tank geometry is obtained and check final geometry for various process changes such as density, concentration and inflow rate.
Step 7	Asses the influence of the settling velocity and sludge density input parameters and repeat steps 3 - 6 if necessary.

TEMPERATURE EFFECT

Settling Velocity Correction Factor

In order to define a correction factor for the settling velocities based on temperature difference, the temperature effect on the zone settling velocity has to be determined. Figure 11 displays graphically the value of the relationship Vs_{T2} / Vs_{T1} and μ_{T2} / μ_{T1} for the data presented in Table 5 at temperatures T_s (summer temperature) and T_w (winter temperature).

Table 5: Settling velocity and dynamic viscosities for summer and winter temperature

SS mg/l	Settling velocity V m/h	Inlet temperature °C	Outlet temperature °C	Dynamic viscosity µ kg/m.s
CFD Calculated at summer temperature				
60	1.5	27.5	27.5	8.5e-04
50	1.7	27.5	27.5	8.6e-04
25	1.83	26	26	8.7e-04
15	2.52	25.4	25.4	8.8e-04
CFD Calculated at cooled temperature				
60	0.95	8	9.2	1.3e-03
50	1.05	6.6	6.8	1.35e-03
25	1.9	7.8	8.8	1.29e-03
15	2.7	7	8.9	1.30e-03

From Figure 11 can be observed that the numerical values of the ratios Vs_{T2} / Vs_{T1} and μ_{T2} / μ_{T1} are very close, suggesting that an easy correction in the zone settling velocity for different temperatures can be made with a correction factor based on the dynamic viscosity of the water at the two temperatures. Figure 12 shows an extended data set indicating the relationships between the ratios Vs_{T2} / Vs_{T1} and μ_{T2} / μ_{T1}.

Fitting a straight line to the data point presented in Figure 13 can find a correction factor for the settling velocities based on temperature

$$V_{ST_2} = V_{ST_1}\left(\frac{10^{\left[\frac{247.8}{T1+133.15}\right]}}{10^{\left[\frac{247.8}{T2+133.15}\right]}}\right) \quad (1)$$

Equation 1 can be applied to correct the settling velocities for difference in temperatures in whichever of the four types of sedimentation, i.e., unflocculated discrete settling, and flocculated discrete settling. Even though equation 1 can be used for a sensitivity analysis on the performance of the model for different seasons, e.g. summer and winter, there is no evidence that the settling properties can be accurately extrapolated from one season to another.

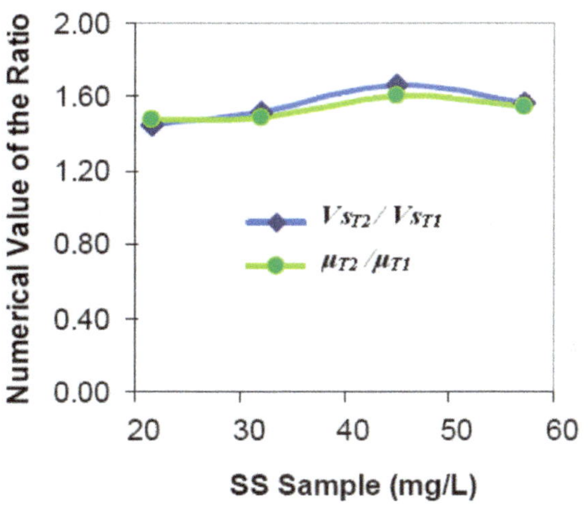

Figure 11: Ratios of Vs_{T2} / Vs_{T1} and μ_{T2} / μ_{T1} for Different suspended solid (SS) concentrations.

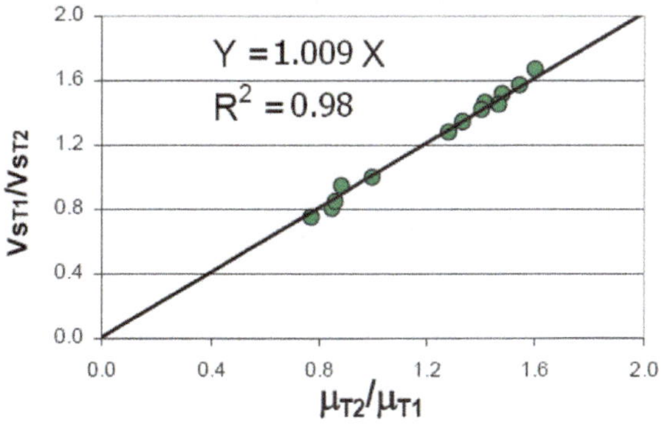

Figure 12: Effect of Temperature on Settling Velocity.

VALIDATION OF THE MODEL

The validation process involves comparing the model response to actual measured data. The model was validated using measured data from the AL-DEWANYIA WTPs.

After the development of the hydrodynamic model, and turbulence model, the ST model was tested. The ESS predicted by the model was tested during seven days (from a 10 day period) showing a very good agreement with the field data. Figure 13 presents a comparison between the experimentally measured and the simulated values of the floc concentration in the effluent of the existing tanks in AL-DEWANYIA. Apparently, there is a good agreement between measured and predicted values.

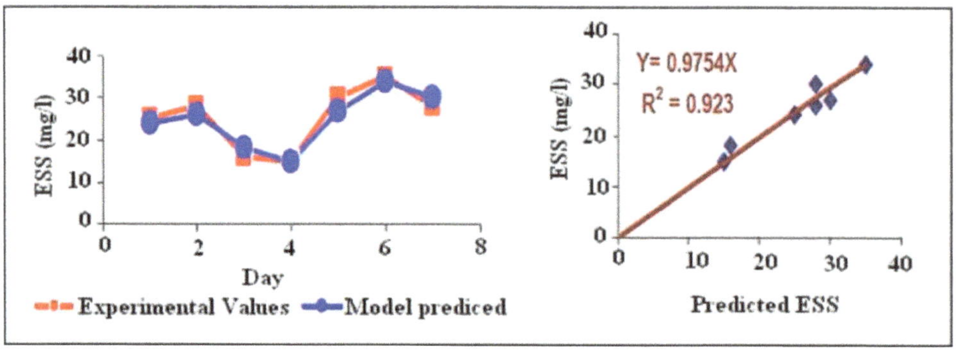

Figure 13: Validation of the ESS Simulated by the Model.

The average values of effluent concentration of improvement are presented in Table 6. The tank operation conditions in the data collection periods (February-April, 2007, (June-August, 2006)) and the CFD model predictions are very close as shown in Table 6. The comparison of model predictions with the subsequent field data indicates that the significantly improvement of tank performance was obtained by using the minor modifications based on the 2-D computer modeling.

Table 6: Comparison of model predictions with field data

	Operation conditions	Effluent concentration (mg/l) and improvement		
	Ave. concentration mg/l	No Modifications	baffle Modifications	Baffle and launder Modifications
Field Data June-August 2006	50	28	-	-
Model Predictions	47	27	6 (+78%)	5 (+82%)

CONCLUSIONS

The introduction of this study made clear that many factors influence the performance of settling tanks. They may be categorised as physico-chemical and hydraulic influences. To account for them in terms of process operation and design, mathematical models may be utilised. In this respect, Computational Fluid Dynamics (CFD) enables the investigation of internal processes, such as local velocities and solids concentrations, to identify process in efficiencies and resolve them. Although these complex models demand for considerable computational power, they may become an option for the study of process operation and control as computer speed increases. Nowadays, they mostly find applications in the world of settling tank design.

The main purpose of this investigation was to develop a CFD ST model capable of simulating the major processes that control the performance of settling tanks, this goal was achieved. The accomplished objectives of this research include: the development of a compound settling model that includes the representation of the settling velocity for the suspended solids usually encountered in this type of tank (horizontal sedimentation tanks) the inclusion of iron removal effects, a flocculation sub-model, and a temperature sub-model.

These types of sub-models have not been previously incorporated in CFD ST models. The model was rigorously tested and validated. The validation process confirms the utilities and accuracy of the model. An important benefit of this research is that it has contributed to a better understanding of the processes in STs. The results presented in this research clarify important points that have been debated by previous researchers. This research may also open the discussion for future research and different ways for improving the performance of existing and new STs. In summary, this research has led to more complete understanding of the processes affecting the performance of settling tanks, and provides a useful tool for the optimization of these corn stone units in water treatment. The major conclusions, general and specifics, obtained from this research are:

- CFD modeling was successfully used to evaluate the performance of settling tank.

- The usually unknown and difficult to be measured particle density is found by matching the theoretical to the easily measured experimental total settling efficiency. The proposed strategy is computationally much more efficient than the corresponding strategies used for the simulation of wastewater treatment.
- Solid removal efficiency can be estimated by calculating solids concentration at effluent.
- High solid removal efficiency was achieved for all cases tested.
- Baffling inlet arrangement succeeded in controlling kinetic energy decay.
- Improved energy dissipation due to an improved inlet configuration.
- Reduced density currents due to an improved inlet configuration.
- Improved sludge removal due to the inlet configuration.
- Troubleshoot existing STs and related process operations.
- The effluent quality can be improved by more than 60% for any cases.
- Evaluated ST design under the specified process conditions.
- Develop reliable retrofit alternatives with the best cost-effectiveness.
- The changes in temperature on STs play an important role on the performance of STs.
- Scrape is important in the settling process and play a big role in changing the flow field.
- In this work we improved the STs guidelines design procedure.
- The fairly good agreement between model predictions and field data.

In general the study demonstrated that CFD could be used in reviewing settling tank design or performance and that the results give valuable insight into how the tanks are working. It can be inferred that CFD could be use to evaluate settling tank designs where the tanks are not functioning properly.

ACKNOWLEDGEMENT

This chapter is supported by the Scientfic Grant Agency of Ministry of Education-VEGA Projekt No. 1/1143/11 with at the Department of Sanitary and Environmental Engineering, Faculty of Civil Engineering, Slovak University of Technology in Bratislava.

REFERENCES

1. Huggins D.L., Piedrahita R.H., & Rumsey T., (2005). Use of computational fluid dynamics (CFD) for aquaculture raceway design to increase settling effectiveness, Aquacult. Eng. 33, 167–180.
2. Dick, R. I. (1982) Sedimentation Since Camp, J Society of Civil Engrs, 68, 199-235.
3. Abdel-Gawad, S. M. & McCorquodale, J. A. (1984b). Hydrodynamics of Circular primary clarifiers. Canadian J. Civil Engr. ,11, 299-307.
4. Adams, E.W. & Rodi, W. (1990). Modelling flow and mixing in sedimentation tanks, Journal of Hydraulic Engineering, 116(7):895-913, Jul.
5. Launder, M. & Spalding, D. (1974). The numerical computation of turbulent flows, Computer methods in applied mechanics and engineering, (3), 269-289.
6. Ghawi, A.H., (2008). A numerical model of flow and settling in sedimentation tanks in potable water treatment plants, Edícia vedeckých prác (Science Works Edition) , Workbook no. 64, STU Bratislava, pp. 150, ISBN 978-80-227-2964-2

CHAPTER 26

A MULTI-SCALE HYBRID CFD-DEM-PBM DESCRIPTION OF A FLUID-BED GRANULATION PROCESS

Maitraye Sen, Dana Barrasso, Ravendra Singh and Rohit Ramachandran

Department of Chemical and Biochemical Engineering, Rutgers, The State University of New Jersey, Piscataway, NJ 08854, USA

ABSTRACT

In this study, a hybrid multi-scale model has been developed for a continuous fluid bed wet granulation process by dynamically coupling computational fluid dynamics (CFD) with a discrete element model (DEM) and population balance model (PBM). In this process, the granules are formed by spraying the liquid binder on the fluidized powder bed. The fluid flow field has been solved implementing CFD principles and the behavior of the solid particles has been modeled using DEM techniques whereas the change in particle size has been quantified with the help of PBM. The liquid binder droplets have been modeled implicitly in DEM. A detailed understanding of the process aids in the development of better design, optimization and control strategies. The model predicts the evolution of important process variables (i.e., average particle diameter, particle size distribution (PSD) and particle liquid content) over time, which have qualitative similarity with experimentally observed trends. The advantage of incorporating the multi-scale approach is that the model can be used to study the distributions of collision frequencies, particle velocity and particle liquid content in different sections of the fluid bed granulator (FBG), in a more mechanistic manner.

INTRODUCTION, MOTIVATION AND OBJECTIVES

Granulation is widely applicable in industries which deal with powder handling processes (e.g., food, pharmaceutical, catalyst, fertilizer industries etc.). During granulation process, fine powder particles form agglomerates which are granules of larger size with improved properties (e.g., flowability, uniform composition etc.). In pharmaceutical industries, wet granulation is an important unit operation in the downstream tablet manufacturing process, since it has a significant effect on the mechanical properties of the tabletting material (e.g., hardness, dissolution rate). As a result, the understanding of the process dynamics of this particular unit operation is critical. An inefficient operation of the granulation process leads to high batch rejection rates (if operated in batch mode) or high recycle ratio (if operated in continuous mode) [1]. Since it is necessary to closely monitor the product quality at every step in powder handling processes, the Food and Drug Administration (FDA) has introduced the principles of Quality by Design (QbD) and Process analytical technology (PAT). These guidelines allow consistent building of product quality at every step of the manufacturing framework and help reduce production of inferior quality products. As the pharmaceutical industries are transitioning towards implementation of QbD and PAT guidelines, in order to improve the processing efficiency, there is a need to develop more science-based models which can be used to capture detailed process dynamics. Such a

Citation: Maitraye Sen, Dana Barrasso, Ravendra Singh and Rohit Ramachandran, A Multi-Scale Hybrid CFD-DEM-PBM Description of a Fluid-Bed Granulation Process, doi:10.3390/pr2010089.

process model can be used for virtual experimentation and optimization, prior to testing in the actual plant. A mechanistic model developed from the first-principles can be more efficient in capturing the dynamics of a process compared to empirical or statistical models [2]. Since fluid bed granulation process consists of both solid and fluid phases, it is desired to model both the phases individually and understand how they interact with each other. The fluid phase can be treated as a continuum and the flow dynamics can be described with the help of continuity equation and equation of motion. On the other hand, solid particles are discrete entities which will require the implementation of DEM techniques in order to capture their flow dynamics. Such types of flow which include both fluid and solid phases can be simulated efficiently with the help of CFD-DEM coupling. CFD will account for fluid flow whereas the same can be done for the solid phase using DEM. Establishing a connection between the flow field of the fluidizing medium and the contact pattern of the particles which in turn will dictate the aggregation rate will provide a detailed process dynamics of the fluid bed granulation process. A multi-scale modeling scheme in this case is required. The advantage of developing a multi-scale model is that it stores the information from various scales (i.e., micro, meso and macro) and couples the information to simulate the process dynamics. Continuum models are associated with macro scale simulation, volume averaged equation over a cell with grid size larger than the particle size is an example of meso scale simulation and in micro scale simulations, the flow of each and every individual particle is considered [3]. For the first time, this work introduces a framework which uses macro scale information from a CFD model (based on continuity equations), meso scale information based on a PBM and particle level information from DEM to simulate the process dynamics.

Objectives

The purpose of this study is to develop a hybrid CFD-DEM-PBM model for fluid bed granulation process with the following objectives:

- Present a hybrid CFD-DEM-PBM framework using dynamic two-way coupling.
- Incorporate multi-scale information such that the model can be used to study the detailed process dynamics.
- Study the heterogenous particle velocity distribution and liquid binder distribution.
- Study the evolution of average particle diameter and particle liquid content with time.

The CFD model has been solved using Fluent™ (version 14.5.0) (ANSYS Inc., Canonsburg, PA, USA), DEM has been solved using EDEMTMTM (version 2.5.1) (DEM Solutions, DEM Solutions Ltd. Edinburgh, UK) and in order to link the PBM with DEM, a user-defined library model has been built using the application programming interface within EDEMTMTM.

BACKGROUND

In order to facilitate agglomeration during granulation process, a liquid binder is added and granulation takes place as a result of three rate processes (1. wetting and nucleation; 2. consolidation and aggregation and 3. breakage and attrition) [4]. Based on these rate processes, a granule can be classified by three internal properties (e.g., size, liquid content and porosity). In a FBG, the particles are fluidized in order to ensure maximum contact between the liquid and the powder particles. At first the particles are rendered wet when they come in contact with the liquid binder and form granule nuclei. As the wet particles collide with one another, they form liquid bridges resulting in aggregation of small sized particles to form a larger sized particle. Similarly, particle breakage occurs because of the presence of shear stresses, compressive and tensile forces within the granulator due to particle-particle interaction or particle-wall (vessel interior wall) interaction, which can break a particle into smaller fragments.

The particle wetting rates in a FBG is highly dependent on the particle flow pattern which in turn is affected by the flow field of the fluidizing gas [5]. The flow behavior of the powder within the granulator also depends on the geometry of the vessel. Several attempts are being made to understand the granular flow pattern but many aspects of it (i.e., a more detailed study of the vessel geometry, fluid flow field, collision frequencies etc.) using sophisticated modeling tools and techniques are yet to be developed [6]. The physical properties of the granules that are important in deciding the compactibility of the granules and which also affect the mechanical strength of the tablets, are particle liquid content, particle size, particle surface area and pore diameter. Porosity relates to the void spaces present in the particle and it is an important particle property because the surface liquid content of a particle will depend on its porosity and the surface liquid content will further decide the particle's aggregation rate. Porous particles can easily deform during compaction to create new bonding surfaces, thus affecting the tablet hardness [7]. Hardness of the tablet also depends on the granule surface area [8]. An appropriate design and operation of the granulation process can help to control the PSD and also help to achieve the desired flow characteristics of the particles. Improved granule characteristics will lead to efficient operation of the other unit operations present in the tablet manufacturing framework [9] and also help to achieve better product quality.

PBMs have been used extensively in order to model granulation process [10,11]. PBMs are used to calculate the rate processes (e.g., aggregation, consolidation and breakage) during granulation, but certain information, used in calculation of mechanistic aggregation and breakage kernels (i.e., effect of particle properties, spatial effect, collision frequency, particle velocity etc.) cannot be determined from a PBM alone. A DEM on the other hand is able to determine these entities. Such a PBM-DEM coupled model for particle aggregation has been reported [12]. A one way coupling consisting of PBM-DEM hybrid model has been reported [13,14] in case of continuous mixing, in which the PBM consists of spatial discretization only (as there is no size change taking place during mixing).

Since a CFD-DEM approach has been used to study several complex particle-fluid flow systems [15,16,17], it is an effective tool to model fluid bed granulation. Interested readers are referred to the review presented by [18] on the discrete particle modeling of fluid beds. A combined CFD-DEM approach was first presented by [19] for simulating plug flow through horizontal pipes. Since then, there have been several works reported in scientific literature [20,21,22,23]. Liquid-solid interactions have been studied in case of gas bubble formation in a gas-liquid-solid system [24,25,26].

Similar studies have been reported in case of gas and liquid fluidized beds in the bubbling regimes where a rigorous two way coupling framework has been introduced to explain the fluid-particle interaction [27,28]. Fernandez et al. used CFD-DEM coupling in case of centrifugal separation systems [29]. Previous studies have been carried out in CFD-DEM coupling in case of granulation and are well documented [5,30]. A direct numerical simulation model for a three phase flow in case of wet granulation, consisting of solid, liquid and gas has been developed by [3]. The liquid-gas flow has been solved using one of the CFD multi-phase solvers, and the solid particle flow has been solved by DEM.

A CFD-PBM framework has been already implemented in order to model size change taking place in a particle-fluid system [31,32] and a PBM derived from kinetic theory of granular flow (KTGF) has been developed by [33] for fluidized melt granulation. Hydrodynamic modeling of wet granulation process in a FBG using CFD principles has been reported by [34]. However there is a major scope still open to study a CFD-DEM-PBM coupled framework, which will capture a detailed interaction between the phases and quantify the size change. Such a multi-scale model formulation has been reported by [35] for agglomerate breakage in fluid bed. The present work aims at capturing the particle aggregation during a granulation process in a FBG. The framework can be further developed in future by adding particle breakage along with the aggregation.

MULTI-SCALE MODEL DEVELOPMENT

This section illustrates the mathematical equations and modeling techniques of each domain of the coupled framework. The CFD model calculates the fluid flow-field and the DEM adds it to the force-field calculation for each particle. The PBM obtains the various information required to calculate the rate processes from DEM to quantify the size change and update the PSD.

CFD Model for the Fluidizing Medium

The fluid phase has been simulated using FLUENT™. The flow model chosen is that of a laminar viscous flow. The governing equations for incompressible flow are given in Equation (1) and Equation (2) [5,36]:

Navier-Stokes Equation

$$\frac{\partial \epsilon \mathbf{u}}{\partial t} + \nabla.(\epsilon \mathbf{u}\mathbf{u}) = -\frac{1}{\rho}\epsilon \nabla p - \frac{\mu}{\rho}\nabla^2.(\epsilon \mathbf{u}) + \mathbf{g} - S_p \quad (1)$$

and Equation of continuity

$$\nabla.(\epsilon \mathbf{u}) = 0 \quad (2)$$

where □ is the solid volume fraction, u is the fluid velocity, ρ is the fluid density, p is flow pressure, μ is the viscosity, S_p is the source term and g is gravity. The fluidizing medium in this case has been considered to be air and the fluid velocity at the inlet has been kept constant at 30 m/s. The fluid velocity at the inlet has been scaled as a function of the particle size in DEM to simulate observed experimental fluidization in a granulator. The inlet velocity has been determined by running several simulation trials such that the particles are fluidized and the effect of gravity can be nullified. In real granulation process, the particle size is in microns. But in the simulation, the initial particle size is in millimeters (i.e., 2 mm diameter), which is much higher than the initial particle size used in real granulation process. As the simulation bed weighs more, therefore a higher air velocity is required to counter the effect of gravity. The solution methods used for calculating the momentum and particle volume fraction is first order upwind [37]. A first order Implicit scheme has been selected for the transient formulation. The boundary conditions have been set as follows:

- Flow near wall is laminar and the velocity varies linearly with the distance from wall.
- A no slip boundary condition has been set at the wall.
- A velocity inlet boundary condition has been used for the air entering the geometry.
- An outlet-vent boundary condition has been used at the geometry exit.

Discrete Element Model

DEM essentially uses Newton's laws of motion (as shown in Equation (3) and Equation (4)) to simulate the particle force fields (namely contact forces and body forces [38]). The contact forces are due to particle-particle or particle-boundary (vessel internal wall) contacts whereas the body forces are any external force fields (i.e., gravity or fluid flow field in case of fluid bed granulation) acting on the particles. The net force F_{total} is calculated for each particle at a time interval which is approximately in the order of 10^{-5} or 10^{-6} s and the new particle state is calculated by numerically solving Newton's law and Euler's equations of rotational motion. The simulation uses a damped Hertzian normal contact model with Mindlin-Deresiewicz/Coulomb friction tangential force model. A detailed discussion on the contact models along with the governing equations have been provided by [38].

$$m_i \frac{dv_i}{dt} = F_{total} \quad (3)$$

$$F_{total} = \sum F_{contact} + \sum F_{body} \quad (4)$$

The DEM has been simulated using EDEM™. The geometry of the FBG has been imported within EDEM™ and the initial PSD has been created. The initial PSD is uniform such that all particles are of fixed size (1 mm in radius). A virtual particle factory plate has been introduced within the geometry in the lower half and the particles are created at a generation rate of 50,000 particles per second. The liquid addition has been captured in EDEM™. The liquid droplets have been modeled as solid particles with similar properties as the initial powder particles. Liquid particles have been created continuously at a rate of 50,000 particles per second from a virtual plate placed in the upper half of the granulator. It should be noted that the liquid addition has been started at time equal to 0.2 s in order to let the particles fluidize first. The feed rate and the number of particles have been set in a way such that the simulation speed is fast and a reasonable granulation is achieved within the simulation time (when compared qualitatively with the experimental results [39,40]). In a FLUENT™-EDEM™ coupled simulation, it is important that the particle size is comparable with the mesh size (i.e., the particle size shouldn't be very large compared to the mesh size) [41]. For example in this work the primary particle diameter is 2 mm and the mesh size is 4 mm. These parameters have been fixed by running several trial simulations.

The change in particle size and new PSD is calculated using the PBM in the subsequent time steps as will be explained later (in Section 3.3). Whenever a liquid particle collides with a powder particle, it is deleted from the system and the liquid content of the particle increases by the liquid particle volume. If two liquid particles collide with each other, a single droplet with volume equal to the total volume of the two particles is formed. It should be noted that as the liquid particles are deleted from the system immediately upon contact, therefore EDEM™ does not consider the contact forces due to these particles in the simulation. The EDEM™ tracks the collisions occurring in between different particles and the information is stored in form of an array, which is made use by the PBM for calculating the new population distribution function at every population balance time step. Table 1 gives the material properties and the other parameters for particle-particle and particle-wall interaction.

Table 1: DEM simulation parameters

Particle properties	
Shear modulus	1×10^6 Nm^{-2}
Poisson's ratio	0.25
Density	1030 kgm^{-3}
Particle-particle interactions	
Coefficient of restitution	0.2
Coefficient of static friction	0.5
Coefficient of rolling friction	0.01
Granulator walls	
Material	Steel
Shear modulus	7.6×10^8 Nm^{-2}
Poisson's ratio	0.29
Density	7800 kgm^{-3}
Particle-wall interactions	
Coefficient of restitution	0.2
Coefficient of static friction	0.5
Coefficient of rolling friction	0.01

The density of the powder material is 1030 kg/m^3 (which is similar to that of a pharmaceutical active ingredient (S)-Ibuprofen [42,43]). Detailed information on the exact range of parameter values (particle-particle and particle-wall interaction parameters of EDEM™) for S-Ibuprofen is not available in literature, hence the values have been adapted from [38], who simulated fines in a continuous mixer, and from the work of [44].

Population Balance Model for FBG

The Population Balance Equation (PBE) can be expressed as shown in Equation (5) [45]:

$$\frac{\partial}{\partial t}F(\mathbf{x},\mathbf{z},t) + \frac{\partial}{\partial \mathbf{x}}\left[F(\mathbf{x},\mathbf{z},t)\frac{d\mathbf{x}}{dt}\right] + \frac{\partial}{\partial \mathbf{z}}\left[F(\mathbf{x},\mathbf{z},t)\frac{d\mathbf{z}}{dt}\right] = \Re_{formation} - \Re_{depletion} \quad (5)$$

Here, F(**x**,**z**,t) is the population distribution function, **x** is the vector of internal co-ordinates used to express the particle size, z is the vector of external co-ordinates used to represent spatial position of the particles and t is the time. The term $\frac{\partial}{\partial \mathbf{x}}\left[F(\mathbf{x},\mathbf{z},t)\frac{d\mathbf{x}}{dt}\right]$ accounts for the rate of change of particle distribution due to change in particle size. The term $\frac{\partial}{\partial \mathbf{z}}\left[F(\mathbf{x},\mathbf{z},t)\frac{d\mathbf{z}}{dt}\right]$ accounts for the rate of change of particle distribution with respect to spatial co-ordinates. $R_{formation}$ and $R_{depletion}$ stand for particles being formed and depleted respectively due to aggregation and breakage.

The present study considers the particle size change only, therefore the internal coordinates (represented using solid and liquid volume) have been retained and the spatial coordinates have been dropped from the PBE. At present only the particle aggregation has been accounted for and breakage has been neglected. Equation (5) has been modified accordingly to obtain Equation (6) as shown below:

$$\frac{\partial}{\partial t}F(s,l,t) + \frac{\partial}{\partial l}\left[F(s,l,t)\frac{dl}{dt}\right] = \Re_{aggregation} \quad (6)$$

The first and second terms of the equation represent evolution of the particle distribution function with respect to time and rate of liquid addition respectively. s and l are the volumes of solid and liquid (per particle), respectively and $\frac{dl}{dt}$ is the liquid addition rate. The liquid addition has been captured implicitly in EDEM™ (as explained in the previous Section 3.2). Therefore the liquid addition term has been omitted from Equation (6) and the PBE has been further modified as shown in Equation (7):

$$\frac{\partial}{\partial t}F(s,l,t) = \Re_{aggregation} \quad (7)$$

The aggregation rate process is defined in Equation (8) to Equation (10)

$$\Re_{aggregation} = \Re_{formation} - \Re_{depletion} \quad (8)$$

where

$$R_{formation} = \frac{1}{2}\int_0^s \int_0^l \beta(s-s',l-l',s',l',t)F(s-s',l-l',t)F(s',l',t)dl'ds' \quad (9)$$

$$R_{depletion} = \int_0^\infty \int_0^\infty \beta(s,l,s',l',t)F(s',l',t)F(s,l,t)dl'ds' \quad (10)$$

β(s,l,s′,l′,t)is the aggregation kernel, as given in Equation (11). The aggregation kernel has been defined as a function of the collision frequency (C) and collision efficiency ψ. The collision frequency (which is a function of particle size [46]) is calculated based on the number of collisions occurring between the particle groups which can be obtained from EDEM™ based on the particle properties.

$$\beta(s,l,s',l',t) = C(s,l,s',l',t)\psi(s,l,s',l') \tag{11}$$

The collision frequency is an important parameter. As previously mentioned, it is already known to be a function of particle size [46], but a study has been conducted by the authors showing collision frequency as a function of the PSD as well. A summary of this study has been provided as an appendix in this manuscript for the benefit of the readers. The effect of the particle size is captured in the collision frequency which in turn controls the aggregation rate. Therefore, the aggregation rate kernel of this model depends on the PSD at any point of time. Collision frequency can be calculated as shown in Equation (12) [46]:

$$C(s,l,s',l',t) = \frac{N_{coll}(s,l,s',l',t)}{F(s,l,t)F(s',l',t)\Delta t} \tag{12}$$

In the above equation, N_{coll} is the total number of collision between two particle types represented by solid and liquid bins $(F(s,l,t)$ and $F(s',l',t))$ during the time interval Δt.

A simple expression (as shown in Equation (13)) has been adapted for collision efficiency based on the works of Biggs et al. [47]:

$$\psi(s,l,s',l') = \begin{cases} \psi_0, & LC(s,l) \geq LC_{min} or LC(s',l') \geq LC_{min} \\ 0, & LC(s,l) < LC_{min} or LC(s',l') < LC_{min} \end{cases} \tag{13}$$

Here LC stands for liquid content of the particles and LC_{min} is the minimum liquid content required for the particle coalescence. The above expression essentially means that if the liquid content of any two particles is greater than or equal to the minimum value (specified as 0.2 in this case), then the particles may aggregate to form a new particle upon collision depending on the value of the collision efficiency (which has been kept at a constant value of 0.01 in this study). The liquid content and collision efficiency values have been chosen such that the process variables show qualitative similarity in trend when compared to experimental results [40].

The PBM has been implemented by creating a custom contact model and custom factory for particle aggregation and liquid addition within the EDEM™ simulation. The initial particles (both liquid and solid) which are created within EDEM™ has a uniform solid and liquid volume respectively. Since a particle consists of both solid and liquid part, therefore the internal coordinate has been discretized linearly based on the particle's solid and liquid volume (also referred to as "bins").

Information Exchange in the Coupling Framework

The information exchanged over the coupled network along with the model assumptions have been summarized in this section. The model assumptions have been listed below:

- The PBM considers aggregation only, breakage and consolidation has not been incorporated since FBG processes are low shear processes with reduced consolidation and breakage (similar approach has been followed by [48]).
- A simple aggregation kernel has been formulated based on collision frequency and collision efficiency (adapted from [46]).
- The collision efficiency in the aggregation kernel is size independent, non-mechanistic and conditional based on the liquid content of the powder particles (adapted from [47]).
- Liquid addition has been captured in EDEM™ by creating particles which get deleted from the system upon contact.

Figure 1 shows a schematic of the main CFD-DEM-PBM coupling framework and the DEM-PBM framework has been magnified to show the ongoing steps within it. The CFD-DEM framework has been developed by coupling EDEM™ and FLUENT™ through the commercially available coupling interface between them. CFD simulates the flow-field of the fluidizing air. When the solution converges, the fluid flow-field is passed to the CFD-DEM coupling interface which then calculates the drag force acting on each particle. The calculated drag force is then transferred to the DEM solver which updates the particle flow-field to obtain the new particle positions (or state). It should be noted that the PBM has been coded within EDEM™ using the Application programming interface. The PBM time step is greater than the DEM time step. Therefore as the DEM solver performs the iteration, PBM waits till the PBM time step is reached. The number of collisions between each pair of solid and liquid bins over the time interval is recorded within the DEM simulation. This information is transferred to the PBM when the PBM time step is reached. The PBM uses this information to calculate the aggregation kernel as given by Equation (8)–Equation (12). Each time the PBM is solved, a new PSD is calculated. This new PSD is then implemented within the DEM simulation for the subsequent DEM time step until the next PBM step is encountered. The DEM solver iterates until the CFD time step ends. The CFD-DEM coupling interface then takes the information of the new particle position (or state) from the DEM solver, updates the solid volume fraction in each fluid cell and passes the information to the CFD solver. The CFD solver again iterates over the next time step until the solution converges and the same steps are repeated. The EDEM™ time step should be always less than the FLUENT™ time step and it is suggested that the EDEMTMTM time step is kept between (1/10)th to (1/100)th of the FLUENT™ time step for stability [41]. In this case, the CFD time step is 2.0×10^{-3} s and the DEM time step is 3.25×10^{-5} s. It means that the DEM solver iterates approximately 61 times within one CFD time step. The PBM time step is 0.25 s (i.e., PBM is solved every 0.25 s). The framework has been simulated for 2 s. Therefore the PBM has been solved 8 times in this simulation time. The PBM time step is an important factor and has been decided after running a few simulation trials. It is important to choose the time-step in a way such that:

- A reasonable number of collisions occur among the particles between any two subsequent time steps.

- The PBM is solved a reasonable number of times such that there is a more consistent distribution of the particle size (as described in Section 4.3)

If the PBM time step is too low, then the model will not be able to capture enough number of collisions which results in difficulty of computational tractability of PBM. A resolution study of the PBM time step and its effect on PSD has been carried out (details have been provided in the results and discussion section (Section 4.3)).

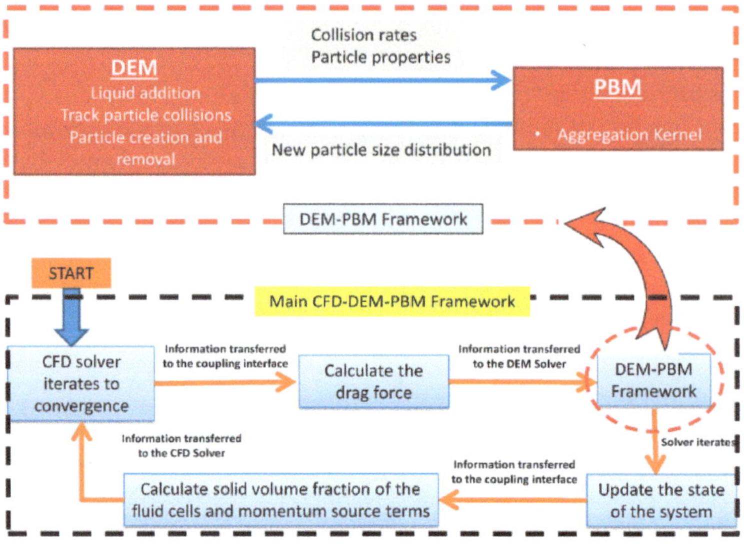

Figure 1: Schematic of the coupled multi-scale framework.

The coupled framework has been run for a time period of 2 s which took approximately 6 h of CPU time, running on an Intel Core i7-2600 CPU processor (3.4 GHz) (Intel Corporation, Santa Clara, CA, USA) with 16 GB of RAM.

Model Outputs

This section presents a brief description of the model outputs which are PSD, average fractional liquid content and average diameter.

The PSD has been obtained by post-processing the simulation data. The diameter of each particle present in the system at any particular time point can be obtained from EDEM™. Several size classes are defined by ranges of the diameter and each bin from the PBM are grouped in these size classes. The total mass of particles in each of these size classes is determined individually and each of them is further normalized by dividing with the overall mass of particles to obtain a mass frequency as seen in Equation (14). The mass frequency can be plotted with respect to the size classes in order to obtain the PSD.

$$\mu_m(t) = \sum_{L_m \le D_{ij} < L_{m+1}} \left(\frac{F(s_i, l_j, t)}{\sum_{i=1}^{ns} \sum_{j=1}^{nl} F(s_i, l_j, t)} \right) \frac{\rho V_m}{\sum_{m=1}^{nm} \rho V_m} \tag{14}$$

where μ_m is the mass frequency of mth size class (which is a function of particle diameter), ρ is the density, ns is the number of solid bins, nl is the number of liquid bins, V_m is the particle volume in m th size class, nm is the number of size classes and D_{ij} is the particle diameter corresponding to the ith solid bin and jth liquid bin.

Similarly the average diameter (D_{avg}) can be calculated as shown in Equation (15):

$$D_{avg}(t) = \frac{\sum_{i=1}^{ns} \sum_{j=1}^{nl} F(s_i, l_j, t)(6(s_i + l_j)/\pi)^{1/3}}{\sum_{i=1}^{ns} \sum_{j=1}^{nl} F(s_i, l_j, t)} \tag{15}$$

The average fractional liquid content (x_{avg}) can be calculated as shown in Equation (16):

$$x_{avg}(t) = \frac{\sum_{i=1}^{ns} \sum_{j=1}^{nl} F(s_i, l_j, t)(l_j/(s_i + l_j)}{\sum_{i=1}^{ns} \sum_{j=1}^{nl} F(s_i, l_j, t)} \tag{16}$$

RESULTS AND DISCUSSION

This section includes the discussion on the model geometry and the results obtained from the coupled framework. A brief section on setting up the coupled model has been included which enumerates the step wise set-up of the simulation.

Simulation Procedure

This section lists the procedure followed while setting up the coupled simulation, in a nutshell.

1. The geometry has been made using ANSYS Design Modeler™.
2. The geometry has been meshed using ICEM-CFD™.
3. The mesh file has been imported within FLUENT™.
4. The mesh has been converted into Polyhedra domain.
5. The gravity is defined in the correct direction and a transient simulation is selected.
6. The flow model has been selected to be viscous lami-

nar.

7. The coupling server has been started.

8. The FLUENT™ is coupled with EDEM™ for the desired fluid domain by selecting the Eulerian-Eulerian option.

9. The coupling server will automatically import the geometry with the specified direction of gravity in EDEM™ and set the source terms in x-momentum, y-momentum and z-momentum calculation. The value of the simulation parameters of the coupling interface has been set as follows:

   - Sample points: The number of points used by FLUENT™ to calculate the volume fraction of the fluid cell. This value has been set at 10, which means that a large particle can transfer its volume between 10 cells. This particular parameter decides the stability and speed of the simulation. A higher value of sample point may increase the stability but decrease the simulation speed.
   - Relaxation factor: The relaxation factors again help with stability and convergence of the solution. Reducing the value helps to increase stability and achieve convergence. Both momentum-MTM-under-relaxation factor and volume under-relaxation factor have been set at 0.7.

10. The inlet fluid velocity has been defined as 30 m/s.

11. The custom contact model and custom factory (for PBM calculation) have been imported within EDEM™.

12. The material properties, particle-particle and particle-wall interaction parameters as given in Table 1 have been set in EDEM™.

13. The initial PSD has been created in EDEM™.

14. The liquid particles have been created in EDEM™ (the liquid addition starts at 0.2 s).

15. Once the EDEM™ simulation is set up, initialize the solution in FLUENT™.

16. Run the calculation.

Model Geometry

The measurement details of the model geometry has been given in Figure 2. Figure 2 also presents the mesh, which has been done in ICEM-CFD™. The geometry has been meshed using the tetra/mixed meshing tools with a grid size of 4mm. The mesh has been converted into polyhedra domain after importing it in FLUENT™. As shown in Figure 2, the geometry has been divided into five fluid domains or cell zones (inlet domain, fluid inlet domain, main fluid domain, fluid outlet domain and outlet domain). The CFD-DEM coupling has been performed for the cell zone (main fluid domain). Every time the simulation leaves Fluent™ and enters EDEM™, all the relevant data in the linked zones are passed to EDEM™. Reducing the number of linked zones will reduce the amount of data transferred between the two softwares and thus increase the simulation speed. Therefore the simulation has been linked for the "main fluid domain" only. Although the coupling has been done for the "main fluid domain", the particles are free to move around throughout the geometry, but only the data from the "main fluid domain" are passed to EDEM™. The results/plots reported in the manuscript are for the "main fluid domain" only. It should be noted that the spatial co-ordinates have not been considered in the PBM. The main idea is to present a CFD-DEM-PBM framework which is able to capture the granulation dynamics. Therefore the assumption is that the "main fluid domain" is representative of the whole geometry.

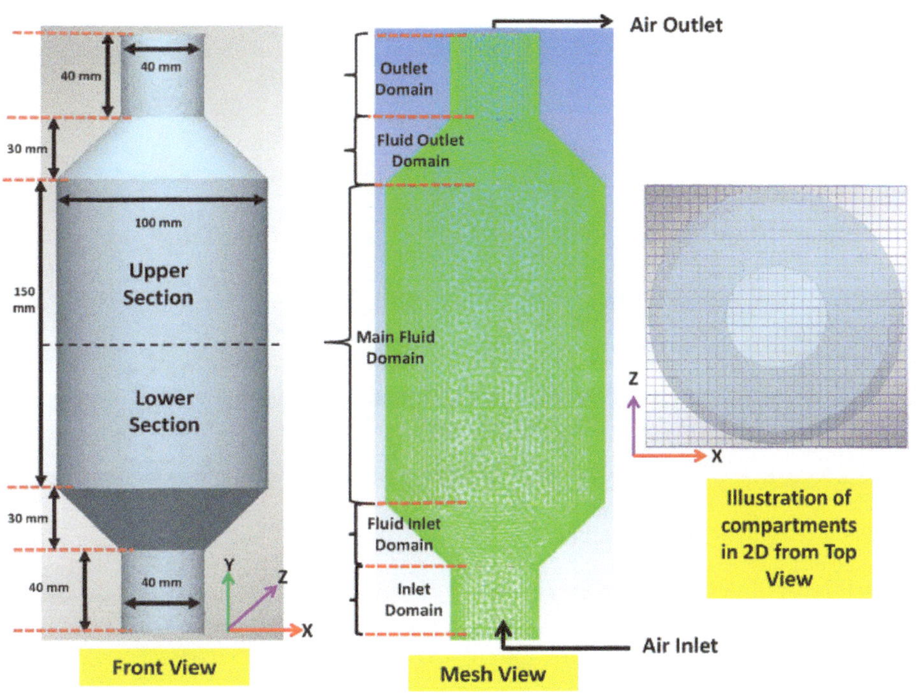

Figure 2: Model Geometry and mesh of the FBG.

Multi-Scale Model Results

The multi-scale model has been simulated to obtain the average diameter and liquid content of the particles as a function of time. This model can be also used to study the distribution of particle velocity and particle liquid content, which will help to understand the flow pattern and liquid content of the solid particles. Since the aggregation rate depends on the liquid content of the particles, therefore it is important to study these aspects. It will also help in understanding the impacts of the equipment design and improve it for a better operational efficiency.

In order to analyze the particle flow pattern and liquid distribution, the "main fluid domain" has been divided equally into two sections (upper half and lower half). Each half has been divided into several compartments and the DEM simulation has been post processed to obtain the particle velocity in each of these compartments. As shown in Figure 2, the geometry has been divided into 30 grids in both x and z direction and 1 grid in y direction, such that there are 30×1×30×1 compartments in total. Figure 2 gives an illustration of the compartments in 2D as seen from the top view. Figure 3 shows the contour plots of the particle velocities (i.e., the resultant velocity of the three components v_x, v_y and v_z) averaged over different time intervals. Figure 3a,b present the particle velocities for both lower and upper halves, averaged over time 0 s–0.5 s and 0.5 s–1 s respectively. It can be seen that the velocity of the particles increases with time. However the lower section shows a more uniform distribution of velocity compared to the upper section. The plot for upper section in Figure 3a shows a non-uniform distribution with high velocity present sporadically in few locations and the velocity values in most of the compartments are closer to zero. This is because the particles are being injected into the system from a virtual plate placed in the lower half and it takes about 0.4 s for the bed to be fully fluidized. Similarly Figure 3c,d present the particle velocity distribution in both the halves averaged over time 1 s–1.5 s and 1.5 s–2 s respectively. These plots also show that the particle velocity increases with time as the bed gets fully fluidized and presents a more uniform velocity distribution in the upper half. It can be also noted that an absolute steady state in the particle velocity is not being realized. This is due to the mechanics of the granulation process (as the partcles interact with each other and the boundary in numerous ways, it is difficult to obtain an absolute steady state).

Figure 4a,b show the particle liquid content distribution averaged over 0 s–0.5 s (initial time of simulation) and 1.5 s–2 s (final time of simulation). Figure 4a shows a non-uniform distribution over a higher range for the upper part, with the values in most of the compartments lying close to zero (similar observation has been made in the velocity distribution as well). The distribution becomes comparatively uniform with time as seen in Figure 4b. The liquid content of the particles in the upper part is more than the liquid content of the particles present in the lower part. This is because the liquid droplets are being injected into the system from a virtual plate located near the upper half of the granulator, hence only a few of them are able to penetrate through the void space and reach the lower half. This shows that the binder liquid distribution is highly dependant on the particle flow pattern, which in turn also affects the aggregation rate (particle liquid content is an important factor in deciding the granulation rate processes).

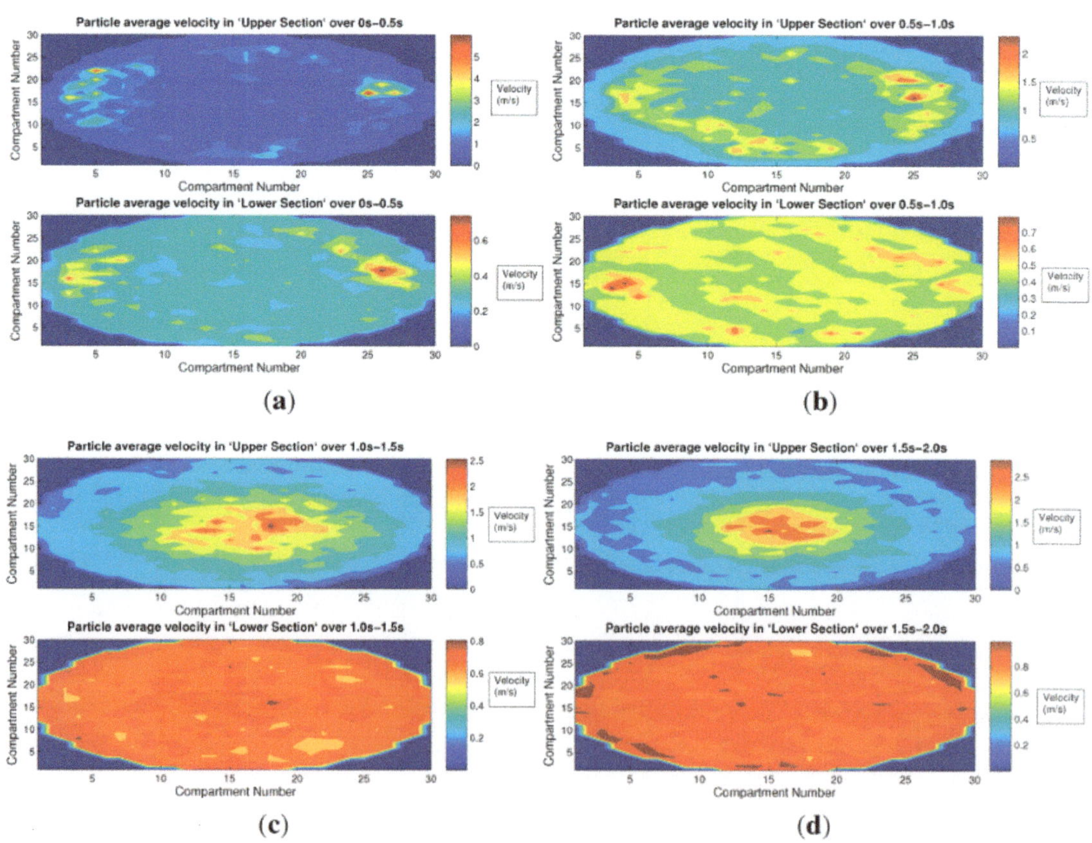

Figure 3: Velocity contour plots. (**a**) Velocity distribution over 0 s–0.5 s; (**b**) Velocity distribution over 0.5 s–1.0 s; (**c**) Velocity distribution over 1.0 s–1.5 s; (**d**) Velocity distribution over 1.5 s–2.0 s.

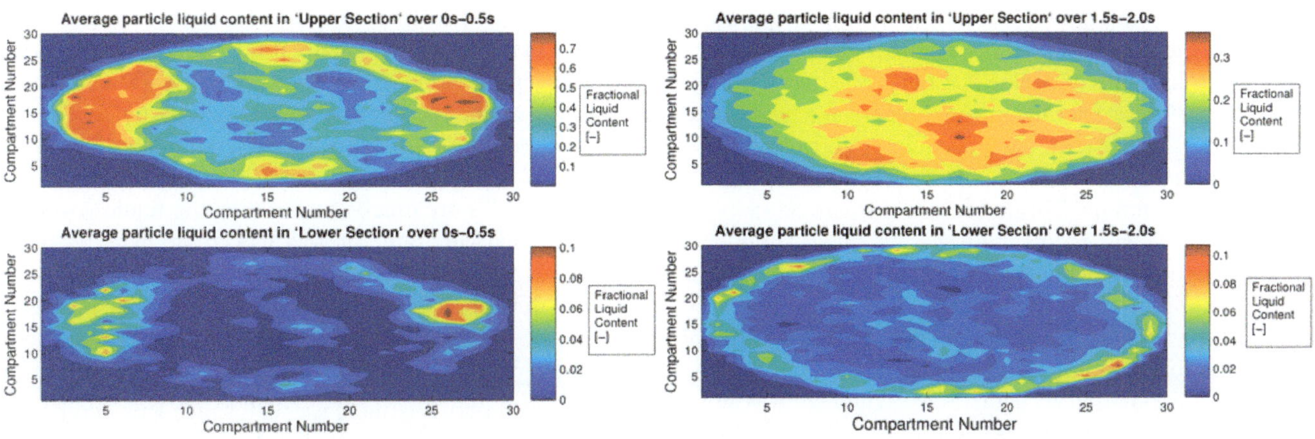

(a) (b)

Figure 4: Plots for particle liquid content distribution. (**a**) Liquid content over 0 s–0.5 s; (**b**) Liquid content over 1.5 s–2.0 s.

Figure 5 presents snapshots of the particles within the granulator taken at different time points based on their liquid content.

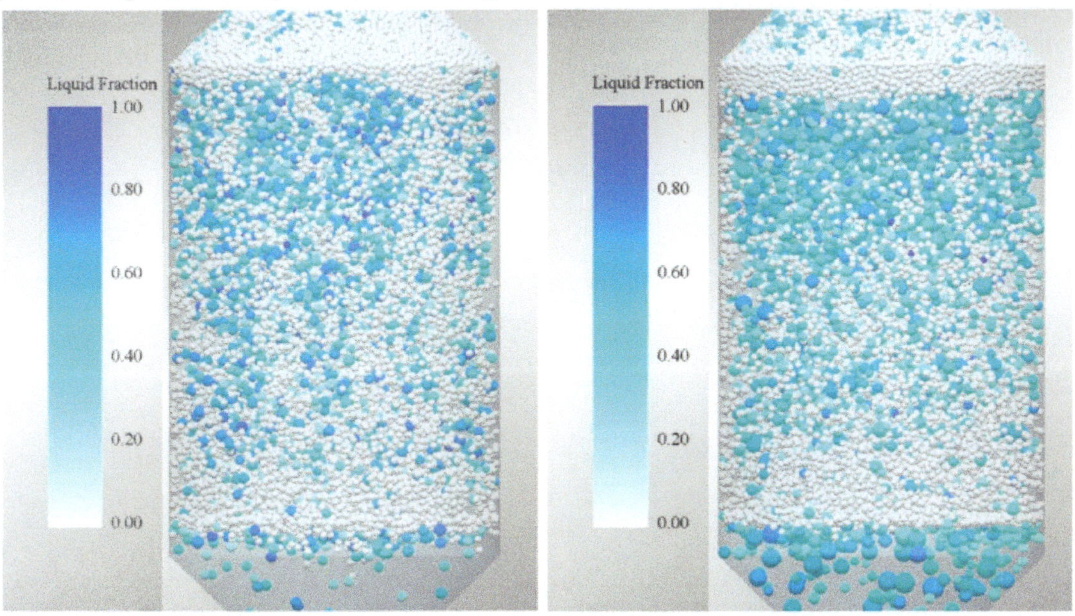

(a) (b)

Figure 5: EDEM™ snapshots of liquid content. (**a**) Particle liquid content at time = 1 s; (**b**) Particle liquid content at time = 2 s.

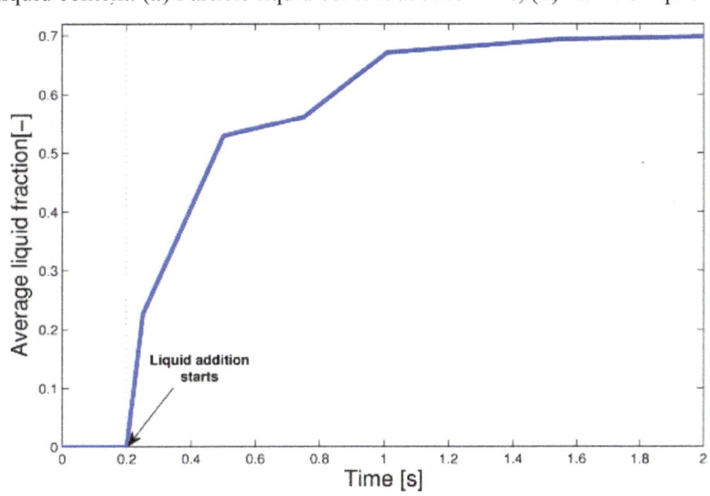

Figure 6: Liquid content vs. time.

As previously mentioned, the amount of liquid present in the particles is an important attribute as it controls the rate process, which in turn will decide the increase in particle size. The snapshots (Figure 5) of the liquid content have been taken at 1 s and 2 s. The particles have been color coded based on their fractional liquid content. It can be seen that more particles are getting wet with time. This is because the particle velocity increases with time and a more uniform velocity distribution is obtained throughout the granulator (also seen from the contour plots) resulting in more contacts between the liquid and solid particles. So the average liquid fraction (i.e., the ratio of liquid volume to the total volume of a particle, where total volume is equal to the summation of liquid volume and solid volume of a particle) increases with time and levels off gradually as can be seen in Figure 6 (plot of particle liquid content versus time). Since the particles are free to move around (in all domains), therefore there are particles exiting the "main fluid domain". Initially there is a steep increase in the liquid content because the bed takes some time to be fully fluidized. As a result the number of particles in the "main fluid domain" is still increasing and more particles are getting wet with time. But as the bed is fully fluidized, an approximate steady state is being realized in the "main fluid domain". It should be noted that in this model, the liquid droplets which have been created are of comparable size to the solid particles. Therefore an assumption has been made that one liquid particle can wet only one solid particle. Under certain circumstances, where the powder particles are much smaller than the liquid particles, it is possible that one liquid droplet wets more than one powder particle (depending on the liquid to solid volume ratio). This feature will be included in the coupled framework in future.

Figure 7 presents the snapshot of the relative particle diameter taken at 1s and 2s. It can be seen that the number of large sized particles increase with time. This can be again explained on the basis of the particle movement pattern. With time, the velocity distribution becomes more uniform and the contact between the particles increases which results in aggregation. Figure 8 is a plot for the average diameter as a function of time. It is seen that the average diameter of the particles increase with time. Figure 9 is a plot for the PSD taken at different time points. It can be seen that a gradual progression towards larger sized particles is being realized during granulation.

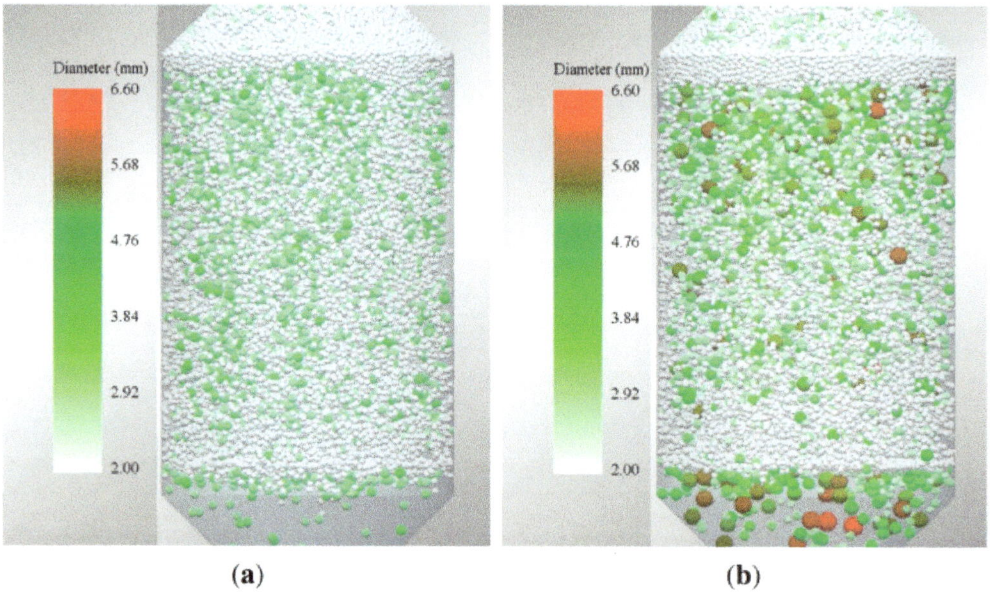

Figure 7: EDEM™ snapshots of particle diameter. (**a**) Particle diameter at time = 1 s; (**b**) Particle diameter at time = 2 s.

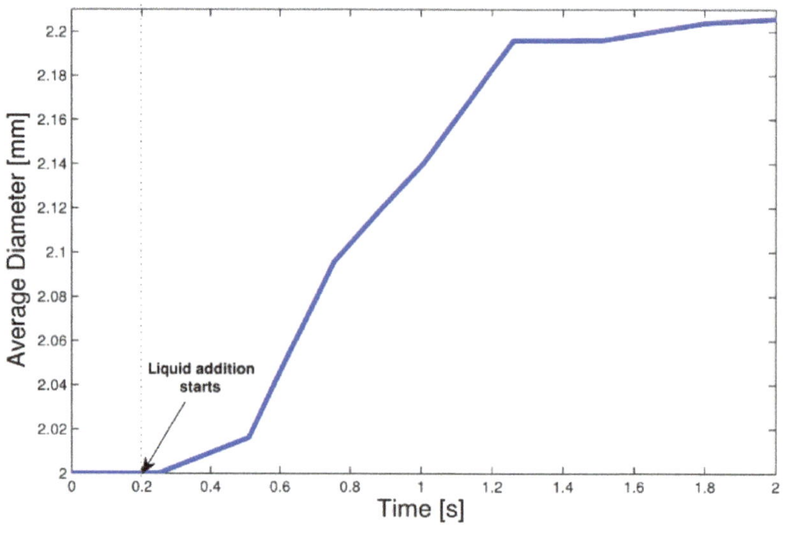

Figure 8: Average diameter vs. time.

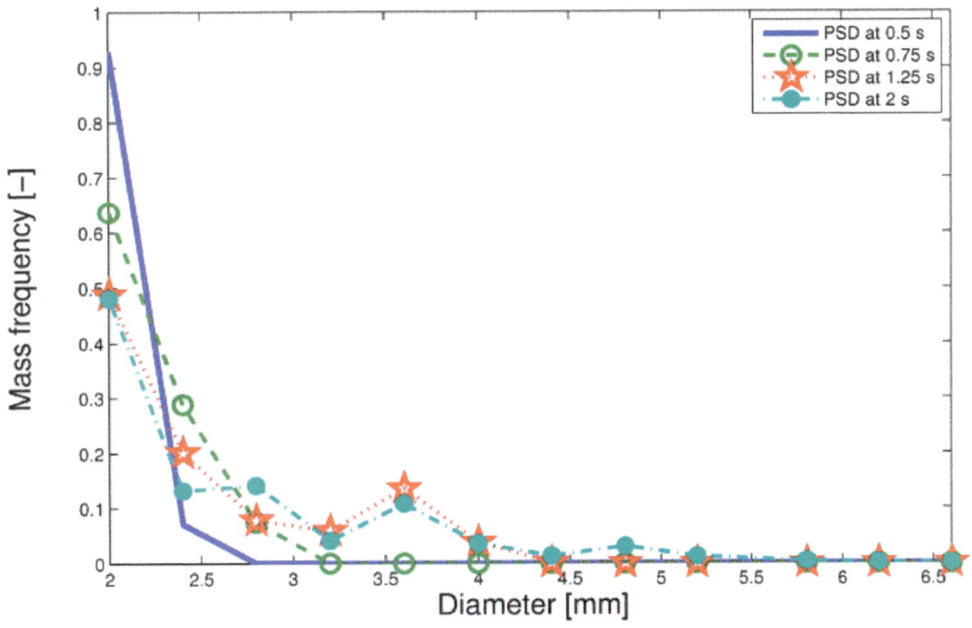

Figure 9: PSD at different time points.

The trends observed in this study are qualitatively consistent with experimental results [39,40]. These studies show that the granulation is being achieved as the particle liquid content increases. They present bimodal particle size distributions that broaden over time by starting with finer primary particles.

In order to show the effect of PBM time step, two more simulatons have been run with PBM time step equal to 0.2 s (PBM has been solved 10 times) and 0.4 s (PBM has been solved 5 times) and compared with the base case (PBM time step equal to 0.25 s, where PBM has been solved 8 times). Figure 10 presents the PSDs obtained at the final time point (time = 2 s) from each simulation. It can be seen that the PSD for the largest PBM time step (0.4 s) has a lower frequency of very large particles (due to the presence of the highest peak). This is because the PBM has been solved only 5 times during the simulation thus limiting the opportunities for large particles to collide and aggregate with each other. On the other hand, PSDs with smaller PBM time step (0.2 s and 0.25 s) show a more consistent distribution suggesting more accuracy. Therefore it is essential to perform a comparative analysis of the model predicted data obtained by running the simulation with different PBM time step, with experimental data.

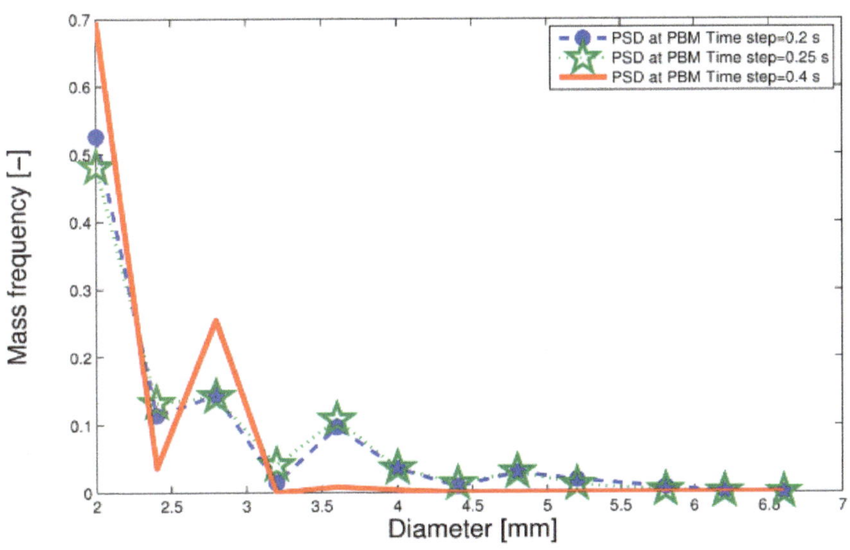

Figure 10: PSDs for different PBM time step.

CONCLUSIONS

This work presents a multi-scale hybrid model for granulation, using information from different scales of CFD, DEM and PBM techniques. Granulation has been extensively modeled following a PBM approach, which groups a lump of particles in different classes (based on size, porosity etc.), but this particular framework tracks the flow-field of each particle with the help of DEM. The CFD aspect helps to add in the drag force acting on the particles due to the flow field of the fluidizing medium. The aggregation kernel is a function of the collision efficiency and collision frequency, which in turn depends on the particle size distribution. An empirical expression, based on the liquid content has been used to determine the collision efficiency. Future effort will be to introduce a mechanistic approach in determining the collision efficiency based on relative velocity of the colliding particles, particle mass and liquid content. The PBM which has been implemented is two-dimensional which can track the change in particle size as well as liquid content of the particles. The liquid addition has been captured implicitly in the DEM and the aggregation kernel has been decided based on the information provided by the DEM on collision frequency and particle liquid content. PBM calculates the new PSD at every PBM time step, which has been implemented in the DEM by removing old particles and creating new ones. The model can be used to study the material flow pattern, equipment geometry (see if dead zones are present) and key performance criteria (i.e., average diameter and liquid content). The present framework considers particle aggregation only, but it will be further extended with the addition of breakage and consolidation processes in future. The model can be updated by addition of spatial co-ordinates as well. This model can be used to study the effect of material properties as well, which can be an interesting future investigation. This coupled model is detailed as it is able to store information from different scales and can be used as an effective tool to understand the process dynamics of a fluid bed granulation process.

ACKNOWLEDGMENTS

This work is supported by the National Science Foundation Engineering Research Center on Structured Organic Particulate Systems, through Grant NSF-ECC 0540855. The authors would like to acknowledge Atul Dubey of Tridiagonal Solutions Pvt Ltd., Pune, India for useful discussion on the DEM simulation and Vijaykrishna Bodla of Krist Gernaey's group at Technical University of Denmark for useful discussions on the CFD simulation.

APPENDIX. EFFECT OF PSD ON COLLISION FREQUENCY

Four PSDs (as shown in Figure A1 have been used in order to demonstrate the effect of the size distribution on the collision frequency. All the four distributions have particles defined by size between 1 mm and 5 mm with varying number of particles of each size. D1 is uniformly sized (i.e., same mass frequency of each particle size). D3 is a normal distribution with a median diameter same as that of D1. D2 and D4 are skewed distributions where D2 has more small sized particles and D4 has more large sized particles. The number of collisions between each particle size has been recorded between 0.5 s to 1.5 s and the collision frequencies have been calculated according to the Equation (12). This study has been performed with DEM only and no PBM has been implemented. Figure A2 shows the collision rates for each distribution. It can be seen, from all the plots that collision between large sized-small sized particles is more frequent than large size-large size particles. The drop in the collision rates along the axis of symmetry shows that the similar sized particles collide with each other less frequently.

D1 and D2 show similar trends of collision frequencies however D3 shows lesser collision frequency among the particles of higher size range and greater collision frequency among the particles of medium size range. D2, which has more number of small sized particles shows lower collision frequencies overall. D4, which has more number of large sized particles shows higher collision frequencies overall. Therefore this study proves that there is a dependance of the collision frequency on the PSD.

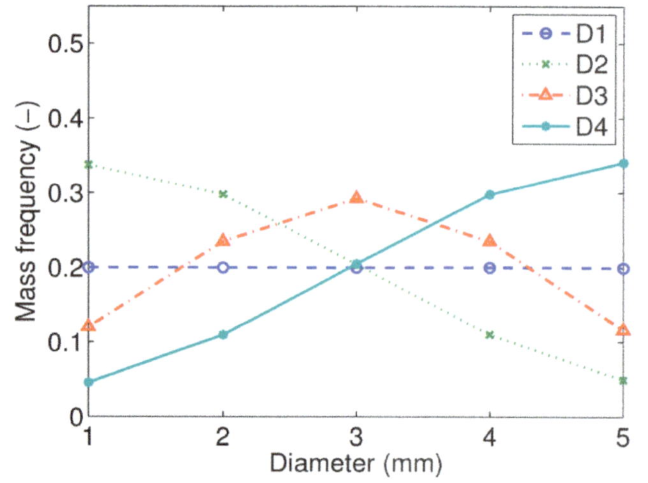

Figure A1: PSD used in DEM simulations to determine collision rates.

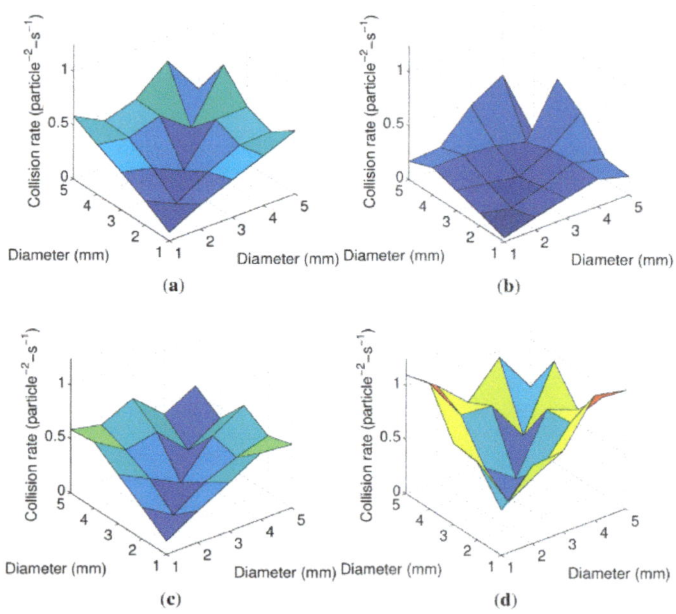

Figure A2: Collision frequency versus particle size for each distribution based on DEM simulations. (**a**) D1; (**b**) D2; (**c**) D3; (**d**) D4.

REFERENCES

1. Wang, F.Y.; Cameron, I.T. Review and future directions in the modelling and control of continuous drum granulation.Powder Technol. 2002, 124, 238–253.
2. Cameron, I.T.; Wang, F.Y.; Immanuel, C.D.; Stepanek, F. Process systems modeling and applications in granulation: A review. Chem. Eng. Sci. 2005, 60, 3723–3750.
3. Washino, K.; Tan, H.S.; Hounslow, M.J.; Salman, A.D. A new capillary force model implemented in micro-scale CFD-DEM coupling for wet granulation. Chem. Eng. Sci. 2013.

4. Iveson, S.M.; Litster, J.D.; Hapgood, K.; Ennis, B.J. Nucleation, growth and breakage phenomena in agitated wet granulation processes: A review. Powder Technol. 2001, 117, 3–39.

5. Fries, L.; Antonyuk, S.; Heinrich, S.; Palzer, S. DEM-CFD modeling of a fluidized bed spray granulator. Chem. Eng. Sci.2011, 66, 2340–2355.

6. Ng, B.H.; Ding, Y.L.; Ghadiri, M. Modelling of dense and complex granular flow in high shear mixer granulator—A CFD approach. Chem. Eng. Sci. 2009, 64, 3622–3632.

7. Habib, Y.; Augsburger, L.L.; Shangraw, R.F. Production of inert cushioning beads: Effect of excipients on the physicomechanical properties of freeze-dried beads containing microcrystalline cellulose produced by extrusion-spheronization. Int. J. Pharm. 2002, 233, 67–83.

8. Cantor, S.L.; Kothari, S.; Koo, O.M.Y. Evaluation of the physical and mechanical properties of high drug load formulations: Wet granulation vs. novel foam granulation. Powder Technol. 2009, 195, 15–24.

9. Boukouvala, F.; Chaudhury, A.; Sen, M.; Zhou, R.; Mioduszewski, L.; Ierapetritou, M.G.; Ramachandran, R. Computer-aided flowsheet simulation of a pharmaceutical tablet manufacturing process incorporating wet granulation. J. Pharm. Innov. 2013, 8, 11–27.

10. Barrasso, D.; Walia, S.; Ramachandran, R. Multi-component population balance modeling of continuous granulation processes: A parametric study and comparison with experimental trends. Powder Technol. 2013, 241, 85–97.

11. Ramachandran, R.; Immanuel, C.D.; Stepanek, F.; Litster, J.D.; Doyle, F.J., III. A mechanistic model for breakage in population balances of granulation: Theoretical kernel development and experimental validation. Chem. Eng. Res. Des.2009, 87, 598–614.

12. Reinhold, A.; Briesen, H. Numerical behavior of a multiscale aggregation model-coupling population balances and discrete element models. Chem. Eng. Sci. 2012, 70, 165–175.

13. Sen, M.; Dubey, A.; Singh, R.; Ramachandran, R. Mathematical development and comparison of a hybrid PBM-DEM description of a continuous powder mixing process. J. Powder Technol. 2013.

14. Sen, M.; Ramachandran, R. A multi-dimensional population balance model approach to continuous powder mixing processes. Adv. Powder Technol. 2013, 24, 51–59.

15. Chu, K.W.; Yu, A.B. Numerical simulation of the gas-solid flow in three dimensional pneumatic conveying bends. Ind. Eng. Chem. Res. 2008, 47, 7058–7071.

16. Ibsen, C.H.; Helland, E.; Hjertager, B.H. Comparison of multifluid and discrete particle modelling in numerical predictions of gas particle flow in circulating fluidised beds. Powder Technol. 2004, 149, 29–41.

17. Rong, D.G.; Horio, M. Behavior of particles and bubbles around immersed tubes fluidized bed at high temperature and pressure: A DEM simulation. Int. J. Multiph. Flow 2001, 27, 89–105.

18. Deen, N.G.; Annaland, M.S.; van der Hoef, M.A.; Kuipers, J.A.M. Review of discrete particle modelling of fluidized beds. Chem. Eng. Sci. 2007, 62, 28–44.

19. Tsuji, Y.; Tanaka, T.; Ishida, T. Lagrangian numerical-simulation of plug flow of cohesionless particles in a horizontal pipe. Powder Technol. 1992, 71, 239–250.

20. Helland, E.; Occelli, R.; Tadrist, L. Numerical study of cluster formation in a gas particle circulating fluidized bed.Powder Technol. 2000, 110, 210–221.

21. Hilton, J.E.; Ying, D.Y.; Cleary, P.W. Modelling spray coating using a combined CFD-DEM and spherical harmonic formulation. Chem. Eng. Sci. 2013, 99, 141–160.

22. Rhodes, M.J.; Wang, X.S.; Nguyen, M.; Stewart, P.; Liffman, K. Use of discrete element method simulation in studying fluidization characteristics: Influence of interparticle force. Chem. Eng. Sci. 2001, 56, 69–76.

23. Zhou, Z.Y.; Kuang, S.B.; Chu, K.W.; Yu, A.B. Assessments of CFD-DEM models in particle fluid flow modelling. J. Fluid Mech. 2010, 661, 482–510.

24. Chen, C.; Fan, L.-S. Discrete simulation of gas liquid bubble column and gas liquid solid fluidized beds. AIChE J. 2004,50, 288–301.

25. Li, Y.; Yang, G.; Zhang, J.; Fan, L.-S. Numerical studies of bubble formation dynamics in gas liquid solid fluidization at high pressures. Powder Technol. 2001, 116, 246–260.

26. Li, Y.; Zhang, J.; Fan, L.-S. Numerical simulation of gas-liquid-solid fluidization systems using a combined CFD-VOF-DPM method: bubble wake behavior. Chem. Eng. Sci. 1999, 54, 5101–5107.

27. Renzo, A.D.; Maio, F.P.D. Homogeneous and bubbling fluidization regimes in DEM-CFD simulations: Hydrodynamic stability of gas and liquid fluidized beds. Chem. Eng. Sci. 2007, 62, 116–130.

28. Tsuji, T.; Yabumoto, K.; Tanaka, T. Spontaneous structures in three-dimensional bubbling gas-fluidized bed by parallel DEM-CFD coupling simulation. Powder Technol. 2008, 185, 132–140.

29. Fernandez, X.R.; Nirschl, H. Simulation of particles and sediment behaviour in centrifugal field by coupling CFD and DEM. Chem. Eng. Sci. 2013.

30. Fries, L.; Antonyuk, S.; Heinrich, S.; Dopfer, D.; Palzer, S. Collision dynamics in fluidised bed granulators: A DEM-CFD study. Chem. Eng. Sci. 2013, 86, 108–123.

31. Drumm, C.; Attarakih, M.M.; Bart, H. Coupling of CFD with DPBM for an RDC extractor. Chem. Eng. Sci. 2009, 64, 721–732.

32. Yan, W.-C.; Luo, Z.-H.; Guo, A.-Y. Coupling of CFD with PBM for a pilot-plant tubular loop polymerization reactor.Chem. Eng. Sci. 2011, 66, 5148–5163.

33. Tan, H.S.; Goldschimdt, M.J.V.; Boerefijin, R.; Hounslow, M.J.; Salman, A.D.; Kuipers, J.A.M. Building population balance model for fluidized bed melt granulation: Lessons from kinetic theory of granular flow. Powder Technol. 2004,142, 103–109.

34. Rajniak, P.; Stepanek, F.; Dhanasekharan, K.; Fan, R.;

Mancinelli, C.; Chern, R.T. A combined experimental and computational study of wet granulation in a wurster fluid bed granulator. Powder Technol. 2009, 189, 190–201.

35. Dosta, M.; Antonyuk, S.; Heinrich, S. Multiscale simulation of agglomerate breakage in fluidized bed. Ind. Eng. Chem. Res. 2013, 70, 165–175.

36. Bird, R.B.; Stewart, W.E.; Lightfoot, E.N. Transport Phenomena; John Wiley & Sons (Asia) Pte. Ltd.: Singapore, 2007.

37. Hirsch, C. Numerical Computation of Internal and External Flows; John Wiley & Sons: Hoboken, NJ, USA, 1990.

38. Dubey, A.; Sarkar, A.; Ierapetritou, M.G.; Wassgren, C.R.; Muzzio, F.J. Computational approaches for studying the granular dynamics of continuous blending processes, 1 DEM based methods. Macromol. Mater. Eng. 2011, 296, 290–307.

39. El Hagrasy, A.S.; Cruise, P.; Jones, I.; Litster, J.D. In-line size monitoring of a twin screw granulation process using high-speed imaging. J. Pharm. Innov. 2013, 8, 90–98.

40. El Hagrasy, A.S.; Hennenkamp, J.R.; Burke, M.D.; Cartwright, J.J.; Litster, J.D. Twin screw wet granulation: Influence of formulation parameters on granule properties and growth behavior. Powder Technol. 2013, 238, 108–115.

41. EDEM CFD Coupling Interface, 2013. EDEM 2.5.1 Documentation. DEM Solutions. Available online: http://www.dem-solutions.com/ (accessed on 17 July 2013).

42. Meade, E.A.; Smith, W.L.; Dewitt, D.L. Differential inhibition of prostaglandin endoperoxide synthase (cyclooxygenase) isozymes by aspirin and other non-steroidal anti-inflammatory drugs. J. Biol. Chem. 1993, 9, 6610–6614.

43. Mehvar, R.; Jamali, F. Pharmacokinetic analysis of the enantiomeric inversion of chiral nonsteroidal antiinflammatory drugs. Pharm. Res. 1988, 5, 76–79. [PubMed]

44. Mangwandi, C.; Cheong, Y.S.; Adams, M.J.; Hounslow, M.J.; Salman, A.D. The coefficient of restitution of different representative types of granules. Chem. Eng. Sci. 2007, 62, 437–450.

45. Ramakrishna, D. Population Balances; Academic Press: San Diego, CA, USA, 2000.

46. Gantt, J.A.; Cameron, I.T.; Litster, J.D.; Gatzke, E.P. Determination of coalescence kernels for high-shear granulation using DEM simulations. Powder Technol. 2006, 170, 53–63.

47. Biggs, C.; Sanders, C.; Scott, A.; Willemse, A.; Hoffman, A.; Instone, T.; Salman, A.; Hounslow, M. Coupling granule properties and granulation rates in high-shear granulation. Powder Technol. 2003, 130, 162–168.

48. Madec, L.; Falk, L.; Plasari, E. Modelling of the agglomeration in suspension process with multidimensional kernels.Powder Technol. 2003, 130, 147–153.

CHAPTER 27

COMPUTATIONAL FLUID DYNAMICS (CFD) MODELING OF PHOTOCHEMICAL REACTORS

Masroor Mohajerani, Mehrab Mehrvar and Farhad Ein-Mozaffari

Department of Chemical Engineering, Ryerson University, Toronto, Ontario Canada

INTRODUCTION

Advanced oxidation processes (AOPs) play an important role in the degradation or the production of a wide range of organic materials. Many organic compounds such as pharmaceuticals, dyes, herbicides, and pesticides have been subjected to degradation and remediation purposes in water and wastewater treatment systems using AOPs. Some of the organic compounds such as drugs, vitamins, or fragrances could be also produced by oxidation processes.

As the standard of living increases, many chemicals such as pharmaceuticals, pesticides, herbicides, and dyes are extensively consumed. Each of these products may cause health issues by their accumulation in aquatic environment. Pharmaceuticals such as antibiotics are partially metabolized and excreted by humans and animals. Improper disposal, dumping, and accidental discharge of drugs lead to the increase of the concentration of compounds such as analgesics, antibiotics, steroids, and hormones in aquatic environment, which cause environmental and health problems. Residual pesticides and herbicides originate from the direct pollutant in production plant, disposal of empty containers, equipment washing, and surface runoff. High levels of these compounds are toxic, mutagenic, carcinogenic, and tumorigenic. Some other wastes such as landfill leachate are subjected to advanced treatment methods. Old landfill leachates (>10 years) are nonbiodegradable in nature due to the existence of organic compounds with high molecular weights. Although the composition of landfill leachates varies widely with respect to the age of the landfill, type of wastes, and climate conditions, they can be categorized into four groups of dissolved organic matter, inorganic macro components, heavy materials, and xenobiotic organic substances. Another type of toxic chemicals which cannot be removed using conventional treatment methods is endocrine disrupting compounds (EDCs). EDCs, especially the steroidal hormones, are well recognized exogenous agents that interfere with the synthesis, action, and/or elimination of natural hormones in the body. Conventional processes are not effective in destruction of these types of organic compounds; therefore, powerful advanced treatment processes are required in order to mineralize them. There are several options for choosing an oxidation process: wet air oxidation, supercritical water oxidation, incineration, and AOPs.

AOPs have been promising in the treatment of contaminated soils and waters. The AOPs could be employed to fully or partially oxidize organic pollutants usually using a combination of different oxidants. In contrary to the

Citation: Masroor Mohajerani, Mehrab Mehrvar and Farhad Ein-Mozaffari, Computational Fluid Dynamics (CFD) Modeling of Photochemical Reactors, ISBN 978-953-51-0271-7.

conventional physical and chemical treatment processes, AOPs do not transfer pollutants from one phase to another, i.e., organic pollutants are completely destroyed. Most AOPs are able to generate hydroxyl radicals. These hydroxyl radicals are active and powerful, capable of reacting with almost all types of organics, including non-biodegradable and recalcitrant compounds. These oxidizing hydroxyl radicals are initiated by photolysis, photocatalysis, sonolysis, radiolysis, and other AOPs alone or in combination in the presence of some reagents such as hydrogen peroxide (H_2O_2), ozone, and homogeneous or heterogeneous catalysts. Hydroxyl radicals are extremely powerful, short-lived oxidizing agent, and non-selective in nature and could react with a wide range of organic chemicals with reaction constant of several orders of magnitude higher than the reaction with molecular ozone under the same conditions. AOPs are used to convert toxic, non-biodegradable, inhibitory, and recalcitrant pollutants into simpler and less harmful intermediates, so they could be treated subsequently in biological processes. Due to the high operating and capital costs of AOPs, the complete mineralization of organic compounds using AOPs is practically impossible; therefore they are mainly combined with other processes such as biological systems to be cost-effective. With sufficient contact time and proper operating conditions and design, the mineralization efficiency of the AOPs is maximized by optimization of the processes in such a way that the total costs of the processes are minimized while the removal rates of organic pollutants are maximized.

Since conducting experiments is time-consuming and expensive or sometimes impossible to do, the modeling of the photoreactors is necessary. Computational fluid dynamics (CFD) is one of the numerical methods to solve and analyze the transport equations. CFD has been used recently in order to find the velocity, radiation intensity, pollutant concentration distribution inside the photoreacotrs (Mohajerani et al., 2010,2011; Qi et al., 2011; Duran et al., 2011a,b, 2010; Vincent et al., 2011; Denny et al., 2009,2010; Elyasi and Taghipour, 2006; Taghipour and Mohseni, 2005; Mohseni and Taghipour, 2004). In this chapter, the velocity distribution in ten different configurations of single lamp and multi-lamp photoreactors in both laminar and turbulent regimes are provided. Moreover, the light intensity distribution inside the photoreactors is also presented.

PHOTOREACTOR DESIGN

The design and analysis of photochemical reactors have attracted researchers' interests for the past four decades. The main advantages of the photoreactors over the conventional thermal excitation reactors are their selectivity and low operating temperature. In spite of these advantages, photoreactors are not widely used in industrial scale. The use of a photoreactor results in higher product costs; therefore, the photochemical process is used when there are no other available feasible alternative conventional (thermal or catalytic) processes. Some other disadvantages such as size limitations, design and construction difficulties, and fouling on lamp walls have restricted industrial applications of these types of reactors. Radiation field is the main characteristic of photoreactors determining the kinetics and the photoreactor performance. The photochemical reaction rate is proportional to the local volumetric rate of energy absorption (LVREA), an important parameter playing a crucial role in the photoreactor design. The LVREA depends on the radiation field presents within the photoreactor. The LVREA is a complex function of the lamp intensity, the concentration of absorbing species, the geometrical characteristics of the photoreactor system, and some other physicochemical properties. The light attenuation caused by the absorbing species makes the LVREA non-uniform. A Photochemical reactor design requires the solution of momentum, energy, and mass balances. The expression for the conservation of momentum is similar to that of the conventional reactor. It is believed that light irradiation does not affect the photochemically reactive fluid flow. The mass balance or continuity equations for each compound in the system should be solved simultaneously to find the concentration profile of each compound inside the photoreactor. The mass and energy conservation equations are directly related to the light irradiation. The energy balance can be divided into two parts: thermal and radiant energy balances. Although for isothermal photochemical reactions, thermal balances are negligible, a complete equation of change for energy balance is required for exothermic and endothermic reactions. The first step in the LVREA evaluation is to state a radiant energy balance at steady state condition in a homogeneous system. A radiation source model is also needed to be considered. Different source models have been developed during the past years to describe the radiant energy field and to calculate the LVREA.

Gaertner and Kent (1958) were among the first group that modeled photochemical reactions. They studied the photolysis of aqueous uranyl oxalate in an annular tubular reactor. The results of their experiments were correlated with a remarkable accuracy by assuming that the rate of photolysis is proportional to the residence time multiplied by the intensity of radiation at a particular radial position. They also proposed a mathematical model without considering the reactant concentration. Therefore, their model was applicable when the conversion was small (less than 12%). The effect of diffusion was also neglected in their model. The modeling of radiation field is classified into two categories: incidence and emission models. In the former approach, a specific radiation distribution exists in the photoreactor, while in the latter approach, the photon emission rate is employed to derive an incidence model. Although incidence models are mathematically simple to implement, there is no way of using this approach without experimentally adjustable parameters. Therefore, this approach cannot be employed as a reasonable systematic method to design photoreactors. Different incidence models have been developed and used for photoreactor modeling including radial incidence model (Gaertner and Kent, 1958; Schechter and Wissler, 1960; Cassano and Smith, 1966, 1967; Cassano et al., 1968; Matsuura and Smith, 1970b, 1971; Santeralli and Smith, 1974; Dolan et al., 1965; Jacob and Dranoff 1969; Williams and Ragonese, 1970), partially diffuse model (Matsuura and Smith, 1970a), and diffuse incidence model (Jacob and Dranoff, 1969; Williams, 1978; Matsuura et al., 1969; Matsuura and Smith, 1970c; Harada et al., 1971; Roger and Villermaux, 1983).

In emission model, a lamp is considered as an emitting line, a surface, or a volume source. Different emission models have been developed including line source with parallel plane emission model (Harris and Dranoff, 1965), line source with spherical emission model (Jacob and Dranoff, 1966, 1968, 1970; Jain et al., 1971; Dworking and Dranoff, 1978; Magelli and Santarelli, 1978; Pasquali and Santarelli, 1978), and line source with diffuse emission model (Akehata and Shirai, 1972). Line and surface models are classified into specular and diffuse emission models. In specular emission, the magnitude of the light intensity vector is not a function of the emission angle. Photoreactors equipped with mercury arc and neon lamps belong to this class, while fluorescent lamps, in which the magnitude of the light intensity vectors depend on the angle of the emission, fall into the diffuse emission. There are some great reviews in which developed models describing the light distribution in photoreactors are presented in details in the open literature (Alfano et al., 1986a,b; Cassano et al., 1995; Alfano and Cassano, 2009).

Photoreactors are generally grouped into homogeneous and heterogeneous classes. Homogeneous reactors operate in a single phase, gas or liquid, in air and water purification systems. In heterogeneous photoreactors, a photocatalyst is added to the system in order to increase the process efficiency. Heterogeneous photoreactors are subcategorized into attached (immobilized) or suspended (slurry) modes. Most of the early photoreactors have employed a titanium dioxide (TiO_2) suspension because it offers a high surface area for the reactions and almost no mass transfer limitation exists in the system. In slurry photoreactors, a recovery step is necessary to separate, regenerate, and reuse the photocatalyst. The efficiency of an immobilized system is less than that of the slurry photoreactor but the photocatalyst is continuously used for a longer period of time.

The disadvantages of the slurry photocatalysis include 1) difficulty and time consuming process of separation or filtration of the photocatalyst after the photocatalytic process; 2) particle aggregation and agglomeration at high photocatalyst concentration; and 3) difficulty of using the suspended photocatalyst in continuous processes (Sopyan et al., 1996). To overcome these drawbacks, immobilized photocatalysts are usually recommended. Photocatalysts could be immobilized on various supports such as glasses (Sabate et al., 1992; Kim et al., 1995; Fernandez et al., 1995; Wang et al., 1998; Piscopo et al., 2000; Sakthivel et al., 2002; Neti et al., 2010), tellerette packings (Mehrvar et al., 2002), silica (Van Grieken et al., 2002; Ding et al., 2001; Xu et al., 1999; Alemany et al., 1997; López-Muñoz et al., 2005; Marugán et al., 2006), polymers (Kasanen et al., 2009), and clays (An et al., 2008). The analysis and design of photochemical reactors have received increasing interests in chemical engineering recently. The photochemical reactors could be used to either produce and synthesize chemicals or destroy water and wastewater contaminants. The photochemical reactors are one of the least well known industrial reactors. As mentioned earlier, the main advantages of the photochemical reactors over the conventional ones are their selectivity and low reaction temperature. These advantages have become the subject of several reviews (Doede and Walker, 1955; Marcus et al., 1962; Cassano et al., 1967; Shirotsuka and Nishiumi, 1971; Roger and Villermaux, 1979).

RADIATION FIELD PROPERTIES

A bundle of rays with a specific energy carrying photons is called spectral specific intensity and radiance. This parameter plays a detrimental role in radiation field inside a photoreactor. The spectral specific intensity is defined as the total amount of radiative energy passing through a unit area perpendicular to the direction of propagation, per unit solid angle about the direction, per unit frequency, and per unit time. Therefore, the spectral specific intensity unit is Einstein per square meter per stradian, per unit frequency, and per second (Cassano et al., 1995):

$$I_v(x,\Omega,t,v) = \lim_{dA,d\Omega,dt,dv\to 0}\left(\frac{dE_v}{dA\cos\theta d\Omega dt dv}\right) \tag{1}$$

Another important photochemical property is the spectral incident radiation that could be calculated by integrating the spectral specific intensity over the entire photoreactor volume as follows:

$$G_v = \int_{\Omega} I_v d\Omega \tag{2}$$

In spherical coordinates, the spectral incident radiation could be written as follows (Cassano et al., 1995):

$$G_v = \int_{\theta_1}^{\theta_2}\int_{\phi_1}^{\phi_2} I_v \sin\theta d\phi d\theta \tag{3}$$

Equation (3) is valid for the monochromatic radiation while for the polychromatic radiation, an extra integration is required as follows (Cassano et al., 1995):

$$G_v = \int_{v_1}^{v_2}\int_{\theta_1}^{\theta_2}\int_{\phi_1}^{\phi_2} I_v \sin\theta d\phi d\theta dv \tag{4}$$

The local volumetric rate of energy absorption (LVREA), the absorbed intensity, is the product of the spectral specific intensity and the absorption coefficient. The rate of a photochemical reaction is the product of the LVREA and the quantum yield. The quantum yield indicates the number of moles of chemicals reacted per moles of photons absorbed.

PHOTON TRANSPORT EQUATION

The photon balance in a photochemical reactor can be written as follows (Ozisik, 1973; Cassano et al., 1995; Alfano and Cassano, 2008):

$$\begin{bmatrix} \text{time rate of change} \\ \text{of } \Omega, v \text{ photons in} \\ \text{the volume} \end{bmatrix} + \begin{bmatrix} \text{net flux of } \Omega, v \text{ photons} \\ \text{leaving the volume} \\ \text{across the surface } A \end{bmatrix} = \begin{bmatrix} \text{net gain of } \Omega, v \text{ photons owning} \\ \text{to emission, absorption, in - scattering,} \\ \text{and out - scattering in the volume} \end{bmatrix} \tag{5}$$

The right hand side of Equation (5) could be divided into two sources (emission and inscattering) and two sink (absorption and out-scattering) terms, therefore, the photon balance equation could be written as follows (Cassano et al., 1995):

$$\frac{1}{c}\frac{\partial I_{\Omega,v}}{\partial t} + \nabla.\left(I_{\Omega,v}\Omega\right) = \left(W_{\Omega,v}^{em} + W_{\Omega,v}^{in-s}\right) - \left(W_{\Omega,v}^{ab} + W_{\Omega,v}^{out-s}\right) \tag{6}$$

The absorption and emission terms are useful for both homogeneous and heterogeneous photochemical reactors. The loss of photons due to the absorption could be calculated using the following expression (Cassano et al., 1995):

$$W^{ab}_{\Omega,\upsilon} = K_\upsilon(x,t)I_\upsilon(x,\Omega,t) \tag{7}$$

where K_v is the absorption coefficient, which shows the fraction of the incident radiation that is absorbed by the molecule per unit length along the path of the beam. The emission term highly depends on temperature as shown in Equation (8). Photolytic and photocatalytic reactors usually operate at low temperatures, therefore, this term can be neglected for photoreactor modeling:

$$W^{em}_{\Omega,\upsilon} = K_\upsilon(x,t)I_{\upsilon,b}\left[T(x,t)\right] \tag{8}$$

The following two equations account for the gain and loss of photons due to in-scattering and out-scattering, respectively (Cassano et al., 1995). In homogeneous photoreactors, these two terms (in-scattering and out-scattering) are neglected due to the absence of semiconductors:

$$W^{s-in}_{\Omega,\upsilon} = \frac{1}{4\pi}\int_{\Omega'=4\pi}\int_{\upsilon'=0}^{\infty} \sigma_\upsilon(x,t)p(\upsilon' \to \upsilon,\Omega' \to \Omega)I_{\upsilon'}(x,\Omega',t)d\upsilon' d\Omega' \tag{9}$$

$$W^{s-out}_{\Omega,\upsilon} = \sigma_\upsilon(x,t)I_\upsilon(x,\Omega,t) \tag{10}$$

PRINCIPLES OF PHOTOCATALYSIS

The photocatalytic process in a heterogeneous photoreactor takes place in the presence of a semiconductor photocatalyst, in which the whole process is divided into five stages as follows (Herrmann, 1999; Fogler, 1998; Qi et al., 2011):

- Mass transfer of the organic compounds in the liquid phase to the photocatalyst surface;
- Adsorption of the pollutants onto the photoactivated semiconductor surface;
- Heterogeneous photocatalytic reaction;
- Desorption of the product(s) from the photocatalyst surface; and
- Mass transfer of the product(s) from the interface region to the bulk solution.

The most common photocatalyst is titanium dioxide (TiO_2) which catalyzes the reaction as a result of the interaction of the electrons and holes generated during the photocatalytic process. Titanium dioxide has a diverse range of applications especially in paint and coatings, plastics, sunscreens, ointments, and toothpaste. The whiteness, brightness, and opacity of titanium dioxide make the powder as a good option for a wide range of applications. Titanium dioxide exists in three different crystalline polymorphs: rutile, anatase, and brookite. TiO2 powders could be employed in solar and UV irradiation systems (photoreactors) because this semiconductor has a high transparency to visible light, high refractive index, and low absorption coefficient.

Wide range of metal oxides and sulfides have been used as photocatalysts including ZnO (Daneshvar et al., 2004; Sakthivel et al., 2003; Kormann et al., 1988; Khodj et al., 2001; Gouvêa et al., 2000; Lizama et al., 2002; Kansal et al., 2007; Chakrabarti and Dutta, 2004), WO_3 (Waldner et al., 2007; Sayama et al., 2010; Saepurahmanet et al., 2010; Cao et al., 2011), WS2 (Jing and Guo, 2007), Fe_2O_3 (Chen et al., 2001; Bandara et al., 2001; Pal and Sharon, 1998), V_2O_5 (Akbarzadeh et al., 2010; Teramura et al., 2004a,b), CeO_2 (Lin and Yu, 1998; Coronado et al., 2002; Ji et al., 2009; Song et al., 2007), CdS (Bessekhouad et al., 2004; Reutergårdh and Langphasuk, 1997; Tang and Huang, 1995), ZnS (Torres-Martínez et al., 2001), and CuO (Lim and Kim, 2004; Sathishkumar et al., 2011; Nezamzadeh-Ejheieg and Hushmandrad, 2010).

The light intensity in homogeneous and heterogeneous photoreactors determines the LVREA which is proportional to the rate of photochemical reaction. The light distribution in homogeneous photoreactors is well established. However, the heterogeneous photoreactor containing semiconductor particles makes the photoreactor modeling extremely difficult. In homogeneous photoreactors, the main solution parameter is the absorption coefficient, while in heterogeneous photoreactors, the absorption and scattering coefficients must be determined simultaneously.

MOMENTUM BALANCE AND COMPUTATIONAL FLUID DYNAMICS

The momentum balance equations are applied for two flow regimes: the laminar and turbulent flow for these two single and multi-lamp photoreactors. The fluid flow in laminar regime is described by Navier-Stokes equation. For the case of the turbulent flow, different known models have been developed. Among them, the k-ε model has shown a great attraction to researchers for modeling turbulent flows. The k-ε model is a two-equation model in which fluctuating velocities and Reynolds stresses have been related to the properties of the turbulent flow itself such as k and ε. These two fluid flow properties are the turbulent kinetic energy per unit mass of the fluctuating components (k) and the turbulent dissipation rate of the kinetic energy (ε), respectively. The continuity equation and steady state momentum balances for an incompressible fluid are described by following equations (Bird et al., 2002):

$$\nabla \cdot V = 0 \tag{11}$$

$$\rho V \cdot \nabla V = -\nabla P + \nabla \cdot \left(\mu + \rho\eta_T\right)\left(\nabla V + \left(\nabla V\right)^T\right) + F \tag{12}$$

In these equations, r, V, and P, represent the density, time-averaged turbulent velocity, and time-averaged pressure, respectively. The μ and η_T are the dynamic and kinematic turbulent viscosities, respectively. F is also the external force on the control volume. The k-ε model depicts that the kinematic turbulent viscosity (η_T) at any point should depend only on k and ε at that point according to the following expression (Bird et al., 2002):

$$\eta_T = C_\mu \frac{k^2}{\varepsilon} \tag{13}$$

where C_μ is an adjustable model constant. The turbulent kinetic energy (k) is the average kinetic energy per unit mass of eddies in the turbulent flow that is produced by the buoyant thermal and mechanically generated eddies based on the following equation (Bird et al., 2002):

$$k=\frac{1}{2}\left(\overline{u'^2}+\overline{v'^2}+\overline{w'^2}\right) \tag{14}$$

Parameters u¢, v¢, and w¢ are the fluctuating velocities in x, y, and z directions, respectively. The steady state equations for the turbulent kinetic energy (k) and the turbulent dissipation rate (ε) are as follows (Wilkes, 2005):

$$\rho V\cdot\nabla k=\nabla\cdot\left[\left(\mu+\rho\frac{C_\mu k^2}{\sigma_k\varepsilon}\right)\nabla k\right]+\rho C_\mu\frac{k^2}{\varepsilon}\left(\nabla V+(\nabla V)^T\right)^2-\rho\varepsilon \tag{15}$$

$$\rho V\cdot\nabla\varepsilon=\nabla\cdot\left[\left(\mu+\rho\frac{C_\mu k^2}{\sigma_\varepsilon\varepsilon}\right)\nabla\varepsilon\right]+\rho C_{\varepsilon_1}C_\mu k\left(\nabla V+(\nabla V)^T\right)^2-\rho C_{\varepsilon_2}\frac{\varepsilon^2}{k} \tag{16}$$

All five model constants have been selected as follows such that the k-ε model gives an estimation that fits reasonably well with the experimental data (Wilkes, 2005):

$$C_\mu=0.09;\quad C_{\varepsilon1}=1.44;\quad C_{\varepsilon2}=1.92;\quad \sigma_k=1.0;\ \text{and}\quad \sigma_\varepsilon=1.3 \tag{17}$$

Equations (15) and (16) along with continuity and momentum Equations (11) and (12) should be solved simultaneously. Boundary conditions for the momentum balance are as follows: Inlet of the photoreactor: The inflow velocity is specified by the ratio of the volumetric flow rate to the surface area. The k and ε are also calculated as follows:

$$k=\left(\frac{3I_T^2}{2}\right)\left(\overline{V}^2\right);\quad \varepsilon=C_\mu^{0.75}\frac{\left[\frac{3I_T^2}{2}\left(\overline{V}^2\right)\right]^{1.5}}{L_T}; \tag{18}$$

where I_T and L_T are the turbulent intensity scale (initial turbulence intensity) and turbulent length scale (eddy length scale). Their magnitudes could be calculated by Equations (19) and (20), respectively (Wilkes, 2005):

$$I_T=0.16Re^{-1/8} \tag{19}$$

$$L_T=0.07L \tag{20}$$

Outlet of the photoreactor: A normal flow is assumed so that the normal stress in the outlet is zero. Therefore, the gradient of k and ε are also calculated as follows (Wilkes, 2005):

$$\nabla k=0\quad;\qquad \nabla\varepsilon=0 \tag{21}$$

Walls of the photoreactor: No slip boundary condition could be assumed for the photoreactors and all the walls of the lamps. However, the logarithmic wall function is used for walls as a modification of the k-ε model. The boundary conditions for the k-ε model at a no-slip wall are obvious, but the near-wall behavior of the model, especially the ε-equation, is not appropriate. Therefore, the model produces poor results when integrated to the wall without modification. In fact, the integration of the k-ε model through the near-wall region and application of the noslip conditions yield unsatisfactory results. Therefore, the logarithmic wall function is used as the boundary condition on the walls of the photoreactor and lamps. In the near-wall region, therefore, the equation is solved for the first grid node away from the wall. The logarithmic wall function for a smooth pipe, therefore, is as follows (Wilkes, 2005):

$$u_z^+=5.5+2.5\ln y^+$$

where

$$y^+=\frac{y\sqrt{\tau_w/\rho}}{\mu/\rho}\quad \text{and}\quad u_z^+=\frac{\overline{u_z}}{\sqrt{\tau_w/\rho}} \tag{22}$$

in which y^+, uz^+, and τ_w are the dimensionless distance from the wall, the dimensionless velocity, and the wall shear stress, respectively. The y and $\overline{u_z}$ are the distance from the wall and time-averaged axial velocity, respectively. Simultaneous solution of Equations (11), (12), (15), and (16) as well as applying boundary conditions give various flow characteristics such as velocity distribution, minimum and maximum velocities, turbulent kinetic viscosity, turbulent kinetic energy, turbulent dissipation rate distribution, and vorticity. Equations (10) to (21) are applicable to both single and multi-lamp photoreactors.

CASE STUDY

Annular photoreactors are the most common continuous photoreactors. These types of photoreactors contain one or more UV lamps with quartz sleeve around them. In this study, 10 different photoreactor configurations were considered for modeling purposes as listed below:

- Single lamp photoreactor, the UV lamp is located at the center of the photoreactor;
- Two lamp photoreactor, each of them located at 0.5R from the center line of the photoreactor;
- Two lamp photoreactor, each of them located at 0.25R from the center line of the photoreactor;
- Two lamp photoreactor, each of them located at 0.75R from the center line of the photoreactor;
- Four lamp photoreactor, each of them located 0.5R from the center line of the photoreactor;
- Four lamp photoreactor, each of them located 0.25R from the center line of the photoreactor;
- Four lamp photoreactor, each of them located 0.75R from the center line of the photoreactor;
- Four lamp photoreactor, two of them at 0.5R and two at 0.25R from the center line of the photoreactor;
- Four lamp photoreactor, two of them at 0.5R and two at 0.75R from the center line of the photoreactor; and
- Four lamp photoreactor, two of them at 0.25R and two at 0.75R from the center line of the photoreactor.

In all cases, the diameter of the quartz sleeve is assumed to be 15 mm with the photoreactor radius (R) of 200 mm. The cross sections of these photoreactors (a to j) are shown in Figures 1, 2, and 3. The flow domains inside these photoreactors were simulated both in laminar and turbulent regimes. The inlet velocities for laminar and turbulent regimes were assumed to be 0.01 and 1 ms^{-1}, respectively. In laminar flow, the 3D Navier-Stokes equations were solved in cylindrical coordinates for single lamp and Cartesian coordinates for multilamp photoreactors. In turbulent regime, the k-ε turbulent model was used. The velocity field determination is important in the photoreactor design. The local velocity determines the residence time of the solution at a specific point. As mentioned earlier, the radiation intensity inside the photoreactor is highly non-uniform; therefore, there must be a trade-off between the residence time and the radiation intensity received at particular location.

Figures 1 and 2 depict the velocity distributions at the outlet of the photoreactors in turbulent and laminar regimes, respectively. From these two figures, three distinct regions could be considered: the region around the UV lamp(s), the region around the photoreactor wall, and the remaining parts. A lower velocity is observed around the UV lamps due to the no-slip boundary condition; therefore, a higher residence time is obtained near the UV lamps. On the other hand, around the UV lamp, a higher LVREA exist. Both higher residence time and greater radiation intensity result in a higher degradation of pollutants in water. The region near the photoreactor wall also has lower velocity and higher residence time, but the LVREA in this region is the lowest compared to those of the other regions. The light intensity was also simulated for these ten photoreactors as shown in Figure 3. In multilamp photoreactors, the light intensity of UV lamps could provide a synergetic effect depending on their location. Although closer UV lamps (in four lamp photoreactor) provide higher light intensity in the photoreactor center, the shadowing effect of the lamps reduces the light intensity in the region near the photoreactor wall. A comparison between Figures 3e and 3f show that the photoreactor in Figure 3f performed a higher light intensity at the center, however, the shadowing effect is high, therefore, Figure 3e shows a good trade-off between the synergetic effect and the shadowing effect.

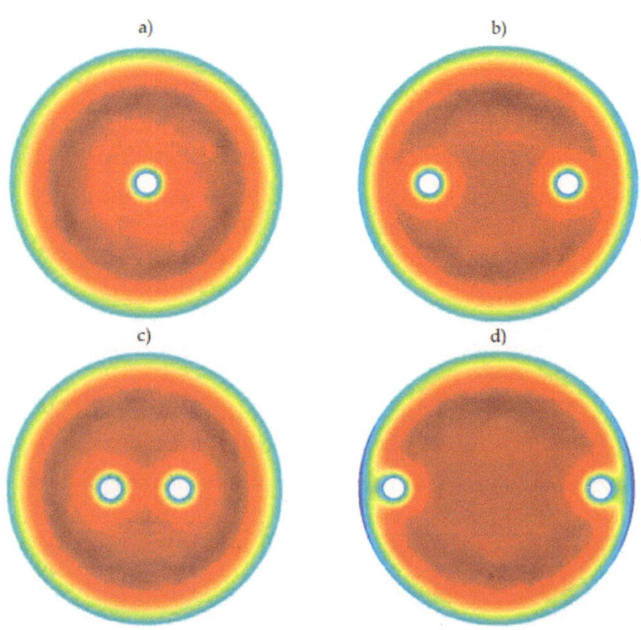

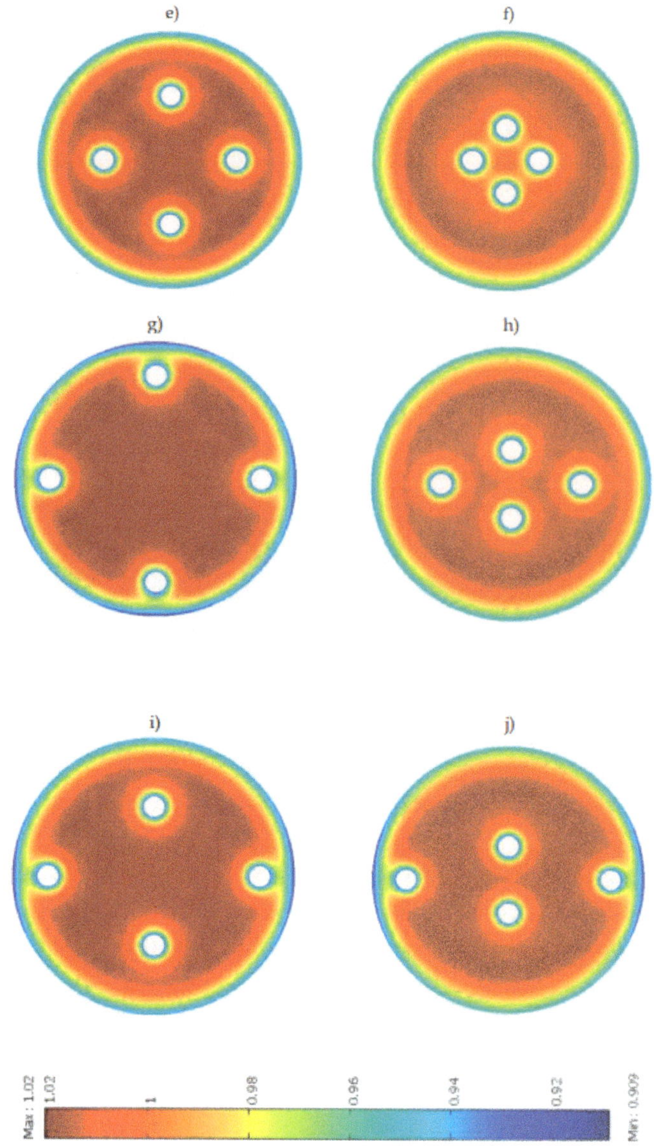

Figure 1: Velocity distribution at the outlet of the photoreactors (turbulent regime). All geometries are defined in Section "Case study".

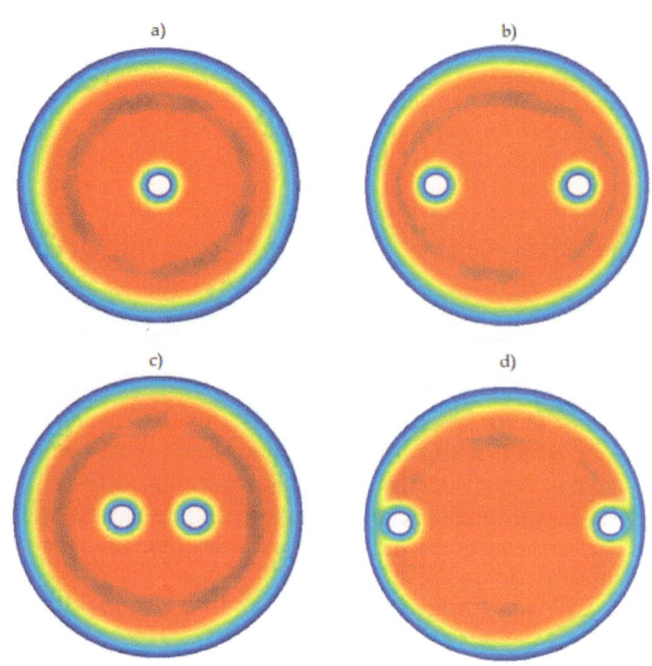

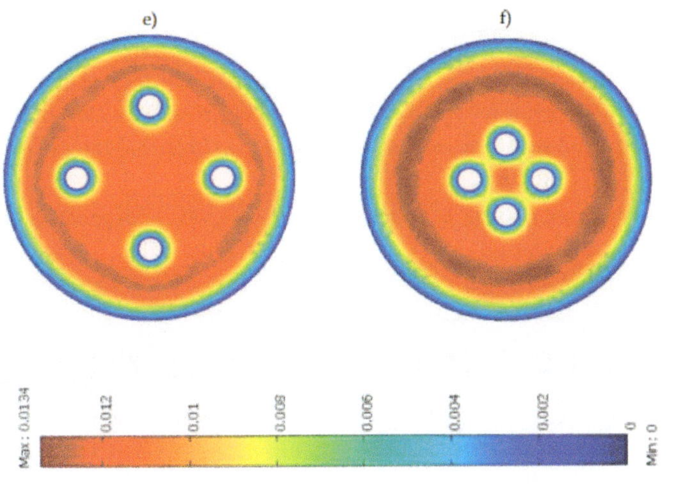

Figure 2: Velocity distribution at the outlet of the photoreactors (laminar regime). All geometries are defined in Section "Case study".

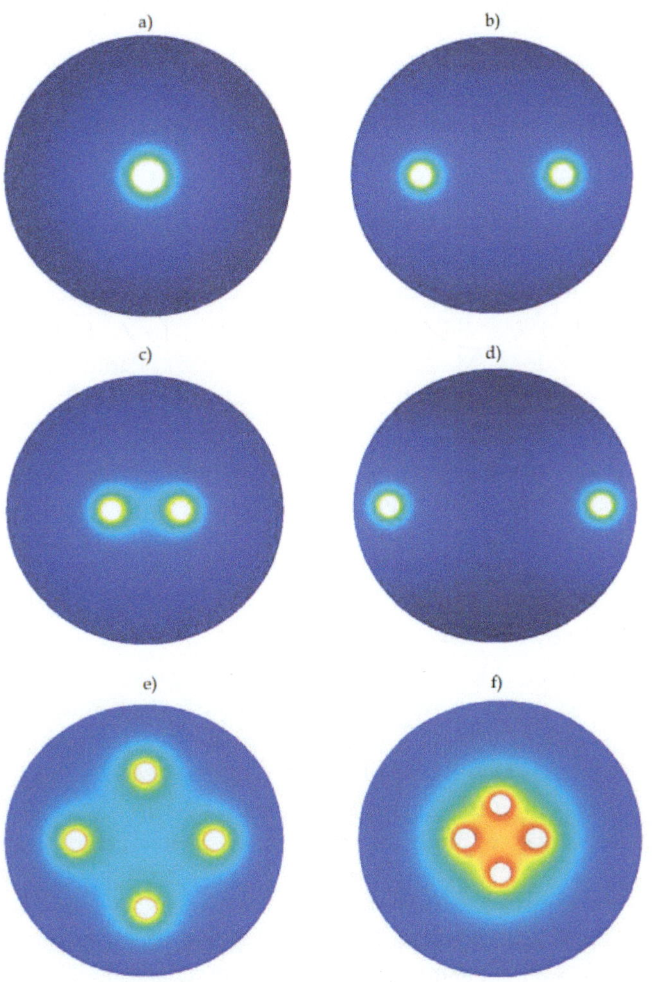

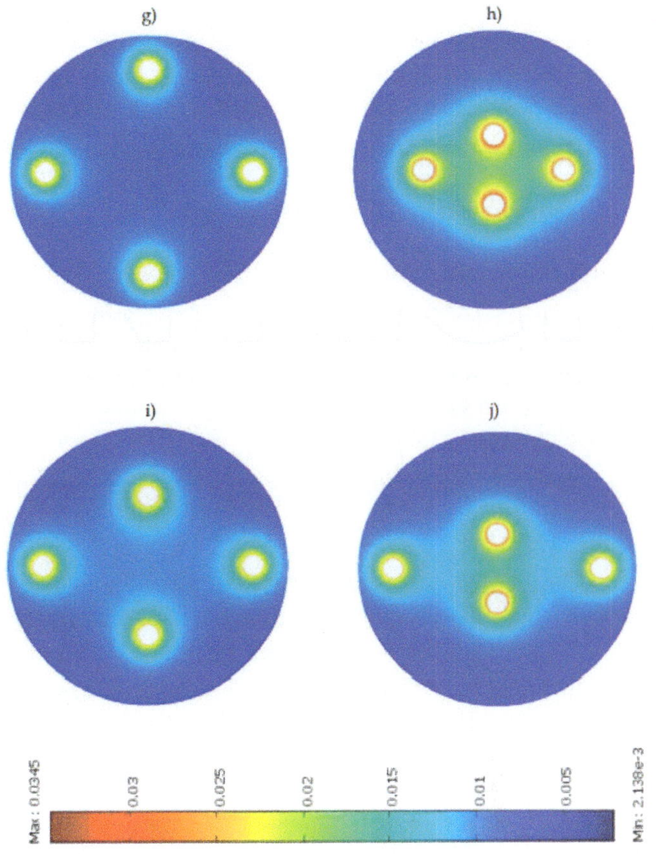

Figure 3: Light intensity distribution at the outlet of the photoreactors for both laminar and turbulent regimes. All geometries are defined in Section "Case study".

CONCLUSIONS

The degradation and mineralization of polluted waters containing non-biodegradable compounds have become increasingly important. CFD modeling helps to design, optimize, and scale-up photochemical reactors. Velocity, light, and pollutant(s) concentration distribution would give better understanding of photoreactors operation for further decisions. Velocity and light intensity distribution in ten different annular photoreactors were described. Velocity profiles were sketched for both laminar and turbulent flows. The effect of number and location of UV lamps was studied to find the enhancement in light distribution in multi-lamp photoreactors. Closer UV lamps performed a higher light intensity in the center of the photoreactor but the closer lamps provided a higher shadowing effect.

ACKNOWLEDGEMENTS

The financial support of Natural Sciences and Engineering Research Council of Canada (NSERC) and Ryerson University is greatly appreciated.

REFERENCES

1. Akbarzadeh, R.; Umbarkar, S.B.; Sonawane, R.S.; Takle, S. & Dongare, M.K. (2010) Vanadiatitania thin films for photocatalytic degradation of formaldehyde in sunlight. Appl. Catal. A: General, Vol. 374 No. 1-2 pp. 103-109.

2. Akehata, T. & Shirai, T. (1972) Effect of light-source characteristics on the performance of circular annular photochemical reactor. J. Chem. Eng. Japan Vol. 5 pp.385-391.
3. Alemany, L.J.; Bañares, M.A.; Pardo, E.; Martin, F.; Galán-Fereres, M. & Blasco, J.M. (1997) Photodegradation of phenol in water using silica-supported titania catalysts. Appl. Catal. B: Environ., Vol. 13, No. 3-4, pp. 289-297.
4. Alfano, O.M.; Romero, R.L. & Cassano, A.E. (1986a) Radiation field modelling in photoreactors-II. Heterogeneous media. Chem. Eng. Sci., Vol. 41 No. 5 pp. 1137-1153.
5. Alfano, O.M.; Romero, R.L. & Cassano, A.E. (1986b) Radiation field modelling in photoreactors-I. Homogeneous media. Chem. Eng. Sci., Vol. 41 No. 5 pp. 421-444.
6. Alfano, O.M. & Cassano, A.E. (2008) Photoreactor modeling: applications to advanced oxidation processes. Int. J. Chem. Reactor Eng., Vol. 6, P2, pp.1-18.
7. Alfano, O.M. & Cassano, A.E. (2009) Scalling-up of photoreactors. Applications to advanced oxidation process. Adv. Chem. Eng., Vol. 36 pp. 229-287.
8. An, T.; Chen, J.; Li, G.; Ding, X.; Sheng, G.; Fu, J.; Mai, B. & O'Shea, K.E. (2008) Characterization and the photocatalytic activity of TiO2 immobilized hydrophobic montmorillonite photocatalysts. Degradation o decabromodiphenyl ether (BDE 209). Catal. Today Vol. 139 No. 1-2, pp. 69-76.
9. Bandara, J.; Mielczarski, J.A.; Lopez, A. & Kiwi, J. (2001) Sensitized degradation of chlorophenols on iron oxides induced by visible light comparison with titanium oxide. Appl. Catal. B: Environ, Vol. 34 No. 4 pp. 321-333.
10. Bessekhouad, Y.; Robert, D. & Weber, J.V. (2004) Bi2S3/TiO2 and CdS/TiO2 heterojunctions as an available configuration for photocatalytic degradation of organic pollutant. J. Photochem. Photobiol. A: Chemistry, Vol. 163 No. 3 pp. 569-580.
11. Bird, R.B.; Steward, W.E. & Lightfood, E.N. Transport Phenomena, 2nd ed. John Wiley & Sons, Inc.: NY, 2002
12. Cao, J.; Luo, B.; Lin, H. & Chen, S. (2011) Photocatalytic activity of novel AgBr/WO3 composite photocatalyst under visible light irradiation for methyl orange degradation. J. Hazard.s Mater., Vol. 190 No. 1-3 pp. 700-706.
13. Cassano, A.E. & Smith, J.M. (1966) Photochlorination in a tubular reactor. AIChE J., Vol. 12 pp. 1124-1133.
14. Cassano, A.E. & Smith, j.M. (1967) Photochlorination of propane. AIChE J., Vol. 13, pp. 915-925.
15. Cassano, A.E.; Silverston, P.L. & Smith, J.M. (1967) Photochemical reaction engineering. Ind. Eng. Chem., Vol. 59, pp. 18-38.
16. Cassano, A. E.; Matsuura, T. & Smith, J.M. (1968) Batch recycle reactor for slow photochemical reactions. Ind. Eng. Chem. Fundam., Vol. 7 pp. 655-660.
17. Cassano, A.E.; Martín, C.A.; Brandi, R.J. & Alfano, O.M. (1995) Photoreactor analysis and design: Fundamentals and applications. Ind. Eng. Chem. Res., Vol. 34, No. 7, pp. 2155-2201.
18. Chakrabarti, S. & Dutta, B.K. (2004) Photocatalytic degradation of model textile dyes in wastewater using ZnO as semiconductor catalyst. J. Hazard. Mater., Vol. 112 No. 3 pp. 269-278
19. Chen, F.; Xie, Y.; Zhao, J. & Lu, G. (2001) Photocatalytic degradation of dyes on a magnetically separated photocatalyst under visible and UV irradiation. Chemosphere, Vol. 44 No. 5 pp. 1159-1168.
20. Coronado, J.M.; Javier Maira, A.; Martínez-Arias, A.; Conesa, J.C. & Soria, J. (2002) EPR study of the radicals formed upon UV irradiation of ceria-based photocatalysts. J. Photochem. Photobiol. A: Chemistry, Vol. 150 No. 1-3 pp. 213-221.
21. Daneshvar, N.; Salari, D. & Khataee, A.R. (2004) Photocatalytic degradation of azo dye acid red 14 in water on ZnO as an alternative catalyst to TiO2 J. Photochem. Photobiol. A: Chemistry, Vol. 162; No. 2-3; pp.317-322.
22. Denny, F.; Scott, J.; Pareek, V.; Peng, G.-D. & Amal, R. (2010) Computational fluid dynamics modelling and optimal configuring of a channeled optical fibre photoreactor. Chem. Eng. Sci., Vol. 65 No. 17 pp. 5029-5040.
23. Ding, Z.; Hu, X.; Yue, P.L.; Lu, G.Q. & Greenfield, P.F. (2001) Synthesis of anatase TiO2 supported on porous solids by chemical vapor deposition. Catal. Today, Vol. 68, No.: 1-3; pp. 173-182.
24. Doede, C.M. & Walker, C.A. (1955). Photochemical engineering. Chemical Engineering, pp. 159-178.
25. Dolan, W.J.; Dimon, C.A. & Dranoff, J.S. (1965) Dimensional analysis in photochemical reactor design. AIChE J., Vol. 11 pp. 1000-1005.
26. Duran, J.E.; Taghipour, F. & Mohseni, M. (2011a) Evaluation of model parameters for simulating TiO2 coated UV reactors. Water Sci. Technol., Vol. 63 No. 7 pp. 1366-1372.
27. Duran, J.E.; Mohseni, M. & Taghipour, F. (2011b) Design improvement of immobilized photocatalytic reactors using a CFD-Taguchi combined method. Ind. Eng. Chem. Res., Vol. 50 No. 2 pp. 824-831.
28. Duran, J.E.; Taghipour, F. & Mohseni, M. (2010) Irradiance modeling in annular photoreactors using the finite-volume method. J. Photochem. Photobiol. A: Chemistry, Vol. 215 No. 1 pp. 81-89.
29. Dworkin, D. & Dranoff, J.S. (1978) Free radical transport in a photochemical reactor. AIChE J., Vol. 24 pp. 1134-1137.
30. Elyasi, S. & Taghipour, F. (2006) Simulation of UV photoreactor for water disinfection in Eulerian Framework. Chem. Eng. Sci., Vol. 61 No. 14 pp. 4741-4749.
31. Elyasi, S. & Taghipour, F. (2010) Simulation of UV photoreactor for degradation of chemical contaminants: Model development and evaluation. Environ. Sci. Technol., Vol. 44 No. 6 pp. 2056-2063.
32. Fernandez, A.; Lassaletta, G.; Jimenez, V.Z.; Justo, A.;

Gonzalez-Elipe, A.R.; Herrmann, J.M.; Tahiri, H. & Ait-Ichou, Y. (1995) Preparation and characterization of TiO2 photocatalysts supported on various rigid supports (glass, quartz and stainless steel). Comparative studies of photocatalytic activity in water purification. Appl. Catal. B: Environ., Vol. 7 No. 1-2 pp. 49-63.

33. Fogler, H.S. Elements of Chemical Reaction Engineering, third ed., Prentic Hall PTR, New Jersey, 1998.

34. Gaertner, R.F. & Kent, J.A. (1958) Conversion in a continuous photochemical reactor. Ind. Eng. Chem. Res., Vol. 50 No. 9, pp. 1223-1226.

35. Gouvêa, C.A.K.; Wypych, F.; Moraes, S.G.; Durán, N.; Nagata, N. & Peralta-Zamora, P. (2000) Semiconductor-assisted photocatalytic degradation of reactive dyes in aqueous solution. Chemosphere, Vol. 40 No. 4 pp. 433-440.

36. Harada, J.; Akehata, T. & Shirai, T. (1971) Light intensity distribution in an elliptical photoreactor. Kagaku Kogaku, Vol. 35 pp. 233-239.

37. Harris, P.R. & Dranoff, J.S. (1965) A study of perfectly mixed photochemical reactors. AIChE J., Vol. 11 pp. 497-502.

38. Herrmann, J.-M. (1999) Heterogeneous photocatalysis: Fundamentals and applications to the removal of various types of aqueous pollutants. Catal. Today, Vol. 53 No. 1 pp. 115-129.

39. Jacob, S.M. & Dranoff, J.S. (1966) Radial scale-up of perfectly mixed photochemical reactors. Chem. Eng. Prog. Symp. Ser., Vol. 62 pp.47-55.

40. Jacob, S.M. & Dranoff, J.S. (1968) Design and analysis of perfectly mixed photochemical reactors. Chem. Eng. Prog. Sym. Ser., Vol. 64 pp. 54-63.

41. Jacob, S.M. & Dranoff, J.S. (1969) Light intensity profiles in an elliptical photoreactor. AIChE J., Vol. 15 pp. 141-144.

42. Jacob, S.M. & Dranoff, J.S. (1970) Light intensity profiles in a perfectly mixed photoreactor. AIChE J., Vol. 16 pp. 359-363.

43. Jain, R.L.; Graessley, W.W. & Dranoff, J.S. (1971) Design and analysis of a photoreactor for styrene polymerization. Ind. Eng. Chem. Prod. Res. Develop., Vol. 10 pp. 293-298.

44. Ji, P.; Zhang, J.; Chen, F. & Anpo, M. (2009) Study of adsorption and degradation of acid orange 7 on the surface of CeO2 under visible light irradiation. Appl. Catal. B: Environment., Vol. 85 No. 3-4 pp. 148-154.

45. Jing, D. & Guo, L. (2007) WS2 sensitized mesoporous TiO2 for efficient photocatalytic hydrogen production from water under visible light irradiation. Catal. Commun., Vol. 8 No. 5 pp. 795-799.

46. Kansal, S.K.; Singh, M. & Sud, D. (2007) Studies on photodegradation of two commercial dyes in aqueous phase using different photocatalysts. J. Hazard. Mater., Vol. 141 No. 3 pp. 581-590.

47. Kasanen, J.; Suvanto, M. & Pakkanen, T.T. (2009) Self-cleaning, titanium dioxide based, multilayer coating fabricated on polymer and glass surfaces. J. Appl. Polym. Sci., Vol. 111 No. 5 pp. 2597-2606.

48. Kim, D.H.; Anderson, M.A. & Zeitner, W.A. (1995) Effects of firing temperature on photocatalytic and photoelectrocatalytic properties of TiO2. J. Environ. Eng., Vol. 121 No. 8 pp. 590-594.

49. Khodja, A.A.; Sehili, T.; Pilichowski, J.-F. & Boule, P. (2001) Photocatalytic degradation of 2- phenylphenol on TiO2 and ZnO in aqueous suspensions. J. Photochem. Photobiol. A: Chemistry, Vol. 141 No. 2-3 pp. 231-239.

50. Kormann, C.; Bahnemann, D.W. & Hoffmann, M.R. (1988) Photocatalytic production of H2O2 and organic peroxides in aqueous suspensions of TiO2, ZnO, and desert sand. Environ. Sci. Technol., Vol. 22 No. 7 pp. 798-806.

51. Lim, T.H. & Kim (2004), S.D. Trichloroethylene degradation by photocatalysis in annular flow and annulus fluidized bed photoreactors. Chemosphere, Vol. 54 No. 3 pp. 305- 312.

52. Lin, J. & Yu, J.C. (1998) An investigation on photocatalytic activities of mixed TiO2-rare earth oxides for the oxidation of acetone in air. J. Photochem. Photobiol. A: Chemistry, Vol. 116 No. 1 pp. 63-67.

53. Lizama, C.; Freer, J.; Baeza, J. & Mansilla, H.D. (2002) Optimized photodegradation of reactive blue 19 on TiO2 and ZnO suspensions. Catalysis Today, Vol. 76 No. 2-4 pp. 235-246.

54. López-Muñoz, M.-J.; Grieken, R.V.; Aguado, J. & Marugán, J. (2005) Role of the support on the activity of silica-supported TiO2 photocatalysts: structures of the TiO2/SBA-15 photocatalysis. Catal. Today, Vol. 101, pp. 307-314.

55. Magelli, F. & Santarelli, F. (1978) The modelling of batch photoreactors. Chem. Eng. Sci., Vol. 33 pp. 611-614

56. Marcus, R.J.; Kent, J.A. & Schenck, G.O. (1962) Industrial photochemistry. Ind. Eng. Chem., Vol. 54, pp. 20-28

57. Marugán, J.; Hufschmidt, D.; López-Muñoz, M.-J.; Selzer, V. & Bahnemass, D. (2006) Photonic efficiency for methanol photooxidation and hydroxyl radical generation on silica-supported TiO2 photocatalysts. Appl. Catal. B: Environ, Vol: 62, No.: 3-4, pp. 201-207.

58. Matsuura, T.; Cassano, A.E. & Smith, J.M. (1969) Acetone photolysis: kinetic studies in a flow reactor. AIChE J., Vol. 15 pp. 495-501.

59. Matsuura, T. & Smith, J.M. (1970a) Light distribution in cylindrical photoreactors. AIChE J., Vol. 16 pp. 321-324.

60. Matsuura, R. & Smith, J.M. (1970b) Kinetics of photocomposition of dodecyl benzene sulfonate. Ind. Eng. Chem. Fund., Vol. 9 pp. 252-260.

61. Matsuura, R. & Smith, J.M. (1970c) Photodecomposition kinetics of formic acid in aqueous solutions. AIChE J., Vol. 16 pp.1064-1071.

62. Matsuura, R. & Smith, J.M. (1971) Photodecomposition kinetics of formic acid in aqueous solutions. Ind. Eng. Chem. Fundam., Vol. 10, pp. 316-318.

63. Mehrvar, M.; Anderson, W.A. & Moo-Young, M. (2002) Preliminary analysis of a tellerette packed-bed

photocatalytic reactor, Adv. Environ. Res. Vol. 6 No. 4, pp. 411–418.

64. Mohajerani, M.; Mehrvar, M. & Ein-Mozaffari, F. (2011) Photoreactor design and CFD modelling of a UV/H2O2 process for distillery wastewater treatment. Can. J. Chem. Eng. (in press).

65. Mohajerani, M.; Mehrvar, M. & Ein-Mozaffari, F. (2010) CFD modeling of metronidazole degradation in water by the UV/H2O2 process in single and multilamp photoreactors. Ind. Eng. Chem. Res., Vol. 49 No. 11 pp. 5367-5382.

66. Mohseni, M. & Taghipour, F. (2004) Experimental and CFD analysis of photocatalytic gas phase vinyl chloride (VC) oxidation. Chem. Eng. Sci., Vol. 59 No. 7 pp. 1601-1609.

67. Neti, N.R.; Parmar, G.R.; Bakardjieva, S. & Subrt, J. (2010) Thick film titania on glass supports for vapour phase photocatalytic degradation of toluene, acetone, and ethanol. Chem. Eng. J., Vol. 163 No. 3 pp. 219-229.

68. Nezamzadeh-Ejhieh, A. & Hushmandrad, S. (2010) Solar photodecolorization of methylene blue by CuO/X zeolite as a heterogeneous catalyst. Appl. Catal. A: General, Vol. 388 No. 1-2 pp. 149-159.

69. Ozisik, M.N. (1973) Radiative transfer and interactions with conduction and convection, Wiley, New York

70. Pal, B. & Sharon, M. (1998) Photocatalytic degradation of salicylic acid by colloidal Fe2O3 particles. J. Chem. Technol. Biotechnol., Vol. 73 No. 3 pp. 269-273.

71. Pasquali, G. & Santarelli, F. (1978) Radiant energy transfer in batch photoreacting media. Chem. Eng. Commun., Vol.2 pp. 271-274.

72. Piscopo, A.; Robert, D.; Marzolin, C. & Weber, J.V. (2000) TiO2 supported on glass fiber for the photocatalytic degradation of benzamide. J. Mater. Sci. Lett., Vol. 19 No. 8 pp. 683-684.

73. Qi, N.; Zhang, H.; Jin, B. & Zhang, K. (2011) CFD modelling of hydrodynamics and degradation kinetics in an annular slurry photocatalytic reactor for wastewater treatment. Chem. Eng. J., (in press).

74. Reuterġardh, L.B. & Langphasuk, M. (1997) Photocatalytic decolourization of reactive azo dye: A comparison between TiO2 and CdS photocatalysis. Chemosphere, Vol. 35 No. 3 pp. 585-596.

75. Roger, M. & Villermaux, J. (1983) Modelling of light absorption in photoreactors. Part II. Density profile and efficiency of light absorption in a cylindrical reactor. Experimental comparison with complex kinetics. Chem. Eng. Sci., Vol. 38 pp. 1593-1605.

76. Roger, M. & Villermaux, J. (1979) Modelling of light absorption in photoreactors. Part I. General formulation based on the laws of photometry. Chem. Eng. J., Vol. 17, pp. 219-226.

77. Sabate, J.; Anderson, M.A.; Aguado, M.A.; Giménez, J.; Cervera-March, S. & Hill Jr., C.G. (1992) Comparison of TiO2 powder suspensions and TiO2 powder suspensions and TiO2 ceramic membrans supported on glass as photocatalytic systems in the reduction of chromium. J. Mol. Catal., Vol. 71 No. 1 pp. 57-68.

78. Saepurahman, Abdullah, M.A. & Chong, F.K. (2010) Preparation and characterization of tungsten-loaded titanium dioxide photocatalyst for enhanced dye degradation. J. Hazard. Mater., Vol. 176 No. 1-3 pp. 451-458.

79. Sakthivel, S.; Shankar, M.V.; Palanichamy, M.; Arabindoo, B. & Murugesan, V. (2002) Photocatalytic decomposition of leather dye. Comparative study of TiO2 supported on alumina and glass beads. J. Photochem. Photobiol. A: Chemistry, Vol. 148 No. 1-3 pp. 153-159.

80. Sakthivel, S.; Neppolian, B.; Shankar, M.V.; Arabindoo, B.; Palanichamy, M. & Murugesan, V. (2003) Solar photocatalytic degradation of azo dye: Comparison of photocatalytic efficiency of ZnO and TiO2. Sol. Energy Mater. Sol. Cells, Vol. 77 No. 1, pp.65-82.

81. Santarelli, F. & Smith, J.M. (1974) Rate of photochlorination of liquid n-heptane. Chem. Eng. Commun., Vol. 1, pp. 297-302.

82. Sathishkumar, P.; Sweena, R. & Wu, J.J. (2011) Anandan, S. Synthesis of CuO-ZnO nanophotocatalyst for visible light assisted degradation of a textile dye in aqueous solution. Chem. Eng. J., Vol. 171 No. 1 pp. 136-14.

83. Sayama, K.; Hayashi, H.; Arai, T.; Yanagida, M.; Gunji, T. & Sugihara, H. (2010) Highly active WO3 semiconductor photocatalyst prepared from amorphous peroxotungstic acid for the degradation of various organic compounds. Appl. Catal. B: Environ., Vol. 94 No. 1-2 pp.150-157.

84. Schechter, R.S. & Wissler, E.H. (1960) Photochemical reactions in an isothermal laminarflow chemical reactor. Appl. Sci. Res., Vol. 9 No. 1, pp. 334-344

85. Shirotsuka, T. & Nishiumi, H. (1971) Theoretical basis for making an apparatus for photochemical reactions. Kagaku Kogaku, Vol. 35, pp. 1329-1338.

86. Song, S.; Xu, L.; He, Z.; Chen, J.; Xiao, X. & Yan, B. (2007) Mechanism of the photocatalytic degradation of C.I. reactive black 5 at pH 12.0 using SrTiO3/CeO2 as the catalyst. Environ. Sci. Technol., Vol. 41 No. 16 pp. 5846-5853.

87. Sopyan, I.; Watanabe, M.; Murasawa, S.; Hashimoto, K. & Fujishima, A. (1996) An efficient TiO2 thin-film photocatalyst: photocatalytic properties in gas-phase acetaldehyde degradation. J. Photochem. Photobiol., Vol. 98 No. 1-2. pp. 79-86.

88. Taghiour, F. & Mohseni, M. (2005) CFD simulation of UV photocatalytic reactors for air treatment. AIChE J., Vol. 51 No. 11 pp. 3039-3047.

89. Tang, W.Z. & Huang, C.P. (1995) Photocatalyzed oxidation pathways of 2,4-dichlorophenol by CdS in basic and acidic aqueous solutions. Water Res., Vol. 29 No. 2 pp. 745-756.

90. Teramura, K.; Tanaka, T.; Kani, M.; Hosokawa, T. & Funabiki, T. (2004a) Selective photooxidation of neat cyclohexane in the liquid phase over V2O5/Al2O3. J. Mol. Catal. A: Chem., Vol. 208 No. 1-2 pp. 299-305

91. Teramura, K.; Tanaka, T.; Hosokawa, T.; Ohuchi, T.; Kani, M. & Funabiki, T. (2004b) Selective photo-oxidation of various hydrocarbons in the liquid phase

over V2O5/Al2O3. Catal. Today, Vol. 96 No. 4 pp. 205-209.

92. Torres-Martínez, C.L.; Kho, R.; Mian, O.I. & Mehra, R.K. (2001) Efficient photocatalytic degradation of environmental pollutants with mass-produced ZnS nanocrystals. J. Colloid Interface Sci., Vol. 240 No. 2 pp. 525-532.

93. Van Grieken, R.; Aguado, J.; López-Muoz, M.J. & Marugán, J. (2002) Synthesis of sizecontrolled silica-supported TiO2 photocatalysts. J. Photochem. Photobiol. A: Chemistry, Vol. 148 No. 1-3, pp. 315-322.

94. Vincent, G.; Schaer, E.; Marquaire, P.-M. & Zahraa, O. (2011) CFD modelling of an annular reactor, application to the photocatalytic degradation of acetone. Process Saf. Environ. Prot., Vol. 89 No. 1 pp. 35-40.

95. Waldner, G.; Brüger, A.; Gaikwad, N.S. & Neumann-Spallart, M. (2007) WO3 thin films for photoelectrochemical purification of water. Chemosphere, Vol. 67 No. 4 pp. 779-784.

96. Wang, K.-H.; Tsai, H.-H. & Hsieg, Y.-H. (1998) The kinetics of photocatalytic degradation of trichloroethylene in gas phase over TiO2 supported on glass beads. Appl. Catal. B: Environ., Vol. 17 No. 1-2 pp. 25-36.

97. Wilkes, J.O. Fluid mechanics for chemical engineers with microfluidics and CFD, 2nd ed. Prentice Hall: NJ, 2005.

98. Williams, J.A. & Ragonese, F.P. (1970) Asymptotic solutions and limits of transport equations for tubular flow photoreactors. Chem. Eng. Sci., Vol. 25 pp. 1751-1759.

99. Williams, J.A. (1978) The radial light intensity profile in cylindrical photoreactors. AIChE J., Vol. 24 pp. 335-338.

100. Xu, Y.; Zheng, W. & Liu, W. (1999) Enhanced photocatalytic activity of supportedTiO2: Dispersing effect of SiO2. J. Photochem. Photobiol. A: Chemistry, Vol. 122, No.: 1, pp. 57-60.

INDEX

H

I

J

L

M

N

O

P

Q

R

S

T

V

W

Z